Geography and Geographers

Geography and Geographers provides a survey of the major debates, key thinkers and schools of thought in human geography in the English-speaking world, setting them within the context of economic, social, cultural and political, as well as intellectual, changes. It focuses on the debates among geographers regarding what their discipline should study and how that should be done, and draws on a wide reading of the geographical literature produced during a seventy-year period characterised by both growth in the number of academic geographers and substantial shifts in conceptions of the discipline's scientific rationale.

The pace and volume of change within the discipline show little sign of diminishing; this seventh edition covers new literature and important developments over the past decade. An insightful and reflective examination of the field from within, *Geography and Geographers* continues to be the most comprehensive and up-to-date single volume overview of the field of human geography.

This seventh edition has also been extensively revised and updated to reflect developments in the ways that geography and its history are understood and taught. Providing a thoroughly contemporary perspective, the book maintains its standing as the essential resource for students and researchers across the field.

Ron Johnston is Professor of Geography at the University of Bristol.

James D. Sidaway is Professor of Political Geography at the National University of Singapore.

WITHDRAWN

1 5 MAR 2024

Praise for the fifth edition of *Geography and Geographers*:

'[*Geography and Geographers*] has probably done more to shape human geographers' collective sense of what geography is and has been about than any other single source.'

Murray Low, Political Geography (2004)

Praise for this edition of *Geography and Geographers*:

'This new edition of *Geography and Geographers* is especially welcome. By providing what the authors call "wider discussions of the contexts" within which geographical endeavour has been located, it shows that the historical geography of geography has come of age. As a working map of the territory, this is a superlative piece of intellectual cartography that no geographer wanting to orientate themselves can afford to be without.'

Professor David N. Livingstone, Queen's University Belfast, Northern Ireland

'*Geography and Geographers* is a living classic. It provides a compelling and subtle narrative of the key intellectual shifts shaping contemporary geography, one that is sensitive to the range of factors that shape academic knowledge. An invaluable resource for scholars from students to professors, and an intellectual achievement in its own right.'

Professor Clive Barnett, Professor of Geography and Social Theory,
University of Exeter, UK

'Tony Blair once said that the Beatles produced the "music that was the background of our lives". *Geography and Geographers* is the constant in the background of the busy and changing life of geography. You can always count on it, and it is a comforting and reassuring presence. Like the Beatles' music you never tire of it, and it seems to get better with age.'

Professor Trevor Barnes, Department of Geography,
University of British Columbia, Canada

Geography and Geographers

Anglo–American human geography since 1945

Seventh Edition

Ron Johnston and James D. Sidaway

Routledge
Taylor & Francis Group

LONDON AND NEW YORK

Seventh edition published 2016
by Routledge
2 Park Square, Milton Park, Abingdon, Oxon OX14 4RN

and by Routledge
711 Third Avenue, New York, NY 10017

Routledge is an imprint of the Taylor & Francis Group, an informa business

First edition published by Edward Arnold 1979
Sixth edition published by Hodder Arnold 2004

British Library Cataloguing in Publication Data
A catalogue record for this book is available from the British Library

Library of Congress Cataloging in Publication Data
Johnston, R. J. (Ronald John)
    Geography & geographers : Anglo-American human geography since 1945 /
    R.J. Johnston and J.D. Sidaway. — Seventh Edition.
        pages cm
    Includes bibliographical references and index.
    1. Human geography—United States—History.   2. Human geography—Great
    Britain—History.   I. Sidaway, James D.   II. Title.   III. Title: Geography and
    geographers.
    GF13.J63 2016
    304.2—dc23   2015027516

ISBN: 978-0-415-82737-9 (hbk)
ISBN: 978-0-340-98510-6 (pbk)
ISBN: 978-0-203-52305-6 (ebk)

Typeset in Joanna
by Keystroke, Station Road, Codsall, Wolverhampton

Printed and bound by CPI Group (UK) Ltd, Croydon, CR0 4YY

For Lorna, Josh and Heather
And commemorating Jasmin Leila and Joseph Trevor

Contents

Figures

Preface
Situating *Geography and Geographers*

I remember being at a party in Minneapolis, sitting on the floor with other guests in a loose circle, taking turns saying what we did. There was a nurse, a medical researcher, a planner . . . When I said I was a graduate student in economic geography, one of them burst into uncontrollable laughter. 'So what do you do, find new places?'

(Trevor Barnes, 2002, p. 9)

I do think that sometimes we have written the history of geography as though it were something quite exotic but harmless, consigned, at the very least, to the last century, if not to oblivion: from our lofty positions as postmoderns, postcolonials or whatever, we detach ourselves from our predecessors without so much as a backward glance. Isn't it about time we took a more modest view? Perhaps that history which we want to transcend is precisely what makes our own geographies possible? Now there's a thought.

(Felix Driver, 2003, p. 230)

Geographers themselves have been very bad at writing their own history.

(Neil Smith, 2003, p. xxii)

the story of . . . geography . . . cannot be narrated in isolation from a much wider global and historical framework.

(David Livingstone, 2003a, p. 13)

The digital age means that many of the correspondences vanish the moment they are sent; even the fax paper we used has disintegrated 20 years on. So we offer this historical narration in order to begin the process of active remembering of the recent past in an attempt to avoid what has plagued the history of geography – its erasure of women and gender.

(Mona Domosh and Liz Bondi, 2014, p. 1068)

Human geography has been through many changes – in terms of its foci, practices, methods and styles of presentation. Reflecting aspects of this, between 1979 and the end of the twentieth century there were five previous editions of this book (1979, 1983, 1987, 1991, 1997), all sole authored by Ron Johnston. When a further, sixth, jointly authored edition appeared in 2004, we explained some of the changes made and the rationales behind these. By way of an entrée to this seventh edition, we first cite from that Preface to the sixth:

Since 1979, the book has expanded considerably. In order to contain it within reasonable margins, in this sixth edition we have cut back some of the details on behavioural geography that appeared in the earlier editions (veteran readers will notice that this no longer has a chapter of its own) and shortened some other sections. These selective cuts have allowed us to accommodate more on poststructuralism, postmodernism, postcolonialism and feminist geographies (through two new chapters derived in part from what hitherto was a single chapter entitled 'The cultural turn'), as well as to rearrange the chapter on radical geography and update others. In all cases we have sought to strike a balance between length and scope. Nor are we trying to write-out episodes of the history of the discipline. More or less everything covered in past editions remains here. We have however – with the benefit of hindsight – shifted the emphases somewhat. As with previous editions, the specific trajectories of significant subdisciplines, such as population, development, medical/health, or political geography (to name but four) are neglected here – on account of our word limit; such reviews appear annually in the journal *Progress in Human Geography* and are a feature of Gaile and Willmott's (1989, 2004) collections. However, we are aware that from such subdisciplinary vantage points, the shifts that *Geography and Geographers* outlines would be illuminated in different ways. In part, this is because the respective subdisciplines have been influenced by changes in the subjects with which they (sometimes uneasily) share affiliation; development geography with development studies and branches of area studies, for example. In part, it is because paradigmatic change in human geography has always been profoundly uneven, with individuals in some places, departments and fields of study setting agendas that only later came to have much wider influence across a heterodox discipline.

The addition of new material on recent trends in the discipline is complemented here by wider discussions of the contexts within which earlier disciplinary debates and developments occurred. The last decade has seen an increase in the amount of research and writing on the history of human geography since 1945 – especially on the decades prior to 1980. This has included not only retrospectives written by those involved and other biographical works, but also by scholarly interpretations of some of the major events and personalities produced by those interested in applying the perspectives and methods of the history of science to geography. As in the previous editions, however, the book depends almost entirely for its source material on the published record, on 'the written statements of others' as it was described in the original Preface (on which approach see also Keylock, 2003, and Lorimer, 2003). But that published record now includes a substantial volume of historical and interpretative material, which means that this current edition is less a history created from the contemporary debates and substantive contributions and more reliant – particularly in the earlier chapters – on biographical and autobiographical recollections of the conditions of their production.

The growth of historiographical work on recent human geography is in itself indicative of the importance that many geographers give to an appreciation of their discipline's past. Some authors contend that this is unnecessary: Barnett (1995, p. 417), for example, has 'doubts about the value of expending energy studying the history of geography as a means of throwing light on the state of the discipline today' (see also Thrift, 2002). We are not so sure. Appreciation of a discipline's history – of where it has come from, and why it has its current contemporary form – is not only valuable when, as Barnett accepts, it is deployed in a 'political' context to sustain its professional identity, but also for an understanding of the discipline's present condition. To paraphrase Marx, 'geographers have made – and continue to make – geography, but not in conditions of their own choosing'. Geographers are socialised into intellectual communities with accepted bodies of knowledge and practices, and their orientations are constrained, though not determined, by the nature of the communities they join – as argued by one of the theories (of structuration . . .) which a number of them have adopted to guide their work. Furthermore, the practice of geography itself illustrates one of the discipline's major contributions to the deployment of Marx's dictum: the 'conditions not of their own choosing' within which geographers operate are themselves spatially variable. Studying the history of geography – especially the recent history – is valuable not only for the light that it sheds on what geographers are and do today, but also for the case study it provides of a wider concern – the geography of knowledge production and the importance of context therein.

Finally, what of the 'Anglo-American' in the subtitle? In recent years, the Anglo-Americanness of human geography has come under increased scrutiny. In the first place, several geographers have pointed to the continued (indeed in some ways increasing) relative dominance of English language geography vis-à-vis geography written in other languages (Desbiens, 2002; Garcia-Ramon, 2003; Gutiérrez and López-Nieva, 2001; Minca 2000; Short *et al.*, 2001; see also Rodríguez-Pose, 2004). Thus Vaiou (2003, p. 136), writing from Greece, notes that: 'the power-knowledge system, in which Anglophone scholarship can posit itself as international and even universal has not (yet?) totally subsumed local initiatives'.

Others have questioned what is excluded and included when the term Anglo-American is used (Samers and Sidaway, 2000; Minca, 2003). Not only are the USA and the UK (and geographers working in these countries) far from being exclusively 'Anglo', but geography produced there contains many 'foreign' influences – as will become clear in the chapters that follow. The way that other predominantly anglophone contexts and geographers (for example, Australia, South Africa and New Zealand) are marginalised has also been critiqued (Berg and Kearns, 1998). The fact that one of the co-authors of this new edition is based in neither the UK nor North America, but in Singapore (although he was educated in the UK and has held academic posts there), and that some of his own geographical research has been conducted in Portuguese- and Spanish-speaking milieux has sharpened our sense of this (as Gregory, 1994, has also written regarding the impact of his move from the UK to Canada on his 'positionality'); the other author spent the first eleven years of his career in Australia and New Zealand, to which he has frequently returned. Nevertheless, we have retained the subtitle 'Anglo-American Human Geography since 1945'. But the book should be read with the caveat that it remains a survey of key debates within English language human geography in the past sixty years, with a primary focus on the UK and anglophone North America. Even within this particular geographical tradition, we cannot adequately cover everything, however. More than twenty years ago, in a debate on the place of Marxism within radical geography (one of our concerns in Chapter 6), Walmsley and Sorenson (1980, p. 137) reminded readers of the UK-based journal *Area* that: 'To speak of a "Marxist approach" [in geography] is therefore to distil a highest common factor from a variety of writings.' Our distillations here mean that this is not the story of anglophone human geography (which is itself is certainly not the story of human geography tout court over the past sixty-odd years). To pretend otherwise would take us into the realm of the mythological map described by the Argentinian Jose Luis Borges in one of his extraordinary fables. In Borges's story, the cartographers of an imaginary empire draw up a map so detailed that it ends up exactly covering the territory so that the 'real' territory underneath the map is obscured. The people of this empire came to relate more closely to this map than they did to the original territory underneath (they live, work and play on it, etc.) When eventually the map frays and disintegrates, the people are unnerved and nostalgic for the map they once knew and lived upon, feeling that they have lost something of their reality.

Our account of *Geography and Geographers* makes no such claims to be a replacement for an encounter with the original writings that are précised here. Moreover, readers should not be unnerved if they find – as they venture out into the broader fields of human geography – that some things are not quite as neat or clear as they might have been led to believe from this text alone.

(Johnston and Sidaway, 2004, pp. x–xii)

This conundrum crops up in the prefaces or introductions to many other textbooks or edited volumes on geographic thought. Most recently, the introduction to an edited *Handbook of Human Geography* (one of the multiplying quantity of such multi-author compilations) invites readers 'to engage with the current structure and contents of the text through critical reflection on extant closures and absences, distortions and reflection on the ways in which these limitations might be addressed' (R. Lee *et al.*, 2014, p. ix). Such invitations perform some of the function that warning readers of incompleteness and compression does in prior textbooks, such as James Bird's (1993, p. viii) advisory that 'Where this guide tries to summarise the arguments of whole books and papers, there is a danger that compression will have telescoped stages of those arguments'. An introduction to *Philosophy and Human Geography* (Johnston, 1983a), which one of us wrote as a companion to the second edition of *Geography and Geographers*, similarly begins with a declaration that 'All that is attempted is an introduction to the nature of the philosophies which human geographers are now exploring; any student wishing to join the quest will need to read a great deal more' (p. vii). The challenge multiplies along with the caveats that preface other recent textbooks. Thus, Nayak and Jeffrey (2011, p. xv) claim to be providing 'an aperture with no attempt at closure', and Cresswell (2013, p. 3) wants to:

> make no claims to completeness. Geographers, like practitioners of many other disciplines, are constantly arguing about ideas. Often it is the people who are supposed to be in agreement that are doing the arguing. We are used to the idea of advocates of competing ideas clashing with each other. In these arguments large numbers of people are lumped together as 'positivists' or 'Marxists' for instance. But if we look closely we find that these groups are constantly arguing with each other too, over what it means to be a positivist or a Marxist. A book like this cannot hope to recount each and every one of those arguments. Such a book would be an encyclopedia of many volumes . . . This is a road map and there are many small towns and hamlets and even some major cities that these roads do not connect. You will have to go off road occasionally to find them.

In providing our own fairly fine-grained 'road map', we also realise that at times *Geography and Geographers* risks becoming a crammed and wearisome atlas, especially for those 'off-roaders'. Some readers might prefer either a detailed map of their particular area of interests or a simpler sketch of the whole of disciplinary territory. For those who do want a (portable and printable) disciplinary atlas, however, we do our best to provide a serviceable one here: within the limits afforded by a single volume. However, the sheer impossibility of doing justice to the entire substance of anglophone human geography looms even larger more than a decade on from the last edition of our book. Many of the debates and changes that we flagged in that 2004 Preface have continued to rage. New historical work on the discipline has appeared, opening up the range of and contexts to what counts as geographical scholarship. When the first edition was published in 1979, the breadth and depth of work on the history of ideas and debate in human geography was much smaller. The Preface to an earlier survey, first published in the UK in 1961, but later republished in Chicago (1962) and subsequently reprinted several times in the 1960s, could begin with a note that 'The history of geography is not an over-tilled field . . .' (Freeman, 1962, p. 9). That relative paucity of such scholarship meant that *Geography and Geographers* found a ready market in 1979. The author of what would later become a key work on *The Geographical Tradition* (Livingstone, 1992) has noted how the attempt to sample and describe the evolving field of human geography registered in the first edition of this book was a source of 'consolation' to someone interested in geography's history:

> When *Geography and geographers* first appeared in 1979, I was still a graduate student working in a rather isolated situation on the history of geographical knowledge in nineteenth century America. I well remember the sense of excitement I felt when the book came out that matters of history and theory were beginning to be taken with much greater seriousness in the discipline. *Geography and geographers*, together with one or two other recently published items, were opening the door to a greater critical spirit, to a desire to connect the subject with wider currents of intellectual moment, and to a concern to locate human geography's coordinates in a broader scholarly landscape.
>
>   (p. 45)

Fifteen years later, he added:

> The appearance of Ron Johnston's *Geography and geographers* in 1979 did an enormous service in recovering and imposing coherence on human geography's collective memory, and thus its identity, since 1945. It had a major role in reawakening the subject's historical consciousness and injecting new life into the study of its narrative.
>
> (Livingstone, 2007, p. 43)

The task of recovering and imposing that coherence has increased exponentially over the thirty-seven years since the appearance of the first edition, reflecting not just the expansions in human geographical scholarship and the ever-broadening canvass of geographical research, but also a growing volume of biographical and autobiographical works, providing new insights to specific events in disciplinary history and much changed (and still changing) economic, political and institutional contexts in which those practices were situated. When the lecture course on which the first edition was based was first given in 1975, subsequent twists in 'the intellectual diversity of the discipline, the contingencies of its historical trajectories, and the stresses and strains of its institutional architectures' (Livingstone, 2007, p. 43) could hardly have been imagined.

Reflecting on those previous editions, one of us (Johnston, 2007a, p. 48) indicated that not only was he reluctant to produce the sixth but also that having done so he was even more convinced that there should not be a seventh – not least because commentators had indicated that after thirty years there needed to be 'a profound recasting of the venture' and 'a more compelling and expressly geographical argument about human geography' (Johnston, 2007a, p. 50). He agreed, being not only 'very unsure whether such a major project is feasible' but also even more certain that, as a senior, he no longer had a broad and deep enough appreciation of the multifaceted discipline to undertake it: it was 'time for somebody else to take up the challenge' (p. 50). Yet here he is, having been convinced by James that a further edition was both desirable and feasible, with some restructuring. *Geography and Geographers* is, after all, just two authors' take on seventy years of disciplinary histories: others take different positions – and our appreciation of history advances by bringing them into contact: out of different theses a new synthesis might emerge, which in turn will stimulate further anti-theses. And so the book's structure has remained largely unchanged, but what became clear in its revision was that one of us is much more comfortable revisiting the earlier decades and the continuation of certain types of work in the light of more recent publications while the other has kept pace with changes in the last decade. Thus the overarching message – as transmitted in the final chapter – endures (though not unaltered in the light of events), but the material on which it is based continues to broaden and deepen. However pessimistic we may be about the state of the world and the challenges universities face as contexts for geographical scholarship since the sixth edition was written in 2003, we remain enthusiastic about and impressed by human geography's continued flowering and vitality. And we are glad that others join us in arguing that

> teaching the history of geography matters to the disciplinary present – not because it offers a neat and Whiggish explanation of how things came to be as they are now, but because it functions as a whetstone against which students' critical thinking is honed. At its most effective, such teaching offers insight into the complex making and breaking of ideas, into the complex social circumstances which govern the production, circulation, and reception of knowledge, and into the ways in which questions of race, gender, and class permeate and inform all we do. It is precisely the messy, nonlinear and conflicted nature of geography's history that provides its most valuable and transferable lessons.
>
> (Keighren *et al.*, 2015, p. 3)

Today, there are several new books that tackle the same or similar ground as *Geography and Geographers* (e.g. Couper, 2015; Cox, 2014; Cresswell, 2013; Nayak and Jeffrey, 2011; Holt-Jensen, 2009, a fourth edition of another that is translated from Norwegian) adding to the crop that appeared (sometimes as new editions) in the 1990s or early 2000s (Bird, 1993; Cloke *et al.*, 1991; Martin, 2005; Peet, 1998; T. Unwin, 1992 – we have been unable to incorporate the many insights offered by Geoffrey Martin's (2015) magnum opus, published at about the time we submitted this manuscript, but we are flattered by his chosen title), and others that add to our appreciation of the discipline's recent history (such as DeVivo, 2015, and Pacione, 2014), not to mention the increasing number of departmental histories, many either privately published or available on the internet (http://old.geog.psu.edu/hog/dept_ histories.html provides a listing of those available in North America). Recently, too, the histories of key moments and trends in human geography, especially the quantitative revolution and advent of spatial science (the focus of Chapters 3 and 4) are subjects of fresh scholarship and debate regarding their causes (e.g. Barnes and Farish, 2006; Barnes, 2008a, 2008b; Johnston *et al.*, 2008). More recently, there has also been deeper historical scrutiny of the emergence of radical and critical geography (our focus in Chapter 6 as well as some of the discussion of applied geography in Chapter 9) from the 1960s and 'until the mid-1980s, when key decisions were taken that moved radical/critical geography into the mainstream of the discipline' (Peake and Sheppard, 2014, p. 305). Moreover, among the implications of feminism in geography (our focus in Chapter 8) have been significant challenges to the way the history of the subject is recounted, who is included in such accounts, and what and who has tended to be systematically neglected. In the words of a key book that challenges exclusions, by focusing not only on what are taken to be the major debates and figures, but also 'minor' figures, including those who may not have published much, but whose teaching and impacts have to be recovered from archives such as university records, oral histories or even obituaries:

> Those who held office in geographical societies, those who taught at universities and those who wrote influential texts will be represented here, but so too will the work of school teachers, teacher trainers, non-academic authors and those who might be called public servants, both 'major' and 'minor' figures.
>
> (Maddrell, 2009, p. 12)

All this needs to be reckoned with. And it challenges us (and we trust our readers) to reconsider the histories that *Geography and Geographers* recounts. In addition, some new (or rather renewed) cross-cutting fields, around nature and the environment for example, have returned to a more central position in disciplinary debates. Yet we cannot refuse the challenge of producing an updated survey. The caveat is that we recognise more than ever that this cannot be comprehensive. However, we are confident in claiming that the basic structure we decided on for the sixth edition still serves useful purposes. Chief among these is encouraging readers to find paths into and through the realms of human geography; to offer them a sense of how the parts, which they might otherwise not see far beyond, might be fitted together and relate to those of the past.

Moreover, reviews of the sixth edition were encouraging in seeing value in the venture. Rogers (2006, p. 546) notes how 'Like the discipline it seeks to understand, the book shows no sign of slowing down or losing direction', while also complaining that 'By being so thorough and authoritative, there is a sense in which the book crowds out alternative readings, or at least discourages curiosity' (p. 547). That is not our intention. We don't see a contradiction between the arguments that we advance here and other ways into the discipline. Or, to be more exact, where there might be, we encourage readers to mix and match and pick up other texts and the original sources. In similar terms, Mark Boyle (2005, p. 161) declares that:

> *Geography and Geographers* has undoubtedly been the formative text for many teachers currently charged with the responsibility of delivering courses on the history and philosophy of the

discipline. Nevertheless, the addition of a whole series of new textbooks has opened up fresh opportunities for those keen to deliver material in more innovative ways.

Those opportunities, he notes, might include:

1   Contextualist approaches, more explicitly locating geography in wider social, political and economic backdrops. *Geography and Geographers* has always done some of that: in particular, the account of the structure of disciplines and the regulation of universities has been a key to these material backdrops. Blending this with sufficient account of wider social shifts remains a challenge, given the primary focus here on how these were mediated in debates in the discipline.

2   Thematic approaches, which introduce how geographic thought impacts on and is read through an object, such as globalization, or a region, such as Europe, or a theme, such as nature–society or scale, that transcends a single subdiscipline and has been the subject of changing approaches (such as Castree, 2005; Herod, 2009; Hinchliffe, 2007; Zimmerer, 2010). Such texts and accounts are not new. But *Geography and Geographers* works through a coarser frame of reference, with the account organised according to the macro-framework of approach, rather than the objects of the analysis per se.

3   Subdisciplinary approaches – accounts of, say, political, health, economic, cultural or environmental geography as mirrors to the wider course of the discipline. There are very many of these subdivisions, and they offer distinctive vantage points both on the wider trajectory of geography, and on overlaps with and differences from other disciplines. For example, investigation of these has formed the subject of a book-length study of the relationship between history and historical geography, for example, in which Baker (2003, p. 206) documents how 'The academic battlefields of geography and history are littered with aphorisms about each other, as well as about their "mysterious" offspring, historical geography'.

4   Speculating on the future rather than the past. In practice these date quickly (one of us edited one, Johnston, 1985a) but nevertheless become useful mirrors to the issues at a particular moment (as in Dixon and Jones, 2004).

5   Focus on the wider performance of geography, rather than its canon of writings or concepts (on which a conference was held in Oxford in June 2012 which contributes to a special issue of the *Journal of Historical Geography*: R. C. Powell, 2015; see also the commentaries introduced by Keighren *et al.*, 2012). To some extent, Maddrell's (2009) important account, as noted earlier, moves in this direction. Others, notably Philo (1998), have pushed this in other directions, charting student writings in a geographical magazine as a mirror to the consumption and reproduction of the discipline. And for Sullivan (2011, pp. 2–3) geography is:

> fraught with a perpetual identity crisis and thus overtly vulnerable to pre-emptive acts of performance as geographers attempt to stake out geography as being determined by this or that methodology or philosophical perspective. The history of geography can be viewed as one long performative struggle, as ideographic and nomographic, qualitative and quantitative, cultural and physical agonists feint and parry their way to some imagined dominance of the profession, suing speech acts and other assorted performative acts (articles, presentations, as well as hiring and tenure practices) as their weapons of choice.

6.   Biography as an approach (such as Hubbard *et al.*, 2004; and Lorimer and Withers 2007, 2014 – the latter is the 33rd volume in a superb series of *Biobibliographical Studies* of geographers launched by the International Geographical Union's Commission on the History of Geographical Thought), sometimes supplemented by a focus on key texts (Hubbard *et al.*, 2008).

These do not exhaust the possibilities. Matthew Gerike's (2012) Ph.D. thesis on 'Explorations in Historiographies of Geographical Knowledges' draws on biography and autobiography, but also on correspondence (both published and unpublished), symposia and content analysis (of textbooks and journals). Greater attention to the historical geography of geography has emerged, along with further perspectives on international connections and flows (Best, 2014), though Seemann (2015, p. 16) still laments the way that 'The teaching of geography wrongly emphasises time over space.' Similarly, in reviewing the range of ways that geographical theory can be represented, Cresswell (2013, p. 12) suggests that:

> It could be written through places where theories were developed: German geography in the late nineteenth and early twentieth centuries, radical theories in Clark University, Massachusetts, spatial science at the University of Washington, or even the new cultural geography in Lampeter, Wales.

In similar spirit, Trevor Barnes (2014, p. 204) wants to:

> stress the importance of the geography of geography's own knowledge, which is bound inextricably to the history of the discipline. Geography's geography is not mere background colour, a gazetteer-like list of places, but must be conceptualized, theoretically connecting the history of the discipline's ideas with where they were produced and travelled.

Barnes joins those who are critical of this book's chronological approach:

> If their goal is to know what geographers 'are and do' [citing from p. xi of the 2004 edition of *Geography and Geographers*] – that is, their work now – why do we need to study history? Wouldn't it be better to learn about geographers as they are currently, focusing on what they do in the present?
>
> (p. 205)

Furthermore, others who also approach geographic thought through narrating its historical shifts (or, if preferred, $\Delta t_1 - t_2$) have classified the structure and trajectory of geographical thought in different ways from that developed in *Geography and Geographers*, such as the 'four traditions' ('spatial', 'area studies', 'man-land' and 'earth science') identified by Pattison (1964) and subsequently revised by J. Lewis Robinson (1976). We touch upon and briefly return to some of these alternative ways of studying the history of geography in the opening and concluding chapters. At this point, however, we should reiterate that the focus of *Geography and Geographers* remains (as it has been since 1979) on the overall corpus of published writings in the field, and on claims and debates about how geography should be practised rather than the subject matter of its practices. The book is broadly chronological, though with inevitable jumps, discontinuities and interconnections. Moreover reviewers have pushed us to reconsider and supplement the meta-framework inherited from earlier editions – the account of paradigms and disciplinary structures/change that forms a large part of Chapter 1 and which we also re-evaluate in the last chapter.

Our overall challenge is both enhanced and enabled by the continued vitality of human geography in the last few decades. While the discipline has faced significant challenges regarding its identity (in the UK, for example, increasingly ending up being merged with other disciplines into university departments or Schools of the Environment, or Environmental and Earth Sciences, and in some countries, such as Australia, almost disappearing as a single subject discipline), the number of undergraduate students taking degrees in the subject has grown (in tandem with wider growth in student numbers). The volume and range of specialist and generalist journals devoted to geography has likewise increased (new, entirely online ones, have appeared, such as *ACME: An International E-Journal for Critical Geographies* and *Geography Compass*), as have the number of text-books, readers, encyclopaedias and companions offering guides to the field (see the accounts by

Demeritt, 2008, and Johnston, 2010a). The vast archive of 'Geographers on Film', compiled between 1970 and 2004 by Maynard W. Dow, is also being digitised and made more readily available; it is likely to be a valuable research and teaching resource (G. Martin, 2013). We believe, however, that there is still considerable value in this single-volume survey of the evolution of human geography, paying attention to where it has come from. Such texts can be influential (Johnston, 2006) and will never please all readers. But to abandon an attempt to survey human geography through the means and methods of *Geography and Geographers* seems to us to miss an opportunity of enriching debate, including or inculcating new voices – hence this seventh edition. In short, we judge that the original rationale (as well as the limits) of such a survey are still valid, approaching four decades on from the first edition. These were described in the Preface to that first edition (that full Preface and all subsequent Prefaces also appear in the fifth edition) as follows:

Most students reading for a degree or similar qualification are required to pay some attention to their chosen discipline's academic history. The history presented to them commonly ends some time before the present. This has advantages for the historian, because the past is often better interpreted from the detachment of a little distance and there is less chance of hurting scholars still alive. But there is a major disadvantage for the students. In virtually every other component of their courses they will be dealing with the discipline's current literature, and so if the history ends some decades ago, they are presented with the contemporary substance but not with the contemporary framework, except where this is very clearly derivative of the historical context.

This state of affairs is unfortunate. Students need a conspectus of the current practice in their chosen discipline and should encounter a relevant overview which describes, and perhaps explains too, what scholars believe the philosophy and methodology of the discipline presently are and should be. Such an overview will allow the substantive courses comprising the rest of the degree to be placed in context and appreciated as examples of the disciplinary belief system as well as ends in themselves.

As a discipline expands, so the need for a course in its 'contemporary history' will grow too. In the last few decades, for example, most disciplines, and certainly human geography, have expanded greatly, with expansion measured by the number of active participants and their volume of published work. And the more active members there are, almost certainly the greater the variety of work undertaken, making it difficult for individual students to provide their own conspectus of the discipline from their own reading. Hence the need for a 'contemporary history' course at the present time.

The present book is the outcome of teaching such a course for several years (in the 1970s–1980s and then, briefly, the late 1990s), and is offered as a guide for others, both teachers and students. As with all texts, it has many idiosyncrasies. The course on which it is based is taught to final-year students reading for honours degrees in geography, and is probably best used by people at that level since it assumes familiarity with the concepts and language of human geography. Further, my view is that students probably benefit most from a framework after experiencing some of its contents; this book provides the matrix for organizing the individual parts rather than a series of slots into which parts can be placed later, which would be the case if the course were taught early in the degree.

A series of constraints circumscribes the contents and approach of this contemporary history. First, it deals only with human geography, for several reasons. The most important is that I find the links between physical and human geography tenuous, as those disciplines are currently practised. The major link between them is a sharing of techniques and research procedures, but these are shared with other disciplines too, and are insufficient foundation for a unified discipline. (What price a department of factorial ecology?) Further, my own competence, work and interests lie wholly within human geography and although I have been trained with and by and have worked among physical geographers, and have obtained stimuli from this, I am incompetent to write about physical geography. And finally, much of the human geography discussed here is of North American origin, and many geographers there, especially in the USA, encounter no physical geography as it is understood in UK universities. To a considerable extent, therefore, human and physical are separate, if not independent, disciplines. Throughout the book, I use the terms 'geography' and 'human geography' interchangeably.

The second constraint is cultural; the subject matter of the book is Anglo-American human geography. Most of the work discussed emanates from either the USA or the UK: there are some contributions from workers in Australia, Canada and New Zealand, but the efforts of geographers in the rest of the world are largely ignored. (A partial exception is Sweden, which has major academic links with the Anglo-American tradition; much Swedish geographical work is published in English.) Such academic parochialism in part reflects personal linguistic deficiencies, but it is not entirely an idiosyncratic decision. Contacts between Anglo-American human geography, on one hand and, say, that of France and of Germany, on the other, have been few in recent decades, so to concentrate on the former is not to commit a major error in separating a part from an integrated whole.

The final constraint is temporal, for the book is concerned with Anglo-American human geography during the decades since the Second World War only. Again, this is in part a reflection of personal competence, for I have been personally involved in academic geography for the last twenty years. But the Second World War was a major watershed in so many aspects of history, not least academic practice, and much of the methodology and philosophy currently taught in human geography has been initiated since then.

(Johnston, 1979, pp. ix–xi)

We close this Preface with another extract from the one written thirty-six years ago. In acknowledgement of how supportive personal and professional 'contexts' had aided its production, the Preface of the first edition of *Geography and Geographers* had the following to say: 'I make it clear in Chapter 1 that the progress of any individual's academic career depends considerably on the actions (and sometimes inactions) of others. No academic is an island' (Johnston 1979, p. xii). Very many have assisted Ron Johnston over the years and are named in previous editions. Ron has continued to benefit greatly from interactions with his colleagues at Bristol, with Derek Gregory and the co-editors of the *Dictionary of Human Geography*, his many co-editors of *Environment and Planning A* and *Progress in Human Geography* over a quarter-century of ever-increasing workloads, and Charles Pattie, Dave Rossiter, Dave Cutts, Kelvyn Jones, Rich Harris, David Manley, Mike Poulsen, Jim Forrest, the late Andy Trlin and others who have kept his nose to the ever-enjoyable research grindstone. James D. Sidaway would also like to thank Robina Mohammad for her encouragement and feedback arising from using the sixth edition to teach an honours-year module on geographic thought at the University of Strathclyde. He also thanks those he has worked with over the last two decades in the UK and Amsterdam as well as his co-editors at *Political Geography*, and once again thanks colleagues and students at the Department of Geography of the National University of Singapore, a stimulating and relatively diverse intellectual environment (see Olds, 2001) from which to reflect on the dynamics of Anglo-American human geography since 1945. Both of us are also grateful to the publishers of the past editions (Arnold) for commissioning a new edition and providing us with anonymous reviews by other academics who had used the sixth edition in their teaching. At Routledge (who have purchased the rights to publish *Geography and Geographers* from Arnold), we also thank Sarah Gilkes and Andrew Mould for their patience while we finalised the seventh edition, emailing numerous drafts between the port cities of Bristol and Singapore.

Ron Johnston and James D. Sidaway
Bristol and Singapore

Acknowledgements

The authors and publisher are grateful to the following for permission to reproduce figures included in this and previous editions of the book:

Allan Pred and the Department of Social and Economic Geography, University of Lund, for Figure 4.5.
Association of American Geographers for Figures 3.2 and 3.3.
Barbara Kennedy and Prentice-Hall for Figures 4.2 and 4.3.
Chris Philo for Figure 10.3.
David Harvey for Figures 3.1 and 7.1.
Derek Gregory for Figures 6.1 and 6.3.
Geraldine Pratt for Figure 8.1.
Olavi Gräno for Figure 10.1.
Oxford University Press and Taylor & Francis Group Ltd for Figure 4.6.
Peter Haggett for Figure 4.1.
Peters Edition Limited, London, for Figure 7.2.
Reg Golledge and Guilford Publications for Figure 4.7.
Richard Peet for Figure 10.2.
Royal Statistical Society for Figure 3.4.
Taylor & Francis Group Ltd for Figure 6.2.

Every effort has been made to contact copyright holders for their permission to reprint material in this book. The publishers would be grateful to hear from any copyright holder who is not hereby acknowledged and will undertake to rectify any errors or omissions in future editions of this book.

Chapter 1

The nature of an academic discipline

[M]ost effective academic communities are not that much larger than most peasant villages and just about as ingrown . . .

(Clifford Geertz, 1983, p. 157)

The history of geography involves many stories that start at different times in different places: they slowly become intertwined, and their narrative threads can be unpicked and rewoven into many different designs.

(Derek Gregory and Noel Castree, 2012, p. xxv)

[Th]e distinctive cultures within academic communities . . . refer to sets of taken-for-granted values, attitudes and ways of behaving, which are articulated through and reinforced by recurrent practices among a group of people in a given context.

(Tony Becher and Paul Trowler, 2001, p. 23)

[Geographers] appear to live in an intellectual world characterized by groups of people plowing their own theoretical furrows, with little outright objection to others doing their own thing.

(Tim Cresswell, 2013, p. 196)

This book is a study of part of an academic discipline, particularly of changes in its approaches and content. Approaches and content cannot be appreciated fully without understanding context, however, which this chapter provides. To study an academic discipline is to study a miniature society, which has a stratification system, power structures, a set of rewards and sanctions, and a series of bureaucracies, not to mention occasional interpersonal conflicts (some academic, others not). An outsider may perceive academic work as objective, but many subjective decisions must be taken: what to study and how; whether to publish the results; where to publish them and in what form; what to teach; whether to question the work of others publicly, and so on.

Studying an academic discipline involves studying a society within a society; both set the constraints to individual and group activity. Two questions are focused on here: how is academic life organised; and how does that academic work, basically its research, proceed? Use of the term 'society within a society' strongly implies that academic life does not proceed independently in its own closed system but rather is open to the influences and commands of the encompassing wider society. A third necessary question, therefore, is: what is the nature of the society which provides the environment for the academic discipline being studied and how do the two interact? Answering these three questions is set within a framework of studying the ways in which academic life is institutionalised. These are the basic material frameworks within which disciplines evolve. Much of this is not particular to human geography, although its relative novelty as a discipline (a theme considered in more detail in Chapter 2) and relatively small size in terms of overall numbers of students and teaching faculty (especially when compared with disciplines like English literature, history or physics, for example) does reflect back on the structure and perhaps reinforces the sense of community/networks. We therefore start with some details about how these, and attendant hierarchies, are configured in all academic life.

Academic life: the occupational structure

Pursuit of an academic discipline in modern society is part of a career, undertaken for financial and other gains; most of its practitioners see their career as a profession, complete with entry rules and behavioural norms. In the initial development of almost all current academic disciplines, some of the innovators were amateurs, perhaps financing their activities from individual wealth. There are virtually no such amateurs now; very few of the research publications in human geography are by other than either a professional academic trained in the discipline or a member of a related discipline with interests in some aspects of geography – indeed, there has been a recent trend for scientists in a number of other disciplines to become increasingly interested in research topics that many professional geographers tend to consider 'their own' (see, for example, Bettencourt and West, 2010; Clauset *et al.*, 2009). The profession is not geography, however. Rather, geography is the discipline professed by individuals who probably state their occupation as university teacher, professor or some similar term. Indeed, the great majority of academic geographers are teachers in universities or comparable institutions of higher education; 'university' is used here as a generic term. Academic geographers are distinguished from other professional geographers (many of whom are also teachers) by their commitment to all three of the basic canons of a university: to propagate, preserve and advance knowledge. The advancement of knowledge – the conduct of fundamental research and the publication of its original findings – identifies an academic discipline; the nature of its teaching reflects the nature of its research.

The academic career structure

University academic staff members, termed faculty members in North America – hereafter termed academics here – follow their chosen occupation within a well-defined career structure, of which two variants are relevant to this discussion: the British model and the North American model (Figure 1.1), although they have tended to converge in recent years (so the title associate professor is increasingly tending to replace reader and senior lecturer grades in the UK). Entry to both is by the same route. With rare exceptions, the individual must have been a successful student as an undergraduate and, more especially, as a postgraduate. The latter involves pursuing original research, guided by one or more supervisors (in North America, an advisory committee with a chair plays this role) who are expert in the relevant specialised field. The research results are almost invariably presented as a thesis for a research degree (usually the Ph.D.), which is examined by relevant experts; possession of a Ph.D. is now an almost obligatory entrance ticket for both models. Along the way, publication of research findings in specialist journals is expected.

While pursuing the research degree, most postgraduate students obtain experience in teaching undergraduates. Some universities finance many research students through such teaching activities, notably those operating the North American model. The individual may then proceed to further research experience (such as a postdoctoral fellowship or as a research worker on a project directed by a more senior academic), or may gain appointment to a limited-tenure university teaching post, offering an 'apprenticeship' in the teaching aspects of the profession while either the research degree is completed or the individual's research expertise is consolidated.

Beyond these limited-tenure positions lie the permanent teaching posts, which is where the two models deviate. In the British model, the first level of the career structure is the lectureship. For the first years of that appointment (usually three) the lecturer is on probation: training

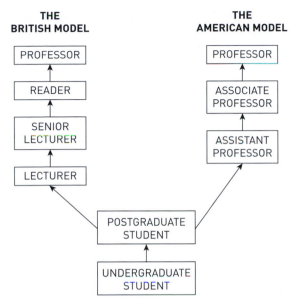

Figure 1.1 The academic career ladder. Note that whereas in the American model it is very unusual not to progress up through the stages in an orderly sequence, in the British model missing steps is quite common (i.e. some lecturers move direct to readerships, even to professorships, without first being senior lecturers – there are no senior lecturers in the 'Oxbridge' universities – and many professors are recruited from the senior lecturer rather than the reader grade). In recent years, many British universities have adopted the terminology of the American model.

in teaching and related activities is provided (with, increasingly, satisfactory performance on accredited programmes required), annual reports are made on progress as teacher, researcher and administrator, and advice is offered on adaptation to the demands of the profession. At the end of the probationary period, the appointment is either terminated or confirmed.

Appointment in this model is almost always to a department, which in many cases is named for the discipline which its staff members profess; others may be in a broader school or division – of social sciences, say. The prescribed duties involve undertaking research and such teaching and administration duties as directed by the head of the department, though frequently negotiated through a variety of committees. The lecturers are on a salary scale and receive an annual increment; accelerated promotion may be possible. In the UK there is an 'efficiency bar' after a certain number of years' service; 'crossing the bar' involves a promotion, and is determined by an assessment of the lecturer's research, teaching and administrative activities.

Beyond the lectureship are further grades into which the individual can be promoted. In the universities that were established before the 1990s, the first was the senior lectureship for which there is no allocation to departments but competition across the entire institution. Entry is based again on assessments of the lecturer's conduct in the three main areas of academic life, and in many universities these include evaluations of research performance and potential from outside experts. (The universities of Oxford and Cambridge are exceptions to this; they do not have a senior lecturer grade, and their salary scale for lecturers is much longer.) The next level is the readership, a position generally reserved for scholars with established research records. The criteria for promotion to it require excellence in research (as demonstrated by their publications and influence) alongside satisfactory performance (at least) on the other criteria. In many universities it is now possible for promotion directly from lectureship to readership. In the post-1990 universities (most of them originally established as polytechnics), the promotional grades were principal and senior lecturer, with readerships only very rarely created. In both, the North American category (see below) of associate professor has increasingly been adopted since the turn of the millennium.

The final grade (associated with academic departments as against administration of the university as a whole) is the professorship. Although superior in status to the others, this may not be a promotional category; until the last few decades most professors were appointed from open competition preceded by public advertisements. Initially, the posts of professor and head of department were synonymous, so a professor was appointed (from a field of applicants after public advertisement) as an administrative head to provide both managerial and academic leadership. With growing departmental sizes and restructurings, however, and the development of specialised subfields within disciplines, each requiring separate leadership, it has become common for departments to have several professors. In most, the headship is a position independent of the professoriate to which other grades may aspire, although many heads are also professors. The post-1990 universities had very few professorships while they were polytechnics, but have since converged with the older universities in terms of these structures. Finally, in the British model, promotion is possible to non-advertised, personal professorships. These positions became increasingly common in the 1990s onwards; in most cases, a personal professorship is awarded for excellence in research and scholarship.

The system under the North American model is simpler. There are three permanent staff (or 'faculty' as they are known there) categories: assistant professor, associate professor and professor. The first contains two subcategories – those staff on probation who do not have 'tenure' and those with security of tenure. Each category has its own salary range but no automatic annual increments. (In the USA, these scales vary from university to university; the UK has a national salary scale that applies in all universities with only slight local modifications, although individual universities have different scales above the agreed minimum for professors.) Salary adjustments result from personal bargaining based on academic activity in the three areas already listed, and salaries for the three categories may overlap within a department. Movement from one category to another is promotional, as recognition of academic excellence. Professorships are simply the highest promotional grade and do not carry obligatory major administrative tasks. Heads (more usually termed 'chairs') of departments in the American model are separately appointed or elected, usually for a limited period only, and they need not be professors (usually referred to as 'full professors'). Elsewhere in the world, the structures may be some combination of these models, invariably mediated through local norms. And, as we have mentioned, there has also been considerable hybridisation of the original templates in recent years, with a few universities in the UK and elsewhere now referring to those of their academics who would hitherto have been designated as lecturers, senior lecturers or readers as assistant or associate professors.

The nature of academic work

Academic work has three main components: research, teaching and administration (the latter is more often termed 'service' by USA-based academics). Entrants to the profession have little or no experience of academic administration and their teaching has probably been as assistants to others. Thus, it is very largely on proven and potential research ability that an aspirant's initial potential for an academic career is judged, especially in universities that present themselves as 'research driven' and where, on average, staff members spend less time on teaching than is the case in other institutions. Research ability can be partly equated with probable teaching and administrative competence, since all require some of the same personal qualities – enthusiasm, organisation, motivation, incisive thinking and ability to communicate orally and in writing: most universities provide appointees with training in teaching and administrative skills. Increasingly, in the British system such formal training has become compulsory and is completed (among many other tasks and much 'learning by doing' in the first few years of an academic appointment). To a considerable extent, however, academics gain their first position on faith in their potential (as demonstrated by their letters of application and curricula vitae, references, and their performance at formal and informal evaluation processes, such as interviews and seminars, plus their initial publications), hence the usual probationary periods before tenure is granted.

Once admitted to the profession, academics undertake all three types of work, so that promotional prospects can be more widely assessed. Teaching and administration have traditionally carried less weight with those responsible for promotions than research. This is partly because of perceived difficulties in assessing performance in the first two, and partly because of a general academic ethos which gives prime place to research activity in peer evaluation and wider recognition.

Although not necessarily progressive in terms of increasing level of difficulty, administrative tasks tend to be more complex and demanding of political and personal judgement and skills as the academic becomes more senior, although not necessarily less onerous in terms of time required. A person's ability to undertake a certain task can often only be fully assessed after appointment or promotion to the relevant position (this has been called the 'Peter principle': Peter and Hull, 1969), so decisions must frequently be based on perceived potential; although it is possible to point to somebody whose administrative skills are insubstantial, it is not always easy to assess who will be able to cope with the more demanding tasks.

It has long been argued that assessment of teaching ability is difficult (although clear inability is often very apparent). Student and peer evaluations are increasingly used, as is external scrutiny of students' work. But criteria for judging teaching performance at university level are ill-defined, and expectations of a lecturer, tutor or seminar leader often vary quite considerably within even a small group of students, as well as among external assessors. Thus the majority of teachers are usually accepted as competent if undistinguished. Indeed, the 'inspections' of teaching sessions – most of them lectures – undertaken during the 1995 quality assessment of teaching in geography in UK universities invariably rated the majority as 'satisfactory', with a substantial minority being deemed 'excellent' and very few – none in most cases – 'unsatisfactory' (Johnston, 1996a). In most universities now, promotion exercises require proven excellence in at least two of the three criteria generally considered – research, teaching and administration; some add a fourth criterion – service to the wider community.

Research performance is a major criterion for promotion, therefore, although a relatively undistinguished record in this area can be compensated by excellence elsewhere. Some argue that teaching competence is not only undervalued (despite the promotion of geographical pedagogy, notably through the *Journal of Geography in Higher Education*), but that as a consequence of the increasing emphasis on research and its financing in most universities, it has received less attention from the 1980s onwards (Jenkins, 1995; Sidaway, 1997). How is research ability judged? Details of how research is undertaken are considered in the next section; here the concern is with its assessment rather than with stimulus and substance.

Successful research involves making an original contribution to a field of knowledge. It may involve the collection, presentation and analysis of new information within an accepted framework; it may be the development of new ways of collecting, analysing and presenting material; it may comprise promoting a new way of ordering facts – a new theory or hypothesis, or simply new interpretations of and insights into existing material; or it may be some combination of all three (on the definitions of research and scholarship, see Collini, 2012). Its originality is judged through its acceptance by those of proven expertise in the particular field. The generally accepted validation procedure is publication, hence adages such as 'unpublished research isn't research' and 'publish or perish'.

The main publishing outlets for research findings in most disciplines are their scholarly journals, which operate fairly standard procedures for scrutinising submitted contributions. Some of the major journals are published by academic learned societies and others by commercial publishers (on the shifting relative status of these in geography, see Bosman, 2009), with many learned societies now contracting out the production and marketing of their journals to commercial publishers while retaining full editorial control (Luke, 2000; Johnston, 2003a). Manuscripts are submitted to the editor, who seeks the advice of qualified academics on the merits of the contribution; these referees will recommend either publication or rejection, or revision and resubmission. When accepted, a manuscript will enter the publication queue, which may be up to two years long, although increasingly journals are publishing accepted papers on their websites as soon as they have been accepted for publication.

Although widely accepted, this procedure is subjective because it is operated by human decision-makers. The opinions of both editor (on the manuscript and on the choice of referees) and referees may be biased or partial, so a paper can be rejected by one journal but accepted by another, even without alteration. (On the canons of editorship see, for example, Hart, 1990, and Taylor, 1990a; for a radical alternative, see Symanski and Picard, 1996.) Most disciplines have an informal prestige ranking of journals, and it is considered more desirable to publish in some rather than others, because of the stringency of their review procedures, their circulation, readership and hence visibility, and their generally-assumed status (Luke, 2000; Johnston, 2003a). In addition, over recent decades much effort has been put into the development of citation indices to rate journals, individual articles in journals (plus books) and authors. A corporation (Thompson Reuters) scans all the items referenced in a large selection of journals. Originally known as the Institute for Scientific Information (ISI), the index compiled by Thompson Reuters has since been joined by others: Scopus (owned by the publisher Reed Elsevier) and others whose datasets are free to access for all, such as Google Scholar and Publish or Perish (set up in 1999 by the then Melbourne based academic, Ann-Wil Harzing).

Collation of this material allows these organisations to provide data on the number of times each author is cited by others – taken as evidence of the impact and importance of her/his work; the number of times each item is cited – taken as evidence of its importance in advancing knowledge; and the number of times each journal is cited – taken as evidence of the importance and impact of the items it publishes. (Yeung, 2002, details how these citation counts and impact factors are calculated and Paasi, 2005, examines their influence on academics' publishing practices. Several geographers have deployed them to identify the most influential geographers according to their citation frequency – e.g. Bodman, 1991, 1992, 2010; Foster et al., 2007; Whitehand, 1985; and Belgian geographers have debated – in English – the impacts of these English-dominated citations metrics across the linguistic divides in Belgium: Derudder, 2011; Meeus et al., 2011; Schuermans et al., 2010.) Citation counts for individuals and the journals in which they publish are often taken into consideration by appointment and promotion committees. The issue of the standing of journals and the use of citation counts became crucial in the UK after the mid-1980s because of the use of publication data as a proxy indicator for the periodic national assessments, which not only rate every department but also determine its future funding (the higher rated received more money, see Johnston, 1993b). The relationship between citation numbers and quality and impact of scholarship is, however, rather complex and, as we will describe, contested.

Some research results are published in book form, rather than as journal articles. Most academic books in geography are texts, however, published by commercial companies whose main interest is marketability among the large student population. The textbook may be innovative in the way that it orders and presents material, and can be beneficial to its author's reputation (and bank balance), but it is not usually a vehicle for demonstrating research ability, although, as we detail below, textbooks can significantly shape fields of scholarship, by virtue of how they chose to present things to their student readers from whom future generations of potential researchers are drawn (on types of books published, see Johnston, 1986a, pp. 165–6). Many companies, including university presses, publish research monographs, however, presenting the results of major research projects or theoretical elaborations to relatively small academic markets. Their decisions on whether to publish are made on academic as well as commercial grounds, usually with the aid of academic referees, and their output is validated through the journal book-review columns. (Some concern has been expressed on the future of monograph publishing in geography, generating considerable debate: see Harvey, 2006 and Ward *et al.*, 2009. On the relative standing of books in discussions of a discipline's influential outputs, see Johnston, 2009a.)

Processes of promotion and appointment: patronage

Whatever the weighting given by appointment and promotion committees to the three main academic activities (others are pursued by some, such as consulting for outside bodies and contributions to the work of learned societies), how they make assessments is frequently contentious. One of the main sets of 'objective' information that can be presented to them is lists of publications, most of which have been validated by academic journals. But how is such information evaluated?

Two modes of assessment are widely used and relied upon: the written opinion of a third party (either a referee, selected by the applicant, or an assessor, selected by the appointment/promotion committee) and the interview, sometimes associated with presentation of a lecture or seminar. The British model places considerable weight on the former. Applicants must supply a list of referees who will provide opinions on their suitability for the post: applicants are likely to ask people, especially senior people they have worked with, and are believed to be favourably inclined towards them, to act in this capacity. As appointment committees tend to be influenced by such evaluations, especially when preparing a shortlist of candidates to be interviewed, the opinions of well-respected referees are often crucial. Thus, certain leaders in a subject often find it easier to get their candidates appointed to posts than do others: there is a considerable element of patronage involved in obtaining a university post, especially a first university post. Increasingly, too, it seems that appointment committees for junior posts are influenced by where candidates obtained their postgraduate training: attendance at a highly rated graduate school is considered an advantage.

Promotions under the British model are frequently strongly influenced by the head of the candidate's department, although in almost all places now she or he is expected to consult senior colleagues before making a recommendation. Reports must be made on each lecturer: annually during the probation period; at the confirmation stage after probation; on reaching the 'efficiency bar'; and for either accelerated promotion within the lecturer scale (or beyond it, to 'discretionary points'), or promotion to senior lecturer. This also contains a strong element of patronage, although constraints are built into the committee system to promote fairness for all, including the right either to present one's own case or to appeal against a decision. Many universities use outside referees and assessors when considering promotions to senior lecturer; promotions to readerships and professorships almost invariably involve such consultation.

Referees, assessors and interviews are also used in the appointment of professors. Referees nominated by the candidates provide a confidential report on their potential for the job in question. Assessors nominated by the university evaluate the list of candidates, suggest other worthy candidates who might be approached, and in many cases attend the formal interviews: their potential patronage power is great.

Procedures are slightly different under the American model. For appointments, more weight is usually given to an enlarged interview, with candidates giving seminars to the department and meeting with its existing members (such procedures are increasingly used in UK universities). The reference is still important, however, especially for aspirants to first teaching positions, and a letter from a respected leader in a field, especially one working in a prestigious graduate school, can be very influential in gaining an appointment for a former student (Morin and Rothenberg, 2011; Liu and Zhan, 2011). The individual candidate is more active in this system than in the British, however, perhaps canvassing for interviews during the annual conferences of scholarly societies, for example. For promotions and for salary rises there is considerable bargaining between individual, department chair and university administrators; external referees' opinions may be sought when tenure is being confirmed, or promotion to full professor proposed.

In all these procedures the applicants depend to a considerable extent on the opinions of senior academic colleagues in evaluating their potential, performance and prospects. Some opinions carry more weight than others, so it is important when developing a career to identify potentially influential individuals, to keep them informed of your work and to enlist their support for your advancement. Because of the lack of truly objective criteria for measuring research, teaching and administrative ability, such patronage is crucial.

Other rewards and the sources of status

The tangible rewards of an academic career are the salary and the relative autonomy of the work (which beyond teaching and administrative commitments enjoys fairly flexible hours), the relative security of employment plus the occasional extra earnings that are possible – for examining, writing, lecturing and consulting. These are experienced unevenly; by no means do all academics experience job security and for many the relative autonomy of academic work has been eroded by growing pressures and measures of performance (on which more follows). And in many places part-time academics and those on temporary contracts perform significant roles, but without the same levels of tangible rewards or security as others (Binnie *et al.*, 2005). Nonetheless, for many the hours are still relatively flexible, the possibilities for travel are often considerable and the constraints on when, where and how work is done are relatively few compared with most other professions.

Other, less tangible, rewards include involvement with the intellectual development of a discipline and its students, or the excitement of research and attraction of sustained immersion in intellectual ideas, practical challenges or both. There is also the charisma associated with recognised excellent teachers, and even more so for leading researchers, whose publications are widely read, whose invitations to give outside lectures are many, and whose opinions as examiners, referees and reviewers are widely canvassed; a few attain the status of 'public intellectual' with their views published widely through various media (Castree, 2006a; Ward, 2007). The conduct of research brings its own rewards, apart from the charisma; the satisfaction from having identified and solved a significant problem is often considerable, as is that from publishing a widely – and positively – cited article or book. And for a few, a successful research career is recognised by honours such as the annual prizes and medals awarded by learned societies such as the Association of American Geographers and the Royal Geographical Society (RGS) (with the Institute of British Geographers); international recognition such as the Prix Vautrin Lud, awarded annually by the Festival International de Géographie, the Anders Retzius Gold Medal of the Swedish Society for Anthropology and Geography, and the Laureat d'Honneur of the International Geographical Union; and election to national honorary learned societies such as the British Academy and Academy of the Social Sciences and, in the USA, the National Academy of Sciences and the American Academy of Arts and Sciences.

Academic life also offers a reward common to many social systems: power. Patronage is power, as is work as examiner, referee or reviewer. Others' careers are being affected, and exercise of this power can bring with it the loyalty and respect of those who benefit. Because the academic system is so dependent on individuals' opinions and some individuals' opinions are more valued than others' are, and because power is a 'commodity' widely desired in most societies, many academics seek positions of influence or service. These may include journal editorships or other similar broker roles that in turn shape the structure and trajectory of disciplines (on one aspect of power within the academy, see Hammett, 2012).

Among the most influential positions in an academic discipline are the administrative headships of departments. Their holders can instruct other staff members (usually after consultation) in their teaching and administrative duties; they are frequently used by staff and students as referees for job and other (such as research grant) applications, and their reports are crucial in the promotions procedure. The departmental organisation of universities is a bureaucratic device which makes for relative ease in administering what may be a very large institution. (It also tends to fossilise disciplinary boundaries, as discussed later.) Departmental heads not only have power over members of their own discipline's staff, they also participate in their university's administration, and they treat within its committee system for departmental resources. As in all bureaucracies, there is a tendency for the status and power of the various departmental heads to be a function of the size of their 'empires' (Tullock, 1976; large departments in UK universities certainly get the highest research ratings: see Johnston *et al.*, 1995). Heads of large departments, especially those which are also growing (growth being widely considered a 'good thing') and those attracting large numbers of students and research grants/contracts, are very often the most influential within a bureaucracy. Their heads have the incentive to build up their departments, which usually means increasing student numbers, since universities tend to allocate at least a major portion of their resources according to the numbers taught. This brings power and prestige, both within the university and beyond; it is an added reward for the academic bureaucrat, and the power over resources which it involves usually benefits the whole department. The successful department head may find that it leads to appointments to more senior positions within the university, such as a deanship, overseeing the work of a group of departments.

Finally, the academic bureaucrat can gain power beyond the home university through, for example, positions on committees. These may be concerned with the subject through professional bodies, with the allocation of research moneys by public or private foundations, including government-funded Research Councils, or with a wide range of public duties. Again, the status and power obtained can spill over to others, since patronage is important in all of those roles.

Disciplines and institutionalisation

By far the most important contemporary identifier for an academic discipline is that it is taught in universities – it has departments named for it, degrees are awarded that carry its name, and the teaching for those degrees is undertaken by individuals who have been trained in and identify with the discipline. This identifier is related to what Morrell (1990) associates with the 'professionalization of science': scholars in full-time paid positions; specialist qualifications; formal training programmes; specialised publications; disciplinary solidarity and self-consciousness; and professional reward systems. Golinski (1998) added a firm boundary between amateurs and professionals (see also Barton, 2003). Geography exists as an academic discipline because there are people who call themselves geographers, who have geography degrees and who teach on geography degree programmes. These programmes may not, however, always be in departments of geography. Especially in recent years, many departments have been merged into larger ones (most combining geography with earth or environmental sciences: see Holmes, 2002, on the situation in Australia, and for the UK see Wainwright *et al.*, 2014, and Hall *et al.*, 2015). This has raised questions about the future prospects for the discipline and what strategies will be required in the context of a rapidly changing university system, especially one under pressures to be relevant, accountable, and make broader social and economic impacts. Reflecting on this from the UK, Castree (2011, p. 298) reflects that:

> A century ago, a small number of university geographers in England and elsewhere worked hard to create a subject that is, today, far larger and more buoyant than they could possibly have imagined. Is this one of those critical moments when concerted and coordinated action of the sort we're nowadays unaccustomed to taking is warranted?

A university presence has long been an indicator of a discipline's existence: it is the key element in its institutionalisation (Johnston, 2003b). It indicates that the discipline meets a felt need, in particular that there are students who want a higher education in geography and scholars who wish to teach them. This, however, for geography, as with many other disciplines, is a relatively recent phenomenon: it dates only from the late nineteenth century (as illustrated by the chapters in Dunbar, 2002). That doesn't mean that geography didn't exist before the first courses were launched in universities: as some of the work briefly discussed in Chapter 2 indicates, it has a long and sometimes distinguished tradition as a form of enquiry and practice (Withers and Mayhew, 2002). But it was only from the 1870s on that it was formalised as an academic discipline, separate from the science and humanities study that characterised the traditional universities, significantly supported in the British case by the RGS (who were able to fund a number of geography lecturerships in UK universities; Powell, 2011). Moreover, as Chapter 2 describes in more detail, until expansion commenced in the early twentieth century, there was only a handful of geography departments in the UK.

Most disciplines are established in universities to teach a body of knowledge: they are identified by their subject matter. Their existence indicates a demand for instruction in that material, by people who are expert in it. Being an expert almost invariably means playing some part in producing and reproducing that body of knowledge, which involves undertaking original research in some aspects of the discipline and publishing the findings to inform other researchers and their students. Academic disciplines in universities involve teacher-researchers, people interested not only in the propagation of knowledge but also in its advancement; indeed, the production of new knowledge is considered an essential component of university teacher-researchers.

Although the creation of departments for training students by experts is a necessary component of the institutionalisation of a discipline, it is far from sufficient. In order to promote both their discipline and their own work, academics need to be able to interact with colleagues interested in similar issues, most of whom will work at other universities. Some organisational structure is needed to facilitate this. It is invariably done by academic or learned societies, bodies established to promote a discipline through a variety of means, of which the most important include the holding of meetings at which research issues and findings can be discussed and to facilitate the publication of research results, through books, journals and other media. All academic disciplines have such bodies, most of which are nationally organised; in some countries there may be several. Some of those societies may predate the creation of the academic discipline, having been inaugurated to promote its study before it achieved a separate identity and a body of professional scholars associated with it. Indeed, in many cases those societies were involved in promoting the discipline and lobbying for its presence in universities – perhaps even paying for the creation of teaching posts. Others were established after the formal institutionalisation of the discipline. Most are open for anybody to join if they are prepared to subscribe to the discipline's advancement, but in effect membership is very largely confined to university staff and (postgraduate) students – especially if there is no named profession linked to the academic discipline which retains the practitioners' affiliation (as with medicine and architecture, for example). Some societies have closed memberships, limited to those who meet certain criteria (such as a postgraduate degree in the discipline); others have several membership categories which separately distinguish the professional academics from those with a non-professional interest in the subject.

Those societies have been central to the development of disciplines as individual academics create their own research agenda. Their conferences and seminars, journals and other publications have been the main media until relatively recently (when commercial journal publication on a large scale took off), whereby knowledge has been disseminated and debates about its production and interpretation conducted. They also undertake other important functions, one of which is the promotion and defence of the discipline as a whole. Even when established in a country's universities, a discipline is not necessarily secure. There are struggles for wider recognition, for greater resources, for expansion into institutions where the discipline has yet to be established, and so forth. The learned societies play crucial roles in these tasks, acting as lobby groups for their discipline and seeking to ensure its health by influencing public and private decision-makers concerning its utility and value. And above them are even broader societies, representing a range of cognate disciplines, such as the Royal Society and the British Academy in the UK, and the National Academy of Sciences and the American Academy of Arts and Sciences in the USA; these play a wider role in advancing the interests of their subject groupings.

The institutionalisation of an academic discipline involves creating an identity, a subject with its own niche within the academic division of labour with which individuals are affiliated. In some cases, the discipline – like geography and history – has a name whose interpretation extends beyond academia, its vernacular definition (Johnston, 1986b). This may create some difficulties in establishing an academic identity, because while the two definitions and/or interpretations (or understandings) almost certainly overlap, they may not coincide (see, for example, Harvey, 2005a, on the various forms of 'geographical imagination'). This is certainly the case with geography. Much popular interpretation of the term is associated with the acquisition of basic information about the earth – what is where (what has sometimes been referred to as 'capes and bays geography') – and not with the contemporary concerns of academic researchers. Such interpretation is sustained by magazines such as the widely circulated *National Geographic* (Lutz and Collins, 1993; Schulten, 2001; Rothenberg, 2007), which carries much informative material about the earth's environments and peoples, but little that reflects geographers' current research. Similarly, periodicals such as *New Zealand Geographic* are closer to natural history than to geography; its definition and coverage of the subject has very little in common with the academic discipline whose practices are represented in the *New Zealand Geographer* (Johnston 2009b, 2009c). One of us has suggested that it is therefore analytically useful to distinguish between these forms of geography:

> There is a strong thread not only in the USA but also in the UK of geographical imaginations that have exerted . . . and continue to exert . . . very considerable popular and public influence, shaping and re-shaping views of the world . . . But their links to the academic discipline are, at best, relatively weak. The various worlds of geography seem not to intertwine – indeed some stand rather aloof from each other: the *National Geographic* has virtually no links with academic geography in the USA, for example. Magazines such as *Africa Geographic*, *Australian Geographic*, *Canadian Geographic* and *New Zealand Geographic* hardly recognise the existence of the academic discipline in their respective countries . . . The current existence of separate geographical imaginations raises a number of issues . . . such separate imaginations have existed in the past and . . . their relative strengths in different places and at different times may have been crucial influences on variations in the nature of geography as it is currently taught in schools and practised in universities. Those who write about the history of geography may recognise the existence of those separate imaginations as foundations upon which an academic discipline was built but then, once that discipline has been founded, focus almost entirely on it alone, very largely ignoring the other imaginations . . .
>
> (Johnston, 2013, p. 90)

Despite these issues of representation, boundaries and appreciation, however, geography is a substantial academic discipline in the countries that we focus on in this book. In the UK, for example, by the end of the twentieth century there were about 1,600 staff in university and college departments offering geography programmes, with some 7,000 places available on undergraduate degree programmes in geography each year, plus over 170 taught postgraduate programmes with nearly 3,000 places available; in addition, there were more than 800 full-time students registered for postgraduate research degrees, and a further 640 registered as part-time students. In the early 2000s, an increasing number of British geography departments were merged with allied subjects, usually environmental science, geology or earth science, but the overall numbers studying for a degree in geography have not declined. In the USA, the majority of universities lacked a geography department and there was a graduate programme in only a minority, including just one (Dartmouth) of the prestigious network comprising some of the older East Coast universities known as the 'Ivy League'. Richard Wright and Natalie Koch (2009) chart the history of geography at these eight Ivy League universities, noting how in the first decades of the twentieth century geography was part of the curricula at them all, but then lost ground owing to the relatively low status of the discipline (on the closure of Harvard's geography department, see N. Smith, 1987a; Cohen, 1988; Martin, 1988). The exodus of academics from the closure of their programmes, however, meant that some capable geographers (such as Edward L. Ullman, see p. 70 in this book) either moved into public universities elsewhere where geography remained, or moved to planning or allied programmes at the Ivy League:

> Increased demand for practical education and the attack on environmental determinism was devastating for geography in the Ivy League universities; these institutions increasingly emphasized the importance of theory and held technical instruction in low esteem. In contrast, geography departments in the new land-grant colleges of the Midwest prospered in this environment . . . designed in part to support . . . the Midwestern agricultural economy and serve the broader public, they welcomed the applied elements of geography.
>
> (p. 618)

Despite this uncertain start, today a large number of geographers are active in the USA: the Association of American Geographers (the AAG) has over 10,000 members (although not all are based in the USA, for the AAG promotes membership internationally) and its annual meeting regularly attracts several thousand participants, including hundreds from other countries. For many human geographers in North America, including Ph.D. students, participation at the AAG meetings, which may mean presenting papers, acting as discussant re other papers or organising and chairing panels of papers and the like, is an important site of networking and recognition. Other 'national' conferences, such as that of the London-based RGS (with the Institute of British Geographers), and similar events in other countries are also significant venues for this kind of professional interaction, but the sheer scale of the AAG means that many geographers will travel far to attend – it is held in a different American city each early spring (Hsu and Sidaway, 2009). (The AAG was very much a small, closed elite organisation with high barriers to entry before 1950, when it merged with the American Society for Professional Geographers – see James and Martin, 1978; Miller, 1993.) This large number of practising academics is part of a buoyant discipline whose interests are wide-ranging and whose contributions to knowledge advancement are substantial (Thrift, 2002). The nature of those contributions over a period of some seven decades is the core of this book's concerns.

The academic working environment: disciplines and the academic division of labour

The continuing goal of an academic discipline is the advancement of knowledge, pursued within its own particular areas of study. Its individual members contribute by conducting research and reporting their findings, by integrating material into the disciplinary corpus, and by pedagogical activities aimed at informing others, promoting and reproducing the discipline; they may also argue for the discipline's 'relevance' to society at large. But there is no fixed set of disciplines, nor any one correct division of academia according to subject matter. As David Livingstone (1992, pp. 28–9) puts it:

> The idea that there is some eternal metaphysical core to geography independent of historical circumstances will simply have to go . . . geography has meant different things to different

people in different places and thus the 'nature' of geography is always negotiated. The task of geography's historians, at least in part, is thus to ascertain how and why particular practices and procedures come to be accounted geographically legitimate and hence normative at different moments in time and in different spatial settings.

Those disciplines currently in existence are contained within boundaries established by earlier communities of scholars. The boundaries are porous, so disciplines interact. Occasionally the boundaries are changed, usually through the establishment of a new discipline that occupies an enclave within the pre-existing division of academic space. Moreover, the boundaries are not necessarily the same in all comparable institutions.

Just as there is no immutable set of disciplinary boundaries, so there is also no right way of undertaking research nor, in many cases, any exact criteria for determining whether research findings, and even more their interpretations, are correct. The 'right' and 'wrong' ways of doing research, the 'correct' and 'incorrect' interpretations of research findings, and the 'proper' ways of presenting knowledge and training students are all the product of decisions by academics themselves within the culture that they are constantly reproducing. Thus, not surprisingly, there is considerable debate within disciplines over these issues. At any one time, there may be consensus within a disciplinary community regarding both its subject matter and its research methods. But controversy is just as likely, as academics discuss the relevance of particular research findings, the validity of certain research methodologies, and so on. Indeed, there is controversy over the nature and definition of knowledge itself.

The study of the controversies and consensuses that characterise academic disciplines is the function of historians of science. (Science is used here as a very broad term encompassing the entire range of academic disciplines in the natural and social sciences: the Oxford English Dictionary defines science as 'an organized body of knowledge on a subject'.) To the outsider, the workings of a science are generally mysterious – especially regarding disciplines that require a great degree of prior training before original research is undertaken and whose literature is virtually impenetrable to the untrained. It is generally believed, however, that science is an objective activity undertaken within very strict rules, involving the continuous excitement of the search for new discoveries. Indeed, scientists frequently present themselves in this light. Certain values are universally subscribed to within academia, it is argued, with the main five being:

1 The norm of originality: academics strive to advance knowledge, and conduct original research designed to discover and account for aspects of the world as yet not fully understood.
2 The norm of communality: all information is shared within the academic community – it is transferred through accepted channels (notably the research journals) and its provenance is always recognised when it is being used.
3 The norm of disinterestedness: academics are devoted to their subject, and their main reward is the satisfaction of participating in the advancement of knowledge – a reward that may bring with it charisma and promotion.
4 The norm of universalism: judgements are made on impartial criteria, which take into account the academic merits of work only and make no reference to the personalities of the researchers.
5 The norm of organised scepticism: knowledge is furthered by a continuing process of constructive criticism, in which academics are always reconsidering both their own work and that of others.

(Mulkay, 1975, p. 510)

According to these five norms, therefore, academic work is carried out neutrally; there is a complete lack of partiality, self-seeking, secrecy and intellectual prejudice. Objective assessment criteria are assumed, as are high levels of ability and humility on the part of members of the academic community.

Against this ideal view of science and scientists, which many of the latter promote, are the results of studies in the history of science (most of which refer to the physical sciences: Mulkay, 1975; Mulkay et al., 1975; Latour, 1987, 1999; Latour and Woolgar, 1979). These show that the procedures adopted are seldom straightforwardly objective or neutral, and they present a picture of disciplines growing 'by gathering more detail in areas already investigated, and by stumbling across new sets of facts in areas of experience never previously investigated' (Barnes, 1974, p. 5). Science is a culture, with each discipline comprising one or more subculture(s). It has its own rules and procedures which are open to change and hence are the subject of debate, even conflict. Furthermore (as suggested in more detail below), scientific cultures are parts of wider cultures, and although to some extent scientists – because of their (self-imposed) expert status – can impose their own views of themselves on their host societies, they are also subject to external influences (see Barnes, 1974).

Each scientific discipline is a separate academic community, therefore; many are groupings of several related communities. Their goal is the advancement of knowledge, but the definitions of advancement and of knowledge itself are influenced, if not determined, by the community's members. Thus, study of a discipline's history is not simply a chronology of its successes. It is an investigation of the sociology of a community, of its debates, deliberations and decisions, as well as of its findings. (See James Watson, 1968, for an example of this; for a fictional account, see William Cooper, 1952; an excellent introductory history is in Gribbin, 2002. Livingstone, 1992, p. 30, suggests that 'it might be helpful if we were to think of geography as a tradition that evolves like a species over time'.) Most of those communities – including geography – are divided into a substantial number of subcommunities involving individuals working on a particular subject matter, and/or set of problems, and/or particular methodology. Indeed, Geertz (1983, p. 157) argues that:

> most effective academic communities are not that much larger than most peasant villages and just about as ingrown . . . From such units, intellectual communities if you will, convergent data can be gathered, for the relations among the inhabitants are typically not merely intellectual, but political, moral, and broadly personal (these days, increasingly, marital) as well. Laboratories and research institutes, scholarly societies, axial university departments, literary and artistic cliques, intellectual factions, all fit the same pattern: communities of multiply connected individuals in which something you find out about A tells you something about B as well, because, having known each other too long and too well, they are characters in one another's biographies.

Untangling such biographies, individual and collective, is the key to understanding geography's institutionalisation and contemporary practices: its history indicates where it has come from and why it currently takes the forms that it does. Furthermore, the history of most of those communities involves both conflict with others (within a university, for example: Johnston, 1998) and within the communities (Johnston, 2000a) over the allocation of resources and rewards. Such communities are frequently territorial, in the defence of their portion of the academic division of labour, and sometimes aggressively so in strategies to expand their portion of the whole.

But once it is established, why and how does the content of a discipline change? What accounts for this, beyond clashes of interest and the ideas or biographies of individual academics? What forms do such shifts take? What triggers them? In the first half of the period that we consider here, most of the models available for explaining and interpreting changes were developed for, and tested on, the physical sciences, revolving around the idea of paradigms. Their relevance to human geography was soon suggested, however (e.g. Haggett and Chorley, 1967; Harvey, 1973), and they are presented below as a backcloth for a later evaluation of the substantive material arranged in Chapters 2–9. The final chapter of the book will return to debates about interpreting the evolution of human geography.

Kuhn, paradigms, normal science and revolutions

A framework for studying the history of science that has received much attention (indeed, the book became highly cited by social scientists) is provided by T. S. Kuhn's (1962, 1970a) *The Structure of Scientific Revolutions*. Reviewing its impacts fifty years on from publication, Gordon (2012, p. 73) notes how:

> Anyone who works at the interstices of intellectual history and philosophy and the history and philosophy of science will be quick to rank Thomas Kuhn's *The Structure of Scientific Revolutions* amongst the most influential works of the last half-century. But its influence extends well beyond these disciplines as well. First published in 1962 as a contribution to the *International Encyclopedia of Unified Science*, over the last fifty years it has enjoyed a rich afterlife, leaving in its wake an immense if contested inventory of ideas whose significance has transcended the well-policed boundaries that often separate the natural sciences from the social sciences and the humanities. Even more surprising for a book of its academic character, it has enjoyed a reception in popular discourse that exceeds its disciplinary bailiwick. Its trademark terms – not only the celebrated ideas of a *paradigm* and a *paradigm shift* but also more technical themes such as *normal science*, *incommensurability*, and *anomaly* – have been naturalized into mundane English with a degree of success that puts to shame just about any other work of recent scholarship. Paraphrasing one of its characteristic claims, one may be tempted to observe that, since the publication of Kuhn's *Structure*, we all live in a different world.

The fact that Kuhn's book appeared just as human geography was on the verge of significant changes probably encouraged those charting and advocating the changes to adopt Kuhn's language of disciplinary revolutions, thus describing human geography as undergoing a 'quantitative' or 'scientific' 'revolution' – a series of shifts and debates that we describe in Chapters 2 and 3.

Yet as Kuhn (1970b) and many commentators have pointed out, his book is a study in the sociology of science: it is a positive interpretation – presenting what scientists actually do – rather than a normative programme – an argument for what scientists ought to do. It generalises from its observations, identifying common elements in disciplinary histories: Kuhn's goal was to identify 'what science, scientific research as it is actually practised, is really like' (Barnes, 1982, p. 1). Following Kuhn's death in 1996 a number of books on his life and work have been published and provide valuable introductions to, and critiques of, his work (e.g. Bird, 2000; Conant and Haugeland, 2000; Fuller, 2000; Sharrock and Read, 2002; Nickles, 2003).

Kuhn argued that scientists work in communities of researchers and teachers who share a common approach. They operate within an agreed philosophy, concur on the theoretical focus of their work and use accepted methodological procedures. They deploy those procedures to solve problems identified within the theoretical framework, thereby adding to knowledge (the store of problems solved) and extending the range of their theory. The framework, its procedures and its empirical substance are codified in their textbooks. What they share is termed a paradigm: 'a scientific community consists of men [sic] who share a paradigm' (Kuhn, 1970a, p. 176). As Popper (1959) expresses it, this means that once scientists are socialised into a research field they can proceed directly to its unsolved problems. The existing framework defines these problems, providing both a context and a methodology for tackling them. The researcher does not set out to tackle a problem de novo, therefore, but based on what is already known in the chosen field. A paradigm is thus 'an accepted problem-solution' (Barnes, 1982, p. xiv), which by its very nature poses the next problem. Progress in science is achieved by problem solving, by the current generation of scientists agreeing that 'If I have seen further it is by standing on the shoulders of giants' who preceded them – a phrase used by Isaac Newton.

To undertake research within a paradigm requires understanding and acceptance of its philosophy and methodology. This involves a period of training, during which the potential new researcher is socialised into the paradigmatic culture – that is, its way of thinking about its scientific problems – its language, literature, and methods of research, result presentation and argumentation. The key is provided by the paradigm's textbooks, the summaries of its literature which set out what is known and how further knowledge can be obtained. The training prepares individuals for work within an accepted mould:

> Scientific training is dogmatic and authoritarian . . . [it] does not generate or encourage traits such as creativity or logical rigour; rather it equips scientists so that it is possible for them to be creative, or rigorous, or whatever else, in the context of a specific form of culture.
>
> (Barnes, 1982, pp. 1, 17)

Having been integrated into the paradigm, the scientist joins a research community. Such communities, sometimes termed 'invisible colleges', operate through close interaction involving attendance at specialised conferences and the private circulation of research papers in pre-publication form: their existence is reflected in citations to each other's work (Crane, 1972). Increasingly, electronic communications media allow community members to be in frequent, almost constant, contact and collaborative work without regular meetings is now common (Johnston, 2004a). Individual members receive recognition for the value of their work and attract the patronage of leaders who may assist their career advancement; some communities have only a few patrons, perhaps even one, and may be identified as a particular school of thought – perhaps focused on a single leading individual.

Research within a paradigm involves filling in the gaps; the researcher 'has to make the unknown into an instance of the known, into another routine case' (Barnes, 1982, p. 49). The paradigm provides the needed resources – guidance but not direction. Success, which brings the rewards of recognition, patronage and status, involves conforming to the paradigmatic norms (Mulkay, 1975):

> It is clear that the quality or significance of a scientist's work is judged in relation to the existing set of scientific assumptions and expectations. Thus, whereas radical departures from a well-defined framework are unlikely to be granted recognition early under normal circumstances, original contributions which conform to established preconceptions will be quickly rewarded.
>
> (p. 515)

Thus, the dominant norm of academic life is not one of the five listed earlier (p. 13), but rather conformity or convergence. Science is not the constant search for novel discoveries but rather the careful application of agreed procedures to the solution of problems in order to extend existing well-structured bodies of knowledge. Judgements about relevance and importance are being made all the time, but within an academic environment carefully structured by the training process. Science progresses through filling the gaps in a predefined framework.

This operation of a paradigm is known as normal science:

> Perhaps the most striking feature of the normal research problems . . . is how little they aim to produce major novelties, conceptual or phenomenal. Sometimes . . . everything but the most esoteric detail of the result is known in advance . . . the range of anticipated, and thus assimilable, results is always small compared with the range that imaginations can conceive . . . the aim of normal science is not major substantive novelties . . . the results gained in normal research are

significant because they add to the scope and precision with which the paradigm can be applied. . . . Though its outcome can be anticipated, often in detail so great that what remains to be known is itself uninteresting, the way to achieve that outcome remains very much in doubt. . . . The man who succeeds proves himself an expert problem solver.

(Kuhn, 1962, pp. 35–6)

Within normal science, therefore, the researcher has available:

1 an accepted body of knowledge, ordered and interpreted in a particular way;
2 an indication of the puzzles that remain to be solved; and
3 a set of procedures for puzzle-solving.

Training within a paradigm provides the tools for extending the paradigmatic body of knowledge. The result is 'conventional, routinised practice' (Barnes, 1982, p. 11). This does not imply that normal scientific activity is necessarily one of 'long periods of dreary conformity' (p. 13), however, because extending and developing knowledge are not simply 'a matter of following instructions or rules. Rather, normal science is a test of ingenuity and imagination, with paradigms figuring largely among the cultural resources of the scientist' (p. 13). Solving problems is rarely easy; Kuhn (1970c, pp. 36–9) uses chess-playing as an example, arguing that much ingenuity as well as existing knowledge (practice) must be brought to bear if the problems posed in individual games are to be solved successfully.

Scientists are not omniscient. They do not understand everything – even within a particular paradigm – so their predictions occasionally prove inaccurate. While the process of normal science continues, therefore, slowly accumulating extra knowledge as problems are solved, it sometimes throws up anomalies, findings that are not in accord with the paradigm's assumptions. These must be accounted for:

Puzzle-solving activity frequently attempts to show that what is prima facie anomalous is either the spurious product of bad equipment or technique, or a familiar phenomenon in disguise. And most anomalies are successfully assimilated in this way.

(Barnes, 1982, p. 53)

Either the work was badly done or the researcher interpreted the results wrongly. Minor adaptation of the paradigm may be needed, but the general process of normal science continues.

Some anomalies cannot readily be accounted for, however, and they may continue to worry a few scientists, whose persistence leads to work on alternative paradigms, potential new frameworks that structure knowledge so that there are no anomalies. This is extraordinary research, conducted outside the bounds of the accepted paradigm. When it is successful, a 'revolutionary episode' is in progress. Regarding such research, Kuhn notes that:

Almost always the men who achieve these fundamental inventions of a new paradigm have been either very young or very new to the field whose paradigm they change . . . obviously these are the men who, being little committed by prior practice to the traditional rules of normal science, are particularly likely to see that those rules no longer define a playable game and to conceive another set that can replace them.

(1962, pp. 89–90)

The result is an alternative paradigm. Adherents of the current normal science are then asked to choose between two competing views of their subject: either the accepted mode of working is to be maintained, despite the anomalies, or a new subculture is to be adopted. The invitation is to discard existing authorities and habits and to take up new ones, which are claimed to be superior, because they are better predictors of that aspect of the world being studied. If the need for change is accepted, then a revolution in scientific practice occurs; one paradigm is replaced by another.

That choice between competing paradigms is an extremely difficult one because of their incommensurability; there are no common standards against which both can be compared. As Kuhn (1970a, p. 148) points out, 'the proponents of competing paradigms will often disagree about the list of problems that any candidate for paradigm must resolve. Their standards or their definitions of science are not the same.' The new paradigm will almost certainly use some of the language and procedures of the one it is seeking to replace, but in slightly different ways, giving rise to considerable misunderstanding in discussions between the rival paradigms' protagonists. More importantly, however, the two groups of scientists may be looking at the world in very different ways:

> Both are looking at the world, and what they look at has not changed. But in some areas they see different things, and they see them in different relations one to the other. That is why a law that cannot even be demonstrated to one group of scientists may occasionally seem intuitively obvious to another.
>
> (p. 150)

Thus, any switch from one paradigm to another is not forced simply by logic. It is what Kuhn calls a 'gestalt switch', a decision to abandon one way of viewing the world and replace it by another on intuitive grounds, rather than through the application of strict scientific criteria; a subjective decision has to be made that one is better than the other. Not all scientists may come to the same intuitive decision, therefore, so that some continue to work in the context of a paradigm that others have discredited. One of the often-cited examples of such a paradigm shift – on which Kuhn, 1957, wrote his first book – was the Copernican revolution in astronomy, as promoted by Galileo. Likewise, within Western geography, Columbus's 'discovery of America' forced a revolution in world cartography.

Kuhn represents scientific activity as researchers trained to employ a proven mode of looking at their subject matter using an accepted methodology for solving problems. They proceed in a steady, cumulative manner, adding to the store of knowledge: small modifications may be needed to accommodate minor anomalies encountered en route. Occasionally, however, anomalies are detected which cannot be either explained away or accommodated. Some researchers may focus on these, developing new paradigms to account for them, as well as everything else that was already known. When this is achieved, the new paradigm is presented for approval. A revolution is invited, for the alternative paradigms are incommensurable: the community is asked to redirect its work. Science proceeds in a steady fashion along well-trodden paths, therefore, with occasional major breaks in its continuity marked by important changes in the organisation of its material, in the definition of its problems, and in its techniques for problem-solving.

Criticisms of Kuhn's approach, and his response

Kuhn's work quickly stimulated a great deal of debate among philosophers of science because it challenged orthodox views of scientific progress and implied, especially in the concept of revolutions, that some scientific decision-making was 'irrational' (see Watkins, 1970, on Kuhn's analogy between scientific and religious communities, and Barry Barnes, 1982, for a wider evaluation of Kuhn's impacts). Indeed, the introduction and treatment of the concept of the revolutionary episode attracted much of the attention (see Stegmuller, 1976), because most of the views that Kuhn was challenging were normative rather than positive; they prescribed what science should be like, whereas Kuhn claimed to describe what it was actually like.

A major problem with Kuhn's original presentation for many commentators was the varying use of the term 'paradigm'. Masterman (1970, p. 61) identified no less than 21 different usages, which 'makes paradigm elucidation genuinely difficult for the superficial reader'. From these, she distilled three main groups of definitions:

1 the metaparadigms (or metaphysical paradigms), which can be equated with 'world views', or general organising principles;
2 the sociological paradigms, which are the concrete scientific achievements of a community defining their working habits; and
3 the artefact or construct paradigms, the classic works (exemplars) that provide the tools for future work.

The second provides the structure within which individual scientists work, whereas the third presents their means for puzzle-solving; the first comprises their overall view of the nature of science and its objects of study. Kuhn accepted Masterman's representation of his work, and later writings clarified his views, focusing almost entirely on the second and third of Masterman's definitions. He indicated that if he rewrote the original book he would give primacy not to paradigms but to scientific communities (Kuhn, 1970a). These operate at various scales: the global community of natural scientists, for example; the main professional disciplinary groups (physicists, chemists, etc.); and intradisciplinary groups working on particular empirical problems (such as nuclear physicists). He focused on the last: 'Communities of this sort are the units that this book has presented as the producers and validators of scientific knowledge. Paradigms are something shared by members of such groups' (p. 178). With regard to the use of the term 'paradigm' in this context, Kuhn (1977, p. 460) suggested a two-part definition, the second component fitting within the first: 'One sense of "paradigm" is global, embracing all the shared commitments of a scientific group; the other isolates a particularly important sort of commitment and is thus a subset of the first.'

The global, or sociological, definition is of a disciplinary matrix: 'what the members of a scientific community, and they alone, share' (p. 460). It comprises (Kuhn, 1970c, pp. 152 ff.): the accepted generalisations, shared commitments to particular models, or guiding frameworks for the construction and validation of theories.

Revolutions can therefore involve:

1 the replacement of one exemplar by another;
2 modification of the existing set of exemplars to accommodate new material; or
3 the replacement of the disciplinary matrix.

The last, a revolution in the sociological paradigm, is likely to be a major event in the history of a science; the first and second can occur without affecting the fundamentals of the disciplinary matrix. Others, such as Lakatos (1978a, 1978b) and Popper (1970) responded to Kuhn. Writing about Kuhn years later, Alexander Bird (2000, p. 50) suggests a continuum of changes.

Lakatos not only recognised but also required theoretical pluralism (which led to criticism from Barry Barnes, 1982, that his is a normative view of science, not a positive one like Kuhn's). Within this pluralist situation, 'Criticism of a programme is a long and often frustrating process and one must treat budding programmes leniently' (Lakatos, 1978a, p. 92). Although Lakatos denies Kuhn's concept of normal science as one of paradigm dominance in a discipline, he accepts that progressive research programmes comprise relatively routine extensions to knowledge through the testing of new hypotheses derived from their positive heuristic. Refutations are rare; the aim is verification and progress.

Popper argued against this, claiming that 'science is essentially critical' (1970, p. 55) and characterised not by normal science but by extraordinary research. To him, the normal scientist is really an applied scientist:

> 'Normal' science . . . is the activity of the non-revolutionary, or more precisely, the not-too-critical professional: of the science student who accepts the ruling dogma of the day; who does not wish to challenge it; and who accepts a new revolutionary theory only if almost everybody else is ready to accept it – if it becomes fashionable by a kind of bandwagon effect. To resist a new fashion needs perhaps as much courage as was needed to bring it about . . . The success of the 'normal' scientist consists, entirely, in showing that the ruling theory can be properly and satisfactorily applied in order to reach a solution of the puzzle in question.
>
> (Popper, 1970, pp. 52–3)

For Popper, science consists of bold conjectures and the conduct of empirical research (he had experimental science in mind) designed to refute them. In his view of science (Popper, 1970, p. 77) hypotheses can never be wholly verified, only falsified. It is the testability of any concept that matters. The ability of concepts to endure attempts to disprove them is at the heart of what science ought to be. Kuhn's (1970a, p. 243) response was that '[Popper] and his group argue that the scientist should try at all times to be a critic and a proliferator of alternate theories. I urge the desirability of an alternate strategy which reserves such behaviour for special occasions'.

Popper's arguments were extended by Feyerabend (1975), who claimed that science is a series of accidents – which is how it should be. Science should be allowed to evolve in that way, since this is the best means of ensuring progress. Most discoveries have been made by individuals who either deliberately or unwittingly flouted the rules, but modern scientific education tries to prevent this by constraining its practitioners into myopic paradigms. Feyerabend (1975, p. 187) claims that the only rule in science should be 'anything goes', in a scientific anarchy whose hallmark, like a political anarchy's, is 'opposition to the established order of things: to the state; its institutions, the ideologies that support and glorify those institutions'. Like Popper, however, Feyerabend presents a normative model, concerned with major scientific developments and not with their everyday extensions. Regarding the latter, if Lakatos rather than Kuhn is correct, revolutions take a long time, so a discipline could be characterised by dissensus rather than consensus over much of its history. Is it likely that a body of scholars (especially a large body) will agree on ends and means, and will not differ at all on fundamental interpretations, except very occasionally and then only for fairly short periods? Branches may break away, however, establishing new communities – by 'quiet revolutions' (Johnston, 1978a). Such breaks may not be favoured by the parent community on political grounds, for they are likely to be against the interests of the disciplinary bureaucracy; separate communities competing for students and research funds, yet covering similar subject-matter, are unpopular. Thus, communities may be prepared to accommodate dissenting groups, even potential revolutionaries, rather than risk breakaway success.

Such dissent is often contained, however, and may be marginalised or ignored (see Lichtenberger, 1984, for an example from within geography). The results are frequently complex, comprising interconnections, disputes, disjunctures and alliances which can be approached from a variety of perspectives – for example, along the different lines indicated by Mark Boyle (2005), that were described in the Preface. Reviewing a similar range, including analysis of the social relations between individuals within scholarly institutions, the development of the subject in universities, patterns of citation, biography and the one taken here, that of the history of ideas and themes, Stoddart (1986, p. 27) noted how: 'Each of these approaches has its own validity and its own strengths. I claim here only that the history of geography is an infinitely richer and more varied landscape than acquaintance with the standard works on the subject would suggest.'

Stoddart is among those who critiqued the way that the paradigm concept was being applied in telling the history of geography, or being used to bolster the case for new ways of doing human geography. We will return to this issue in the last chapter of the book. But it should be noted here that he has since been joined by others – drawing on a variety of other frameworks of interpretation of the history and structure of disciplines.

Other frames: Bourdieu and Foucault, and other interpretations

All of Kuhn's examples come from the natural sciences (as Kuhn, 1977, himself stressed). Indeed, Steve Fuller notes that:

> Kuhn observed that social scientists could never agree on what counted as an exemplary piece of research, and so could never establish a common frame of reference for anchoring their disputes ... Accordingly, Kuhn concluded that the history of the social sciences has not witnessed a clear succession of paradigms because social scientists have been unable to agree on research exemplars to underwrite the activity of normal science. . . . the peculiar character of the social sciences can be traced historically to the social scientists' having been guided by larger, conceptually unwieldy social problems that cannot be reduced to well-defined puzzles. Consequently historians of each of the social sciences have portrayed their disciplines' trajectory as pulled in three distinct directions that can be characterized independently of any intrinsic concern for the nature of social reality:
>
> 1. From *above* (in an administrative or managerial capacity, as the trustees of state or business);
> 2. From *within* (in an ethical capacity, as the secular successor of pastoral theology);
> 3. From *below* (in a rhetorical capacity, as the voice of politically disenfranchised groups).
>
> (2000, p. 228)

Working more on social sciences, humanities and professions (such as law) which have a basis in university disciplines, the French sociologist Pierre Bourdieu (1988) proposed that they comprise semi-autonomous 'fields'. These are structured social spaces, with power structures that relate to but have a relative autonomy from other class or social structures. Bourdieu's book on these, focusing on academia in France, with most of the material drawn from his own discipline of sociology, appeared in French in 1984 and was translated into English in 1988. It thus entered English language debates about disciplinary change at a time when Kuhn's had already been debated for a quarter of a century. Bourdieu has since been seen as a rich entrée to thinking about power and structures of the academy (Swartz, 1997; and for a reading of his work into geography, Sidaway, 1997, 2000b). But his work is less of a template for studying changes in a discipline than a mirror to its structures that links them to contests for authority, distinction and power. Swartz (1997, p. 273) notes that for Bourdieu, 'it is the material and symbolic reward structures, not the normative ideals, that more decisively shape scientific behaviour'. In other words, the focus on ideas and explanations that Kuhn foregrounds needs to be supplemented with nuanced accounts of power, influence, connections and status.

Changes in geography have also been interpreted in terms of intergenerational struggle (whereby one paradigm becomes associated with a particular cohort, who seek to advance their careers in opposition to their forebears), as in Peter Taylor's (1976) account of debates about quantification or John Mohan's (2004) claim that change in human geography resembles a product cycle, with new theories quickly replacing older ones, in a succession of trends. We return to these and other debates about the evolution of the discipline in the last chapter. At this point, however, it is important to note that another potential framework for analysing a social science discipline without the particular world view of Kuhn and his critics is outlined in Foucault's (1972) *The*

Archaeology of Knowledge. This offers an analysis of the 'pure description' of discourses, unified systems of statements (however expressed) that can only be understood within their context (Foucault, 1972, p. 97): 'One cannot say a sentence, one cannot transform it into a statement, unless a collateral space is brought into operation.' Anything said or written can only be understood by those privy to the context of the discourse; as with a language, one can only comprehend the words, and the sequence in which they are used, because one understands the rules governing their use.

According to this view, the history of ideas is the history of discourses, of systems (somewhat akin to languages) used for the discussion of subjects that are defined within the discourses. There is, for example, no fixed definition of madness; each discourse that discusses it has its own definition and any discussion of madness is particular to that definition and the discourse in which it is set. The nature of a discourse is deposited in a collection of texts and practices, and it is the function of archaeology to reconstitute the discourse from such 'archives', to describe and understand what was being done in a particular configuration (Foucault, 1972, pp. 138–40, 157–65). As Foucault describes it, nineteenth-century medical science:

> was characterized not so much by its objects or concepts as by a certain style, a certain constant manner of statement . . . medicine no longer consisted of a group of traditions, observations and heterogeneous practices, but of a corpus of knowledge that presupposed the same way of looking at things . . . [in addition it was] a group of hypotheses about life and death, of ethical choices, of therapeutic decisions, of institutional regulations, of teaching models, as a group of descriptions . . . if there is a unity, its principle is not therefore a determined form of statements; is it not rather the group of rules, which, simultaneously or in turn, have made possible purely perceptual descriptions? What one must characterize and individualize is the coexistence of these dispersed and heterogeneous statements; the system that governs their division, the degree to which they depend upon one another, the way in which they interlock or exclude one another, the transformation that they undergo, and the play of their location, arrangement, and replacement.
>
> (1972, pp. 33–4)

Discourses are therefore sets of mutually agreed rules that govern description and discussion among the members who agree those rules, and are similar to the scientific communities given central place in Kuhn's (1970a) later statements. They are not independent of wider conditions, however. Archaeology may identify separate discursive formations, but these must be related to what Foucault calls the episteme:

> something like a world-view, a slice of history common to all branches of knowledge, which imposes on each one the same norms and postulates, a general stage of reason, a certain structure of thought that the men [sic] of a particular period cannot escape.
>
> (1972, p. 19)

This sets out how a particular world view (metaparadigm?) predominates during a period and influences the contents and determines the parameters of 'individual' discourses (the idea of an individual is itself the product of a particular discourse, for example). Presumably, too, there must be periods of competition between epistemes (which may involve a similar process to that outlined by Lakatos). Foucault is less interested in how such change occurs, however.

Foucault's ideas set the study of scientific disciplines more firmly in the context of the wider environment in which (and for which) they are practised than is the case with the other approaches discussed here, all of which very largely abstract the study of science from its social milieu. (As Berdoulay (1981, p. 9) expresses it, 'little interest is paid to historical contexts or intellectual climates since the focus is placed on the internal evolution of each science'.) Such abstraction is especially unfortunate for the social sciences, the disciplines that investigate and interact with their milieu and whose contents, in the broadest sense if not in detail, are likely to be strongly influenced by that context. Just how social science and society interact is challenging to map in detail (as, of course, is the interaction between natural science and society because the contents of the former, too, are clearly influenced by the environmental constraints), and the reception of Foucault's proposals of shifting epistemes has generated much debate.

That human geography should be studied in its social context appears an irrefutable claim, however. The latter section of this chapter outlines that context.

Daily life, texts and authority in the (social) sciences

Many of the major approaches to the history of academic disciplines, including those summarised here, focus in particular on changes in their practices rather than the everyday pursuit of research within agreed frameworks. This – what Kuhn saw as 'normal science' – occupies most academics for most of their time; they conduct research within an established matrix of practices, and rarely become involved in more general debates about the nature of those practices and whether their discipline or subdiscipline is pursuing a 'correct path'. Revolutionary calls are rare; revolutions are even rarer.

That matrix of accepted disciplinary practices may be very diverse, comprising a substantial number of the 'villages' that Geertz (1983) identified with each group's members undertaking their original investigations of specific topics and/or subject areas in a particular way. These are not independent entities – many of the 'villages' have no specific identity – but their interdependence is often limited. In their day-to-day teaching activities and interactions with students and departmental colleagues, academics encounter other approaches, but many of their formal interactions tend to be with colleagues in other institutions, who they meet only rarely at conferences and similar gatherings but with whom – given the benefits of electronic media – they can be in regular impersonal contact, sharing ideas and manuscripts. And many of their formal interactions will be directed at relatively limited audiences; although the general journals published by the major learned societies are among the most prestigious for placing one's research findings, there is now a plethora of journals directed at specific subfields and which are the main destinations for specialised research findings.

The presentation of 'normal science' suggests that for substantial periods of time academic disciplines are relatively untroubled by debates about what, why and how their members' research should proceed. Their academic space is divided up into a series of overlapping specialisms in which practitioners pursue new knowledge without challenges regarding the means of production that they deploy. And as disciplines have become larger with the expansion of higher education, so that archipelago of academic villages has become more extensive – and pluralistic in its orientations.

This is not to say that there are no debates about disciplinary means and ends and challenges to the status quo. In part this is because of the contest for resources within a discipline, especially in the locales where it is mainly practised – university departments. With few exceptions these will be relatively catholic bodies in a discipline where a range of paradigms (disciplinary matrices and exemplars, if not world views) prevail, and although the representatives of each may co-exist in reasonable harmony for much of the time, there are invariably tensions over resources. If a teaching position is to be filled, for example, different research groups may compete for it to be allied to their specialism – and there will be similar debates over degree curricula, the allocation of research funds, and so forth. Political contestation if not conflict is thus common, as academics with different views on what their discipline's dominant practices should be seek a significant – if not dominant – place for them within the department's activities.

Such politics are local and may have no implications beyond the particular institution, with the result that university departments in the same discipline often differ substantially in the particular aspects of the discipline that they promote – and also in those that they (relatively at least) ignore. But occasionally the politics becomes a large-scale activity, because groups want to promote changes across the discipline more generally, for a variety of possible reasons: they may have identified what they consider a superior set of practices to that currently deployed, for example. Such challenges to the status quo almost invariably involve collective action and, as Frickel and Gross (2005) argue, are necessarily contentious and political: groups are competing for the 'heart and soul' of a discipline.

According to a model of such intradisciplinary contestation developed by Latour (1999; for its application to the history of regional science (p. 83 of this book), see Trevor Barnes, 2003a, 2004a), four interrelated, largely sequential, processes are involved in mounting such a challenge:

1   mobilisation, which involves the announcement of a new agenda for the discipline, stressing its novelty and claimed superiority over (at least some of) current disciplinary practices;
2   autonomisation, involving strategies to promote the new agenda to various audiences, seeking to attract a broad support base for the project;
3   building alliances by winning space for the project within the discipline, such as in teaching programmes, through which support from students – and thus the next generation of researchers and academics – can be gained; and
4   public representation – whereas building alliances involves intradisciplinary politics wider support in the academy and beyond can help those promoting the new agenda (if, for example, research funding bodies accept the new approach as part of the discipline's portfolio).

The result of such a challenge is usually that room is found in the discipline for the new approach, and its size and influence spread over time as those it seeks to replace slowly wither. (All members of a discipline are unlikely to be converted – although this may be the case in the natural sciences where a proposed change leads to a fundamentally new paradigm because those it challenges are totally discredited. Given the subjectivity that underpins most of the social sciences, that is unlikely to happen there.)

In the challenge for intradisciplinary status it may be that the logic of the new agenda wins over adherents in an entirely uncontroversial way. That is rare, however, because any challenge to the status quo raises doubts about not only current practices but also the standing of their practitioners. Thus a variety of strategies may have to be deployed in the politics of potential disciplinary change, including:

1   the politics of denigration, whereby the promoters of the new agenda argue that the practices they wish to displace are unworthy of a place in the disciplinary portfolio;
2   the politics of critique, which are less assertive than the first strategy and seek to win over converts by argument and demonstration rather than assertion;
3   the politics of dismissal, in which previous practices are simply dismissed as no longer viable;
4   the politics of silence in which dismissal is assumed because the 'to be replaced' practices are simply ignored in the writings and arguments of the new agenda's promoters – a strategy particularly valuable in the autonomisation stage of the process involving students: if they are not made aware of particular practices their choice set will be constrained and their own practices hopefully focused on the new agenda;
5   the politics of accommodation, whereby advocates – having mobilised considerable support – seek a place alongside established practices (which they hope will subsequently wither) during the building alliances and public representation stages; and

6   the politics of unity which involves arguing that the discipline, and its ability to attract resources, will be stronger if the commonality of various approaches is presented outside the discipline, rather than a view of a fissiparous discipline with no clear trajectory.

The tactics employed in these strategies will vary according to the local context. While much of the initial debate over alternatives may take place in the relative privacy of seminar rooms and similar locales, if a major shift in a discipline's orientation is to be achieved, those debates will almost certainly become public – in the discipline's journals and other publications. In the politics of silence, for example, textbooks may be the vehicle deployed for mobilising student support; if certain practices are not presented in the volumes that tell students what their chosen subject studies and how, then they are, in effect, captured by those approaches that are represented to them – until such time as they may begin to question the status quo (see Johnston, 2006). The extent and intentionality of such textbook omissions/foci in propagating geography's agendas is contested (Hubbard and Kitchin, 2007; and the response by Johnston, 2007b). However, the fact that a leading disciplinary journal (*Progress in Human Geography*) regularly contains an article on classic 'text-books that shaped generations' indicates that many academic geographers recall, retain and are sometimes still responding to those texts that inspired them when students.

Geographers, Kuhn and contexts

Soon after the appearance of Kuhn's (1962) classic volume, the concept of a paradigm was introduced to the geographic literature by Haggett and Chorley (1967). Since then it has been widely used in discussions of geography's history, not least in previous editions of this book (on which see Mair, 1986). Its notion of normal science punctuated by revolutions was soon discarded, however, as it was realised that several incommensurate (or entangled) paradigms were co-existing within the discipline, and more seemed to be added on a regular basis without others being dropped; the discipline was seen as what Alexander Bird (2000, p. 31), following Hoyningen-Huene (1993), terms an 'immature science': a 'mature science' is one wherein normal science is practised within a single paradigm, whereas an 'immature science' has 'several competing paradigm-like schools'. Maturity only comes when there is consensus, when:

> one school scores a victory over its competitors by producing a signal achievement. That achievement wins defectors from other schools and attracts the favour of younger scientists. Support for the competitors dwindles and they eventually die out.

This suggests 'political' decision-making, at least in part. It may be that all members of a community accept the 'achievement' (with those who do not perhaps defecting to other disciplines), or it may be that those with power within the discipline impose it, by declining to make any appointments of people who challenge it, and by stifling dissent from within. According to Bird (2000, p. 32):

> Immature science is that initial period in the history of a science that occurs when for the first time sufficient interest in a phenomenon or set of related phenomena crystallizes distinct groups or schools around particular theories or approaches; the schools not only pursue research on the basis of their favoured ideas but also compete with one another for intellectual, social and professional supremacy.

This book clearly demonstrates that on those terms geography has been an immature science throughout the period surveyed. It also, at least implicitly, poses the question whether it will ever reach maturity, as defined by Bird.

Indeed, unless such 'maturity' is imposed on a discipline by preventing dissent and debate, it is doubtful whether any discipline can be termed a mature science in a strict sense. That certainly may be the case with the social sciences. As Bird (2000, p. 267) noted, Kuhn's book 'had little direct influence on the functioning of the natural sciences but its impact on social science was enormous'. That impact included the provision of 'a template whereby the histories of the social sciences, almost entirely ignored in *Structure* itself, could be described'. We return to the uses of Kuhn in human geography in this book's final chapter; at this stage we merely note that too much emphasis in geographers' discussions of Kuhn has focused on the concept of cycles of normal science punctuated by revolutions and less on that of communities, their struggle for dominance and their promotion of alternative discourses, which Derek Gregory (1994, p. 11) identifies as 'all the ways in which we communicate with one another . . . that cast networks of signs, symbols and practices through which we make our world(s) meaningful to ourselves and to others' in local contexts, 'the contexts and casements that shape our local knowledge'.

Nor do we seek to resolve all these issues in *Geography and Geographers*. Our contribution to the task of charting the history of Anglo-American human geography over the last seven decades involves identifying the major features in the various debates. That is the role of the next eight chapters; they are followed by one which evaluates those debates, not by trying to reconcile the various positions, but rather by setting them in context and seeking to appreciate why and when they occurred and were settled (to the extent that they were). Moreover, these frameworks for interpreting the evolution of geography have relatively little to say about how it has been mediated through places, its geography. While recognising the unevenness and power structures that accompany it (Paasi, 2005), some have argued that national trends or national 'schools' in geography are less important than hitherto in an increasingly globalised academic arena, dominated by the English language (Claval, 2009). Even within anglophone universities and predominantly anglophone countries, however, there have been key nodes, centres and relative peripheries. Writing from one of the latter, Andrew Wilson and Matthew Henry:

> argue that the evolution of geography in New Zealand cannot be simply read off from geography as conceived and practised in what Johnston and Sidaway [2004] identify as the centres of Anglophonic geography. Rather, it has been constituted between the rhythms of a wider Anglophonic tradition within which geographers in New Zealand have been firmly embedded and the situated pressures of an evolving context in which geography has often been framed by a colonial legacy and emerging projects of national development.
>
> (2011, p. 116)

A fuller account of these, and other, *geographies* of geography and geographers are beyond our scope here (though we will return to them in the final chapter), for the focus remains those broad rhythms, which in turn are themselves situated, not least among wider social, political, cultural and economic environments.

The external environment

The discussion so far has suggested that scientific disciplines (and/or discourses) comprise communities and subcommunities, small societies which are microcosms of their containing social systems. As such, the proper study of how they operate is sociological, although philosophy may provide a normative framework.

Sociological studies of academic disciplines accept that such communities are not autonomous. Their members are not isolated and they need the support of a wider society in order to exist: society employs academics to teach and research – either collectively, because universities receive public funding to teach and undertake research, or through market systems, whereby students pay for their education and sponsors pay for research. Whereas scientists may to some extent be able to impose their own priorities over what type of work is done, they are strongly influenced by external factors:

> Social, technical and economic determinants routinely affect the rate and direction of scientific growth. . . . It is true that much scientific change occurs despite, rather than because of, external direction or financial control. . . . Progress in the disinterested study [of certain] . . . areas has probably occurred just that bit more rapidly because of their relevance to other matters.
>
> (Barnes, 1982, pp. 102–3)

The study of a discipline must be set in its societal context. It must not be assumed that members of academic communities fully accept that context and its directives and impulses. They may wish to counter them, and use their academic base as a focus for discontent and promotion of alternatives. But the (potential) limits to that discontent are substantial. Most university academic communities are dependent (indirectly if not directly) for their existence on public funds disbursed by governments which may use their financial power to influence, if not direct, what is taught and researched. Some universities (notably in the USA) are dependent on private sources of finance, so they must convince their sponsors and students that their work is relevant to current societal concerns (as Taylor, 1985a, suggests).

In that context, the Second World War is more than a convenient period from which to commence this chapter on the history of human geography; it marked a major watershed in the development of the societies which are the prime focus of the book – the UK and the USA. It cannot be considered in isolation, however. Just as important for the present discussion are the worldwide economic depression which preceded it and the Cold War, the economic boom, and then the recession and restructuring which followed. For the first time, a major international conflict was not determined solely by sacrifices in battles on land and sea, although there were many of them during the Second World War. And the extra dimensions of this war did not just involve the development of air space as a further arena for conflict – as well as restructuring views of spatial relationships (Schulten, 2001; Abrahamson, 2010). The war was fought not only between military forces with guns and bombs but also between scientists, and victory was hastened, if not ensured, by the scientific superiority of the Allied Powers, most obviously at Hiroshima and Nagasaki, as well as in the cracking of secret codes. Science and technology had long been major elements in the developing industrialisation of the Western world, but their dominance was established between 1939 and 1945, and there was to be no retreat from the many technological advances made by the researchers involved in the military effort. Thus, the war heralded the predominance of technology and 'science' in human affairs, not least in the militarisation of academic research during the subsequent Cold War, in which, according to Trevor Barnes (2008b), some geographers were implicated.

Associated with this growth in scientific activity, and its consequent prestige (with governments and with society at large), was a parallel development of social engineering. The major economic depression of the 1930s, inaugurated by the Wall Street crash of 1929, had a massive impact on governments and stimulated many measures aimed at relieving poverty and deprivation, simultaneously assuaging the liberal conscience. In the USA, this was represented by the New Deal legislation of President Roosevelt's governments, which aimed at relief and encouragement to industry and, through the Social Security Act, providing public support for those who, by no fault of their own, were indigent. Similar measures were introduced by the national government in the UK; more were foreshadowed by plans formulated during the war, such as those promoted in the Beveridge Report on social security, and the Labour Party's landslide election victory in 1945 heralded the introduction of many social democratic policies which gave government a much greater peacetime role in the organisation of the economy and society than previously envisaged, let alone achieved.

This development of social engineering was associated with a rising status for the social sciences, and a great expansion in their activities. Economics achieved prestige first, notably through the contributions of Keynes and others to solving problems of the depression, the organisation of economies during wartime, and participation in the planning of a new world economic order after the war. Others followed. Social psychological research was widely used in the evaluation of personnel by the armed forces, and after the war opinion surveys proved valuable to governments and political parties while market research was increasingly used by industry and commerce (along with psychology in its advertising efforts). All these fields adopted the 'scientific methods' of the more prestigious hard sciences, and their successes were envied by other disciplines, such as sociology, social administration and, later, geography. To be scientific was to be respectable and useful.

The war years saw the end of the extreme economic deprivations of the depression, as manufacturing output was boosted to provide the machinery of war. Afterwards, there were many years of doing without to be compensated for, and with full employment, government direction of, and increasing involvement in, economic affairs, plus the need to re-equip industries, the two decades following the war were characterised by an economic boom in the Western world. Technological developments which had emerged in the 1920s–1930s, and which had been finessed in wartime, found wider applications in mass markets and continued to be fostered by Cold War research and development. Apart from the greater government involvement, this era in industrial development was marked by another major new characteristic, the advance of the giant firm, including the multinational corporation. The concentration and centralisation of capital proceeded rapidly; the average size of firms and factories increased and the economy of the world became dominated by a relatively small number of concerns. The decline and eclipse of the overseas empires in the context of the rise of the American and Soviet superpowers and the birth of the 'Third World' out of the debris of war and imperialism, all reinforced the sense of a fast-changing 'new world'.

Rebuilding the ravaged war arenas placed new demands on societies, and the planning profession emerged from earlier obscurity to take on a major role in preparing the blueprint for a new social order. The need for such action in the UK was realised during the war with the preparation of a series of reports concerned with future land-use patterns, and with the spatial distribution of economic activity, at local and regional scales. Cities were to be rebuilt; new towns were to be constructed; a more balanced interregional distribution of industry and employment was to be ensured; agricultural land was to be protected and residential environments were to be improved: all of this made great demands on social scientists, as well as engineers (an opportunity recognised by some geographers: Johnston, 2002a, 2003b, 2003c). The greater degree of commitment to the private ownership of abundant land in the USA led to slower acceptance of the need for spatial planning there, but its heyday came with the rapid growth of problems involved in catering for the upsurge of automobile ownership and use: transport planning and engineering soon became major activities, allied with the automobile industry and the companies which constructed the major highways (many of which were designed for 'strategic' defence purposes in the Cold War). Suburbanisation (and later gentrification) and other new socio-spatial structures of mobility changed landscapes in much of North America and Western Europe in striking ways. A baby boom after the war and new migration flows (e.g. among black and Hispanic communities in the USA, and from the former Caribbean, Asian or African colonies and parts of the Balkans to Western Europe) shifted the social, 'racial' and cultural dynamics of many cities. Many of these trends had begun before the war, but they accelerated and some of their vectors shifted afterwards.

All the associated tasks of governance – related to economic growth and planning, spatial planning, social administration, technological change, management, etc. – generated a need for educated personnel, and the universities received unprecedented demands for their graduates to serve the new needs of society. Existing universities and colleges expanded and many more were founded. Science and social science departments grew in numbers and size to meet student demand. The additional academics were involved in research, which became a salient component of academic life and so increased the tempo of paradigm development and questioning. Rather than places where a small elite was educated for the professions and a few privileged individuals followed their research interests, the universities became centres of society's development – the 'white-hot technological revolution' which Harold Wilson promised the British in the early 1960s. Research projects became bigger, supported by large grants from outside bodies (including government-established research councils) and carried out by specially employed graduates; the rate and volume of publication increased exponentially (Stoddart, 1967a; Johnston, 1995a, 1995b, 1996b).

The years from 1945 to about 1965 were a period of scientific and technological dominance, therefore. It was argued that the problems of production had been solved, because enough goods and services could be provided to satisfy all (and such benefits – it was argued – would soon accrue to the 'developing world' too). The problems of distribution were still being tackled, however, for as yet there was inequality of provision at all spatial scales. But these could be handled, it was argued, through a combination of market processes and state provision, and the prospect of a prosperous and healthy life for all was widely canvassed. Academic disciplines were contributing substantially to this problem-solving by their own – scientific – progress. Advances in the natural sciences and technol-ogies were solving the problems of production – of food, housing and consumer goods – as well as of ill-health. Meanwhile advances in the social sciences were aiding in the management of success and the delivery of well-being. Investment in education was thus investment in social progress (as well as an investment in the life chances of the individuals involved).

The deprivations of the 1920s and 1930s produced political responses two decades later, char-acterised by a determination not only to ensure no return to economic depression but also to provide permanent protection to those who suffered short- or long-term deprivation. The Beveridge Report in the UK identified five causes for concern: want, disease, ignorance, squalor and idleness. Policies aimed at their removal led to the creation of the major edifice of the Welfare State, which offered basic minimal standards of living for all, through such mechanisms as a National Health Service that was free at the point of demand; a universal education service for all aged between 5 and 14 (later 15 and then 16), plus subsidised further and higher education systems; subsidised state housing; universal child allowances; unemployment benefits; benefits and care for the sick and disabled; and universal pensions. Other countries created different structures, but the overall pattern in the 'developed world' was of a major state role in sustaining its citizens (see Johnston, 1992).

Growth of the 'interventionist state' involved a major change to the nature of polity as well as society. A much wider range of activities came within government orbits, and the public bureau-cracies grew accordingly – with consequent benefits in both employment opportunities for graduates and outlets for applying scientific and social scientific research. The decades after the Second World War saw the dominance of the welfare–corporate state in public affairs, with policy advisers and analysts provided by many professional groups (including scientists and social scientists).

Despite the successes of this initial postwar period, some doubted the desirability of such an all-embracing state. These concerns were growing during the 1960s, and by the early 1970s were having a major impact on the world scene. The seeds of doubt were many. Initially they focused, especially in the USA, on the problems of nuclear weapon development and of war, particularly the increasingly unpopular intervention in Vietnam and surrounding countries where technology was not carrying all before it against strong popular resistance; the casualty rates, destruction and corruption were increasingly deplored. Alongside were a number of major humanitarian concerns, including doubts about the inequalities that continued, both on a world scale and within the 'successful' capitalist societies. Poverty was not being alleviated; if anything, the disparities between rich and poor were being extended (and continued to be experienced through 'race' and racism), and absolute standards of living remained appallingly low in many parts of the world. The prospects for solving these problems were much less rosy than they had been a few years earlier, it seemed.

Furthermore, it was increasingly realised that success was being bought at considerable cost. Scientific and technological advancement required the dominance and discipline of the machine and the large factory. Work for large proportions of the labour force was being made more repetitive and boring as skilled tasks were taken over by automated production and assembly lines. Alienation of the individual from society was increasing. Particular groups suffered more than others did, as the result of prejudice and discrimination. The status of black people and women became the foci of emerging civil and human rights movements, which extended later to other oppressed groups within Western society. Finally, interest was kindled in the growing degradation of the environment to fuel the production goals of advanced capitalist societies.

The problems of the dignity of the individual, the repression of minorities, the quality of life and the depletion of environmental resources were not new in the late 1960s. What was realised then, however, was that the form of social 'progress' advanced during the previous decades was in many cases exacerbating and not solving such problems. As the realisation grew, so the proposed solutions varied. To some, the problems could be solved by greater state involvement, on international and national scales. Human and civil rights must be protected; greater equality must be achieved through the redistribution of wealth; the environment must be conserved, and where necessary preserved. In the language of the previous section, the research programme was maintained, but major efforts were made within its positive heuristic to solve the many anomalies. To others, this solution was insufficient. It would simply generate new anomalies because, according to the developing critique of capitalist systems, these are necessary to social 'progress'; capitalism, it was argued, necessarily survives on inequalities, on alienation and on the degradation of the environment.

Universities were the focus of much of this developing concern. There were major confrontations between students and the authorities in the late 1960s, in particular, for example, at Berkeley, Chicago, London and Paris, and the student body was at the forefront of the anti-war movements. Some academics, notably in the social sciences, focused on the 'managerial' issues involved in producing a 'better, more equal' society, and these were considered more crucial than the 'production' issues covered by the natural sciences and technologies. A threefold division developed. In one, the need for social scientists to become more active in developing solutions to the problems of distribution and environmental depletion was advanced: social science must become more 'relevant', more 'policy-oriented', within the constraints of the proposed societal 'research programme'. A second argued for greater concentration on the problems of individual alienation; people should be released from the overbearing dominance of the machine and the big organisation and encouraged to take a much greater part in creating their own lives. The individual needed protection against the increasingly distrusted expert. Finally, a third group developed a critique of capitalist society, seeking to show that while specific problems may be soluble, this would merely lead to others, while the general problems of inequality and injustice (Dorling, 2010) would remain because they are endemic to that mode of social organisation. The call was for major social reform − from some, revolution (the vanguard of which was often identified with the Vietnam War and other Third World liberation struggles) − as the only long-term solution to the problems of human dignity and inequality. Feminist and civil rights movements in Western societies emerged out of similar radical commitments. These would influence and inspire other movements − for gay rights/liberation and native/indigenous/first people's rights or advocating self-determination for small/colonised nations or linguistic minorities, for example. Much talk of freedom and revolution was in the air. Changes in popular culture (music, fashion and art) and academia soon embodied these changes. Indeed, academia would be at the forefront of ensuing contests.

The force of these arguments can be identified in a variety of ways, not least the declining popularity of scientific and technological subjects among students, and the growing demand for places in the social sciences and humanities in the 1960s and 1970s. A further major problem then arose, because Europe and North America spent substantial periods of the 1970s in economic recession, punctuated by periods of increasingly speculative and inflation-prone growth. This was accompanied by shifting geographies of production and consumption as lower-end service and many industrial or extractive industries moved to cheaper sites of production overseas (Dicken, 2011). These shifts would be heightened in subsequent decades. The long boom of the 40s–70s was over. And countervailing tendencies to the leftist radicalism of the 1960s developed – for example, right-wing evangelistic forms of Christianity in the USA (themselves linked to the decline of organised labour and the socio-spatial shifts, as dissected by Moreton, 2009) and Islamic fundamentalism elsewhere (for example, the Iranian revolution of 1979).

The recessions were argued as reflecting the failure of state policies of demand management employing Keynesian principles, and the search for an alternative saw a growing divergence between political parties. In the UK, for example, the relatively high degree of cross-party agreement over demand management broke down after 1974, with the Conservative Party promoting a greater emphasis on markets and a reduction in the direct role of the state. The postwar political consensus (in which Labour and Conservative policies had converged) was thus breaking down and the sense of crisis and change was reinforced by the revival of violent conflict over Northern Ireland and of nationalist claims in Scotland and Wales, all of which contested the established arrangements of the British state (Nairn, 1977). This growing polarisation added to problems in the UK, some claim, because of the uncertainty it engendered about the future; a change of government would bring a major shift in policy. In the USA, similarly, there was a rise in support for 'new right' policies promoted within the Republican Party, which countered the liberal policies of the Democrats in the 1960s.

To some social theorists, recession and its major impacts (notably unemployment) should accelerate the demands for reform and revolution. Both the UK and the USA at the end of the 1970s elected right-wing governments dedicated to programmes of economic regeneration by a reduction in public expenditure and a liberalisation (or rather a deepening) of those capitalist forces that produced the inequalities so widely condemned only a decade earlier. The role and size of the state were to be significantly reduced (Gamble, 1988); its task was to enable a buoyant economy (particularly through sound management of public finances and the removal of restrictive practices – notably, though not only, by trade unions). 'Enterprise' was to be encouraged by a system which rewarded successful initiatives and risk-taking, and the result, it was claimed, would be a wealthier society from which all would benefit, while the state provided protection only for those in proven need, whose 'rights' for such assistance were balanced by state-created 'obligations' (Johnston, 1992). Elsewhere, communist and socialist regimes either collapsed (as in the Soviet Union, and many regimes it supported in Eastern Europe and some elsewhere, such as in Afghanistan and South Yemen), abandoned all socialist commitments (as cases like Angola, Mozambique and Ethiopia) in the face of crisis and armed resistance, or embraced many aspects of capitalism and became much more integrated into a globalising economy (as in China and, to an extent, in Cuba and Vietnam). These developments continued (under governments of a variety of political stances, but all committed to continue the shifts unleashed in the 1980s) until the latter part of the twenty-first century's first decade, when, after a period of financial and real-estate speculation, the failures in the banking system and the inability of a number of states to sustain their very large budget deficits stimulated a major crisis and political response (Hendrikse and Sidaway, 2010). In many countries, including the USA and the UK, the response involved a fresh range of restructuring and marketisation (policies that have been termed 'neoliberalism' – see Chapter 6), alongside attempts to stimulate a private-sector-led economic recovery out of recession. Meanwhile, those cuts meant that few universities were recruiting additional academics (Franklin and Ketchum, 2013), and the funding and governance of universities underwent further shifts.

The contemporary situation

Education and research has been subject to significant policy changes since the 1970s because of these deep shifts in the political context. For the first time for more than two decades, expenditure on higher education in the UK was cut in 1981, for example, and the numbers of students undertaking undergraduate degrees and postgraduate training reduced. The cuts were selective, with relative protection for science and technology – growth in which was seen as necessary for economic progress – alongside substantial reductions in the social sciences, believed by many politicians to be the homes of left-wing radicals and the fomenters of discontent. Research funds were similarly cut and redistributed. The reaction to these policies took the form, in part, of attempts to defend academic freedom and independence and the need for a 'critical conscience' within society (involving curricula designed to develop critical intellects as well as foster 'transferable skills' to be deployed by graduates in the labour market) but there was also a desire to reorient work within disciplines and make them more 'relevant' to current societal concerns.

The educational system within which the research components of academic disciplines are located benefited from the boom years of the first three post-1945 decades. Expansion was rapid and academic activity was considerable. It was then cut back, especially in the UK, as – to some decision-makers at least – expansion, especially expansion in the arts and social sciences (excluding business studies and management), was seen as an unaffordable luxury. Economic progress did not require large numbers of students and potential researchers being trained in disciplines with little relevance to perceived societal needs and working on topics that were critical of societal structure. Disciplines and scholars had to prove their relevance and sell their skills in the market-place. Academic freedom was not entirely removed, but was to be curtailed, simply by denying it resources. The result is that after participating gleefully in the booms of the 1950s and 1960s, academia went into deep depression by the late 1970s, suffering internal and external crises of confidence and subject to considerable political direction.

There was a reversal in political attitudes towards higher education at the end of the 1980s, however, if not also in the funding provided. In the UK, for example, there was a growing realisation that national competition in the restructured global economy required developing human resources to their full potential, through a substantial increase in the participation rate of the 18–25-year-olds in further and higher education and an expansion of continuing education for older people. It was argued that the UK had a lower percentage of its population in higher education and obtaining relevant qualifications compared with competitor, more successful, economies such as the USA and much of continental Europe. More qualified students were needed and universities should be allowed to expand, though in a more rigorous market system. (For an evaluation of these policies, see Wolf, 2002; Palfreyman and Tapper, 2014.) When a Labour government was returned to power in 1997 after eighteen years in opposition, it built on the expansion of student numbers during the preceding decade with a goal of half of all school-leavers proceeding to a university education. (The then percentage attending universities was in the high 30s, compared to less than 10 per cent four decades earlier.) During its period of office, however, it ended the tradition of that university education being free: students were required to pay an increasing proportion of their tuition costs, with universities being given some flexibility in the amounts charged, though with a fixed maximum (which most adopted).

Higher education was implicated in the post-2008 restructuring of public finances after the banking and financial crises. Although a well-educated society was seen as necessary to national economic prosperity in a world where many manufacturing and routine clerical jobs had been lost to emerging economies, for example, the UK government decided that most university students would have to pay fees towards their degree courses through publicly funded loans (later sold off by the government to private firms) to be repaid once they passed an earnings threshold. Although it was recognised that there was a public as well as a private gain from education, further direct public money (in the form of subsidies to the costs of degree courses) would only be made available for those studying science, technology, engineering and mathematics (the STEM disciplines); there would be no subsidy to universities for social science and humanities students. Research funding was also reduced, especially in the non-STEM subjects, and there was an increased emphasis in grant-making on the work's potential impact, especially on 'wealth creation' and 'wider impacts' (see Collini, 2012). Since geography frequently bridged the science–social science/humanities divides (given physical geography), the immediate negative consequences for the subject were somewhat contained, but the longer-term ones are uncertain.

Thus, despite a demographic downturn (which particularly affected the size of the teenage population in the socioeconomic classes from which students traditionally moved into higher education), universities, polytechnics and colleges (most of them later elevated to university status) in the 1990s were expected to expand their provision and numbers. However, they were expected to become more efficient in doing their teaching (i.e. to take more students without additional resources – Jenkins and Smith, 1993), and to obtain more income (especially for research and continuing education) from sources other than the state, including their students (in many countries overseas students are charged more than the standard fee for 'home' students). Further, there was an increased emphasis on the 'customer pays' principle (for long the norm in North America), with students meeting more of the costs of their education (in part, it was argued, because they would benefit from higher incomes as a consequence of their qualifications and expertise), so being more concerned with getting 'value for money'; courses with a clear vocational orientation became more popular. Again, geography fared relatively well in this respect, since the mixture of qualitative and quantitative skills developed by its graduates conferred advantages in a competitive labour market

The UK government also became increasingly concerned over accountability for the funds provided to the universities for research. These were thought to be too widely spread and their use subject to insufficient evaluation. From the late 1980s on, therefore, funding for research in each discipline was to be concentrated on those institutions where work was evaluated as of high quality internationally, and a series of regular Research Assessment Exercises (RAEs) was instituted which grade all departments on the quality of their research outputs and environment; the higher the grade achieved, the more money provided, and the greater the amount of time and resources available for research. At the same time, the Research Councils, which provide funding for individual projects, were restructured to promote 'user interests' and were directed to focus their funding on areas of national economic importance identified through a regular Technology Foresight exercise. Financial pressure was thus used to promote a particular orientation to research (and teaching) within universities, one that was sympathetic to the ideology of the 'enterprise economy' (Johnston 1995a, 1995b). In addition, the ways in which universities were funded meant that whereas success in the RAEs brought substantial financial rewards, teaching activities did not; indeed, student numbers in universities were capped for some years, so that no extra income could be obtained, and research – both income from external sponsors and the public money allocated according to RAE grade – was the main source for new income. Not surprisingly, research occupied an increasingly important place in academic life, putting pressures on individuals, which many believed was to the detriment of their teaching roles (Sidaway, 1997).

There have now been seven RAEs (later renamed Research Excellence Framework, REF) in the UK. After each, the distribution of the rewards has become increasingly concentrated on the highest graded departments. This increased the pressure on departments and their members to be successful according to the criteria employed by the RAE/REF panels. These are supposed to recognise research quality, both retrospective and prospective, but the main 'evidence' used in those judgements, especially in disciplines like geography, is publications. There were claims that papers take precedence over books and that papers in certain journals are considered of higher quality than others (e.g. Short, 2002). Certainly, some journals have been cited much more than others among the four publications that staff members are required to submit as indicative of their best work (Johnston, 2003a), although those involved in the assessment for geography in 2008 refuted a straightforward correlation between impact or reputation of the journal and the quality they ascribed to submissions (Richards *et al.*, 2009). In addition, other 'indicators' are deployed to assist in the subjective judgements, such as research income and studentships gained, and citation indices. The ability of the procedure to produce a result significantly different from a purely quantitative audit has been debated (see Kelly and Burrows, 2011, on the case of sociology). As well as the long-standing evaluations of excellence based on peer review of staff research publications, the quality of the research environment and various indicators of esteem, a substantial portion of the 2014 REF evaluation (20–25 per cent) was based on an evaluation of 'research impact', on indications that the work undertaken has achieved 'demonstrable benefits to the wider economy and society', which excludes its impact on other academic researchers. 'Research users' therefore played an important role on the assessment panels, seeking to evaluate how academia is generating substantial returns beyond those purely scholarly (on, for example, 'the take-up or application of new products, policy advice, medical interventions, and so on': Collini, 2012, p. 170; Sayer, 2015). The implications for much work in the social sciences and humanities, its future funding, are unclear. Debate in human geography about impact, however, took place in the context of earlier discussions about the merits and nature of applied geography – our focus in Chapter 9.

In part, as a response to the claims that the RAEs were over-emphasising research at the expense of teaching, a parallel external evaluation of teaching quality was established in the mid-1990s, though those departments that were highly rated received no rewards other than the status of the perceived distinction, and very few indeed got the 'unsatisfactory' grades that threatened their ability to recruit students (Johnston, 1996a). Furthermore, although some of the evaluation involved external assessors observing and grading actual teaching sessions, much of it focused on other aspects of curriculum design, student success and so forth. The assessment procedure was as much a bureaucratic inspection of bureaucratic practices as it was an evaluation of the quality of the student learning experience. Its impact has been to further increase the pressures on staff, alongside the increased pressure to obtain research grants and contracts, and publish high-quality research to win high RAE grades. This has occurred during a period of very significant reductions in the resourcing of universities on a per capita (student) basis; research and teaching productivity has been increased, at substantial, if unmeasurable, costs (Johnston 1994a, 1995a, 1995b).

This audit and accountability culture in UK universities (Johnston, 1994b) has spread, with similar assessment procedures introduced in Australia and New Zealand, and elsewhere in Europe – for example, Castree *et al.* (2006) consider their consequences for the production of geographical knowledge. In the USA the decentralisation of responsibility for higher education, the existence of a large number of private universities and the market-orientation of education provision has precluded the introduction of centralised evaluations. Nevertheless, as Luke (2000) demonstrated with regard to political science there, there are similar pressures to win grants, produce high-quality research papers in the 'right' journals and gain high citation indices as universities compete for the best staff and graduate students, and the reputations that go with them. And although there aren't centralised bureaucratic evaluations as in the UK, there is a variety of reputational exercises that result in rankings of departments and institutions in 'league tables' which are widely used to promote institutional prestige. As Luke puts it, these exercises, and the inputs on which they are based, involve 'arithmetical economies of professional correctness [that] drive or stall careers' on the basis of which 'universities and disciplines . . . track their various cases of success and failure, growth and decline, winners and losers in this complex symbolic economy' every bit as much as do companies listed on the Stock Exchange (Luke, 2000, pp. 212–13). It is the outcome of his description of academic life as:

an existence pegged to perpetual examinations: seminar discussions, research papers, disser-
tation defenses, conference papers, journal submissions, book contracts, teaching evaluations,
committee assignments, tenure hearings, academic promotions, annual reports.

(p. 217)

These are now routine components of the academic profession. Its members are compelled
to 'trade actively in the sign-value economies of professional correctness' and forced 'to adhere
correctly to very clear disciplinary expectations in order to succeed' (p. 226). Academics research
and publish in order to succeed in their chosen careers: the results of what they do – their
publications – form the store of knowledge that represents their discipline's achievements. What
they have chosen to write about, and why, determines the contemporary nature of their science.

An example of a USA assessment exercise is the National Research Council's (2010) data-based
evaluation of doctoral programmes across 62 disciplines and 212 of the country's leading
universities. (For a long-term analysis of geography doctoral programmes and their success in
placing graduates in leading universities, see Liu and Zhan, 2011.) Its results, like those of many
others, including the UK's RAEs, have been widely publicised in the form of 'university league
tables'. Indeed, the production of such tables has become a feature of the specialist media (such as
the UK's *Times Higher Education*) as well as some newspapers and universities and other institutions,
using a wide range of indicators – designed, it is claimed, to assist students in their choice of
university. Those tables – and the politics of their construction – are now a focus of considerable
attention: universities and their departments enjoy the charisma of high rankings, whatever the
quality of their production (on which, see Foley and Goldstein, 2012).

Departments of geography and their individual members have had to respond to these political,
institutional and potential market changes, therefore, and to restructure their course and research
offerings accordingly. The nature of the science that they practised and taught was necessarily
always strongly influenced by the 'culture of the times' (which does not mean that all conformed
to it). As Livingstone (1992, p. 347) expresses it, this is because, 'geography changes as society
changes, and . . . the best way to understand the tradition to which geographers belong is to get a
handle on the different social and intellectual environments within which geography has been
practised'. As we seek to appreciate the history of geography over recent decades, we have to do this
in the context of the constraints within which it has been practised.

Conclusions

The thesis of this chapter is that the history of an academic discipline must be set in a context
comprising three elements: the occupational structure, the organisational framework for research,
and the societal environment. These three interact in a complex variety of ways. The occupational
structure is very much enabled and constrained by the societal environment – for example, as
indicated by the expansion in student and hence numbers of academics to teach them in the 1960s,
negligible opportunities for promotion in UK universities in the late 1970s and early 1980s because
of the cuts in educational funding, and the uncertainty that prevails today, after another period of
relative expansion (albeit not with a commensurate increase in funding) in the 1990s and the first
decade of the twenty-first century. Similarly, the framework for research, although established by
and for academics, is subject to societal support. Some frameworks are much more acceptable than
others and so are more likely to receive the needed public finance. The governance of academics and
academia through metrics that seek to quantify their achievements and qualities, however, has fed
into a system that was already structured by hierarchy and more informal reputational judgements.
These metrics mimic markets (and combine with more conventional markets in the hiring
of academics or the charging of student fees, for example). Writing about the UK case, one
critical observer traces how 'their performative character contributes to the generation of the

[commercialised and quantified] structures of feeling that have come to increasingly define academic life' (Burrows, 2012, p. 357).

In the following chapters, the content of human geography in the UK and North America since the Second World War is reviewed, within the context set by this discussion. The pre-war contexts and making of the modern discipline (as well as some of its antecedents) are also briefly documented in Chapter 2. In all cases, however, the emphasis is on 'extraordinary research' rather than the cumulative achievements of what Kuhn termed 'normal science', stressing the debates over how research should be done in human geography. No attempt is made to test the models outlined in the section on 'the academic working environment', although the ideas outlined there have clearly influenced the organisation of the book. The main purpose is to present a reasoned summary of debates in human geography (the discipline being defined as comprising that which is claimed as human geography), based principally on the published record. The relationship between this summary and the models of disciplinary change only resurfaces in the final chapter.

Chapter 2

Foundations

the central core of our discipline – why we call ourselves geographers and why what we do has so much in common – has been essentially European in derivation and characteristics ... developed over a remarkably short space of time at the end of the eighteenth and beginning of the nineteenth centuries ... these characteristics have persisted as enduring concerns through the series of 'revolutions' to which the discipline is said to have been subjected ...

(David Stoddart, 1986, p. 28)

any analysis of how nature is understood in geography is necessarily one about the nature of geography.

(Noel Castree, 2005, p. xix)

geographical imagination A sensitivity towards the significance of PLACE and SPACE, LANDSCAPE and NATURE in the constitution and conduct of life on earth. As such, a geographical imagination is by no means the exclusive preserve of the academic discipline of GEOGRAPHY.

(Derek Gregory, 2009, p. 282)

For all the battles which geography has witnessed since the nineteenth century ... none has been anyway near as profound or thoroughgoing as that which destroyed the empirical descriptive geography of the early modern age and replaced it with the connective causal discipline of modern geography. Modern geographers are all, to ape A. N. Whitehead on Plato, merely a series of footnotes to that moment.

(Robert Mayhew, 2011a, p. 44)

Although this book is mostly about human geography since 1945, a brief outline of the nature of the discipline in preceding decades is necessary, for several reasons. First, although 1945 was a watershed year in many aspects of the social, economic and intellectual life of the countries considered, it was not a significant divide in the views on geographical philosophy and methodology. Not surprisingly, the war years were not a major period of intradisciplinary academic debate. Most academics spent the time either on active service or in associated intelligence activities: although some of the latter retained their teaching commitments, for most the everyday activities of teaching, pure research and administration were replaced by commitment to the war effort. (On UK geographers' war activities, see Balchin, 1987; Clout and Gosme, 2003; and Maddrell, 2008, who focuses on the roles on women geographers engaged in cartographic work in wartime; on the USA, see Stone, 1979; Hohn, 1994, focuses on one geographer's specific role; on European geographers, see Clayton and Barnes, 2015.) After 1945 it took a few years for academic life to return to something like normality, to assimilate the new staff needed to replace war losses, to teach the student backlog and to react to new social and economic environments.

A second reason for retrospection relates to the processes of change discussed in Chapter 1. New practices are responses to perceived failings of those currently in favour, not inventions produced in an intellectual vacuum. Thus, post-1945 changes were reactions to the philosophies and methodologies developed and taught in the preceding decades about which some knowledge is necessary.

Finally, academic changes are not instantaneous. New research programmes may take years to mature, while experimentation with alternatives takes place, the programmatic statements are written and converts are won over. Meanwhile, the current practices prevail. Adherents continue in their accepted ways, researching, publishing and teaching undergraduates according to the conventional wisdom. Even when a new paradigm has been crystallised, it may co-exist with its predecessors for several years, while competing for support; it is quite feasible for several world views to have adherents at the same time, quite possibly in the same academic department, especially in the social sciences and humanities – indeed, that could well be a virtually permanent state.

Geography in the modern period

The hallmark of an academic discipline is an educational organisation which provides specialist training in the subject. James (1972) dates the beginning of such an organisation for geography around 1874, when the first ten university geography departments were established in German universities (Wardenga, 1999; Schelhaas and Hönsch, 2002; see also Taylor, 1985a). Claval notes how:

> Towards the 1860s the economic development of Germany had been accompanied by the development of applied sciences, and the accompanying development of the universities itself demanded that they redefined their functions. . . . it was the presence of geography in primary education that was the main stimulus to the growth in numbers of university teachers of geography after 1874. . . . Certainly the new situation created by the unification of Germany between 1864 and 1871 imposed new demands in terms of texts and geography courses at the primary and secondary level. To this must be added the requirements of German imperialism . . . the formation of the German empire in 1871 and the acquisition of colonies from 1884–85 (after the Congress of Berlin) gave birth to a new politics which was the impulse to the creation of new university posts in and departments of geography.
>
> (1981, pp. 96–7, our translation)

Berdoulay thus describes the intellectual influence of the German example (which also stimulated responses in France and then elsewhere in Europe):

> German and French geographic thought and achievements played a foundational role in the development of geography as a discipline in European and American academic institutions during the late nineteenth century and the first decades of the twentieth.
>
> (2011, p. 74)

Previously, geography was investigated either by amateurs or by scientists trained in other fields. In the UK, for example, geography was a recognised area of scholarly work for several centuries, even though it had no separate disciplinary or departmental identity in the small number of universities then operating. Map-making, exploration and chorography (the description of the characteristics of different areas) were all taught at English and Scottish universities by the end of the sixteenth century, for example (Cormack, 1997; Withers and Mayhew, 2002; Withers, 2001;

Mayhew, 2011a). Geographical knowledge was important to traders who, along with bankers and private individuals, invested in expeditions in the expectation of benefiting commercially from the new knowledge obtained – about new lands and potential resources, for example, as well as new and quicker routes to known destinations. Increasingly, exploitation of such resources called for state support and protection for their activities and contributions to the national wealth, which stimulated state-sponsored settlement, colonisation and the growth of imperial naval power. The practice of geography was central to this imperial ethos of much of nineteenth-century Europe (Driver, 1998; Butlin 2009; Bell *et al.*, 1995; Clayton, 2011).

Promotion of geography in this context involved the establishment of geographical societies in many national capitals and major trading cities. Among the first were those founded in Paris (1821), Berlin (1828), London (1830) and St Petersburg (1845), with later creations in Manchester (1884), Newcastle upon Tyne (1887), Liverpool (1891) and Southampton (1897) (see Johnston 2003b; Heffernan, 2003; McKendrick, 1995; Butlin, 2009). Many of these societies had royal patronage and were strongly supported by members of the mercantile, diplomatic and military classes as well as audiences drawn from the upper and middle classes who attended their lectures. The societies' main roles involved collating and publishing information, sponsoring expeditions and fostering relevant scientific developments, as in navigation and cartography. The American Geographical Society was similarly established in New York in 1851 as 'a merchant's information bureau' (Koelsch, 2002, p. 253; Wright, 1952; Morin, 2011; Crampton *et al.*, 2012; Johnston, 2013).

Some of these societies became involved in educational activities, promoting geography as a subject in their countries' schools and universities. In the UK, the RGS lobbied for geography's presence in the evolving school curricula (Keltie, 1886; Ploszajka, 1999; Wise, 1986). It then turned its attention, in the late nineteenth century, to the universities, realising that to sustain the subject's presence in schools required trained teachers who had a university qualification in the discipline while its presence in those institutions gave it status. The society's attentions focused on the universities of Cambridge and Oxford, on the argument that geography would only be recognised as an academic discipline if it were taught in England's 'ancient universities'; posts were established there, with RGS financial support for nearly forty years, in the late 1880s, although chairs were not filled in either until the 1930s (Stoddart, 1975a; Scargill, 1976; Pawson, 2009.) Other universities took the lead in establishing full departments and degree programmes (Slater, 1988) – the first was at Liverpool in 1917 – as a result of varying pressures for geography teaching. Some of this pressure came from cognate disciplines, notably geology and economics, which required geography courses within their syllabuses; the first permanent chair was held by L. W. Lyde at University College London, for example, who was appointed to teach geography for economics students in 1903 (see Dickinson, 1976; Clout, 2011a), whereas at Glasgow a newly appointed professor of geology pressed strongly, and successfully, for geography to be taught there too. The growth was hence fast, as charted in Figure 2.1. By the Second World War, there was a geography department (albeit invariably small) in virtually every UK university and university college, plus many of those in the British Empire. (For more detail on this aspect of the discipline's institutionalisation there, see Johnston 2003b. The essays in Steel, 1987, discuss British geography in the 1930s.)

There was no national pressure for geographical education in the USA, largely because of the decentralised nature of educational systems there: the American Geographical Society showed very little interest in geographical education in the latter decades of the twentieth century (Wright, 1952; Morin, 2011), and the same was largely true of the National Geographic Society until late into the twentieth century (Schulten, 2001). The establishment of geography teaching in some of the country's universities resulted from perceived needs in other degree programmes (many of these can be traced through the increasing number of departmental histories available online at http://old.geog.psu.edu/hog/dept_histories.html). One was geology, and some of geography's

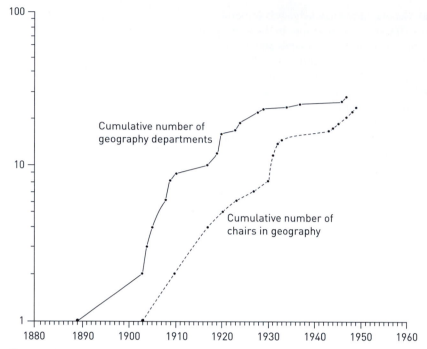

Figure 2.1 The foundation of departments of geography and of chairs in geography in British universities

Source: Stoddart, 1986, p. 46.

early protagonists – such as William Morris Davis (p. 43, this book) at Harvard – were trained geologists. Departments of economics and schools of business and commerce also identified the need for geography teaching (Fellmann, 1986); indeed, the first fully fledged department of geography – at the University of California Berkeley campus – emerged from such an origin, in 1898 (Dunbar, 1981). Many of those courses lapsed by 1920, but by then geography had been established elsewhere in a number of universities, although to nothing like the same relative extent as in the UK.

Most of the initial teachers on those geography programmes were trained in other disciplines and were attracted to geography by its core interest in the relationships between people and their environments. Many of their stimuli came from early work in Germany and France, where a number went to study (Berdoulay, 2011). By virtue of their primacy in establishing geography departments and the intellectual heritage of Alexander von Humboldt and Karl Ritter (see p. 46, this book), French and especially German influences were strong in the original formulation of geographers' core agenda (Claval, 1981). Others were trained by the discipline's pioneers – such as the many who took the certificates and attended the summer schools offered at the University of Oxford by Mackinder and Herbertson (who, like many of the other pioneers, also wrote innovative school textbooks); Mackinder's (1887) paper 'On the scope and methods of geography' was an early, influential essay in defining a particular niche for geography within British academia (see also Kearns, 2009): as Withers (2010) discusses, it was important for British geographers at that time to establish their 'scientific' credentials in order to win recognition (and respectability) for their discipline's prospectus. The late nineteenth and early twentieth centuries thus saw the transition from what James (1972) termed geography's classical age into its modern period; he termed post-1945 its contemporary period. James's modern period is virtually co-extensive with the decades surveyed in Freeman's (1961) *A Hundred Years of Geography*, which identified six interrelated trends in the geographical literature:

1   The *encyclopaedic trend* covered the collection of information about the world, particularly areas little known to Western Europeans and North Americans. Although the great age of discovery was over, and by the late nineteenth century much of the world had been visited by European explorers, there remained vast tracts, notably in Africa, which if not *terra incognita* were almost empty on contemporary Western maps.

2   The *educational trend*, in which the nascent academic discipline was establishing its role and relevance in educational systems, thereby ensuring its reproduction. (On the role of the RGS in this in the UK, see Freeman, 1980a, 1980b; on the Geographical Association, founded in 1893, see Balchin, 1993, and Walford, 2001. Johnston, 2003b, overviews the discipline's institutionalisation in UK universities.)

3   The *colonial trend* reflected a major preoccupation during the modern period's early decades, especially in the UK (Driver, 1998; M. Bell *et al.*, 1995). Organisation of the imperial world required information, whose provision became a major task of geographical research while its propagation was the keystone of geographical education. Furthermore, part of the creation of an imperial national identity and ethos involved educational activities, which gave children a sense of their identity in the world – and, by implication at least, of their superiority and that of their environment. (On school texts in the period, see Ploszajska, 1999; J. M. Smith, 2001; Kearns, 2009. Buttimer and Fahy, 1999, p. 179, argue that school geography in Ireland has been taught 'as an *instrumentum regni*, to transmit political, economic, and social attitudes of prevailing regimes ... transmitting images rather than evoking curiosity'. On geography's wider role as popularly understood in the creation of world views, see Schulten, 2001.) Geography was centrally involved in the creation of this imperial identity in the UK. In the USA, however, geographers were less involved at this stage, and a particular notion of American identity and destiny was created by historians such as Frederick Jackson Turner, who associated the American nation with its expanding frontiers (Kearns, 1984). The entanglements between USA imperialism and geography would come later (Morin, 2011).

4   The *generalising trend* describes the use to which geographical data were increasingly employed. Academic study involved more than collecting and collating facts; these had to be interpreted, and the methods and aims of such interpretation defined the early paradigms of the discipline's development, as will be discussed.

5   The *political trend* was reflected in the contemporary uses made of geographical expertise. For example, Isaiah Bowman, trained at Harvard under Davis, became Director of the American Geographical Society in 1915, and was selected as an adviser to Woodrow Wilson at the conferences (notably at Versailles near Paris) which redrew the map of the world after the First World War (Martin, 1980; N. Smith, 1994). Bowman subsequently wrote one of the first books on political geography (Bowman, 1921). According to N. Smith (2003, p. 183):

> World War I forever transformed U.S. geographical research, and no one sensed this more acutely than the chief territorial specialist of the U.S. delegation at Paris. The new world that confronted everyone after 1919, Bowman understood, required a new geography ... human rather than physical fashioning of the world's landscapes was now preeminent. For a U.S. geographical tradition heavily modeled on German geography and umbilically connected to geology, this was a dramatic paradigm shift. It was not his physical but his political geography that was most exploited in Paris.

In the UK, the first appointee to the School of Geography at the University of Oxford, H. J. (later Sir Halford) Mackinder also wrote on political geography, which later informed a wealth of writing around the theme of geopolitics (though Mackinder himself did not use that term). His heartland model (Mackinder, 1890, 1904; Kearns, 2009) presented the Eurasian continent as the fulcrum of world power, hence the 'geographical pivot of history' as encapsulated in the triplet:

> Who rules East Europe commands the Heartland;
> Who rules the Heartland commands the World-Island;
> Who rules the World-Island commands the World.

Mackinder later became an MP and then was involved in diplomacy, but continued to write on geopolitical issues until the 1940s; his *Democratic Ideals and Reality* (Mackinder, 1919) summarised his views (on his career, see Blouet, 1987; W. H. Parker, 1982; Kearns, 2009).

6   The *specialisation trend* was a reaction to the growth of knowledge and the inability of any one individual to master it all, even within the single discipline of geography. Prior to the modern period, many scientists and other academics had catholic interests and expertise. As the volume of research literature increased and the techniques of investigation demanded longer and more rigorous training so it became necessary to specialise first, in this context, as a geographer and then within geography.

Clearly, these trends overlapped. However, three major paradigms (as disciplinary matrices) characterise human geography's modern period. The development of each was strongly influenced by a few individuals who left lasting impressions on the discipline. Until after the Second World War there were only small numbers of geographers working in universities, and many of the pioneers had no training in the discipline (Johnston, 2005). Their main role was as teachers, and many did relatively little research or publication of original work. A small number took up leadership roles, however, defining geography's academic agenda and strongly influencing what the discipline studied and its methods (Mackinder – e.g. Mackinder, 1887; Kearns, 2009 – and Davis, 1906, were especially influential in the UK and the USA respectively). Almost all of these early definitional essays focused on geography as a whole; human geography emerged as a separate subdiscipline relatively late in the modern period (Johnston, 2010a). Those who occupied powerful positions within the discipline – as professors and departmental heads, for example – were able, through combinations of the force of their personalities, the strength of their arguments and carefully chosen appointments to their own and other departments (including their post-graduate students), to play crucial roles in setting the discipline's academic directions. There were differences and debates, but in small communities only a few ideas generally prevailed – a very different situation from the pluralism that developed in the second half of the twentieth century as geographers became much more numerous and key individuals, though still important, were less dominant.

Paradigms in the modern period

Exploration

This first approach was carried over from the classical period; exploration was the major activity popularly recognised as geography through most of the nineteenth century. The collection and classification of information about 'unknown' parts of the earth (unknown, that is, to Western Europeans and North Americans) was undertaken by explorers and navigators, many of whose expeditions were sponsored by geographical societies. Information gained was used to enhance cartographic knowledge and disseminated widely through lectures and books. The map of the world was completed and filled in at an increasing pace.

The importance of exploration within geography declined in the late nineteenth century, although soon after his appointment at Oxford, Mackinder (who had been trained as a biologist and historian) felt it necessary to establish his geographical credentials by becoming the first recorded person to climb Mt Kenya (Kearns, 2009). Much *terra incognita* remained on European maps of the rest of the world, however, and the geographical societies maintained their interest in and sponsorship of expeditions throughout the period. The exploration tradition is maintained in the USA by the National Geographical Society (NGS) and its popular publication, *National Geographic* magazine. Like other magazines, such as *New Zealand Geographic* and the *Geographical* magazine, *National Geographic* concentrates, though not exclusively, on material illustrating the relationships between people and their environments (with copious high-quality photographic material and maps). In such media, the boundaries of geography and natural history are imprecise. (The NGS also has a TV channel devoted to similar material, and has given large sums of money in recent years to develop geographical programmes for schools in the USA, after surveys revealed considerable ignorance about what is where in the world: Johnston, 2009d.)

Although most of it was not strictly exploration, the work summarised by Freeman under the colonial and encyclopaedic trends can be included here, since its aims were the collection, collation and dissemination of information. Much of the material was about commercial activities and infrastructure, as in volumes such as G. G. Chisholm's *Handbook of Commercial Geography* (first edition, 1899) and *Gazetteer of the World* (1895), which were aimed at the world of commerce, with companion volumes for schools (Wise, 1975; Barnes, 2000, 2001a). Their content comprised statistics and descriptions of production and trade, and a training in this type of geography involved the assimilation of large bodies of factual knowledge ('capes and bays' geography). Similar texts were produced in the USA (Barnes, 2001a; Fellmann, 1986; Lawton and Miller, 2001; Johnston, 2010a).

The value of such geographical information and expertise was widely recognised, and was called on during both world wars when geographers were recruited into intelligence services. In the UK, for example, they were responsible for the preparation of reports about areas in which Allied troops were likely to be engaged, and their Second World War British Admiralty Handbooks (edited at Oxford and Cambridge under the leadership of Kenneth Mason and Clifford Darby respectively: Clout, 2003a; Clout and Gosme, 2003) were put on sale afterwards. The comparable volumes produced in the USA by the Office of Strategic Services (of which Richard Hartshorne was Deputy Head) – the Joint Army-Navy Intelligence Studies – remained classified documents, however (Barnes, 2006a).

Environmental determinism and possibilism

Environmental determinism and possibilism represented the first attempts at generalisation by geographers during the modern period. Rather than just present information in an organised manner, either topically or by area, geographers sought explanations for the patterns of human occupation of the earth's surface. Their major initial source for explanations was the physical environment, with a general belief that the nature of human activity was controlled by the parameters of the physical world within which it was set.

The origins of this environmental determinism lie in Charles Darwin's landmark *On the Origin of Species* (first published in 1859) which influenced many scientists (Livingstone, 1992; Armstrong, 1999, argues that Darwin was influenced by the pioneer geographer–naturalist von Humboldt). Darwin's notions regarding evolution were taken up by an American geographer, William Morris Davis, in his cycle-of-erosion model of landform development (Chorley *et al.*, 1973; Beckinsale 1976; Vale, 2002). Ideas of natural selection and adaptation formed the basis of statements regarding environmental determinism, including Davis's (1906) programmatic paper identifying the core of geography as the relationship between the physical environment as the control and human behaviour as the response (Stoddart, 1966; Martin, 1981; Campbell and Livingstone, 1983; and Livingstone, 1984, discuss the related influence of Lamarckism in the development of geography; see also Peet, 1985a).

Chief among the late nineteenth-century speculations on environment-society was the work of the German scholar, Freidrich Ratzel, who was trained in zoology before taking an interest in

geology and ethnological (the classification and study of 'races'/peoples) research on the diffusion and distribution of peoples. Ratzel began lecturing on geography in Munich in the 1870s, then
become professor of geography at Leipzig (1886–1904). Before Ratzel, geography had largely been
seen as a natural (physical) science. According to Berdoulay (2011, p. 76), Ratzel's 'breakthrough'
was key 'in setting geography within a scientific-evolutionary discourse'. His *Anthropogeographie*
(Ratzel, 1882–91) related the course of history to the earth's physical features (illustrating organic
notions that had been stimulated by Darwin); and his *Politische Geographie* (Ratzel, 1897) adapted
Darwinian arguments to states, which he treated as organisms that struggle for land (*Lebensraum*,
or living space), with the strongest states able to expand territorially. Although others had also
elaborated the idea prior to Ratzel (see Halas, 2014), these arguments were later taken up by a
Swedish conservative political scientist (Rudolf Kjellen) who coined the term 'geopolitics' to signify
the codification and study of this mode of political geography. In the 1920s and 1930s, a German
military officer and geographer (Karl Haushofer) expounded such geopolitics to provide an
element of the putatively 'scientific' underpinning for the 1930s–1940s Nazi policy policies of
territorial expansion, although sometimes in tension with the 'racial' ideas about space and power
that became central to Nazism (Kost, 1989; Natter 2003; Parker, 1985; Barnes and Minca, 2013;
Barnes and Abrahamsson, 2015). In turn, however, geopolitics soon acquired a life of its own with
work in many European countries (from Italy and Portugal to Denmark, the Netherlands and
Romania) and Japan, as part of a wider conversation about empires, sovereignty, space and power,
in which geographers played a central (though not exclusive) role (Dodds and Atkinson, 2000).
After the Second World War, such classical geopolitics continued in right-wing military circles from
Turkey to Portugal and in many South American countries, and it has more recently been adopted
in post-communist Russia as well as in parts of Central Asia.

To return to the development of anglophone human geography, however, Ratzel's ideas were
promoted in the USA by Ellen Churchill Semple, who opened her book *Influences of Geographic
Environment* (1911) with the statement that 'Man is the product of the earth's surface.' Semple, who
was Davis's student, had travelled to Germany to consult Ratzel. In Europe, she connected also with
British geographers via seminars at the RGS and RSGS. Indeed, there was a transatlantic commerce
in these debates about environmental determinism. Ellsworth Huntington advanced theories
relating the course of civilisation to climate and climatic change (Huntington, 1915, 1945). In
some hands, the environmental influences adduced were gross, and with hindsight it is hard to
believe that they were taken seriously; Tatham (1953), for example, illustrates the extent to which
authors were prepared to credit all aspects of human behaviour with an environmental cause.
The reception of Semple was complex, however; the wide range of debates that her work triggered
has been dissected by Keighren (2006, 2010; see also the survey conducted by John K. Wright,
1962, to identify the extent of her influence on other geographers in the first part of the twentieth
century). There were frequent links with discourses about 'race'. Though not all those who took
part were explicitly racist, it is evident that the wider framework in which debates evolved was
shaped by racialised assumptions about categories, lands and peoples that owe much to the wider
imperial moment (as in various essays and books by T. G. Taylor – e.g. 1927, 1937; on Taylor, see
Strange and Bashford, 2008). Livingstone (1992, p. 221) notes how 'The idea that climatic regions
on both local and global scales implied an ethnic moral topography was an idea that weaves its way
through the corpus of nineteenth- and early-twentieth-century writings.'

The debates lingered in the form of 'tropical geography' into the postwar era, when the
prospects for the development of tropical regions (mostly colonies and former colonies) re-entered
the debate. Indeed, there was a wider conversation – resting on translation – between anglophone
and francophone geographers writing about these themes (Power and Sidaway, 2004) of tropical
geography.

Such debates were informed by reaction to the extreme generalisations of the environmental
determinists which had led to a counter-thesis, that of possibilism. This presented individuals as

active rather than passive agents. Led by French geographers, themselves followers of the *Annales*-school historian Lucien Febvre, the possibilist argument had people perceiving the range of alternative uses to which they could put an environment and selecting that which best fitted their cultural dispositions. Taken to extremes, this approach could be as ludicrous as that which it opposed, but possibilists generally recognised the limits to action which environments set, and avoided the great generalisations which characterised their antagonists.

Debate over environmental determinism and possibilism continued into the 1960s (Lewthwaite, 1966; Spate, 1957, for example, proposed a middle ground with the concept of 'probabilism'). And it was not only in tropical geography that they were registered. The doughtiest advocate of the determinist cause was T. Griffith Taylor, foundation professor of geography at the University of Sydney, whose views so angered politicians interested in the further white settlement of outback Australia that he was virtually hounded out of the country (Powell, 1980a). He argued that possibilists had developed their ideas in temperate environments which offer several viable alternative forms of human occupance. But such environments are rare: in most of the world, as in Australia, the environment is much more extreme and its control over human activity accordingly much greater. He coined the phrase 'stop-and-go' determinism to describe his views. In the short term, people might attempt whatever they wished with regard to their environment, but in the long term, nature's plan would ensure that the environment won the battle and forced a compromise out of its human occupants (Taylor, 1958; Sanderson, 1988; Strange and Bashford, 2008).

Many debates begin as two opposing, extreme views, and end with a compromise accepted by all but the most fervent devotees of each polar position. Thus, the lengthy discussion among geographers about whether people are free agents in their use of the earth or whether there is a 'nature's plan' slowly dissolved as the antagonists realised the merits in each case. Some geographers studied human–environment interactions outside the confines of these debates (see Fleure, 1919; the debate's deep foundations are the subject of Glacken's (1967) magnum opus, *Traces on the Rhodian Shore*). While some geographers strongly promoted environmental determinism, however, respect for their discipline was relatively low in the wider academic community. As a consequence, geography's next focus, which nevertheless had some roots in environmental determinism, was very much an introspective and conservative one, alongside other attempts to develop an alternative paradigm, which lacked extensive support, such as Barrows's (1923) presentation of geography as human ecology.

The region and regional geography

This third approach dominated British and American geography for much of the middle twentieth century. Like environmental determinism, it was an attempt at generalisation, but it lacked structured explanation and so was of a very different type from the increasingly discredited law-making attempts of the previous writings. This drew heavily on contemporary developments in Germany and France. Of great importance in Germany was the work of two individuals who many identify as the founders of modern geography. Both died in 1859, one having established the modern roots of what became known as systematic geography and the other having performed a similar task for regional geography. Both accepted the definition of geography set out by the Enlightenment philosopher, Immanuel Kant (1724–1804), who lectured on physical geography at the University of Königsberg (Harvey, 2009). He argued that knowledge can be organised and classified in two ways: because of similarities in origin wherever they occur (the logical classification, which is the basis of the various sciences); or because of similarities in when or where they occur – the discipline that looks at similarities in terms of time is history, whereas that which looks at similarities in terms of place of occurrence is geography (J. A. May, 1970).

Alexander von Humboldt (1769–1859) was a naturalist with very broad scientific interests. He spent five years travelling in Central and South America, amassing information about the environment and its exploitation by humans. He assembled this material (over a 20-year period in Paris) to show how environments varied, with differences in agricultural practices and patterns of human settlement, for example, reflecting interactions among altitude, temperature and vegetation. This approach to the discipline emphasised field collection of data and its synthesis through maps, leading to inductive generalisations. Sachs (2007, p. 49) describes von Humboldt's

> effort to see new things deeply, in context, and in connection to everything else he's seen and learned. The juxtaposition of his literal acts of observation and his *philosophical* observations, his adventures and his scientific 'tangents' leaves us with a sense of swirling intellectual currents.

The remainder of his career was spent assembling similar materials from a wide range of sources, which were synthesised in the five-volume *Kosmos* (von Humboldt, 1845–62). The first volume (1845) provided an overview of the universe, which was followed by one (1847) on representations of the earth in art and literature and a history of scientific writing on the earth since Egyptian times. The third (1850) was largely concerned with astronomy; the fourth (1858) turned to human interactions with the earth, with a vast range of observational material described and used to derive generalisations about those interrelationships; and the final volume (published posthumously in 1862) dealt with geology and volcanology. (On von Humboldt, see G. Martin and James, 1993; Bowen, 1970, 1981; Rupke, 2005; Sachs, 2007.)

Karl Ritter (1779–1859) was trained in a variety of disciplines – during which he met, and was impressed by, von Humboldt – and by 1811 had published a two-volume work on the geography of Germany. He was appointed to a chair of geography in Berlin in 1820 (a year after a similar appointment in history in Frankfurt). After early field work in Italy and Switzerland, as well as Germany, he travelled little and his teaching was largely based on secondary sources. He focused on the connections between phenomena in places – of 'unity in diversity', which he believed represented 'God's plan'. This involved defining regions, separate areas of the earth's surface with distinct assemblages of phenomena. The material accumulated was used to produce his 19-volume (unfinished) *Erdkunde* (Ritter, 1817–59).

Regional geography was also the main focus of an influential French geographer, Paul Vidal de la Blache (1845–1918) who, unlike von Humboldt and Ritter, obtained a higher degree in geography after initial training in ancient history and literature. He was appointed to a post at the Sorbonne in 1898, where he maintained close links with the *Annales* school of historians. French geography had strong early connections to both mapping and history (Claval, 1999), which were sustained by Vidal's concentration on defining and describing regions, relatively small homogeneous areas (*pays*) whose distinctive *genres de vie* resulted from the interactions of people with their physical milieux. Unlike some German contemporaries, however, he did not see those inter-actions as predominantly determined by the physical environment. He was attracted to possibilism (which he first encountered in Ratzel's *Anthropogeographie*). The environment offers people a range of options, and they choose how to modify nature according to their cultural and technological inheritances – as Lucien Febvre put it, 'nowhere necessities . . . everywhere possibilities'. Vidal's major contributions were his *Tableau de la Géographie de la France* (1903), an introduction to the multi-volume *Histoire de la France* (see Ozouf-Marignier and Robic, 1999), and the 15-volume *Géographie Universelle*, finished in 1948. Many of his students wrote dissertations on individual *pays* (Clout, 2009); they dominated French geography through the first half of the twentieth century (Buttimer, 1971) and fitted into the conservative and nationalist milieu of interwar France amid territorial threats from Germany (see Gregory, 1994; Heffernan, 2001). French regional geography thus emphasised France's essential 'unity in diversity', blending climatic and cultural influences from the Mediterranean and the Atlantic into an essential *élan français*.

The early development of geography in Germany and France relative to the UK and the USA meant that as the academic discipline was being established in the latter countries, some of those attracted to it sought inspiration and training from French and German sources. There was considerable personal interaction, and it was the norm at both UK and USA universities during the first half of the twentieth century for would-be geographers to be fluent in either if not both German and French. Ideas about the nature of geography thus infiltrated the English-speaking world from continental Europe. They were modified to fit local circumstances, but dominated contemporary thinking for several decades. (On a major institutionalised interaction between American and continental European geographers – the 1912 American transcontinental excursion, in which a small number of British geographers participated – see Clout, 2003b, 2004; Clout and Stevenson, 2004.)

Hartshorne and American views

The ideas and methods of regional geography were taken up in the USA after environmental determinism had been largely rejected. In the late 1930s, two non-geographers published a major survey of American regionalism (Odum and Moore, 1938) and in 1939 the Association of American Geographers published a monograph – Richard Hartshorne's *The Nature of Geography: A Critical survey of current thought in the light of the past* – which was rapidly established as the definitive statement of the current orthodoxy (Stoddart, 1990; Martin, 2015). As Hartshorne (1948, 1979) later made clear, there was much debate among American geographers during the 1930s (most of it apparently unpublished, though see James and Mather, 1977) about the nature of their discipline. He was concerned about both the tone and the content of that debate (particularly in Leighley, 1937), and in 1938 submitted a paper to the *Annals of the Association of American Geographers* as a contribution to the philosophical discussions. He then proceeded to Europe for political geography fieldwork on boundary problems. This was frustrated by the political situation, and he spent his time reading further European, mainly German, work on the nature of geography. He used this to extend his 1938 paper, adding the crucial subtitle; the result was a 'paper' of some 230,000 words which became the major philosophical and methodological contribution to the literature of geography in English then available.

A synopsis of Hartshorne's book, and his interpretations of others' works, notably Hettner's (Harvey and Wardenga, 2006), is not possible in a few paragraphs, and only the main conclusions can be highlighted. Hartshorne's statements were positive ones – of what geography is. They were only normative in the sense of him saying that geography should be what others (notably Hettner, whose approach to geography reflected a line of thinking through Kant and von Humboldt to his own mentor, Richtofen) have said that it is. Thus, Livingstone (1992, p. 306) describes Hartshorne's project as seeking 'to determine the nature of geography from scrutinizing its history', and Lukermann (1990, p. 58) claimed that the *Nature* was 'a search for authority to validate the conclusions drawn from selected premises – largely formulated by Hettner, who had philosophical associations and leanings rather than historical associates'. Butzer (1990) argued that Hartshorne was selective in his use of Hettner's material, and Derek Gregory (1994, p. 51) claimed that 'Hartshorne's views were developed through a highly selective exegesis of a German intellectual tradition. His approval of Hettner (in particular) was unrestrained, but the regional geography that he constructed was purged of both the physic–ecological and the cultural–historical implications that were indelibly present in Hettner'. In sum, Hartshorne transmitted to an American audience his interpretation of a particular German argument as to the nature of geography.

Hartshorne argued forcefully that the focus of geography is areal differentiation, the mosaic of separate landscapes on the earth's surface (see Agnew, 1990, on the representation of Hartshorne's focus as 'areal variation' rather than 'areal differentiation'). It is defined as:

> a science that interprets the realities of areal differentiation of the world as they are found, not only in terms of the differences in certain things from place to place, but also

in terms of the total combination of phenomena in each place, different from those at every other place.

(Hartshorne, 1939, p. 462)

The discipline 'is concerned to provide accurate, orderly and rational descriptions and interpretations of the variable character of the earth's surface' (p. 21) and

seeks to acquire a complete knowledge of the areal differentiation of the world, and therefore discriminates among the phenomena that vary in different parts of the world only in terms of their geographic significance – i.e. their relation to the total differentiation of areas. Phenomena significant to areal differentiation have areal expression – not necessarily in terms of physical extent over the ground, but as a characteristic of an area of more or less definite extent.

(p. 463)

The principal purpose of geographical scholarship is thus synthesis, the integration of material on relevant characteristics to provide a total description of a place, or region, which is identifiable by its peculiar combination of those characteristics. Hartshorne identified a close (Kantian) analogy between geography and history; the latter provides a synthesis for 'temporal sections of reality', whereas the former performs a similar task for 'spatial sections of the earth's surface' (p. 460). From this separation of their roles, Hartshorne concluded that there was no need for geographers to study change, since that was the province of historians.

To Hartshorne, 'the ultimate purpose of geography, the study of areal differentiation of the world, is most clearly expressed in regional geography', so that the discipline's research methods had to focus on regional definition and depiction. Regions are characterised by their homogeneity on prescribed characteristics, selected for their salience in highlighting areal differences. Identification of such regions 'depends first and fundamentally on the comparison of maps depicting the areal expression of individual phenomena, or of interrelated phenomena . . . geography is represented in the world of knowledge primarily by its technique of map use' (pp. 462–4). Others followed his lead, as in the editorial chapters in a later overview of American geography (James and Jones, 1954).

Hartshorne emphasised map use. Although it is valuable for geographers to appreciate the methods of map construction, the sciences of surveying and map projections are of only secondary interest to them; their prime task is map interpretation. Much information to be interpreted may have been placed on the maps by geographers during fieldwork, whose role and nature were of considerable interest to his contemporaries (Johnston, 2010a); they established detailed methods for mapping land use as the bases for regional delimitation, developed at a series of intensive 'field camps' (see Whittlesey, 1954).

Preparation of a regional synthesis required materials from both other sciences specialising in certain phenomena (though usually not their areal patterning) and the emerging topical systematic specialisms within geography which complemented, but were eventually subsidiary to, regional geography. Physical, economic and political were the main systematic subdivisions recognised at the time Hartshorne wrote (Johnston, 2010a), although a later survey, set firmly within the regional paradigm, identified many other 'adjectival geographies', including population, settlement, urban, resources, marketing, recreation, agricultural, mineral production, manufacturing, transportation, soils, plant, animal, medical and military, plus climatology and geomorphology (James and Jones, 1954). A number were of only minor importance, however, so that despite the apparent diversity of interests among geographers of the time, the 'classic' regional study usually followed a sequence comprising physical features, climate, vegetation, agriculture, industries, population and the like (Freeman, 1961, p. 142), and was summarised by a synthesis of the individual maps to produce a set of formal regions.

To most geographers of the period spanning the Second World War, regional geography was at the forefront of their discipline's scholarship and systematic studies were the providers of information for that enterprise. To James, 'Regional geography in the traditional sense seeks to bring together in an areal setting various matters which are treated separately in topical geography' (1954, p. 9). Urban geographers studied towns because they 'constitute distinctive areas' (Mayer, 1954, p. 143), in line with the regional concept; political geographers studied the functions and structures of an area 'as a region homogeneous in political organisation, heterogeneous in other respects' (Hartshorne, 1954a, p. 174); and in defining a 'new' field of social geography, J. Wreford Watson (1953, p. 482) saw it 'as the identification of different regions of the earth's surface according to associations of social phenomena related to the total environment' (see also Johnston, 1993b). Each topical specialism produced its own regionalisation and each had links (although often weak) with the relevant systematic sciences. The key differentiating factor between geography and other, systematic, disciplines was the geographer's focus on the region. For Hartshorne, every geographer should have not only a substantive specialism (he considered himself a political geographer and wrote the chapter on that in *American Geography: Inventory and prospect* – Hartshorne, 1954a. He also made an early, prescient, apparently original – since it made no reference to any of the work of early location theorists discussed in the next chapter – contribution to the study of industrial location – Hartshorne, 1927) but also a regional specialism. It was the latter that distinguished a geographer from scholars in other disciplines.

There was a tension within this promotion of regional geography, however. On one hand, there was the presentation of regions – following Vidal's example – as small (usually rural) homogeneous areas, with many countries comprising a patchwork mosaic of such separate districts, even though in many cases their boundaries were relatively indeterminate. In this view of the region, the geographer's field of study was inherently local, save in areas of little physical variation over large distances. On the other hand, the argument that geographers should be regional specialists was made at a much larger scale – an individual country, perhaps, but more commonly an entire continent or major subcontinental area. In this context, geographers presented themselves as specialists on, say, Latin America or the Indian subcontinent. Their teaching was at this scale even if their research and field knowledge was on specific parts only and textbooks were structured at the larger scale. There was thus a hierarchy of regions: the large areas used to structure considerable sections of many degree programmes (courses on the regional geography of 'x'), and the myriad small areas which had their own unity and whose identification was the focus of much geographical enquiry and writing (following Vidal's emphasis on small regional units with distinct physical characteristics, notably in soils and drainage, and associated agricultural specialisms – Buttimer, 1971, 1978a).

British views

Examining the period between its nascent establishment as a university discipline in the 1880s and the middle of the twentieth century, Stoddart (1986, p. 51) describes it as 'pragmatic, concerned with practical issues', as well as 'pedagogic: all were deeply connected with education in the schools and with the training of teachers'. He also noted that

> they brought to bear on the problems selected for study an almost bewildering range of formal training and interest, obtained before geography had itself become established as a formal discipline, but unified through a shared belief in geographical methods and objectives.

Stoddart adds that there was a strong emphasis on fieldwork and planning issues in the UK.

Hence, British geographers were less concerned with philosophical and methodological debate than were their American counterparts during the 1920s, 1930s and 1940s (though see the exchange in the *Scottish Geographical Magazine* during the late 1930s, initiated by Crowe, 1938). They were apparently more pragmatic, less prone to contemplate the nature of their subject and more prepared, perhaps, to adopt the well-used adage that 'Geography is what geographers do'. But they too generally accepted that geography's *raison d'être* was synthesis, integrating the findings of various systematic studies, but with a strong emphasis on genesis distinguishing their approach from their American contemporaries', as in the studies of geomorphology and historical geography (Darby, 1953; K. J. Gregory, 2003). According to Wooldridge and East (1958), 'geography . . . fuses the results, if not the methods, of a host of other subjects . . . [it] is not a science but merely an aggregate of sciences' (p. 14): 'its raison d'être and intellectual attraction arise in large part from the shortcomings of the uncoordinated intellectual world bequeathed us by the specialists' (pp. 25–6) and 'in its simplest essence the geographical problem is how and why does one part of the earth's surface differ from another' (p. 28).

All these statements indicate some transatlantic common body of opinion (Stoddart (1990) discusses Hartshorne's influence on Woodridge) although, despite a statement that 'The purpose of regional geography is simply the better understanding of a complex whole by the study of its constituent parts' (p. 159), Wooldridge and East did not elevate the regional doctrine as much as their American counterparts (nor were they carried to excesses of environmental determinism in earlier decades). Nevertheless, Wooldridge (1956, p. 53) wrote in 1951 that:

> the aim of regional geography . . . is to gather up the disparate strands of the systematic studies, the geographical aspects of other disciplines, into a coherent and focused unity, to see nature and nurture, physique and personality as closely related and interdependent elements in specific regions.

He argued that in any university department of geography each staff member should be committed to the study of a major region (p. 64; Mead, 1963, 2007).

As in the USA, much early development in the UK involved work at two scales (Freeman, 1961, p. 84; Johnston, 1984a). The large scale is exemplified by Herbertson's (1905) exercises dividing the earth into major natural regions, usually based on climatic parameters and thus having some links with the earlier determinism. At the smaller scale:

> The fundamental idea was that the small area would legitimately be expected to show some distinct individuality, if not necessarily entire homogeneity, through a study of all its geographical features – structure, climate, soils, vegetation, agriculture, mineral and industrial resources, communications, settlement and distribution of population. All these, it has often been said, are united in the visible landscape, linked into one whole and dependent one on another. And more, every area, save those few never occupied by man, has been influenced, developed and altered by human activity, and therefore the landscape is an end-product, moulded to its present aspect by successive generations of people. The practice has therefore been to take an evolutionary view and . . . to attempt to reconstruct the landscape as it was a hundred, or a thousand years ago.
>
> (Freeman, 1961, p. 85)

The delineation of regions at these two scales also included attempts to devise hierarchies (or 'orders'), whereby smaller regions are grouped into larger units – with the various orders given separate names, such as 'tracts' and 'stows' (Unstead, 1933; Johnston, 1984a).

Although much regional definition and description was undertaken for pedagogical purposes, the practical value of appreciating regional divisions was also pressed. Notable was the work of L. D. (later Sir Dudley) Stamp. A University of London geology graduate, Stamp made the translation to geography when working as Professor of Geology and Geography at the University of

Rangoon (his initial employment in Burma was as an oil geologist). He returned to the UK in 1926, to the London School of Economics, where he developed wide interests and published a large number of school textbooks on all parts of the world, as well as a much revised text on *The British Isles* (Stamp and Beaver, 1947) and several revisions of G. G. Chisholm's *Handbook of Commercial Geography* (1895). In 1930, Stamp launched what was then by far the largest research project undertaken by British geographers – the Land Utilisation Survey of Britain. Over four years he mobilised some 250,000 students at about 10,000 schools to map land use over the entire country (a later survey was organised in Northern Ireland) at the scale of 6 inches to the mile. These provided the data for maps of land use at the scale of 1 inch to the mile, a series of county reports, many written by academic geographers and including regional summaries, and for his summary volume on *The Land of Britain* (Stamp, 1946a; see Rycroft and Cosgrove, 1999; Wise, 1968). Stamp promoted the survey in particular, and geographical skills in general, as valuable for planning land use (as in Stamp, 1934, 1946b, 1949), and this was one of the main foundations for his later advocacy of applied geography (Stamp, 1948, 1960). His expertise led to him either serving on or advising a number of important government commissions regarding land use and related issues during the Second World War and after (such as the Royal Commission on Land Utilisation in Rural Areas, of which he was Vice-Chairman, and the Royal Commission for the Common Land – Buchanan, 1968). We reconsider Stamp's contributions again in Chapter 8 where his arguments for an applied geography are evaluated.

In addition to the definition of what are generally termed formal (or uniform) regions – areas of any scale that are relatively homogeneous on the selected phenomenon or phenomena – there was also interest in functional regions. These are also homogeneous areas on criteria regarding interactions between places: the unity of a functional region is provided by links to a common dominant node. Functional regions were introduced to British geography by Fawcett (1919), who suggested that the main cities' hinterlands should be the territorial framework for regional governments. One of his students identified the hinterlands of both large cities – Leeds and Bradford (Dickinson, 1930) – and market towns – in East Anglia (Dickinson, 1933). Dickinson travelled widely in France, Germany and the USA during the 1930s, assembling material on this aspect of urban geography and developing his argument – crystallised in his postwar text, *City Region and Regionalism* (Dickinson, 1947) – that functional regions should be the basis for dividing up a country for the purposes of public administration (Johnston, 2000c, 2002a). Dickinson was a strong proponent of the regional approach and saw no tension between the formal and functional regional concepts, arguing that the view of the region as developed by what he termed the 'landscape purists' (Dickinson, 1938, p. 12) was 'the "objective manifestations" of economic circulation . . . men and things in movement'.

A significant difference between British and American geographers by the 1950s was in attitudes to physical geography. Both groups had strong traditions of work in this field, and many geographers had academic roots in geology. But this tradition had slowly dissolved in North America (the USA much more than Canada) and interest in the physical environment waned, particularly its understanding as against its description (Leighley, 1955). This may have been a consequence of the excesses of environmental determinism, with a subsequent desire to remove all traces of that connection and to see society as the formative agent of landscape patterns and change. Thus, with regard to geomorphology – the science of landform genesis – Peltier wrote:

> the geographer needs precise, factual information about particular places. What landforms actually exist in a given area? How do they differ? Where are they? What are their distribution patterns? The geomorphologist may concern himself with questions of structure, process, and stage, but the geographer wants specific answers to the questions: what? where? and how much?
> (1954, p. 375)

Geographers, according to this view, were only interested in the geography of landforms: geomorphology, the genetic study of landforms, was a part of geology and deemed outside the geographical enterprise.

Similar reactions saw reduction, if not removal, of material from climatology and biogeography from American geographical curricula, and their replacement by introductory physical geography courses that described landforms, climates and plant assemblages – usually in a regional context – but paid little or no attention to their origins. A substantial revival of physical geography in the USA after 1970 reflected its perceived relevance to understanding and resolving environmental problems, realising the potential of technological advances in remote sensing and associated technologies (see Marcus, 1979, and the essays in Gaile and Willmott, 1989, 2004). British geographers did not follow this American trend. According to Wooldridge and East:

> To treat geography too literally as an affair of the 'quasi-static present' is to make both it and its students seem foolish and superficial. It is true that our primary aim is to describe the present landscape; but it is also to interpret it. . . . Our study has therefore always to be evolutionary. . . . It is unscholarly to take either landforms or human societies as 'given' and static facts, though we must not let temporal sequences obscure spatial patterns.
>
> (1958, p. 47)

Geography students at UK universities in the 1950s rarely specialised in either physical or human geography, except perhaps in the final year of their course. Both were considered essential parts of a geographical education (Johnston and Gregory, 1984; Cosgrove, 1989a). By that time, however, most British geographers were research specialists in either physical or human geography (though rarely exclusively so), although most also practised a regional specialism in which they 'integrated' studies from 'both sides' of their subject, as in the regional textbooks of the period. The 'dogma of regional synthesis' (Darby, 1983b, p. 25) was being softened, however, and geographers were increasingly turning their attention from regions to systematic studies, identifying themselves as either physical or human geographers.

Historical geography

One systematic specialism which stood slightly apart from the others was historical geography. Two separate approaches operated from the late 1920s on; indeed one (predominantly American) was not presented as historical geography.

The first approach was closely associated with the work of H. C. (later Sir Clifford) Darby in the UK (Perry, 1969; Darby, 2002). Darby was very much influenced by Cambridge historians in the 1920s, and his Ph.D. (the first to be awarded in geography at Cambridge) was on the medieval Fenland (Darby, 1940a, 1940b). He then turned his attention to the 1086 Domesday Book, editing a multi-authored set of regional volumes depicting the geography it disclosed, county by county, with a summary volume (Darby, 1977; Perry, 1979). He also enunciated a broad approach to historical geography which combined detailed cross-sectional analyses of particular times – selected according to the availability of source materials – with linking narratives describing the intervening changes (as in two edited volumes: Darby, 1936, 1973). Darby's approach was set out in major essays (1953, 1962) and posthumously in a volume assembled from his surviving lecture notes (Darby, 2002; he also left notes on his course on English landscape change – see Darby, 1951—but without the illustrative material, which prevented these also being published in book form). During his career he supervised a large number of research students, establishing a strong presence for historical geography within the discipline (Prince, 2000); indeed, he restructured the department at University College London after his appointment as head in 1949 to reflect the importance he gave to historical geography (Clout, 2003c).

Largely contemporary with Darby's was an approach centred on the works of Carl Sauer (on whom, see Williams, 2014) and his associates. Sauer was a Chicago geography graduate and strongly influenced by German work but, unlike Hartshorne, he did not elevate the regional focus and concentrated on change rather than pattern. (On the differences between Hartshorne and Sauer, see Lukermann, 1990, and Butzer, 1990; for critical assessments of Sauer, see Stoddart, 1997b, and Symanski, 2002.) Some of Sauer's early work has environmental determinist tinges (e.g. Sauer, 1918, pp. 421–2), but after his move to Berkeley in 1923 he focused on the cultural processes leading to change as reflected in the landscape, beginning at the prehuman stage of occupance (Mikesell, 1969); most of this work was conducted either outside the USA (particularly in Latin America) or in its less industrialised parts.

Sauer's (1925) first methodological statement constrained geographical endeavour closely to the generic study of landscapes, emphasising their cultural features (although work was also done on the borderlands between geography and botany, his main interdisciplinary links were with anthropologists – see Kroeber, 1952); there was no glorification of the region, however. In his later 'sermons' – as he called his methodological and philosophical statements – Sauer (1941, 1956a) encouraged research over a much wider field, but emphasised the study of cultural landscapes, through field-based studies (Speth, 1999). It was a creative art-form whose hallmark was that it was not prescribed by pattern or method: the human geographer is obliged 'to make cultural processes the base of his thinking and observation' (Sauer, 1941, p. 24).

This genre of work involved neither detailed reconstruction of past geographies nor close consideration of regional boundaries. Instead, it led to a catholic historical geography ('The elements of the landscape that have a cultural origin . . . cannot be understood rationally, but only historically' – Leighley, 1937, p. 141) whose rationale was that 'through its study we may be able to find more complete and better answers to the problems of interpretation of the world both as it is now and as it has been at different times in the past' (A. H. Clark, 1954, p. 95). Not all American historical geographers followed this lead – Brown (1943), for example, worked on detailed reconstructions of past periods (Meinig, 1989) – but the 'Berkeley school', which Sauer founded and led for almost five decades, had many followers, and a particular point of view focused on a single iconoclast (Hooson, 1981). Sauer's influence was continued by his students, notably Leighley, Parsons, Kniffen and Clark (Bushong, 1981; M. Williams, 1983, 2014).

A major statement consistent with Sauer's approach was a 1955 Wenner-Gren symposium 'Man's Role in Changing the Face of the Earth' (see Glacken, 1983) which brought UK and USA practitioners together and resulted in a substantial publication (52 chapters plus discussions: 1,193 pages in all – Thomas, 1956). Sauer identified its theme as:

> the capacity of man to alter his natural environment, the manner of his so doing, and the virtue of his actions. It is concerned with historically cumulative effects, with the physical and biologic processes that man sets in motion, inhibits, or deflects, and with the differences in cultural conduct that distinguish one human group from another.
>
> (1956b, p. 49)

It presented no grand methodology or general findings – indeed, Sauer's closing statement (1956c) criticised the tendency of American authors who 'have an inclination to universalize ourselves' (p. 1133). The volume stressed cultural differences as the basis for both diversity in human response to environments and its impacts on them. Mumford's (1956, p. 1142) conclusion has affinities with that advanced in the 1980s by adherents of structuration theory (see p. 221, this book): 'the future is not a blank page; and neither is it an open book'. The symposium was updated some 30 years later (Turner et al., 1990).

One Berkeley school contributor to that symposium was Glacken (1956), whose *magnum opus Traces on the Rhodian Shore* (Glacken, 1967) surveys interpretations of nature and demonstrates 'how all-pervading teleology has been in the history of Western interpretation of nature' (Glacken, 1983, p. 32). Like the 1955 symposium publication, this book is widely recognised as a classic on society–nature interrelationships. But by then that topic was receiving diminishing attention among geographers; its impact was less than might otherwise have occurred, as was Tuan's (1968) monograph on vernacular interpretations of *The Hydrological Cycle*.

The body of work stimulated by Sauer is set in a wider frame of studies of human–environment interactions concerned with anthropomorphically generated change rather than the influence of the environment on people which characterised the early twentieth century (Turner, 2002). Geographers' basic concerns became the environments that we encounter and the changes that we create in order to make them more comfortable for our desired lifestyles (Sack, 2001). Increasingly, concern with the deleterious consequences of those changes focused attention on conservation (Manners and Mikesell, 1974), for which one stimulus was George Perkins Marsh's *Man and Nature* (1864), widely recognised as a pioneer statement on environmental abuse by humans and the needs for conservation (Lowenthal, 2001).

Geography in the early 1950s

Two early 1950s edited volumes characterise geographical practices at the end of the modern period. *American Geography: Inventory and Prospect* (James and Jones, 1954) was a 'semi-official' compilation celebrating the Association of American Geographers' 50th anniversary with 'a progress report on the objectives and procedures of geographic research' (p. vii). Although each chapter has a named author, all had been commented on in draft by others, with some collaborating in their production. All but two chapters deal either with a systematic field within geography (urban, political etc.) or with geographical methods (field survey, air photo interpretation and cartography). But the first two – on 'The field of geography' and 'The regional concept and the regional method' (by Preston James and Derwent Whittlesey respectively) – dominate the book.

Geography is defined as 'concerned with the arrangement of things on the face of the earth, and with the association of things that give character to particular places' (James, 1954, p. 4); the region is 'the geographic generalization of phenomena associated in area' (p. 9) and the map is 'the fundamental implement of geographic research'. Whittlesey (1954, p. 21) claimed that 'Geographers are in general agreement that regional study is an essential part of their craft' and concluded that regional study 'underlies and is applicable to all aspects of geography' (p. 65). These views permeate the entire volume, covering subject matter as diverse as political, agricultural, geomorphic, soil and zoogeographical regions, and regions in marketing geography. (On the dissolution of this core to geographical study provided by regions and maps, see Johnston, 2004b.)

Geography in the Twentieth Century: A Study of Growth, Fields, Techniques, Aims and Trends (T. G. Taylor, 1957 – this was an expanded version of one published in 1951) had a wider range of authors, from Canada (6; T. G. Taylor – see p. 45, this book – was then a Professor at Toronto), the USA (6), England (8, including Darby, Stamp and Wooldridge), and one each from Czechoslovakia and Poland. T. Griffith Taylor defines geography as 'the discussion of the causes of patterns of distribution' (p. v), and introduces the book as

> an attempt to answer questions which are engaging the attention of all geographers. What are the salient features of modern geography? What are we trying to accomplish? How have our ideas as to what are the important fields of our discipline changed during the last fifty years? How do our studies touch the fields of allied disciplines? Have America, Britain, France, Germany, and the Slav nations and even Canada produced special contributions? . . . Broadly speaking, are there different schools of geographic thought which cut across national boundaries to some degree?
>
> (p. 3)

The book has three parts, with five further introductory chapters on the history of geography, ten on 'the environment as a factor', and thirteen on 'special fields of geography'. This structure – particularly the ordering of the last two sections – reflects Taylor's position as a 'stop-and-go determinist', and although there are no chapters on regional geography or the regional concept per se (there is one on regionalism as regional administration) and the term 'region' does not appear in the book's index, Taylor states that 'Most geographers accept regional geography as the core of our discipline' (p. 8).

These two books clearly illustrate the transitional situation of Anglo-American geography at the beginning of the post-1945 period. Systematic studies were coming to dominate its internal organisation, but there was still a very strong tie to the regional concept and the belief in regional synthesis as the core of geographical knowledge and, for some at least as exemplified by Taylor's book, the remnants of environmental determinism in the primacy given to the 'environmental factor'. Later work on environmental geography would trace some of its lineages to these earlier human–environment debates, while losing the baggage of determinism/possibilism. Indeed, as subsequent chapters will chart, nature, landscape and environment have continued to be key themes in human geography, albeit from very different vantage points (or paradigms if you prefer) from those we have considered here.

Conclusions

This chapter has presented a brief outline of geography during its 'modern period', as a foundation for studying what followed. Three major approaches have been identified, although others also attracted attention (Taylor, 1937). All three lasted into the contemporary period, but the regional dominated in the years before and just after the Second World War. Its main focus was areal differentiation, the varying character of the earth's surface; its portrayal of that variation was built up from parallel topical studies of different aspects of physical and human patterns. Kevin Cox (2014, p. 21) judges 'human geography in the first half of the twentieth century' as 'an extremely conservative subdiscipline':

> There was an odd disinterest in modern urban society, apart from a few visionaries like Fawcett and Dickinson. The notion of methodological or theoretical debate was utterly alien. There was no sense of forward movement. And lacking a strong sense of the social, human geography found itself closeted off from the other human sciences.

Yet by the 1950s, initially in the USA and then in the UK too, there was growing disillusionment with the empiricist philosophy of regional geography. The topical specialisms slowly came to dominate disciplinary practice and regional synthesis was increasingly ignored.

In both countries, those developments were produced by only a small number of active geographers. In the UK, for example, there was no separate academic learned society organising conferences for geographers alone until the mid-1930s (Johnston, 2003b; Withers, 2010), and by 1946 there were only about 120 employed in the country's universities – many lacking any formal training in the discipline. Darby (1983b, p. 17) cites from an unsigned 1934 obituary that 'geography is far from having consolidated a definite position, British geography particularly so'. In the USA, the discipline was larger (in absolute terms). The Association of American Geographers (which was small and difficult to join without an established research reputation) had been founded in 1904; in 1949 it merged with the Association of Professional Geographers and ended its restrictive entry requirements (the RGS was the focus of considerable debate for a number of years before it was finally agreed to allow women to become Fellows – Bell and McEwan, 1996). Nevertheless, the discipline was not strong in the country's universities with American academics, according to Schulten (2001, p. 125), finding 'little incentive to join the fight for their subject's independence even as they bemoaned its position within the social studies'. Writing about Oxbridge (and by extension, the longer established Scottish universities, comprising Aberdeen, Edinburgh, Glasgow and St Andrews) in the 1920s, Hodges (2014, p. 77) claims that 'there was a great deal about the ancient universities which had less to do with learning than with social status, with courses in geography and estate management for those of a less academic turn of mind'. While according to Cox:

> For the most part, complacency ruled. In Great Britain, this was absolutely the case. In the United States, though, there had been some rude shocks challenging that self-satisfaction and reflecting a view in some universities that geography was marginal to their intellectual purpose. The most notable of these was the closure of the department at Harvard, the most prestigious of American universities, in 1948.
>
> (2014, p. 21)

Thus, the changes discussed in subsequent chapters of this book were built on fairly shallow and sometimes shaky roots, but when they came they coincided with a massive growth in the size (both absolute and, in the case of the UK, relative) of the academic geographic enterprise.

The modern period saw the institutionalisation of geography as an academic discipline, with a particular focus. The nature of the subject matter appropriated to that discipline, and the associated research and pedagogic practices, formed what Harvey (2005a) terms a particular 'geographical imagination' (the term draws from Mills's (1959) earlier discussion of a 'sociological imagination'; see also Gregory, 1994). The nascent academic discipline focused on a particular view of 'geographical knowledge', one that served the interests of states, nations and empires and associated commercial interest groups during the apex of imperial capitalism, so that the academics involved became 'tacit agents of state power, captive of a particular national and geopolitical vision' (Harvey, 2005a, p. 222); as different countries had different geopolitical and military agendas, particular 'national schools' emerged. But other geographical imaginations were emerging alongside these, reflecting alternative deployments of geographical knowledge. These included the growing tourist and travel industries that promoted a 'particular kind of geographical understanding marked by adventure, engagement with difference (and hence liberation from self), exoticism (the romancing of the other), eroticism (the themes of "sea, sun, sex and sand" are widespread) and culture (presented as unique and authentic to some locality)' (p. 232). Associated with this are the imaginations promoted by the media – increasingly by the cinema and TV, alongside geographical magazines (many of which promote the exotic, especially in wildlife and local culture – Johnston, 2009d) and the many local knowledges generated as part of daily life. The type of geographical knowledge produced by academic geographers, and disseminated not only by them but also by school teachers, thus co-exists alongside a number of others that use the name geography – and which some academics might think improper (Harvey, 2005a, p. 240). By the beginning of the end of the modern period in the 1950s, however, there was general appreciation that all types of geographical knowledge were concerned with 'what is where?' or 'what is it like there?'. With the movement into the contemporary period and the explosion of a variety of academic geographies, as described in subsequent chapters, the distancing between what many – arguably most – academic geographers do and vernacular understandings of their discipline increased.

Chapter 3

Growth of systematic studies and the adoption of scientific method

Some carried the idea as far as to believe that the 'core of geography' lay in 'regional synthesis', in which the facts of geology, climate, agriculture, industry and so on could, by some artistry, be fused into a delineation of 'the personality of a region' . . . one must remember that much so-called regional geography was in no sense a synthesis, but simply a way of handling the facts of, say, agriculture or morphology.

(H. Clifford Darby, 1983b, p. 15)

The fact that one can do little with the unique except contemplate its uniqueness, has led to the present unsatisfactory position . . .

(Peter Haggett, 1965c, p. 3)

The realization dawned that where a qualitative and descriptive approach had previously sufficed, it was now possible to press analysis much further than had previously seemed conceivable. Although the roots of this change can be traced back to the early decades of this century and even earlier, the adoption process undoubtedly accelerated during the 1950s and 1960s.

(Michael Chisholm, 1975a, p. 172)

The aim of geographical research is to provide laws and theories about the spatial structure of the earth. The task is formidable and modern efforts to develop geographic generalis- ations have a relatively recent origin. Progress has been slow . . . there was a solid, but rapidly declining, aversion to theory-building and mathematical methods within the ranks of geographers . . . there has been a process of trial-and-error, using methods developed in other disciplines . . .

(Maurice Daly, 1972, p. 98)

Every beginning is difficult, holds in all science.

(Karl Marx, 1867 Preface to the first German edition of *Das Kapital*)

Dating the origin of change in a discipline's orientation and practices, or even a part of it, is difficult. Several pieces which contain the kernel of the new ideas can usually be found in its literature, but often these are derivative of the earlier teachings of others, whose views may never have been published but only disseminated to their students; others may promote the ideas that are widely adopted later, but have no impact on those developments for a variety of reasons (see Johnston, 1993b, 1996b). Change can also emanate contemporaneously from several separate, though usually not entirely independent, nodes. An attempt to locate the first stirrings against the regional paradigm would be a futile exercise, therefore. Instead, the present chapter identifies the most influential statements, published by geographers.

The critique of regional geography

Major change within a discipline involves both dissatisfaction with existing approaches and the promotion of an acceptable alternative disciplinary matrix (if not world view, see p. 19, this book). The 1950s saw widespread dissatisfaction with the regional approach, as indicated by Freeman:

> disappointment with the work of regional geographers has led many to wonder if the regional approach can ever be academically satisfying and to turn to specialization or some systematic branch of the subject.
>
> (1961, p. 141)

He suggested three reasons for this. First, much regional classification was naive, particularly on the broader scale; its generalisations contained too many discrepancies. The second, and perhaps most important, was the 'weary succession' of physical and human activity 'facts' which characterised much regional writing: 'The trouble has perhaps been that many regional geographers have tried to include too much' (Freeman, 1961, p. 143). Third, one of the most influential models of regional writing, derived from work on the French *pays*, suggested that the whole of the earth's surface could be divided into clearly identifiable regions, each with its own character; that this proved false was reflected by many pedestrian studies of areas lacking such 'personality'. A satirical essay highlighting the poverty of much regional geography was published anonymously in an early issue of the IBG *Newsletter* (Anon., 1968). The author's name was given as Llwynog Llwyd, 'an exiled Celt . . . [working in] an Antipodean University'– it was Keith Buchanan (Johnston, 1999). Buchanan's critique emerged in two book reviews (Buchanan, 1958a, 1958b). He concluded that Harrison Church's *West Africa: A study of the environment and man's use of it* (1957), 'fails to capture anything of the quality of West African life or the vividness of the West African scene. These are lost, veiled in a grey harmattan haze of geological details and climatic statistics, of catalogues of towns and production data' – K. Buchanan, 1958a, p. 278. (For his own quest to portray the 'vividness of the scene', see Buchanan and Pugh, 1958; Buchanan, 1966.)

An earlier critique of regional geography as practised in the UK was Kimble's (1951) essay on 'The inadequacy of the regional concept'. He claimed that most geographers would 'contend that the highest form of geographical enquiry was a kind of "hunt-the-region" game, in which (provided we were offered enough clues) we were bound to discover that life in a given land or continent resolves itself into a neat pattern of cultural entities we call regions' (p. 151). But geographers had no agreed definition of a region (he had found no less than a hundred in the literature) and so they were unable to present a united front regarding their discipline to the wider academic world. He concluded that not only was there no agreement on how to define regions but also the concept was largely obsolete in the contemporary world:

> From the air it is the links in the landscapes, the rivers, roads, railways, canals, pipe-lines, electric cables, rather than the breaks that impress the aviator . . . regional geographers may be trying to put boundaries that do not exist around areas that do not matter.
>
> (p. 159)

His later reference to the situation in Germany that 'If the only type of regionalised unity possessed by modern . . . Germany was its circulation pattern, then the day of the region is nearly at the end, since mobility is the catalyst of regional diversity, and must sooner or later dissolve the whole compartmental structure of our civilization' (p. 168) makes clear that he was referring only to formal regions (see p. 89, this book); functional regions may well exist. (The reference to Germany was based on Dickinson's work: see Johnston, 2000c.) Formal regions were of little contemporary interest, however, even in work on different parts of the globe:

> Whatever the pattern of the new age may be, we can be sure there will be no independent, discrete units within it – no 'worlds within worlds'. There will be no neatly demarcated 'regions' where geographers . . . can study a 'fossil' community. Man's 'region' is now the world. This does not render superfluous the continued organization of geographical studies on a systematic areal basis. . . . But it does mean that we should be well advised, when making these studies, to refrain from searching for 'unitary patterns of living', 'entities of distribution', and from assuming, in the manner of determinists, that 'regional unity' is the goal towards which civilized society is moving. . . . At best a regional study can be only a personal work of art, not an impersonal work of science – a portrait rather than a blueprint. As such, it can have substantial value, but its value will lie in the realm of illumination and suggestion rather than of definitive analysis and synthesis.
>
> (p. 173)

Alongside these critics, in the USA Ackerman (1945) forcefully argued that insistence on the primacy of regional geography was undermining the associated systematic studies. Drawing on his experience of working in the wartime intelligence services (American Association of Geographers, 1946; Harris, 1997a), Ackerman identified two major failings among professional geographers there: their inability to handle foreign languages and the weakness of their topical specialisms. Regarding the latter, he categorised much geographical work as undertaken by scholars who were 'more or less amateurs in the subject on which they published' (p. 124), so that when called upon to provide intelligence material for wartime interpretation what they produced was extremely thin in its content. Regional geographers could provide only superficial analyses, and the division of labour within the discipline whereby people specialised in different areas of the earth was both inefficient and ineffective. (Gould, 1979, p.140, called the geography of the 50 years prior to 1950 'shabby, parochial and unintelligent . . . bumbling amateurism and antiquarianism'.)

Ackerman suggested that rectifying this major deficiency required much more research and training in the systematic specialisms. This was not necessarily contrary to giving primacy to regional synthesis, he claimed, since more detailed systematic studies could lead to greater depth in regional interpretations. There is little evidence that his paper had an immediate impact, however, as illustrated by the abstracts of papers presented at the Association of American Geographers' annual meetings (published then in the *Annals* each year). These indicated that no major shift in the orientation of academic work with the return to postwar 'normality', save in the case of the abstracts presented by Garrison and McCarty for the Cleveland meeting in 1953, which were clearly based on a different methodology to that widely used (see Garrison, 2002, p. 108, and also p. 72, this book). The systematic fields had undoubtedly been gaining in importance prior to Ackerman's statement, and continued to do so, as indicated by their extensive treatment in the review volume edited by Preston James and Clarence Jones (1954). But it was not until the mid-1950s that this was matched by widespread changes in disciplinary methodologies and philosophy. A few geographers who had shared Ackerman's wartime Washington experiences, such as Chauncy Harris and Edward Ullman (Eyre, 1978; Boyce, 1980), did follow this lead and were among the pioneers of the 'new geography' launched in the early 1950s.

Schaefer's (1953) paper and the response

A substantial (and ultimately successful) revolution against the regional paradigm got under way in the USA in the 1950s, initiated through debates over philosophy and methods. The first major contribution was a posthumously published paper by Schaefer (1953), often identified as one of the origins of the 'quantitative and theoretical revolutions' (on its impact, see Cox, 1995, and Getis, 1993). Schaefer trained as an economist: he joined the group of geographers teaching in the economics department at the University of Iowa after his escape from Nazi Germany (Bunge, 1979).

Schaefer claimed that his paper was the first to challenge Hartshorne's interpretation of the works of Hettner and others. He criticised Hartshorne's exceptionalist claims for regional geography and presented an alternative case for geography, adopting the philosophy and methods of the positivist school of science (Martin, 1990, p. 72). He first outlined the nature of a science and then defined the peculiar characteristics of geography as a social science. He argued that a claim for geography as the integrating science which put together the findings of the individual systematic sciences was arrogant, and that its products were somewhat lacking in 'startlingly new and deeper insights' (p. 227). A science is characterised by its explanations, and explanations require laws: 'To explain the phenomena one has described means always to recognise them as instances of laws.' Geography's major regularities refer to spatial patterns: 'Hence geography has to be conceived as the science concerned with the formulation of the laws governing the spatial distribution of certain features on the surface of the earth' (p. 227). These spatial arrangements, not the phenomena themselves, should be the subject of geographers' search for law-like statements. Their procedures would be the same as those employed in other natural and social sciences: observation would lead to a hypothesis which would then be tested against large numbers of cases, providing the basis for a law if it were verified.

Schaefer criticised Hartshorne's exceptionalist position that geography does not share the methodology of other sciences because of the peculiar nature of its subject matter – the study of unique places or regions; this puts geography in a similar situation as history, which studies unique periods of time. Using analogies from physics and economics, Schaefer argued that geography is not peculiar in focusing on unique phenomena. All sciences deal with unique events which can only be accounted for by integrating laws from various systematic sciences in particular circumstances, but this does not prevent the development of laws, although undoubtedly making it more difficult: 'It is, therefore, absurd to maintain that the geographers are distinguished among the scientists through the integration of heterogeneous phenomena which they achieve. There is nothing extraordinary about geography in that respect' (p. 231).

Schaefer traced the exceptionalist view in geography back to an analogy drawn by Kant between geography and history, an analogy repeated by both Hettner and Hartshorne (Harvey, 2009). Kant argued in his *Physische Geographie* (Vol. I, p. 8) that, 'Geography and history together fill up the entire area of our perception: geography that of space and history that of time' (Schaefer, 1953, p. 233; on Kant's geography more generally, see May, 1970; Elden and Mendieta, 2011; Harvey, 2009). But when Kant was working, Schaefer claims, history and geography were cosmologies, not sciences, and a cosmology is 'not rational science but at best thoughtful contemplation of the universe' (p. 232). Hettner followed Kant's views and developed geography as a cosmology, arguing that both history and geography deal with the unique, and thus do not apply the methods of science. Schaefer argued that this is a false position, for in explaining what happened in a certain time period historians must employ the laws of the social sciences. Time periods, like places, are unique assemblages of phenomena, but this does not preclude the use of laws in unravelling and explaining them. History and geography can both be sciences, for 'What scientists do is . . . They apply to each concrete situation jointly all the laws that involve the variables they have reason to believe are relevant' (p. 239). Schaefer further argued that Hartshorne disregarded one aspect of Hettner's writing which was nomothetic in its orientation, and in so doing misled American geographers. (Müller-Wille, 1978, p. 55, claims that Hettner predated Christaller in the development of ideas regarding central place theory – see p. 72, this book; Hartshorne made no reference to the paper by Hettner cited by Müller-Wille. On the same point, see Butzer, 1990, and N. Smith, 1990.)

The final part of Schaefer's paper reviewed some problems of applying a nomothetic (law-producing) philosophy to geography as a spatial, social science. He recognised difficulties of experimentation and quantification, for example, and suggested a methodology based on cartographic correlations (a topic explored by his Iowa colleagues: McCarty *et al.*, 1956). He argued that geographical laws are morphological (i.e. concerned with spatial forms), whereas those from other, 'maturer' social sciences are concerned with processes (hence morphogenetic). In order fully to comprehend the phenomena assemblages described in morphological laws, therefore, it is necessary to deploy process laws from other social sciences, a procedure which requires team work (the last point was also made by Ackerman). Thus, geography, according to Schaefer, is the source of the laws on location, which may be used to differentiate the regions of the earth's surface.

Hartshorne's responses

Schaefer's paper did not produce much direct reaction in print, despite claims that it was a major stimulus to later work in his proposed genre (Bunge, 1962). It drew considerable response from Hartshorne, however, first in a letter to the editor of the *Annals* (Hartshorne, 1954b) and later in three substantive pieces (Hartshorne, 1955, 1958, 1959). The last was another major book which, although not as influential as the 1939 volume, showed Hartshorne's continued importance to American geographers as an interpreter of their subject's methodology and philosophy.

The purpose of Hartshorne's first (1955) paper (which subsumed the earlier letter) was to indicate flaws in Schaefer's scholarship (see also D. Gregory, 1978a, p. 31). He begins with a discussion of the mores of methodological debate (Hartshorne, 1948). Schaefer was limited in his references, drew unsupportable conclusions, and misrepresented the views of others, so that 'In every paragraph, in nearly every sentence of this third section, there is serious falsification, either by commission or by omission, of the views of the writer discussed' (p. 236: this statement refers to the third part of Schaefer's paper, which focused on Hartshorne's interpretations of Hettner). More generally, Hartshorne claimed that Schaefer's paper 'ignores the normal standards of critical scholarship and in effect offers nothing more than personal opinion, thinly disguised as literary and historical analysis' (p. 244). Since Hartshorne himself (1959, p. 8) strongly believed that 'geography is what geographers have made it', to him all methodological and philosophical statements should be based on a close and careful analysis of others' published works.

Although most of the 1955 paper examined Schaefer's 'evidence', in the final section Hartshorne turned to the anti-exceptionalist argument. He pointed out that in concluding that geography should take process laws from the systematic sciences and use them to produce morphological laws, Schaefer came very close to preaching the sort of exceptionalist claim that he sought to destroy, so that his critique 'is a total fraud' (p. 237). Schaefer's position is summarised as: 'geography must be a science, science is the search for laws, and all phenomena of nature and human life are subject to such laws and completely determinable by them' (p. 242). Such scientific determinism is opposed to the summary of what geographers do set out in *The nature of geography*, which was treated in a most cavalier way by Schaefer.

Hartshorne's second paper (1958) addressed Schaefer's claim that Kant was the source of the exceptionalist view. Literary analysis suggests that both von Humboldt and Hettner reached the same position independently, being unaware of Kant's views when they were writing. May (1970, p. 9) suggests that both Hartshorne and Schaefer could have misunderstood Kant's conception of a science, however, and of geography's status as a science, although he confirms Hartshorne's dismissal of Schaefer's interpretation of the source of Kant's ideas (see the later exchange between Hartshorne, 1972, and May, 1972).

The third and most substantial piece in Hartshorne's rebuttal was a further monograph (*Perspective on the Nature of Geography*, Hartshorne, 1959) whose production was stimulated by requests from colleagues that he respond in detail to Schaefer's argument, but was also used to discuss other issues raised during the two decades since publication of the original statement (Hartshorne, 1939). The discussion used a framework of ten separate questions/topics; the aim was to provide a methodology by which geography could meet its need for 'new conceptual approaches and more effective ways of measuring the interrelationships of phenomena' (p. 9), which could only develop out of an understanding and acceptance of the subject's 'essential character'.

The first set of questions concerned the meaning of areal differentiation, the definition of the earth's surface, the particular geographical interest in the integration of phenomena – 'the total reality [that] is there for study, and geography is the name of the section of empirical knowledge which has always been called upon to study that reality' (p. 33) – and the determination of what is significant for geographical study. His answers led Hartshorne to define geography as 'that discipline that seeks to describe and interpret the variable character from place to place of the earth as the world of man' (p. 47). Human and natural factors do not have to be identified separately – any prior insistence on this was a function of environmental determinist arguments – and a division into human and physical geography is unfortunate, because it limits the range of possible integrations in the study of reality.

Turning to processes operating over time, Hartshorne argued that geographers need only study proximate genesis, since classification by form of appearance rather than by provenance is central to investigations of areal differentiation. As most landforms are stable, or virtually so, from the point of view of human occupancy, for example, then study of their change is irrelevant to the aims of geography (see p. 51, this book). According to this argument, geomorphology, insofar as it is the study of landform genesis, is not part of geography, but the study of landforms is. With regard to cultural features in the landscape, Hartshorne drew an important distinction between expository description and explanatory description: 'geography is primarily concerned to describe ... the variable character of areas as formed by existing features in interrelationships ... explanatory description of features in the past must be kept subordinate to the primary purpose' (p. 99). Thus, historical geography should be the expository description of the historical present, 'but the purpose of such dips into the past is not to trace developments or seek origins but to facilitate comprehension of the present' (p. 106); studies of causal development and genesis are the prerogative of the systematic sciences. By this statement, Hartshorne at least partly bridged the gap between his and Sauer's views of the discipline perhaps reflecting that one of Sauer's students, Andrew Clark (whose Ph.D. was on the European colonisation of New Zealand (Clark, 1949)) had joined Hartshorne on the staff at the University of Wisconsin-Madison.

In answering the question 'Is geography divided between systematic and regional geography?', Hartshorne modified his position from that in *The Nature of Geography*. He accepted that studies of interrelationships could be arranged along a continuum 'from those which analyse the most elementary complexes in areal variation over the world to those which analyse the most complex integrations in areal variation within small areas' (p. 121). The former are topical studies and the latter regional studies, but whereas 'every truly geographical study involves the use of both the topical and the regional approach' (p. 122), there is no argument that the latter is superior and that to which all geographers should aspire. Hartshorne thereby somewhat downgraded his earlier view regarding the centrality of regional synthesis in the geographical enterprise.

Regarding Schaefer's important question 'Does geography seek to formulate scientific laws or to describe individual cases?', Hartshorne argued for the latter, largely by pointing out the difficulties of establishing laws through geographical investigations. He did not argue against geographers seeking and using general laws for understanding individual cases, however: it is an 'erroneous presumption that to focus on studies of individual places and to focus on generic concepts are opposing alternatives mutually exclusive' (Hartshorne, 1984, p. 429). Scientific laws must be based on large numbers of cases, but geographers study complex integrations in unique places. Scientific laws can best be established in laboratory experiments which allow only a few potential influences to vary, but such work is rarely possible in human geography. Interpretation of such experiments requires skills in the systematic sciences which are beyond the capability of geographers, and scientific laws suggest some kind of determinism, which is inappropriate to the human motivations that are in part the causes of landscape variations. For all these reasons, he argued, the search for laws is irrelevant to geography. Laws are not the only means to the scientific end of comprehending reality, however, instead:

> Geography seeks (1) on the basis of empirical observation as independent as possible of the person of the observer, to describe phenomena with the maximum degree of accuracy and certainty; (2) on this basis, to classify the phenomena, as far as reality permits, in terms of generic concepts or universals; (3) through rational consideration of the facts thus secured and by logical processes of analysis and synthesis, including the construction and use wherever possible of general principles or laws of generic relationships, to attain the maximum comprehension of the scientific interrelationships of phenomena; and (4) to arrange these findings in orderly systems so that what is known leads directly to the margin of the unknown.
>
> (pp. 169–70)

This, he says, is a perfectly respectable scientific goal. Its empiricist base is very similar to the overall goal of positivist work and is the reason why several commentators see very little difference in ends, if not means, between Hartshorne's work and that of spatial scientists.

Finally, in discussing geography's position within the classification of sciences, Hartshorne returned to the Hettnerian analogy of geography as a chorological science with history as a chronological science. This is valid, he argues, because it describes the way in which geographers have worked, on both topical and regional subjects (this view was revived by Harris, 1971; see Earle, 1996).

Reconciliations?

The major basis of their methodological and philosophical differences was that Hartshorne had a positive view of geography – geography is what geographers have made it – whereas Schaefer's was normative, of what geography should be, irrespective of what it had been. Over the decade after Hartshorne published his *Perspective*, the view associated with Schaefer prevailed with many geographers on both sides of the Atlantic, although his personal influence via the 1953 paper was probably very slight and the real iconoclasts of the 'revolution' were those discussed in the next section. (According to N. Smith, 1990, Hartshorne's view that geography should be what (he said) it had always been in effect 'turned the discourse of geography into a museum and [Hartshorne] appointed himself curator' – D. Gregory, 1994, p. 285.) Indeed, in the UK, although Hartshorne's two books were clearly widely read and referenced, Schaefer's paper was apparently not. It receives no mention in Freeman's (1961, 1980a) books, none in Chorley and Haggett's (1965b) trail-blazing *Frontiers in Geographical Teaching*, and only one in their major edited collection *Models in Geography* (Chorley and Haggett, 1967) – in the chapter by Stoddart (see, however, Stoddart, 1990). Thus, it is not surprising that relatively little attention has been paid elsewhere in the geographical literature to the Schaefer/Hartshorne debate (Gregory, 1978a, p. 32: see also Martin, 1951 and E. Jones, 1956, for a separate, British, debate). Interestingly, Schaefer is not in the index of authors referred to in the encyclopaedic *Geography in America* either (Gaile and Willmott, 1989). Schaefer and Hartshorne's importance to the discipline's history is not so much the influence of the two principals themselves as the clarification they brought to what became a major debate over disciplinary means and ends.

Guelke (1977a, 1978) argued that Hartshorne's and Schaefer's views were not as antagonistic as they suggested (see also Gregory, 1978a, p. 31, and Entrikin, 1981, 1990). Hartshorne generally supported use of the scientific method as later defined by other geographers, but created problems for this method's application in geography by his view on uniqueness. Schaefer, on the other hand, not only accepted the full position of what geographers came to term 'scientific method', but also showed that uniqueness was a general problem of science, and not a peculiar characteristic of geography. Thus:

> In extending the idea of uniqueness to everything, Schaefer effectively removed a major logical objection to the possibility of a law-seeking geography and demonstrated that Hartshorne's view of uniqueness as a special problem was untenable for anyone who accepted the scientific model of explanation.
>
> (Guelke, 1977a, p. 380)

Furthermore, Guelke argued, Hartshorne's distinction between idiographic (understanding the individual case) and nomothetic (law-seeking) approaches was misleading; both he and Schaefer ignored the possibility of geographers being major 'law-consumers'. To Hartshorne, the alternatives were either law-making or the description of unique places, while to Schaefer geographers had to develop morphological laws, and ignore the interest in process laws which characterises the systematic sciences.

Guelke (1977a, p. 348) claimed that Schaefer's insistence on the need for geographers to develop laws 'created a major crisis within the discipline'. Within a decade of Schaefer's paper being published, however, many human geographers had adopted at least part of his manifesto with their growing concern for quantification and law-making. Presented with a choice between such activity and the sort of contemplation of the unique advocated by Hartshorne then, as Guelke (1977a, p. 385) points out, 'Not surprisingly, most geographers opted for geography as a law-seeking science' because at that time (Guelke, 1978, p. 45):

> Universities were expected to produce problem-solvers or social-technologists to run increasingly complex economies, and geographers were not slow in adopting new positions appropriate to the new conditions. Statistics and models were ideal tools for monitoring and planning in complex industrial societies. The work of the new geographers, however, often lacked a truly intellectual dimension. Many geographers were asking: 'Are our methods rigorous?', 'What are the planning implications of this model?', and not 'How much insight does this study give us?', 'Is my understanding of this phenomenon enhanced?', 'Does this study contribute to geography?'. The last-mentioned question was considered of little consequence. Yet it should have been asked, because one of the weaknesses of the new geography was a lack of coherence.

Developments in systematic geography in the USA

Whether because of or independent from the statements by Ackerman (1945), Schaefer (1953) and Ullman (1953), systematic studies became much more important in the research and under-graduate and graduate teaching in many American (and a few Canadian) university departments in the 1950s. (On Hartshorne's influence on systematic studies, see Butzer, 1990.) This did not mean a total departure from Hartshorne's contemporary views, since by 1959 he no longer gave primacy to regional studies, but the trend towards the scientific method proposed by Schaefer marked a clear break with the Hartshornian tradition.

The growing popularity of topical specialisms is illustrated in the review chapters in the collection edited by James and Jones (1954). Very few of the investigations reported there aimed at generating laws, however. Indeed, some could almost be categorised under the exploration paradigm, since their major purpose was the provision of new factual material; such work is best described as empiricist – it wants 'the facts' to speak for themselves.

Fundamental to scientific progress in Schaefer's approach is the development of theory. Geographers were aware of theory pre-Schaefer, as shown by their references to theoretical work on space and the economics of location by non-geographers (on the part of August Lösch first published in German 1940, intended as a contribution to economics, see Gregory, 1994, pp. 56–7; Leslie Curry wrote his MA thesis in 1950 on the utility of location theory in economic geography: see Johnston, 2010b) and a few – notably Chauncy Harris (1945) and Edward Ullman (1941) – wrote review papers summarising that material (see C. D. Harris, 1978, 1997b; Ullman, 1962). Stoddart (1986, p. 13) also points to a foreshadowing of the commitment to science in a paper by James (1952), which 'considered the problem of pattern, process and scale, and attempted to reconcile them with the prevailing theme of geography as the study of areal differentiation'. Although James was 'a leader of the old paradigm', according to Stoddart, the issues flagged in his 1952 paper had a considerable vintage, appearing in technical debates about the terminology and approaches of regional description that had been published in the *Annals* in the 1930s and 1940s. In Stoddart's (1986, p. 13) view, therefore, 'the revolution, if there was one, had long and respectable antecedents: the precursors themselves were central features of the old paradigm'. As indicated below, James was not alone.

Yet, despite these precedents, relatively little work was done in human geography prior to the mid-1950s which explicitly followed the dictates of the 'scientific approach'; there were pockets of innovation and wider awareness of relevant work on spatial arrangements being undertaken by other scholars, but relatively few geographers were utilising such approaches in their own research.

Once a new idea gains circulation through the professional journals it is available to be taken up by all. Initial development is usually concentrated in a few places only, however, where pioneer teachers encourage students to conduct research within the new framework. Thus, most of the methodological changes in systematic studies in geography during the 1950s can be traced to a few centres in the USA. These were connected through the movements of key individuals, and the dissemination of texts and technologies on location theory, produced by Walter Christaller and August Lösch and others, originally written in German in the 1930s–1940s, were translated into English by the 1950s too, making them more accessible to those seeking alternatives to the established regional tradition. The changes in American geography were largely concerned with method, and their scientific underpinning was at first little stressed, although law-seeking was the clear goal. Certainly, methods dominated the writings of both the pros and cons; many early contributions by the former group were in relatively fugitive, departmental publications, because of difficulties in getting such 'new' material accepted by the small number of mainline journals (as described by Berry, 1993; see Barnes, 2001b; Johnston, 2004a).

The Iowa school

Although Schaefer was at Iowa until his death in 1953, he had little apparent influence on changes promoted by geographers there, who for a number of years were members of the economics department and thus exposed to the approaches deployed in that more established social science. The group's leader was Harold McCarty, author of a major text on American economic geography (McCarty, 1940); associated with him were J. C. Hook, D. S. Knos, H. A. Stafford and, later, J. B. Lindberg, E. N. Thomas and L. J. King (McCarty, 1979; King, 1979a).

McCarty and his co-workers wanted to establish the degree of correspondence between two or more geographical patterns – to make map comparison more rigorous and generate laws of association akin to the morphological laws of accordance discussed by Schaefer. Interestingly, none of their publications refer to Schaefer's paper, although they do refer to works and assistance given by Gustav Bergmann, a positivist philosopher of the Vienna school who also strongly influenced Schaefer and read the proofs of his 1953 paper (Davies, 1972a, p. 134; L. King, 1979a; Golledge, 1983; see also G. Martin, 1990). These laws were to be embedded in a theory, thus:

> If we are to accept the idea that economic geography is becoming the branch of human knowledge whose function is to account for the location of economic activities on the various portions of the earth's surface, it seems reasonable to expect the discipline to develop a body of theory to facilitate the performance of this task.
>
> (McCarty, 1954, p. 96)

Such theory could be either topically or areally focused, and in its early stages of development would probably be restricted, both in its areal coverage and in the topics whose spatial interrelationships it considered.

The purpose of theory is to provide explanations, of which McCarty recognised two types. The first involved searching for the causes of observed locational patterns, which 'can never produce an adequate body of theory for use in economic geography. . . . Variables became so numerous that they were not manageable, and, in consequence, solutions to locational problems were not obtainable' (p. 96). The second, and preferred, type focuses on associations:

> Its proponents take the pragmatic view that if one knew that two phenomena always appear together in space and never appear independently, the needs of geographic science would be satisfied, and there would be scant additional virtue in knowing that the location of one phenomenon caused the location of another.
>
> (p. 97)

Such laws of association are built up in a series of stages: (1) a statement of the problem and the necessary operational definitions; (2) measurement of the phenomena (with consideration of attendant problems of sampling in time and space); and (3) a statement of the findings, in tabular or graphical form. These three descriptive stages precede analysis which seeks correlations between distributions:

> the nub of the problem of research procedure seems to lie in finding the best techniques for discovering a, b and c in the 'where a, b, c, there x' hypothesis in order to give direction to the analysis. But where shall we search for its components? . . . One source of . . . clues lies in the findings of the systematic sciences. The other source lies in the observations of trained workers in the field or in the library.
>
> (p. 100)

In seeking morphological laws, geography is thus to a considerable extent a consumer of the laws generated in other disciplines, which may be theoretically rather than empirically derived. According to the causal or process approach to explanation:

> Models may be created showing optimal locations for any type of economic activity for which adequate cost data may be obtained. These models may then be used (as hypotheses) for the comparison of hypothetical locations with actual locations. Divergences of pattern may then be noted and the hypothesis altered to allow for them (often by inclusion of factors not ordinarily associated with monetary costs). Ultimately the hypothesis becomes generally applicable and thus takes on the status of a principle.
>
> (McCarty, 1953, p. 184)

This statement, although not referenced as such, very faithfully reflects accepted views on how science progresses by the continual modification of its hypotheses, so as better to represent reality.

McCarty *et al.* (1956) discussed several statistical procedures for measuring spatial association and adopted multiple regression and correlation, used previously among geographers by John Kerr Rose (1936) and Weaver (1943), both apparently after contacts with agricultural economists. McCarty's empirical context was the location patterns of manufacturing industries in the USA and Japan; other studies by the group included Hook's (1955) on rural population densities, Knos's (1968) on intra-urban land-value patterns, and Leslie King's (1961) on the spacing of urban settlements. Edwin N. Thomas (1960) had used similar procedures in his study of population growth in suburban Chicago, presented as a Ph.D. thesis to Northwestern University (where he was a contemporary of Garrison); he extended the methodology with a paper on the use of residuals from regression for identifying where the putative laws of association do not apply fully, thereby suggesting further hypotheses for areal associations. (This last paper developed on an earlier one by McCarty, 1952, which was not widely circulated.) McCarty (1958) later expressed some doubts about the statistical validity of the procedure, but the method that he and his associates pioneered, with its focus on the testing of simple hypotheses derived either from observation or from theoretical deductions, became an exemplar for much research in the ensuing decades.

Wisconsin

The Department of Geography at the University of Wisconsin, Madison, had a long tradition of research with a quantitative bent; its early products included John Weaver's Ph.D. thesis on the geography of American barley production, which included a major section – published in his 1943 paper with no supporting methodological argument – using multiple correlation and regression to identify the influences of climatic variables on barley yields. Weaver later taught at the University of Minnesota, where he developed a widely adopted statistical procedure for defining agricultural regions (Weaver, 1954). Other work at Madison focused on the quantitative description of population patterns (e.g. J. Alexander and Zahorchak, 1943). A combination of these two interests was furthered by a group led by Arthur H. Robinson, whose main interests were in cartography; cartographic correlations were introduced to him by his research supervisor at Ohio State University, Guy-Harold Smith (S. E. Brown, 1978). Robinson also worked with R. A. Bryson, of the university's Department of Meteorology, who was a source of statistical ideas and expertise. (Cartographic work was for a long time called 'mathematical geography' by some.)

Robinson, like McCarty, wanted to develop statistical methods for map comparison, as indicated by the title of an early paper – 'A method for describing quantitatively the correspondence of geographical distributions' (Robinson and Bryson, 1957). J. K. Rose and Weaver's lead was followed with the adoption of correlation and regression procedures. Particular attention was paid to the problems of representing areal data by points (Robinson *et al.*, 1961) and of using correlation methods in comparing isarithmic maps (Robinson, 1962). Two such map types began to appear more in geographic papers and books in the late 1950s and 1960s: isometric maps, based on a point-occurring data, such as precipitation and temperature, and isoplethic maps, based on data occurring over space, especially within geographical areas called spatial units. The latter were normally represented by rate/ratio variables such as population density and crop yield per acre.

Like McCarty (1958), Robinson, too, was aware of difficulties in applying classical statistical procedures to areal data, and he proposed a procedure to circumvent one of these (Robinson, 1956). Edwin Thomas and David Anderson (1965) later found this wanting, as it dealt with a special case only and not with the more general problems. Interestingly, however, the main early work on this topic was published by a group of Chicago sociologists, under the title *Statistical Geography* (Duncan *et al.*, 1961); perhaps even more interestingly, this work was virtually ignored by geographers, even though Brian Berry (1993, p. 439) developed a close working relationship with Otis Dudley Duncan in the late 1950s and persuaded him to give the book that title.

The social physics school

This group's work was initiated and developed independently from the other three, and its early publications preceded Schaefer's paper by more than a decade. The leader was J. Q. Stewart, an astronomer at Princeton University who traced the origins of social physics in the work of a number of natural scientists who applied their methods to social data (Stewart, 1947). His investigations began when he noted certain regularities in various aspects of distributions that were akin to the laws of physics, such as a tendency for the number of students attending a university to decline with increasing distance of their home addresses from its campus. These observations led to his ideas on social physics, which he defined as:

> that the dimensions of society are analogous to the physical dimensions and include numbers of people, distance, and time. Social physics deals with observations, processes and relations in these terms. The distinction between it and mathematical statistics is no more difficult to draw than for certain other phases of physics. The distinction between social physics and sociology is the avoidance of subjective descriptions in the former.
>
> (Stewart, 1956, p. 245)

Stewart established a laboratory at Princeton to investigate a wide range of such regularities.

Stewart introduced his ideas to geographers through a paper in *The Geographical Review* (Stewart, 1947). Four empirical rules were adduced:

1   The rank-size rule for cities showed that in the USA the population of a city multiplied by its rank (from 1 for the largest to n for the smallest) and standardised by a constant, equalled the population of the largest city, New York (Carroll, 1982).
2   At various dates the number of cities in the country with populations exceeding 2,500 was very closely related to the proportion of the population living in such places.
3   The distribution of a population could be described by the population potential at a series of points, in the same manner as the potential in a magnetic field is described in Newtonian physics.
4   There was a close relationship between this population potential and the density of rural population in the USA.

From these regularities, Stewart claimed that 'There is no longer any excuse for anyone to ignore the fact that human beings, on the average and at least in certain circumstances, obey mathematical rules resembling in a general way some of the primitive "laws" of physics' (p. 485). No reasons were given why this should be so (Curry, 1967, p. 285, called this a 'deliberate shunning of plausible argument'); the rules were presented as empirical regularities which had some similarity to the basic laws of physics. Causal hypotheses were not even postulated, let alone tested.

Stewart's main collaborator was William Warntz, a graduate of the University of Pennsylvania who was later employed by the American Geographical Society as a research associate, working on what he termed 'the investigation of distance as one of the basic dimensions of society' (Warntz, 1959a, p. 449; see also Warntz, 1984, where he remarks that he was introduced to Stewart's ideas via a book on *Coasts, Waves and Weather* (Stewart, 1945) that he was reading when working as a weather observer during the Second World War. The book was 'prepared primarily to explain to marine and air navigators the physical environment . . . [but] Stewart could not resist the temptation to include an exotic chapter describing potential of population and its sociological importance'.) The wide range of empirical regularities which they observed (see Stewart and Warntz, 1958, 1959) was used to develop their concept of macrogeography (Warntz, 1959a, 1959b). Warntz (1959a) claimed that geographical work was dominated by micro-studies: 'The tendency of American geographers to be preoccupied with the unique, the exceptional, the immediate, the microscopic, the demonstrably utilitarian, and often the obvious is at once a strength and a weakness' (p. 447) because 'the assembly of more and more area studies involving an increase in the quantity of detail does not mean per se a shift from the microscopic to the macroscopic' (p. 449). Geographers were in danger of being unable to perceive general patterns within their welter of local detail, and to counter this Stewart and Warntz suggested the search for 'regularities in the aggregate'.

As examples of that search, Stewart's concept of population potential was used to describe general distributions, and was related to a large number of other patterns in the economic and social geography of the USA. Although these findings were recognised as only empirical regularities, they could be used as the basis for theory development (Stewart and Warntz, 1958, p. 172): geography needed theory which, according to Warntz (1959b, p. 58), 'has as its aim the establishment and coordination of areal relations among observed phenomena. General laws are sought that will serve to unify the individual, apparently unique, isolated facts so laboriously collected'. Their approach to theory was inductive (see Figure 3.1) rather than deductive, although there was a clear underlying belief in the importance of distance and accessibility as influences on individual behaviour. The work of German geographer Walther Christaller was key for Warntz as well as Garrison, Ullman and others (see Bunge, 1968).

These macroscopic measures, particularly population potential, were used in a variety of contexts, as in Warntz's (1959c) *Toward a Geography of Price*, which established strong relationships between the prices of agricultural commodities in the USA and measures of supply and demand potential; an early textbook in economic geography (Huntington *et al.*, 1933) included a short section of geographic variations in the price of and profits from wheat in the USA; and Chauncy Harris (1954) had used the potential measure in studies of industrial-location patterns, having been influenced by agricultural economist Colin Clark when he was studying at Oxford, noting that the measure he adopted from Clark's work was similar to J. Q. Stewart's (Harris also acknowledged help from Garrison and Isard); it was later taken up by a Chicago student (Pred, 1965a). The 'macrogeographers' also did much work on various distance-decay functions (see Chapter 4, p. 101, this book) and extended early Russian work on centrographic measures (Sviatlovsky and Eels, 1937; Neft, 1966); this was done both at the American Geographical Society (where its pioneering, mathematical nature ran contrary to the Society's general research interests) and later, under Warntz, at the Graduate School of Design's Laboratory for Computer Graphics and Spatial Analysis at Harvard University, where it was the forerunner of later developments in computerised cartography and GIS (see p. 158, this book).

These lines of work contrasted markedly with that of the other three groups reviewed here, in a variety of ways. First was the issue of scale; Stewart and Warntz perhaps conformed more than any others to calls for a scientific approach which aimed at a high level of generality. Second, there was their approach to theory: macrogeography was largely inductive in its search for regularity rather than testing deductive hypotheses. Finally, the analogies sought for human geography were in a natural science – physics – and not in the other social sciences.

W. L. Garrison and the Washington school

By far the largest volume of work in the spirit of Schaefer's and McCarty's proposals published during the 1950s came from the University of Washington, Seattle. The group of workers there was led by W. L. Garrison, whose Ph.D. was from Northwestern University, where he was associated with Thomas (the two returned to Northwestern in the early 1960s; Taaffe, 1979). According to Bunge (1966a, p. ix), Garrison had been influenced by Schaefer's paper, although his earliest

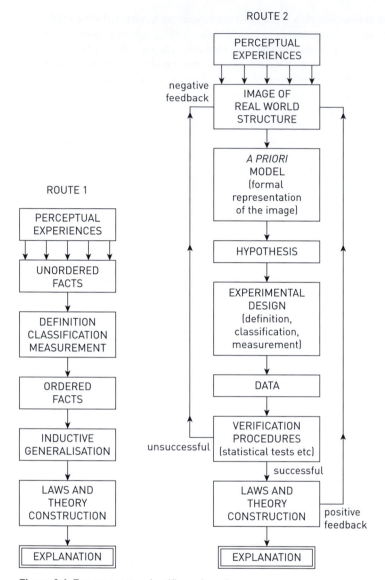

Figure 3.1 Two routes to scientific explanation

Source: Harvey, 1969a, p. 34.

publications indicate that he was applying the proposed methods to systematic studies in human geography before 1953; his 1950 Ph.D. thesis was on the shopping centres of northern Chicago, following the exemplar set by Proudfoot's (1937) work, which identified a hierarchy of five shopping centre types in a comparative study of several cities (Garrison, 2002). Also involved was Ullman, who moved to Seattle from the closing department at Harvard in 1951 (he was previously at Chicago, as was Chauncy Harris: Harris, 1977, 1978, 1990, 1997b). Ullman had already done pioneering research on urban location patterns and transport geography (see Hay, 1979b; Morrill, 1984). Ullman was strongly critical of regional geography as traditionally practised and committed to a 'scientific approach', as illustrated by a commissioned paper published in 1953. He argued that spatial interaction should be the focus of work in economic geography rather than regional definition:

> Geographers conventionally spend much time mapping the characteristics and use of the land. This is important to many problems, but sometimes I feel that we do this simply because it is tangible and distinctive, in fact by now almost instinctive.
>
> (Ullman, 1953, p. 57)

and:

> areal differentiation [is] the current catholic definition . . . I cannot accept areal differentiation as a short definition for outsiders because it implies that we are not seeking principles or generalizations or similarities, the goal of all science.
>
> (p. 60)

Ullman's (1959) early work on commodity flows in the USA was seen as 'mere description' by the graduate students who went to Seattle to work with him (Morrill, 1978) but his earlier paper in a relatively obscure source (Ullman, 1954a; see also Ullman, 1954b; Hay, 1979b) provided a theoretical basis for analysing the observed patterns based on the concepts of complementarity and intervening opportunities as well as distance (the 1954 papers included references to Zipf and the gravity model).

Soon after Garrison and Ullman were recruited to Seattle a new chair of the department, G. Donald Hudson, was appointed from Northwestern. He shifted its emphasis from teaching to research, focused on four areas – cartography, economic geography, the Far East and the Soviet Union, and Anglo-America (Velikonja, 1994). He supported the work of Garrison and his students against some opposition from other colleagues (Berry, 2002a, notes that although Ullman, as an established name, attracted many of the graduate students to Seattle, most of them chose to work with the more approachable Garrison, who joined the department just before Ullman went on leave and undertook his teaching, which included statistical methods). Hudson also worked hard to place them in other prestigious universities after graduation, thus aiding the spread of their pioneering ideas (80 MAs and 40 Ph.Ds were completed during his headship – 1950–63). When Hudson retired, the Washington department was placed fourth in national rankings of graduate programmes.

Garrison (2002, p. 105) records that a 'fork in the road' for him came in 1952 when he held a visiting post at the University of Pennsylvania. Encounters beyond geography introduced him to work on the spatial organisation of markets and he decided to take some courses in statistics. He returned to Washington and recast his course in transportation geography. A large group of graduate students soon joined him, with several becoming leaders in the new methodology during the subsequent decade, including B. J. L. Berry, W. Bunge, M. F. Dacey, A. Getis, D. F. Marble, R. L. Morrill, J. D. Nystuen and W. R. Tobler (Garrison, 1979, 2002). The group also benefited from a 1957 visit by Leslie Curry, then a climatologist at the University of Auckland but with a background of studying location theory (Johnston, 2010b) who later switched his interests to spatial theory, another in 1959 by a Swedish geographer, Torsten Hägerstrand, who was developing methods for generalising spatial patterns and processes (see p. 107, this book), and from Garrison's contacts with the university's business and engineering schools (Halvorson and Stave, 1978). Berry (1993) and Getis (1993) both provide accounts of the intellectual activity at Seattle during the mid-1950s: Getis stresses Donald Hudson's role, and his 'inferiority complex that comes from being remote' (p. 530), which led him to place Washington graduates as faculty members in several of the main Midwestern graduate schools (such as Chicago, Northwestern and Michigan Universities) in order to establish his department's intellectual reputation. This was also facilitated by their access to the department's reprographic facilities, which allowed them to circulate draft papers very widely (Velikonja, 1994; Morrill, 2002; Johnston, 2004a).

Garrison and his co-workers had catholic interests in urban and economic geography. Much of their research was grounded in theory gleaned from other disciplines, notably economics, although they were much attracted to Christaller's central place theory, which Bunge (1968, p. 133) described as 'geography's finest intellectual product'. Their efforts were directed towards both testing those theories and applying them to planning problems. In the former activity they drew on a much stronger mathematical base than was the case at Iowa and Madison. They also searched widely for relevant statistical tests for analysing point and line patterns – the biological sciences provided several, such as nearest-neighbour analysis of point patterns (Dacey, 1962), whereas others, for grouping, classifying and regionalising, were derived from psychology (Berry, 1968). Garrison (1956a, p. 429) argued that 'there is ample evidence that present tools are adequate to our present state of development. No type of problem has been proposed that could not be treated with available tools', which contradicted a claim by Reynolds (1956), although Garrison (1956b) also criticised some uses of standard techniques. The dominant thrust of the group's work therefore involved importing relevant normative theories plus mathematical methods and statistical procedures from other systematic sciences, with which to develop morphological laws.

The wealth of the work done at Seattle is illustrated by their major publications. Garrison (1959a, 1959b, 1960a), for example, contributed a three-part review article on the state of location theory. The first part reviewed six recent books – none by geographers – addressing the question 'What determines the spatial arrangement (structure, pattern or location) of economic activity?' (Garrison, 1959a, p. 232). Each incorporated locational considerations into traditional economic analysis and offered valuable economic insights to traditional geographical problems.

Central place theory was the dominant location theory on which the group worked, however. This had several independent origins (Ullman, 1941; see also C. Harris, 1977, and Freeman, 1961, p. 201, who notes that findings akin to those in central place theory were reported by the 1851 Census Commissioners of Great Britain). Christaller's (1966) thesis attracted most attention, however (see Müller-Wille, 1978). Working in Germany in the 1930s, Christaller developed ideas regarding the ideal distribution of settlements of different sizes acting as the marketing centres of functional regions, within constraining assumptions relating to the physical environment and the goals of both entrepreneurs and customers; a translation by C. W. Baskin became available in the late 1950s and was formally published in 1966. Studies of functional regions were not novel, of course (see p. 51, this book), and other geographers had tested Christaller's notions regarding a hierarchical organisation of settlements distributed on a hexagonal lattice (e.g. the work of Brush, 1953, at Wisconsin). Related, more inductive, work on 'principles of areal organization' was reported by Philbrick (1957). Dacey wanted to make the analysis of these spatial hierarchies more rigorous mathematically (Dacey, 1962), whereas Berry focused on empirical investigations of the settlement pattern north of Seattle and the retail centres in the city of Spokane (Berry and Garrison, 1958a, 1958b; Berry, 1959a).

The second part of Garrison's (1959b) review article dealt with possible geographical applications of the mathematical procedures of linear programming, which identify optimal solutions to problems of resource allocation in constrained situations. He illustrated how neoclassical economic analytic procedures could be adapted for investigations of ideal spatial solutions to six problems of where to locate economic activities and how to organise flows of goods:

1   The transportation problem, which takes a set of points, some with a given supply of a good and some with a given demand for it, plus costs of movement, and determines the most efficient flow pattern of the good from supply to demand points which minimises the expenditure on transport.

2   The spatial price-equilibrium problem, which takes the same information as the transportation problem, but determines prices as well as flows.
3   The warehouse-location problem, which determines the best location for a set of supply points, given a geography of demand.
4   The industrial-location problem, which determines the optimum location for factories from knowledge of the sources of their raw materials and the destinations for their products.
5   The interdependencies problem, which locates linked plants so as to maximise their joint profits.
6   The boundary-drawing problem, which determines the most efficient set of boundaries (i.e. that which minimises total expenditure on transport), as with school catchment areas.

If these are used to investigate actual patterns and not as the bases for future plans, the purpose, as Lösch (1954) put it, is to see whether reality is 'rational', whether decision-makers have acted in ways that would produce the most efficient solutions: efficiency is defined as cost-minimisation, particularly transport-cost minimisation. Relevant investigations by the Washington group included studies of interregional trade (Morrill and Garrison, 1960) and the optimal regional location of agricultural activities in the USA (Garrison and Marble, 1957).

In the final part of his review article, Garrison (1960a) dealt with four further books on locational analysis, which were empirical in orientation and shared a common interest in the agglomeration economies reaped by industrial clusters. Several topics and techniques were discussed, such as the use of input–output matrices to represent industrial systems, and Garrison concluded by stressing the need for geographers to investigate location patterns as systems of interrelated activities.

The empirical work with a planning orientation undertaken by the Seattle group is illustrated by their research on the impact of highway developments on land use and other patterns (Garrison et al., 1959). This book included four studies: Berry's on the spatial pattern of central places within urban areas; Marble's on the residential pattern of the city (as indexed by property values) and relationships between household characteristics, including location, and their movement patterns; Nystuen's on movements by customers to central places; and Morrill's on the locations of physicians' offices, both actual and the most efficient. In addition, Garrison worked on the impacts of highway improvements and devised indices of accessibility based on graph theory (Garrison, 1960b; the work was continued by Kansky, 1963, after Garrison and Berry moved to the Chicago area). He also used the simulation procedures developed by Hägerstrand (1968) to investigate urban growth processes (Garrison, 1962), a topic extended by Morrill (1965).

Somewhat separate from the work of the others in the group, although aligned with their general purpose, was Bunge's thesis on *Theoretical Geography* (1962, reissued in enlarged form in 1966a: see Cox and Macmillan, 2001). This displays a catholic view of geography, together with an acknowledged debt to Schaefer (Bunge worked at Iowa for a short period, and also at Madison, where he clashed with Hartshorne). Its basic theme is that geography is the science of spatial relations and interrelations; geometry is the mathematics of space; hence, geometry is the language of geography. The early chapters establish geography's scientific credentials in a debate with Hartshorne's published statements, especially those concerning uniqueness and predictability. Bunge claimed that Hartshorne confused uniqueness and singularity (on which, see P. W. Lewis, 1965); he opposed the claim that geography cannot formulate laws because of its paucity of cases by arguing for even more general laws, and countered the argument that geographical phenomena are not predictable with the claim that science 'does not strive for complete accuracy but compromises its accuracy for generality' (1966a, p. 12).

Having established geography's scientific credibility to his satisfaction, Bunge then investigated its language. An intriguing discussion of cartography led him to conclude that descriptive mathematics is preferable to cartography as a more precise language. The remainder of the book looked at aspects of the substantive content of the science of geography, beginning with 'a general theory of movement' and then a chapter on central place theory: 'If it were not for the existence of central place theory, it would not be possible to be so emphatic about the existence of a theoretical geography ... central place theory is geography's finest intellectual product' (1966a, p. 133). Problems testing the theory suggested the need for map transformations (see also Getis, 1963), and the final chapter of the first edition clarified the links between geography and geometry: 'Now that the science of space is maturing so rapidly, the mathematics of space – geometry – should be utilised with an efficiency never achieved by other sciences' (1966a, p. 201). Bunge's *Theoretical Geography* is peppered with maps and geometrical visualisations of spatial relationships.

The richness of the work done by this group during the mid- and late 1950s continued in various locations after it broke up (only Ullman remained, although Morrill later joined the staff at Seattle). Berry was one of the most prolific and seminal, not only in his original field of central place theory (Berry, 1967) but also over a wide range of other topics in economic and social geography. His work always had a very strong empirical and utilitarian base (Yeates, 2001), whereas Dacey continued to work on the mathematical representation of spatial, especially point, patterns (e.g. Dacey, 1973) and Tobler (1995) moved into computer cartography (on Tobler's background, see Barnes, 2004c, 2008b). In total, the work of this significant group of scholars influenced the research and teaching of a whole generation of human geographers throughout the world. In subsequent decades, the Washington department also provided a major lead in regional economic analysis. Led by Morgan Thomas, a number of graduate students – some of whom were later appointed to the Washington faculty – produced innovative work on topics such as growth-pole theory, regional input–output analysis, and the spatial strategies of firms and corporations, and in turn several of them became leaders in this area of locational analysis (Velikonja, 1994).

In a series of essays on the history of geography's 'quantitative revolution' in the USA (e.g. T. Barnes, 2003b, 2004b), Trevor Barnes has associated that shift not only to the experience of Ackerman, Ullman and others interacting with other social scientists in the Office of Strategic Services during the Second World War (Barnes, 2006a; Barnes and Farish, 2006), but also to the Cold War interests of the 'military-industrial complex' which focused on the 'received view' regarding scientific method and its use of models as 'instruments to think about the world, and instruments to alter it' (Barnes, 2008b, p. 6). The very centre of geographers' discipline (T. Barnes used the examples of Garrison, Tobler and geomorphologist Strahler) was 'displaced, shifted, and realigned, mangled, in part because of its connections to the military, its connections to the Cold War' (p. 15; see, however, Johnston *et al.*'s, 2008, rejoinder, which points out that the parallel developments in the UK were not linked to the Cold War and the military-industrial complex). (Barnes and Minca, 2013, similarly illustrate the Nazi regime's use of Christaller's central place theory to create a replica of the German homeland in the eastern territories captured in the first years of the Second World War.)

From Seattle and elsewhere to Chicago and further

The developments outlined earlier marked the beginning of major changes in the field of human geography, which were rapidly taken up by others, within and beyond the USA. Although the focus was on theory and measurement and the development of 'geographical laws', the work did not deviate too far from Hartshorne's expanded (1959) definition of the nature of geography (see p. 62, this book); indeed, in a major programmatic statement, Berry (1964a) used the regional concept as the core for a revised framework of geographical research. The main difference between the new work, with its focus on systematic studies, and its regional predecessor was the greater faith of 'new geographers' in their ability to produce laws, to work within the canons of accepted scientific method, to master and apply relevant mathematical and statistical procedures, and to move out of their self-imposed academic isolation (Ackerman, 1945, 1963).

Diffusion of the new approach was rapid, involving a number of graduates from the four centres identified, especially Seattle, plus other willing change-agents. Ned Taaffe, for example, trained in journalism and meteorology and, after experience teaching economics and statistics, was involved in the development of spatial analysis at Northwestern University (on which, see also Hanson, 1993) and then moved to Ohio State University 'with a mandate to build a research department'. Taaffe had studied urban hinterlands using air traffic data for his Chicago Ph.D. as exemplars of functional regions and following the stimulus of his mentor, Robert Platt; his supervisor was Harold Mayer. He then taught economic geography and statistics for four years in the economics department at Loyola University (Gauthier, 2002), before taking up a joint appointment with the Department of Geography and the Transportation Center at Northwestern in 1956 (Garrison was appointed to the latter in 1960). The graduate students that Taaffe worked with (and to whom he promoted the 'spatial view' which won converts to geography's role from other social sciences; Taaffe, 1974) included, almost contemporaneously with the Washington department, a number who were to become leaders in the field, such as Emilio Casetti, Ed Conkling, Lawrence (Larry) Brown, Barry Garner, Howard Gauthier, Peter Gould, Frank Horton, David Reynolds and Maurice Yeates (Gauthier, 2002, p. 582).

At Ohio State, Taaffe appointed 'a group of geographers, each of whom influenced the development of spatial analysis in geography in a somewhat different way' (Taaffe, 1993, p. 423). He names Howard Gauthier, Les King, George Demko, Kevin Cox, John Rayner, Emilio Casetti, Larry Brown, Reg Golledge, John Arnfield and Harold Moellering. Golledge, King – Taaffe's 'most influential appointment' according to Getis (1993, p. 522) – Rayner and Bill Clark were all at the University of Canterbury, New Zealand when Harold McCarty was a visiting professor there in the late 1950s (see the essays in King, 2008). Taaffe was also involved in several summer workshops conducted at Northwestern University with funding under the National Defense Education Act, which introduced many other geographers to quantitative methods (Taaffe, 1979, p. 137, lists some of them).

A further major centre to which the 'new geography' was initially transferred, and where a great deal was accomplished in the 1960s and 1970s, was the University of Chicago, which Brian Berry joined in 1958. (Indeed, several examples of early Chicago work, much of it inspired by Robert Platt, Charles Colby and later Chauncy Harris – 1990 – foreshadowed the later developments.) Already on the staff when Berry arrived were Chauncy Harris, who had published pioneering papers on location theory and had used macrogeographic ideas in his research (see p. 68, this book), and Harold Mayer, a leading urban geographer whose collection of essays (Mayer and Kohn, 1959) was a pioneering early urban text (including in New Zealand, which Mayer visited in 1961: Clark, 2002) and whose contents Berry lightly influenced (Berry, 2002a). The research that Berry stimulated (some of it involving staff and students at nearby Northwestern University) was carried forward by a large number of graduate students. It focused initially on central place studies, extending his Seattle work, but was soon expanded into a wide range of studies including the internal structure of cities, ethnic segregation and housing allocation, and the definition of urban built-up areas and their functional hinterlands. This work, much of which was undertaken for external sponsors – thus demonstrating the applied relevance of the 'new geography' – was widely cited, and the large numbers involved became nodes in what Yeates (2001, p. 524) has termed the 'continental and global network of professionals' associated with a further 'Chicago school' that very strongly influenced the development of human geography during that period. That influence spread beyond the fields of urban and economic geography on which Berry and his students specialised. Other research groups there were shaped by this, such as Ginsburg's on patterns of economic development (Ginsburg, 1961), especially in Asia, and the work by Gilbert White and his associates (such as Ian Burton, Robert Kates, Thomas Saarinen) on issues of resource management (see p. 126, this book).

Other centres were soon established. Two Washington graduates – John Nystuen and Waldo Tobler – were appointed to the University of Michigan at Ann Arbor, for example, and William Bunge to Wayne State University in Detroit. Together with others at Michigan State University, they established an informal inter-university seminar group which met at a tavern midway between the three centres at which they and invited visitors gave papers. Several of these were 'published' as the Michigan Inter-University Community of Mathematical Geographers (MICMOG) discussion papers, which were widely circulated, and which Tobler (2002, p. 309) reports were critical in attracting Gunnar Olsson to Michigan. Their publication was halted when the journal *Geographical Analysis* was founded, but several of them (such as Gould's original paper on mental maps – see p. 129, this book) had by then become influential 'mini-classics'.

The role of certain individuals and institutions in the promotion of change was thus crucial (on which, see the exchange between Morrill and Johnston: Morrill, 1993, 1994; Johnston, 1994c). Their publications influenced geographers, especially young geographers, around the world (Adams, 2002; Clark, 2002). The work that they promoted succeeded, according to Berry (2002a, p. 561), because (1) it was scientifically conducted, and thus replicable; (2) theory was continually being improved, through the interplay of speculation and empirical analysis; (3) it was multidisciplinary; and (4) it was continually applied, creating a successful tension between theory and practice – to the benefit of both. They didn't 'take over' the discipline – as James Wheeler's (2002a) analysis of articles published in major journals during the 1960s shows – but they were highly influential in setting its research and training agenda for several decades.

Scientific method in human geography

A sense of a revolution?
The changes just described heralded a potential major reorientation in the nature of much geographical research. This was not focused on any single blueprint, grand design or programmatic statement, however. No paper or book published in either the 1950s or the early to mid-1960s provided either a philosophy for the new approach or detailed how research should be conducted within that framework. Schaefer's paper said nothing about how geographical laws were to be derived, and although McCarty and his co-workers discussed methods in both general (McCarty, 1954) and specific (McCarty *et al.*, 1956) contexts, they provided no programmatic statement for how the discipline should be practised. This exemplifies what Derek Gregory (1978a, p. 47) notes, that 'geography has (with some notable exceptions) paid scant attention to its epistemological foundations'. Livingstone (1992, p. 328) suggests that 'Geography's confrontation with the vocabulary of logical positivism [in the 1950s] . . . was a post hoc means of rationalizing its attempt to reconstitute itself as spatial science'. What was promoted was very largely a technical revolution which only later sought philosophical legitimation.

One piece postdating many of the early essays, but which was then widely quoted in the 1960s as the new ideas spread, was Ackerman's (1958) essay on *Geography as a Fundamental Research Discipline*. This was mainly an analysis of research organisation. He argued that the ultimate goal must be interdisciplinary integration, and that 'If any one theme may be used to characterise this period, that theme would be one of illuminating covariant relations among earth features' (p. 7). As a science, even one which is eventually an idiographic science since it deals with unique places, geography, according to Ackerman, needed to strive for 'an increasingly nomothetic component'. Its fundamental research

> need not necessarily be law-giving . . . Much fundamental research in geography has not been
> law-giving in the strict sense but it has been concerned with a high level of generalization, and it

has given meaning to other research efforts which succeeded it. In this sense it has a block-building characteristic.

(p. 17)

Such fundamental research 'is likely to rest on quantification ... accurate study depends on quantification' (p. 30) and should 'furnish a theoretical framework with capacity to illuminate actually observed distributional patterns and space relations' (p. 28).

Ackerman issued a clarion call for theory development, the application of quantitative methods and a focus on laws and generalisations to form the building-blocks for further nomothetic research. But his essay lacked any detailed discussion of how such research should be undertaken. Seven years later, a report of a National Academy of Sciences-National Research Council (NAS-NRC, 1965, p.1) committee on *The Science of Geography* set out 'geography's problem and method' in the statement that

Geographers believe that correlations of spatial distributions, considered both statistically and dynamically, may be the most ready keys to understanding existing or developing life systems, social systems, or environmental changes. In the past ... progress was gradual, however, because geographers were few, rigorous methods for analysing multivariate problems and systems concepts were developed only recently.

(p. 9)

This, too, was a general statement on research orientation with no accompanying detail on research procedures. And yet a paper published in 1963 had claimed that an intellectual revolution – the quantitative and theoretical revolution – had been completed in geography: 'The revolution is over, in that once-revolutionary ideas are now conventional' (Burton, 1963, p. 156). Something had become conventional, but nobody had yet written a full formulation for the discipline of what that something was. The groups of researchers who proposed changes in the nature of human geography undoubtedly had a clear, if implicit, rationale for their arguments, but these were not yet discussed in detail in print. If their citations in the published works are any lead, many did not research deeply into the philosophy that they were adopting (though a few, such as Peter Gould, regularly attended courses in philosophy throughout their subsequent careers: see Gould, 1999). The exceptions are the references to Bergmann in the Iowa group's papers (see Golledge, 1983) and in Bunge's (1962) thesis. Texts on statistical and mathematical procedures were widely quoted and, although many relied on those written for other disciplines (e.g. Blalock, 1960; Krumbein and Graybill, 1965; Garrison worked with Krumbein at Northwestern in the 1960s: Garrison, 2002, p. 110), several were produced by and for geographers (e.g. S. Gregory, 1963; L. King, 1969b). But there was very little discussion of either epistemology – the theory of what constitutes valid knowledge – or of ontology – the theory of what can be known; these were largely ignored (or considered 'common sense') and instead the emphasis was on methods.

Explanation in geography

The first major work on the philosophy of the 'new geography' was not published until 1969, by a British geographer (who moved to the USA at about the time his book, entitled *Explanation in Geography*, appeared), based on several years teaching the epistemology and ontology of science to geography undergraduates at the University of Bristol in order to give a philosophical foundation to the technical procedures (Harvey, 1969a; a parallel, but briefer, statement is Moss, 1970). (On Harvey's early work, see Gregory, 2006, and Barnes, 2006b.)

Harvey identified two routes to explanation (Figure 3.1). The first, the 'Baconian' or inductive route, derives generalisations from observations: a pattern is observed and an explanation developed from and for it. This involves a dangerous form of generalising from the particular case, however, because acceptance of the interpretations depends too much on the unproven representativeness of the case(s) discussed and, perhaps, the charisma of the scholar involved (Moss, 1970). Thus, the second route in Figure 3.1 is much preferred (though see Bennett, 1985a and T. Barnes, 1996). This also begins with observers perceiving patterns in the world; they then formulate experiments, or some other kind of test, to test the veracity of the explanations offered for those patterns. Only when ideas have been tested successfully against data other than those from which they were derived can a generalisation be produced.

Scientific knowledge obtained via the second route is 'a kind of controlled speculation' (Harvey, 1969a, p. 85). Its philosophy, known as positivism, was developed by a group of philosophers working in Vienna during the 1920s and 1930s based on procedures developed a century earlier in France (Gregory, 1978a; Guelke, 1978). It is based on a conception of an objective world in which there is order waiting to be discovered. Because that order exists – spatial patterns of variation and co-variation in the case of geography – it cannot be contaminated by the observer (a position known as empiricist). A neutral observer, acting on personal observations or reading others' research reports, will derive a hypothesis (a speculative law) about some aspect of reality and then test it; verification of the hypothesis translates the speculative law into an accepted one. (For a more nuanced discussion of the search for order, see Symanski and Agnew, 1981.)

A key tenet of this philosophy is that laws must be proven through objective and replicable procedures: 'the plausibility or intuitive reality of a theory is not a valid basis for judging a theory' (Bunge, 1962, p. 3). A valid law must predict successfully, so that having developed an idea about certain patterns the researcher must formulate it into a testable hypothesis – 'a proposition whose truth or falsity is capable of being asserted' (Harvey, 1969a, p. 100) – for which an experiment can be designed. Data are then collected and the hypothesis's validity evaluated. If the test results do not match the predictions, then either the observations on which the hypothesis was derived or the deductions on which it was based are thrown into doubt. There is thus negative feedback (Figure 3.1) and the image of the world has to be revised, creating a new hypothesis.

According to Harvey (1969a, p. 105):

> A scientific law may be interpreted as a generalization which is empirically universally true, and one which is also an integral part of a theoretical system in which we have supreme confidence. Such a rigid interpretation would probably mean that scientific laws would be non-existent in all of the sciences. Scientists therefore relax their criteria to some degree in their practical application of the term.

After sufficient (undefined) successful tests, therefore, a hypothesis may be accorded lawlike status, and fed into a body of theory, which comprises a series of related laws. There are two types of statement within a full theory: the axioms, or givens – statements that are taken to be true, such as laws; and the deductions, or theorems, from those initial conditions, which are derived consequences from agreed facts – the next round of hypotheses. There is a positive feedback from the theory stage to the world view, therefore (Figure 3.1), so that the whole scientific enterprise is a cyclical procedure whereby the successes of one set of experiments become the building blocks for thinking about the next. (Haggett, 1965c, had already used this argument in his pioneering textbook.)

One stage in Figure 3.1 so far ignored is the model, a widely used term with a variety of meanings (Chorley, 1964). Models have two basic functions: as representations of the real world, such as a scale model, a map, a series of equations and some other analogue (Morgan, 1967); and as ideal types, representations of the world under certain constrained conditions. Both are used in the positivist method to operationalise a theory as a guide to the derivation of testable hypotheses.

Quantification is central to this scientific method. Mathematics is particularly useful in developing models, as in the linear programming procedures discussed by Garrison. Relatively few geographers had strong backgrounds in mathematics, however (this was especially true in the 1950s), so little work involved representing the real world as sets of equations. Instead, the central role was given to statistics. Two types of statistics are available: descriptive statistics represent a pattern or relationship; inductive statistics are used for making generalisations about a carefully defined population from a properly selected random sample. They use the same procedures; the difference lies in how they are deployed. Many geographical researchers confused the two. Inductive statistics employ significance tests to show whether what has been observed in the sample probably also occurs in the parent population, so that if the data analysed are not a sample, such tests are irrelevant (some disagreement was expressed over 'what is a sample?' – see Meyer, 1972, and Court, 1972 – and debates continue: see Johnston *et al.*, 2014a). Many geographers use inductive methods in a descriptive manner only, however, using the significance tests as measures of the validity of their findings, asking what is the likelihood of a relationship that they have discovered arising by chance given the dataset's contents (as argued in Hay, 1985b).

The main attraction of statistics to many early adherents of the 'new geography' was their precision and lack of ambiguity – compared to the English language – in description. This was expressed by Cole (1969), who annotated a quotation from a well-known textbook, *The British Isles*, in which the text is Stamp and Beaver's (1947, pp. 164–5) and the annotations in parentheses, to show the ambiguities, are Cole's:

> The present distribution of wheat cultivation in the British Isles (space) raises the conception of two different types of limit. Broadly speaking (vague), it may be said that the possible (vague) limits (limit) of cultivation of any crop are determined by geographical (vague), primarily by climatic conditions. The limits so determined (how?) may be described (definition) as the ultimate (vague) or the geographical (vague) limits.
>
> (p. 160)

Cole argued that the full quotation (only part is reproduced here) is so full of ambiguities that it could refer to about one million possible combinations of some forty counties and it is impossible to reconstruct a map from that description:

> the correlations suggested are so tentative and imprecise that they leave the reader still wondering why wheat is grown where it is. The application of a standard correlation procedure . . . in itself would give a more precise appreciation of the relationship.
>
> (p. 162)

Similar views became widely held during the 1960s, and quantification became the sine qua non of training in research methods (LaValle *et al.*, 1967). But they were only a means to the end, as Gould (1979, p. 140) expressed in a later retrospective:

> It was not the numbers that were important, but a whole new way of looking at things geographic that can be summed up in Whitehead's definition of scientific thought, 'To see what is general in what is particular and what is permanent in what is transitory.'

The scientific method increasingly adopted by geographers was a procedure for testing ideas, therefore, but a highly formalised one, about which there has been a great deal of debate among philosophers of science and others (Harvey, 1969a). Although many aspects of the method were used by geographers, their citations indicate relatively little training or exploration in depth into the full philosophy of positivism which, as used here, refers to what is often known as 'scientific method'. It is embraced by the philosophy of logical positivism, which claims that only scientifically obtained knowledge is valid knowledge (Johnston, 1986a, 1986b).

An alternative to positivism is critical rationalism, a view of science associated with the philosopher Karl Popper. He argued that any hypothesis can be found wanting by a single falsification. Harvey (1969a, p. 39) gave only eight lines to this approach, however, preferring to argue that only 'severe failure'– which he does not define – discredits a hypothesis totally (see, however, Moss, 1977; Bird, 1975; Petch and Haines-Young, 1980; Haines-Young and Petch, 1985). Hay (1985a), Marshall (1985) and James Bird (1989) all present cases for human geography adopting Popper's critical rationalism. The goal is the same as in positivism – to develop comprehensive theories which allow predictions with high degrees of certainty. The two differ on means, not ends. Critical rationalists believe that hypotheses can never be comprehensively verified, only falsified. If the researcher is a good observer and a logical thinker, falsification of hypotheses should be rare. If the test is successful, then the speculation in the hypothesis becomes an acceptable generalisation. One successful test will not turn it into a universal law, however, since, according to Popper, all knowledge is provisional, subject to continued testing through critical experiments (Symanski and Agnew, 1981). Knowledge accumulates through the falsification of invalid hypotheses.

Reactions to scientific method

Despite (or perhaps because of) the lack of a clear programmatic statement of the 'new Theology' (Stamp, 1966, p. 18), at least until the appearance of Harvey's (1969a) book, reactions to the developments were many and varied. (James, 1965, p. 35, called the debate 'continued, bitter and uncompromising warfare'.) Two related issues were the main foci of contention: whether quantification was sensible in geographical research, and whether law-making was possible. As Peter Taylor (1976) points out, to some extent the debate was intergenerational: some of the 'old guard' thought the proposals were just not geography and should be banished to another corner of academia.

The quantification issue was the less important and few spoke out against it entirely, although its extent was criticised. Spate (1960a, p. 387) recognised quantification as 'an essential element':

> This is, like it or not, the Quantified Age. The stance of King Canute is not very helpful or realistic; better to ride the waves, if one has sufficient finesse, than to stake attitudes of humanistic defiance and end, in Toynbee's phrase, in the dustbin of history.
>
> (p. 391)

He identified three dangers, however. The first was a confusion of ends and means; some protagonists wanted to quantify everything (after Lord Kelvin, 'when you cannot express it in numbers, your knowledge is of a meagre and unsatisfactory kind': Spate, 1960b), but some things, like the positions of Madrid and Barcelona in Spanish thought, cannot be treated in that way. Second, there was the dogged analysis of trivia, producing platitudinous findings, a fault which Spate recognised as part of all academia, and especially its revolutions: 'Quantified or not, the trivial we will always have with us' (Spate, 1960a, p. 389), and the problem is usually the extreme positions taken up – Arthur H. Robinson's (1961) perks (the hyperquantifiers) and pokes (the hypoquantifiers). Finally, there was the quantifiers' vaunting ambition and belief that solution of the world's problems lay just around the corner.

Spate was more generous than many critics. Burton (1963) identified five types of criticism:

1 geography was being led in the wrong direction;
2 geographers should stick with their proven tool – the map;

3   quantification was suitable for certain tasks only;

4   means were being elevated over ends, with too much research on methods for methods' sake; and

5   objections were not to quantification per se but to the quantifiers' attitudes.

He believed that quantification had already proved more than a fad or fashion, however. Geography would soon proceed beyond a stage of testing relatively trivial hypotheses with its new tools so that 'The development of theoretical, model-building geography is likely to be the major consequence of the quantitative revolution' (p. 156).

One episode regarding the reaction to quantification concerns Stamp's response to Haggett's (1964) early paper on land use change in Brazil: Stamp suggested that Haggett's quantitative analysis could be likened to using 'an enormous steam hammer . . . in the cracking of nuts'. These comments are included in the published record of the discussion of Haggett's paper (Haggett, 1964, p. 380) and end with an exclamation mark, with the implication that these may not have been Stamp's own views, though his call for those using quantitative techniques to 'tear up some of their results' suggests a negative attitude to at least some of the 'new lines of thought'. Haggett had defended his approach – prior to Stamp's summing-up – by noting that 'While method has loomed large in this paper I hope that it has not blotted out the fact that . . . traditional problems of forest depletion are reaching critical levels. Statistical analysis can help to clarify some of the problems and provide some checks on intuitive solutions' (p. 380). It is also recorded that a few days after the paper was given, Haggett was told by the head of department at Cambridge that he was 'bringing the subject of geography into disrepute by applying such mathematical methods' (Chorley, 1995, p. 360).

More critical to many geographers than quantification was the issue of theory, and in particular the role of laws in geography. For some, this continued the debate over environmental determinism, which was still active in the UK (Clark, 1950; Martin, 1951; Montefiore and Williams, 1955; Jones, 1956; see Johnston, 2008). Emrys Jones, for example, extended the debate to cover scientific determinism and its implications for human free will. Martin (1951, p. 6) had argued that possibilism is 'not merely wrong but is mischievous' because all human actions are determined in some way, so that in human geography:

> Unless we can assume the existence of laws or necessary conditions similar in stringency to those of physical science, there can be no human geography nor social sciences worth the name, but only a series of unexplainable statements of bare events. . . . such laws cannot differ, except in respect of . . . far greater complication, from those of physical science.
>
> (pp. 9–10)

Emrys Jones (1956) indicated the impossibility of discovering universal laws about human behaviour and pointed to the use of two types of law in physics: the determinate laws of classical physics, which apply macroscopically, and the probabilistic quantum laws, which refer to the behaviour of individual particles. The latter allow for the exercise of free will within prescribed constraints, and their application in human geography would at least allow answers to be offered to the question 'how?' if not to 'why?'. But the question of causality clearly worried many, as indicated by Peter Lewis's (1965) counter-argument that 'it is erroneously assumed that causes compel their effects in some way in which effects do not compel their causes' (p. 26).

Golledge and Amedeo (1968, p. 760) addressed this same problem, pointing out that critics of law-seeking in human geography defined a law as a universal postulate which brooked of no exception: 'not all are aware [even among those who accept the nomothetic goal] of the nature, types, and relationships of scientific laws developed by philosophers of science'. Scientists use several types of law, however, and the veracity of a lawlike statement can never be finally proven, since it cannot be tested against all instances, at all times and in all places. Four types were identified with relevance for human geographers:

1   Cross-sectional laws describe functional relationships (as between two maps) but show no causal connection, although they may suggest one.
2   Equilibrium laws state what will be observed if certain criteria are met.
3   Dynamic laws incorporate notions of change, with the alteration in one variable being followed by (and perhaps causing) an alteration in another. They may be either historical – for example, showing that B would have been preceded by A and followed by C, or developmental, in which B would be followed by C, D, E, etc.
4   Statistical laws are probabilistic statements of the likelihood of B happening, given that A exists; all laws of the other three categories may be either deterministic or statistical, with the latter almost certainly the case with phenomena studied by geographers.

Having illustrated all four types, they concluded with regard to the existence of situations where laws seem not to fit that 'the idea of laws being strict universals no longer dominates in the interpretation of the concept' (p. 774); the goal is to advance knowledge by continually reassessing the validity of assumed laws.

This discussion was not part of an ongoing, published, debate on quantification and theory-building; it was a reaction to widely held and discussed attitudes rather than to published critiques. There were none for several years in the UK, for example (Taylor, 1976), but there was one debate in the American literature, initiated by Lukermann's (1958) reaction to Warntz's views on macrogeography (see p. 68, this book) and a paper by Ballabon (1957). The latter claimed that economic geography lacked general principles; it was 'short on theory and long on facts' (p. 218). Ballabon stressed the utility of location theory being developed by economists as a source for hypotheses. Lukermann responded that the main problem in Ballabon's and Warntz's arguments lay in the assumptions behind their hypotheses (Warntz's analogies with physics and Ballabon's with economics) which did not conform to his view of geography as an empirical science. Statistical regularities and isomorphisms with other subject matter do not provide explanations, so that hypotheses derived from such models test only the models themselves and (Lukermann, 1958, p. 9; see also Moss, 1970) 'the hypotheses to be tested are neither statistically nor rationally derived; that is, they are derived neither from empirical observation nor from deductions of previous knowledge in the social, economic or geographic fields'.

Berry (1959b) countered with the contention that models, for all their simplifications and unreal assumptions, offer insights towards understanding the real world: 'A theory or model, when tested and validated, provides a miniature of reality and therefore a key to many descriptions. There is a single master-key instead of the loaded key ring' (p. 12). Lukermann (1960a) was not convinced that models based on assumptions of perfect knowledge and competition, for example, could help towards understanding if they were not empirically derived: 'the crucial problem is the construction of hypotheses from the empirical realities of economic geography . . . more light is shed and less truth is sophisticated through inventory than through hunches' (p. 2). Leslie King (1960) claimed that all laws are really only hypotheses and that deviation of observed from expected values in their testing indicate where the assumptions are invalid. Lukermann responded three times. The first paper showed the lack of consensus in 'explanations' of the geography of cement production in the USA (Lukermann, 1960b) because economic analyses ignored 'Historical inertia, geographical momentum, and the human condition' (p. 5). Second, in response to King, he presaged some of the arguments developed later by Sack, who worked with him at Minnesota (see p. 121, this book), and pointed out that much of the theory being introduced to economic geography (such as Lösch's) was not based on providing understanding of, and explanation for, reality (Lukermann, 1961). Finally, a discussion of several aspects of the debate concluded with the statement that:

> Thus, we see scientific explanation as far removed from the context within which the macroscopic geographers would have us put it – the end product of geographic research. Science does not explain reality, it explains the consequences of its hypotheses.
>
> (Lukermann, 1965, p. 194)

He called for explanations in geography to be based on observations and not on imported analogies which cannot offer explanations, only unreal assumptions. Lukermann's basic point, never fully tackled by his critics, was that tests showing conformity between empirical reality and a model were tests of the model only, and could not indicate how empirical reality was created if the assumptions on which the model was constructed were not themselves grounded in reality. (See T. Barnes, 1996, Chapter 9, for an evaluation of Lukermann's work.)

This difference of opinion was over the way in which geographers should seek explanations and not about the positivist scientific method itself; it was concerned with the inputs to the images of the real-world structure (Figure 3.1). It is doubtful whether papers such as those of Jones, Lewis, and Golledge and Amedeo quieted the fears of others unconvinced by the quantifiers' arguments, any more than Berry and King convinced Lukermann. But the differences soon became a non-issue, at least in the published papers resulting from the research activities of geographers in many topical specialisms. As Burton claimed, by the mid-1960s the changes seem to have been widely accepted, and the regional approach had certainly been ousted from its prime position in the publications of American human geographers (see J. Wheeler, 2002a). Increasingly, quantitative and theoretical material came to dominate not only the more obvious journals, such as *Geographical Analysis* (a 'journal of theoretical geography' founded in 1969, to a considerable extent because those adopting the new approach found difficulty in getting their work accepted by the established journals: Berry, 1993; Barnes, 2001b), but also the prestigious general journals, notably *Economic Geography* (especially after a change of editor) and the *Annals of the Association of American Geographers*. (*The Geographical Review* was an early partial 'convert' through the American Geographical Society's sponsorship of the macrogeographers, although Berry states that it rejected his early papers with Garrison as 'not geography': Halvorson and Stave, 1978; Berry, 1993; Barnes, 2001b.) Most of the work contributed little to theory, however. It was quantitative testing of theory- or model-derived hypotheses in some cases, but with little indication of how good the results were. In others, it was quantitative description that could inform theory and model development, but in itself was merely a series of 'factual reports'. By the 1970s, textbooks were being published which began with discussions of scientific method and quantification before proceeding to the substantive content of the 'empirical science' (Abler *et al.*, 1971; Amedeo and Golledge, 1975); the 'revolution' was apparently becoming the orthodoxy.

Spread of the scientific method

The initial development of systematic studies using the positivist scientific method in the USA was very largely focused on economic geography and associated economic aspects of urban geography. This reflects the relative sophistication of economics within the social sciences at that time and the existence of several approaches to 'location theory' (see p. 72), providing models for geographers to emulate, not only to advance their discipline but also to promote its cause in the search for utility to the worlds of business and government. The long tradition of empirical work in human geography meant, however, that with few exceptions research in the systematic areas mainly comprised the statistical testing of relatively simple hypotheses, with little mathematical modelling or writing of formal theory.

Contemporaneous with, and an important stimulus to, these developments in human geography was the emergence of a new discipline in the USA – regional science (see Berry, 1995, on regional science as a stimulus). This was very much the product of one iconoclast scholar

– Walter Isard, an economist who built spatial components into his models to provide a stronger theoretical basis for urban and regional planning than had existed previously. In general terms, regional science is economics with a spatial emphasis, as illustrated by Isard's (1956a, 1960) two early texts, but the Regional Science Association attracted relatively more practising geographers than economists. To some, regional science and economic geography are hard to distinguish: the former can be separately characterised by its greater focus on mathematical modelling and economic theorising, however, whereas geographical work has remained more empirical and less dependent on formal languages. (Initially Isard, 1956b, saw geographers as doing the empirical tests of the regional scientists' models.) Over time, the interests of regional scientists broadened (Isard, 1975), but the strong theoretical base has remained. (The history of regional science, forty years after the foundation of the Regional Science Association, is discussed in two issues – volume 17, numbers 2 and 3, 1995 – of the *International Regional Science Review*; see, in particular, Isserman, 1995. Subsequently, Isard, 2003, published his own history; see also Donaghy, 2014.)

The nascent discipline did not create a substantial niche within American academia, however. (Garrison, 1995, likened it to Moses' forty years wandering in the wilderness!) Nor did it have a lasting, substantial impact on geography and geographers, although the latter remain a considerable proportion of the association's membership. Berry (1995) argues that, for him, relative disillusion with regional science was engendered by Isard's categorisation of geographers as the empirical workers who tested the regional scientists' theories, which implied a subservient position for geographers in a two-class academic society and would hinder the reformation of economic geography that Berry sought. (This has an interesting parallel to the debates within geography on the division of labour between 'theorisers' and 'empirical testers'.)

The Regional Science Association has flourished internationally, however, and Isserman (1995, p. 261) claims that, 'Regional science has become a mainstream group within geography and geography departments ... [and] ... so successful within geography that it became worthy of caricature.' He then notes major tensions between those geographers who remain committed to the goals of regional science (almost certainly a relatively small number within the profession) and those whose interests lie elsewhere. Warf (1995, p. 192) points to these – 'Class and gender, historical sensitivity, the politics of the state, the recovery of the living subject and everyday life, the unintended reproduction of social worlds' – and invites a rapprochement between regional science and a geography inspired by what is now known as social theory. (Interestingly, the 1990s saw the recreation of a form of spatial economics – termed 'new economic geography' by its adherents – that has many parallels to regional science; on its relationships with geography, see Barnes, 2003a; Martin, 1999; Schoenberger, 2001; Clark et al., 2000; and Bosker et al., 2010; a recent text in the field is Brakman et al., 2009.)

The emphasis on statistical methods in so much of the new work in American human geography led to its partial rapprochement with physical geography. (Several of the leading 'quantitative geographers' of the 1960s were introduced to the approach during their wartime training, as essays in Gould and Pitts, 2002, illustrate.) More physical geography papers were published in the leading journals, more physical geographers were appointed to university departments, geologists such as Krumbein, Leopold, Schumm and Wolman were major sources of quantitative ideas, and there was a common interest in the training of graduate students (LaValle et al., 1967). This shared concern over procedures was illustrated at a 1960 conference on quantitative geography, which resulted in two volumes on methodological developments, one for human geography and the other for physical geography (Garrison and Marble, 1967a, 1967b). Other conferences and summer schools to train geographers in quantitative techniques were held at this time (on their impact, see Gould, 1969; Taaffe, 1979), and American geographers were to the forefront in launching an International Geographical Union Commission on Quantitative Methods. Nevertheless, the rapprochement was far from full, as Marcus (1979, p. 522), himself a physical geographer, complained in his presidential address to the AAG which was given just after the appearance of a special issue of the *Annals* to commemorate seventy-five years of American geography: 'not one article [of 30 in the issue] focused on physical geography, its roots, its evolution, its practitioners' (even though 42 per cent of all students enrolled in geography courses at USA universities were studying physical geography and thirty-five of the AAG's presidents had been practising physical geographers at the time that they held the post). There was a need for greater awareness and cooperation, with much hard work in order to generate a 'rational geography that fulfils the promise we hold for the study of earth and humankind together' (p. 532).

The quantitative revolution and geography's position in the academic division of labour

The launch of the 'quantitative and theoretical revolutions' identified by Burton (1963) was concentrated on a few topical specialisms within American human geography only, so an early task for the 'revolutionaries' was to spread their ideas through the discipline, convincing others of the benefits which quantification and the associated scientific method could bring to their special interests. A major piece of advocacy was the NAS-NRC (1965) report on *The Science of Geography* which was prepared in order to chart research priorities within the discipline and establish a place for geography as a social science within the American National Academy (Johnston, 2000b). The case was presented for more 'theoretical-deductive' work to balance the earlier emphasis on 'empirical inductive analysis', the detailed argument being based on four premises:

(a) Scientific progress and social progress are closely correlated, if not equated. (b) Full understanding of the world-wide system comprising man and his natural environment is one of the four or five great overriding problems in all science. (c) The social need for knowledge of space relations of man and natural environment rises, not declines, as the world becomes more settled and more complex, and may reach a crisis stage in the near future. Last, (d) progress in any branch of science concerns all branches, because science as a whole is epigenetic. The social need for knowledge of space relations means an imminent practical need. As the population density rises and the land-use intensity increases, the need for efficient management of space will become even more urgent.

(NAS-NRC, 1965, p. 10)

The committee members (E. A. Ackerman, B. J. L. Berry, R. A. Bryson, S. B. Cohen, E. J. Taaffe, W. L. Thomas Jr and M. G. Wolman) defined geography as 'the study of spatial distributions on the earth's surface' (p. 8) so it followed for them that, 'Geographic studies will be irreplaceable components of the scientific support for efficient space management' (p. 10). The positivist scientific method was being sold to geographers and at the same time geography was being sold to the scientific establishment as a positivist science, from whom financial research support could then be sought. (Until that time, geography had little access to major funding sources for large research programmes, though the Office of Naval Research — surprisingly, given its title — did provide some: Pruitt, 1979.)

The committee chose four problem areas to illustrate geography's potentials as a 'useful science'. The first was physical geography. The second was cultural geography, which studies 'differences from place to place in the ways of life of human communities and their creation of man-made or modified features' (p. 23), with a major focus on landscape development and the diffusion of specific cultural features over space and time: 'applying modern techniques to studying the nature and rate of diffusion of key cultural elements and establishing the evolving spatial patterns of culture complexes' (p. 24) was identified as a profitable avenue for development. The third was political geography, with proposed work on boundaries and resource management. Finally, the committee recognised location theory studies, an amalgam of work in economic, urban and transport geography in which the 'dialogue' between the empirical and the theoretical had gone furthest, 'revealing the potential power of a balanced approach when applied to other geographical problem areas' (p. 44). Location theory involved work on spatial patterns, the links and flows between places in such patterns, the dynamics of the patterns, and the preparation of alternative patterns through model-building exercises which identify efficient solutions.

In the development of the science of geography:

> A major opportunity seen by workers in the location theory problem area is that of integrating their work more closely with other geographers as they begin to deal with spatial systems of political, cultural, and physical phenomena. . . . This could be achieved . . . by the accelerated diffusion of techniques and concepts to other geographers, and communication on the definition of research problems. The result would be to hasten the confrontation of empirical-inductive studies by theoretical-deductive approaches throughout geography. . . . Testing the theory in a variety of empirical contexts should aid in the overall development and refinement of viable theories. It should also serve to connect geographic progress to local problems more rapidly and more effectively.
>
> (pp. 50–1)

The deductive-theoretical scientific methodology was central to the committee's blueprint for the advancement of geographical research, therefore. All geographers would have a role to play in this movement forward, for:

> Geographers have one other asset that should be capitalized on. Those who have been interested in the study of a specific part of the earth (regional geography) develop competences for interpreting the physical-cultural complexes of the regions that they study. Students of the way a particular part of the earth has evolved (historical geography) have other competences for interpreting the historical development and modification of a region. These two groups have students that are particularly qualified to undertake the field observation and field study of problems recognized in a more systematic way and to conduct field tests of generalizations arrived at through systematic study. . . . The regional or historical geography specialist who has mastered the technique of field observation and historical study thoroughly . . . can make himself indispensable if he understands the direction in which the generalizing clusters are headed and relates his work closely to their growing edges.
>
> (p. 61)

A clear division of labour was being suggested, comprising theoretical-deductive 'thinkers' and empirical-inductive 'workers', a division which was apparently unequal in status and was resented by some as such (James, 1965; Thoman, 1965). Berry (1995; p. 84, this book) reports a similar reaction to a proposed division of labour between regional scientists and geographers (all of which suggests parallels with a nineteenth-century division of scientists – reported in Colin Russell, 1977 – into the original, whose time is taken up with their own ideas; the deputed: 'men of little or no originality of mind but who are well trained and competent to work out problems suggested by others'; and the drudges: 'men who can make analyses, calculate constants, collect data, etc.' – the quotes are from Barton, 2003, p. 105). Although not all of the committee's recommendations were taken up, it did win institutional recognition for geography within the NRC with the establishment of a geography and regional science programme; one of its main achievements, when under the direction of Ron Abler, was the establishment of a national centre for Geographical Information Systems (GIS) in the 1980s (see p. 158, this book).

A somewhat similar report was prepared a few years later for the Committee on Science and Public Policy of the National Academy of Sciences and the Problems and Policy Committee of the

Social Science Research Council (Taaffe, 1970). Also the product of a committee (E. J. Taaffe, I. Burton, N. Ginsburg, P. R. Gould, F. Lukermann and P. L. Wagner), this was a difficult task because (as pointed out by Gauthier, 2002, p. 577) there had been considerable debate over the inclusion of geography because members of the other panels (for economics, sociology, psychology, anthropology and political science) did not consider geography as a social science. The report had to convince them otherwise – it succeeded and 'changed people's minds by making a strong argument for geography as a social science that made contributions to society and the development of public policy'. That strong case stressed human geography as 'the study of spatial organization expressed as patterns and processes' (pp. 5–6), incorporating people–environment relationships and cultural landscapes and stressing relevance to planning and other policy issues. Much of the report promoted human geography's cause that:

> there are many opportunities for the expansion and improvement of geographic research. If geography is to have a strong and beneficial impact on the constantly changing patterns of spatial organization of American society, it will be necessary to continue this development.
>
> (p. 131)

This led to six conclusions regarding the discipline's needs: (1) greater collaboration among the social sciences; (2) alleviation of geography's manpower shortage; (3) establishment of centres for cartographic training and research; (4) development of remote sensing and related data bases; (5) greater support for foreign area study; and (6) programmes established 'to strengthen the mathematical training of geographers'. Human geography was presented as an integral component of the social sciences, increasingly sophisticated in its analytical tools, focusing on spatial organisation, and offering particular skills in mapping and data acquisition.

There is little evidence that these two volumes, plus the efforts of successive directors of the NRC's geography and regional science programme and the growing volume of work on the geographical-regional science interface, had a major impact, however, and geography's hold on its place in the American scientific academy was tenuous. Thus, in 1997 a further report was published by a committee of sixteen. It was commissioned because of a 'well-documented growing perception (external to geography as a discipline) that geography is useful, perhaps even necessary, in meeting certain societal needs' (NAS-NRC, 1997, p. ix). To provide a showcase for geography's strengths as 'good science and socially relevant science' (p. vii), the committee was asked to:

1   identify critical issues and constraints for the discipline;
2   clarify teaching and research priorities;
3   link developments in geography to national educational needs;
4   increase the appreciation of geography within the country's scientific community; and
5   communicate with that community about the discipline's future directions.

(p. 2)

In doing that, it focused almost entirely on a role for geography, and of types of geographical research, that was very similar to that laid out in the reports of thirty years previously (on which, see Johnston, 1997a, 2000b). The message had to be repeated since the scientific credentials of geography (both human and physical) were still not widely appreciated within the American academic community; one result of its impact within the NRC was its establishment of a Committee on Geography (Turner, 2002, p. 66).

Little more than a decade later, a further report essayed the same general task. A committee was established by the National Academies at the request of the National Science Foundation, the US Geological Survey, the National Geographic Society and the Association of American Geographers, which sought to persuade the American scientific academy 'how the *geographical sciences* can best contribute to science and society in the next decade through research initiatives aimed at advancing understanding of *major issues facing Earth in the early 21st century*' (NRC, 2010, 2, our italics). The terminology indicates that the coverage is wider than the institutionalised discipline of geography, but the chosen adjective indicates that a political point was being made regarding the discipline's centrality to the more amorphously defined 'spatial sciences' (Johnston, 2011); interestingly, the report's authors identify archaeologists, economists, astrophysicists, epidemiologists, biologists, geologists, landscape architects and computer scientists among the 'geographical scientists' but omit sociologists, political scientists and a wide range of scholars in the humanities disciplines that have experienced a 'spatial turn' in recent decades (Warf and Arias, 2008; Withers, 2009). Its brief was wider than geography *sensu stricto* – as indicated by the reference to the 'social sciences' (though relatively little social science was covered) but its emphasis – as in the previous volumes – was strongly on techniques and technologies, reflecting the recent rapid growth of Geographical Information Science (see p. 157, this book). Much stress was placed on the need for a research infrastructure (called a cyberinfrastructure) to enable spatial analysis of large and complex datasets, and the need for training in the use and development of that infrastructure in order to realise the potential. Although geography was by then a very much broader discipline, this report, like its predecessor, continued to portray geographers' role within the sciences as largely technical (as argued in a symposium on the report published in *The Professional Geographer*: Sui, 2011).

Geography's subdisciplinary fields

The argument advanced by proponents of the new methodology focused on a common set of procedures to tackle geographical problems. Berry (1964a) argued that the geographer's viewpoint emphasises space, with regard to distributions, integration, interactions, organisation and processes. Its data can be categorised in a single matrix (Figure 3.2) in which 'places' form the rows and 'characteristics' the columns; each cell defines a 'geographic fact'. Berry recognised five types of geographical study which focused on different elements of this matrix: study of either a single row

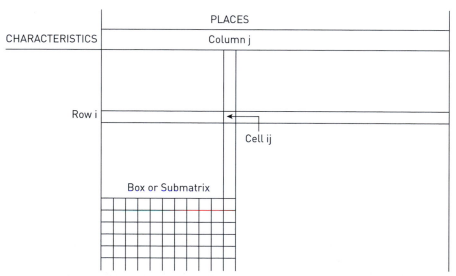

Figure 3.2 The geography matrix: each cell – ij – contains a 'geographic fact', the value of characteristic i at location j

Source: Berry, 1964b, p. 6.

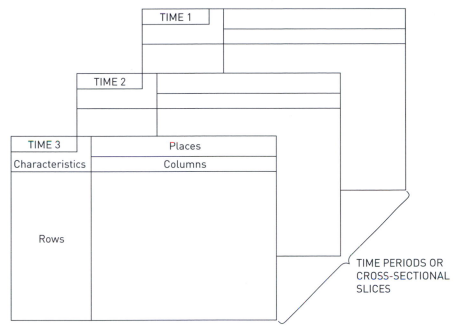

Figure 3.3 A third dimension to the geographic matrix: each cell – ijt – contains a 'geographic fact', the value of characteristic i at location j at time t

Source: Berry, 1964b, p. 7.

(a place) or column (a characteristic); comparison of two or more rows (places) or columns (characteristics); or study of rows and columns together (regions). Adding further matrices, one for each time-period (Figure 3.3), allows five further types of study, based on the earlier five but concentrating on changes over time. Thus, he concluded, systematic and regional geography are part of the same enterprise – a repetition of Hartshorne's (1959) arguments – with neither sufficient in itself.

Berry's matrices referred only to the characteristics of places; further matrices (Figure 3.4) show flows between places, with one matrix for each flow category in each time-period (Clark *et al.*, 1974). Berry used this extension, though he did not formalise it, in a fusion of the procedures for formal and functional regionalisation (see p. 59, this book) to produce a general theory of spatial behaviour – a field theory, which he applied in a large study of the spatial organisation of India (Berry, 1966, 1968). The techniques employed became widely used in the 1960s, as access to high-speed computers became easier. They were given the umbrella term 'factorial ecology' (Berry, 1971; Davies, 1984), and were widely applied in several geographic subfields, resulting in a methodological unity which was previously unknown across the various systematic specialisms.

One systematic area colonised early by such new methods was that part of urban geography which dealt with the internal spatial structure of cities. Until the 1960s, little work had been done on this topic except with regard to commercial land uses (particularly the Central Business District and the relationship of the pattern of suburban shopping centres to the postulates of central place theory); almost no attention was paid to the human content of residential areas, perhaps because geography was seen as the science of places, not of people. Recognition that 'people live in cities' (Johnston, 1969) generated interest in residential areas, which gained much stimulus from the work of the urban ecology school of sociologists at Chicago (some of their works had been

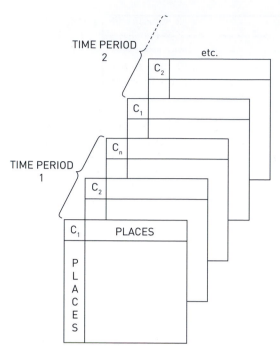

Figure 3.4 Geographic flow matrices – one per commodity (C_1 – C_n) per time period

Source: after Clark *et al.*, 1974.

introduced to geographers earlier – Harris and Ullman, 1945; Dickinson, 1947 – but with little impact for a decade or so – Agnew, 1997; Lichtenberger, 1997; Harris, 1990, 1992, 1997b). Social area analysis, and then factorial ecology, gained in popularity as a methodology (Berry, 1964b), with the allocation of housing among competing groups being modelled, mapped and statistically analysed (Johnston, 1971; Timms, 1971).

By the mid-1960s use of statistical methods to test hypotheses was common across many of human geography's specialisms. They were united by their methods, remaining very separate identifiable branches within the geographical enterprise in terms of their substantive interests. The decline of interest in regional syntheses (see Taaffe, 1974) meant that human geography experienced a centrifugal trend with regard to substance, contemporaneous with a centripetal one with regard to procedures. Since positivist scientific method, and the statistical techniques with which it was associated by many geographers, was used much more widely than in human geography alone, the former trend was probably the most important.

Transatlantic translations

By the early 1960s the quantitative and theoretical revolutions were having considerable impact beyond the USA, as the result of two agencies. The first was the publication of work by the American iconoclasts in the major journals. Second, and probably more importantly, during the 1950s and 1960s a number of British geographers went to the USA, either as postgraduate students or as visiting staff members (see Haggett, 1990); during the 1960s and 1970s many British academics were recruited to teach on the intensive summer-school courses for part-time students. Some of those visitors encountered the new ideas and disseminated them 'back home', in part via the newly established (1962) Study Group in Quantitative Methods of the Institute of British Geographers

(Gregory, 1976). Several, including Brian Berry, stayed in North America after completing their postgraduate training, and others were attracted to move there (on the attraction of America, see Berry, 1993, 2002b).

Some UK geographers attracted to the 'new approach' had a local base on which to build. For most, this was in physical geography, arising out of the early use of statistics by climatologists (e.g. Crowe, 1936); as a result, perhaps surprisingly, the first undergraduate text in statistics for geographers was written by an English academic, Stan Gregory (1963), in which the positivist methodology is only implicit, however. His Preface speaks of the geographer's raw material 'becoming progressively more of a quantitative nature' and 'the need to present both data and conclusions in sound quantitative terms' (pp. xiii–iv); statistical techniques are needed, but their use in hypothesis testing is not set out. In addition, there was some interest in location theory – it was taught at University College London in the early 1950s, for example (Halvorson and Stave, 1978; Berry, 1993; Barnes, 2001a, 2001b), and at the London School of Economics (Johnston, 2003c). Some worked on industrial location (W. Smith, 1949; Rawstron, 1958, though see Rawstron, 2002) and settlement patterns was influenced by theoretical work, too (Dickinson, 1933; Smailes, 1946; note, however, the unwelcoming anonymous review of Christaller's thesis on central place theory in the *Scottish Geographical Magazine*, 1934; on Dickinson and Christaller, see Johnston, 2002a), but this was far from normal.

Although statistics courses were introduced to several UK university departments of geography by the mid-1960s, and aspects of the scientific methodology were taught in at least a few (Whitehand, 1970), the main focus for the introduction of the 'new geography' to the UK during the early years of the decade was the University of Cambridge. The leaders were R. J. Chorley (an Oxford graduate and geomorphologist who had studied geology in the USA: Beckinsale, 1997; Stoddart, 1997a) and P. Haggett (a Cambridge-trained human geographer, although his early published work was in biogeography; he had also visited the USA and experienced the developments there: Haggett, 1965c, p. vi, 1990; see also Haggett and Chorley, 1989. Haggett's interests in human geography – and in particular on spatial diffusion – were stimulated by a footnote in Lösch's classic book, but he was only able to initiate that research when he moved to Bristol in the mid-1960s.) Their impact on British geography was considerable, through innovative research and teaching (Gregory, 1976); they supervised a substantial number of graduate students who proceeded to academic posts elsewhere and 'spread the message' in the same way that the 'space cadets' did in the USA (Hanson, 1993).

Chorley and Haggett worked on the adaptation of certain statistical techniques to geographical (both physical and human) problems (Chorley and Haggett, 1965a; Haggett, 1964; Haggett and Chorley, 1969), but their most lasting initial contribution was probably editing two collections of essays which emerged from extramural courses that they directed, aimed at introducing the 'new geography' to teachers. (Note that Barnes, 1996, emphasises Haggett's role as a change-agent within British geography, but in so doing overlooks the substantial behind-the-scenes impacts of others, notably Stan Gregory who was very influential not only in establishing the Study Group on Quantitative Methods within the IBG, but also within the Geographical Association and the Joint Matriculation Board, whose examinations were used for entrance to most UK universities.)

The first of these books – *Frontiers in Geographical Teaching* (Chorley and Haggett, 1965b) – was based on a 1963 course designed 'to bring teachers and like persons into the University, there to encounter and discuss recent developments and advances in their subjects' (p. xi; see also Haggett, 2015). In it E. A. Wrigley (1965) discussed the changing philosophy of geography, identifying the increasing use of statistical techniques as the contemporary development 'of singular importance' (p. 15). He pointed out that techniques of themselves do not form a methodology and that 'Geography writing and research work have in recent years lacked any general accepted, overall view of the subject even though techniques have proliferated' (p. 17). He offered no outline of such a view, however, arguing that eclecticism in mode of analysis was likely to be most productive: 'the best sign of health is the production of good research work rather than the manufacture of general methodologies' (p. 17) and 'The final test of the value of any intellectual labour is its ability to help men to understand questions in which they are interested' (p. 19).

Many of the other chapters interpret geography as if the 'revolutions' had not occurred in the USA, however. C. T. Smith's (1965) chapter on historical geography, for example, is an excellent British companion to the American statement published a decade earlier (Clark, 1954). Elsewhere, Pahl (1965) introduced the models of the Chicago school of urban sociologists and suggested a social geography in which the prime factor is distance (p. 108), but only the chapters by Haggett and Timms introduced much of the transatlantic turmoil. Haggett (1965a) wrote on models in economic geography, both those based on simple views of the world, such as developments on von Thünen's (Chisholm, 1962), and those derived from observations of particular cases (e.g. Taaffe *et al.*, 1963). Haggett noted that:

> Perhaps the biggest barrier that model builders in economic geography will have to face in the immediate future is an emotional one. It is difficult to accept without some justifiable scepticism that the complexities of a mobile, infinitely variable landscape system will ever be reduced to the most sophisticated model, but still more difficult to accept that as individuals we suffer the indignity of following mathematical patterns in our behaviour.
>
> (1965a, p. 109)

He introduced the notion of indeterminacy at the individual level (see also Jones, 1956; p. 81, this book) and showed how random variables must be introduced to make models operational; his chapter on scale problems (Haggett, 1965b) illustrated methods of sampling and of map generalisation from samples.

Timms (1965) demonstrated the use of statistical techniques for analyses of social patterns within cities (based on Shevky and Bell's, 1955, social area analysis, and developed independently of Berry's work on this topic (see p. 89, this book). Later work in this mould by Robson, 1969, followed a period spent working with Berry at Chicago, but earlier developments at Keele, also based on Shevky and Bell (Williams and Herbert, 1962) illustrated that some at least of the changes in UK geography were largely independent of those occurring in the USA, depending instead on accessing the relevant literature in other disciplines.) Timms pointed out that:

> The sciences concerned with the study of social variation have as yet produced few models which can stand comparison with the observed patterns or which can be used to predict those patterns. . . . Prediction rests on accurate knowledge of the degree and direction of the interrelationships between phenomena. This can only be attained by the use of techniques of description and analysis which are amenable to statistical comparison and manipulation. If the goal of geographical studies be accepted as the formulation of laws of areal arrangement and of prediction based on those laws, then it is inevitable that their techniques must become considerably more objective and more quantitative than heretofore.
>
> (1965, p. 262)

The majority of the contributors to *Frontiers in Geographical Teaching*, almost all of whom were associated with the Department of Geography at the University of Cambridge, were not as committed to the 'new geography' as Timms (who later, like Pahl, became a professor of sociology; E. A. (Tony) Wrigley became an economics historian/demographer, and President of the British Academy). This cannot be levelled at the editors, whose epilogue presented a strong case for the 'theoretical revolution':

We cannot but recognize the importance of the construction of theoretical models, wherein aspects of 'geographic reality' are presented together in some organic structural relationship, the juxtaposition of which leads one to comprehend, at least, more than might appear from the information presented piecemeal and, at most, to apprehend general principles which may have much wider application than merely to the information from which they were derived. Geographical teaching has been remarkably barren of such models. . . . This reticence stems largely, one suspects, from a misconception of the nature of model thinking. . . . Models are subjective frameworks . . . like discardable cartons, very important and productive receptacles for advantageously presenting selected aspects of reality.

(Haggett and Chorley, 1965, pp. 360–1)

This view dominated their next, and substantially more influential, volume *Models in Geography* (Chorley and Haggett, 1967), most of whose contributors were also linked with the Cambridge department. This was a major synthesis of most of the work completed before the mid-1960s by adherents to the 'quantitative and theoretical revolutions'. Individual authors were asked 'to discuss the role of model-building within their own special fields of research' (Haggett and Chorley, 1967, p. 19), which resulted in a series of substantive review essays, some dealing with topical specialisms (urban geography and settlement location; industrial location; agricultural activity – there were similar reviews for physical geography), some with particular themes ranging across several specialisms (economic development; regions; maps; organisms and ecosystems; the evolution of spatial patterns), and some with methods and approaches (demographic models; sociological models; network models). A catholic use of the term 'model' was deployed, as a synonym for a theory, a law, a hypothesis, or any other form of structured idea (see Moss, 1970). The approach was strongly nomothetic, however:

the student of history and geography is faced with two alternatives. He can either bury his head, ostrich-like, in the sand grains of an idiographic human history, conducted over unique geographic space, scowl upon broad generalization, and produce a masterly descriptive thesis on what happened when, where. Or he can become a scientist and attempt, by the normal procedures of scientific investigation, to verify, reject, or modify, the stimulating and exciting ideas which his predecessors presented him with.

(Harvey, 1967a, p. 551)

All the contributors had chosen the latter course: their focus was on models – on generalisations of reality – and methods were very much secondary.

The orientation of this significant volume is given by the editors' introduction. (Its significance lay in its two uses: first, as a synthesis and argument, the volume was widely read and used by researchers and teachers as a guide; second, as a series of major reviews, when republished in separate paperback volumes, the book was extensively employed as an undergraduate text.) Haggett and Chorley (1967, p. 24) presented the model as 'a bridge between the observational and theoretical levels . . . concerned with simplification, reduction, concretisation, experimentation, action, extension, globalisation, theory formation and explanation'. It can be descriptive or normative, static or dynamic, experimental or theoretical (see also Chorley, 1964). It forms the basis for their proposed new paradigm, which made no attempt 'to alter the basic Hartshorne definition of Geography's prime task' (p. 38) but offered hope for much greater progress:

the new paradigm . . . is based on faith in the new rather than its proven ability . . . There is good reason to think that those subjects which have modelled their forms on mathematics and physics . . . have climbed considerably more rapidly than those which have attempted to build internal or idiographic structures.

(p. 38)

Models in Geography stands as a statement of that faith. Although the editors and contributors comprised many of the early active participants in the move to change British geography towards a 'more scientific' approach, others involved are not directly represented. Notable among them was a group who graduated at the University of Cambridge in the 1950s, having been tutored by A. A. L. Caesar. (Chisholm and Manners, 1973, p. xi, credit his role, one which continued until the end of the 1970s, with a steady stream of productive research workers from St Catharine's College.) It included Michael Chisholm, Peter Hall, Gerald Manners and Kenneth Warren as well as Haggett, who was the only one who worked within the 'quantitative revolution'. Chisholm, for example, focused on theoretical developments in economic geography (e.g. Chisholm, 1962, 1966, 1971a) but with relatively little quantitative analysis (though see Chisholm and O'Sullivan, 1973), and both Hall and Manners were more concerned, as was Caesar, with geographical analyses of contemporary policy issues, although in some cases the analyses led to theory-derivation (Hall, 1981a, 1998).

Chorley and Haggett's editing, and their joint work on technical developments, such as trend surface analysis (Chorley and Haggett, 1965a) and network analysis (Haggett and Chorley, 1969), reflected a belief in the unity of physical and human geography (further demonstrated in Haggett's (1972, 2001) major undergraduate text). This assumed that a shared interest in methods and techniques could unite the two – Haggett (1967, p. 664) writes on 'the basic proposition that a wide range of different geographical networks may be usefully analysed in terms of their common geometrical characteristics'. As in North America, if geometry (i.e. spatial form) was the focus of geographical analysis, then physical and human geographers could find common cause (Woldenberg and Berry, 1967). Explanation of the geometry required the study of very different processes, however, and the two soon separated again.

The two books edited by Chorley and Haggett were the first major indication of change within British geography – and they appeared too late to assist geography's cause during the first stages of a major expansion of universities there. By the end of the 1950s, there was a geography department in just about every UK university and many of them were expanding to meet growing demand from students. But as a discipline, geography did not appear to have a high standing within British academia, and when a series of seven universities was founded in England in the late 1950s and early 1960s, only one (the University of Sussex) included geography within its course offerings from the outset (as a social science though including physical geography; there was no separate geography department, however, only a 'Geography Laboratory'). Two others included geographers within schools of environmental studies, as did two further universities established a little later in Scotland and Northern Ireland, but only one – the University of Lancaster – eventually established a Department of Geography, because it was having difficulties recruiting students for other disciplines. Thus geography, because of its image as not being closely linked with the burgeoning social sciences and the lack of an influential pressure group, failed to gain a place in a new group of universities, many of which soon became major research centres in the social and environmental sciences. (On this episode, see Johnston 2003b, 2004c.)

This marginalisation of geography within the UK was exacerbated in the mid-1960s when a Social Science Research Council (SSRC; it was later renamed the Economic and Social Research Council – ESRC) was established to fund postgraduate and research work, and from which geography was initially excluded. In response to this, a number of geographers (led by Robert Steel, an Oxford graduate, then Professor of Geography at the University of Liverpool and later to be Principal of University College Swansea) took the initiative of approaching the SSRC chairman to promote their discipline's cause. He invited them to make a detailed case, a task undertaken for the group by Michael Chisholm (then a lecturer in the Department of Geography at the University of Bristol). His detailed document and case – not accepted by all departmental heads – was successful, and a Human Geography and Planning Committee was established within the SSRC (Chisholm 2001; Johnston, 2003c). Chisholm later developed his case for geography as a social science into two books (Chisholm, 1971c, 1975a) and edited a third, which illustrated it with a series of major case studies (Chisholm and Rodgers, 1973). The first of these (written specifically as a review conducted for the SSRC) argued in its introduction that the real change in geography associated with quantification was 'a substantial tightening up in logical rigour' (Chisholm, 1971c, p. 2) involving three main themes (pp. 5–6):

1   All phenomena occur in a spatial context.
2   Economic and social processes require a built infrastructure (buildings and transport media).
3   Human activity takes place within environmental contexts.

There were only five chapters: one considered static patterns (points, surfaces, regions and their correlation), another considered dynamic patterns (processes such as urbanisation and economic development) and a third theories of spatial organisation. The other two were concerned with research organisation and priorities with seven of the latter identified (by the SSRC's Human Geography Committee and not just Chisholm): perception studies; simulation models; forecasting; regional taxonomy; environmental standards; population and migration; and processes of regional economic and social development. (These brief chapters were the core of the later, more extended treatment: Chisholm, 1975a.)

The relatively untouched: historical and cultural geography

The NAS-NRC (1965) report (see p. 85, this book) identified two main systematic specialisms within human geography that had been largely untouched by the developments outlined here: cultural and historical geography (see also Darby, 1983a). In addition, despite Berry's (1964a) attempt to reframe it, regional geography remained largely apart from the changes in methodological emphasis. Furthermore, political geography was described by Berry, 1969, as a 'moribund backwater' (though Muir, 1975, sought to apply quantitative methods to the field). Not all cultural, historical and regional geographers ignored the changes occurring elsewhere, and some were in the 'revolutionary' vanguard: two of the chapters in *Models in Geography*, for example, were written by David Grigg and David Harvey, whose empirical research for their Ph.Ds was on historical topics (e.g. Harvey, 1985c). But in general terms, the NAS-NRC report was correct; there is little evidence of early success in winning cultural, historical and regional geographers over to the new methodology.

Of the three groups, historical geographers were probably most concerned about their apparent isolation within the discipline. This was summarised by Baker in terms of the approaches which historical geographers need to consider in greater detail:

> An assumption is necessary here: that methodologically the main advances can be expected from an increased awareness of developments in other disciplines, from a greater use of statistical methods, from the development, application and testing of theory, and from exploitation of behavioural approaches and sources. . . . Rethinking becomes necessary because orthodox doctrines have ceased to carry conviction. As far as historical geography is concerned, this involves a questioning of the adequacy of its traditional methods and techniques.
>
> (1972, p. 13)

All of these would have to be followed with care, and the potentials of the methodological developments assessed cautiously, but Baker clearly believed there was considerable scope for change as had already been shown in economic and social history, and perhaps even more so in archaeology (see Hodder and Orton, 1976; Renfrew, 1981). Particular areas of historical geography, including those relating to urban settlements (e.g. D. Ward, 1971), were more open to such changes, if for no other reason than the better quality and quantity of available data (e.g. Whitehand and Patten, 1977; Johnson and Pooley, 1982; Dennis, 1984). There were possible implications in such work, however, as Baker noted:

> Studies in, for example, 'historical agricultural geography', 'historical urban geography' and 'historical economic geography' seem to offer possibilities of fundamental development, particularly in terms of a better understanding of the processes by which geographical change through time may take place. Such an organization of the subject would view historical geography as a means towards an end rather than as an end in itself.
>
> (1972, p. 28)

In the early discussions of the relationships between historical geography and the 'theoretical and quantitative revolutions', most attention focused on quantification rather than theory. Vance (1978) pointed out that the development of theory does not have to involve 'quantitative abstraction', as his own work showed (Vance, 1970; see also Pred, 1977b, and Conzen, 1981; Meinig's (1986–2004) later four volume *magnum opus* on *The Shaping of America* was built around graphical models). Available data can be manipulated to test theories regarding past spatial patterns (e.g. Goheen, 1970), but Radford (1981, p. 257) argued that theory is the more important: 'In the cities of the nineteenth-century United States, a set of principles . . . was taken to something approaching a logical conclusion.' This assumes that theory is possible in historical geography. As illustrated in Chapter 5, some dispute this – for geography as a whole and not just for historical geography. The positivist method implies objectivity, but the geographer in describing a landscape is subjective:

> In describing a landscape, is he not committed by his past training and his past experiences – by his prejudices, if you will? Just as the portrait an artist paints will tell you much about the artist as well as his sitter, so the description of a countryside will tell you a great deal about the writer.
>
> (Darby, 1962, p. 4)

Darby portrayed geography as both a science and an art:

> [it] is a science in the sense that what facts we perceive must be examined, and perhaps measured, with care and accuracy. It is an art in that any presentation (let alone any perception) of those facts must be selective, and so involve choice, and taste, and judgement.
>
> (p. 6)

Examining the subsequent evolution of historical geography, Baker (2003) was especially concerned with debates about its relationship with history. Reviewing the impacts of the approaches that would transform wider human geography after the 1960s, he claimed that:

> the more explicit adoption of spatial and locational concepts in historical geography (and in history) does not require us, Mao-like, to reject our traditions and embrace a brave 'new' world of geographical history. The more energetic prosecution of a new geographical history in the guise of spatial and locational histories is to be welcomed, but it does not of necessity involve forsaking all other geographical perspectives upon history.
>
> (pp. 70–1)

Cultural geographers were less concerned about their apparent drift away from the mainstream of geographical activity than historical geographers, perhaps because of the lack of any parallel developments to those affecting geography in anthropology, the discipline with which many contemporary cultural geographers had most contact. As Mikesell (1978, p. 1) expressed it, 'Stubborn individualism and a seeming indifference to academic fashion are well-known character-istics' of cultural geographers, whose preferences are for a historical orientation; a focus on the role of human agency in environmental change, on material culture, and on rural areas; links with anthropology; an individualistic perspective; and field work. In the USA, the Berkeley 'school', founded and led for several decades by Carl Sauer, remained an influential site of resistance to the new geography, continuing to stress research on the *Morphology of Landscape* (Sauer, 1925). This had rejected environmental determinism for a more open approach, examining for example, human intervention in landscape evolution through plant and animal domestication, the use of fire, the diffusion of ideas and artefacts, and the creation of settlements (all themes reviewed in W. L. Thomas, 1956, and maintained by Sauer's disciples, such as Kniffen: see Mathewson, 1993). Sauer was departmental head at Berkeley from 1923 to 1955 and 27 post-graduates completed Ph.Ds under his supervision in those years. His student, Jim Parsons, succeeded him and more students continued to work on the ethos of 'Sauer's project of elucidating the cultural landscape' (Williams, 2009, p. 303, 2014). More traditional regional geography continued to be written by others too; even if it could never recover the relative centrality it enjoyed in anglophone geography before the mid-1950s. The field was surveyed by Paterson (1974), whose essay had two main sections – 'On the problems of writing regional geography' and 'Is progress possible in regional geography?'. The first investigated six problems, including the growing shortage of subordinate materials (micro-regional studies), and the increasing submergence of regional distinctiveness, though:

> only a certain amount of innovation is possible if the regional geographer is to perform his appointed task, which is to convey to his reader the essentials of his region; to illuminate the landscape with analytical light. Landforms and climate are common to all terrestrial landscapes . . . and human activities to most of them: how shall repetition be avoided?
>
> (p. 8)

> So long as contrasts between region and region [remain], and no matter to what they are attributable, there is work for the geographer to do.
>
> (p. 16)

Paterson did not conclude that there is no possibility of progress, however, despite the constraints: regional geography can advance on two criteria – content and insight. Reference to Wilbur Zelinsky's (1973a) book illustrated the increased range of content currently being introduced; discussion of Meinig's work (e.g. Meinig, 1972) showed the ability of regional geographers to produce fresh spatial insights, although Paterson concludes that 'Adventurousness is not a quality that most of us associate with regional geography' (p. 9). Thus:

> The way is open for regional studies which are less bound by old formulae; less obliged to tell all about the region; more experimental and, in a proper sense of the word, more imaginative than in the past, and covering a broader range of perceptions, either popular or specialist.
>
> (p. 23)

> Regional Geography['s] . . . goals are general rather than specific; it is not primarily problem-orientated but concerned to provide balanced coverage, and its aims are popular and educational rather than practical or narrowly professional. Such relevance as it possesses it gains by its appeal . . . to the two universal human responses of wonder and concern. . . . One may recall Medawar's assertion that in science we are being progressively relieved of the burden of singular

instances, the tyranny of the particular, and in turn assert that there is a frame of mind on which the particular exercises no tyranny, but a strong fascination.

(p. 21)

Elsewhere, Mead (1980, p. 297: see also Mead, 2004) strongly defended geographers who 'adopt other lands . . . share other cultures . . . [and] make a contribution to the store of knowledge about them'. Similarly, though with greater intensity, Hart (1982a, p. 1) claimed in his presidential address to the AAG that: 'Society has allocated responsibility for the study of areas to geography: this responsibility is the justification for our existence as a scholarly discipline'.

This responsibility involves responding to people's curiosity about places, promoting an 'understanding [of] the meaning of an area [which] cannot be reduced to a formal process' (Hart, 1982a, p. 2). Unfortunately, geographers were very insecure with that mission and so had retreated into 'scientism' which involved studying 'people abstracted from their real world contexts' (p. 16). The quantitative revolution was giving priority to 'techniques and procedures rather than places and people' (p. 17) and diverting attention from geographers' obligation to answer the key questions of 'how much of what is where and why it's there, about the where and why of places and people, about the land and how people have used and abused it' (p. 19) – about which there is so much 'woeful ignorance'. Thus, to Hart (p. 29) an important element in the 'highest form of the geographers' art' (his term for regional geography) requires individuals to 'adopt a region, to immerse themselves in its culture, to acquire a specialist understanding of it'. Ten geographers from the University of California at Santa Barbara responded by arguing that just as Hart claimed that 'We cannot allow ourselves to be intimidated by those who flaunt the banner of science' (Hart, 1982a, p. 5) so 'We equally cannot allow ourselves to be intimidated by those who flaunt the banner of anti-science, those who would reject all that is scientific about the discipline, and those who would urge a return to the descriptive morass from which we have recently emerged' (Golledge et al., 1982, p. 558), especially at a time when geography was under threat in some American universities. Hart's (1982b, p. 559) response was that he was critical of shoddy work of any type, did not want geography to be forced into any particular mould, and that 'we should be most unwise to jettison so blithely one of our grand traditions merely because some of our colleagues do not understand, appreciate, or know how to practise it' (see also Clout, 2003a, and Murphy's, 2013, argument for geographers to produce 'grand regional narratives' as a way of improving [non-academic] geographical understanding).

The implication is that there was a strongly perceived difference between, on one hand, many historical geographers and, on the other, most cultural and regional geographers with regard to the degree to which they felt 'left behind' or 'relatively untouched' by the changes that occurred during the 1950s and 1960s in other branches of human geography. Historical geographers, at least, seem to have been impelled to consider the possibility of making methodological changes, whereas the other two groups continued to work within their established tradition (see also Mikesell, 1973). Not all historical geographers would agree with Baker's analogy from systems theory that simply 'Historical geography has a long relaxation time' (p. 11), however, and Chapter 5 illustrates the degree to which they mounted attacks against the positivist approach.

Conclusions

Critiques of the established regional approach began to crystallise in the USA during the mid-1950s. The aims of research in human geography had been resolved around regions and description until then, and relatively traditional definitions of the field were still observed: the main issues concerned means and methods. Some of these debates were among those who would contribute to the new agendas in the 1950s. The innovations of that period involved the strengthening of the systematic and topical geographies, and their release from a largely subservient relationship to regional geography, by attempts to develop laws and theories of spatial patterns, using models of various kinds for illumination, and applying mathematical and, especially, statistical procedures to facilitate the search for generalisations. Whereas regional geographers saw their discipline as, at most, law-consuming, those of the new persuasion aimed at producing their own laws, which could be used to explain particular regional outcomes.

These changed means to the geographic end were rapidly accepted in many branches of human geography, particularly in those topical specialisms dealing with economic aspects of contemporary life. They were soon accepted in the growing field of contemporary social geography, but were relatively ignored in historical geography and almost completely shunned in cultural and regional investigations, which, it was claimed, focus on unique characteristics of unique places. They also spread into the corresponding fields of study across the Atlantic (and across the Pacific, too: King, 2008), and within little more than a decade British geographers produced a major review volume (*Models in Geography*) containing 816 pages of testimony to the innovators' enthusiasm and their links (largely one-way) with other social sciences. It is evident that in the UK there was some fertile ground for the reception and dissemination of this 'new geography'; location theory and quantitative methods were already being taught in some geography departments in the 1940s–1950s.

Everywhere those who promoted these shifts evidently judged the existing regional geography framework inadequate. What, larger, contextual, forces prompted them has more recently become the subject of debate, with Trevor Barnes (2008a, 2008b) pointing to the key role, in the USA, of the wider scientific ethos of the moment linked to the Cold War. This had been shaped by the Allied and especially American techno–military–science in the Second World War and would be further bolstered by the Cold War. Others (Johnston *et al.*, 2008) signal multiple causes and diverse points of origin, especially in the UK, citing Stan Gregory's (1976) identification of three groups who became key to the development of the quantitative revolution there: those who had been to North America (usually as postgraduate students), a few introduced to quantification elsewhere, and the 'home grown' advocates. Moreover, methods are insufficient to sustain an academic revolution unless they can be applied to a coherent substantive core. The quests for such a core and some of their ongoing trajectories are the subject of the next chapter.

Chapter 4

Human geography as spatial science

[O]ne of the greatest legacies that our discipline is endowed with is the frontier spirit that accompanied the quantitative revolution.

(Trisalyn Nelson, 2012, p. 92)

Undoubtedly the 1960s have witnessed a major transformation in the subject; today, geographers are probably at least as numerate, if not more so, than most other social scientists. However, the movement to quantification should not be regarded as an end in itself; it is but a means to an end, a symptom of something deeper that is going on within the subject.

(Michael Chisholm, 1975a, p. 48)

[T]hese developments have been both exciting and disturbing. Exciting because they promised to breathe life into traditional university geography which – with a few exceptions – remained wedded to conservative regional or commodity lines or split-up into confusing sub-varieties like 'resource' geography or 'medical' geography. Disturbing because they increasingly used mathematical methods and were scattered with unfamiliar and unwelcome incognitas like Eigen values or Beta coefficients.

(Peter Haggett, 1965c, p. v)

Our images – the maps and models of the world we carry around with us – need larger and more relevant information inputs.

(Peter Gould and Rodney White, 1974, p. 192)

By our theories you shall know us.

(David Harvey, 1969a, p. 486)

The changes in human geography that emanated from several centres in the USA during the 1950s were very much concerned with methods of investigation. Systematic studies were in the ascendant, however, and the implicit intention was to develop laws and theories within an, often unstated, positivist framework.

In this context, human geographers promoting the changes increasingly sought a clear identity for their discipline within the social sciences, alongside economics, sociology and political science. Disciplines are sustained by their content rather than their methods, however, since many of the latter (especially those involving statistics) were common to several disciplines – or so it was believed. Geographers argued that their discipline provided a particular viewpoint and contribution to the overall social science goal of understanding society. Their focus was on spatial variables and the study of spatial systems, so that within their own discipline they elevated the concept of space over that of place, while also redirecting its focus away from the idiographic and towards the nomothetic. Other concerns and approaches, from the late 1960s and into the

1970s, 1980s and beyond soon yielded different ways of doing geography that were linked to a series of adjectives – notably, humanistic, radical (and sometimes anarchist and Marxist), postmodern and feminist. These form the foci of Chapters 5–8. However, while those discussions developed, the adoption of scientific methods that we recounted in Chapter 3 was also carried forward through a range of methods, debate and technologies. This chapter describes these, which include ideas of spatial and systems theory, notions of behavioural geography and time-geography, geostatistics and geocomputation and, arguably most influential of all, the advent of Geographical Information Systems (GIS) and Science (GISci).

Spatial variables and spatial systems

Geography is a discipline in distance, according to a Scottish professor's inaugural lecture (Watson, 1955; Johnston, 1993b), with the relative location of people and places as its central theme. Kevin Cox (1976) later argued that the contemporary importance of relative location within society stemmed from alterations in societal structure consequent upon technical change. The main interactions in less developed societies are between relatively isolated groups of individuals and their physical environments, creating vertical relationships between societies and 'a spatially differentiated nature' (p. 192) as an obvious focus for geographical scholarship. With technological advancement, however, the main links are among individuals. Interdependence within and between societies increases with the more complex division of labour, so that the most important aspects of modern human existence relate to spatially differentiated societies, not to a spatially differentiated nature. This 'horizontal' interdependence between groups living in different places creates the patterns of human occupance on the surface of the earth and provides the basic subject matter for human geographers.

The approach clarified: three pioneering textbooks

As noted earlier (p. 15, this book), textbooks play a significant role in the transmission of paradigms within disciplines so that a clear statement of a major shift occurring should be provided when new textbooks appear which codify a 'new approach'. For human geography, this occurred in the mid- and late 1960s. Three texts promoted the new approach, focusing on spatial arrangements (the areal differentiation in human activities and the spatial interactions which this produces) and on the role of distance as a crucial variable influencing the nature of those arrangements.

The pioneer among such texts was Haggett's (1965c) *Locational Analysis in Human Geography*. It presented material that he had been teaching to undergraduates for several years at the University of Cambridge. (On the stimuli to Haggett's work, see Haggett, 1990, and Chorley, 1995; on Chorley, see Stoddart, 1997a.) Its depiction of pattern and order in spatial structures was placed within a decomposition of nodal regions into five geometrical elements; a sixth was added in the second edition (Haggett *et al.*, 1977). The book contained little programmatic material either 'selling' or 'justifying' the new approach; he noted at the outset that any book reflects its author's biases and made clear that his were a preference for quantitative over qualitative analysis and a search for 'order, locational order, shown by the phenomena studied traditionally as *human geography*' (p. 2, original italics).

Haggett's presentation of spatial systems was based on a simple schema comprising its major elements (Figure 4.1). This assumed a spatially differentiated society within which there is a desire for interaction within regional systems, and which results in patterns of movement – of goods, people, money, ideas, and so on – between places. The first stage in analysing nodal regions involves their representation. Most movement is channelled along routes, so the second stage involves characterisation of those channels as networks comprising edges and vertices; in a transport system, many of the latter are the nodes, the organisational nexus. Their spatial arrangement forms the

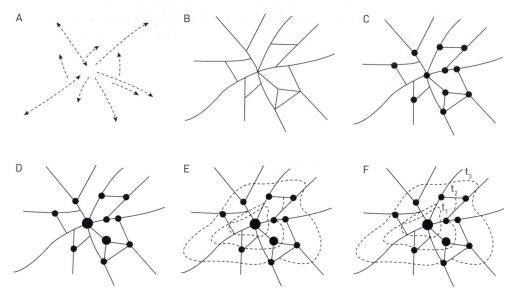

Figure 4.1 The elements in Haggett's schema for studying spatial systems: A: movement; B: channels; C: nodes; D: hierarchies; E: surfaces; and F diffusion

Source: Haggett *et al.*, 1977, p. 7.

schema's third element, and the fourth investigates their organisation into hierarchies which define the importance of places within the settlement framework. Finally, in the original scheme, there are the surfaces, the areas of land within the skeleton of nodes (settlements) and networks (routes) that are occupied by land uses of various types and intensities.

Patterns of human occupation of the earth change frequently and the spatial order to such changes forms the sixth element in Haggett's revised schema. Change does not occur uniformly over space in most circumstances; it usually originates at one or a few nodes and spreads to others (as with the 'new' geography in the USA discussed in the previous chapter), along the movement channels, through the nodes, across the surfaces and down the hierarchies. The processes of change over space and time thus involve spatial diffusion.

Haggett stressed his conception of geography as a science of distributions, emphasising the regularities in various elements of these distributions: the first half of his book – entitled 'Models of locational structure' – allocated a chapter to each of the schema's elements. The first edition (1965c) was very reliant on work in other disciplines; the second (Haggett *et al.*, 1977) illustrated the substantial amount of work done by geographers in the intervening decade. In his collection of semi-autobiographical essays, Haggett (1990) frequently reiterated his view of geography's 'emphasis on space and on geometry' (p. 5), which he expressed as personal, 'though I hope not wholly ... quirky'. Maps, as demonstrations of spatial structure, form the foundation of his geography, as illustrated by essays on 'Levels of resolution', 'The art of the mappable', 'Regional synthesis' and 'The arrows of space'. Thus, although he recognised that 'there are other delights in geography' (p. 184), he retained a 'simple delight in the beauty of geographical structures and the challenges posed in finding them and mapping them' (see also Gould, 1993, and D. Gregory, 1994, for more reflections on the aesthetics of spatial science). Although *Locational Analysis* covered human geography only, Haggett believed in the indissolubility of physical and human geography, as argued in his biographical essays (Haggett, 1990) and introductory texts (Haggett, 1972, 2001). In 1991 he wrote that:

Geography is now too large, too multicultural and, hopefully, too mature, to be tempted by any single orthodoxy. Yet even a braiding stream may sometimes converge on a few major channels and I hope geography may ever delight in rediscovering how much it has to gain from allowing environmental and societal themes to merge together.

(Haggett, 1991, p. 302)

Other texts produced in that period took a similar approach, although differing in emphases. Morrill's (1970a) title – *The Spatial Organization of Society* – clearly emphasised his view of the role of geographical analysis in the larger task of the social sciences: human geography's core elements are 'Space, space relations, and change in space – how physical space is structured, how men relate through space, how man has organized his society in space, and how our conception and use of space change' (p. 3). Space has five qualities relevant to the understanding of human behaviour: (1) distance, the spatial dimension of separation; (2) accessibility; (3) agglomeration; (4) size; and (5) relative location, which can be put together to build theories:

Virtually all theory of spatial organization assumes that the structure of space is based on the principles of minimizing distance and maximizing the utility of points and areas within the structure, without taking the environment, or variable content of space, into account. Although the differential quality of area is interesting and its effect on location and interaction is great, most of the observable regularity of structure in space results from the principles of efficiently using territory of uniform character. The theoretical structures for agricultural location, location of urban centres, and the internal patterns of the city are all derived from the principle of minimizing distance on a uniform plane.

(Morrill, 1970a, p. 15)

Thus, all land use and location decisions are taken to minimise movement costs. The spatial approach (i.e. geography) claims that one particular variable on which it concentrates – distance – is a significant influence on human behaviour and seeks accounts for observed spatial patterns within this framework. And so in Morrill's book:

the explanation of spatial structure proceeds from the deductive – what would occur under the simplest conditions – to the inductive – how local factors distort this 'pure' structure. To begin with, all the local variation may introduce is a risk of missing the underlying structure. Modern theory of location therefore stresses the spatial factors – above all, distance – which interact to bring about the regular and repetitive patterns.

(p. 20)

Morrill's summarising 'theory' of spatial structures proceeds as follows:

1   Societies operate to achieve two spatial efficiency goals:

(a) to use every piece of land to the greatest profit and utility; and
(b) to achieve the greatest possible volume of interaction at the least possible cost.

2   Pursuit of these goals involves four types of location decisions:

(a) the substitution of land for transport costs when seeking accessibility;
(b) substituting production costs at sites for transport costs when seeking markets;
(c) substituting agglomeration benefits for transport costs; and
(d) substituting self-sufficiency (higher production costs) and trade (higher transport costs).

3   The spatial structures resulting from these decisions include:

(a) spatial land-use gradients; and
(b) a spatial hierarchy of regions.

These are somewhat distorted by environmental variations to produce:

(c)　more irregular but predictable patterns of location.

Whereas over time distortion may result from:

(d)　non-optimal location decisions; and

(e)　change through processes of spatial diffusion.

Like Haggett, therefore, Morrill focused on the geometry of human organisation of activities on the earth's surface. Whereas Haggett emphasised pattern, however, Morrill paid more attention to decision-making processes which would produce the most 'efficient' patterns, as foundations for appreciating the imperfect examples of such patterns which are observed in the 'real world'. The importance of economic and spatial variables is less dominant in the third edition of the book, however (Morrill and Dormitzer, 1979, p. 7). Location theory is presented as providing:

> fairly simple models that permit us to highlight some essential principles and factors of human location. The real world, of course, does not correspond very closely to the patterns projected by location theory because the human landscape is the complex product of many different forces – historical, physical, cultural, political, and behavioral, as well as economic (spatial).

Nevertheless, location theory is presented as revealing, 'the basic spatial order that underlies our often confusing world . . . the landscape around us makes sense when viewed as the outcome of human decisions to use land rationally in the context of a particular culture, level of technology, and physical environment' (p. xix). Later, however, in a paper on 'Some important geographic questions', Morrill (1985) returned to a strong focus on spatial variables.

A third, highly influential text in this mould was produced by Abler *et al.* (1971; see Palm, 2003; an important difference is that whereas Haggett's book was written for UK undergraduate audiences, based on a course he had developed at Cambridge over the previous eight years, the Abler *et al.* book was based on an introductory graduate-level course at Penn State University: Abler, 1993b, p. 66). *Spatial Organization* has a stronger focus on positivist methodology (see also Gould, 1977, 1978). Like Haggett, its authors emphasised the search for order, noting that this was the prime goal of sciences, that in their view human geography is a social and behavioural science, and all societies 'impose order on our experience continual . . . order . . . is a fundamental requirement of human welfare. The need for order must be fulfilled, even if the order must be created where none can be discovered' (Abler *et al.*, 1971, pp. 6–7). Order can be produced in a variety of ways. They identify four: theological, aesthetic and emotional, common sense, and scientific; for them, the scientific is the most important because it is 'adaptable whereas the others are rigid. Its verification and replication procedures enable [it] to remain viable in the rapidly changing world it creates. Far more than the others, it is a supra-individual, empirical ordering system' (p. 19 – i.e. it is 'objective').

For Abler *et al.* (1971, p. 20) geography's 'one basic question [is] "Why are spatial distributions structured the way they are?"' (p. 20). Like Haggett, tackling that question needs a textbook that not only synthesises the relevant available knowledge, but also sets out the methodology for such a task. Haggett has 'Methods in locational analysis' as the second part of his book, introduced by the statement that 'Further development of the locational models . . . depends largely on our ability to test them against existing geographical patterns' (Haggett, 1965b, p. 185); that ability is set out in four chapters covering data collection and description, region building and hypothesis testing. For Abler *et al.*, however, methods take precedence. The first two sections of their book are on 'Order, science and geography' and 'Measurement, relationship and classification'; only then do they turn to the substantive material, with sections on 'Location and spatial interaction', 'Spatial diffusion' and 'Spatial organization and the decision process'. (Palm, 2003, notes that this emphasis on the nature of geography as a science reflects Abler's and John Adams' experiences as graduate students at the University of Minnesota, where their adviser was Fred Lukermann, who wrote on that issue at the time: Lukermann, 1964a, 1964b. To Lukermann −1964b, p. 167 − geographers should focus on interactions within spatial systems, with their discipline presented as 'a catalogue of questions, and the questions − not the phenomena, not the facts, not the method − are geographic'.) Morrill focused on the substantive material alone, and in a slightly later book Amedeo and Golledge (1975) used a similar organisation to Abler *et al.* (1992a) to introduce geographers to scientific reasoning, especially theory construction.

These textbooks illustrate the centrality of space and distance as the major focus of geographical interest in the 1960s. An emphasis on observed patterns rather than abstract theoretical arrangements was noted in King's (1969b, p. 574) review of 'The geographer's approach to the analysis of spatial form . . . the mathematics which are used and the geometrical frameworks which are favored'. He focused on descriptive mathematics, which represent 'what is' rather than 'what should be', realising that, 'when they are pursued to their extremes in very formal terms these studies [of 'what should be'] run the risk of appearing as seemingly sterile exercises in pure geometry' (p. 593; see also King, 1976, 1979b). Geographers had by then not proceeded very far in providing process theories which would account for observed spatial patterns; however, they were working backwards, identifying order that called for explanation rather than deducing what the world should be like from knowledge of human behaviour.

In almost all of this work, space was treated as a continuous variable; there was very little on its discontinuous nature in some aspects of human organisation. Bounded spaces, at a great variety of scales, from the largest nation-state through to the 'bubble of personal space' around each individual, were largely ignored by geographers promoting their discipline as spatial science, perhaps because they found generalisations more difficult to identify than in work on continuous space. The study of territories had been advanced by a French political geographer (Gottmann, 1951, 1952), whose work had little impact on his American contemporaries (Johnston, 1996c, 2004d); despite his periods of working there, he was 'outside the project'. Interestingly, some forty years later Cutter *et al.* (2002, p. 307) had asked 'Is there a deeply held human need to organize space by creating arbitrary borders, boundaries and districts?' as the second of their 'big questions in geography' (on which, see also Sack, 1986; Elden, 2013).

Spatial theory?

Just as the methodological developments reviewed in the previous chapter lacked clear guidelines from programmatic statements, the growth of the spatial science viewpoint similarly lacked a manifesto, at least until Harvey's (1996a) *Explanation in Geography*. J. Wreford Watson's (1955) paper was not widely referenced (Johnston, 1993b) and the only attempt to provide a lead − apart from general statements about geography and geometry, such as Bunge's (1962) − was a paper by Nystuen in 1963 which reached a wider audience when reprinted in 1968. His objective was 'to consider how many independent concepts constitute a basis for the spatial point of view, that is, the geographical point of view' (Nystuen, 1968, p. 35). Rather than analyse the 'real world' with its many distorting tendencies, he sought clarity in considering abstract geographies.

To illustrate his basic concepts, Nystuen used the analogy of a mosque completely lacking furniture (i.e. an isotropic plain) in which a teacher chooses a location at random. The students then distribute themselves so that they can see and hear the teacher; their likely arrangement is in semicircular, staggered rows facing and as close as possible to the teacher's position. This arrangement has three characteristic features:

1 directional orientation − all students face the teacher, to perceive expressions and to hear better;

2   distance – the students cluster around the teacher, because audibility diminishes with distance; and

3   connectiveness – they arrange themselves in rows, organised so that each has a direct line of sight to the teacher.

The third is in part a function of distance and direction, but not entirely so:

> A map of the United States may be stretched and twisted, but so long as each state remains connected with its neighbors, relative position does not change. Connectiveness is independent of distance and direction – all these properties are needed to establish a complete geographical point of view.
>
> (Nystuen, 1968, p. 39)

In addition:

> Connections need not be adjacent boundaries or physical links. They may be defined as functional associations. Functional associations of spatially separate elements are best revealed by the exchanges which take place between the elements. The exchanges may often be measured by the flows of people, goods, or communications.

Within the mosque, therefore, the connectivity between teacher and student involves not only a direct line of sight between them but also a direct flow from one to the other, in this case of ideas.

Nystuen presented these three concepts as both necessary and (probably) sufficient for the construction of an abstract geography, grounded in the study of sites (abstract places) rather than of locations (real places):

> The terms which seem to me to contain the concepts of a geographical point of view are direction or orientation, distance, and connection or relative position. Operational definitions of these words are the axioms of the spatial point of view. Other words, such as pattern, accessibility, neighborhood, circulation, etc., are compounds of the basic terms. For abstract models, the existence of these elements and their properties must be specified.
>
> (p. 41)

Nystuen was unsure, however, whether these three comprised the full set of necessary and sufficient concepts for a geographical argument (see also Papageorgiou, 1969); boundary might be a primitive concept too rather than derivative of the basic three, but bounded space was of much less interest to spatial theorists than continuous space (albeit often transformed: Haggett, 1990; Cliff and Haggett, 1998).

Nystuen's general case, that arguments in human geography could be based on a small number of foundational concepts, was implicitly accepted by much of the work in 'the spatial tradition', although rarely explicitly referred to. Haynes (1975) suggested an alternative approach to writing theory in human geography, based on the mathematical procedures of dimensional analysis. Five basic dimensions – mass, length, time, population size, and value – were defined and manipulated to indicate the validity of functional relationships, such as distance-decay equations (see p. 108, this book), by checking their internal consistency. His defence of this approach is that:

> Although most quantitative geographers would probably claim to be engaged in the discovery of relationships, it appears that geography has not passed the first stage [in the development of a science] with any degree of rigor. . . . With no clear idea of which variables are relevant and which

particular characteristics in a system should be isolated, it is pragmatic to define our measurement scales with regard to a particular set of observations rather than the other way round. The method of physical science ... is a superior system, as measurements can be interpreted exactly, different results compared, and experiments replicated.

(p. 66)

This deductive approach is not as closely tied to a 'spatial view' as Nystuen's but has the same goal – the derivation of a set of fundamental concepts which can form the basis for writing geographical theory to be tested in the 'real world' (see also Haynes, 1978, 1982).

These attempts at isolating human geography's primary concepts differed from most contemporary efforts at producing geographic theory which, according to Harvey (1967b, p. 212), were 'either very poorly formulated or else derivative'. Central place theory, for example, was based entirely on economic postulates about how people behave as 'rational economic actors' with regard to the costs of travel, producing a theory about the size and spacing of settlements. Its attraction for many was that geographers could contribute their own basic concept to theory development, without depending totally on other disciplines. Harvey (1970) identified a group of concepts which could comprise geographical elements in building integrated social science theories – location, nearness, distance, pattern, morphology; most are compounds of Nystuen's basic terms.

An example of theory development involving both derivative (from other disciplines) and indigenous (geographical) concepts was research on spatial diffusion, which received considerable attention in the late 1960s (Brown, 1968). The basic behavioural postulate, taken from sociological research, was that word of mouth is the most effective form of communication about innovations. Geographers introduced the effect of distance on this contact; much interpersonal communication is between neighbours, so information about innovations should spread outwards in an orderly fashion from the initial adopters' locations. Pioneer work on this hypothesis was conducted in Sweden by Hägerstrand, who introduced it to the Washington school in the 1950s, where it was taken up by Morrill (1965, 1968, 2002) whose doctoral research was undertaken in Sweden. Hägerstrand's major work was made available in an English translation in 1968. He was in many ways the pioneer of spatial analysis, working almost entirely independently in Sweden for more than a decade before his ideas were taken up in North America and the UK (see, for example, the essays on him by Buttimer and Mels, 2006; Persson and Ellegård, 2011); his studies of migration fields were very early examples of Monte Carlo simulations, for example, and he was one of the first to realise the potential of computers for spatial analytic research: Hägerstrand, 1967; Morrill, 2005; Buttimer, 2007.

Much subsequent research has been done on diffusion processes and, even more, on patterns of spatial spread which are assumed to result from diffusion processes (see Abler et al., 1971, Chapter 11; Brown, 1981; also p. 102, this book), in which the notions of spatial spread across space and down settlement hierarchies is valid because of the role of direct human contact in the transmission of many diseases (see Thomas, 1992). The lead in much of this work was provided by Haggett and Cliff, and included classic spatial epidemiological investigations of influenza, measles and AIDS, as well as more general studies (e.g. Cliff and Haggett, 1988; Haggett, 1994, 2000; Smallman-Raynor et al., 2004; Cliff et al., 2009). Changes in communications technology, among others, have had substantial impacts on these flows, introducing new technical issues (Cliff and Haggett, 1995).

Social physics and spatial science

Two of Nystuen's three basic concepts – distance and connectivity – received much more attention from those advocating geography as a spatial science than the other. Direction was relatively ignored, except for some work on migration patterns (Wolpert, 1967; Adams, 1969). A substantial volume of work in spatial science followed the agenda established by the social physics school (p. 68, this book). The relationship between distance and various types of interaction – migrations, information flows, movements of goods, etc. – had been identified by several nineteenth-century workers, such as Carey (1858), Ravenstein (1885; Grigg, 1977; Tobler, 1995) and Herbert Spencer (1892). Their impact is unclear: McKinney (1968) suggested that Stewart and others were unaware of the seeds of the 'gravity model' and 'population potential' ideas in Carey and Spencer's writings, and claimed that 'current geographers could learn much' (p. 105) from their publications; Warntz (1968) retorted that Stewart was well aware, although McKinney's rebuttal pointed out that Stewart did not refer to them in his 1940s' papers, only in the 1950s. Ravenstein's papers, on the other hand, were very influential on later research into migration patterns. Interestingly, the pioneering work on distance-decay and the gravity model, as with much of the rest of spatial science, was done outside geography; as Tocalis (1978, p. 124) expressed it, 'geographers' contributions to the theoretical evolution of the gravity concept were minor', although one major contributor – Alan Wilson – later 'became a geographer' (Wilson, 1984a).

Apart from Stewart's work, one of the major social physics' influences on geographers was George Zipf's (1949) 'principle of least effort'. If individuals organise their lives to minimise the amount of work undertaken and movement involves work, then movement-minimisation is part of the general principle of least effort. To explain this, Zipf expanded Stewart's finding that with increasing distance from Princeton, fewer students from each state attended that university. Two aspects of work are involved: (1) that needed in acquiring information about the university; and (2) that needed in travelling there. Thus, the greater the distance between potential students' homes and Princeton, the less they are likely to know about the university and the less prepared they will be to travel there. The validity of this argument was tested against many datasets; material on the contents of newspapers and on their circulation illustrated the expected distance-decay trend in information flows, for example, and data on movements between places showed that the greater the distance separating them, the smaller the volume of inter-place contact.

Zipf called this regularity the $P_1 P_2/D$ relationship, and Stewart noted the analogy between it and the Newtonian gravity formula, hence its popularisation as the 'gravity model'. It was fitted to flow data in many research projects, as shown in Carrothers' (1956) and Olsson's (1965) reviews. To achieve reasonable statistical fits, the various elements of the equation had to be weighted ($P_1 P_2$ – the measures of mass at the origin and destination respectively, indicators of interaction-generation potential; and D – the distance between the two places). Because different weights were determined empirically in almost all studies, Olsson (1965, p. 48) claimed that the gravity model of interaction was 'an empirical regularity to which it has not yet been possible to furnish any theoretical explanation'. (Wilson, 1967, 1970, later offered a statistical explanation and developed a whole family of interaction models based on statistical mechanics.) Empirical studies showed that the strength of the influence of distance varied from place to place, from population to population, and from context to context, but it was invariably related to the volume of interaction to some extent; the influence of distance on flow volumes seemed virtually universal, but no available theory could account for the variability in the strength of its impact.

It was not only in social physics that distance was receiving attention at this time. Economists and sociologists were becoming increasingly aware of its influence on behaviour (Pooler, 1977), the former in location theories and Isard's development of regional science, and the latter in studies by the Chicago school on urban residential patterns (Johnston, 1971, 1980a). Thus:

> a number of geographers became aware of the spatial enquiries that were being undertaken in a social science context outside their own discipline and, upon realizing their relevance to geography, proceeded to emulate them. The appearance of the spatial tradition was prompted, not by discoveries from within geography, but by an awareness and acceptance of investigations external to the discipline. The space-centred scientific enquiries of other social sciences became

paradigms for geographers, simply because those enquiries, being spatial, were seen to be of relevance by some practitioners.

(Pooler, 1977, p. 69)

This fitted both the philosophical framework outlined by Schaefer (1953) and the 'quantitative revolution'. In adopting spatial viewpoints pioneered in other disciplines, however, human geographers were frequently selective in what they imported. Work on urban residential patterns concentrated on certain aspects of the urban sociology of the Chicago school, for example. Robert Park's dictum relating social distance to spatial distance stimulated much work on residential segregation (Peach and Smith, 1981), but the social Darwinism and ecological theory underlying this dictum was largely ignored (though see Robson, 1969; Entrikin, 1980) and the humanistic concerns in Park's work were only identified later when spatial science was losing popularity (Jackson and Smith, 1981, 1984).

In their spatial analyses some social scientists looked not only at the influence of distance but also at its meaning and measurement. Stouffer (1940), for example, established a relationship in which migration between places X and Y was accounted for not so much by the distance between them, but rather by the number of intervening opportunities. He measured distance in terms of opportunities; the greater the number of opportunities available locally, the less work that has to be expended in moving to one. (Berry, 1993, refers to contacts with Stouffer when he was a student at the University of Washington.) Others followed this flexible approach to measurement of the basic variable. Ullman (1956), for example, developed a schema for analysing commodity flows in which the amount of movement between two places was related to three factors:

1   complementarity – the degree to which there is a supply of a commodity at one place and a demand for it at the other;
2   intervening opportunity – the degree to which either the potential destination can obtain similar commodities from a nearer, and presumably cheaper, source or the potential source can sell its commodities to a nearer market; and
3   transferability – for complementarity to be capitalised on, movement between the two places must be feasible, given channel, time and cost constraints.

This was not as easily tested statistically as the 'gravity model' (Hay, 1979b). Fitting such models also requires accepting that the influence of distance, measured as either time or cost, varies from place to place and from time to time, producing time – and cost – distance maps that are transformations of the traditional grid (Abler, 1971; Forer, 1974; Janelle, 1968, 1969). This issue was taken up by later work on 'local statistics' (Fotheringham et al., 2002; p. 153, this book).

Analyses of movement patterns were stimulated not only by their obvious relevance to geography's growing spatial science focus and its development of location theories, but also by their applicability in forecasting contexts. Planning land-use patterns and transport (especially road) systems became increasingly sophisticated technically during the 1950s and 1960s. Data analyses showed both the traffic-generating power of various land uses and the patterns of interaction between different parts of an urban area, with the gravity model being used to describe the latter. Future land-use configurations were then designed, their traffic-generating potential derived, and the gravity model used to predict flow patterns and identify needed road systems. Later models, most based on Ira Lowry's, assessed different land-use configurations in terms of traffic flows, thereby suggesting the 'best' directions for future urban growth (see Batty, 1978; on the extent of their application, see Batty, 1989).

The demands for sophisticated planning devices stimulated much research using gravity and Lowry models. American economists initiated this but it was later taken up by British workers, including Alan Wilson, who was appointed to a professorship of geography at the University of

Leeds in 1970. (He was trained as a physicist and became interested in traffic forecasting when a city councillor in Oxford: Wilson, 1984a.) Wilson (1967) derived the gravity model mathematically, thereby giving it a stronger theoretical base, and extended Lowry's model into a more general suite concerned with location, allocation and movement in space (Birkin *et al.*, 1990: see also p. 72, this book; these later became the foundation of a major commercial operation at the University of Leeds–GMAP Ltd).

Collaboration between academic geographers and practising planners on these modelling exercises led to developments which paralleled those in regional science in the USA. Two new journals in the UK catered for these research areas – *Regional Studies* and *Environment and Planning* (now *Environment and Planning A*); both attracted contributions and readers from other social sciences (Johnston and Thrift, 1993; Johnston, 2003a). The new geographical methodology was proving of considerable applied value, therefore. Careers as planners became extremely popular among geography graduates in the 1960s and 1970s, and academic geographers' growing desire for their discipline to equal that of other social sciences in its policy-making relevance seemed well on the way to fulfilment. Geographers' analyses of spatial patterns and interactions were resulting in applicable models which could be used to identify feasible future locational and flow patterns.

Systems

The study of systems was introduced to the geographical literature by Garrison (1960a) and Chorley (1962), although Foote and Greer-Wootten (1968) claimed that systems analysis was promoted in Sauer's (1925) programmatic statement *The Morphology of Landscape* with the words 'objects which exist together in the landscape exist in interrelation'. More generally, the notion of a system has a long history, as Bennett and Chorley (1978, pp. 11–14) point out: teleological traditions, for example, postulate the world as 'a vast system of signs through which God teaches man how to behave' (p. 12), whereas functionalism links observed phenomena together as 'instances of repeatable and predictable regularities' of form.

However, the field of modern systems analysis as applied to geography in the 1960s was a derivation from a Cold War project of managing complex military systems. Developed first at the American Air Force Research and Development (RAND) 'think tank' in Santa Monica, California, systems analysis embodied postwar American scientific application across a spectrum of science, engineering, economics, political science and sociology (Collins, 2002; Hughes, 2000). Its impacts in the late 1950s and early 1960s on both military thinking (warfare strategies and nuclear weapons systems) and public policy (the growth of centralised top-down decision-making and technocratic 'expertise') were profound. Its appeal lay in the bold claims it made to predict trends and outcomes and the sense of rational order that it claimed to bring to bear on complex problems (at the same time as moving them out of the arenas of political and public debate). The keystones of such study of systems were connectivity and prediction. Although geographers made little or no contribution to developments at RAND or to the status of general systems theory, geography would inevitably be influenced. Thus, Harvey's (1969a) *Explanation in Geography* adopts key aspects of the logic of systems analysis. He points out that reality is infinitely complex in its links between variables, but systems analysis provides a convenient abstraction of that complexity in a form which maintains the major connections (p. 448).

A system comprises three components:

1   a set of elements;
2   a set of links (relationships) between those elements; and
3   a set of links between the system and its environment.

The last component may be absent, creating closed systems. These are extremely rare in reality (because 'everything is related to everything else'), but are frequently created for analytical purposes either experimentally or, more usually in human geography, by imposing artificial boundaries in order to isolate a system's salient features. Thus, just as an internal combustion engine comprises a set of linked elements which receives energy from, and returns spent fuel to, its environment, so a set of settlements linked by communications networks forms a spatial system, with links to other settlements outside the defined area of the system providing the contacts with the environment. A system's elements have volumetric quantities and material flows along the links; as the system operates, the various quantities may change.

Systems terminology was widely adopted by human geographers in the 1960s, but much of the early literature on systems analysis was programmatic rather than applied; it suggested how the terminology might be applied, often reinterpreting old material (McDaniel and Eliot Hurst, 1968). Relatively few applications were reported, and more than a decade later much of the literature assessed in a major review was written by other scientists (Bennett and Chorley, 1978). Nevertheless, Harvey wrote that:

> If we abandon the concept of the system we abandon one of the most powerful devices yet invented for deriving satisfactory answers to questions that we pose regarding the complex world that surrounds us. The question is not, therefore, whether or not we should use systems analysis or systems concepts in geography, but rather one of examining how we can use such concepts and such modes of analysis to our maximum advantage.
>
> (1969a, p. 479)

In seeking answers to this question, two variants on the systems theme have been employed. The first is systems analysis, which in turn triggered debate about information and the estimation of order and disorder (under the term entropy); the second is general systems theory, which was an attempt (no longer considered effective) to provide a more unified science than current disciplinary boundaries allow.

Systems analysis

Several typologies have been suggested for studying geographical systems (Harvey, 1969a, pp. 445–9). Chorley and Kennedy (1971) identified four types (Figure 4.2):

1   Morphological systems are statements of static relationships – of links between elements: they may be maps showing places joined by roads, or equations describing the functional relationships between variables. Much of the spatial analysis outlined earlier in this chapter described such morphological systems.

2   Cascading systems contain links along which energy passes between elements: factories are cascading systems, for example, with the output from one part forming the input for another. Each element may be a system itself, producing a nesting hierarchy of cascading systems, as with Haggett's (1965c) nodal regions and the input–output matrix representation of an economy (Isard, 1960). Berry (1966) linked these two examples of cascading systems in his interregional input–output study of the Indian economy. Within each element, the material flowing through is manipulated in some way (the industrial process in a factory, for example). The nature of the manipulative process may be ignored entirely in the investigation, with focus on the inputs and outputs only; such a representation of the element is termed a black box. White-box studies investigate the transformation process, whereas grey-box analyses make a partial attempt at their description.

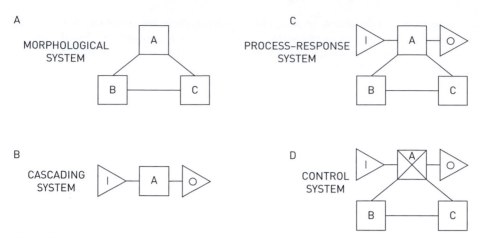

Figure 4.2 Types of system: A, B and C indicate system elements, I represents input, O represents output, and in the control system A is a value

Source: Chorley and Kennedy, 1971, p. 4.

3   Process-response systems are characterised by studies of the effects of linked elements on each other. Instead of focusing on form, as in the first two types, these are studies of processes, of causal interrelationships. In systems terms these may involve, for example, the effects of one variable, X, on another, Y; in the analysis of spatial systems they could involve the effect of variable X in place a on variable Y in place b, as with the transmission of a disease from one area of a country to another.

4   Control systems are special cases of process-response systems, having the additional characteristic of one or more key elements (valves) which regulate the system's operation and may be used to control it.

Attention focused on process-response and control systems. Langton (1972), for example, suggested that the former provide an excellent framework for studying change in human geography. He identified two subtypes. Simple-action systems are unidirectional in their nature: a stimulus in X produces a response in Y, which in turn may act as a stimulus to a further variable, Z. Such a causal chain merely reformulates 'the characteristic cause-and-effect relation with which traditional science has dealt' (Harvey, 1969a, p. 455); in another language it is a process law. The second subtype comprises feedback systems. According to Chorley and Kennedy (1971, pp. 13–14), 'Feedback is the property of a system or subsystem such that, when change is introduced via one of the system variables, its transmission through the structure leads the effect of the change back to the initial variable, to give a circularity of action'. Feedback may be either direct – A influences B, which in turn influences A (Figure 4.3a) – or it may be indirect, with the impulse from A returning to it via a chain of other variables (Figure 4.3b). With negative feedback the system is maintained in a steady state by self-regulation processes termed homeostatic or morphostatic: 'A classic example is provided by the process of competition in space which leads to a progressive reduction in excess profits until the spatial system is in equilibrium' (Harvey, 1969a, p. 460). But with positive feedback the system is characterised as morphogenetic, changing its characteristics as the effect of B on C leads to further changes in B, via D (Figure 4.3d).

The concept of feedback, with the associated notions of homeostasis and morphogenesis, provides 'the nuclei of the systems theory of change' (Langton, 1972, p. 145). In many spatial systems, feedback may be uncontrolled; others may include a regulator, such as a planning policy

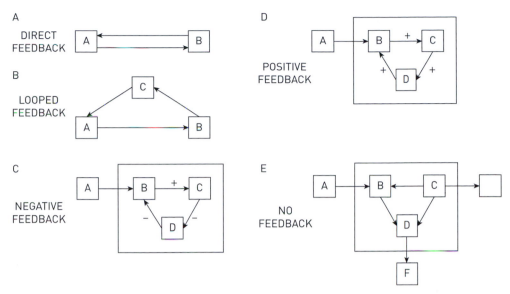

Figure 4.3 Various types of feedback relationships in systems

Source: Chorley and Kennedy, 1971, p. 14.

(Bennett and Chorley, 1978). There were few geographical studies of such feedback processes, however. For homeostatic systems, Langton cited investigations of central place dynamics in which the pattern of service centres is adjusted as the population distribution changes, to reproduce the previous balance between supply and demand factors (Badcock, 1970). Morphogenetic systems are illustrated by Pred's (1965b) model of the process of urban growth, in which expansion in a sector generates further growth there, as in Myrdal's (1957) more general theory of cumulative causation. Such systems modelling has been used to predict urban futures (Forrester, 1969). In most of these studies the input from systems theory was relatively slight, however, leading Langton (1972, pp. 159–60) to the paradoxical conclusions that:

> First, there is little correlation between the extent of the penetration of the terminology of systems theory and the rigorous application of its concepts. The empty use of terminology, which is typified by the use of the term feedback as an explanatory device rather than as a description of a fundamental research problem, must be counter-productive in a situation in which the terms themselves may be given many subtle different shades of meaning. . . . Second, some-what paradoxically, many of the concepts of systems theory are already used in geography without the attendant jargon and without apparently drawing direct inspiration from the litera-ture of systems theory.

One of the most substantial attempts to apply systems theory to a problem in human geography was Bennett's (1975) study of the dynamics of location and growth in north-west England. Having represented the system – its elements, links and feedback relationships – he estimated the influence of various external (i.e. national) events on the regional system's parameters, isolated the effects of a government policy on the system's structure, and forecast the region's future spatiotemporal morphology. He developed the forecasting aspects of this methodology in later papers (Bennett, 1978a, 1979), suggested how an optimal distribution of government grants could be achieved Bennett, 1981a), and outlined the likely spatial variation in the impact of a new tax (the 'poll tax': Bennett, 1989a; see also Hepple, 1989).

One area of investigation which firmly adopted the systems approach covers the intersection of human and physical geography. The ecosystem is a process-response system involving energy flows through biological environments, most of which include, or are affected by, people. It is also a control system whose living components regulate the energy flows: 'they further represent a major point at which human control systems must intersect with the natural world' (Chorley and Kennedy, 1971, p. 330). Most naturally occurring ecosystems are homeostatic for much of the time (see Chapman, 1977, Chapter 7), but human 'interference' often transforms them into morphogenetic systems, with potentially catastrophic effects (Johnston, 1989).

Stoddart (1965, 1967b) argued that the ecosystem should be employed as a basic geographical principle, as did other programmatic statements (e.g. Clarkson, 1970), including two based on the associated concept of community (Morgan and Moss, 1965; Moss and Morgan, 1967). But Langton's conclusion about relatively little substantive research in human geography based on systems' thinking remained valid (see Grossman, 1977). Similar attempts, but involving less consideration of the biotic environment, were made in allied disciplines, and were occasionally imported into the geographical literature. Sociologists' human ecosystem models (e.g. Duncan, 1959; Duncan and Schnore, 1959) were used as frameworks for investigating migration (Urlich-Cloher, 1972) and urbanisation (Urlich-Cloher, 1975), for example, and the operational research techniques of economists, with their important feedback mechanisms, stimulated work in transport geography (e.g. Sinclair and Kissling, 1971).

The most comprehensive early attempt to forge a systems approach to geographical study was Bennett and Chorley's book (1978), written to provide 'a unified multi-disciplinary approach to the interfacing of "man" with "nature"' (p. 21), with three major aims:

> First, it is desired to explore the capacity of the systems approach to provide an inter-disciplinary focus on environmental structures and techniques. Secondly, we wish to examine the manner in which a systems approach aids in developing the interfacing of social and economic theory, on the one hand, with physical and biological theory, on the other. A third aim is to explore the implications of this interfacing in relation to the response of man to his current environmental dilemmas. . . . It is hoped to show that the systems approach provides a powerful vehicle for the statement of environmental situations of ever-growing temporal and spatial magnitude, and for reducing the areas of uncertainty in our increasingly complex decision-making arenas.

This task involved elucidating not only the 'hard systems' of physical and biological sciences but also the 'soft systems' characteristic of the social sciences. The latter cover a large and fertile literature concerned with, first, the cognitive systems describing people as thinking beings and their decision-making systems (as individuals and in groups) and, second, the socioeconomic systems made up of very many of these interacting individuals and groups. They then attempt to interface the two types, because:

> in large-scale man-environment systems the symbiosis of man as part of the environment of the system he wishes to control introduces all the indeterminacies of socio-economic control objectives. . . . In particular, we need to ask what are the political and social implications of control, and for whom and by whom is control intended?
>
> (p. 539)

Their essay ends with a discussion of the substantial problems involved in such interfacing, many of which are firmly tackled in the text.

Use of systems analysis in human geography assumes (usually implicitly) that valid analogies can be drawn between human societies on one hand and both natural phenomena-complexes and machines on the other. Individual system elements have predetermined roles, and can only act and change in set ways, depending on the system's structure and its interrelationships with the environment. As a descriptive device, this analogy allows the structure and operation of society and its components to be portrayed and analysed, providing a source of ideas from which hypotheses can be generated (see Coffey, 1981). And once a system has been defined and modelled, systems analysis can predict the likely nature of the elements and links following certain environmental changes (such as the introduction of new elements and/or links, as in the classic Lowry model used to predict the impact of new land-use configurations on traffic flows: Batty, 1978).

The potential fertility of this analogy was examined in Wilson's (1981a) discussion of methods for analysing environmental systems. He defined environment as 'natural, man-nature, and manufactured "systems of interest"' (p. xi), in all of which 'the main concern is with complicated systems whose components exhibit high degrees of interdependence. The behaviour of the "whole" system is then usually something very much more than the sum of the parts' (p. 3). He argued that moorland ecosystems, water resource systems and cities (his three initial examples) can all be studied in the same way and set out available methods for that; a further book (Wilson, 1981b) presented the mathematics for studying systems in which the rates and directions of change suddenly alter, and he returned to the topic in a later summary volume (Wilson, 2000). Huggett (1980) similarly argued that systems analysis has wide applications in both human and physical geography, plus the interface where people and environment interact (see also Huggett and Thomas, 1980). Once a system has been successfully modelled, it can be manipulated using control theory: 'a dynamic optimization technique . . . [that] permits optimal allocation over long time horizons . . . [and] shifts emphasis from mere model construction to model use' (Chorley and Bennett, 1981, p. 219). Such a combination of models describing systems with a theory of system control has a wide range of potential applications, according to Chorley and Bennett, in such fields as pollution control, catchment management, inter-area resource allocation and urban planning. It suggests a commonality of interest, focused on methods, between applied physical and applied human geography.

Systems theory, information and entropy

A system has been defined here so far as a series of linked elements interacting to form an operational whole. This was challenged by Chapman who opened his book with:

> I do not think that the concept of a system will have any great operational consequences in geography for a long time yet. It represents an ideal that the real world does not fully approach. On the other hand, in conceptual terms I think the concept is extremely important and useful, and that it has a great and immediate role to play for those who are about to plan the strategy of their research. As a framework for analysis, it has no current peers.
>
> (1977, p. 6)

For Chapman, a system comprises a series of elements which can take alternative states, and his definition – following Rothstein (1958) – is:

> A system is a set of objects where each object is associated with a set of feasible alternative states: and where the actual state of any object selected from this set is dependent in part or completely upon its membership of the system. An object that has no alternative states is not a functioning part but a static cog.
>
> (p. 80)

An example is a number of farms, each comprising a series of fields: every farmer has to decide how to use those fields. Each decision will in part reflect the farm's general operations and the uses to which all other fields are put; in part, too, it will be a function of the external market and the decisions made by other farmers regarding use of their own fields. Thus, there is a large number of possible states for the system of fields – different configurations of land uses. According to Chapman, systems analysis should involve investigating those configurations and placing the observed pattern within the context of the alternatives:

> to theorize merely about what does exist is not very useful. If we restrict ourselves to that alone, all explanation will be merely historical accidental. At all stages it is most important to include consideration of what else could have been. The definition of organization in a system even explicitly requires the assessment of what else could have been.
>
> (pp. 120–1)

Taylor and Gudgin (1976) argued similarly in a different context: instead of simply asking 'is there a bias [towards one political party] in the electoral districting of a borough?', they ask 'what is the likelihood of a bias occurring, given the system constraints?' (the number of different ways that the districts could be constructed). This allowed them to then set the study of electoral districting, the aggregation of spatial units into constituencies and its consequences, on a strong statistical footing (Gudgin and Taylor, 1979; Johnston *et al.*, 2001; Johnston, 2012).

Such analyses focus on one configuration of the system as a sample from a set of possible configurations, and among their key concepts is entropy. In the second law of thermodynamics an increase in system entropy involves an increase in uncertainty, as with the introduction of a layer of hot water to a body of cold water. Initially, the two are separate, and one can be completely certain about the location of the hot molecules, for example. But with no external influence the two slowly mix, until all are at the same temperature. As the mixing proceeds, so the entropy increases.

Social scientists have drawn their usage of entropy from two separate, though linked, definitions. Thermodynamic entropy relates to the most probable configuration of the elements within the system constraints, as just illustrated. In information theory, entropy refers to the distribution of the elements across a set of possible states, and is an index of element dispersion. One can be completely certain about a distribution, in terms of predicting where one element will be, if all elements are in the same state; conversely, one will be most uncertain when elements are equally distributed through all possible states, so that prediction of the location of any one element is most difficult. (See Webber, 1977, and Thomas, 1982, on the relationship between entropy and uncertainty.)

In its simplest form, the information-theory measure of entropy is another descriptive index, but it can be developed in a variety of ways. Chapman (1977) illustrates three uses:

1   as a series of indices of variations in population distributions;
2   as an index of redundancy in a landscape, where redundancy is defined as relating to a regular sequence so that it is possible to predict the land use at place 'a' from knowledge of the land uses at neighbouring places; and
3   as a series of measures of reactions to situations in states of uncertainty.

In general terms, these continue the tradition of Stewart and Warntz's macrogeography (p. 68, this book): the aim is to describe a pattern rather than to explain it, although the nature of the constraints used to derive the entropy measures provides an input to explanatory modes of analysis (Webber, 1977).

The use of entropy as developed in statistical mechanics was introduced to the geographical literature by Wilson's (1970) extensions to the gravity model. His initial example was a flow matrix. The number of trips originating in each of a series of residential areas is known, as is the number ending in each of a series of workplace areas, but the entries in the cells of the matrix – which people move from which residential area to which workplace area (the flow pattern) – are unknown. What is the most likely flow pattern? Even with only a few areas and relatively small numbers of commuters, the number of possibilities is very large. Wilson defines three system states:

1   The macro-state comprises the number of commuters at each origin and the number of jobs at each destination.
2   The meso-state consists of a particular flow pattern: five people may go from zone A to zone X, for example, and three from the same origin to zone Y, but it is not known which five are in the first category and which three in the second.
3   Finally, a micro-state is a particular example of a flow pattern – one of the many possible configurations of eight people moving from zone A, five to X and three to Y. Entropy-maximising procedures find that meso-state associated with the largest number of micro-states; the most probable distribution is that with the greatest number of micro-states giving rise to it. (For illustrations, see Johnston and Pattie, 2000, 2003.)

The most likely (maximum likelihood) distribution corresponds to the position where we are most uncertain about the micro-state of the system, as this has the largest possible number of such states and we have no grounds for choosing between them. This approach, too, follows the macrogeography tradition. It is not an attempt at explanation and Wilson sees his work illustrating:

> the application of the concept of entropy in urban and regional modelling; that is, in hypothesis development, or theory building. ['Model' and 'hypothesis' are used synonymously, and a theory is a well-tested hypothesis.] . . . the entropy-maximizing procedure enables us to handle extremely complex situations in a consistent way.
>
> (pp. 10–11)

In reviewing Wilson's presentation, Gould (1972, p. 689) termed it 'the most difficult I have ever read in geography' but continued, 'he has planted a number of those rare and deep concepts whose understanding provides a fresh and sharply different view of the world'.

The hypothesis that the entries in the flow matrix conform to the most likely distribution can be tested against 'real' data. If it is falsified, either entirely or in part, it can be refined by building in more constraints. Wilson does this with intra-urban transport models, for example, by introducing travel-cost constraints, different types of commuters (class, age, etc.), different types of jobs, and so on. The aim is to describe the most likely system structure from a given amount of information, which is incomplete. He subsequently developed both the theory of his modelling and the substantive applications (see Wilson, 1981a). A general text (Wilson, 1974) introduced a whole family of models that can be used to represent, and then forecast, the various components of a complex spatial system such as an urban region. Some of these models were expanded (e.g. Rees and Wilson, 1977, on demographic accounts) and applied, with varying success, to the West Yorkshire region (Wilson et al., 1977).

Wilson's work was applied and developed by others (see Batty, 1976, 1978, for example), though rarely in other contexts. Michael J. Webber (1977) extended Wilson's argument that the purpose of entropy-maximising models is to draw conclusions from a dataset that are 'natural' in that they are functions of that dataset alone and contain no interpreter bias. For him, they provide a convenient way of organising thinking about a complex world, and he identifies an 'entropy-maximizing paradigm' (p. 262). At Leeds, Wilson's work on the entropy-maximising approach was extended into the field of micro-simulation (Ballas et al., 2002; Birkin and Clarke, 1988, 1989; Clarke and Holm, 1988; Clarke, 1996; Williamson et al., 1998) which produces statistically reliable estimates of micro-level characteristics of a population from macro-data. Given the number of people in each age group in each of a number of small areas, for example, and, separately, the number with cancer and the cross-tabulation of age by cancer for a larger area comprising those small areas, it lists the individuals in each area in each age group who are estimated to have cancer; as with entropy-maximising, these are the maximum-likelihood estimates given the constraints (the number of people in an age group with cancer in all of the small areas must equal the sum for the larger area).

One of the few examples of the application of entropy-maximising models to other research contexts is in electoral geography. It may be, for example, that information on the number of votes for each party in each district at each of two elections suggests that the number of voters who changed their choice between the elections varied across districts, but no district-by-district data are available to see if this is the case. Johnston (1985b) adapted Wilson's method to address this issue using national survey data on party-switching as a constraint to enable estimates of the volume and direction of switching in each district. Tests using available data have shown that the estimates are generally very accurate representations of the actual patterns (Johnston and Pattie, 2000, 2003), and the method has been employed in a large number of applications to demonstrate that there are significant spatial variations in various form of electoral behaviour (Johnston and Pattie, 2001). They also linked it to other attempts to tackle one of the major statistical issues in quantitative social science – the solution of the ecological inference problem. This relates to how individual relationships and behaviour can be inferred from aggregate data, such as differences between ethnic groups in their turnout at elections and spatial variations in those differences, when there are no data on turnout by ethnic group only on the ethnic composition and turnout rates of different areas (King, 1997; Johnston and Pattie, 2001).

In emphasising aggregate patterns, this work is macrogeographic; according to Webber (1977, p. 265), 'The entropy-maximizing paradigm asserts . . . that, though the study of individual behavior may be of interest, it is not necessary for the study of aggregate social relations.' The patterns predicted by the models are functions of the constraints (the information provided at the meso-state), so that knowledge of them means that 'the entropy-maximizing paradigm is capable of yielding meaningful answers to short-run operational problems' (p. 266) of immense value for immediate planning purposes. But:

> in the longer run, much of the economic system is variable: the constraints and the spatial form of the urban region may change . . . the research task facing entropists is (1) to identify the constraints which operate upon urban systems, which is partly an economic problem; (2) to deduce some facets of the economic relations among the individuals within the system from the use of the formalism; and (3) to construct a theory which explains the origins of the constraints. Only when the third task has been attempted may the paradigm be adequately judged.
>
> (p. 266)

The entropy-maximising model acts not only as a 'black-box' forecasting device (see p. 111, this book), therefore, but also as a hypothesis: if a system's operation is to be understood, the axioms – the constraints – must themselves be explained. Given the nature of the constraints (in Wilson's initial example, why people live where they do, why people work where they do, and why they spend a certain amount of time, money and energy on transport), the task is a major one: entropy-maximising models aim to clarify it and indicate the most fruitful avenues for investigation.

Dynamic processes

The study of systems allows dynamic processes to be incorporated within geographical analyses, instead of focusing on static patterns that are the outcomes of such processes. Thus, much of the mathematical development reviewed has been concerned with such processes, to advance understanding of change and the ability to forecast it. (On the use of the term 'process' in this context, see the critique by Hay and Johnston, 1983.) Dynamical systems theory and analysis were the focus of a major research programme conducted at the University of Leeds by Alan Wilson and his associates from the mid-1970s, for example, during which a number of major advances were made; in particular, attention was directed away from relatively straightforward linear modelling to complex representations in which change is presented as discontinuous and not necessarily unidirectional – often termed chaotic (see Wilson, 1981b).

Many geographical analyses of dynamical systems have focused on static spatial patterns or structures, representing these as equilibrium or steady-state situations within the ongoing dynamic processes. Change is then handled, as Clarke and Wilson (1985, p. 429) describe it, 'by forecasting (in some other theory or model) the independent variables associated with a system and then calculating the new equilibrium or steady state'. (There are parallels here with Darby's approach to historical geography, linking cross-sections through narratives of change, see p. 96, this book.) Their applications of dynamical systems analysis suggest that this approach is not tenable, however, because in complex systems 'There are too many possibilities of transition to different kinds of equilibrium or non-equilibrium states' (Clarke and Wilson, 1985, p. 431), which casts doubts on the validity of the traditional approach to forecasting and hence the contribution of geographical modelling to planning. The nature of that contribution must be rethought, therefore, because conditional forecasting is of little value – 'there are simply too many possible futures for this to be useful' (p. 446). The new contributions, they argue, could involve the following:

> First, it may often be possible to recognize the 'nearness' of some instability or structural shift to an undesirable state. Policy can then be focused on conservation. Secondly, it may be possible to see how to bring about a shift to a new desired state by changing policy in order to move a parameter through some critical value. Thirdly, both the ideas of dynamical analysis and the capabilities of modern computer technology lend themselves to the construction of planning systems focused on information retrieval . . . and monitoring, so that at least planners, and policy makers are in a position to respond more rapidly when difficulties are identified.
>
> (p. 446)

Geographers thereby come more to terms with the inherent complexity, and hence unpredictability, of the world.

The theory of systems and general systems theory

Like regional science and the impact of Walter Isard, the development of General Systems Theory (GST) was very much tied up with the work of one person – Ludwig von Bertalanffy (1950). He focused on isomorphisms, the common features among the systems studied in different disciplines: GST's 'subject matter is the formulation and derivation of those principles which are common for systems in general' (Walmsley, 1972, p. 23). The goal was to write meta-theories with rules that apply in a variety of contexts; application was usually by analogy from one discipline to another (as in Chappell and Webber's, 1970, use of an electrical analogue of spatial diffusion processes and in work on artificial intelligence using computer modelling: Couclelis, 1986a). GST was presented as offering geographers an organising framework, an empirical exercise using inductive procedures to fashion general theories out of the findings of particular disciplines (Coffey, 1981).

Although it was claimed that there have been no advances in either the theoretical base or the empirical application of GST (Greer-Wootten, 1972), some advances were made using models of one system as metaphors for understanding another. Woldenberg and Berry (1967) drew analogies between the hierarchical organisation of rivers and central-place systems, for example; Berry (1964b) argued that cities are open systems in a steady state, as exemplified by the stability of their behaviour-describing equations; several authors (e.g. Ray et al., 1974) applied the concept of allometry – that the growth rate of a component of an organism is proportional to the growth of the whole; and Haggett (1965c, p. 22; 1990) was attracted to D'Arcy Thompson's 1917 work On Growth and Form, which 'illustrates how many subjects find common ground in the study of morphology; there is inspiration still to be found in his treatment of crystal structures or honeycomb formations' (see also Werritty, 2010).

Some analysts were attracted to the concept of fractals (Goodchild and Mark, 1987, p. 265). If a line is measured at two scales, the second larger than the first, then we might expect its length to increase as a ratio of the two scales. In many areas of geography, however, this is unlikely to be the case: as the scale of a map increases (i.e. the representative fraction falls and more detail is shown) the length of a line – such as a stretch of coast – will probably increase by more than the ratio of the scales, because extra detail is added. Fractals are the geometrical objects that operate in this manner; even though the two lines may be generated by the same process (i.e. they are technically self-similar so that, for example, any part has the characteristic features of the whole), at different scales more detail can be shown. The concept was extensively used by Batty and Longley (1994) in analyses of urban form – both the shapes of entire built-up areas and those of their component parts; identification of fractal regularities allowed them to simulate urban growth and change (see also Batty, 2007).

According to its proponents, GST's advantages for human geography lie in its interdisciplinary approach, its high level of generalisation and its concept of the steady state of an open system (Greer-Wootten, 1972; Walmsley, 1972), but they also contended that geography's strong empirical tradition means that it has more to contribute than to take from GST. One critic, however, claimed that 'General systems theory seems to be an irrelevant distraction' (Chisholm, 1967, p. 51). Chisholm summarised the case for GST as follows:

1   there is a need to study systems rather than isolated phenomena;
2   there is a need to identify the basic principles governing systems;
3   there is value in arguing from analogies with other subject matter; and
4   there is a need for general principles to cover various systems.

He argued that something as grand as a meta-theory is unnecessary in order to convince people of the need to understand what they study, of the value of interdisciplinary contact, and of the potential fertility of arguing by analogy. His case was generally accepted, and references to GST soon disappeared from the discipline's literature.

Yet although a considerable number of programmatic statements were written about the benefits of adopting a system approach in human geography, relatively few applications were reported outside a small number of topical areas: the study of land use and transport patterns, as in the many analyses by Wilson and his co-workers already discussed; the study of input–output matrices (for examples, see Isard, 1975); and the study of population changes. With the last of these, for example, Robert Woods (1982) promoted a more theoretical approach to population geography than had previously characterised the subdiscipline, with an emphasis on systems, and with Rees he edited a pioneering collection of essays launching an approach termed 'spatial demography' (Woods and Rees, 1986). Many of the essays in the latter treated the three main components of a population system – birth, death and migration – in a multiregional spatial systems framework, which addressed the issue:

> Given that the characteristics of the population vary in interesting ways across space and that their time paths of development may be very different . . . how do we represent the processes of demographic change in formal models which will make the predictions about the future path of development possible?
>
> (Rees, 1986, p. 97)

The nature of the proposed accounting systems for interregional population changes was set out in detail in Rees and Wilson (1977).

A further area in which systems thinking was suggested concerns the interrelationships between society and nature, including ecology. There were some attempts to integrate the two through input–output and related models (as in Bennett and Chorley, 1978) and a number of authors presented flow diagrams of human–nature interrelationships that portrayed them as systems (as in Blaikie, 1985; Cooke and Reeves, 1976; Goudie, 1986a; Huggett, 1980; and Wilson, 1981a). In general, however, the concept of a system has been applied much more by physical than by human geographers. A small group of the latter has continued to work on the formal representation of spatial systems. Wilson (2000, p. vii) claimed some thirty years after his first essays into the area that such modelling of urban and regional systems remained underdeveloped, yet had enormous potential for policy-making and planning; for him, geography's 'classical theorists' remained von Thünen, Weber, Palander, Hoover, Hotelling, Christaller and Lösch, Burgess, Hoyt, Chauncy Harris and Ullman, plus the original formulators of the gravity model.

Spatial separatism?

Opposition to arguments for human geography as a spatial science developed largely as counters to claims for its separate status within the social sciences, with distance as the key variable. Crowe (1970), for example, portrayed using the spatial variable in nomothetic studies as naive spatial determinism.

The most sustained argument against geography as a spatial science – what he called the 'spatial separatist' theme – came in a series of papers by Sack, a former associate of Lukermann at the University of Minnesota (see p. 82, this book). Reality has three dimensions – space, time and matter – and geography, according to the spatial separatist view, is the science of the first. Sack contended that space, time and matter cannot be separated analytically in an empirical science, however, and that geometry is not an acceptable language for such a science (Sack, 1972). Geometry is a branch of pure mathematics that is not concerned with empirical facts; its laws are static, with no reference to time, and they are not derivable from any dynamic or process laws.

Geographic facts have geometric properties (e.g. locations) but if, as Schaefer (1953) proposed, geographical laws are just concerned with the geometries of facts, they will provide only incomplete explanations for them. To illustrate this contention, Sack used an analogy of chopping wood. If the answer to the question 'Why are you chopping wood?' is 'because the force of the axe on impact splits the wood', then it is a static, geometric law; but if the answer is 'to provide fuel to produce heat', then it is an instance of a process law, which incorporates the geometric law. In this analogy, a process law is equated with the intention behind an action. The static laws of geometry are sufficient to explain and predict spatial patterns, so that if geography aimed only to analyse points, lines and shapes on maps it could be an independent science using geometry as its language. But 'We do not accept the description of the changes of its shape as an explanation of the growth of a city' (p. 72), so that 'Geometry alone, then, cannot answer geographic questions' (p. 72), leading to the conclusion that (p. 77):

> To explain requires laws and laws (if they are valid) explain events. Since the definition of an event implies the delimitation of some geometric properties (all events occur in space), the explanation of any event is in principle an explanation of some geometric properties of events.
>
> Thus geography will be closely allied with geometry in its emphasis on the spatial aspects of events (the instances of laws), but geometry alone is insufficient as a basis for explanation and prediction since no processes are involved in the derivation of geometries.

Bunge's (1973a) response claimed that spatial prediction was quite possible with reference to the geometry alone, as instanced by central place theory and Thünian analysis of land use patterns. Such geometries provide 'classic beauty', and 'purging geometry from geography reduces our trade to no apparent gain' (p. 568). Sack (1973a) replied that the static laws espoused by Bunge are only special cases of dynamic laws having antecedent and consequent conditions, and that:

> Although the laws of geometry are unequivocally static, purely spatial, non-deducible from dynamic laws, and explain and predict physical geometric properties of events, they do not answer the questions about the geometric properties of events that geographers raise and they do not make statements about process.
>
> (p. 569)

Sack did not argue, as Bunge supposed, that geometry should be purged from geography, only that space should not be considered independently from time and matter. He further contended that:

> for a concept of physical space to conform to the rules of concept formation and be useful in a science of geography every instance of the geometric or spatial terms must be connected or related to one or more instances of non-geometric terms (to be called substance terms).
>
> (1973b, p. 17)

Thus, physical distance is not a concept in itself; it is necessary to know the terrain that a road crosses, for example, in order to assess the significance of its length in a gravity model – geometry alone is not enough. Since there is no such thing as empty physical space, there are no frictions of distance per se. There are frictions which demand work in crossing a substance, but the substance itself and the context in which it is being crossed create the frictions (such as transport costs), not simply the distance: 'There are frictions and there are distances, but there is no friction of distance' (Sack, 1973b, p. 22). Geography, according to Sack, is concerned to explain events and so requires substantive laws; such laws may contain geometric terms, such as the frictions of crossing a substance, but these terms alone are insufficient to provide explanations.

The spatial separatist approach proposes an independent position for geography within the social sciences using geometry as its language, but Sack (1974b, p. 446) contended that 'The spatial position's aim of prying apart a subject matter from the systematic sciences by arguing for spatial questions and spatial laws does not seem viable'. Instead, two types of law relevant to geographical work must be identified (Sack, 1974a):

1   Congruent substance laws are independent of location: statements of 'if A, then B' are universals which require no spatial referent.
2   Overlapping substance laws, on the other hand, involve spatial terms: 'if A, then B at C' contains a specific reference to location.

Both are relevant and necessary in answering geographical questions, so no case can be made for a necessary 'spatialness' to human geography's substance laws.

Extending his argument, Sack (1981, pp. 3–4) contended that 'Space is an essential framework of all modes of thought' but 'geographic space is seen and evaluated in different ways at different times and in different cultures'. His book illustrated this by examining several approaches to the study of space: that of the social scientist concerned with objective meanings of space; that of the social scientist concerned with subjective meanings; the practical view of people who live in and learn about space (he used children as his example); the mythical and magical views of space; and the societal conceptions within which organisations and institutions structure and use space. Only the first of these is the concern of 'spatial separatists', whose approach takes space out of its relational context, with the consequence that 'Ignoring spatial relations or conceiving them non-relationally will hinder the discovery and confirmation of social science generalizations' (p. 85).

J. A. May (1970, p. 188) also argued against Schaefer's claim that geography is the study of spatial relations:

> If we extend Schaefer's argument to include time, and assign the study of temporal sequences or relations to the historian, then the only conclusion respecting this matter that can be drawn is that economic, social, political, and other relations must be non-spatial and non-temporal. Hence economics, sociology, political science, etc. are non-spatial, non-temporal sciences. But this is absurd . . . insofar as economics qualifies as a science possessing empirical warranty, then its generalizations must apply to given spatio-temporal situations.

If all sciences have a spatial content, what defines a separate discipline of geography? May lists five possibilities:

1   Geography is a 'super-science' of spatial relations, 'a generalizing science of spatial relations, interactions and distributions' (p. 194) drawing on the findings of other sciences. This would leave the latter truncated and their studies unfinished. Such an approach had already proved unsuccessful; 'the issue of the conception of geography as a generalizing or law-finding science that somehow stands above the social sciences and history is not even appropriately debatable' (p. 195).

2   Geography is a lower-level science of spatial relations, applying in empirical contexts the laws of higher-level, generally more abstract sciences. (This may be a valid description of much of the geography produced in the 1960s.) This again truncates the latter sciences and raises the question: 'What differentiates economic geography from economics, and vice versa?'.

3   Geography is the study of geographical spatial relations. This implies that there are spatial phenomena not studied in other social science disciplines and which can therefore be claimed as geography's; May could conceive of no objects which are purely geographical (or, in the parallel argument, purely historical either).

4   Geography is the study of 'things in reality' spatially. Yet again, this abstracts from other sciences; May admits that certain 'bits and pieces' are not studied elsewhere, but argues that they do not offer a satisfactory empirical foundation for a separate discipline.

5   'Geography is not a generalizing or law-finding science of spatial relations' (p. 203).

The first four indicate that, because of the analytical indissolubility of time, space and matter, all social sciences are concerned with spatial relations, so that May, like Sack, argued that geography cannot claim an independent status on the basis of the spatial variable and the geometrical aspects of space alone. R. P. Moss (1970, p. 27) reached a similar conclusion:

> geometrical relationships must be assigned economic, social, physical, or biological meaning before they can in any sense become explanatory . . . though geometries may be important tools in geographical study and research, they cannot be a source of theory since their analogy with geographical phenomena is simply through particular logical structures, and not through explanatory deduction . . . such an application implies that space, area, distance, etc., are important in and of themselves, quite independently of any implications they may have in terms of diffusion, of cost, of time, or of process. This is manifestly false.
>
> (p. 27)

Derek Gregory (1978b, 1980) also criticised the extremely narrow, even superficial, view of spatial processes identified in many spatial scientists' work. Their claims are instrumentalist, involving theories which cannot be validated conclusively but can only be evaluated pragmatically against the real world. He argued that Bennett (1974), for example, accepted that his models did not mirror actual processes but assumed that they did in order to allow policy formulation (and therefore produced self-fulfilling prophecies – as Lukermann had argued: see p. 82, this book); because they can postdict the world as it presently is, it is assumed that they explain it and can be used to predict the future.

The arguments reviewed in this section were critical of much of the work undertaken by geographers in the paradigm debate over which dominated the 1960s, impelled by the 'Victorian myth of the supremacy of the natural sciences' (Gregory, 1978a, p. 21). The case for spatial analysis continued to be argued, however. Gatrell (1983), for example, countered Sack's critique of spatial science – while accepting his case that the separation of space from substance is untenable. He did not confine spatial analysis to a positivist philosophy, however, stating that:

> My response to both structuralists and humanists is that, since they too deal with relations (among individuals or social groups, or between man and his environment), they cannot avoid the notion of space, since any relation defines a space. Moreover, because every relation has a geometry associated with it . . . they cannot avoid the fact that geometry underlies much of what they deal with. Structures . . . are intrinsically spatial, but not in any simple geographical sense.
>
> (p. 5)

Gatrell's definition of space is much broader than simply distance, therefore, and he promoted spatial analysis not as a separate paradigm of geography, but as an arsenal of tools to be used in all empirical research. Much of that research, he argued elsewhere (Gatrell, 1985, p. 191), involves portraying objects arranged in space, investigating the role of distance as a constraint on human spatial organisation and achieving efficiency in locational arrangements.

Behavioural geography

Although not apparently directly influenced by the arguments of Sack and others regarding geography and geometry, by the end of the 1960s some critiques of 'quantitative and theoretical geography' were being articulated on the grounds that testing of models of spatial organisation and behaviour based on rational decision-making dominated by monetary criteria (and hence, in geography, minimising transport costs) were not substantially aiding understanding. They were too simplistic in their assumptions about how people make decisions and behave. Instead, there were calls for more 'realistic' modelling based on observed rather than assumed decision-making processes.

These proposed modifications to the spatial-science approach stimulated work that became known as 'behavioural geography', whose birth was announced in a key collection of essays (Cox and Golledge, 1969) that would be revisited twelve years later in a companion volume (Cox and Golledge, 1981), and which was further synthesised in two editions of a major textbook (Golledge and Stimson, 1987, 1997). Its essential ingredients, as set out by Golledge and Timmermans (1988), were:

1   a search for models which were alternatives to those of normative location theory based on economically and spatially rational beings;
2   a search to define environments other than objective physical reality as the milieux in which human decision-making and action took place;

3 an emphasis on process rather than structural explanations of human activity and the physical environment;

4 an interest in unpacking the spatial dimensions of psychological, social and other theories of human decision-making and behaviour;

5 a change in emphasis from aggregate populations to the disaggregate scale of individuals and small groups;

6 a need to develop new data sources other than the generalised mass-produced aggregate statistics of government agencies which obscured and over-generalised decision-making processes and consequent behaviour;

7 a search for methods other than those of traditional mathematics and inferential statistics that could aid in uncovering latent structures in data, and which could handle datasets that were less powerful than the traditionally used interval and ratio data; and

8 a desire to merge geographic research into the ever broadening stream of cross-disciplinary investigation into theory-building and problem-solving.

Towards a behaviouristic spatial science

The main ground for disillusion with 'spatial science' was thus a growing realisation that many of the models being propounded and tested provided poor descriptions of reality; progress towards the development of geographical theory was painfully slow and its predictive powers consequently weak. The large body of work based on central place theory, for example, was built on axioms regarding human behaviour regarding choice between spatial alternatives, from which a settlement pattern was deduced. But the deductions were often only weakly reflected in settlement morphologies. The theory suggested how the world would look under certain circumstances of economic rationality in decision-making on an isotropic plain; that those circumstances did not prevail suggested that the world should be looked at in other ways in order to understand how people structure spatial organisation. As Brookfield put it with regard to the models then popular:

> We may thus feel that we have proceeded far enough in answering our questions when, by examination of a sufficient number of cases, we can make assertions such as the following: population density diminishes regularly away from metropolitan centres in all directions; crop yields diminish beyond a certain walking distance from the centres of habitation; air-traffic centres lying in the shadow of major centres do not command the traffic that their populations would lead us to expect. . . . Such answers, which represent the mean result of large numbers of observations, whether statistically controlled or otherwise, are valuable in themselves, and sufficient for many purposes. But each is also an observation demanding explanations which may seem self-evident, or which may in fact be very elusive. Furthermore, there will be exceptions to each generalization, and in many cases, there are also limits to the range of territory over which they hold true. Both the exceptions and the limits demand explanation.
>
> (1964, p. 285)

The concern was, therefore, not with the basic goal of establishing generalisations and theories, but rather the route being taken to that goal. Better models than those of 'rational economic man' were needed, and their search took a more inductive route than previously followed. Rather than base hypotheses on assumptions about behaviour, models were to be constructed that replicated observed behaviour.

Rationality in land-use decisions

Some of the earliest attempts by geographers to explore behaviour inductively, as a prelude to later modelling, were investigations of human responses to floods and other environmental hazards initiated at the University of Chicago during the late 1950s and early 1960s. Their director was Gilbert White, whose own thesis on human adjustment to floods was published in 1945, and who has been described as 'the outstanding geographer in the man–land [sic] tradition, in the study of natural resources and hazards, and the study of the human environment' (Kates and Burton, 1986b, p. xi). (White is one of the most honoured of geographers, having been awarded both the Public Welfare Medal of the National Academy of Sciences, its highest award, in 2000 and the National Medal of Science in 2001: Turner, 2002, p. 68; these aren't mentioned in White's 2002 autobiographical essay.) Water resources were his main focus, in which 'he found himself a leader in the newly developing geography of perception, the world inside people's minds' (p. xi; Kates and Burton, 1986a, provides a collection of his writings and of valedictory essays; see also Hinshaw, 2006; Macdonald *et al.*, 2012).

Gilbert White's associates developed a behaviourist approach for studying reactions to hazards based on Herbert Simon's (1957) theories of decision-making. Roder (1961), for example, categorised Topeka residents according to their attitudes to the probability of future floods there, concluding that:

> Flood danger is only one of the variables affecting the choices of the flood-plain dweller, and many considerations operate to discourage a resident from leaving the flood plain, even when he is aware of the exact hazard of remaining.
>
> (p. 83)

Such behaviour did not fit easily into notions of profit-maximising decision-making. Alternative theories were needed.

A major exponent of this behaviouristic approach was Kates (1962, p. 1), whose study of flood-plain management began: 'The way men [sic] view the risks and opportunities of their uncertain environments plays a significant role in their decisions as to resource management'. Kates developed a schema for studying such decision-making based on four assumptions:

1   People are rational when making decisions. Such an assumption may be either normative/ prescriptive – describing how people should behave – or descriptive of actual behaviour. The latter seemed most fruitful, both for understanding past decisions and for predicting those yet to be made. Kates suggested adoption of Simon's (1957) concept of bounded rationality. Decisions are made on a rational basis, but in relation to the environment as it is perceived by the decision-maker, which may be quite different from either 'objective reality' or the world as seen by the researcher. Rational decision-making is constrained, therefore, and is not necessarily the same as the maximum rationality assumed in the neoclassical normative models discussed earlier; people make decisions in the context of the world as they observe and interpret it, which may differ from others' perceptions (including those of geographers studying them).

2   People make choices. Many decisions are either trivial or habitual and accorded little or no thought immediately before they are made (Kahneman, 2012). Some major decisions regarding environmental use may also be habitual, but such behaviour usually develops only after a series of conscious choices has been made, which may generate a stereotyped response to future similar situations.

3   Choices are made on the basis of knowledge, which is usually only partial. Only very rarely can decision-makers bring together all the information relevant to their task, and they are frequently unable to assimilate and use all that is available.

4   Information is evaluated according to predetermined criteria. In habitual choice the dominant criterion is what was done before, but in conscious choice the information must be weighed according to certain rules (individually determined, though usually in a social context). Some

normative theory prescribes maximising criteria (of profits, for example); descriptive theory may use Simon's notion of satisficing behaviour, involving decision-makers who seek a satisfactory outcome only (a given level of profit, perhaps, as in Rawstron's, 1958, industrial location theory: see Smith, 1981).

Models based on such behavioural axioms are likely to differ very substantially from those which assume not only rationality but also complete information, perfect decision-making ability and common goals. As Kates (1962, p. 16) describes his model, 'men [sic] bounded by inherent computational disabilities, products of their time and place, seek to wrest from their environments those elements that might make a more satisfactory life for themselves and their fellows'. As a consequence:

> Thus, a descriptive theory of choice must deal with the well informed and the poorly informed and the choices that men make under certainty, risk or uncertainty . . . such a theory must deal with the eventuality that not only do the conditions of knowledge vary, but the personal perception of the same information differs.
>
> (p. 19)

People behave rationally, but within constraints, many imposed by their context – the local cultures in which they have been socialised to make decisions.

The results of decision-making that do not match the predictions from the theories employed by normatively inclined spatial scientists do not imply irrational behaviour, therefore. Most decisions are made rationally on the basis of, probably non-random, selections of information, are intended to satisfy a goal which does not imply making a perfect decision, and are based on criteria which vary somewhat from individual to individual. Having learned a satisfactory solution to a given class of problems, decision-makers apply it every time such a problem occurs, unless changed circumstances require a re-evaluation.

Kates wanted to understand why people chose to live in areas that are prone to flooding. Their information was based on their knowledge and experience, and they varied in the certainty of their perceptions regarding further floods. In justifying their decisions, most were boundedly rational; they made conscious choices to satisfy certain objectives. Similar findings were reported by others of White's co-workers and had impacts on public policy formulation internationally (White, 1973, 2002), covering a wide range of environmental hazards such as floods, water management and desertification (Burton et al., 1978). Their initial impact on the wider geographical enterprise was not great, however, especially in the early years of their work, perhaps because they were operating close to the boundaries between human and physical geography. Later work in other fields brought Simon's ideas more forcefully before geographical audiences, however. These articulate with the question of the boundaries of human and physical geography – a theme we return to in Chapter 10.

The decision process in a spatial context

In the 1960s human geographers were introduced to the behaviourist approach by Julian Wolpert's (1964) paper, based on his Wisconsin Ph.D. thesis with field data from Sweden. The normative theory then espoused by many geographers assumed a rational economic decision-maker who:

> is free from the multiplicity of goals and imperfect knowledge which introduce complexity into our own decision behavior. Economic Man has a single profit-maximizing goal, omniscient powers of perception, reasoning, and computation, and is blessed with perfect predictive abilities . . . the outcome of his actions can be known with perfect surety.
>
> (p. 537)

In the study of spatial patterns, however, 'Allowance must be made for man's finite abilities to perceive and store information, to compute optimal solutions, and to predict the outcome and future events, even if profit were his only goal' (p. 537). Farmers face an uncertain environment – both physical and economic – when making the decisions that in aggregate result in a land-use map. Wolpert suggested that differences between these decisions and those that would be made by 'economic man' reflect aspects of the farmers' economic and social environments.

Comparing the observed labour productivity of farms in an area of Sweden with what could have been achieved under optimising decision-making, Wolpert decided that the farmers were probably satisficers, although such a conclusion is difficult to verify without detailed knowledge of aspiration levels. How they acted was undoubtedly contingent upon their available information, and he argued that spatial variations in the levels of potential productivity achieved were the consequences of parallel spatial variations in knowledge. Only conspicuous alternatives are considered, it was suggested, and the result is rational behaviour, adapted to an uncertain environment. Gould (1963, 1965) approached a similar subject – land-use decisions in Tanganyika – using a different approach based on the mathematical theory of games. (See also Chapman, 1982, who uses games to illustrate decision-making in uncertain situations.)

Wolpert (1965) continued his theme regarding the role of information with studies modelling the decision-making behind observed aggregate patterns of migration. He contended that the gravity model is inadequate to represent such flow patterns; indeed, 'Plots of migration distances defy the persistence of the most tenacious of curve fitters' (p. 159; see also Taylor, 1971). Rational individuals make sequential decisions: first, whether to move, and second, where to, doing both on the basis of place utilities – their evaluations of the degree to which each potential location, including that currently occupied, meets defined needs. The information on which these utilities are computed is invariably far from complete; indeed, for many places people have none. Each individual has an action space – 'the set of place utilities which the individual perceives and to which he responds' (Wolpert, 1965, p. 163) – whose contents may deviate considerably from that portion of the 'real world' which it purports to represent. Once the first decision – to migrate – has been made, then the action space may be changed as the would-be mover searches for potential satisfactory destinations and, if necessary, extends the space if no suitable solution to the search can be found (see Brown and Moore, 1970).

Wolpert's papers heralded the development of behavioural geography (Cox and Golledge, 1969, pp. 1–3), an approach 'united by a concern for the building of geographic theory based on postulates regarding human behaviour . . . upon social and psychological mechanisms which have explicit spatial correlates and/or spatial structural implications'. Early work focused on topics related to decision-making in spatial contexts, much of it involving researchers associated with the Ohio State University. Golledge (1969, 1970), for example, looked at models of learning about space and of habitual behaviour therein, and with Brown investigated methods of spatial search (Golledge and Brown, 1967). Others researched the information flows on which decisions are based, indicating the influence of local context on behaviour (Cox, 1969), and Brown and Moore (1970) extended Wolpert's place utility and action space concepts for the study of intra-urban migration.

The fundamental aim of behavioural geography, according to a review by Golledge *et al.* (1972, p. 59), is to derive alternative theories to those based on 'economic man', 'more concerned with understanding why certain activities take place rather than what patterns they produce in space'. This involves 'the researcher using the real world from a perspective of those individuals whose decisions affect locational or distributional patterns, and . . . trying to derive sets of empirically and theoretically sound statements about individual, small group, or mass behaviors'. Individuals are active decision-makers, not passive reactors to institutionally created stimuli (Cox and Golledge, 1981).

In evaluating such behaviouristic endeavours, Golledge *et al.* (1972) indicated the seminal influence of Hägerstrand's (1968) use of the concept of a mean information field (analogous to the action space of a place's residents) to model migration flows and the adoption of innovations.

Further developments included a series of pioneering papers by Wolpert and his associates on political decision-making. Regarding the distribution of certain artefacts in the landscape, Wolpert (1970, p. 220) pointed out that the location of, for example, a public facility in an urban area is frequently the product of policy compromise:

> Sometimes the location finally chosen for a new development, or the site chosen for a relocation of an existing facility, comes out to be the site around which the least protest can be generated by those displaced. Rather than being an optimal, a rational, or even a satisfactory locational decision produced by the resolution of conflicting judgements, the decision is perhaps merely the expression of rejection by elements powerful enough to enforce their decision that another location must not be used. . . . These artefacts are rarely 'the most efficient solutions', and frequently not even satisfactory neither for those responsible for their creation nor for their users.

(This argument avoids considering any definition of optimal, in either economic or political terms.) Such decisions involve what Wolpert terms 'maladaptive behaviour'. Kates and others had suggested that decisions are adaptively rational, within the constraints of uncertainty, utility and problem-solving ability. Coping strategies under the mutual exchange of threats between interested parties can involve decision-making which does not involve the careful and methodical investigation of alternatives until a satisfactory solution is found, however. Instead, decisions are the consequences of conflict between groups with different attitudes and motivations, and are not the result of joint application of criteria on whose relevance there is a consensus.

> This formulation lays the framework for the interpretation of locational decisions which appear to be more the product of pressurised responses than the end result of a dispassionate and considered selection of alternatives posited by the classical normative approaches or even the Simon scheme of bounded rationality.
>
> (Wolpert, 1970, p. 224)

The approach was applied in a variety of contexts, such as the routes for intra-urban freeways and the siting of community mental-health facilities (Wolpert et al., 1975). Many of the outcomes of such decision-making processes reflected the uneven distribution of power among the various groups of participants.

Mental maps

One concept enthusiastically adopted by a number of workers was that of a mental map of the environment, which guides a decision-maker's deliberations. The term was not new to the geographical literature, having been used in Wooldridge's (1956) descriptions of the perceived environments within which farmers make land-use decisions. Gould (1973) revived it in 1966, in a seminal, widely circulated discussion paper, which included his guiding belief that:

> If we grant that spatial behavior is our concern, then the mental images that men hold of the space around them may provide a key to some of the structures, patterns and processes of Man's work on the face of the earth.
>
> (p. 182)

Increasingly, he argued, location decisions involve perceived environmental quality, so it is necessary to know how people evaluate environments and whether their views are shared by their contemporaries. To investigate such questions, Gould asked respondents in various countries to rank-order locations according to their preferences for them as places in which to live, and these rankings were analysed to identify their common elements – the group mental maps (Gould and White, 1974, 1986). Such maps, he argued, are useful not only in the analysis of spatial behaviour but also in planning social investments, such as offering differential salaries to attract people to less desirable areas (Gould and Ola, 1970).

Those who followed Gould's lead investigated a range of methods for identifying and analysing spatial preferences (Pocock and Hudson, 1978). Their results made little impact on theoretical development, however, and Downs (1970, p. 67) wrote that:

> Even the most fervent proponent of the current view (that human spatial behaviour patterns can partially be explained by a study of perception) would admit that the resultant investigations have not yet made a significant contribution to the development of geographic theory.

Apart from Gould's rank-ordering procedures, Downs identified two other major approaches to the study of environmental images. The structural approach enquires into the nature of the spatial information stored in people's minds and which they use in their everyday lives – Kevin Lynch's (1960) book was a model for such work: his five-element categorisation of urban environments based on map-drawing exercises comprised paths, boundaries, districts, nodes and landmarks as the main components of the urban scene identified by users. The evaluative approach addresses the question: 'What factors do people consider important about their environment, and how, having estimated the relative importance of these factors, do they employ them in their decision-making activities?' (Downs, 1970, p. 80).

With this latter approach geographers moved into the wider field of cognitive mapping – 'a construct which encompasses those cognitive processes which enable people to acquire, code, store, recall and manipulate information about the nature of their spatial environment' (Downs and Stea, 1973, p. xiv). Although medical technologies (based on brain scanning) would only be applied decades later, some of the pioneers of behavioural geography worked alongside both psychologists, who were becoming increasingly interested in the individual's relationship to a wider area than the proximate environment and the development of relevant non-experimental research techniques, and designers concerned with creating more 'liveable' environments. The journal *Environment and Behavior* was launched in 1969 to cater for this interdisciplinary market, but despite some interest (e.g. Tuan, 1975a; Downs and Stea, 1977; Pocock and Hudson, 1978; Porteous, 1977), the general field has not made a major impact within human geography. Gold (1992) argued that although a small number of scholars have continued working in this area, the challenges from other types of work (see Chapters 5–7 below) meant not only that it failed to develop into a large component of the discipline, but also that it was marginalised, as both passé and tainted with the generally perceived problems of positivism. Gold quotes Cloke *et al.* (1991) as presenting cognitive-behavioural geography as forming 'something of a "bridge" leading from the "peopleless" landscapes of spatial science through to the "peopled" landscapes of humanistic geography' (Gold, 1992, p. 242). The impression he gives is of a small specialism 'retaining a precarious autonomy on the fringes of human geography' (p. 246).

The concept of 'mental map' and the associated process of 'cognitive mapping' – which 'seems to imply the evocation of visual images which possess the kinds of structural properties that we are familiar with in "real" cartographic maps' (Boyle and Robinson, 1979, p. 60) – became the centre of considerable debate both among behavioural geographers and between them and outside critics. Gould's initial work, and that which it stimulated, was criticised as the study of space preferences only (Golledge, 1981a; see also Golledge, 1980, 1981b; Guelke, 1981a; Robinson, 1982). But, as Downs and Meyer (1978, p. 60) made clear, 'perceptual geography' – 'the belief that human behaviour is, in large part, a function of the perceived world' – extends much further than the elicitation and mapping of space preferences.

The basic arguments of behavioural geography are that:

1 people have environmental images;
2 those images can be identified accurately (i.e. quantitatively) by researchers (as in landscape evaluation: Penning-Rowsell, 1981); and
3 'there is a strong relationship between environmental images and actual behaviour' (Saarinen, 1979, p. 465).

The nature of those images – whether they are maps in the generally understood sense of that word and whether they can be apprehended by researchers – remained a problem; the concept of a 'mental map' may be a red herring but, to behavioural geographers, the argument on which it is based is not. (For an early review of research into cognitive mapping, see Golledge and Rushton, 1984.) Similarly, the concepts of cognitive mapping were increasingly used in work on the history of cartography (on which, see Lewis, 1998).

Time-geography

One area of work sometimes identified as behavioural, but which extends into the humanistic and realist philosophies discussed in later chapters, was developed by Hägerstrand from the late 1960s on and introduced to a wider audience by Pred (1973, 1977a; Persson and Ellegård, 2011; Sui, 2012). As interpreted by Carlstein *et al.* (1978), time and space are resources that constrain activity. Any behaviour requiring movement involves individuals tracing a path simultaneously through space and time, as depicted in Figure 4.4, where movements along the horizontal axis indicate spatial traverses and those along the vertical signify the passage of time. All journeys, or lifelines, involve movement along both and are displayed by lines that are neither vertical nor horizontal; vertical lines indicate remaining in one place; horizontal lines are not possible for people, though they are (or virtually so) for the transmission of messages (see Adams, 1995).

Movement in space and time is constrained in three ways, according to Hägerstrand:

1 capability constraints include the biological need for about eight hours sleep in every twenty-four, and movement across space is constrained by the available means of transport;

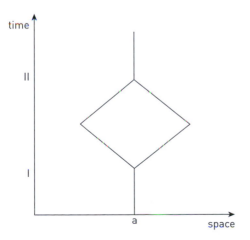

Figure 4.4 The time–space prism. In this simple example, a person cannot leave place a until time I and must return there by time II; the prism between those two times indicates the maximum available spatial range of travel

2   coupling constraints require certain individuals and groups to be in particular places at stated times (teachers and pupils in schools, for example) and thus limit the range of mobility during 'free time'; and

3   authority constraints may preclude individuals from being in stated places at defined times.

Together, these three define the time–space prism (Figure 4.4) which contains all the possible lifelines available to an individual who starts at a particular location and has to return there by a given time.

Pred (1977a, p. 207) claimed that time-geography 'has the potential for shedding new light on some of the very different kinds of questions customarily posed by "old-fashioned" regional and historical geographers, as well as "modern" human geographers' because:

> It is . . . a great challenge . . . to cease taking distance itself so seriously . . . to accept that space and time are universally and inseparably wed to one another; to realize that questions pertaining to human organisation of the earth's surface, human ecology, and landscape evolution cannot divorce the finitudes of space and time . . . it is a challenge to turn to the 'choreography' of individual and collective existence – to reject the excesses of inter- and intradisciplinary specialization for a concern with collateral processes.
>
> (p. 218)

Thus, time-geography was in part a critique of spatial analysis as currently practised by human geographers. For Hägerstrand it was much more, however, because he was addressing problems concerned with the quality-of-life implications of packing people together in space and time; for example, we are enabled to

> see existing together what we otherwise have chosen to see apart. Neighbour meets neighbour in some sort of association. My concern now will be with how their presence together has come about, how they go along together and what is going to happen ahead in time.
>
> (Hägerstrand, 1984, p. 375)

To explore what Pred (1977a, p. 213) calls the 'principle of togetherness', Hägerstrand employed three concepts:

1   Paths (or lifelines in the earlier terminology) are 'successions of situations' (Hägerstrand, 1982, p. 323) traced by individuals. To study paths alone is just to map outcomes (in the same way that geographers map migrations or shopping trips); investigation of the path alone may reveal very little about the purpose and meaning behind the events that it incorporates, because it tells nothing about the 'living body subject, endowed with memories, feelings, knowledge, imagination and goals – in other words capabilities too rich for any conceivable kind of symbolic representation but decisive for the direction of paths' (p. 324).

2   Projects, goals and strategies for project fulfilment are the activities which generate paths, whose intersections produce situations, particular moments in the flow of history in particular places.

3   Dioramas. Geographical study of paths and projects has traditionally involved the concept of landscape, devised to represent 'the momentary thereness and relative location of all continuants' (p. 325). Hägerstrand claimed that this concept insufficiently incorporates 'the human body subjects, the keepers of memories, feelings, thoughts and intentions and initiators of projects' (p. 320) and preferred the concept of diorama, normally used to denote static museum displays which depict people and animals in their usual environments. The concept implied to Hägerstrand that 'All sorts of entities are in touch with each other in a mixture

produced by history, whether visible or not ... [we] appreciate how situations evolve as an aggregate outcome quite apart from the specific intentions actors might have had when they conceived and launched projects out of their different positions' (p. 320).

Hägerstrand illustrated his concepts with a description of his childhood home, demonstrating the importance of coupling, authority and capability constraints on the flow of daily life and the dioramas (particular situations) that resulted; Pred (1979) applied the concepts to his academic career. The study of paths alone is insufficient if time-geography is adequately to portray the 'real life of real people' (Hägerstrand, 1982, p. 338), however; thus autobiographical material (as in Buttimer, 1983; Billinge *et al.*, 1984) provides an important source on projects and dioramas because 'Only one's own experience is able to provide the kind of intimate detail which can bring the study of project and situation into any real depth. He is after all an expert on his own networks of meanings' (Hägerstrand, 1982, p. 338).

Most interpretations and applications of Hägerstrand's initial work focused on the study of paths, and were set in the positivist mould of other behavioural geography enterprises. Indeed, this aspect of time-geography was the only one presented in an introductory review by Thrift (1977), who emphasised that a '"physicalist" approach is the backbone' (p. 4). With it, for example, Parkes and Thrift (1980) sought 'to place time firmly in the minds of human geographers' (p. xi) in an approach they termed 'chronogeographical', arguing that 'Together, territories, societies and times are the principal components of urban and social geography' (p. 34); this focus on the meaning and construction of time in historical contexts was very substantially extended by Glennie and Thrift (1996).

The concept of diorama, and the 'togetherness' of a situation that it denotes, were absent from these early presentations, apart from passing references such as Thrift's (1977, p. 7), quoting Hägerstrand (1975), that 'every situation is inevitably rooted in past situations'. Projects were also mentioned, as in Thrift's (1977, p. 7) statement that 'All human beings have goals. To attain these they must have projects, series of tasks which act as a vehicle for goal attainment and which, when added up, form a project', but it was implicit that those projects can be treated in the same sort of aggregate data analysis and policy prescription mould that characterise most behavioural geography (see Palm and Pred, 1978). Van Paassen (1981) argued that Hägerstrand's work is essentially humanistic, however, aiming 'to provide insight into what is specifically human in man's nature and ... [to] elucidate the specific human situation' (p. 18), a case accepted by Hägerstrand (1982) in his reference to human intentions (Pred, 2005; see also Thrift, 2005).

The concept of diorama later led authors to associate Hägerstrand's work not so much with the humanistic approaches outlined in the next chapter, but rather with the realist proposals discussed in Chapter 6. The social theorist Giddens (1984), for example, drew heavily on Hägerstrand in his arguments for a structurationist approach (see p. 221, this book). (Note, however, Gregson's, 1986, critique of Giddens's attempts to incorporate time-geography into structuration theory, but also Gregory's, 1985a, p. 329, contention that time-geography 'must draw on substantive theories of structuration for its explications of contingency and necessity'.) In all these applications, however, authority constraints and their associations with bounded spaces have been largely ignored, with continuous time–space the dominant concern.

Most criticisms of time-geography focused on the physicalist description of paths. Thus, Baker argued that:

> While space and time may usefully be considered as resources whose competitive allocation gives rise to patterns of use which may be observed empirically and modelled theoretically, the nature of that human struggle to control and structure time and space – the process underlying the form – should be of paramount concern rather than descriptions of temporal or spatial organisation.
>
> (1979, p. 563)

As with other aspects of spatial science, description of outcomes, however sophisticated, cannot elucidate the processes involved in their production. (Processes here are defined as mechanisms, the products of human agency, and not just sequences: see Boots and Getis, 1978; Haining, 1981; Hay and Johnston, 1983.) Thrift and Pred responded to Baker by denying his physicalist interpretation; to them:

> time-geography is much more than that. It is a discipline-transcending and still evolving perspective on the everyday workings of society and the biographies of individuals. It is a highly flexible and growing language, a way of thinking about the world at large as well as the events and experiences, or content, of one's own life.
>
> (1981, p. 277)

They argued that 'time-geographers' are necessarily concerned with underlying processes, with the ideological uses of time and space as devices to channel individual paths, and with the crucial role of human agency in the production of particular situations. They tied time-geography into Giddens's ideas on structuration and also with developing Marxist humanism, concluding that:

> Some see the graphs used in time-geography as just neat pieces of art but others, in turn, are able to internalize the perspective represented by the graphs and use the path and project language as a way of thinking about themselves and the world. This will we believe be the lasting legacy of time-geography.
>
> (p. 284)

Baker's (1981, p. 440) response was largely positive, accepting that time-geography could be of value in a reorientation of geographical work: 'we should be examining the social organisation of space and time, not the spatial and temporal organisation of society for this is to put the cart before the horse'. Gregory (1985a) was more cautious, however, arguing that Hägerstrand too readily focused on paths rather than on the people whose projects fashion those paths, thereby failing to explore the meanings that are hidden beneath the tasks that produce the biographies. (Elsewhere, Gregory, 1994, implied that the fault is more with those applying Hägerstrand's ideas than with Hägerstrand himself; others – e.g. Ellegård and Svedin, 2012 – have illustrated the importance of Hägerstrand's concepts to the study of certain types of behaviour, as in transport geography.)

Hägerstrand's papers were written before the 'information technology revolution' which enabled the almost instantaneous transmission of information to multiple sites around the globe and promised virtual reality, whereby people in one place could operate as if in another. This led Paul Adams (1995) to promote a reconsideration of what he termed 'personal boundaries in space-time', using Janelle's (1973) concept of 'personal extensibility', which he defines as 'the ability of a person (or group) to overcome the friction of distance through transportation or communication' (P. Adams, 1995, p. 267). As that ability increases and 'distant connections become easier to maintain' (p. 268), so interaction patterns alter, to the extent that 'Some believe that time-space and cost-space convergence have reached the point that one's location is of little or no significance for an increasing variety of interactions.' Extensibility promotes the transcendence of place, breaking down the crucial role of spatially bounded locales (see p. 213, this book) as interaction containers. Thus, people are both point-entities and extensible persons. They comprise:

> A) a body rooted in a particular place at any given time, bounded in knowledge gathering by the range of unaided sensory perception and, in action, by the range of the unaided voice and grasp; and B) any number of fluctuating, dendritic, extensions which actively engage with social and natural phenomena at varying distances. This dynamic entity depends on media.
>
> (p. 269)

Acceptance of this argument requires a reworking of the importance of the presence/absence distinction in much analysis of human behaviour, and a new form of social geography (heralded, Adams argues, by Melvin M. Webber in 1964). It implies the need to distinguish between the geography of the body from that of the person: the former is grounded in space–time, but technology removes that constraint for the latter, with many implications for the exercise of power and responsibility.

Methods in behavioural geography

Whereas early theory- and model-builders of the spatial-science school of human geographers derived much of their stimulus from neoclassical economics, in some cases via regional science, the alliance for behavioural geography was mainly with other social sciences, notably psychology and sociology. The behaviouristic approach is largely inductive, aiming to build general statements out of observations of ongoing processes. The areas studied were very much determined by the work in the spatial-science school, however. As Brookfield's statement quoted earlier suggests (p. 125, this book), the spatial-scientists' models and theories, such as central place theory, raised many of the queries which the behaviourists followed up, stimulated by observations of the theories' failings when matched against the 'real world'. In terms of the accepted route to scientific explanation (see Figure 3.1), therefore, behavioural geography involved moving outside the accepted cyclic procedure to input new sets of observations on which superior theories might be based. In doing this, it did not move far from the spatial-science ethos, and many of the methods were those used in that work.

Somewhat away from this general orientation of behavioural work, Pred (1967, 1969) presented an ambitious alternative to theory-building based on 'economic men' in a two-volume work *Behavior and Location*. His critique of location theory was based on three groups of objections concerned with:

1   logical inconsistency – it is impossible for competing decision-makers to arrive at optimal location decisions simultaneously;
2.   motives – maximising versus satisficing behaviour; and
3.   capacity – human ability to collect, assimilate, manipulate and use all possible information.

To Pred (1967, p. 17):

> Bunge's theoretical geography is easily distinguished from geographical location theory because its optimal final goals are disassociated from the interpretation of real-world economic phenomena . . . and because these same goals can only yield a body of theory that for all intents and purposes is totally abehavioral and static rather than dynamic.

Location and land-use decisions are made with imperfect knowledge by fallible individuals and as a consequence there is bound to be disorder in the ensuing spatial patterns. (For another critique, see Barnes, 1988.)

As an alternative, Pred proposed a behavioural matrix (Figure 4.5), whose axes are quantity and quality of available information and ability to use that information; completely informed rational decision-makers occupy the bottom right-hand corner. Because of the nature and importance of information flows, decision-makers' positions on the first axis partly depend on their spatial locations; positions on the second reflect aspiration levels, experience and the norms of any groups to which the individuals belong (many of which are spatially bounded). Different people in the same matrix position could vary in their decisions, therefore, if they acted on different information and in the pursuit of different goals, even if their stores of knowledge and ability to

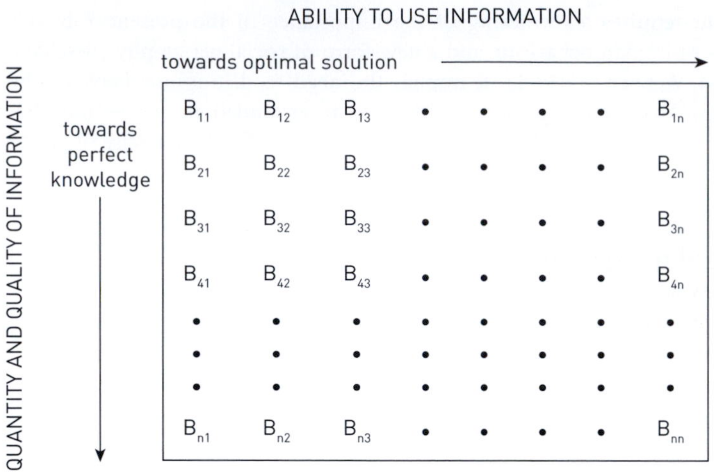

ABILITY TO USE INFORMATION

Figure 4.5 The behavioural matrix

Source: Pred, 1967, p. 25.

use them were volumetrically commensurate. Similarly, people occupying different positions could make the same decision, based on different information and/or different ways of using it.

Individuals do not stay at the same position in the matrix; as they learn and react, so their positions change. Pred's (1969) second volume introduced a dynamic element, by shifting individuals through the matrix; as they make decisions, they change the environment in which they and others operate. As people learn, they may both acquire more and better information, and become more skilled in its use; they shift towards the bottom right-hand corner of the matrix, some of them in advance of others who benefit from the decision-leaders' experience. The unsuccessful are gradually eliminated, so that a concentration of 'good' decision-makers close to the optimum position evolves, although new entrants to the matrix will probably not be located there. Changes in the external environment (such as market prices for farmers or the closure of some outlets for shoppers) generate parametric shocks, however, which result in decision-makers becoming less informed and less certain; as a consequence, they are shifted back towards the upper left-hand corner and another learning cycle begins. As long as parametric shocks occur relatively frequently, then optimal location patterns will never emerge, except perhaps by chance; the environment will have changed before all people reach the bottom right-hand corner of the matrix.

Pred (1969) presented the behavioural matrix as a 'gross first approximation' (p. 91), arguing that any theory is better than none (p. 139), even if the model itself is literally untestable (p. 141; see also Pred and Kibel, 1970). Harvey (1969b) treated it to a scathing review, however, calling the two dimensions vaguely defined, ambiguous, unoperational and an oversimplification of the complex nature of the factors influencing behaviour. Indeed, Harvey (1969c) was generally sceptical about the potential of any behavioural location theory, a view shared by Olsson (1969) who pointed out the difficulties of studying processes and demonstrated that much behavioural geography involved only inferring processes from aggregated data on individual behaviour.

Others argued that such inference could be very strong, however. Rushton (1969), for example, accepted that any one pattern of behaviour – what he termed behaviour in space – was largely a function of the spatial structure within which it occurred (the choice of shopping centre to patronise, for example), but general rules of spatial behaviour could be deduced from examining the types of preferences displayed within particular patterns: 'To say that these preferences do not exist independently of the environment where the decision is made is to argue that environments could exist about which the person would be unable to reach a decision' (p. 393). Thus a distance-decay pattern reflects the details of the environment in which it is observed, and its production involves decision-making based on certain rules, which Rushton claimed his analytical procedure could isolate. Its validity was queried, however (Pirie, 1976), and, like Pred's behavioural matrix, it did not attract many disciples. (Rushton's arguments suggest activity in a 'taken-for-granted world': see p. 185, this book.) Few others took up the explicit notion of the behavioural matrix, but many of the ideas that it was founded on were explicit in other studies, such as those of industrial geographers who studied the spread of 'best practice' methods.

Harvey (1969c) suggested two alternatives to behavioural location theory – further development of normative theory and the construction of a stochastic location theory; he claimed that both offered more immediate pay-offs for understanding spatial patterns than behavioural theory, because of the latter's conceptual and measurement problems. A stochastic approach was also favoured by Leslie Curry (1967), who argued that large-scale patterns are the outcomes of small-scale indeterminacy; individual choices may be random within certain constraints, but when very many of them are aggregated they may display considerable order. Similarly, Webber (1972) attempted to model locational decision-making processes in states of uncertainty using normative approaches, concluding that 'uncertainty increases agglomeration economies' (p. 279), thereby leading to greater concentration of economic activities and people into cities than predicted by models based on 'economic men' (see also H. Richardson, 1973; Johnston, 1976a; Scott, 1988). Game theory is a mathematical procedure developed to handle decision-making in uncertainty that has received little attention from human geographers (Gould, 1963); the most ambitious attempt to use it, which had very little impact, is in Isard et al. (1969).

For some critics, therefore, micro-analysis at the level of the individual is one or more of impossible, unnecessary or misleading; macro-analysis, and the morphological laws that it produces, provides sufficient insight to the behaviour that generates aggregate patterns. Both are needed according to Mary Watson (1978), however; macro-analysis provides an initial overview, posing questions which can only be answered by behavioural study. (See also Andrew Sayer, 1984, 1992a, on extensive and intensive research; p. 220, this book.) The behavioural approach did not stimulate a major revolution away from the spatial-science focus within human geography, therefore, but rather generated an extension to it – which over time became increasingly the norm as the assumptions underlying the original models were rejected as unrealistic. Whereas those normative approaches started with simplifying assumptions about human behaviour and deduced what spatial patterns follow from the axioms, behaviourists employed inductive procedures to identify the rules of behaviour, using these to predict (and therefore 'explain') spatial patterns (Gale and Golledge, 1982). Their approach involved a sequence of interrelated investigations. An individual is faced with a decision, with either a direct spatial input or spatial consequences. To make it, criteria are established and information is collected to be evaluated against the criteria. That may lead to a decision; alternatively, no satisfactory outcome could be identified so the individual may then alter the criteria, collect more information or both. (This pragmatic approach to problem-solving is similar to the procedures of the scientific method outlined in Figure 3.1.)

Whereas much of the initial work done in the spatial-science mould could be conducted using published data sources (such as censuses) or relatively small field-collection exercises, the behavioural approach generally requires specific data collection from the individual decision-makers. The need for social surveys of various kinds furthered the growing links between geographers and sociologists, psychologists and, to a lesser extent, political scientists, and led to an expansion in the data-handling procedures necessary for the training of geographers. One of the problems in their use is that many of the topics studied in human geography involve very large numbers of individual decisions – as in migrations, journeys to work and to shop, voting decisions, and so on. Very large sample surveys may be necessary to produce valid generalisations about such behaviour, especially across a range of places, but resource limitations have meant concentration on both small selections and only limited segments of the full behavioural sequence.

The behavioural sequence has also been applied in branches of human geography that deal with topics involving relatively few decision-makers. In the study of diffusion, for example, Larry Brown (1975) attempted to divert attention away from overall patterns of spatial spread and the reasons for adoption (or not) to the decision-making which brings certain innovations to places; most of these innovations involve selling products, which he terms 'consumer innovations' (see also Brown, 1981). Similarly, a number of industrial geographers have moved away from investigations of aggregate patterns, which could be compared to those predicted by application of neoclassical economic analysis (D. Smith, 1971), to the study of decision-making behaviour within firms (e.g. Hamilton, 1974; Carr, 1983; Hayter and Watts, 1983; Schoenberger, 1993). Not all of this work involved large datasets whose analysis called for sophisticated quantitative analysis, but the overall approach was similar – seeking to establish general patterns of behaviour as revealed by available data, as in Dicken's (2003) classic work on globalisation.

The analytical procedures employed within behavioural geography became increasingly sophisticated during the 1980s, paralleling developments in statistical procedures for analysing categorical data. Most of the research involved analysing behaviour within the framework of the spatial-science approach and its general positivist orientation. The goal was to explain, using mathematical modelling and statistical analysis, variations in aspects of behaviour (choice of travel mode for the journey to work, for example) in terms of variations in a number of independent variables (such as the characteristics of the decision-makers and the milieux in which their decisions were made: see Kitchin, 1996). Data for such studies are usually categorical in form, involving classifications (which travel mode was used, for example, or the gender of the person concerned) rather than variables measured on interval or ratio scales. Neil Wrigley (1985) provided a major overview of the relevant statistical procedures for analysing such data; Richard Davies and Andrew Pickles (1985) explored the important issue of inferring secular trends from cross-sectional data, while others promoted the analysis of longitudinal data comprising repeated interviews with the same individuals (Dale, 1993). Methods for quantifying attitudes and other aspects of human characteristics and behaviour have been explored, involving both increasingly sophisticated survey instruments for investigating how people cognise and learn about their spatial environments (Golledge and Timmermans, 1990; Timmermans et al., 2001) and technical procedures for representing those cognitions quantitatively (see Aitken et al., 1989).

Many of the choice models underpinning work in behavioural geography are based on theories of utility maximisation. This approach was heralded by Cadwallader (1975), exemplified in fuller detail by Neil Wrigley and Paul Longley (1984), and became the basis for a burgeoning literature (as displayed in great detail in Golledge and Timmermans, 1988, and reviewed in Timmermans and Golledge, 1990; Timmermans et al., 2001). Such theories propose that decision-makers select from within the choice sets presented to (and/or perceived by) them, according to the utility which they allocate to each of the alternatives evaluated; as Timmermans and Golledge (1990, p. 312) express it, all 'are based on (variants of) a conceptual model that explicitly relates choice behaviour to the environment through consideration of perceptions, preference formation and decision-making' (utility being a measure of preference). A full analysis of discrete choices thus involves knowing:

1   the available choice set;
2   the elements of the choice set considered by each individual;
3   the criteria on which each member of the choice set was evaluated by individual decision-makers; and
4   the relative importance they attached to each criterion.

James Bird (1989), following Desbarats (1983), illustrated this (Figure 4.6). The process starts with a listing of all possible opportunities. Some are discarded to form an 'objective choice set'; evaluation of its elements results in some being discarded as unacceptable, producing an 'effective

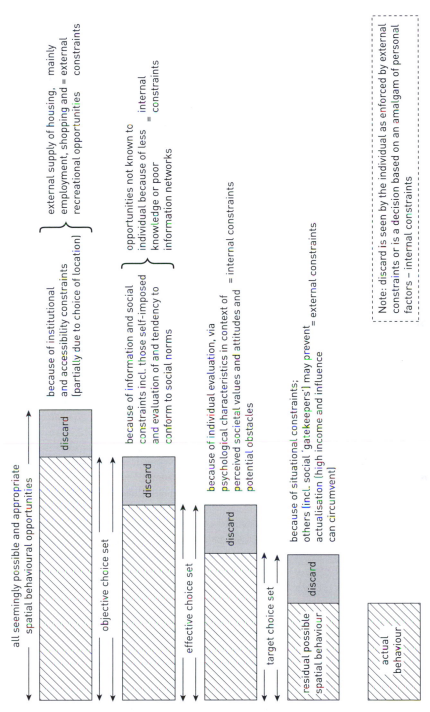

all seemingly possible and appropriate spatial behavioural opportunities

because of institutional and accessibility constraints (partially due to choice of location)

external supply of housing, mainly employment, shopping and = external recreational opportunities constraints

objective choice set

because of information and social constraints incl. those self-imposed and evaluation of and tendency to conform to social norms

opportunities not known to individual because of less = internal knowledge or poor constraints information networks

effective choice set

because of individual evaluation, via psychological characteristics in context of = internal constraints perceived societal values and attitudes and potential obstacles

target choice set

because of situational constraints; others (incl. social 'gatekeepers') may prevent = external constraints actualisation (high income and influence can circumvent)

residual possible spatial behaviour

discard

discard

discard

discard

Note: discard is seen by the individual as enforced by external constraints or is a decision based on an amalgam of personal factors – internal constraints

actual behaviour

Figure 4.6 Choice sets in the spatial decision-making process

Source: Bird, 1989, p. 145, after Desbarats.

choice set'; and further evaluation leads to its reduction to a 'target choice set' from which the final selection is made. A potential house-buyer in a town 'discards' many of the available homes through ignorance, for example, and then whittles down those perceived as viable purchases to a final list from which the selection is made (with, in some cases, a recursive loop to identify further viable buys if none in the final choice set either proves acceptable or can be bought).

Other work within this general framework combines categorical data with other forms, such as indicators of the local context within which decision-makers operate. Studies of voting behaviour, for example, have investigated the decision of which party to support according to both the characteristics of the individuals concerned and various perceived and objective indicators of their local milieux (see, for example, Pattie and Johnston, 1995; Fieldhouse et al., 1996; MacAllister et al., 2001; Johnston and Pattie, 2006).

Behavioural geography, understanding and explanation

Behavioural geography emerged in the slipstream of the wider new geographies that had been opened up with the arrival of commitments to spatial science and positivism. Some of it seeks to account for spatial patterns by establishing generalisations about people–environment interrelationships, which may then be used to stimulate change through environmental planning activities that 'modify the stimuli which affect the spatial behavior of ourselves and others' (Porteous, 1977, p. 12). Initially, it was hoped, a 'powerful new theory' (Cox and Golledge, 1969) would emerge and a decade later Golledge (1981a, pp. 1338–9) claimed that substantial advances had already been made through studying what he called process variables – 'individual preferences, opinions, attitudes, cognitions, cognitive maps, perceptions, and so on'.

Apart from technical issues, the positivist foundation of most behavioural geography has been subject to considerable debate, as part of a wider consideration of the relevance of that philosophy to human geography. Golledge made major contributions, arguing that positivism is a constraining philosophy, and that (Couclelis and Golledge, 1983, pp. 333–4) 'behavioural geography has evolved by gradually shedding the tenets of the philosophy out of which it was born, as these have come to be seen as barriers to further progress and understanding'. He prefers 'analytical research' to describe his approach, which is 'not so much a philosophy as a distinct mode of discourse, a space of possibilities for theoretical languages that meet the criteria of clarity, coherence, intersubjective validity, and a concern never to lose sight of experience' (p. 334). Nevertheless, he retains one of the central tenets of the positivist programme, as illustrated by references to 'a search for generalizations' (Golledge and Couclelis, 1984, p. 181), 'significant generalizations ... about particular sub-groups' (Golledge, 1980, p. 16) and the need to shift research away from 'the more narrow perspective of behaviour in space' towards 'a general understanding of spatial behaviour' (Golledge and Rushton, 1984, p. 30; see also p. 137, this book). In his AAG presidential address, Golledge (2002, p. 12) argued that geographers have a 'common set of primitives and concepts' which define the discipline, that from these they have developed a 'huge array of empirically verified factual data, spatial theories and models, innovative methods of spatial analysis, unique modes of representation, and practical usefulness for decision-making and policy formulation'. Those primitives include 'place-specific identity, location, magnitude, time, boundary and distance' (p. 7), with which geographical information is both informally and formally acquired: behavioural geographers study the informal processes through their own formal procedures. (On Golledge's work, see Golledge, 2007; Loomis et al., 2010; Gold, 2010; Stimson, 2012.)

Golledge and Stimson claimed that analytical behavioural geography is based on 'what is truly positive in positivist thought' (Golledge and Stimson, 1987, p. 9). They rejected the 'classic positivist separation of value and fact' and argued for a positivist position which can 'interpret values and beliefs in a scientific manner'. This they term a transactional or interactionist position, characterised by:

(a) the importance of logical and mathematical thinking; (b) the need for public verifiability of results; (c) the search for generalization; (d) the emphasis on analytic languages for researching and expressing knowledge structures; and (e) the importance of hypotheses testing and the importance of selecting the most appropriate bases for generalization or theorizing.

The goal is to develop quantitatively verified theory (Macmillan, 1989b).

This continued emphasis on key elements of the positivist philosophy links to a second, much less substantial, area of work. Pipkin (1981) noted that much effort had been expended on identifying behaviour patterns but little on exploring the underpinning mental constructs. Understanding why people behave as they do has not been advanced very far, therefore; as Greenberg (1984, p. 193) expressed it, 'For the most part, the intention of behavioural-perceptual geographers has not been to explain the spatial organisation of society, but to illuminate the spatial behavior of individuals.'

Artificial Intelligence (AI) was presented as offering much to the study of cognitive processes (T. R. Smith, 1984), using computer-modelling procedures to represent the decision-making processes and thereby gaining insights (by analogy) to the nature of the human brain. Coulclelis (1986a, p. 2) indicates that this involves the 'human computer' metaphor according to which, 'cognitive functions such as problem-solving, pattern-recognition, decision-making, learning and natural language understanding are investigated by means of computer programs that purport to replicate the corresponding mental processes'. The ability to replicate processes by predicting their outcomes (as in Smith *et al.*, 1984) cannot be equated with understanding them, however. Coulclelis (1986a, p. 111) concluded a detailed discussion of AI's utility in the study of human behaviour by noting that 'reliable predictions can be made about intentional systems even by theories which . . . are totally vacuous psychologically'. Whether one wants to know how or why people behave is crucial; AI can reproduce the former it is claimed, but can it assist in the latter? Nystuen (1984, p. 358) doubted even the former claim, however, arguing that he could:

> see little potential in AI methods available today in addressing problems considered important in the spatial decision-making literature, such as the decision to migrate. . . . These processes would require elaborate models of spatial cognition and tradeoff behavior whereas even the simplest model of a child's wayfinding is complex and contains major unresolved methodological problems.

If the claim were sustained, however, it would only provide a constructed explanation (see the discussion of Lukermann's arguments on p. 82, this book), but:

> I am struck by the fact that careful empirical analysis by biologists describing the anatomy and behavior of bats did not lead to the discovery of how bats navigated in the dark. The explanation was beyond imagination until a purely human system (radar, followed by sonar) was invented and by analogy applied to the behavior and anatomy of the bat. Then all the things fell into place.
> (Nystuen, 1984, p. 359)

In other words, if you can reproduce a process then you may well gain some appreciation of it or, as Nystuen put it:

> If a constructed computer program can repeatedly resolve an issue under varying spatial conditions in a way that is considered useful to geographers, then one might say that we understand the issue. This is a sufficient claim: the problem has been solved by whatever logic or capacities the program has at hand. There is no need to claim that this is necessarily the way human spatial decision-making works.
> (p. 359)

Others reject such an instrumentalist argument, however (e.g. Gregory, 1980: see the discussion of automated geography, p. 160, this book.) Openshaw (1995), on the other hand, argued for the power of AI software for understanding and simulating human behaviour (see also Openshaw and Openshaw, 1997).

Thrift argued that by the end of the 1970s:

> The halcyon days of behavioural geography are long gone. With them have passed the days when behavioural geographers made inflated claims for the explanatory power of their subject area. But the subject area still has its place in human geography.
>
> (1981, p. 359)

He recognised two criticisms of behavioural work: those which perceived it presenting the individual decision-maker as little more than an automaton responding to stimuli in a programmed way; and those which claimed that it ignored the characteristics of society as a whole which are greater than the sum of its individual parts. He argued that behavioural geography might be presented as 'half-blind': 'But to say that behavioural geography is therefore half-blind is not to say that it can see nothing at all. Its explanations may be limited. That does not mean that they are therefore non-existent' (p. 359).

While a substantial volume of work continued during the 1980s (summarised in reviews such as Golledge and Rushton, 1984; Golledge and Timmermans, 1990; Timmermans and Golledge, 1990; Timmermans et al., 2001), it could also be claimed that behavioural geography by then had become (or at least was perceived by most in the discipline) a specialist interest. It was increasingly isolated within human geography (Gold, 1992), because of both the theoretical position and analytical sophistication achieved by its leading practitioners and what Cloke et al. (1991, p. 67) call its 'partial treatment of people' (which they contrast with their 'complete neglect in spatial science'). Many human geographers have accepted the need to collect and analyse individual data through questionnaires and similar methods, to portray how people learn about, represent and behave in space. Others have rejected that approach entirely, because it cannot provide telling insights into individuals – thus, according to Cloke et al., making it 'somewhat limiting', 'dehumanising', 'pallid' and 'horribly reductionist'. Behavioural geography became one of many components of an increasingly diverse discipline: it made a considerable general impact upon geographical practice in its early years and has since been advanced by a small group of specialists somewhat apart from the mainstream (as discussed in Gold, 1992).

Against those claims are arguments – closely aligned to Golledge's distinction between 'positivist law-seeking' and 'analytical research' – which see the methods and approaches developed within behavioural geography as necessary components of understanding central areas of geography's research agenda (Johnston, 2000a). Many millions of people migrate every year; many millions vote in national elections; very many millions commute daily, and so forth. Although full understanding of these events requires appreciation of why individuals acted as they did, that is virtually impossible because of the scale of the task. General patterns of behaviour can only be uncovered through combinations of:

1　analyses of the aggregate situation – the number of migrations between given origins and destinations, for example;
2　analyses of general relationships within such situations – the proportion of migrants in different age and educational groups, for example; and
3　analyses of the reasons for individual decisions.

The first two can only be achieved through work – usually quantitative – on large datasets, which indicate the size and nature of the issue, give an overall picture of general patterns that call for further investigation, and identify some of the relationships that might assist in developing explanations for the observed patterns (Hepple, 2008). Such explanations need not, in the strict positivist mould, imply general laws of cause and effect; they may be simply accounts of a particular situation which can be assessed against theoretical models of the behaviour patterns involved and generate ideas to be tested in new situations. In mass societies, behavioural geography provides insights to mass behaviour. Further, some of the methods developed for studying behaviour have been adapted for particular purposes, such as wayfinding by visually impaired individuals, and applied in a number of contexts (Golledge, 1993; Golledge's work – he was himself visually impaired – was strongly critiqued for its interpretations of disablement; see Imrie, 1996; Gleeson, 1996; and Golledge, 1996).

The continued vitality of the group of specialists working on cognitive-behavioural topics was illustrated by Golledge and Stimson's (1987) text on *Analytical Behavioural Geography* and its 1997 successor *Spatial Behavior: A geographic perspective*. They introduced the approach (Golledge and Stimson, 1987, p. 1), and its origins in the 1960s, as a recognition that:

> in order to exist in and to comprehend any given environment, people had to learn to organise critical subsets of information from the mass of experiences open to them. They sense, store, record, organise and use bits of information for the ultimate purpose of coping with the everyday task of living. In doing this, they create knowledge structures based on information selected from the mass of 'to whom it may concern' messages emanating from the world in which we live. Different elements from these various environments are given different meanings and have different values attached to them. It was the explicit recognition of the relationship between cognition, environment and behaviour that initially helped to develop behavioural research in geography.

They illustrated how the work developed, and structured approaches through a diagram which places individual decision-makers at the interface between environment and behaviour, learning and acting within the environment and changing it as they act (Figure 4.7). The organisation of their book, covering both the methods used by behavioural geographers and their substantive achievements in various topical areas, provides a clear view of a vital subdiscipline, as does the size

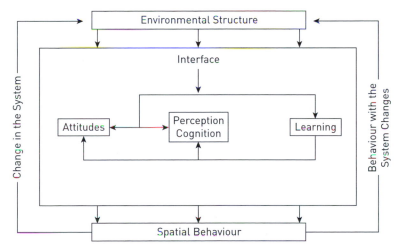

Figure 4.7 The society–environment interface

Source: Golledge and Stimson, 1987, p. 11.

of their bibliography, along with the large number of other pieces listed in the review papers on which parts of this chapter have drawn.

A much-revised second edition extended the coverage of material that illustrated not only the continued volume of geographic work but also the strong links that were being forged with psychology. Behavioural geography was defined as:

> characterized by a concern both for scientific rigor or experimental realism and for phenomenological and anthropomorphic understanding of human-environment systems and relationships.
>
> (Golledge and Stimson, 1997, p. 29, original italics)

Its concerns covered 'philosophical' topics such as experimental design; data collection procedures; the search for validity and reliability; the selection of appropriate analytical methods; the choice of a 'modern, analytical, epistemological basis'; innovative models of behaviour and environments; recognition of socially constraining dimensions; and differing viewpoints based on 'bottom-up' or 'top-down' approaches. Six main subject areas were identified – cognitive mapping and spatial behaviour; attitudes, utility, choice, preference, search and learning; consumer behaviour; location decision-making; wayfinding, mode choice and travel behaviour; and mobility and migration – on the grounds that these 'lend themselves more to the process of measurement, model building, abstraction, generalization, and theory building than do other areas of human geography' (p. 30). The result – as stressed by Golledge (2002, p. 126) – was 'an extra explanatory dimension to more traditional geographic research' examining cognitive processes in a wide range of settings. The use of new tools of functional magnetic resonance imaging (fMRI) to measure cognitive responses within sections of the brain and their relationships to spatial reasoning occasioned by use of maps has also offered fresh directions. Work in this field has indicated how maps are different from other geometric objects, so that:

> even with a relatively simple map, the map influence [as they are orientated and perceived] is significant enough to elicit distinctly different brain activation. We do not offer this study as one of methodology, although we recognize that fMRI is far outside the methodological norm in geography. Instead, we offer this study as one that empirically supports the underlying premise in many definitions and discussions of maps – that they are tools designed for a specific use that distinguishes them from other types of graphics.
>
> (Lobben et al., 2014, p. 110)

Yet one of the intriguing aspects of work in behavioural geography is that those who wrote about it, such as Golledge and Stimson (1997), tended to restrict it to the study of certain types of behaviour. The substantive chapters of their book deal with choice of transport mode, choice of shopping centre to patronise, migration, residential mobility within cities and the problems of specific groups within the population. Other forms of decision-making, such as those related to the location of production and service facilities, were apparently outside the remit of behavioural geographers. And yet a great amount of work was done on such decision-making processes and their outcomes, which involves collecting data from the individuals making the location decisions at a great range of scales from the global, through the national and regional to the local – albeit in many cases within theoretical frameworks that have few links to the models that stimulated much of human geography's urban and quantitative revolution. The theoretical frameworks within which such authors deal (as with Dicken's, 2003, work on globalisation, and Schoenberger's, 1997, work on corporate behaviour) came from political economy and concepts of explanation that are discussed in the next chapter. (The journal *Regional Studies* carries a great deal of this work.) But many of the data collection methods are behavioural, even if there is little complex statistical analysis, and the goals are, as in so much else discussed in this chapter, to identify and account for general patterns of behaviour in space. According to Scott (2000, p. 24), this separation was because behavioural geography 'became more or less the preserve of a few highly specialized researchers who turned increasingly to experimental psychology for their inspiration'. Indeed, as long ago as 1976, Cullen (1976, p. 400) had argued that although behavioural geography should have a great deal to offer for those interested in macro-patterns, it was failing to do so, and was instead an 'intellectual backwater', 'every bit as mechanistic as the neoclassical economic geography it was supposed to replace' (Scott, 2000, p. 24). As with so much in contemporary geography, it is difficult to draw boundaries between different sections of the subject.

Spatial (or geo-)statistics

The application of various approaches to 'geography as spatial science' over the last half-century has involved much attention being given to quantitative data analysis. Initially, most of the researchers involved followed Garrison (1956a) in accepting that statistical procedures developed for other fields of enquiry could be adopted for geographical investigations without difficulty. Some suggested necessary modifications – as in spatial sampling (Berry and Baker, 1968) – but, despite Robinson's (1956; see p. 67, this book) early work, it was generally assumed that no technical problems prevented the application of standard procedures to spatial datasets. A number of textbooks appeared, especially in the late 1960s, which made little reference to any peculiarities of geographical data; they differed from the texts produced by other social scientists only in the nature of their examples. And even in the 1980s, few contributors to a book on *Recent Developments in Spatial Data Analysis* (Bahrenberg *et al.*, 1984) discussed spatial data per se; most of their work referred to the use of standard statistical procedures in geographical applications. (This was also the case in a major review of quantitative geography in the UK: Neil Wrigley and Robert Bennett, 1981, and an earlier special issue of *The Statistician* – Peter Taylor and John Goddard, 1974 – in which geographers introduced statisticians to their quantitative applications, with only one paper – David Unwin and Les Hepple, 1974 – dealing with what Taylor and Goddard, p. 151, refer to as 'the peculiar nature of geographical data, particularly the lack of spatial independence'.) Others simply adapted standard procedures to geographical problems, as with Chorley and Haggett's (1965a) application of polynomial regression to the analysis of spatial patterns – what they termed 'trend surface analysis'.

Indeed, relatively few texts have focused on techniques for spatial (or map) analysis, although some general texts on statistical analysis for geographers (such as Taylor, 1977) have included such material. David Unwin (1981), for example, dealt with point, line, area and surface patterns, along with map comparison, providing a clear emphasis on spatial statistical analysis (see also O'Sullivan and Unwin, 2003); Peter Lewis (1977) covered point and line patterns for a more advanced audience; Graham Upton (a statistician) combined with Bernard Fingleton (a geographer) in an advanced two-volume treatment of aspects of spatial data analysis (Upton and Fingleton, 1985, 1989); Gatrell, also in collaboration with a statistician, published a text with worked examples on a CD-ROM (Bailey and Gatrell, 1995); and Fotheringham *et al.* (2000) brought out a book as 'a statement on the vitality of modern quantitative geography . . . where the [recent] emphasis has been on the development of techniques explicitly for *spatial* data analysis' (pp. xi–xii, original italics). Similarly, relatively few followed the paths set out in two editions of a book on mathematical models (Wilson and Kirkby, 1975; Wilson and Bennett, 1985). To some extent, spatial statistical analysis – as against statistical analysis in geography – was a minority activity.

In this context, there has been infiltration of human geography by technical developments elsewhere. One such was work on classification (a procedure common to the life sciences) and its links to regionalisation. Berry used these widely in his ideas on regional analysis (Berry, 1964a: see

p. 88, this book) in studies of, for example, urban residential areas (or factorial ecology: Berry, 1971) and others have deployed variants on them for the task of defining electoral areas such as parliamentary constituencies and congressional districts (Johnston, 2003d): methods of classification have been substantially advanced in recent years using high-speed computers for the large dataset manipulations involved in the analysis of remotely sensed imagery (Mather, 1999; Tso and Mather, 2001). Elsewhere, Wrigley (1973) pointed out the problems involved in the use of percentage and proportional data in linear regression and related methods and in a series of publications (Wrigley, 1976, 1985) introduced the use of logit and other forms of categorical data modelling. These are now widely used in work with such data – for example, voting studies in which the values are binary (either yes, the individual voted for a given party, or no, he/she did not: Pattie and Johnston, 2000; see also Lloyd et al., 2012).

Kelvyn Jones (1991) introduced methods for exploring varying relationships from work in education. It may be hypothesised, for example, that a student's performance reflects not only her/his innate ability but also home background, school type, the quality of teaching provided, and so forth. Thus, for example, a student from a supportive home background may perform better in a set of tests than one with the same innate ability but from a less supportive background, but the better the teaching provided to the two, the smaller the gap between their performance. This suggests an interaction effect between the independent variables – when operating together they have a greater impact than when operating separately. Such interaction effects can be modelled within standard general linear models (e.g. Andrew Russell, 1997) but Jones has demonstrated that multi-level models developed by educational researchers are more efficient ways of estimating those interactions in complex datasets and has applied them to a range of geographical problems such as differing voting patterns across regions (Jones et al., 1992) and disease patterns (Jones and Duncan, 1996, 1998); they also have the advantage that they explicitly take into account and correct for the influence of any spatial autocorrelation (see below) that may be inherent in data, such as those collected using cluster sampling which have autocorrelation effects in-built.

Various aspects of spatial interaction have continued to be a focus of formal modelling work, as in major studies of the spatial diffusion of diseases (much of it summarised in Haggett, 2000). Bennett and Haining (1985; Bennett et al., 1985) placed such work at the forefront of methodological developments in the discipline. Much of it was summarised by Wilson and Bennett's (1985) volume on *Mathematical Methods in Human Geography and Planning* in a *Handbook of Applicable Mathematics* series.

Others sought to ensure proper use of the standard statistical procedures (e.g. Jones, 1984; Wrigley, 1984), with the use of statistical significance tests forming the basis of further debate (Summerfield, 1983; Hay, 1985b). Better ways of testing hypotheses were also expounded, not only with regard to the use of categorical (or nominal) data (e.g. Fingleton, 1984; Wrigley, 1984), but also in terms of pattern analysis (Upton and Fingleton, 1985) and representing relationships through structural models (e.g. Cadwallader, 1986). Much of this work has been criticised for its reliance on methods that impose structures on data rather than 'letting the data speak for themselves' (Gould, 1981a); the mathematics of q-analysis was proposed to meet the latter criterion (Gould, 1980; Beaumont and Gatrell, 1982; Gatrell, 1983; see also the essays edited by Macgill, 1983) but was not taken up. More interest was shown in exploratory data analysis – the inductive interrogation of datasets unconstrained by prior hypotheses and assumptions such as those underpinning the general linear model (Cox and Jones, 1981; Wrigley, 1983). By assuming less, it was argued, you may discover more.

Increasingly, however, the assumption that all geographers needed was an appreciation of standard procedures was considered invalid. Instead, it was contended that geographical data of various forms required specific procedures because of their particular nature. Geographers thus became involved – in some cases in collaboration with statisticians – in the development of those procedures, as illustrated in the next subsections. Interestingly, some of the earliest work on these issues was done by non-geographers (e.g. Duncan et al., 1961). There have been substantial advances in the development and adaptation of sophisticated modelling and analytical techniques: some, however, thought these advances, although 'scholarship of the highest order', may reflect 'a misdirection of effort' (King, 1979b, p. 157) because they had not illuminated spatial organisation as much as they might. New techniques continue to be advanced, assisted by massive technological advances in data handling. A 'paradigm' in the sense of a community with a clear world view was established within geography in the 1950s and 1960s, and has remained strong and vibrant within the discipline ever since. Its nature has changed considerably since then, with the formal models of spatial structures being largely forgotten and replaced by new approaches. The methods applied have changed too, with greater technological power and sophistication.

Spatial autocorrelation

During the late 1960s a group of researchers at the University of Bristol began to question the widespread assumption about the relevance of standard statistical procedures for analyses of spatial data. Statisticians had for long recognised difficulties in applying the general linear model – notably, in its regression and correlation form – to the time-series data used in economic analysis and forecasting because of what was termed autocorrelation. One of the model's assumptions is that all observations are conditionally independent of each other; the magnitude of one reading on a variable should in no way influence that of any other. Such autocorrelation clearly exists in most time series; the value of the retail price index at one date, for example, strongly influences its value at nearby later dates, and is itself influenced by preceding values. (An early example of this being brought to geographers' attention was in the discussion in Chisholm et al., 1971, p. 465.) Because of such interdependence between adjacent observations, conventional regression methods could not be used; autocorrelation leads to inefficient regression coefficients because of underestimated standard errors and casts doubt on the validity of any forecasts using those models, as well as overestimating the goodness-of-fit measure (usually the correlation coefficient). The estimated impact of the independent variable on the dependent was reliable, but because the standard error of that estimate was overstated with autocorrelated data the strength of the relationship appears statistically less significant than would have been the case without autocorrelation.

The Bristol group (led by Peter Haggett, Andrew Cliff and Keith Ord, who is a statistician) argued not only that this autocorrelation problem applied to spatial datasets too, but also that it was much less tractable than with temporal series. Time proceeds in one direction only, but space is two-dimensional, and the independence requirement can be violated in all directions around a single point, and not necessarily to the same extent. Spatial autocorrelation can therefore involve all neighbours influencing all others with regard to the values of a particular variable, as recognised by some statisticians (e.g. Geary, 1954) and hinted at by geographers such as Dacey (1968), whose work stimulated the Bristol investigations. (Andrew Cliff worked as a graduate student with Dacey at Northwestern University in the early 1960s.) Considerable effort was expended on this problem over a number of years (Cliff and Ord, 1973, 1981), covering a wide range of spatial applications, including the well-known gravity model (Curry, 1972; Johnston, 1976b; Sheppard, 1979; Fotheringham, 1981; Tiefelsdorf, 2003).

The spatial autocorrelation problem severely constrains the application of conventional statistical procedures in geographical analyses if the statistical significance of regression estimates is a focus of attention. Several people thus argued that application of regression methods, and others based on the general linear model such as principal components and factor analyses, was invalid with spatial datasets (Haining, 1980, p. 24). As Harvey put it:

> The choice of the product-moment correlation coefficient for regionalization problems appears singularly inappropriate, since one of the technical requirements of this statistic is independence

in the observations. Since the aim of such regionalization is to produce contiguous regions which are internally relatively homogeneous, it seems almost certain that this condition of independence in the observations will be violated.

(1969a, p. 347)

Gould (1970a) noted that spatial autocorrelation reflects the order that geographers were seeking to establish with their laws and theories, and it was somewhat paradoxical that its presence prevented its identification through accepted technical procedures. As Tobler (1970, p. 236) expressed it more prosaically, the 'first law of geography' is that 'everything is related to everything else, but near things are more related than distant things' – in other words, spatial autocorrelation is the core of geography.

On realising the force of the case, however, some accepted that conventional statistical procedures could not be applied in their work (e.g. Berry, 1973a) and the Bristol group omitted them from their rewrite of Haggett's major text (Haggett *et al.*, 1977). The second part of the book – 'Locational models' – was much expanded and rewritten. In the first edition (Haggett, 1965c), the chapter on 'Testing' included a section on 'Testing via statistical methods' which included standard linear regression analysis. The comparable chapter in the second edition omitted that material, however, on the grounds not only that:

Since the first edition of this book appeared, the use of conventional statistical methods and tests of significance . . . has become commonplace in human geography. It is not the purpose of this chapter to outline the procedures for carrying out such analyses.

(Haggett *et al.*, 1977, p. 329)

but also that:

the aim of this chapter is to examine those properties of geographical data which make their analysis using statistical methods of the sort cited more difficult than might at first appear.

For them, the lesson to be learned from appreciation of the problems of spatial dependence and autocorrelation in geographical data was that 'If geographers continue to apply the usual forms of many of the basic statistical models to spatially autocorrelated data, then a very severe risk is run of reaching misleading conclusions' (Haggett *et al.*, 1977, p. 336). Others have carried the argument forward, developing methods – termed spatial econometrics by Anselin (1988: Anselin studied regional science with Isard – see p. 24, this book) – for modelling with spatially interdependent datasets. (Such software is now generally available through the World Wide Web: Anselin, 2012; Anselin and Rey, 2012.)

Many continued to apply the methods, however, either in ignorance of the autocorrelation case or on the grounds that the biased-coefficients problem refers only to the use of the general linear model in forecasting and prediction (and not in any situation that involves statistical significance testing), and does not affect use of the procedures for either description (see Johnston, 1978b) or certain types of ecological analysis (Johnston, 1982a, 1984b).

Spatial forecasting

One of the Bristol group's substantive interests was in spatial forecasting, involving the development of procedures for estimating how trends – in the spread of a disease, for example, and of prices and unemployment – would proceed through time and over space (Haggett, 1973; Cliff *et al.*, 1975). They organised a major symposium on this (Chisholm *et al.*, 1971) and stimulated a considerable volume of work, much of it technical, on the problems of identifying and forecasting spatio-temporal trends (e.g. Bennett, 1978b). Hägerstrand's work on diffusion patterns provided the foundation for much of this research, which focused on the patterns of spread rather than on their generating processes, and spawned a literature on technical aspects of forecasting (e.g. Bennett, 1979) and deducing 'spatial processes' from mapped patterns (Haining, 1981, 1990). Haggett (1994, p. 6) continued to maintain that 'prediction plays an essential part in the building of models of geographical systems', although he increasingly illustrated this with reference only to epidemiology. (Haggett was initially introduced to ideas about diffusion and spatial spread by reading the English translation of Lösch when he was a postgraduate in Cambridge in 1954. He was in particular stimulated by some of the ideas in Lösch's footnotes – e.g. note 139 on p. 195, and note 38 on p. 415 which was concerned with functional regions – as well as by the figure on p. 496 which showed spatiotemporal trends in the spread of depression in Iowa, 1929–1931. Although Haggett was inspired by these ideas in the mid-19590s it was only in the later 1960s, after he moved to the University of Bristol and obtained one of the first-ever grants to geographers from the SSRC that he began a programme of research into spatiotemporal trends first in economic activity and then, after he had attended a World Health Organization seminar in Geneva, diseases.) Gregory (1994, p. 59) argues that Haggett's 'main concerns have always been descriptive-predictive', and that he has little interest in 'theoretical questions' and explanation. Haggett himself agreed with this in a published interview (Browning, 1982, p. 54); in response to a statement that 'It seems to me that you are naturally inclined towards models but not necessarily theory' he replied, 'Yes . . . I think that the broader theories of the subject are not something I can cope with easily.'

Hay's (1978) review of this work challenged what he saw as its fundamental assumption that phenomena behave coherently in both time and space. He questioned the assumed stability of interplace interrelationships over time in analyses such as Ron Martin and J. E. Oeppen's (1975) work on market-price variations, and argued instead for consideration of the value of catastrophe theory, wherein small changes in the control variable(s) can stimulate major changes in the dependent variable being studied (such major changes are the catastrophes). If catastrophes occur, then the linear extrapolations typical of the Bristol group's work have clear limitations as forecasting procedures. There were relatively few geographical applications of this relatively new area of mathematics, however (Wilson, 1976a, 1981b), which received more attention among physical geographers (Thornes, 1989a, 1989b).

The modifiable areal unit problem and ecological inference

A further problem characteristic of much geographical data is what became termed the modifiable areal unit problem (MAUP). There are many ways in which a large area can be subdivided into smaller units for data collection, reporting and analysis, and geographers showed that different divisions of the same area could result in different analytical outcomes; the relationship between old age and long-term illness may differ in its strength (and even direction) depending on which particular set of areas was used in the analysis. The importance of this issue became apparent in studies of electoral redistricting (Gudgin and Taylor, 1979), as in the reapportionment exercises undertaken after every decennial census in the USA when the goal is to produce Congressional (and other) districts as equal in their populations as possible. Each state is allocated a number of districts. Maps of proposed districts – subdivisions of the state – are created using population data for much smaller areas than the districts, so that each district is an amalgamation of many of those small areas, which should be contiguous. Analysts using computers could identify a very large number of possible ways of producing a set of districts in any state, each of which would have its own electoral impact; it could be possible, for example, for a state that has been allocated twelve districts to have a majority of its seats won by either the Democratic or the Republican Party, depending on how the map is drawn. The implications of the MAUP, therefore, are that one's findings apply only to the particular map configuration used; they may not be more generally applicable. Different maps may result in different geographical conclusions. (The original examples of this were given by Morrill's (1973, 1981) work on redistricting in Washington state, and Openshaw and Taylor's (1979) on the impact of different redistrictings in Iowa. Rossiter and Johnston, 1981, extended Gudgin and Taylor's 1979 classic work on the statistical underpinnings of the geography of redistricting to develop a computer program which identifies all of the possible outcomes; see also Cirincione et al., 2000.)

The MAUP issue was first noted by Gehlke and Biehl (1934) and further generalised by Yule and Kendall (1950). Gehlke and Biehl had data on numbers of juvenile delinquents and median household incomes in the 252 census tracts in the Cleveland metropolitan area; the correlation between those two variables was −0.502. But as the tracts were aggregated into larger units (of approximately similar size), so the correlation increased; with 175 units, it was −0.580; with 125, −0.662; with 100, −0.667; and with just 25 it was −0.763. Using data on wheat and potato yields in English counties, Yule and Kendall (1950) found even greater variation according to the number of units: using all 48 English counties, the correlation was only 0.22; with 24 groups of counties, 0.30; with 12, 0.58; with 6, 0.76; and with just 3 counties, 0.99. In these studies, just one example of each aggregation (e.g. into six groups of English counties) was employed, but in almost all cases there are many different ways in which n units can be aggregated into k regions, where k is smaller than n. Each regionalisation might produce a different correlation coefficient, as demonstrated by Openshaw and Taylor (1979). They had data on the percentage of the population aged over 60 and the percentage of the votes cast in the 1968 presidential election obtained by the Republican candidate for Iowa's 99 counties. The correlation between these two was 0.346. With 30 zones (i.e. aggregations of contiguous counties), the average correlation was 0.33, but with a standard deviation 0.11; as the number of zones decreased, the standard deviation increased. And even this figure concealed the extent of the variation: with 30 zones, for example, it was possible to get a correlation of 0.98 between the two variables at one extreme, and of −0.73 at the other. In effect, as Openshaw (1984a) later demonstrated, it is virtually possible to produce a regionalisation of the 99 Iowa counties to produce any desired OLS relationship between the two variables – both the correlation and the slope coefficient.

Openshaw (1984a, 1984b) showed how different spatial aggregations of the same dataset (counties into regions, perhaps, or census enumeration districts into social areas) can produce very different correlations between two variables as a result of the interaction of two influences – scale and aggregation; indeed, his simulations suggest that with many datasets any correlation between +1.0 and −1.0 might be obtained, so that the particular aggregation employed needs careful justification (see Wrigley, 1995). Openshaw and Taylor (1981) explored three responses to such findings:

1　it is an insoluble problem, and so can only be ignored;
2　it is a problem that can be assumed away, with the results obtained from the particular available dataset being accepted as 'the real ones'; or
3　it is a very powerful analytical device for exploring various aspects of geography and spatial variations, since alternative regionalisations can be produced – this allows the creation of frequency distributions with which one regionalisation can be compared (as with the example from Cirincione et al., 2000, cited earlier), and of optimal regionalisations for particular purposes.

Most geographers and other spatial analysts have at least implicitly adopted the first of these (and perhaps even the second) – in part because they lack the available data which would allow the third response to be adopted. (Flowerdew, 2011, concludes that in most cases MAUP makes little or no difference to the statistical results using English census data.)

Associated with this work on ecological inference and the MAUP has been the development of inductive procedures for classification and regionalisation, using AI (Openshaw and Openshaw,

1997). For example, a region-building exercise may begin with a 'representative sample' of the n building blocks being selected as the k 'cores' for the regions and then a genetic algorithm allocates the remaining (n–k) building blocks to those cores through a 'learning process' until homogeneous regions are created. The algorithm is rerun a large number of times, and the 'best' available regionalisation then selected on a predetermined criterion. Such procedures are often termed 'unsupervised' because the cores are randomly selected. An alternative – 'supervised' – procedure involves creating 'ideal type' cores as the nodes for the regions, rather than using samples of the building blocks; there must therefore be a predetermined number of regions. The cores have a given 'population mix', which may be derived either a priori or empirically, from that or another dataset. Having created those cores, the building blocks are then regionalised using neural net methods, with the nature of the cores changing during the learning process (for an example, see Mitchell et al., 1998). Openshaw et al. (1995) used such a procedure to classify Britain's residential areas. Using census data (85 variables), 145,716 census enumeration districts (EDs, population average c.500), were the building blocks for a clustering into 60 types, themselves aggregated into 17 super-clusters.

These analytical advances have been used to develop regionalisations for practical applications, as in the design of reporting (or output) areas for censuses. Early work on functional regions around cities using flow data (notably for commuting) was used to define metropolitan regions in the USA and travel-to-work areas in the UK (e.g. Berry et al., 1969; Coombes et al., 1982). At smaller scales, in the UK census from 1951 to 1991, for example, the smallest data-reporting units were the EDs – small areas used for the administration of the census collection and tabulation procedures. They were defined for administrative convenience only, and bore no necessary relationship to the underlying geography they were portraying. Many users – private and public sector as well as academic researchers – argued the desirability of these being defined so that they were relatively homogeneous on defined characteristics, within prescribed constraints, notably relating to their size (e.g. population or household minima or maxima). Using data for even smaller areas than EDs (postcode sectors), Martin et al. (2001) used one of Openshaw's classification algorithms (AZP: Openshaw and Rao, 1995) to produce a set of Output Areas (OAs) after the census data had been collected, which met certain criteria – a target population (i.e. average population) with given population minima and homogeneity on specified other variables (such as housing tenure) plus a shape constraint. The result was a geography for publishing data from the 2001 UK census with reporting units that were much more homogeneous than heretofore (Martin et al., 2001); the same procedure was used again in 2011 (Cockings et al., 2011).

The classifications of areas according to their population characteristics using large public and private datasets has not only stimulated much geographical analysis – of the geography of family names, for example (Longley et al., 2007), and what they indicate about migration patterns – but also the development of geodemographics in a wide range of marketing and other applications (Harris et al., 2005). Classifying small areas according to their residents' characteristics allows not only firms but also other organisations, such as political parties, to target likely 'customers' for their products on the assumption that similar people, living in similar areas, have similar tastes – an assumption they can test using data collected on individual buying-habits by stores. The conduct of such classifications has been the focus of much debate, especially with regard to the concept of residential segregation, especially that of ethnic groups (e.g. Poulsen et al., 2011).

Ecological fallacies and inference

Linked to this work on the MAUP was an increasing appreciation of the problem of ecological inference. Much statistical work involves the analysis of data about population aggregates taken from censuses, such as the relationships between a population's age and its health status, from which inferences about individual behaviour are then drawn; if there is a positive relationship between the average age of an area's population and the proportion of its members suffering long-term illnesses, it may be inferred that older people are more likely to suffer in that way. But, as classic works in sociology and statistics had shown (and as reviewed in Gary King, 1997), such an inference may be unsustainable when data on individuals are available. (The classic study of this – by William S. Robinson (1950) – had a major influence, although geographers – Subramanian *et al.* (2008) – have shown using more sophisticated data analyses that his example was flawed.)

The response to that case was twofold: care in the interpretation of what became known as 'ecological correlations' (i.e. analyses based on aggregate data, such as those derived from censuses for administrative areas, but interpreted as showing a relationship at the individual level); and attempts to resolve the problem. Regarding the former, given the available data for tackling many spatial problems, geographers had little alternative but to continue making such inferences – all the while being aware of both the nature of various inferential problems involved (Alker, 1969), and the extra issues raised by Openshaw's work on the MAUP. Only in the late 1990s was a procedure developed which its author claimed resolved the ecological inference problem, providing a method of accurately estimating individual behaviour from aggregate data (King, 1997). Its relevance to geographical work was assessed a few years later (Sui *et al.*, 2000), and a few applications have been undertaken. At the same time, other ways of solving the problem have been advanced, such as the entropy-maximising and micro-simulation approaches already discussed (Johnston and Pattie, 2001). In all of these, the goal is to derive valid statements about individual behaviour from aggregate data, as an input to analyses of possible causal relationships.

Point pattern analysis

The analysis of point patterns was one of the earliest focuses of quantitative interest among geographers, in large part because central place theory and similar models suggested regular patterns for punctiform data: nearest neighbour analysis was adapted from ecology for their analysis (e.g. Dacey, 1962; King, 1962). As such, models attracted diminishing attention, however, so point pattern analysis declined in its importance in human geography. It was revived and much extended, however, in a series of publications by Openshaw.

In studies of the nuclear industry, for example, he evaluated the potential impact of bombing different targets and of selecting different locations for nuclear power stations and hazardous waste disposal by analysing the distribution of population relative to selected points (Openshaw *et al.*, 1983; Openshaw, 1986; Openshaw *et al.*, 1989). The latter work linked with his later studies of the clustering of rare events round certain points – such as particular cancers linked to possible radiation sources – inductively suggesting cause-and-effect relationships; if significant clusters are identified, this provides strong circumstantial evidence that there is a local cause of the disease, worthy of closer investigation. He began this work by building a Geographical Analysis Machine (GAM), which inductively sought all possible clusters in a point pattern, such as one for cases of childhood leukaemia (Openshaw *et al.*, 1988). This involved using massive computing power to identify those clusters – groups of points which were unlikely to occur by chance.

Critiques of Openshaw's approach focused on details rather than the core argument – that inductive searches for patterns can stimulate hypothesis development. With spatial autocorrelation, however, and the problems of underestimated error terms, the significance of clusters may be overemphasised, leading to developments of tests which allow for that autocorrelation – as in Markov Chain Monte Carlo methods (Ward and Gleditsch, 2002; alternative approaches are set out in Elliott *et al.*, 1996; Elliot *et al.*, 2001; and Wakefield *et al.*, 2001).

Local statistics

As stressed repeatedly in this book, one of the basic tenets of geographical work is that places differ. If that is the case, then the nature of statistical relationships may vary across space – as suggested by varying gravity model exponents and the application of multi-level models in geographical research. Furthermore, following from that observation, if the nature of a relationship varies across space, why is that the case: what other variables are related to such variation? To answer such questions, geographers have been involved in the development of what are known as local statistics, as applied to point patterns, univariate and multivariate relationships, and flow patterns (Fotheringham, 1997; Unwin, 1996). The issue was raised early by Leslie King:

> There are no general laws in social science that are consistent over time and independent of the context in which they are imbedded. The search for law-like generalization of cause and effect relationship is an illusion. Any particular patterning of events will not remain stable for very long, and generalization about them cannot provide a firm theoretical basis for intervention.
>
> (1976, p. 304)

John Paul Jones and Robert Hanham (1995) suggested that the reference to context means that relationships vary over space as well as time (and thereby linked their work on local statistics to the realist critique of positivism discussed on p. 219, this book).

A substantial component of the literature on certain types of local statistics has been reviewed by Fotheringham *et al.* (2002, p. 241, original italics), whose book 'is based on the premise that relationships between variables at different points in space *might* not be constant over space'. Because of that, an analysis which does not allow for such spatial variation commits what is known as Simpson's paradox, which is the 'reversal of results when groups of data are analysed separately and then combined' (Fotheringham *et al.* 2002, p. 7). In the former case, as they illustrate, spatial disaggregation may show two negative relationships between a pair of variables, but when they are aggregated into a single dataset the relationship is positive.

A number of procedures have been explored to test for the existence of this paradox and, where necessary, identify the separate relationships between different sets of observations within a population. One (as discussed on p. 146, this book) is multi-level modelling, which takes a linear relationship between two variables and enquires whether its parameters (the intercept and slope coefficients) vary according to subsets of the population, which might be spatially defined: 'Is the relationship between economic deprivation and the propensity to vote for left-wing political parties the same across all regions of a country?', for example (Jones *et al.*, 1992). Another, developed by geographers – the Spatial Expansion Method (Jones and Casetti, 1992) – related the parameters of a regression equation to other variables. Most of the applications were aspatial – they simply tested for differences between subsets of the population, however defined; it was possible to produce a spatial model so that, for example, the strength of a relationship (between distance from the CBD and population density, say) varied by sector of the city. This is a rather limited approach, however, and, as with multi-level modelling, operates with predetermined subsets of the population rather than inductively, searching for variations in relationships – as with Openshaw's searches for clusters of points. (Jones and Bullen, 1994, report on a comparison of the two methods.)

To extend the search process, Fotheringham *et al.* (2002) developed a technique which they term Geographically Weighted Regression (GWR). It tests whether there is spatial variation in the strength, intensity and direction of a relationship inductively through analysing separate subsets of the population of data points, defined by their proximity to a randomly selected core at different scales. This GWR procedure is similar to Openshaw's GAM just discussed, and Openshaw (1998) himself produced a similar Geographical Explanations Machine to undertake similar inductive explorations (and was extended to other statistical procedures – e.g. Harris *et al.*, 2011). With such methods it is possible to explore whether relationships – such as that between school performance and the characteristics of its catchment area – vary spatially.

One aspect of the study of local statistics is related to the issue of spatial autocorrelation (see p. 147, this book). For example, one of the standard spatial autocorrelation statistics may indicate spatial clustering of a variable across a map of areas for which a variable has been measured, but not whether this involves a single core of high values for the relevant variable or a series of local clusters, or 'hot spots'. Geographers, such as Anselin (1995) and Getis (Ord and Getis, 1995), have developed measures which can identify such clusters – such as a G* measure which identifies the degree of clustering of similar values around each area and bearing correlograms indicate whether trends in a variable are stronger in some directions than others; wombling methods identify whether certain boundaries (e.g. between two adjacent areas) act as barriers to such a trend, blocking their development. (For examples of an application of these methods, see O'Loughlin, 2002; Johnston et al., 2009. More generally, Nelson, 2012, reviews recent developments in spatial statistics.)

Spatial modelling

By the 1980s, spatial analysis would no longer occupy the central place within the discipline's literature, as was the case in the 1960s and 1970s. But it remained a foundation of much geographical analysis and has continued to be significant. Indeed early in the 2000s, according to Golledge:

> Today we have all the components deemed necessary to define and justify the existence of a scientific discipline – a huge array of empirically-verified factual data, spatial theories and models, innovative methods of spatial analysis, unique modes of representation, and practical usefulness for decision-making and policy formulation.
>
> (2002, p. 12)

According to Wilson (1989a, p. 29), modelling within human geography had led to 'dramatic advances' and has 'a substantial contribution to make in the long term' (Clarke and Wilson, 1989, p. 30). Wilson (1989a, p. 30) accepted the limitations of the approach – models 'have relatively little to offer in relation to individual behaviour directly', but are important tools which can 'help to handle complexity in a variety of situations'. His review focused on work (much of it his own) which sought to extend the 'classics' of location theory (von Thünen et al.; see also Clarke and Wilson, 1985; Wilson, 2000), and which he considers has provided, through its mathematical sophistication, substantial advances in understanding. (Wilson (1989a, p. 41) claimed that the use of mathematics to model complexity distanced his work from positivism; statistics are needed to calibrate and test models, but 'a purely statistical approach to theory and model-building is limited in scope and is also more directly connected to positivism'.)

The modellers' achievements were similarly lauded by Macmillan (1989a, 1989b), who equates their activity with 'quantitative theory construction'; to him theories are empirically testable statements of 'universal empirical propositions, law-like generalizations' (1989a, p. 93), so that modelling involves the development of theories which contain 'quantifiable propositions' that can be tested. It has already 'helped geography to achieve widespread and significant advances' (1989b, p. 292), and the feelings within the modelling community are of 'frustrated optimism rather than pessimism' (p. 291).

The sort of modelling undertaken – and still promoted by – Alan Wilson (e.g. Wilson, 2011; Dearden and Wilson, 2011; Pagliara and Wilson, 2010) is now a minority interest within 'quantitative geography'. Much less attention is now played to geometry and spatial order than in the early years of the 'quantitative revolution', and although distance remains a key concept in some work – notably of flow patterns, such as migrations and the spread of diseases – it is no longer the key variable in a large number of analyses; geography is no longer perceived and portrayed as very largely, 'a discipline in distance'. Nevertheless, some important work in this area continues, as in Batty's (2013) book on *The New Science of Cities*, in which he revisits some of the 'traditional' concerns of urban geography as a spatial science arguing that 'to understand place we must understand flows, and to understand flows we must understand networks' (p. 1). His modelling leads him to conclude that 'there is an intrinsic order to the number, size and shape of the various attributes of networks and thus, in turn, of spaces and places that depend on them' (p. 2).

Instead, much more attention is paid to local context, to the links between people and their environs – physical, built and, especially, social – as influences on behaviour. In effect, space has been downplayed, and place has become a much bigger leitmotif in much spatial analysis (as argued in Johnston et al., 2014a, 2014b). As Kelvyn Jones (2010, p. 203) expresses it, the goal of such work is to identify and then to understand regularities in behaviour, not to seek potential laws, as some geographers argued for in the 1950s–1960s, but rather to identify 'potentially causal generative processes that operate in particular historical, local, or institutional contexts to produce particular patterns of outcomes'. The development of methods for such analyses continues (e.g. Fotheringham and Rogerson, 2009; Fischer and Getis, 2009; the contents of these books illustrates that such analyses are undertaken in a wide range of disciplines and not just geography).

A considerable amount of the criticism of 'quantitative geography' (discussed in detail in later chapters: see Cloke et al., 1991; Cloke, 2004; Cresswell, 2013) is based on an appreciation of that work as concentrating on the search for geometric order and laws, subordinating the individuality of subjects to mechanistic cause-and-effect, stimulus-response models (hence its omission from textbooks introducing contemporary human geography: Cloke et al., 1999, 2005). Much of the work undertaken in the last three decades is not of that nature, however. Some of the concepts studied by geographers are necessarily aggregate in nature – unemployment levels, indicators of health – and are necessarily studied as such; many public policies, including those aimed at ameliorating if not removing inequalities and injustices, are similarly aimed at population aggregates, including populations characterised by where they live, rather than at individuals. Understanding contemporary society calls for analysis of those aggregates, searching for order. The underlying philosophy, as Jones (2010) argues, is critical realism (see p. 219, this book); the goal is neither generalisable laws nor simply to interpret and appreciate the lived experience of social actors but rather to accept that there are truths, knowledge of which is 'more than just some undetermined socially constructed language system' (p. 203). That knowledge – the observed 'truths' comprising patterns and relationships – lies in the empirical world which the analyst seeks to describe in a rigorous and replicable way, accepting that the structures, mechanisms and tendencies that stimulate the production of those patterns and relationships cannot be 'discovered' in analyses of the empirical world; they have to be determined separately, providing convincing – 'theoretical' – accounts for the worlds that we live in, and frequently want to change.

Geocomputation

By the late 1980s, technological developments had emerged that would significantly deepen and broaden the capacities to tabulate and visualise geographical data. In turn, their adoption soon led to lively and heated debates about the merits of a geography configured around these technologies.

Very few of the statistical and mathematical developments discussed could have occurred without rapid cumulative changes in technology which have allowed massive volumes of required computation to be undertaken increasingly easily and quickly, at decreasing relative prices (Rhind, 1989; Longley et al., 2001, pp. 445–6). As the technology advanced in the last years of the twentieth century many different 'geofutures' could be visualised (Clarke, 1998; Armstrong, 2000).

Computers became generally available for research in universities in the 1960s, and geographers were soon using them. The first paper on the potential application of computers in cartography appeared in 1959 (Tobler, 1959). There was much initial collaboration in writing, developing and operating programs, as illustrated by an early catalogue (Tarrant, 1968), with texts on computer use by and for geographers (Dawson and Unwin, 1976; Unwin and Dawson, 1985). There were two major areas of use from the outset – multivariate statistical analyses and computer mapping (Hägerstrand, 1967; Haggett, 1969). With the former, most geographers increasingly relied on packages developed for wide use – such as SPSS, MICROTAB and MINITAB – although developments in specialised fields of spatial analysis (such as spatial autocorrelation) required custom-built programs. For most human geographers, programming was no longer a required skill by the 1980s since many of their requirements could be met by standard packages or others developed for geographical/spatial analysis (Anselin, 1988; Anselin et al. , 2004; Anselin and Rey, 2010).

This increasing reliance on large computers led to the creation of an approach termed geocomputation, which incorporates such topics as data visualisation and transformation, representing and modelling spatial interactions, and generalisation regarding spatial processes (Longley, 1998). Furthermore, according to Macmillan (1989c, p. 257), such was the computing power available by the late twentieth century that it allowed more than just a step change in the geographer's ability to manipulate information to answer traditional questions:

> Technical advances have been so advanced that the problems we can think about have changed. Moreover, there has been a transformation in the kinds of thoughts we can have [by analogy with computing power expansion and mathematics]. As it became possible to find solutions to mathematical problems computationally, so it became possible to think about new classes of problems that could be solved.

Geocomputation is but one part of a wider area of technical development, which has been termed geoinformatics and incorporates not only geocomputation but also geodesy, cartography, photogrammetry, global positioning systems, remote sensing, geostatistics and geographical information systems (Curran, 2001b). Since then this has been supplemented by the combination of geographically referenced data with the mobile internet, producing a geoweb.

A number of these – notably remote sensing – received considerable attention within physical geography (Cooke and Harris, 1970). The analysis of remotely sensed images – initially from planes, increasingly from satellites – assumed considerable importance because the images provided immediate, regular and frequent information on parts of the world which were difficult to access physically, allowing not only detailed mapping but also estimations of environmental conditions (such as biomass volume, soil wetness and river sediment loads) and assessment of changes, especially short term. Furthermore, such images are the only sources which provide data at the global scale, increasingly important for modelling environmental changes. They have attracted relatively little use in human geography, however, except in work on land use and defining boundaries of different types of built-up area (Longley and Mesev, 2000; Longley, 2002; Mesev, 2003).

Openshaw (1994, 1995) has argued that the computational power available could transform the practice of human geography, provided that (1994, p. 504) 'you are neither blinded by past prejudices about science, not scared of the words computational human geography, nor too hyped up by an overenthusiasm for AI or infected by the neural net virus' (on which, see Fischer and Gopal, 1993). His approach was empiricist and inductive, calling for 'greater intelligence by first becoming less sophisticated in our analysing and modelling technology and then more than compensating by computational intensity' (p. 503). He rejected (1995, p. 161) 'a qualitative, linguistically defined understanding of how a whole system works in vague and general terms' and promoted 'a detailed and precise computer model of one small part of it', using fuzzy set logic as 'a new scientific paradigm for doing geography' (Openshaw, 1996). He called for a HPC (high performance computing) culture within geography focusing on HSM (human system modelling), and the University of Leeds established a Centre for Geographical Computation to advance this work (as illustrated in Openshaw and Rao, 1995); the book by Openshaw and Openshaw (1997) illustrates some of the potential uses of AI within this strategy.

Whereas massive computing power has enabled major advances in modelling and data analysis, it also stimulated improvements in basic analytical procedures. In particular, the development of visualisation methodologies, such as the ability to graph relationships on computer screens and to explore maps (Unwin, 1994), aided exploratory data analyses, such as enquiring whether there is a linear relationship between two variables (something that was often assumed because of the time-consuming procedures involved in graphical procedures: Jones and Almond, 1992; Jones and Wrigley, 1995), as well as allowing spatial exploratory data analyses (Anselin, 1998). Thus, computing power has enabled not only massive advances in spatial analysis but also improvements in standard applications. These include mapping. By the mid-1990s, the availability of computer software mapping packages made much pen-and-ink cartography obsolescent, and allows exploratory maps to be drawn and rapidly displayed on screen. As Dorling and Fairbairn (1997, p. 4) expressed it, 'the craft skills required to undertake map-making are diminishing in importance'; map-making has moved from the field and the drawing board to the laboratory and the keyboard (Rhind, 2003) – atlases are relatively straightforward to produce, given the plethora of data and readily available software, and are frequently deployed as descriptive devices (e.g. Dorling et al., 2010).

One of the contemporary biggest challenges to spatial data analysis is that posed by what are generally known as 'big data' – large, often geocoded, datasets, many of them collated by commercial firms from a variety of sources and 'mined' to provide inputs to a wide range of (including locational) decision-making processes. The manipulation and analysis of these data – characterised by Kitchin (2013, p. 262) as 'high volume, velocity, variety, exhaustivity, resolution and indexicality, relationality and flexibility' – present not only computational issues but also ethical and interpretative (as set out in Graham and Shelton, 2013; Wyly, 2014); traditional spatial analytic procedures based on frequentist models with statistical significance testing are no longer relevant for such datasets, for example (Brunsdon, 2014).

Geographical Information Systems to Geographical Information Science

The arrival of GIS

Spatial data capture, storage, integration, display and analysis have been advanced since the early 1980s by the development of Geographical Information Systems (GIS: Curran, 1984; Green et al., 1985; Chrisman et al., 1989; Foresman, 1998; Longley, Goodchild et al., 2005; Chrisman, 2006; Steinitz, 2014), which brought together rapid advances in computer cartography and data collection in dedicated machines (comprising both hardware and customised software) for the analysis of the three main types of spatial data: those referring to points (such as mapped locations), areas (such as towns and fields) and lines (transport routes, for example). Increasingly, data produced by a wide variety of agencies are geocoded (i.e. given unique spatial references which facilitate spatial analysis), and technological advances are leading to the introduction of computers with which it is possible to gather, store, display and analyse spatial data. Different geocoded datasets (i.e. different maps) can be integrated within a GIS, which performs the technically demanding tasks of transforming different coordinate systems to a common structure and interpolates unknown values from various point and line three-dimensional data (Longley et al., 2011, is an example of the many textbooks in this area). GIS are, according to Longley (2000, p. 157), 'fundamentally a facilitating and applications-led technology, which transparently assesses the importance of space, and as such should be central to our geographical understanding of the world'. Advances in GIS, as he notes (Longley, 2004), have been such in recent years as to blur the distinction between iconic and symbolic modelling; whereas initially GIS stored maps in their iconic form, increasingly they are able to store symbolic (e.g. formulaic) representations of environments and their process systems, in four dimensions.

Development of this technology in the UK was heralded and promoted by a major project initiated to mark the 900th anniversary of the 1086 Domesday Book census. A GIS to provide a modern equivalent was developed and marketed as a teaching aid for schools and other educational institutions (Goddard and Armstrong, 1986). Its massive data base could be used to display a wide range of maps (including the Ordnance Survey 1:50,000 series for the whole of Great Britain) and pictures, and could also be analysed interactively (Openshaw et al., 1986); the technology on which it was based is now obsolete. Other major stimuli included early work on computer cartography at the Experimental Cartography Unit in London, the Department of Geography at Durham and Edinburgh universities, and the Laboratory for Computer Graphics and Spatial Analysis in the Graduate School of Design at Harvard University (for histories of GIS, see Foresman, 1998; Mark et al., 1996; Chrisman, 2006).

GIS allow many problems to be addressed that previously could not be afforded in terms of the time and resources necessary for data collection, analysis and display: as such, they are the basis for much applied research work (see p. 338, this book). The UK's Economic and Social Research Council established a network of Regional Research Laboratories to explore their potential as research tools in the late 1980s, most of them located in university departments of geography which offered postgraduate degree courses in GIS (the October 1988 issue of the ESRC Newsletter, was devoted to the topic 'Working with geographical information systems') and the USA's National Science Foundation invested US$5.5 million in a National Center for Geographic Information and Analysis, based at three sites (see the announcements in the Association of American Geographers' Newsletter for August 1987 and October 1988, and Fotheringham and MacKinnon, 1989); a key individual in winning funding for the latter was Ron Abler, then responsible for geography in the NSF and later Executive Director of the Association of American Geographers (see p. 86, this book; Mao, 2014; and Skupin, 2014; Abler's successor at the AAG, Doug Richardson, also had a background in GIS, having established a successful firm, GeoResearch). Their use was promoted in teaching (Maguire, 1989; Fisher, 1989a); an Association for Geographic Information was established in the UK, following a major report by a House of Lords Select Committee for which a geographer (Rhind) acted as scientific advisor (Rhind, 1986; Rhind and Mounsey, 1989); several specialist journals were launched (such as the *International Journal of Geographical Information Systems*; in 1997 Systems was replaced by Science); a bibliography of over 1,000 items was published in early 1990 (Bracken et al., 1990); and a major encyclopaedic survey was published in the following year (Maguire et al., 1991; a second, much revised and rewritten edition, appeared eight years later; Longley et al., 1999a; for a review of work in urban geography alone, see Sui, 1994). This movement also saw a large number of textbooks published to cater for teaching GIS, alongside specialist technical volumes, and others made explicit the role of GIS in spatial data analysis (O'Sullivan and Unwin, 2003).

In his 1989 presidential address to the Canadian Association of Geographers, Roger Tomlinson (see also Tomlinson, 2007), a pioneer in the development of GIS from the 1960s on, referred to their growing use in the following terms:

> Geographers have a crucial role to play in integrating a wide variety of technologies into new forms of 'earth description' which will act as a foundation for geographical methodology and open the way to richer forms of spatial analysis and geographical understanding.
>
> (1989, p. 298)

Spatial problems will multiply and become more complex in the future, and geographers' ability to handle and analyse large bodies of data in the search for solutions using GIS should demonstrate the strength of their 'integrating science', as part of a 'data are good ethic' (p. 292). Whereas many of those involved in the development of GIS have largely focused on the resolution of the technical issues involved with integrating data bases and their statistical manipulation, and cautioned against over-optimism regarding their potential (see Goodchild, 1995b), some of its promoters have made grand claims, as illustrated by Stan Openshaw's (1991) reaction to Peter Taylor's (1990b) separation of geographical information systems from 'geographical knowledge systems' (see the discussion, p. 156, this book). At its core, however, the use of GIS was characterised, according to one of its leading practitioners, Michael Goodchild, as using 'geographic location as a common key to link data' (Schuurman, 1999, p. 10); the more that is done using such data, the more central 'geography' becomes to a wide range of practices. GIS technology has enabled many of the advances in spatial analysis discussed. To appreciate the importance of place – locale, spatial context – as an influence on behaviour, means have been developed using address files to link individual data (most of them derived from surveys) with information on their local area (much of it derived from censuses); indeed, in some studies 'bespoke neighbourhoods' around each survey respondent's home have been designed, at a variety of scales, to provide the best available approximations of each person's local context (Johnston et al., 2005a; Johnston et al., 2005b; Boyle et al., 2004; Östh et al., 2015). In other studies, such as the UK Millennium Birth Cohort Study, the samples have been deliberately spatially structured to enable analyses of the influence of local geography (such as areas with high levels of child poverty and concentrations of ethnic minorities: Hansen, 2010). Such designs allow for efficiency in data collection; initially, because of the potential impact of the clustering on statistical parameters (the problem of spatial autocorrelation), they were seen to create a 'nuisance' for quantitative analyses, but appreciation of the important role of local context means that such datasets are particularly valuable for exploring the importance of place as a behavioural influence (Jones, 1997), although such procedures have their limitations, as illustrated by what Kwan (2012) terms the 'uncertain geographic context problem'.

The use of GIS has not only expanded exponentially in research over recent decades – across a wide range of the spatial sciences (as illustrated in NRC, 2011) – but has become the basis for major commercial developments in the private and public sectors of most countries, along with the associated use of Geographical Positional Systems (GPS) in many of those applications. As such, it has been promoted by geographers and for geography in many contexts (as in Cutter et al., 2003), and as a major element of a geographical education providing graduates with much-demanded specialist skills is a wide range of employment situations.

Critical responses and debate

GIS soon became the subject of critical debate (see Curry, 1994, and several of the essays in Pickles, 1995a). Pickles (1995b, p. 2), for example, argues that GIS are much more than 'merely more efficient counting machines' providing more accurate descriptions and visual images; their 'virtual representations will produce illusions that will be so powerful it will not be possible to tell what is real and what is not real' (p. 10; see the discussion at p. 82, this book, on Lukermann's arguments regarding models). For others, their use, whether in sophisticated marketing procedures (i.e. geodemographics: see Batey and Brown, 1995; Birkin, 1995; for a critique, see Goss, 1995) or military applications (N. Smith, 1992), raised moral and ethical issues (none of which are specific to GIS but apply to any application of 'knowledge'). To Michael Curry (1995), GIS involve ethical inconsistencies which are necessarily built into systems and can be used to advance surveillance (whether by the state or by other bodies) very substantially. (These issues are discussed in Pickles, 1995a; Longley et al., 1999a; on the wider issue of representation, see Cosgrove, 2001.)

In responding to some of these critiques, Daniel Sui (1994) categorised them into those which stress one or more of the following:

1   ontological inadequacy – GIS present a limited, Cartesian representation of reality, ignoring social and cultural representations;
2   epistemological insufficiency – GIS are deeply rooted in positivist modes of thinking;
3   methodological insufficiency – GIS applications lack coherent theory and are biased by their use of secondary data; and
4   ethical inconsistency – the values embedded in GIS are ignored, along with any recognition of subjective differences in representations of reality.

He claimed that the critiques have 'at best, been refuted by the GIS community or, at worst, have been totally dismissed as anti-progress and anti-science nuisances' (Sui, 1994, p. 268). Examples are presented of GIS researchers who have addressed each of the four issues, showing that:

> the GIS community has realized that the implementation of GIS should go beyond mere technical decisions justified by matters of efficiency and give the ethical use of this information a serious consideration. . . . GIS, if not a fertile ground for common search, at least has initiated a search for common ground.
>
> (p. 271)

(The last sentence draws on the title of Golledge et al.'s, 1988, collection.) Nevertheless, he concluded by noting that 'so far, GIS enthusiasts and GIS opponents have been mutually hostile or at least dismissive of each other's views' (Sui, 1994, p. 272) and suggested that closer cooperation is needed in which 'GIS enthusiasts must avoid the imperialistic claim that science is the only guarantor of objective truth. Postmodernists (see Chapter 7, this book) must relinquish the playful cynicism in their critiques on the scientific chauvinism of GIS'. Schuurman (2000) argued that such cooperation was now developing. She suggests that critiques published during the 1990s drew responses from GIS specialists, which led to the establishment of an NCGIA initiative on the social impacts of GIS. Nevertheless, she feared that the lack of a common vocabulary could reduce the impact of any cooperation, with the critics using approaches based in the sociology of science that have little in common with the scientific and technological languages employed in GIS. For Goodchild, many of the critiques – which focused on GIS use in a range of non-academic contexts – had considerable validity. To him, a 'GIS is a machine that is capable of a wide range of operations and these in turn are compatible with a wide range of philosophical positions' (Schuurman, 1999, p. 5); it can be used both to identify unique features in a dataset and to search for 'the general processes that operate on those unique conditions' (p. 6).

The advances in computer technology, including GIS, have had very substantial impacts on the practice of cartography and other forms of graphic information display. Only a small number of geographers are involved in research on cartographic methods (see MacEachren, 1995), but their work on visualisation has allowed major advances in the computerised presentation of large and complex datasets (see Dorling, 1995). Dobson (1983a) argued that dealing with these massive datasets will require the development of automated geography, an 'integrated systems approach to geographic problem solving in which the problem is defined, the appropriate methods are chosen, and the tools are selected from a broad repertoire of automated and manual techniques' (p. 136). This would be highly dependent on computer hardware and software developments (sufficient of which were already available, though see Cowen, 1983) and would facilitate participation in public policy research on a large scale. Most commentators on Dobson's paper (in the August 1983 issue of The Professional Geographer) argued that automated geography was a misnomer, since the main benefits of the hardware and software that he described related to the scale and speed of data manipulation only. They preferred less emotive terms such as 'computer-assisted geography' or 'computer-assisted geographic systems', and some, like Poiker (1983), argued that the case was no more than one for another set of tools, just as the case for quantification turned out to be decades earlier: 'our tool seems to attract an inappropriate number of prophets' (p. 349). Dobson (1983b) disagreed, however, arguing that 'what I am talking about is not a system or collection of systems. It is a discipline which uses human and electronic cybernetic systems to further understanding of physical and social systems' (p. 351), with some decisions (e.g. with regard to how data are displayed) taken by the computers, not by people. Thus, computers should be programmed so that they incorporate 'the character of our discipline', presenting the results of geographical analysis without the detailed intervention of geographers (see Couclelis, 1986b; Fisher, 1989b; Haines-Young, 1989).

Dobson's arguments were revisited a decade later. He claimed that:

GIS has become a sine qua non for geographic analysis and research in government, business and academia. This advance has been hailed widely as a technological revolution, but I proposed a more exciting prospect beyond technology. The strength of GIS innovation and diffusion suggests that science and society are in the beginning stage of a technological, scientific and intellectual revolution as profound as earlier revolutions brought on by the printing press or the computer.

(1993, p. 431)

Nevertheless, he expressed concerns over the amount of geographical research done using GIS, other than in mapping and survey (p. 432), suggesting that 'the GIS community is still searching for a greater sense of purpose' (p. 433), and claiming that 'the current state of GIS is woefully deficient when compared to our predecessors' ultimate goal of representing landscapes as three-dimensional geographic units with interactions of multiple phenomena in space and time, with order' (p. 434). He ended with a call for geographers to become more involved in what was clearly, to him, an unfulfilled revolution:

The technological revolution, understandably, may be led by private companies and government laboratories, and technical leadership may come from many fields. But academic geographers are essential to the scientific and intellectual revolutions. Geographers are needed as consummate leaders in conceptual design and in geographic analysis employing GIS.

(p. 438)

Others agreed, while noting that hardware and software developments, although substantial, were still far from those needed for completing Dobson's revolution (Armstrong, 1993; Cromley, 1993). But whereas some were optimistic about the growing use of computers in all aspects of geography (and not just those involving numerical applications: Monmonier, 1993), others were pessimistic. Marble and Peuquet (1993, p. 446), for example, argued that since 1983 'the impact of the computer upon geographic research appears to have been even less than our most pessimistic estimates', in part because of 'a substantial turning away from those activities that GIS can do best' (p. 447). Goodchild (1993, p. 445) appeared to agree: 'Whether or not GIS and automated geography have anything to offer to geographic research, we cannot escape the fact that GIS has had a significant impact on many kinds of human activity.' This was reflected in the subtitle of Sheppard's (1993) contribution – 'What kind of geography for what kind of society?' – and in Pickles's (1993) rehearsal of arguments regarding the impact of technology on geographical and other practices (see p. 159, this book).

The technical advances have not been associated with parallel developments in substantive geographic theories it is argued, therefore, in large part because (as discussed in later chapters) spatial analysis has been somewhat marginalised within the discipline. In some of the continuing work, the 'traditional' models associated with the developments of the 1950s/1960s have provided the basic underpinning, focusing on optimum distributions of points and lines, and on flows between those points along the lines. Wilson (1984b) argued that mathematical development enables us to build new models by combining old problems (for example, von Thünen's and Weber's) with new methods, claiming that comprehensive models can now be built combining macro-population backcloth, macro-economic backcloth, spatial interaction, the location of population activities, the location of economic activities, and the development of economic infrastructure into representations of spatial pattern or settlement structure. Two papers 'demonstrate the ability of models of spatial interaction and structure to reproduce the results of classical industrial location theory' (Birkin and Wilson, 1986, p. 305) and they are the foundation for the applied work undertaken at Leeds (Birkin *et al.*, 1995: see p. 351), but relatively few workers now concentrated on such modelling. Even more prefer non-positivistic approaches, leaving the substantial GIS community somewhat isolated within geography as a whole – notwithstanding calls to overcome such divides (Elwood, 2006; Kwan, 2004; Kwan and Schwanen, 2009a, 2009b; Schwanen and Kwan, 2009; Zolnik, 2009), through hybrid, critical or participatory GIS. One of the proponents of deconstructing the divides via a 'critical GIS' argued that by the mid-2000s there had been 'three distinct waves' (Schuurman, 2006, p. 727) of critiques of GIS:

> The first wave, between 1990 and 1994, was characterized by the ferocity of the criticism that human geographers reaped on technical geography and spatial analysis . . . the tone of the papers was frequently scolding and paternalistic . . .
>
>   The second wave of critiques took place between 1995 and 1998. By 1995, more than forty papers had been published by human geographers worried about the implications of a pervasive GIS. . . . Concern was expressed that GIS best served large corporations and public agencies rather than the disenfranchised. . . . Moreover, critics were skeptical about a technology that they deemed a relic of the quantitative revolution . . .

For Schuurman, a third wave of critique followed at the end of the 1990s, but this saw the birth of 'critical GIS', which moved from antagonistic attacks on GIS 'to a means of positively affecting a technology that was being widely adopted in other disciplines and in the commercial sphere'. Yet, during the same decade, Obermeyer (1994) had claimed that GIS now meets all the criteria for identification as a separate profession. At the same time, too, arguments were being advanced that the last term in GIS should be changed to Science.

GISci

Goodchild (1992, p. 31) made the case for GIS to be renamed as Geographical Information Science, arguing that while 'spatial data handling may describe what we do' and that many of the advances in GIS have been technology-driven, there were two questions that suggested the change of name. The first concerned GIS as science, addressing a particular set of scientific questions that are generic rather than restricted to particular applications; the second concerned GIS for science, with wide applications across many sciences. Regarding the first set, he argued that because geographical (or spatial) data have unique properties, then this leads to a set of generic issues that such a science would cover: these relate to data collection, measurement and capture, spatial statistics, data modelling and theories of spatial data, algorithms for representing data, data display, and analytical tools. In each, he claims, there are 'clearly challenging scientific questions [in which] . . . the spatial context appears to be distinctive' (p. 40). Addressing such key questions is the core of his nascent geographic information science (GISci), which focuses on technological solutions to the many issues relating to spatial data handling, whose relevance is widespread across other sciences and in myriad public and private sector applications. (On GISci as a science, see Sui, 2004a; Reitsma, 2013.)

Several years later, Goodchild reaffirmed the value of the term, noting that:

> I wanted to respond [in the 1992 paper] to comments that seemed to marginalize GIS as 'a mere tool'.... The history of science provides ample evidence that there is nothing 'mere' about many scientific tools. The microscope and telescope both produced an outpouring of new science and new ways of thinking about the world. By comparison, however, the effects of the digital computer have been immeasurably greater ... I wanted to suggest that when a researcher in the GIS community used the acronym they were probbaly referring to something more than software ...
>
> (1995a, p. 1)

Fifteen years later, he reviewed the first two decades of GISci (Goodchild, 2010), identifying not only major research achievements which had generated widely adopted, user-friendly, data-handling and associated scientific breakthroughs, but also institutional accomplishments: GIScientists, like Waldo Tobler and David Rhind, had been elected to national science academies, for example, and – perhaps like regional science before it but a much greater size – a growing professionalisation of the field reflected in its conferences and journals (he identified sixteen of the latter). Five issues for the next decade were then discussed: 'knowing where everything is, all the time'; the role of the citizen in what is sometimes known as 'participatory GIS' or neogeography (Turner, 2006; Leszczynski, 2014); the use of GIS in dynamic, 'real time', data collection, analysis and modelling; introducing third (time) and higher dimensions to what is largely a two-dimensional enterprise; and GIS education, leading to the conclusion that:

> the domain addressed by GIScience is well-defined and persistent. There is no danger that GIScience will be absorbed into one of its intersecting disciplines: geography, computer science, or information science. On the other hand a greater interaction with a broader domain of spatial information science seems both logical and desirable.
>
> (p. 16)

(The latter point is also central to the NRC, 2011, report.)

GIS has certainly transformed a substantial proportion of geographic research in recent decades (in physical as well as human geography, as illustrated by papers in the relevant specialised journals, and the birth of new ones in the 1990s such as *Annals of GIS, Transactions in GIS,* and the *International Journal of Geographical Information Science*, which celebrated its 25th anniversary in 2011, with a free access virtual issue containing 25 'classic' papers that shaped the field: www.tandf.co.uk/journals/pdf/freeaccess/TGIS25th.pdf). The sheer weight of publications led to those analysing their content to describe a 'paradigm of GIScience' (Parr and Lu, 2010). In addition, GIS has become widely used commercially and by public agencies for the storage, display and analysis of myriad types of spatial data. As such, skills with GIS are seen as important for future generations of graduates, and geographers have argued that GIS should form a central component of degree programmes, providing graduates with 'transferable skills' that they can apply in a range of careers. Indeed, GIS has been presented by some as a 'saviour' for the discipline, providing it with a raison d'être within the educational structure and a component of the discipline which should enable it to attract students. In surveys of graduate employment and academic job opportunities in the mid-1990s, for example, the Association of American Geographers found that more programmes were preparing graduates for careers in GIS and remote sensing than any other specialism, that GIS was the modal employment category for new graduates, and that GIS specialists were in most demand for posts in USA geography departments – with many more to be hired over the next decade than were expected to retire, resulting in a rebalancing of the profession. (The reports are reproduced in NAS-NRC, 1997; see also p. 87, this book.)

In reporting on a 'renaissance' in geography during the 1990s, the NAS-NRC (1997) volume *Rediscovering Geography* paid particular attention to spatial analysis and GIS as indicative of the discipline's relevance to contemporary social problems. GIS, geographic visualisation and spatial statistics get much of the coverage in the chapter on 'Geography's techniques'– indeed, that chapter makes no reference to the methodologies deployed in other areas of geography (such as those discussed in the next chapters). The case for promoting geography in this way is clarified in the chapter's conclusion:

> Current trends in geography's techniques suggest a future in which researchers, students, business people and public policy makers will explore a world of shared spatial data from their desktops. . . . The users of the future will also bring different world views and theoretical perspectives to these tools. They will challenge the adequacy of current techniques for analysing and understanding geographic phenomena, posing challenges that must be taken up by developers when designing next-generation tools and theories.
>
> (p. 69)

Similarly, chapters on 'Geography's contributions to scientific understanding' and 'Geography's contributions to decision making' focus substantially on spatial analysis and GIS applications, including a 'national spatial data infrastructure'. Not surprisingly, therefore, when it comes to the report's recommendations for 'Strengthening geography's foundations' with regard to research and teaching priorities, the section on 'Geographic competency among college and university students' calls for:

> efforts to ensure that . . . [they] have access to geography courses and perspectives that go beyond a concern with 'where things are' to provide a basic conceptual and analytic grounding in the spatial and environmental dimensions of human and physical processes.
>
> (pp. 150–1)

Furthermore:

> The discipline of geography is in a period of transition from a past where most geographic information was transmitted in the form of paper maps to a future in which most geographic information will be transmitted through digital information systems. . . . Geographers need to learn about these spatial information sources and understand how to use them appropriately in their work. Academic geographers also need to consider how new technologies and data sources will create new opportunities for collaborations with the private sector.
>
> If geography is to meet the challenges of rapid technological change, steps must be taken to familiarize students with new technologies for data analysis and display. . . . Technology now plays a pre-eminent role in a wide range of geographic research. . . . Continued advances of this sort will require a substantial coterie of geographers who understand new technologies and can use them effectively.
>
> (p. 160)

As far as the report is concerned, the future of American geography lies very much with spatial analysis and its enabling technologies, such as GIS. Certainly, GIS was then rapidly expanding as a technology, enabling not only the relatively easy resolution of previous difficulties regarding the integration of spatially coded datasets, but also both a wide range of commercial/public sector applications and the ability to pose questions previously unconsidered. For Longley *et al.* (1999b, p. 1) GIS is a 'generic term denoting the use of computers to create and depict digital representations of the Earth's surface', thereby enabling a wide range of geographical questions to be addressed. By

1997 it was being taught in over 1,500 universities and there were some 500,000 regular users in an industry worth US$12 billion annually. GIS has its own learned societies and literature, and a number of major annual conferences, but the adjective in its title is more than symbolic: geography is the academic discipline most closely connected with developments in the application and teaching of GIS, if not in the enabling hardware developments, with potentials that extend well beyond the narrowly technical (Gilbert, 1995).

Since then, GIS applications have multiplied, as reflected in the NRC (2011) report on *Understanding the Changing Planet*, which has a chapter on 'How might we better observe, analyze and visualize a changing world' which discusses those developments and the future potential of GIS and GISci with the development of what it terms a cyberinfrastructure. The analysis of such data 'requires a set of highly distinct and specialized techniques collectively known as spatial data analysis. The assumptions, data structures, and techniques of spatial data analysis are very different from those of standard statistical analysis and require specially designed software packages commonly known as GIS' (pp. 162–3); training in those specialised techniques is thus central to its programme for the expansion of 'geographical science'.

GIS has led to a reappraisal of the prior moves to quantification, spatial science and speculation on laws in geography. This includes some of the themes that animated behavioural geography, in the form of cognitive models of geographic space (Mark *et al.*, 1999) as well as fundamental questions about geographical representation and spatial analysis (Miller and Wentz, 2003). They have included re-evaluation of Tobler's (1970) proposal for a first law of geography (see p. 148, this book, and Miller, 2004). Sui (2004b, p. 269) notes that this, which he calls 'TFL' for short, had 'remained buried in the literature without much notice throughout the 1970s and 1980s', until GIS in the 1990s 'gradually brought new popularity to TFL both within and outside geography.' Although others have questioned its veracity, universality and meaning (Barnes, 2004b; J. M. Smith, 2004), for Harvey Miller:

> To a large degree, TFL distinguishes geography from other fields of inquiry. Using TFL as a core principal, spatial analysis and geographic information science continue to develop sophisticated techniques for extracting explanatory and predictive power . . .
>
> (2004, p. 288)

GIS has also proved a vehicle for reintroducing maps to geography, albeit in a very different form from previous situations (Kain and Delano-Smith, 2003; Rhind, 2003). In the 1950s, the map was still being promoted as the foundational geographical tool (James, 1954, p. 9; Robinson, 1954). Fifty years later, editors were bemoaning their absence from many geographical papers (Martin, 1998; J. Wheeler, 1998; see also Johnston, 2004b). Dorling (1998), however, argued that mapping is becoming much more common beyond the discipline, in a wide range of contexts:

> Maps have become a very common currency. People in western societies now consult maps so often that they often do not remember using them. These images have become commonplace, from bus maps to maps of offices, to maps in newspapers, in books, and on the news. They are so common that we often fail to question their authority, fail to ask why they are drawn as they are.
>
> (p. 279)

Many of those maps are used to change the way in which people think about the world (as happened in the Second World War when new projections were used to demonstrate America's proximity to Europe with regard to potential aerial attack: Schulten, 2001). 'Not all maps simplify, subvert, collude or conceal' (Dorling, 1998, p. 287), but their use as a means of informing and educating should remain a focus of geographers' concerns, albeit sometimes in new media (Crampton, 2009). For Sui, GIS could therefore be the basis for a:

third culture [that] stresses the cross-fertilization of creative ideas from the arts and sciences via state-of-the-art technology. The recent history of GIS and cartography should be viewed in the context of the third culture, which has the potential to fuse traditional branches of knowledge in truly creative ways ... whether geography is going to become a more vibrant discipline depends on to what extent we can revive geographic imaginations in this computer age via GIScience to address issues of great societal concern.

(2004a, p. 63)

As we chart in Chapter 9, claims about the ways that GIS could bridge social concern, theory, maps and technology have continued in the decade since.

Conclusions

As stressed in the previous chapter, much of the force of the developments in human geography during the 1950s and 1960s concerned methodology and the approach to traditional geographical questions. In addition, however, attempts were made to inaugurate and press a particular geographical point of view, with two main themes. The first – the spatial-science viewpoint – was fairly widely and rapidly accepted, and many geographers placed spatial variables at the centre of their research efforts; as will be indicated in Chapter 5, their work came under increasing attack from the late 1960s. The second – the systems approach – has received much less detailed attention, despite frequent approving gestures, except in studies of environmental and demographic systems. Compared with the spatial-science view, which could be rapidly assimilated within the developing statistical methodology, the systems approach was technically much more demanding, and perhaps for that reason attracted fewer active researchers (though see Coffey's, 1981, defence).

Although there were some antecedents, human geography as spatial science was largely inaugurated in North America in the 1950s. By the end of the 1960s it was providing much of the material for many of the journals published throughout the English-speaking world (Wheeler, 2002a). Much research was positivist in its tone, if not its detail, seeking not only to describe patterns of spatial organisation but also to account for (explain) these as consequences of the influence of distance on human behaviour. A great deal of it used quantitative methods and it contributed to bodies of theory about spatial organisation in general or certain aspects of it in particular. It has influenced other academic disciplines (even though, as Laponce, 1980, expressed it, geography had less impact on the other social sciences than vice versa) and the planning profession.

Since the 1970s, positivist spatial science has been under considerable attack among anglophone human geographers. This attack had little initial effect on the volume of work done in this paradigm, as illustrated by a major review of *Quantitative Geography* by workers in the UK (Wrigley and Bennett, 1981); this work has been stimulated and aided by advances in data-handling technology and display. The approach was also fostered outside the Anglo-American realm, especially in Western Europe where the reaction to quantification was initially stronger than in the UK and North America (see Bennett, 1981b). Many of the models and theories that acted as the foundations to the pioneering work, such as central place theory, are now largely consigned to disciplinary history, as stimuli to work that eventually proved relatively unproductive but generated interest in a wider range of issues that remain central to a geographical agenda, along with the methods needed to address that agenda.

Work in the spatial-science mould continues to thrive, however (though Kevin Cox, 2014, has expressed surprise at the degree to which it has been marginalised within geography). The field of GIS, for example, has spawned its own suite of outlets and professional societies, much like remote sensing before it. This is part of the specialisation and fragmentation that came to characterise human geography (Johnston, 1996b, 2004b) and, given the large volume of work done in other parts of the discipline, means that much of what is being reported can readily go unnoticed by other geographers. For a while it appeared for many that quantitative analyses had been marginalised within geography, while supporters of geocomputation and GIS had promoted their saleable expertise:

> It is clear that geographers specializing in geographical information systems will survive only by combining considerable technical expertise with a taste and talent for the Realpolitik of grants, contracts, committees, administrators, and entrepreneurs. A canny awareness of commercial and political realities will be as necessary as any particular intellectual qualities.
>
> (N. J. Cox, 1989, p. 207)

For Nicholas Cox, this carried the danger that 'preoccupation with data will omit imagination and creativity' (p. 208) but, as the later chapter in this book on applied geography shows, many see it as a major way to protect the discipline in a context where utility and saleability are the crucial indicators of the worth of its work. This great volume of commercial and applied work suggests that there has been a mode of 'progress' within the positivist approach. But there has also been a series of debates about its merits and latterly more engagements between GISci and wider theoretical concerns. For some authors, spatial analysis retained – or should retain – a central position within geography. Casetti (1999, p. 339), for example, identified three modes of enquiry – literary, academic and mathematical. His preference was clearly for the last of these, arguing that 'In geography in general, and in human geography in particular, the mathematical mode of inquiry is progressing much too slowly . . . [to have the required] positive influence on . . . the evolution of the social SCIENCE called geography' (original capitals). Like others involved with the quantitative revolution (e.g. Getis, 1993), he feared that the spatial analytic viewpoint was being marginalised within geography. It had 'saved geography from extinction as a serious university discipline, by attracting and training good students, by writing articles and books that developed theory and method, by gaining a foothold in science at large, and by applying these methods and theories to contemporary social problems' (Morrill, 1993, p. 442), but these gains were at risk as geography becomes increasingly 'anti-science' (p. 446). As stressed by Johnston et al. (2014a), however, there has been a major change of emphasis within contemporary spatial analysis, with studies of place becoming dominant and those of space declining in their volume.

Much of the remainder of this book is concerned with alternative approaches to human geography from those outlined in this chapter; many were initially based on critiques of 'positivist spatial science', and some have attracted large numbers of adherents. They have certainly involved critiques of the idea of human geography as science – at least in the terms that the later came to be understood by geographers in the 1960s – if not always being 'anti-science'.

Yet if the paradigm model of revolutions and normal science discussed in Chapter 1 is relevant to human geography, then as a consequence – given the later developments discussed in the following chapters – interest in the approaches set out here should have waned in the 1970s and 1980s, if not disappeared totally. In relative terms there has undoubtedly been some decline, as the alternative world views on offer (discussed later in this book) have attracted substantial attention. But a substantial amount of work is still being done which is firmly based within the world views that came to the fore in geography in the 1960s, and there continue to be substantial advances in the methods and applications of spatial analysis. This was illustrated in O'Kelly's (1999) listing of some of the most influential papers in *Geographical Analysis* over its first 30 years of publication. Thirteen years later, Nelson reviewed trends in spatial statistics, identifying new opportunities in the better integration of GIS and spatial analysis, the development of methods for large spatial data sets, overcoming the limitations of local statistics, improving how spatial sciences are communicated, and retaining the home of spatial analysis within geography.

(2012, p. 87)

Some of its adherents have sought to accommodate the critiques, while others have counter-attacked and argued for the validity and vitality of their point of view. There have been significant calls too for hybrid approaches that transcend these divides. There is a general consensus among the counter-attackers that modelling, with associated quantitative analysis, is a viable approach to human geography, though there is considerable variability in emphasis as to the way forwards. Moreover, as computers moved from being tools that could be marshalled for geographical analysis to becoming part of the everyday socio-technical worlds that human geography reckons with, so new questions about the relationships between geocodes, technology, society and space have arisen (Kitchin and Dodge, 2011; Pickles, 2004). Latterly, these include the geospatial web or geoweb, where cloud computing, internet and georeferenced data and mobile devices are interacting in ways that shift subjectivities and social relations (Elwood, 2010; Elwood and Leszcynski, 2011; Yang et al, 2011). Haklay *et al.* (2008, p. 2011) thus described how, in everyday life, 'New techniques are being used and new terms have been invented and entered the lexicon such as: mash-ups, crowdsourcing, neogeography and geostack.' We will return to these evolving forms of applied geography in Chapter 9.

Meanwhile, in the decades while those technologies and subjectivities were in formation, other approaches to geographical study have been introduced and debated, many of which have few links with those introduced here; in many cases, the subject matter addressed does not call for the forms of analysis discussed in this chapter. They are discussed in the following chapters, exemplifying the rich variety in geographical scholarship which emerged, in various ways, from the 'revolutions' inaugurated in the decades after the Second World War.

Chapter 5

Humanistic geography

An impartial spectator at the time would likely have bet on the eventual victor of the positivists, who were young, coordinated, and zealous; but a change in the zeitgeist was about to favor humanistic geography for close to 20 years.

(Jonathan M. Smith, 2009, p. 241)

Abstraction and fragmentation in human geography are related facets of a broader cultural condition, and as such they should not be confused with a law of nature.

(David Ley, 1989, p. 244)

Maps can be misleading because they don't – and can't – convincingly represent the seamier side of life. Consider United States Geological Survey Topographic Sheets. Looking through them, one might well conclude that all is well with America. What can be amiss when forests are green, rivers and lakes are blue, and settlements are either a eye-catching yellow or a healthy pink, depending on the scale? Conveying rot and degradation cartographically is difficult, perhaps impossible, because maps use conventional signs, which exhibit pattern, and pattern is by its nature aesthetic, whether it represents tulips or sewage pits.

(Yi-FuTuan, 2002a, p. 129)

In 1985, I sat in a course called humanistic geography. It was a second year course at University College London taught by Peter Jackson and Jacquie Burgess. It was a revelation to me. It made me decide I wanted to be a geographer. Here we learned about Tuan and Relph and Buttimer. We also learned to go out into London and talk to people, to record our impressions, to be creative – to practice a form of ethnography.

(Tim Cresswell, 2010, p. 169)

Although the behavioural geography that was considered in Chapter 4 had involved a reorientation of the work undertaken within the spatial-science approach, it maintained the generally positivist framework and overall commitments to scientific methods. Alongside it from the early 1970s, however, though with roots extending much further back, emerged a fundamental critique of the entire positivist approach. Much of this also argued for a focus on the individual as decision-maker, but denied the goal of explanation and prediction that was inherent to the behavioural approach; its goal was fuller appreciation and understanding of human values, meanings and interpretations. It rebuffed the idea of human geography as a science in favour of it as a discipline aligned with the arts and humanities.

Critics were vocal in their opposition to what had been promoted as the 'new [scientific] geography' (as illustrated in Cosgrove's 1989a, 1989b, cogent presentations of that case). Indeed, J. M. Smith notes how:

> the harbingers of a new mode of inquiry . . . spatial scientists assertively promoted a new 'posi-
> tivist' geography, and in so doing frightened a large number of regional, historical, and cultural
> geographers, as well as some old style economic geographers, into self-conscious awareness of
> their identity as methodological humanists.
>
> (2009, p. 241)

Be it out of fear, or mere bemusement, that opposition to spatial science was clarified by Entrikin's definition of what became known as humanistic geography as:

> a reaction against what they believe to be an overly objective, narrow, mechanistic and
> deterministic view of [the human being] presented in much of the contemporary research in the
> human sciences. Humanist geographers argue that their approach deserves the appellation
> 'humanistic' in that they study the aspects of [people] which are most distinctively 'human':
> meaning, values, goals and purposes.
>
> (1976, p. 616)

A few years later, Meinig (1983) advocated 'geography as an art', reviewing themes such as 'geography and literature', 'geography and the humanities' and the 'geographer as a creator of literature'. Meinig was keen to link human geography more closely to the humanities and arts, rather than see it as a science. He mobilised a range of past currents in geography in this task:

> one can pick out a few bold threads and trace connections . . . but essentially it *is* a new literature,
> for while we have long had a 'human geography' we have never before had an explicitly
> 'humanistic geography' with such a self-conscious drive to connect with that special body of
> knowledge, reflection, and substance about human experience and human expression, about
> what it means to be a human being on this earth.
>
> (p. 315)

Reviewing the variety of similar calls for humanistic geography of the 1970s and 1980s, Cloke *et al.* (1991, p. 69) also noted the way that they emerged from a critique of spatial science:

> It almost goes without saying that the geographers referenced here were unhappy about the
> tendency of spatial science to treat people as little more than dots on a map, statistics on a graph
> or numbers in an equation, since the impression being conveyed was of human beings 'whizzing'
> around in space – travelling from place X to place Y; shopping in centre X rather than in centre Y;
> selling produce in market X rather than at market Y – in a fashion little different from the
> 'behaviour' of stones on a slope, particles in a river or atoms in a gas. Indeed, it was complained
> that such exercises . . . effectively converted human beings into 'dehumanised' entities drained
> of the very 'stuff' (the meanings, values and so on) that made humans into humans as opposed
> to other things living or non-living.

Seventeen years later, one of those authors similarly characterised humanistic geography as:

> driven by a belief that spatial science failed to take seriously the lived experiences, decisions and
> struggles of human beings as human beings (not as dots on maps, numbers in tables or flow-
> lines in diagrams). To some degree, such a worry emerged within spatial science, occasioning a
> behavioural move . . . inquiring into the bases of people's environmental perception, cognition
> and decision-making . . . For other scholars, however, there could be no truck with spatial

science because its very philosophical and methodological precepts disqualified it from embracing humanity in its profounder, meaningful and even spiritual dimensions.

(Philo, 2008, pp. xxix–xx)

The nature of those criticisms, the alternatives advanced and the developing practices of humanistic geography are the subject of this chapter.

Humanistic geography's antecedents in cultural and historical geography

As indicated in earlier chapters, relatively few cultural and historical geographers were attracted by the so-called quantitative and theoretical revolutions. There was some application of statistical procedures to make work in these fields appear more contemporary or 'modern', but few substantive changes were adopted by adherents working within the traditional approaches (despite the efforts of the ad hoc committees established to chart ways forward for the 'new geography': see p. 85, this book). By the 1970s, however, some historical and cultural geographers had taken the initiative and were proposing alternative philosophies to positivism which were humanistic in their orientation: according to Hugill and Foote (1994, p. 12), prior to the Second World War 'cultural geography was more advanced in its theories than was the rest of human geography', but the challenge of the 'revolutions' outlined in the previous two chapters meant that 'Cultural geography ceased to be the most theoretically informed of the subdisciplines and embarked on a long search for theory that is still underway' (p. 14).

The beginnings of these changes can be traced to two papers, one of which had more impact on the geographical discipline at large. In the first, John K. Wright (1947, p. 12) introduced the term 'geosophy', defined as the study of geographical knowledge:

it covers the geographical ideas, both true and false, of all manners of people – not only geographers, but farmers and fishermen, business executives and poets, novelists and painters, Bedouins and Hottentots – and for this reason it necessarily has to do with subjective conceptions.

(p. 12)

Wright conceded that study of such subjective ideas could not employ the strict scientific principles of physical geography, but claimed that it provided indispensable background and perspective to geographical work:

geographical knowledge of one kind or another is universal among men, and is in no sense a monopoly of geographers . . . such knowledge is acquired in the first instance through observations of many kinds. . . . Its acquisition, in turn, is conditioned by the complex interplay of cultural and psychological factors . . . nearly every important activity in which man engages, from hoeing in a field or writing a book or conducting a business to spreading a gospel or waging a war, is to some extent affected by the geographical knowledge at his disposal.

(pp. 13–14)

These words could well have heralded an earlier start to behavioural geography than that chronicled above, but the lack of reference to them in later published works suggests that they had little impact until taken up by Wright's colleague at the American Geographical Society, David Lowenthal (1961, p. 259), in a widely cited paper 'concerned with all geographic thought, scientific and other: how it is acquired, transmitted, altered and integrated into conceptual systems'. Lowenthal argued in a wide-ranging survey that the world of each individual's experience is intensely parochial and covers but a small fraction of the total available. There are consensus views about many aspects of the world, but individuals will often mistakenly assume that their view is the consensus. We all live in personal worlds, which are 'both more and less inclusive than the common realm' (p. 248). Our perceptions of these worlds are personal too; they are not fantasies, being firmly rooted in reality, but because 'we elect to see certain aspects of the world and to avoid others' (p. 251) behaviour based on such perceptions must have its unique elements. Different cultures have their own shared stereotypes, however, which are often reflected in language, and attempts are made to create environments fitting into these stereotypes:

> The surface of the earth is shaped for each person by refraction through cultural and personal lenses of custom and fancy. We are all artists and landscape architects, creating order and organizing space, time, and causality in accordance with our apperceptions and predilections.
>
> (p. 260)

His ideas were implemented in papers concerned with the interpretation of landscapes as reflections of societal norms and tastes (e.g. Lowenthal, 1968; Lowenthal and Prince, 1965), thereby belatedly bringing Wright's ideas before a wider, and perhaps more readily appreciative, audience.

The second of the original papers was by a British geographer, although it was published in India (Kirk, 1951; it is reprinted in Boal and Livingstone, 1989; see also Johnston, 1993c); the main arguments were reiterated in a later article (Kirk, 1963). Kirk stressed that the environment is not simply a 'thing' but rather a whole with 'shape, cohesiveness and meaning added to it by the act of human perception' (Kirk, 1963, p. 365); once this meaning has been ascribed, it tends to be passed to later generations. Thus, he recognised two, separate but not independent, environments: a phenomenal environment, which is the totality of the earth's surface, and a behavioural environment, which is the perceived and interpreted portion of the phenomenal environment: 'Facts which exist in the Phenomenal Environment but do not enter the Behavioural Environment of a society have no relevance to rational, spatial behaviour and consequently do not enter into problems of the Geographical Environment' (p. 367). Since much geography is concerned with decision-making and its consequences, appreciation of the behavioural environment should be central to its study. Indeed, according to Ley (1977a), one cannot proceed without such awareness of what is in the behavioural environment; even an apparently neutral statement such as 'Pittsburgh is a steel town' is, he argued, a value-laden view of a geographer-outsider, which may not accord with the perceptions of the resident-insiders, so that:

> Too often there is the danger that our geography reflects our own concerns, and not the meanings of the people and places we write of. . . . The geographical fact is as thoroughly a social product as the landscape to which it is attached.
>
> (Ley, 1977a, p. 12)

Disciplines, as indicated in Chapter 1, usually comprise a number of 'invisible colleges', groups of scholars working on the same topic who refer to each other's publications. Wright, Lowenthal and Kirk were not members of any substantial 'college' during the early 1960s, and so had little impact on the first phase of behavioural work identified in the previous chapter. None of the three is in Pred's (1967, 1969) reviews of the field, for example, nor referred to by Wolpert's (1964, 1965, 1967, 1970) work through the decade. Golledge et al. (1972, p. 75) make only a passing reference to Lowenthal's work in their overview of behavioural approaches in geography:

> Pursued by insightful researchers, the analysis of literary and other artistic data of past and present can have strong explanatory power. The subjective element in these attempts to assess

the impact of spatial perception is acknowledged, but its presence in many other studies is more subtle and potentially damaging.

Kirk (1978, p. 388) referred to his influence on non-geographers who adopted the concept of the behavioural environment, however, and Spate (1989, p. xix) later described Kirk's initial paper as 'the Catalytic Crystal in the Saturated Solution', a trigger for the 'crystallization' of a humanistic geography (for a detailed exegesis of Kirk's work, see J. Campbell, 1989). The behavioural work of the mid-1960s was in the positivist mould; that of Wright, Lowenthal and Kirk just discussed was not, however, and so they were largely ignored by the former, who were in the ascendancy.

Several cultural and historical geographers did take up the concept of the behavioural or perceived environments though. One of the leaders was Brookfield, a British geographer with field experience, by the mid-1960s, of South Africa, Mauritius, New Guinea, and several Pacific Islands. In reviewing work by cultural geographers on these societies, he noted:

> A difference of approach is apparent between those who have an overtly chorographic purpose, who scarcely ever seek explanations in matters such as human behaviour, attitudes and beliefs, social organization, and the characteristics and interrelationships of human groups, and those whose inquiries are not primarily chorographic, and who are more inclined to undertake a search for processes as a means of reaching explanation.
>
> (Brookfield, 1964, p. 283)

Social organisation provides the key to many explanations, he argued, so that:

> when an individual human geographer is sitting down in one small corner of a foreign land, and seeks to interpret the geography of that small corner, then it is difficult for him to do so without trying to comprehend the perception of environment among the inhabitants.
>
> (p. 287)

Geographers had largely failed to delve into such details of social organisation, however, because of the broad areal scale at which they tended to work, their concern with distributions rather than with processes, and their avoidance of what he terms 'micro-geography'. Enquiry in human geography should involve three stages:

1 general statements about areal patterns and interrelationships;
2 detailed local inquiries which follow up the questions about processes raised by these general statements; and
3 organisation of the general and local material to produce explanatory generalisations.

Brookfield argued for more microgeographical studies at the second of these stages, to provide the foundation for the development of comparative methods with which generalisations could be forged. Brookfield's (1969, p. 53) later literature survey showed that 'decision-makers operating on an environment base their decisions on the environment as they perceive it, not as it is. The action resulting from decision, on the other hand, is played out in a real environment.' Referring to the 'modern' behavioural work, as well as studies by cultural geographers, he pointed out the great problems involved in isolating the perceived environment – something which is 'complex, monistic, distorted and discontinuous, unstable and full of unwoven irrelevancies' (p. 74) – and of building it into an analytical methodology. Further data are needed, too, on such topics as work organisation, time allocation, and budget allocation, on the meaning of consumption and of distance – all necessary tasks for the full understanding of people-environment systems.

The concept of the perceived environment has a considerable pedigree in historical–geographical scholarship, though without the scientistic terminology adopted by behavioural geographers. A key example was Glacken's (1956, 1967) survey of societal attitudes to environments. Historical geographers, Prince (1971a) suggested, must study a trilogy of worlds:

1   the real world, as recorded in the landscape;
2   the abstract world, as depicted by general models of spatial order in the past; and
3   the perceived world: 'Past worlds, seen through the eyes of contemporaries, perceived according to their culturally acquired preferences and prejudices, shaped in the images of their assumed worlds.'

(p. 4)

Using all three, it may be possible to explain landscape changes, which cannot be done by assuming processes from investigations of continua of data over time (see also Moodie and Lehr, 1976). Thus, 'it [is] the province of the intellect to observe the facts, to reduce them to order and to discover relationships among them, but it [is] the imagination which [gives] them meaning through the exercise of judgement and insight' (Prince, 1961–2, p. 21). Reconstruction of past environments is extremely difficult, however, for it involves seeing the written record through the cultural lens of the writer:

> A study of past behavioural environments provides a key to understanding past actions, explaining why changes were made in the landscape. We must understand man and his cultures before we can understand landscapes; we must understand what limits of physical and mental strain his body will bear; we must learn what choices his culture makes available to him and what sanctions his fellows impose upon him to deter him from transgressing and to encourage him to conform.
>
> (Prince, 1971a, p. 44)

Perhaps the enormity of such a task explains why most successful reconstructions concerned the relatively recent past, such as the perceptions which guided the settlement of the American West (e.g. Lewis, 1966), although Wright (1925) essayed a similar task for Europe at the time of the Crusades. Not all perceived worlds refer to either past or present; some landscapes have been fashioned out of Utopian views of the future (Porter and Lukermann, 1975; Powell, 1971). In general – however, and whether of past, present or future, geosophy was not a popular field of study, despite some intriguing essays (Lowenthal and Bowden, 1975). But directly or indirectly, it led to arguments for alternative approaches in human geography to those of the positivists and spatial scientists, or the behavioural geographers and those arguments form the material for the rest of this chapter.

Alternatives to spatial science

From the early 1970s on, some cultural and historical geographers attacked the positivism of spatial-science. To replace it, a variety of humanistic approaches was proposed, focusing on decision-makers and their perceived worlds, and denying the existence of a straight-forward objective world which can be studied by positivist methods. The intent was to reorient human geography towards a more humanistic stance, to resurrect its synthetic character, and to re-emphasise the importance of studying unique events rather than the spuriously general.

Critics and defenders of positivism

The various humanistic approaches have much in common, but they can be separated into different proposals in the present context. The first discussed here is associated with two scholars – Richard Colebrook (Cole) Harris and Leonard Guelke – who were together at the University of Toronto at the end of the 1960s; both are historical geographers.

The basic theme of Harris's (1971) initial paper was that geography is a synthetic discipline, concerned with particular assemblages of phenomena and not with the science of spatial relations (see Cosgrove, 1996). Thus:

> When the history of North American geography in the 1950s and 1960s is written, a paradox with which it will have to deal is how, with little argued, logical justification, so many geographers came to see their subject as a science of spatial relations.
>
> (p. 157)

With May and Sack (p. 121, this book), Harris saw the spatial perspective producing a dismemberment of geography, as specialists communicate more with their contemporaries in other disciplines than with other geographers, and develop theories that are descriptions of how the world might operate under certain conditions, rather than of how it actually works:

> The difficulty in conceiving of geographical theory comes down to this. The development of theory is necessarily an exercise in abstraction and simplification in which the complexities of particular situations are eliminated to the point at which common characteristics become apparent. But if geography is thought to have a particular subject matter, it is certainly not individual phenomena or categories of phenomena which other fields do not study. Rather it is a whole complex of phenomena, many or all of which may be studied individually by other fields but which are not studied elsewhere in their complex interactions.
>
> (p. 162)

The clear parallel is with history, for 'Few historians would attempt to develop a general theory of revolutions. In so doing they would lose grasp of the type of insight that characterizes good historical synthesis' (p. 163). The goal of both history and geography is synthesis, therefore. In developing syntheses, positivist methods may be applicable: historians may be law-consumers, applying the generalisations of other social scientists to particular events, and geographers could operate likewise. Alternatively, both historians and geographers could apply the idealist method, arguing that all activity is based on personal theories. Thus, Harris (1978, p. 126) – who gained his Ph.D. at the University of Wisconsin where he was supervised by Andrew Clark (see p. 62, this book), and was also clearly influenced by Hartshorne, wrote of a 'historical mind':

> Such a mind is contextual, not law-finding. Sometimes it is thought of as law-applying but, characteristically, the historical mind is dubious that there are overarching laws to explain the general patterns of human life.

It is open and eclectic, Clark argued, uses no formal research procedures, sees things in context, is sensitive to motives and values, excludes little and is wary of sweeping generalisations. Its goal is understanding, not planning, and this should be the case with the 'geographical mind' too. To understand an event is to appreciate why it took place, which is the humanistic goal; to explain an event is to predict it, as an instance of a general law or suite of such laws, and that is the positivist goal. (Interestingly, few others have addressed this issue of a 'geographical mind'. Two have been attracted to a parallel concept of 'geographical imagination'. To Harvey (1990; see also Harvey, 2005a) the geographical imagination focuses on 'the historical geography of social conceptions of space and time', with the goal of stimulating critical reflection on the spatial and temporal components of alternative societies to capitalist; to Gregory (1994, p. 11) it involves the roles of space and time, plus nature, in everyday life, just as Mills's (1959) sociological imagination involves the dialogue of biography and history.)

Developing his theme of the parallel between history and geography, Harris (1971, p. 167) identified four points of agreement concerning the nature of history:

1   its primary concern is with the particular;
2   explanation may take into account the thoughts of relevant individuals;
3   explanation may make use of general laws; and
4   'Explanation in history relies heavily on the reflective judgement of individual historians.'

From these, he concluded that 'If geography aims to describe and explain not so much particular events or peoples, as particular parts of the surface of the earth, then these points of agreement about history also apply to geography' (p. 167). The landscape results from actions; behind those actions lie thoughts; study of thoughts allows understanding of landscape. Thus, synthesis is crucial, since

> the idea of synthesis itself becomes more important as it becomes obvious that our larger problems transcend narrow subject-matter fields . . . integration . . . in a larger understanding is still achieved, however aided by statistical methods and computers, by the judgement of wise men who have cultivated the habit of seeing things together.
>
> (p. 170)

Geographers are presumably to be those wise individuals. This was not an original claim for, according to Buttimer the basis of Paul Vidal de la Blache's work was that:

> The task which no other discipline with the possible exception of history claims is to examine how diverse phenomena and forces interweave and connect with the finite horizons of particular settings. Temporality and spatiality are universal features of life so historical and geographical study belong together.
>
> (1978a, p. 73)

Positivist work seeks the same end, interweaving parts of a whole, but its parts are instances of general laws, not unique events.

Many of Harris's arguments were extended by his Toronto graduate student, Leonard Guelke, whose first paper was a strong criticism of the 'narrowly conceived scientific approach' (Guelke, 1971, p. 38) to geography using the positivist method. He argued against geography as a law-seeking activity by asking the proponents of the positivist approach to indicate how their laws would meet the basic standards of scientific acceptability, particularly with regard to prediction. Whereas they might be able to produce generalisations concerning the phenomena which they actually studied, he felt it very unlikely that they could generate laws applicable to all examples of the relevant phenomena. Statistical regularities are not laws and

> Until the new geographers have shown that the laws that might conceivably be discovered in geography will be more than generalizations, which describe common but non-essential connections between phenomena, their claims must be treated cautiously . . . there is little cause for optimism, especially as the statistical methods widely employed by geographers cannot be considered appropriate law-finding procedures.
>
> (p. 42)

With regard to the idea of geography as a law-applying science, Guelke argued that laws of human behaviour are virtually impossible to conceive in anything but the most generalised form because so much behaviour is culturally specific; an a priori statement of the determining conditions for their operation is thus not feasible so that 'Human geographers cannot consider themselves to be law-applying scientists . . . because they have no laws to apply' (p. 45).

Turning to the use of theories and models in geography, Guelke pointed out that for them to serve a valid purpose in the pursuit of understanding, criteria must be specified which indicate how such devices are testable against reality (see also Newman (1973) on the vague use of the term 'hypothesis'). Such criteria have not been, and cannot be, stated, Guelke claimed; for example, studies purporting to test central place theory seemed to operate on the rule that 'one counts one's hits but not one's misses' (p. 48; see also Guelke, 1978, p. 50). Too often, failure to reproduce reality was explained by claims that the test environment was not entirely suitable, and ad hoc hypotheses were frequently adduced to account for observed disparities. Models and theories may have heuristic value for human geographers, clarifying certain aspects, therefore, but they can have no explanatory power. Guelke concluded that:

> The new geography . . . has not yet produced any scientific laws and . . . appears unlikely to produce them in the future. . . . The theories and models . . . are not amenable to empirical testing. . . . The new geographers insisted on . . . logical and internally consistent theories and models. Yet, none of their theoretical constructs were ever complex enough to describe the real world accurately. They had achieved internal consistency while losing their grip on reality.
>
> (1971, pp. 50–1)

His offered alternative to that discredited procedure was an 'idealist' approach mentioned by Harris, which is 'a method by which one can rethink the thoughts of those whose actions he seeks to explain' (Guelke, 1974, p. 193). Understanding these should not be constrained by a single geographical theory. Instead, theory comprises 'any system of ideas that man has invented, imposed, or elicited from the raw data of sensation that make connections between the phenomena of the external world' (p. 194). Many such theories are part of the individual's society and culture; they include its religions, myths and traditions. Using them, 'the explanation of an action is complete when the agent's goal and theoretical understanding of his situation have been discovered . . . One must discover what he believed, not why he believed it' (p. 197). Thus, human geographers do not need to develop theories, since the relevant theories, which led to the action being studied, already exist (or existed) in the minds of the actors involved. The analyst's task is to isolate those theories (a task of considerable difficulty, according to Curry (1982a), because people cannot always identify the reasons for their actions). Some of them may be unique to particular individuals, but most are shared in large part by (sometimes great) numbers of actors; they represent the order which people themselves have stamped on the world and do not require further theories in order to be understood.

Guelke's argument was challenged by John E. Chappell (1975), who pointed out that focusing on the individual actor alone omitted any reference to the environmental constraints and influences on that person's actions (see also Gregory, 1978a). Guelke (1976) accepted the existence of such constraints and influences, but claimed that their investigation lay outside the geographer's domain. Study of environmental causes would, he felt, lead into physiology and psychology, and deflect attention from 'the most critical dimension of human behaviour, namely the thought behind it' (p. 169). Chappell (1976) responded that 'to go so far as to say that there is no possible respectable theory to explain man's rational theories and the actions which flow from them' (p. 170) is to be myopic: 'paradigms not only explain facts but they guide the research of whole disciplines' (p. 171). To him, Guelke's contention that the ultimate causes of actions lie outside the scope of human geography places geographers in an inferior position in the academic division of labour.

In a further essay, Guelke (1975) further sought to counter arguments for the approaches and techniques of positivism. With reference to historical geography, he argued that:

> It is obvious that quantitative techniques will often be useful. . . . [but] Statistical methods put in harness with positivist philosophy are a dangerous combination. . . . Historical geographers need to rethink not their techniques but their philosophy. . . . This can best be achieved by moving from problem-solving contemporary applied geography towards the idealist approach widely adopted by historians.
>
> (p. 138)

Stan Gregory (1976) agreed with the first part of this statement, but not with the proposed solution. Like Chappell, he saw the need to investigate individual action within its constraining structures (see Curry, 1982a).

Positivist approaches in human geography were defended by Hay (1979a), who both responded to the criticisms and raised points of contention with the proposed alternative. He argued that Guelke's case is ill-founded and rests on misconceptions of the nature of positivism, such as:

1   that all theory must be both normative and based on conceptions of optimal decision-making;
2   that to be scientific is always to be nomological (that is relating to basic laws); and
3   that prediction is the same as prophecy (rather than simply testing, from the known to the unknown).

He also claimed that Guelke presented an anti-positivist argument by using a positivist test and failed to realise the value of ad hoc hypotheses in the improvement of theory (the basis of Lakatos's concept of a research programme): Guelke should not, according to Hay, ask 'Does this theory explain Y?' but rather 'Does this theory contribute to an understanding of Y?' – again, a pragmatic approach to knowledge (Does it work?). To Guelke however:

> The assumption that thought lies behind human action is not related to the numbers involved. . . . If thousands of people drive motor cars to their places of work the idealist assumes that each of these journeys is a considered action involving thought. In such situations the investigator will not be able to look at each case individually, but he will seek to isolate the general factors involved in typical circumstances . . . [for which he] might well make use of statistical procedures . . . the value of statistical analysis will largely depend on its successful integration in the general interpretation or explanatory thesis being developed.
>
> (1978, p. 55)

These debates continued over the next decade (Curry, 1982b; Downs, 1979; Guelke, 1981b; Mabogunje, 1977; Mercer and Powell, 1972; Pickles, 1986; Rushton 1979; Saarinen, 1979). As part of them, Guelke (1982) restated his position that geography is an ideographic discipline seeking to understand the 'complexity of human activity on the land' (p. 52), drawing the boundaries such that 'The geographer is not concerned with explaining fluctuations in wheat prices or the level of interest rates, but he is concerned with the impact that these factors might have on, say, farming in western Canada' (p. 53). In seeking to understand the reasons for any action, he accepted that stated reasons may not always help and that a 'historical reconstruction of thought' is necessary to explicate the learned response which is 'part of an individual's cultural heritage' (p. 54). Guelke remained committed to this cause, and over a decade later wrote in a review of Derek Gregory's (1994) discussion of European colonisation in *Geographical Imaginations* that:

The geostrategic thinking of both the colonists and the colonized needs to be reconstructed in a way that provides a convincing explanation of how one group subdued others and how colonization itself was sustained within ideologies of white supremacy. This task involves explorations of how people saw themselves in societies and in relation to the earth, and of the contrasts that existed between different systems of knowledge and territorial control and exploitation as they actually articulated themselves in specific geographical circumstances.

(Guelke, 1995, p. 185)

Phenomenology and related approaches

Phenomenology attracted considerable attention from those developing humanistic geography. Declaring that 'Humanistic geography is concerned with geographical experience – what the spatiality of the earth/world means to those whose lives are part of it – in contrast to those who study geography as externalized, objective data', Backhaus (2009, p. 137) also notes how 'phenomenology offers a methodology that can open up the full spectrum of spatial inscriptions of human life, the fundamental concern of human geographers'.

Talk of a methodology might not have attracted all humanistic geographers, but to some this was exactly the kind of 'philosophically articulated alternative to objectivist science' that could provide an alternative framework 'offering methodologies congenial and adaptable to their geographical concern for the experiential dimension' (Backhaus, 2009, p. 138). The first direct advocate for a phenomenological approach was Relph (1970), who was also associated with the Department of Geography at the University of Toronto. Despite a variety of specific interpretations, he noted that phenomenology's basic aim is to present an alternative methodology to the hypothesis-testing and theory-building of positivism, an alternative grounded in people's lived world of experience. Phenomenologists argue that there is no straightforwardly objective world independent of human existence – 'all knowledge proceeds from the world of experience and cannot be independent of that world' (p. 193). Thus, according to Entrikin (1976, p. 617), 'phenomenologists describe, rather than explain, in that explanation is viewed as [an observer's] construction and hence antithetical to the phenomenologist's attempt to "get back" to the meaning of the data of consciousness'. Seamon (1984, p. 4) too defined phenomenology as a 'descriptive science'.

The phenomenological environment's contents are unique to every individual, for each of its elements is the result of an act of intentionality – it is given meaning by the individual, without which it does not exist but through which it influences behaviour. Phenomenology is the study of how such meanings are defined. It involves researchers identifying how individuals structure the environment in an entirely subjective way; the researchers are presuppositionless, using no personal ideas in seeking to understand their subjects' ideas. Subjectivity becomes the focus of study. Phenomenologists may be satisfied with such empathetic understanding, but some seek to go further, however, and identify essences – elements in individual consciousness which control the allocation of meanings (Johnston, 1983a; Pickles, 1988, p. 252, distinguishes between 'transitional essences' and 'invariant and universal structures (understood carefully)'). Phenomenology then studies human appraisals. It works at the individual level, but may search for the common (imprinted not agreed) elements among those appraisals.

Relph's paper was followed by one from another geographer briefly associated with the University of Toronto (preceded by doctoral training in Berkeley and a period in Arizona, and latterly based in the American Midwest – Minnesota, then Wisconsin-Madison – Tuan, 1999). Geography for Yi-Fu Tuan (1971) is a mirror, revealing the essence of human existence and human striving: to know the world is to know oneself, just as careful analysis of a house reveals much about both the designer and the occupant. The study of landscapes is the study of the essences in the societies which mould them, in just the same way that the study of literature and art reveals much of human life. Such geographical study has its foundations in the humanities, rather than the social or physical sciences. Tuan (1974a, 1975b) illustrated it in a number of essays giving, for example, insights into such topics as the sense of place

> Humanistic geography achieves an understanding of the human world by studying people's relations with nature, their geographical behavior as well as their feelings and ideas in regard to space and place. . . . Scientific approaches to the study of man tend to minimize the role of human awareness and knowledge. Humanistic geography, by contrast, specifically tries to understand how geographical activities and phenomena reveal the quality of human awareness.
>
> (Tuan, 1976, pp. 266–7)

Tuan exemplified this with five themes: the nature of geographical knowledge and its role in human survival; the role of territory in human behaviour and the creation of place identities; the interrelationships between crowding and privacy, as mediated by culture; the role of knowledge as an influence on livelihood; and the influence of religion on human activity. Such concerns are best developed in historical and in regional geography; their value to human welfare is that they clarify the nature of the experience (see also Appleton, 1975, 1994). Indeed, Tuan (1978, p. 204) claimed that 'The model for the regional geographers of humanist leaning is . . . the Victorian novelist who strives to achieve a synthesis of the subjective and the objective', and quoted the first two pages of E. M. Forster's *A Passage to India* as a paradigm example. (Alternative examples – from a geographer, but published after Tuan wrote – are Mead, 1993, 2002.)

Tuan's corpus of work has not usually involved philosophical explorations, and he rarely claims allegiance to any particular approach. Rather, he has been involved in a variety of explorations of the interrelationships between people and environments:

> my point of departure is a simple one, namely, that the quality of human experience in an environment (physical and human) is given by people's capacity – mediated through culture – to feel, think and act . . . I have explored the nature of human attachment to place, the component of fear in attitudes to nature and landscape, and the development of subjective world views and self-consciousness in progressively segmented spaces.
>
> (Tuan, 1984, p. ix)

Thus, for example, he shows that fear is both a representation of the environment and an influence on the creation of environments (Tuan, 1979), that the creation of residential segregation reflects the retreat of individuals from wholes to segmented parts (Tuan, 1982), and that the construction of gardens reflects a desire to dominate the environment (Tuan, 1984). Such works are implicitly phenomenological in that they suggest the existence of general essences, or behavioural stimuli, but the term is not in the index to any of his books. Indeed, as Gregory argues:

> One of the most persistent motifs in Tuan's writings is an appeal to common sense and common experience: to a 'we' whose inclusive address presumes a shared human condition to which 'we' have access without theoretical mediation. His humanistic geography is, in essence, a moral-aesthetic discourse; it is contemplative, at once reflective and speculative, and yet – despite the model of the ideal conversation – at best studiously indifferent to the wider conversations that might be made possible through the theoretical.
>
> (1994, pp. 79–80)

These explorations by Tuan have covered a wide range of topics within what Gregory (1994, p. 385) terms 'ways of being-in-the-world'. Tuan, whose first work (including his Ph.D.) was in geomorphology, grounds his work in his desire to answer his wonderings, since childhood, regarding 'the meaning of existence: I want to know what we are doing here, what we want out of life' (Tuan, 1999, p. 94) – what he terms 'the big woozy questions' and which he related to his own individuality – 'my apartness from others' – which aroused in him both pride and anguish, and which resulted in him being comfortable in the open environments of deserts but dreading rainforests. The title of his autobiographical chapter on his research interests is 'Salvation by geography', because the discipline has 'directed my attention to the world, and I have found there, for all the inanities and horrors, much that is good and beautiful' (p. 115). His books and other essays illustrate those discoveries from the point of view of (p. 10): 'a maverick in the discipline to the degree that, unlike most other geographers, my landscapes are "inscapes", as much psychological conditions as material arrangements'. They are thus individual and particular, but assist others in illuminating their own 'inscapes'.

Tuan's 'inscapes' mean that he focuses more on place than space, from that binary pair of major geographical concepts (as in Tuan, 1977), thus to some extent seeking to correct the bias towards space that characterised 'geography as spatial science, and seeks to present them as complementary perspectives'. (For Taylor (1999, p. 10) Tuan's 'generally undervalued' 1977 book, which provides 'explicit treatment of what are arguably geography's two core concepts by unarguably one of the discipline's leading thinkers', has 'featured only marginally in the stories told of geography's recent past'; it was reprinted in 2001.) In a volume dedicated to Tuan and produced to mark his retirement, however, Adams *et al.* (2001) focus on place, something with texture, with a shape and feel which 'provides a glimpse into the processes, structures, spaces and histories that went into its making' (p. xiii), suggesting a weaving of the terms text and context, and which is conveyed through conversations (or communication acts: p. xiv). For them, *Textures of Place* involves highlighting:

> the geographic tradition of trying to understand the meanings and processes of place – their material and symbolic qualities – as well as the range of peoples and social relations that continuously define and create social and spatial contexts.
>
> (p. xiv)

In this mould, knowing the world involves knowing oneself – a goal that they associate with Tuan's (1974b) writings on topophilia. But for Tuan it is more than just 'knowing oneself'. For him:

> Geography is indispensable to survival. All animals, including American students who consistently fail their geography tests, must be competent applied geographers. How else do they get around, find food and mate, avoid dangerous places? History, in striking contrast, is an esoteric field of knowledge, and for that reason it plays only a small, indeed negligible role in the common tasks of life. It is not too far-fetched to call animals geographers, but it is to call them historians. With animals and probably most human beings, knowledge of the past is folded into present time and space – into geography, where the challenges of living lie.
>
> (Tuan, 2002a, p. 123; see also Tuan, 2002b)

Among other advocates for phenomenology, Mercer and Powell (1972, p. 28) argued that the use of positivism in geography 'left the subject with too many technicians and a dearth of scholars'; they claimed that land-use patterns can never be understood 'by the elementary dictates of geometry and cash register' (p. 42) and the world can only be comprehended through people's intentions and attitudes towards it. In a lengthy discussion of phenomenology and its development in other disciplines, notably sociology, they pointed to 'a very real danger of the research worker assuming that concepts which are cognitively organized in his own mind "exist" and are equally clearly organized in the minds of his respondents' (p. 26). They argued instead for research methods involving empathy between researcher and researched, which for geographers means 'that we make every effort to view problems and situations not from our own perspective, but from the actor's frame of reference' (p. 48) – a position they term 'disciplined naiveté'.

Buttimer (1974) made a similar case for geographers studying the values which permeate all aspects of living and thinking. The order, precision and theory produced by positivists are dearly bought – 'we often lose in adequacy to deal with the values and meanings of the everyday world' (p. 3) – and behavioural geography similarly fails to break away from the mechanistic, natural scientific view of humans as preconditioned responders to stimuli. Ley (1981) also identified the positivist elements in behavioural geography and the suggestions of operand conditioning: people are assumed to react in predetermined ways to particular stimuli, and so can be manipulated accordingly. On the other hand, 'An existentially aware geographer is . . . less interested in establishing intellectual control over man through preconceived analytical methods than he is in encountering people and situations in an open, inter-subjective manner' (Buttimer, 1974, p. 24). Such activity results in 'a meditation on life', with geographers providing more comprehensive mirrors than their colleagues from more specialised disciplines can achieve, thereby clarifying the structural dynamics of life. Prediction would be impossible, apart from 'the most routinized aspects of experience' (p. 29), but the deeper appreciation achieved would allow much more vital social action and planning than is possible using spatial science and behavioural approaches.

Buttimer (1976) also directed geographers' attention to the concept of 'lifeworld', an amalgam of the worlds of facts and affairs with those of values in personal experience – 'the pre-reflective, taken-for-granted dimensions of experience, the unquestioned meanings, and routinized determinants of behavior' (p. 281). Positivism should be rejected for analysing the lifeworld because it separates the observer from the object of study and thus constrains appreciation of the human experience. Phenomenology, on the other hand, provides a path to understanding, on which informed planning can be built:

> It helps elucidate how . . . meanings in past experience can influence and shape the present . . . extremely important as preamble not only to scientific procedure, but also as a door to existential awareness. It could elicit a clearer grasp of value issues surrounding one's normal way of life, and an appreciation of the kinds of education and socialization which might be appropriate for persons whose lives may weave through several milieux.
>
> (p. 289)

The result is an understanding of actions as those involved understand them, rather than in the terms of abstract, outsider-imposed models and theories. Having achieved that understanding, human geographers can transmit it, thereby helping the subjects of their investigation to understand themselves better and realise their potential. In this way, the applied geographer acts as a provocateur, stimulating human development but not forcing it (Buttimer, 1979).

Berry backed this phenomenological orientation, calling for

> a view of the world from the vantage of process metageography. By metageography is meant that part of geographic speculation dealing with the principles lying behind perceptions of reality, and transcending them, including such concepts as essence, cause and identity.
>
> (1973a, p. 9)

But not all accepted that phenomenology can entirely replace the positivist approach. Walmsley (1974), for example, accepted the merits of Buttimer's case because so many human decisions are based on 'experiential' rather than 'factual' concepts, but argued that the scale of geographical enquiry and its long tradition of certain types of empirical work required maintaining the positivist orientation. That the perceived world is not necessarily the same as the real world must be realised, but 'logical consistency and empirical truth will remain central to geographical enquiry provided the importance of values is recognized' (p. 106).

Gregory (1978a) criticised both positivism and phenomenology. Those favouring the former were condemned for making 'social science an activity performed on rather than in society, one which portrays society but which is at the same time estranged from it' (p. 51) and for supporting a procedure which, because it so often assumes *ceteris paribus* in testing its models, can never be sure why these fail to replicate reality (p. 66). He recognised the need for humanistic approaches, but argued that these would not be sufficient to provide a satisfactory foundation, because they ignore the 'constraints on social action which are so much part of the taken-for-granted life-world of the actors' (Gregory, 1978b, p. 166): 'A geography of the life-world must therefore determine the connections between social typifications of meaning and space-time rhythms of action and uncover the structures of intentionality which lie beneath them' (Gregory, 1978a, p. 139). But 'A major deficiency . . . is [the] restricted conception of social structure: in particular, it ignores the material imperatives and consequences of social actions and the external constraints which are imposed on and flow from them'. Phenomenology must be incorporated with broader investigations of those imperatives and constraints; such incorporation produces a critical stance, whose nature is discussed in Chapter 6.

Whereas a key feature of the positivist/spatial-science approach was a great numerical superiority of practitioners over preachers, the phenomenological movement was initially characterised by the converse – much preaching but relatively little practice (Relph, 1981a, pointed to the absence of substantive applications of the phenomenological approach in geography, though see Jackson and Smith, 1984, p. 44):

> There is an essential difference between the contemplative intentions of this transcendental philosophy and the practical concerns of a social science, so that it is scarcely surprising that . . . geographers' . . . efforts have been directed towards the destruction of positivism as a philosophy rather than the construction of a phenomenologically sound geography.
>
> (Gregory, 1978a, pp. 125–6)

Tuan's interpretative essays exemplify its use, and Relph, who 'would much prefer to see substantive applications rather than discussions of the possible uses of phenomenology' (1977, p. 178), published his thesis on *Place and Placelessness* in which – implicitly, he says – phenomenological methods are used 'to elucidate the diversity and intensity of our experiences of place' (Relph, 1976, p. i). His essential themes are the sense of place and identity in the human make-up and the destruction of this through the growing placelessness of modern design (see also Porteous, 1988). Other work generally quoted as phenomenological includes pieces on European settlement of the New World. Joe Powell (1972), for example, has written on images of Australia; his major work on the settlement of Victoria's western plains (Powell, 1970) examined the conflict between official and popular environmental appraisals, the dialogue between these, and the learning process which resulted in the final settlement pattern (see also Powell, 1977, where he presents the study of eiconics, or image-making, in the context of colonisation processes). Billinge queried whether such works are really phenomenological, however:

> the idea has spread that since certain branches of our discipline are less susceptible to quantitative reduction (and, so the argument continues, by false extension, to scientific analysis), we can justify our partially formulated hypotheses, exploit the atypicality of our data, cease worrying about the validity of our reconstruction and within some weakly articulated framework label the whole exercise phenomenological.
>
> (1977, p. 64)

But for Billinge, 'phenomenological we have by no means become' (p. 67) since geographers have paid little attention to consciousness.

Geographers' attempts to adopt a phenomenological approach were also trenchantly criticised by Pickles (1985), not because he opposed them in principle but rather because he believed that those undertaken were misconceived. He accepted that, if for no other reason than its attack on positivism, 'It cannot be denied that the founding and guiding intuitions of a phenomenological approach in geography as it exists at the moment are in the main sound and well intentioned' (p. 68), but argued that work such as Buttimer's and Relph's 'is ungrounded method, unfounded claims, and the actual imposition of unexamined propositions' (p. 71). His criticisms led him to contend that 'we now need to move from what passes for phenomenology in the geographical literature, towards what is actually the case in phenomenology itself' (p. 89). Quoting substantially from Husserl's and Heidegger's original works, he argued that phenomenology is not concerned to explicate subjective meanings as an end in itself, but rather 'to be the science of science through explicating the science of beginnings' (p. 97). Its goal is to identify the essences that underpin individual meanings:

> The essential relation between an individual object and its essence – such that to each object there corresponds its essential structure, and to each essential structure there corresponds a series of possible individuals as its factual instances – necessarily leads to a corresponding relationship between sciences of fact and sciences of essence.
>
> (p. 111)

The subjectivity of the lived world is to be explored for the insights that it can provide on those essential structures, the underpinnings of knowledge itself.

According to this interpretation, phenomenology has much in common with certain forms of structuralism in its basic concern with neither empirical appearances nor actual decision-making but rather with the deep structures of consciousness (see p. 207; Johnston, 1986c; Pickles, 1988, p. 252, implies this too). Thus, it does not fit easily into the humanistic concerns of most geographers who have espoused phenomenology, which Pickles identifies as the search for those essences that give rise to the necessity for geography, as the empirical science concerned with particular facets of behaviour. (Each empirical science should be grounded in such an essence.) Thus:

> we seek an ontological, existential understanding of the universal structures characteristic of man's spatiality as the precondition for any understanding of places and spaces as such. That is, we seek to clarify the original experiences on the basis of which geography can articulate and develop its regional ontology if geography as a human science, concerned with man's spatiality, is to be possible at all.
>
> (p. 155)

Spatiality is that essence, he claims, best represented by the German noun *Raum* and verb *raumen*, 'which means to clear away, to free from wilderness or to bring forth into an openness. *Raumen* is thus a clearing away or release of places, a making room for the settling and dwelling of man and things' (p. 167). In this context, space and place are closely related concepts (see also Gould, 1981b).

Phenomenology is closely associated with existentialism, and some geographers have experienced difficulty in separating the two (Entrikin, 1976). Whereas phenomenology assumes the primacy of essence – the allocation of meanings results from the existence of consciousness – for existentialists the basic dictum is 'being before essence – or man makes himself'. The process of defining oneself (creating an essence) involves creating an environment. Thus, Samuels (1978, p. 31) argued that environments can be read as biographies – 'for every landscape or every existential geography there is someone who can be held accountable'. Generalisation may be possible from analysing such landscapes. Appleton (1975), for example, suggested that landscapes reflect two primal needs – prospect (the need to search for the means of survival), and refuge (the need to hide from threatening others) – although individuals and groups may satisfy these in particular ways (as he illustrates from his autobiography: Appleton, 1994). Lowenthal (1975; see also Lowenthal, 1985) argued that individuals rewrite their biographies, and those of their ancestors, by their choices of what to preserve in the landscape.

All of this suggests that humanistic geography is concerned either with the study of individuals and their construction of, plus behaviour in, phenomenal environments (as in Rowles, 1978) or with the analysis of landscapes as repositories of human meaning. As such, it is separate from the subject matter of much human geography, notably behavioural geography and its investigation of everyday activity within environments. But the phenomenological perspective has been adapted to the latter type of work also, in the application of Schutz's writings on the 'taken-for-granted' world (Ley, 1977b). Much everyday behaviour is unconsidered in that it involves no original encounters with new situations. The behaviour is habitual, because all of the stimuli encountered can be processed as examples of particular types (on which, see Kahneman, 2012). Those types are not externally defined for the individual, but are personally created. The phenomenology of the taken-for-granted world is the study of those individually defined typifications – of the unconsidered 'world of social reality' rather than 'a fictional non-existing world constructed by the scientific observer' (Ley, 1980, p. 10, quoting Schutz; see also Michael Curry's, 1982a, p. 38, discussion of 'the ordinary, everyday actions of individuals' which create for the geographer 'a complex world of complex places and actions' in which the role of the individual decision-maker 'can be determined only on an individual basis, case by case'). Interaction within communities may lead to shared typifications, which quantitative methods may be used to identify descriptively.

Humanistic geography is based on a profound critique of positivist work, which it claims makes major unwarranted assumptions about the nature of decision-making, and seeks inductive laws of human behaviour that can be scientifically verified (Ley and Samuels, 1978; Powell, 1980b). Its counter-argument promotes understanding of the individual as a 'living, acting, thinking' being. To some, however, it is foremost a form of criticism – Entrikin argued that:

> humanist geography does not offer a viable alternative to, nor a presuppositionless basis for, scientific geography as it is claimed by some of its proponents. Rather the humanist approach is best understood as a form of criticism. As criticism the humanist approach helps to counter the overly objective and abstractive tendencies of some scientific geographers.
>
> (1976, p. 616)

For others, however, the human condition can only be discerned through humanistic endeavour, for attitudes, impressions and subjective relations to places (the 'sense of place') cannot be revealed by positivist research. As Pickles (1986, p. 42) put it, 'The value of humanism has been its resilience in consistently raising questions which do not fit within other frameworks ... humanism has been the voice of man against reason, against science'. In similar terms, for Buttimer (1979, p. 30), unless tempered by humanistic impulses, geography risks being managerial, manipulating individuals and their environments rather than seeking to advance the process of 'human becoming'. Relph (1981b, pp. 139–41) coined the term 'geographism' to describe 'the view that people should behave rationally in geographical, two-dimensional space ... that cities and industries and transportation routes should be arranged in the most efficient way' which, when used as a basis for planning, 'will diminish the distinctiveness and individuality of ... communities and places. Geographism involves the imposition of generalizations onto specific landscapes; it breeds uniformity and placelessness.' Both Buttimer and Relph argued that planning should emphasise subjectivity and individuality. Scientists, engineers and planners may seek to improve well-being, but in so doing they make people rootless and deny them individuality. Planning must be allied with environmental humility, by which 'places and communities would increasingly become the responsibility of those who live and work in them instead of being objects of professional disinterest' (Relph, 1981b, p. 201). In this context, humanistic geography is not only a reaction against the dehumanising treatment of people in spatial science and behavioural geography, but also an argument against an applied geography which imposes that treatment on the landscape, and for a form of anarchism in which individuals are encouraged to realise what they are and how they can control both themselves and their environments (see also Pickles, 1986, p. 47). Seamon (1984) also argued that phenomenological work could be 'a learning tool which can help us to discover more about ourselves, others, and the world in which we live' (p. 21), and which may then have practical value in environmental design.

The practice of humanistic geography

Much of the practice of humanistic geography has been concerned with exploring and explicating the subjectivity of human action and its base in meanings (both individual and shared), with relatively little concern for the claims of phenomenological, existentialist, and other philosophies. (Curry, 1996, p. 16, for example, claims that although the earliest figures involved in the break from positivism all referred to phenomenology, 'it seems fair to say that none took it very seriously'.) To the extent that such work is explicitly influenced by philosophical and methodological writings, the (albeit usually implicit) stimulus is more often the pragmatism (and symbolic interactionism) developed by the Chicago school of sociologists in the 1920s and 1930s (and largely overlooked by human geographers who paid much more attention to the relatively small amount of spatial analysis undertaken by that school, as typified by the Burgess model of the internal structure of cities: Johnston, 1971). Pragmatism portrays life as a continuous process of experience, experiment and evaluation through which beliefs are continually reconstructed; such reconstruction is a social process, whereby individuals learn and behave in the context of the beliefs of those with whom they interact (hence the term 'interactionism': see Jackson and Smith, 1984, Chapter 4; both S. Smith, 1988, and Jackson, 1988, illustrate this in their own research).

Understanding social life within this broad framework involves participatory field work (as detailed in Evans, 1988), the methodology of which, according to Susan Smith (1984, p. 353), is 'a hallmark of much geographic humanism'. This, she continues:

> requires a commitment to fieldwork, with the aim of securing data lodged in the meanings ascribed to the world by active social subjects. The strength of this strategy derives from the unique insight it offers into 'lay' or 'folk' perceptions and behaviors. True to the pragmatic maxim, the method allows the truth of a social reality to be established in terms of its consequences for those experiencing it.
>
> (pp. 356–7)

How local societies work is thereby explicated, which was the goal of the ethnographic work of the Chicago school (Jackson, 1984, 1985). Ley's (1974) detailed portrayal of a Philadelphia community illustrates this, as does his general textbook on urban social geography (Ley, 1983).

Field studies, involving interaction with residents over long or short terms, provide a rich source of information about how people structure their lives. Although mindful of how ideas about the field and fieldwork are mediated through technologies, academic structures and theoretical reframing, it is not hard to detect influences from humanistic geography (and those earlier sources on which it sought inspiration, such as Sauer and Vidal de la Blache) in the introduction to a special issue of *The Geographical Review* on 'Doing fieldwork':

> We learn about ourselves in the field: we learn about places from doing fieldwork there. Whatever
> our degree of theoretical engagement, it is what we know about a part of the world – or all of it
> – that is our first fact. And there fieldwork commences.
>
> (DeLyser and Starrs, 2001, p. vii)

Humanistic geography was long sensitive to the range of texts that could become sources, along with other repositories of meanings; Pocock (1983) described humanistic strategies ranging 'from library search, to the observational, to the experiential' (p. 356). The landscape, the creation of those who live/have lived in it, is an important text, and Lowenthal (1985) provides a perspective on its study. He portrays it not as a mirror of the past but rather as an insight to the present:

> Traditions and revivals dominate architecture and the arts; schoolchildren delve into local
> history and grandparental recollections; historical romances and tales of olden days deluge
> all the media. The past thus conjured up is, to be sure, largely an artifact of the present.
> However faithfully we preserve, however authentically we restore, however deeply we immerse
> ourselves in bygone times, life back then was based on ways of being and believing
> incommensurable with our own. The past's difference is, indeed, one of its claims: no one would
> yearn for it if it merely replicated the present. But we cannot help but view and celebrate it
> through present-day lenses.
>
> (p. xvi)

The present landscape is a conglomerate of relics from many different periods in most cases, and by preserving only parts of it we bias our representation of the past:

> Every act of recognition alters survivals from the past. Simply to appreciate or protect a relic, let
> alone to embellish or imitate it, affects its form or our impressions. Just as selective recall
> skews memory and subjectivity shapes historical insight, so manipulating antiquities refashions
> their appearance and meaning. Interaction with a heritage continually alters its nature and
> context, whether by choice or by chance.
>
> (p. 263)

Thus, when use of the landscape as a text involves reading the outcome of a long sequence of selective retentions of earlier forms, so that we learn about what parts of their history people wanted to build into their own presents and futures, 'We must concede the ancients their place . . . But their place is not simply back there, in a separate and foreign country; it is assimilated in ourselves, and resurrected into an ever-changing present' (p. 412). The past does not exist independently of those who seek to interpret it (as Taylor, 1988, makes clear in a very different context); the landscape may tell us more about the past which people wanted to preserve than about the past as it was experienced. (It also tells us much about the people who describe it, as Porteous, 1986, illustrates in his discussion of body imagery as a metaphor for landscape description.) Study of the current processes of landscape-creation illustrates this process, as with the creation of various types of 'living museum' and the promotion of places as 'cities of spectacle' (as in Crang, 1994; Jacobs, 1994a).

An increasingly-used text for the explication of meanings as humanistic geography evolved was literature – or 'creative literature' in Paul White's (1985) term; Porteous (1985) refers to 'imaginative literature', with which geographers have been highly selective:

> Plays are not considered, poetry is but occasionally used, the novel reigns supreme. The
> advantages of the novel lie in its length (meaty), its prose form (understandable), its involvement

with the human condition (relevant), and its tendency to contain passages, purple or otherwise, which deal directly with landscapes and places in the form of description (geographical).

(p. 117)

He argued that when using novels as texts, geographers have concentrated on nineteenth-century, rural contexts (as in Darby's, 1948, paper on Hardy's Wessex). To correct this, he proposed a two-variable categorisation of situations according to whether the subject is an insider or an outsider, and whether the place being described is 'home' or 'away'. The 'home-insider' provides material on sense of place, whereas the much less frequently reported 'away-outsider' refers to those experiencing alienation in a placeless world. (White's, 1985, use of novels to describe migrants' situations fits readily into this category.) 'Home-outsiders' are people who fail to develop insider relationships with their milieux, whereas 'away-insiders' are travellers reporting their experiences (as in 'road, tramp, and down-and-out novels'; Porteous, 1985, p. 119). Such distinctions are reflected in a collection of essays about literature and migration (King *et al.*, 1995), which distinguishes between migrants' autobiographical accounts and general fiction about migration (Duffy, 1995). These are used to explore the renegotiation of identity that accompanies migration experiences (White, 1995), but as texts they are subjected to less critical scrutiny than that undertaken by deconstructionists (p. 259, this book). Similarly, science fiction and related literature gives insights to imaginative geographies of the future: it 'maps out possible future spatialities of the postmodern condition, and provides cognitive spaces which are being used by individuals and institutions in conceiving and making future society' (Kitchin and Kneale, 2001, p. 32). And reviewing the trajectory of 'literary geography', Angharad Saunders (2010, p. 436) argued that 'it was possible to identify a literary geographic tradition which stretched back to the early twentieth century and encompassed work of various theoretical and critical hues'.

Much humanistic geography has been written to describe and appreciate the variety of the human condition, as it is experienced. (Eyles's, 1989, essay, in a volume designed to 'introduce some of the most exciting challenges of the contemporary subject to a wider audience', considers place and landscape without reference to philosophy.) It has been relatively unconcerned with philosophical issues, such as the origins of meanings in human consciousness, and has focused almost entirely on the empirical worlds of experience, even if these have to be interpreted from secondary sources. However, J. Wreford Watson (1983), himself a prize-winning poet, claimed that literature is not a secondary source but 'primary source material for the whole world of images' (p. 397) that illustrates the 'soul' of a place. Its relevance, according to Pocock (1983), is that:

it attempts to unravel the nature of being-in-the-world, as it explores the existential significance of place as an integral part of human existence. In short, it is a geographical contribution to the most fundamental of questions, 'what is man?'

(p. 357)

Meinig (1983) expresses this more vigorously:

By limiting ourselves to describing and measuring and analyzing certain aspects of the world as it seems geographers have denied themselves the possibility of probing very deeply into what it all means. Being unable to convey what it means we cannot help shape it toward what it might become.

(p. 325)

Explicating what it means involves practising geography as art – placing the discipline firmly in the humanities as well as the sciences and the social sciences – as illustrated by Meinig (2002) in a biographical essay.

One problem for such work, brought home to geographers in a series of iconoclastic essays (Olsson, 1978, 1979, 1980, 1982, 2007; see also Pred, 1988; Abrahamsson and Gren, 2012), is that the medium of the text is constraining as well as enabling. Olsson (1982) focuses on language as the most frequently used medium, with the constraints that word definitions and usages put on their application, hence the modes of thought that they support. Thus:

> any social scientist is handicapped by the methodological praxis which requires him to be more stupid than he actually is. Thus, in the interests of discipline, verification and communication he relies mainly on the two senses of sight and hearing: what counts is what can be counted: what can be counted is what can be pointed to; what can be pointed to is what can be unequivocally named. Accumulation of knowledge about the nameable is consequently the point of the scientist's game.
>
> (p. 227)

Ambiguity, therefore, is translated into certainty; describing a phenomenon in words allocates it to a category and can oversimplify its complexity. Hence, our ability to think is made possible by the richness of language, yet also constrained by the categorisations that it imposes; the former aids our understanding but the latter can hinder it (as critical theorists and poststructuralists have explored: see Chapter 7).

Geographers' involvement in humanistic work involves more than interpretation of texts, however, since they themselves create new texts in transmitting their appreciations to others, usually in writing but also through maps, lectures and other media. This involves what is sometimes referred to as a double hermeneutic. Initially developed for the exegesis of biblical texts, hermeneutics was extended by Dilthey (Rose 1980, 1981) to embrace all studies which involve scrutinising an author's intentions when evaluating a text. This involves what Dilthey termed *Verstehen* (interpretative understanding), which not only enables students of a text to appreciate its author but also to increase their own self-awareness:

> Understanding a text from a historical period remote from our own . . . or from a culture very different from our own is . . . essentially a creative process in which the observer, through penetrating an alien mode of existence, enriches his own self-knowledge through acquiring knowledge of others.
>
> (Giddens, 1976, p. 17)

Thus, an applied humanistic geography promotes both self-awareness and awareness of others. A double hermeneutic is involved because it is not only 'the job of human geographers to interpret such texts as a spectator in order to make certain statements about actors operating within the texts', but also 'to communicate the meanings of such phenomena as he should deem important back to the actors involved' (Rose, 1981, p. 124) as well as to others who wish to understand those meanings. Thus, the reader of a research paper by a humanistic geographer is only interrogating those meanings through their interpretation by an intermediary – the humanistic geographer. The latter's selection of those items 'deemed important' means that different geographers may transmit different interpretations of the same text, in the same way as two biographers might interpret a novelist's texts differently and two artists may differ in their portrayal of the same landscape. (On hermeneutics in geography, see Livingstone, 2002, who has also written on 'the geography of reading' (Livingstone, 2005) as part of his general argument regarding the situatedness of knowledge production and reproduction: Livingstone, 2003b.)

Humanistic and 'old' and 'new' cultural geography

Detailed debates about philosophy and methodology in geography have engaged only a minority of the discipline's practitioners, so far as the empiricist/positivist approaches are concerned, whereas a majority have conducted their own empirical inquiries informed by those debates but only marginally connected to them. To a considerable extent this has been the case with humanistic geography too, with Hugill and Foote (1994, p. 18) contending that whereas humanistic geographers turned to such philosophies and methodologies as 'ethnomethodology, existentialism, idealism, phenomenology, symbolic interactionism, and transactionalism . . . [although] successful applications of each of these can be found in the geographical literature . . . none captured the imaginations of more than a handful of geographers at a time'.

A small number of writers have been concerned with philosophical underpinnings and the relative merits of different humanistic approaches, but more have accepted the general tenor of the argument, described by Boal and Livingstone as:

> We can give up the need to find direct empirical connections between terms and objects in the world; we can view knowledge not as presenting the world in some correct way, but as just helping us to get along in it, or to change it. Truth, to repeat, has nothing to do with accurately representing, or, as Rorty has it, mirroring reality; it is just, according to [Courtice Rose, 1987], 'what we are well advised, given our present beliefs, to assert'. The purpose of geography, then, is not to tell us about how the world 'really' is: it tells us nothing about regions or landscapes or economic structures or human agency, because these are mere linguistic fictions; it is just the search for the right vocabulary, the right jargon, the best discourse in which to pursue the kinds of account which help us, in the most basic sense, decide what to do.
>
> (1989, pp. 7–8)

Doing that, according to Pickles (1986, p. 29), involves 'archaeology', uncovering the layers of human behaviour to identify the experience that underpins it, which involves the explication of texts – the 'written records, cultural artefacts, urban landscapes or whatever' (Boal and Livingstone, 1989, p. 15) – which are the repositories of human meanings, and whose interpretation is the goal of humanistic geography.

Some of the work undertaken within this humanistic framework, and without the explicit philosophical underpinnings, goes under the rubric of a form of cultural geography, which has a long and distinguished background, notably in the USA, based on the work of Carl Sauer. Others found some inspiration in Vidal de Blache's commitment to regional description and exploration of the spirit of place, or work on landscape work in prior German geography (see the essays in Ley and Samuels, 1978). In other words, the reaction to spatial science may have been shared, but the literatures mobilised to enable humanistic geography were quite diverse. The story is further complicated by the later talk, from British geographers in the 1980s, of a 'new cultural geography' that would be informed not only by humanistic geography, but also by some of the radical and critical currents that are the focus of the next chapter. Cosgrove and Jackson (1987) note this hybrid origin, going on to declare that:

> If we were to define this 'new' cultural geography it would be contemporary as well as historical (but always contextual and theoretically informed); social as well as spatial (but not confined exclusively to narrowly-defined landscape issues); urban as well as rural; and interested in the contingent nature of culture, in dominant ideologies and in forms of resistance to them. It would, moreover, assert the centrality of culture in human affairs. Culture is not a residual category, the surface variation left unaccounted for by more powerful economic analyses; it is the very medium through which social change is experienced, contested and constituted.
>
> (p. 95)

The suggestion of a break between an 'old' and a 'new' cultural geography was denied in the review in *Geography in America*; Rowntree *et al.* (1989, p. 215) argued that 'Although traditional cultural geography has preferred topics with historical depth, there is increasing interest and emphasis on the study of everyday life and landscapes.' They recognised that trends originating in the UK, as promoted by Cosgrove and Jackson (1987), differed substantially from the more traditional concerns (as represented in the chapter on cultural ecology in the same volume: Butzer, 1989; see also Turner, 1989). 'Whether the British trajectory becomes an integral part of North American cultural geography remains to be seen'; quoting Kofman (1988), they suggested that 'the "new" cultural geography is more talked about than done' (p. 209). Hugill and Foote (1994, p. 18), however, saw the introduction of humanistic approaches (as exemplified by Ley and Samuels's, 1978, anthology) as 'an attempt to define a "new" cultural geography' in an 'attempted coup' which failed. Furthermore, they also portrayed a conflict between traditional cultural geography as practised in North America and promoted in the UK:

> a large number of [American cultural] geographers had continued to cultivate the traditional themes of the Berkeley school. . . . These geographers felt little need to accept humanistic geography and resented its new agenda. Humanistic research had little to say about traditional concerns. The unfortunate mismatch between American cultural geography and British social geography compounded problems of assimilation. The contributions of British social geography to recent advances in humanistic cultural geography have been increasing . . . but the overlap is slight.

James Duncan (1994) refers to this debate as:

> a civil war [that] has been going on in cultural geography since the early 1980s. This struggle has largely, though not exclusively, had an intergenerational character, younger cultural geographers trained in the late 1970s and 1980s assailing the positions of an older generation trained in the 1950s and 1960s.
>
> (p. 95)

The main issue, he argued, was theory:

> The younger generation launched its attacks using an arsenal of theory-seeking weapons provided by suppliers in the humanities and social sciences. Some direct hits were scored on the lightly camouflaged theories of the older generation, but such hits proved indecisive in the war for a number of reasons. Many of the older generation were unaware that they had any theories in their camp: others insisted that these were relic theories long since abandoned. Still others claimed that the theoretical targets had in fact been surreptitiously inserted into their camp by the younger generation to discredit them. At any rate, they argued that no meaningful losses had been sustained.

Others (e.g. M. Price and Lewis, 1993a) argued that the older generation was far from homogeneous in its views, however, and that 'traditional cultural geography' was being misrepresented by those advancing an alternative position.

Cosgrove and Jackson (1987) pointed not only to the continued vitality of cultural geography as the interpretation of past and present landscapes and other texts – for which they appropriated the term 'iconography' (as in Cosgrove and Daniels, 1988; see also Powell, 1977, to whom they do not refer) – but also to growing contacts with contemporary social geography. (Soon after they wrote, the Social Geography Study Group of the IBG was renamed the Social and Cultural Geography Study Group.) The field of contemporary cultural studies offered 'alternative ways of theorising culture without specific reference to the landscape concept' (p. 98), they claimed, with its emphasis on contemporary subcultures and their political struggles rather than 'the elitist and antiquarian predilections of traditional cultural studies'. This is illustrated in a special issue of *Society and Space* (Gregory and Ley, 1988), and in Jackson's (1989) text *Maps of Meaning*, which begins with a definition of culture that characterises a 'new' cultural geography:

> This book employs a more expansive definition of culture than that commonly adopted in cultural geography. It looks at the cultures of socially marginal groups as well as the dominant, national culture of the elite. It is interested in popular culture as well as in vernacular or folk styles: in contemporary landscapes as well as relict features of the past.
>
> (p. ix)

The work of Sauer and the Berkeley school (p. 53 and p. 97, this book) is criticised for being relatively narrow; culture is instead linked by Jackson to the contested concept of ideology, and his substantive chapters cover popular culture, gender and sexuality, racism and language (see Chapter 7). An agenda for future work suggests an even wider coverage, into aspects of social relations that impinge upon the economic organisation of society and thus link cultural geography more firmly with aspects of radical geography (as discussed, in Chapter 6). Work in this mould would presumably further the challenge to the content of traditional cultural geography as 'innocent' (Rowntree *et al.*, 1989, p. 214).

Subsequently, the concept of culture has itself been examined. The definition adopted by Duncan and others, in contrast to the 'superorganicism' of the Berkeley school and its followers, is described as 'socially constructed, actively maintained by social actors, and supple in its engagement with other "spheres" of human life and activity' (Mitchell, 1995, p. 102). For those who adopt this position, culture exists as, according to Cosgrove and Jackson (1987, p. 95), 'the very medium through which change is experienced, contested and constituted'. To Don Mitchell (1995) this means that:

> Culture, therefore, can be specified as something which both differentiates the world and provides a concept for understanding that differentiation. Culture itself is a sphere of human life every bit as important as, yet somehow different from, politics, economy and social relations. It is an important ontological category which must be theorized and understood if we hope to understand human differentiation, behaviour, experience and contest.
>
> (p. 103)

Against this, he argues that 'there is no such (ontological) thing as culture. Rather there is only a very powerful idea of culture, an idea that has developed under specific historical conditions and was later broadened as a means of explaining material differences'. As with the superorganic conception, he contended, the concept of culture promoted by the 'new cultural geography' involves reification and its adherents have reached 'something of a dead end' (p. 104): 'While important empirical work exploring the social creation of many aspects of life continues, none of this work has been able adequately to explain what culture *is*. Cultural geography has remained incapable of theorizing its object.'

Attempts to define culture, according to Mitchell, involve an infinite regress, using terms which themselves are neither internally coherent nor inclusive (see the discussion on metaphors on p. 258, this book). The regress is ended by claims that cultures exist and the term is used as part of a strategy to define and control:

> the naming and representation of cultures create partial, yet globalizing, truths. By localizing social interaction into discrete cultures ... contentious activities are abstracted into the partial truth contained in the idea of culture: namely that there are true and deep differences between people.
>
> (p. 109)

This strategy is crucial in many aspects of life, such as geopolitics (see p. 264, this book) and the promotion of consumerism; indeed, as the essays in Roger Lee and Jane Wills (1997) illustrate (see also Thrift and Olds, 1996; Wheeler, 2002b), it is increasingly difficult to separate the cultural from the economic. For Don Mitchell (1995), therefore:

> like 'race', 'culture' is a social imposition on an unruly world. What *does* exist, and very importantly, is the historical development of the *idea* of culture as a means of ordering and defining the world. . . . Culture is an idea that integrates by dividing . . .
>
> (p. 112)

As such, many cultural geographers were deeply involved in developments during the 1990s, characterised in Chapter 7 as 'the cultural turn'.

We return to these issues in Chapter 7. For now, it is sufficient to note that whereas traditional cultural geographers paid little explicit attention to methodological concerns, those linked to the more recent developments addressed a range of issues involved in the collection of 'data', through the interrogation of both texts and people's lived experiences. In this respect, Ian Cook and Mike Crang (1995) equated (a renewed) cultural geography with humanistic geography in terms of the methods adopted. There is a sense that one of the preconditions for such renewal, however, was the legacy of human-orientated research in geography which came to the fore in humanistic geography's critique of positivism and spatial science. 'Doing humanistic, or cultural, geography' involves a variety of methods often characterised as ethnographic and contrasted to the positivist approach (Cook and Crang, 1995, p. 4) as 'reading, doing and writing . . . thoroughly mixed up' during the course of a piece of research rather than the 'conventional read-then-do-then-write sequence' of other approaches.

Cook and Crang noted that one author has identified forty-three different ways of conducting ethnographic research (Tesch, 1990), but focus on just four:

1   participant observation, in which the researcher(s) lives in and/or works among a selected community in order to appreciate its values and ways of life, and then interprets that culture to a wider audience – the process involves researchers and community developing mutual (or intersubjective) understandings;
2   interviewing, which involves conversations between researcher(s) and researched, which can vary in the degree to which they are structured with predetermined questions by the interviewer;
3   focus groups, whereby researchers initiate information-collection by bringing together groups who interact as they discuss the issues raised by the researcher (and perhaps others which lead from them), thereby treating people as members of interacting communities that influence their behaviour rather than as isolated individuals; and
4   filmed approaches, wherein visual material obtained from the research subjects indicate meanings and values which are not readily expressed in words (Rose, 2001).

Preparing for each of these approaches is a major task, as is the follow-up construction of the information obtained. A number of ethical issues is involved and the continued relationship between researcher(s) and the researched is of considerable importance. Although Cook and Crang's (1995) essay was expanded and brought out as a book (Crang and Cook, 2007) by a commercial publisher, it is interesting to note that it originally appeared in a series that began in 1975, published by the quantitative methods group of the Institute of British Geographers (the series is archived at: http://qmrg.org.uk/catmog/). A series that was originally dedicated to the methods required by spatial science ended up publishing what became a key text on qualitative methods (and, along the way, also published a review on humanism;

Pickles, 1986). Perhaps such methods might seem closer to humanistic geography than to the abstractions of spatial science. Yet as Cloke *et al.* (1991, p. 92) put it, 'doing ethnography' lacks neither rigour nor academic merit; it is not, as some imply, a 'soft option' compared to the 'hard' spatial science:

> rather, it is to engage very honestly with both the enchantment and the problems associated with researchers trying to gain an insight into the worlds of other peoples in other places, and it is also to insist that this insight emerges not by supposing (as might a philosophically inclined phenomenologist or existentialist) that these peoples are basically the same as us but by letting them and their version of humanity simply be different.

Conclusions

A common thread links the material discussed in this chapter and the behavioural work that formed part of the preceding chapter; both are concerned with positive (the way things are) rather than normative (the way they should be) investigations, with attempts to uncover how humans behave in the world rather than with contrasts between actual patterns of behaviour in space and those predicted from normative theories. Both were also part of a more general trend towards a human focus within the social sciences, which reflects reorientations in the external environment. From the late-1960s on, there was a significant disillusion with science and technology, especially in comparison to the 1950s and early years of the 1960s. Accordingly, there was a shift in emphasis from study of the aggregate to the individual, an increase in the relative volume of research conducted at the micro-scale, and growing unease about the role of social scientists in planning mechanisms. Both behavioural geography and especially the various humanistic approaches reflected these trends.

The two approaches remained distinctive. Behavioural geography treated people as responders to stimuli and investigated how different individuals respond to particular stimuli (and also how the same individual responds to the same stimulus in different situations); by isolating the correlates of those varying responses it builds models that can account for, and potentially predict, the probable impact of certain stimuli. The end product is input to processes aimed either at providing environments to which people respond in a preferred way or at changing behaviour by changing the stimuli. The other set of approaches – humanistic geography – treated the person as an individual constantly interacting with the environment and with a range of communities, thereby continually changing both self and milieu. It sought to understand that interaction by studying it, as it is represented by the individual and not as an example of some scientifically defined model of behaviour. And then by transmitting that understanding, it sought both to reveal people to themselves, enabling them to develop the interactions in self-fulfilling ways, and to promote their appreciation by others. A chapter entitled 'A brief history of geographic thought', published in 2002, subtitled its account of this period 'Humanistic thought and poetic geographies' (Hubbard *et al.*, 2002, p. 37), noting how these had 'fuelled a geography in which qualitative methodologies were regarded as superior in the production of meaningful knowledge . . . seen as viable ways of teasing out the emotional, aesthetic and symbolic ties that bound people and place' (p. 41). For David Smith (1988a, p. 266), humanistic work involved a 'new movement', and the chapters in the book that he edited with Eyles:

> have provided persuasive accounts of ways of interpreting the geographical world which are capable of challenging if not displacing prevailing orthodoxy or at least of providing a convincing alternative. The positivist and empiricist paradigm fixated by mathematical process modelling

with a promise of policy relevance . . . is already yielding to approaches more sensitive to actual human achievement.

For him, the struggle between paradigms seemed set to continue, with humanistic interpretative work being rich in empirical example as well as strong in philosophical criticism. According to Gregory (1994, p. 78), however, too many humanistic geographers had been reluctant to 'engage more directly with theoretical concerns'. As the following chapters illustrate, setting human behaviour in a sharpened theoretical (and hence, for many, a geographical and political) context became a prime focus of much geographical work from the 1970s on. Entrikin and Tepple (2006, p. 32) could remark that these later trends saw humanistic geography as 'relatively atheoretical and apolitical', and that humanism in geography might now be thought of 'as a very specific moment in the history of the discipline, a moment that has now passed' but which offered a 'voice against the narrowing of the field and an alternative vision of humans as complex intentional agents' (p. 36). Warf (2012) similarly argues that:

> Humanistic thought made great contributions to the discipline, helping to revive cultural geography and forcing researchers to take seriously the complex question of human consciousness. It jettisoned the myth of objective research and made explicit discussion of values and biases an integral part of the process.
>
> (p. 236)

The meaning of those 'values' and the recognition of 'biases' would soon expand and lead many to go beyond humanistic approaches, as the next three chapters detail. Indeed, another recent review (Seamon and Sowers, 2009) notes that is not really possible to tell the story of how existential and phenomenological geographical work have evolved since the 1970s without looking into how they informed, were interpreted by and have to some extent been supplanted by subsequent trends (including the advent of radical geography, and those critical and poststructuralist impulses that form the basis of the subsequent chapters of this book). For Entrikin and Tepple (2006, p. 36), however, the enduring legacy of humanistic geography was both to inform concepts and to sharpen cultural critique, enabling subsequent reworkings of human geography, as well as to inject a heightened sensitivity to morality, emotions and ethics (see too Cloke, 2002; Valentine, 2005):

> The recent interest in ethics has been evident in articles, books, special sections of journals, and relatively new journals such as *Ethics, Place and Environment* and *Philosophy and Geography*. It is within this emerging subfield that the humanist's concerns with the autonomous intentional agent and humans as the creators and interpreters of meaning are most evident.

Such openings to morality and ethics and fundamental human beliefs also led to both renewed disciplinary engagement with religion (A. Cooper, 1992; Kong, 1990; Park, 1994) and 'new age' beliefs (Holloway, 2000). Mapping religion had been part of traditional cultural geography, but it is only after the critics of a detached/objective geography outside emotions, 'beliefs' and feelings was trail-blazed by humanistic geography that advocacy of an explicitly 'Christian contribution' appeared in mainstream journals (M. Clark, 1991).

It is now more than two decades since Paul Rodaway (1994, p. 6) detected 'a revival of interest in humanistic issues and some cross-fertilisation between humanistic and postmodern thinking in geography.' A more recent set of essays on *GeoHumanities* charts more recent engagements between geography and the humanities (including 'Humanities GIS', 'Geocrativity', 'Geohistories' and 'Geopoetics') but notes how 'The seeds of a distinct geohumanities were planted during the 1970s, when human geography was buffeted by a flood of fresh ideas from the social sciences and the humanities' (Dear, 2011, p. 311). In the decades since humanistic geography was proposed as a reaction to positivism and quantification, geographers have repeatedly drawn on its creative impulses. However, subsequent activism and turbulent political debates subsequently gave these impulses a new character for many geographers who would advocate more radical paths beyond spatial science.

Chapter 6

Radical geography

How and why should we bring about a revolution in geographic thought? The quantitative revolution has run its course . . . to tell us less and less about anything of great relevance. . . . In short, our paradigm is not coping well. It is ripe for overthrow.

(David Harvey, 1973, pp. 128–9)

Everyone who studies Marx, it is said, feels compelled to write a book about the experience.

(David Harvey, 1982, p. 1)

[D]uring the 'summer of love' (1967) geography was perhaps the least sexy subject, certainly in the English-speaking world. The influence of the intellectually conservative Richard Hartshorne . . . still weighed heavily on US geography even as a new generation of positivists in Britain and the US were installing themselves as the next new thing. It is difficult to conceive of a discipline more uncool than geography in 1967. And yet, the influence of the anti-war movement in the US, the feminist and environmental movements, the Prague Spring of 1968, the anti-imperialist movement, radicals discovering socialism and Marx – all of these wider social eruptions in the late 1960s and early 1970s completely transformed the discipline. They had a greater effect on geography than on any other social science in the Anglophone world.

(Neil Smith, 2001, p. 6)

[I]t is precisely because patterns of spatial organization continue to have such strategic significance to capital, states, and social forces at all scales that such concerted political strategies are being mobilized to reshape them. The politics of space thus remain as contradictory and contentious as ever, and their consequences for everyday life remain to be fought out in diverse territorial arenas and at a variety of spatial scales.

(Neil Brenner and Stuart Elden, 2009, p. 33)

Antipode [subtitled *A Radical Journal of Geography*] will reassert itself as not just another academic journal. We want it to be a radical intellectual and political space, and a project that represents both a way of seeing the world and a shared commitment to changing that world by 'being the change we wish to see'.

(Paul Chatterton *et al.* [The *Antipode* Editorial Collective], 2011, p. 189)

Since the late 1960s, a group of interrelated politicised approaches has developed within human geography, which is nevertheless hard to classify under an all-embracing single, descriptive adjective. The term initially favoured was 'radical' (see Smith 1971; Peet, 1977, 1985b; Peake and Sheppard, 2014), but this became less popular in the 1980s as that term was being applied to the political movements associated with the 'new right', 'Thatcherism' and 'Reaganomics'. Blunt and Wills (2000, p. xi) used the term 'dissident geographies' for their textbook 'about radical ideas and practices, their geographical origins and manifestations and their implications for geo-graphical thought'. Since then, the more common term has been 'critical geography', although this includes a considerably wider range of work and influences than those of prior radical geographies (see Best, 2009, and Castree, 1999, for more on this relationship). Subsequently, much of the spirit of earlier radical geography has been incorporated into such 'critical', postmodern, postcolonial, poststructuralist and feminist geographies. Anarchism has also had some influence via a radical rediscovery and celebration of the works of Kropotkin and Reclus, both late nineteenth-century Russian and French anarchist geographers (see the summary in Blunt and Wills, 2000; the 'notes towards autonomous geographies' by Pickerill and Chatterton, 2006; and a reappraisal and repositioning of anarchist geography by Springer, 2013, 2014). Since the 1990s, a variety of 'activist' and 'participatory' geographies 'concerned with action, reflection and empowerment (of oneself and others) in order to challenge oppressive power relations' (Routledge, 2009, p. 7) have re-energised radical stances while also reshaping the meanings of applied geography.

In the late 1960s and 1970s, however, radicalism was more clearly associated with the adoption of a Marxist (or Marxian: on the difference, see Harvey, 1973) approach. Thus in Gibson's (1991, p. 75) words:

> It appears that the stature of human geography within the social sciences has changed dramatically over the last 20 years. From its location as a field of study uneasily straddling the earth sciences–social sciences divide and whose contribution to debates central to the study of society were rather defensively limited to spatial impacts and effects, geography has become 'flavour of the month', especially within left social science discourse. This shift has partly been produced by others 'discovering' geography – Giddens and Foucault, for example. But more importantly, I would say, it has been produced by changes within the discipline itself, particularly those wrought by the marrying of aspects of marxist and geographic theory and the incredible burst of vitality and productivity to which this gave rise. The work of geographers such as Harvey, Soja, Massey and Sayer is now almost as well known outside the discipline as within. Many of us have been surprised, and I suppose rather chuffed, by the way in which human geography has become more broadly validated. But even more surprising over this period has been the acceptance into 'mainstream geography' of significant elements of marxist theory.

For Cumbers (2009, p. 461), moreover, 'The adoption of marxism as a theoretical perspective and a political project in the 1970s was the single most important development in the evolution of a critical human geography.'

It was certainly entangled with debates about the relevance and applications of human geography (to the poor or to radical politics, for example), so that it is useful to read this chapter in tandem with Chapter 9 which focuses on applied geography. For a few years – mostly during the 1980s – radical approaches also became entangled with debates about the merits of an approach to research known as 'philosophical realism' or more often 'critical realism' and about the potential for human geography of 'structuration theory' which had been developed by the sociologist Anthony Giddens. Embodying these, Cloke et al.'s (1991) text contained an entire chapter on realist approaches in human geography and another on structuration theory. A few years later, in Peet's (1998) text, these approaches were relegated to a wider chapter on 'Structuration, realism and locality studies'. In more recent textbooks, both realism and structuration crop up, but neither is given prominence. There are 10 references to structuration theory and 16 references to realism in the textbook by Kevin Cox (2014) and 10 for each in Cresswell's (2013). Nayak and Jeffrey's text (2011) has brief reference to structuration theory as part of a section on challenges to humanistic geography on the part of scholars 'who felt it privileged human free will over societal structures' (p. 65), but none to realism.

It is notable that at the end of the 1980s, Walker (1989a, p. 135) went as far as to claim that 'While not every realist is a Marxist as regards theory of society . . . every Marxist must be a realist.' However, not all (indeed not most) of those geographers influenced by Marxism would agree with Walker. Lawson and Staeheli (1991) defend critical realism from those who see it as an 'arcane sect'. For others, however, such as Yeung (1997, p. 52), realism was also a wider philosophy that offered a path through questions of 'what is the relationship between philosophy, ontology, epistemology, theory and method in the social sciences?'. Certainly for some human geographers, realism seemed to offer a way to specify 'the difference that space makes' (S. Duncan, 1989a; A. Sayer, 1985a, 1985b). We will return to the uses made of realism and structuration theory by geographers later in this chapter. But we should note here that the type of questions posed by Yeung (1997) continued to produce a wide range of different answers among geographers. These have been further enlivened by the breadth of debate in radical and critical geography. Geographers' adoptions of Marxism and the influence of other radical currents, such as feminism, anti-racism/critical race theory and postcolonialism, have produced an enduring and heterodox radical presence in geography.

Geography from the left

In his evaluation of radical contributions to American geography, Walker (1989b) wrote of 'Geography from the left', which involved 'bringing the analytic framework and progressive social agenda of Marxism and allied schools of thought into most of the traditional subject areas of the discipline' (p. 619). The goal of achieving a 'more explicitly spatialized theory of capitalist societies' involved some scholars developing Marxist theory more fully, whereas others concentrated on methods and welcomed 'the clarifications that realism, critical theory, and structuration theory might add to the understanding of social processes and how to grasp them' (p. 620); others focused on 'middle level' theories of such topics as local labour markets. Walker argued that:

> Left geographers can be proud of their achievements in a discipline that is not always noted for its explanatory depth or overriding concern with human oppression and liberation. The left can claim a good deal of credit for broadening the intellectual respectability of the geographic enterprise.
>
> (p. 620)

His review clearly illustrates the breadth of enquiry undertaken by 'the left' in geography over the preceding couple of decades. Also embodying these contributions, but no longer favouring the adjective 'radical', Peet and Thrift (1989) selected the term 'political economy':

> to encompass a whole range of perspectives which sometimes differ from one another and yet share common concerns and similar viewpoints. The term does not imply geography as a type of economics. Rather economy is understood in its broad sense as social economy, or way of life, founded in production. . . . Clearly, this definition is influenced by Marxism. . . . But the political-economy approach in geography is not, and never was, confined to Marxism. . . . So while political economy refers to a broad spectrum of ideas, these notions have focus and order: political-economic geographers practise their discipline as part of a general, critical theory emphasizing the social production of existence.
>
> (p. 3)

This group of approaches began, however, as 'radical geography' in the 1960s, largely as a reaction to the crises of capitalism made clear in increased concern about the persistence of deep pockets of poverty within the supposedly affluent Western societies, the domestic and international response to the American war in Vietnam, the American Civil Rights movement (beginning with a call for an improved status for African-Americans, but later encompassing other 'racial', 'ethnic' and 'cultural' minorities), the nascent Green (ecological and conservation) movement, and the wider rise of a 'counter-culture' (perhaps epitomised by the 'hippies' of the late 1960s and early 1970s) who questioned the 'mainstream' way of life. Critical work on imperialism and war (especially the American-led war in Vietnam) soon featured in radical geography (Blaut, 1970; Folke, 1973; Lacoste, 1973), along with a pioneering paper about the geography of anti-war demonstrations in the USA (Akatiff, 1974; for more on this aspect of the making of radical geography, see Power and Sidaway, 2004; Mathewson and Wisner, 2005; Bowd and Clayton, 2013; Peake and Sheppard, 2014). These mobilisations were accompanied by the 'sexual revolution': changing social mores and a sense of increased sexual freedom in Western societies – although it would be a few years more before these mores were further questioned by the women's and gay liberation move-ments (see Chapter 8). The late 1960s and 1970s also saw a period of creative reinterpretation of Marx(ism) in the aftermath of the Stalinist orthodoxy which prevailed in preceding decades. This was both an intellectual movement, led by scholars in Western universities, and a practical politics of revolution and social change, especially in those parts of the Third World (such as Cuba and Vietnam) where national liberation and revolutionary movements gained or sought power. In short, left-wing and revolutionary causes enjoyed a renaissance and were characterised by experimentation and fresh ideas. Geography connected with these new left and counter-cultures which in America was institutionalised by the creation of several short-lived bodies such as the Union of Socialist Geographers (USG) and the Socially and Ecologically Responsible Geographers (SERGE).

Looking back to this moment, which dates from the late 1960s, Harvey (1999) notes how:

> Marx seemed to speak to those of us (always a minority of course) who were seeking a theo-retical basis to understand the chaos and political disruptions surrounding us at the time (the imperialist war in Vietnam, the strikes and urban unrest, the civil rights movement, the student movements of 1968 that shook the world from Paris to Mexico City, the Czech 'spring' and its subsequent repression by the Soviets, the war in the Middle East – just to name a few of the signal events that made it seem as if the world as we knew it was falling apart).
>
> (p. xi)

The Spanish geographer Horacio Capel (1981, pp. 426–7; our translation) thus notes how, in American geography of the late 1960s: 'That which just a few years before [quantitative geography] had been received as the truly scientific method, seemed now clearly unsatisfactory or a mystification.' Capel designates 1969 as a significant moment in the emergence of radical geography:

> Firstly, the 1969 annual meeting of the Association of American Geographers at Ann Arbor saw a critical mass of radical papers and manifestos. Secondly, in 1969 commenced the publication of a journal which, in its very title already expressed its aspiration to situate itself in the antipodes of geography as it was then practised: *Antipode. A Radical Journal of Geography*.

Capel also notes that, at the same time, William Bunge initiated his so-called 'geographical expeditions' to 'marginal urban areas' (see p. 237 and p. 322, this book) and other geographers increasingly participated in political and citizens' movements. Finally, Capel notes that while the actual numbers of radical geographers were never large, radical geography constituted 'an extraordinarily active movement that has extended its critique to very diverse fields of academic geography and forced it to encounter authentically radical alternative trajectories' (p. 427).

That William Bunge and David Harvey who had both been associated with spatial-science just a few years before were now lending their intellect and authority to radical geography helped to foster the latter. But, radical geography's agendas were also being shaped by a new generation, many of them postgraduate students, who established *Antipode* as a radical forum for publication in 1969. Subsequently, as the revolutionary pressures of the late 1960s became a distant memory, the ideas and critical insights of radical geography percolated into and then – to a significant extent – became part of the intellectual mainstream of anglophone human geography. In the process it had become 'more sober and less combative' (Peet and Thrift, 1989, p. 7). There were a number of reasons for this. In the first place, Marxism was subjected to renewed criticism. Second, the recession and restructurings of the 1980s (which included cuts in academic budgets) led to a more sober and cautious intellectual atmosphere than the heady days of the late 1960s when everything seemed possible – and *Antipode* itself was transferred from an in-house publication at Clark University's Graduate School of Geography to a commercial publisher, with ownership vested in an editorial trust (Peake and Sheppard, 2014). At the same time, greater awareness of the problems of state socialist economies and a wider sense that the tide of history had tipped against statist and socialist projects made the prospects of revolutionary or radical change more precarious. Finally, radical geography became more narrowly professional, and some of the 'radical, anti-establishment Young Turks' joined the establishment (see Barnes, 1996, p. 49). Nevertheless, Peet and Thrift conclude that 'The [radically inspired] political-economy approach . . . has survived counterattack, critique, and economic and professional hard times, and has matured into a leading and, for many, the leading school of contemporary geographic thought' (p. 7).

Radical trajectories

Peet (1977) has claimed that the early 'radical' work by geographers in the late 1960s was mostly liberal in its interpretations and aspirations:

> Radicals investigated only the surface aspects of these questions – that is, how social problems were manifested in space. For this, either we found the conventional methodology adequate enough or we proposed only that existing methods of research must be modified to some extent if they are to serve the analytic and reconstructive policies of . . . radical applications (Wisner, 1970, p. 1) . . . we were fitting into an established market . . . we were amenable to established ways of thinking . . . we were useful in providing background ideas for the formulation of pragmatic public policy directions, and so could not, and were not, engaging in radical analysis and practice.
>
> (p. 245).

Ideas of mapping welfare (Smith, 1975) and later a fully-fledged 'welfare geography' (described in more detail in Chapter 9) were broadly in this reformist spirit, part of what Roger Lee and Chris Philo (2009, p. 225) describe as 'an emerging sense that human geographers should be investigating the geographies of inequality, and perhaps the clearest index was a number of papers appearing in *Antipode* . . . during its earliest years'. Initially, therefore, the purpose of much writing in the emerging radical framework was to demonstrate the shortcomings and lacunae of positivist-inspired (and related behavioural) research. Gray (1975), for example, presented a critique of studies of the operations of housing markets which mapped people's residential moves and inferred that these were the outcomes of choice and the expressions of preferences. He argued that such research:

1   assumes that people are free to choose in which homes (including where) they live, when most are constrained to particular types of dwellings in certain parts of urban areas only;
2   accepts that residential patterns are the consequence of a large number of residents' decisions rather than those of a few developers and institutional managers; and

3   implies that the study of consumers provides the key to understanding the structure of urban
     areas.

People are not free to choose, however: 'Instead, many groups are constricted and constrained from choice and pushed into particular housing situations because of their position in the
housing market, and by the individuals and institutions ... controlling the operation
of particular housing systems' (p. 230). Research in line with Gray's critique focused on the
controllers of access to housing, individuals who became known as 'urban managers'.
According to Pahl's (1969) influential paper, urban residential patterns are the consequence
of two basic sets of constraints: the spatial constraints on access to resources and facilities,
which are usually expressed in terms of time/cost distance; and the social constraints – the
bureaucratic rules and procedures operated by gatekeepers – that govern access and reflect
the distribution of power in society. Study of the latter focused on the key managerial groups,
such as building societies and their lending policies (Boddy, 1976; Dingemans, 1979), the
managers of local government housing stock (Gray, 1976; Taylor, 1979), the local authority
officials who influence the redevelopment of private housing (Duncan, 1974a, 1975), and
the real-estate agents who structure access to housing submarkets (Palm, 1979). Subsequently,
therefore, urban geography became a key entry point of radical ideas into the discipline.
This was given further impetus by the evolution of radical perspectives in the neighbouring
field of urban sociology and the inauguration of the multidisciplinary *International Journal of
Urban and Regional Research* which was founded under the auspices of the International Sociological
Association (ISA) research committee on the sociology of regional and urban development
in 1977.
    The reception of Pahl's work is a case in point (he was trained as a geographer, with a Ph.D.
from the LSE, but then worked in a sociology department). In the 1970s, Pahl was widely read and
cited among radical geographers (especially in the UK and Europe). Pahl himself continued to
revise his ideas, pointing out that planners, for example, are merely 'the bailiffs and estate managers
of capitalism, with very little power' (1975, p. 7; see also Pahl, 1979) so that focusing on them and
other gatekeepers tends to:

> view the situation through the eyes of disadvantaged local populations and to attribute more
> control and responsibility to the local official than, say, local employers or the national
> government ... such 'a criticism of local managers of the Caretaking Establishment' and 'of
> the vested interest and archaic methods of the middle dogs' may lead to an uncritical
> accommodation to the national elite and society's master institutions.
>
> (pp. 267–8)

Whereas the managerial approach focused on the agents who interpret and activate the real
mechanisms, producing patterns of residential segregation, Harvey paid much more attention to
the mechanisms themselves (though with reference to particular outcomes: Harvey, 1974e).
Residential separation is one means of reproducing class differences in a capitalist society. It is
produced by those who manage finance capital, interpreting the basic mechanisms of the capitalist
mode of production (i.e. the need to generate profits), and it results in a series of spatially separated
housing submarkets within which individual households may express their preferences, producing
micro-scale migration patterns:

> But there is a scale of action at which the individual loses control of the social conditions of
> existence ... [sensing] their own helplessness in the face of forces that do not appear amenable,
> under given institutions, even to collective political mechanisms of control.
>
> (Harvey, 1975a, p. 368)

In the American context the promotion of suburbanisation was a major means of stimulating consumption (of housing, automobiles, consumer durables, etc.) at a time of dangerous latent overproduction (Harvey, 1975b, 1978; Walker, 1981a; for the particular outcomes, see Johnston, 1984c). Thus, urban geographers who accepted the Marxist argument saw their basic task as integrating discussions of the mechanisms that drive capitalist (and other) societies, the interpretations of those mechanisms by key agents (the managers and gatekeepers), and the empirical outcomes and experiences. They focused on how and why society operates, not on generalisations regarding the empirical outcomes (as in Bassett and Short, 1980; Badcock, 1984; Johnston, 1980a). However, the study of urban residential patterns is just one element of the wider study of urbanisation, which Scott (1985) – in ways that echoed Harvey's developing work on the urbanisation of capital – related to the complex nature of capitalism:

> the mechanisms of production, the interlinkages of firms, and the formation of local labor markets . . . combine to create a process in which the profit-seeking (cost-reducing) proclivities of producers lead directly to the dense spatial massing of units of capital and, as a corollary, of labor.
>
> (p. 481)

The influence of the initial move to radical urban geography has thus been long and enduring. Subsequent work has explored how cities are both labour markets and places for the reproduction of labour, involving what Castells (1977) and others term 'collective consumption' of commodities produced by the state rather than purchased in the market place (see also Pinch, 1985, 1997), although the shift to 'privatise' these has itself become a domain of critical analysis, amid other regulatory shifts (see p. 211, this book). The issue of gentrification, whereby some formerly working-class inner-city neighbourhoods become areas of middle-class and elite housing also became a subject of investigation, following a key paper (N. Smith, 1979b) that designated gentrification as a 'back to the city movement by capital, not people' (p. 538); for a review of the debates that this initiated, see Lees (2000, 2012) and Lees et al. (2010). More than three decades on, the role of capital flows through urban real estate, especially in American cities (so-called 'subprime loans', designated as 'predatory lending' by radicals) in the making of the financial crisis became a focus, especially via capital's articulations with race and class (Harvey, 2011; Wyly et al., 2008; Wyly et al., 2009). Here, however, we are getting ahead of the story of the formation of radical geography in the late 1960s and early 1970s.

The move from liberal interpretation to radical and Marxist critique was also illustrated in Peet's (1971) paper on poverty in the USA, in which, like Morrill and Wohlenberg (see p. 322, this book), he argued for a series of growth centres in the poverty areas to stimulate economic development and job creation, and also by the tenor of the articles in early issues of the radicals' journal, *Antipode*. Morrill (1969), for example, argued:

> against the 'New Left' premise that a revolution is the only route to progress . . . the dreams of revolution are naive . . . the 'New Left' vastly exaggerates potential support . . . a 'revolutionary program' is hopelessly dated and simplistic . . . the 'New Left' underestimates the capacity of our society for change. . . . All revolutions seem to have been betrayed by incompetents who preferred exercising power to executing reform.
>
> (pp. 7–8)

Again, he argued (Morrill, 1970b) that:

> A simple Marxist-type change in the ownership of business from private to a government (or union) bureaucracy would in all probability decrease production, and would not necessarily bring

any improvement in basic conditions. The key is to retain the institution of private property while instituting social control over its exchange and circumscribing its power over people.

(p. 8)

The case for a Marxist approach was first presented formally by Folke (1972) in a critique of Harvey's (1972) paper on ghetto formation and counter-revolutionary theory. Folke represented geography and the other social sciences as:

highly sophisticated, technique-orientated, but largely descriptive disciplines with little relevance for the solution of acute and seemingly chronic societal problems . . . theory has reflected the values and interests of the ruling class.

(p. 13)

Liberal cases like Morrill's were dismissed as unlikely to succeed. Morrill's two papers in *Antipode* argued for change to be brought about by political consensus, producing a capitalist–socialist convergence. But this is the social democrat method practised in Sweden (and then also influential in other left-of-centre parties, such as the French socialists, German Social Democrats and the British Labour Party) where, according to Folke (1972):

it has been shown over and over again that the idea of equal influence for employers and employees is an illusion. After half a century of social-democratic rule injustices and inequalities still prevail. . . . No small group of experts can accomplish anything . . . when it runs counter to the interests of the dominant social forces. These are not interested in equality or justice, but in profit.

(p. 15)

Radical change requires mass mobilisation and so, to Folke, Harvey's call for a new paradigm within geography was insufficient. What was needed was a new paradigm for a unified social science, containing geography, which deals with problems in all their complexity and provides not only theory but also the basis for action: 'Revolutionary theory without revolutionary practice is not only useless, it is inconceivable . . . practice is the ultimate criterion of truth' (p. 7; see also Eliot Hurst, 1980, 1985).

The major contribution to the case for a Marxist-inspired theoretical development within geography was subsequently made by David Harvey, initially in his collection of essays *Social Justice and the City* (Harvey, 1973). The book is presented as autobiographical, illustrating the evolution of Harvey's views towards an acceptance of Marx's analysis:

as a guide to enquiry . . . I do not turn to it out of some a priori sense of its inherent superiority (although I find myself naturally in tune with its general presupposition of and commitment to change), but because I can find no other way of accomplishing what I set out to do or of understanding what has to be understood.

(Folke, 1972, p. 17)

The first part of the book, entitled 'Liberal formulations', comprises essays which analyse problems of inequality within societies in terms of income-allocating mechanisms; the role of accessibility and location in those mechanisms is stressed. This leads into an attempt at defining territorial social justice, which separates the processes allocating incomes from those which produce them. Only in the second part of the book – 'Socialist formulations' – is it:

finally recognized that the definition of income (which is what distributive justice is concerned with) is itself defined by production. . . . The collapse of the distinction between production and distribution, between efficiency and social justice, is a part of that general collapse of all dualisms of this sort accomplished through accepting Marx's approach and technique of analysis.

(p. 15)

Harvey (2000, pp. 77–8) subsequently noted how he had become disillusioned with the evident limits to the socialism of the British Labour party in the late 1960s (the government of Harold Wilson). However, it was his move to the USA (from the University of Bristol to Johns Hopkins University, Baltimore) at the end of the 1960s that sharpened his political commitment. Thus, when he wrote *Explanation in Geography* (Harvey, 1969a) in 1968 (which, as Chapter 4 charts, was a key text setting out the prospects for a positivist and quantitative geography):

> my politics at that time were closer to a Fabian progressivism, which is why I was taken with the ideas of planning, efficiency and rationality. . . . So in my mind, there was no real conflict between a rational scientific approach to geographical issues, and an efficient application of planning to political issues. But I was so absorbed in writing the book that I didn't notice how much was collapsing around me. I turned in my magnum opus to the publishers in May 1968, only to find myself acutely embarrassed by the change of political temperature at large. By then, I was thoroughly disillusioned with Harold Wilson's socialism. Just at that moment, I got a job in the US, arriving in Baltimore a year after much of the city had burnt down in the wake of the assassination of Martin Luther King. In the States, the anti-war movement and the civil rights movement were really fired up; and there was I, having written this neutral tome that seemed somehow or other just not to fit. I realized I had to rethink a lot of things I had taken for granted in the sixties.

(Harvey, 2000, p. 78)

That political stance did not disappear; when he returned to the UK in 1987 to take up a senior post, 'I was forcefully reminded of what I had long ago rebelled against. The smugness of much of Oxford repelled me, and I could not understand how the left in Britain, within geography and without, was taking such wishy-washy positions in relation to Thatcherism' (Harvey, 2002, p. 179).

The transition in Harvey's approach is marked by his paper on ghetto formation (Harvey, 1972). He begins with a critique of Kuhn's model of scientific development (p. 15, this book), asking how anomalies to the current paradigm arise and how they are translated into crises. The problem with Kuhn's analysis, he claims, is the assumption that science is independent of its enveloping material conditions, when in fact it is very much geared to its containing and constraining society. Recognition of this point is important for geographers because:

> the driving force behind paradigm formulation in the social sciences is the desire to manipulate and control human activity in the interest of man. Immediately the question arises as to who is going to control whom, in whose interest is the controlling going to be exercised, and if control is going to be exercised in the interest of all, who is going to take it upon himself to define that public interest?

(Harvey, 1973, p. 125)

To Harvey, Marxist theory provides 'the key to understanding capitalist production from the position of those not in control of the means of production . . . an enormous threat to the power structure of the capitalist world' (p. 127). It not only enhances an understanding of the origins of the present system, with its many-faceted inequalities, but also propounds alternative practices which would avoid such inequalities: 'we become active participants in the social process. The intellectual task is to identify real choices as they are immanent in an existing situation and to devise ways of validating or invalidating these choices through action' (p. 149). In such a context, geography can no longer be merely an academic discipline, isolated in its 'ivory towers'. Its practitioners must become politically aware and active, involved in the creation of a just society which involves not only reform but rather replacement of the present one. The remainder of Harvey's (1973) book does not make this commitment clear, however; it contains one essay on land use and land-value theory, investigating the difficult concept of rent, and another on the nature of urbanism, presenting a Marxist interpretation of the process of urbanisation. His later (1984) historical materialist manifesto is a very clear statement of that goal, however, and his (1982; second edition, 1999) The Limits to Capital is presented as a venture to extend Marx's theory, excavating the centrality of the production of space and geographically uneven development to the reproduction of capitalist society, in a similar spirit to Neil Smith's (1994a) equally ambitious excavations of the essential, but highly complex, relations between capitalist (re)production and Uneven Development. (Interestingly, Harvey (2002, p. 176) claimed that it was 'far easier to bring Marxism into geography than to take geographical perspectives back into Marxism'.)

Harvey's work continued to develop these themes during the 1980s, 1990s and 2000s. Two further volumes of essays extended his analyses of urbanisation under capitalism (Harvey, 1985a, 1985c; see also Harvey, 1996b, and for a critical overview the essays in Gregory and Castree, 2006, including Harvey's own response) followed by an interpretation of the socioeconomic and geographical backdrop to 'postmodern' forms of culture (1989a). Other essays addressed issues of geopolitics (Harvey, 1985b) and relationships with nature (Harvey, 1994). He then returned to issues of justice (Harvey, 1993a, 1996a) and the imperative for and possibility of alternatives to capitalist urbanisation and 'globalization' (Harvey, 2000), while at the same time defending his approach against a variety of critiques. Throughout, he has advanced the case for Marxism (as a creative and open-ended historical dialectical materialism), as a holistic approach that necessarily puts material issues at the core of its analysis (Harvey, 1989b). Over the years, Harvey's understanding of this materiality has broadened (Harvey, 2000, thus has an extended consideration of 'the body' as a site of accumulation) but remained always resolute and ambitious in the face, as Harvey (2000, pp. 87–8) would see it, of 'The ubiquity and volatility of money as the impalpable ground of contemporary existence.'

Peet had also moved from a liberal to a Marxist position, replacing his earlier paper on poverty (Peet, 1971) by a marxist interpretation (Peet, 1975a) based on the assumption that inequality is inherent in the capitalist mode of production. This led to a 'metatheory dealing with the great forces which shape millions of lives' (p. 567) within which 'Environmental, or geographic, theory deals with the mechanisms which perpetuate inequality from the point of view of the individual. It deals with the complex of forces, both stimuli and frictions, which immediately shape the course of a person's life' (pp. 567–8). Environmental resources act as constraints because they define the milieux within which individuals are socialised and are presented with opportunities for participation within the capitalist system (see also Soja, 1980). Redistribution of income through liberal mechanisms, based on taxation policies, will not solve the problems of poverty, therefore; according to Peet, alternative environmental designs, with removal of central bureaucracies and their replacement by anarchistic models of community control are needed and geographers should work towards their creation. (Harvey (1973, p. 93) disagreed with the latter, pointing out that unless resources are equalised among communities and territories, community control will only result in 'the poor controlling their own poverty while the rich grow more affluent from the fruits of their riches'. See, however, Harvey, 2000.)

Later on, responding to the post-9/11 wars led by the USA, Harvey (2003) examined the new forms of imperialism as well as associated forms of neoliberalism and enclosure. For Harvey, neoliberalism, as the ideology and practice of commodification and privatisation, needs to be studied as a geographical project of remaking state and society in the image of capital, restoring the privilege of the ruling class (producing an expanded super-rich but eroding working-class solidarity). This has:

entailed much 'creative destruction', not only of prior institutional frameworks and powers (even challenging traditional forms of state sovereignty) but also of divisions of labour, social relations, welfare provisions, technological mixes, ways of life and thought, reproductive activities, attachments to the land and habits of the heart.

(Harvey, 2005b, p. 3)

Harvey and other geographers influenced by Marxism's commitment to expose the dynamics of the capitalist system (such as Massey's, 1984a, *Spatial Divisions of Labour* and subsequent research on the geography of work, such as Herod, 2001, as well as other material reviewed elsewhere in this chapter), thus both developed a Marxist (or Marxian) geography and a geographical approach to reading Marx's *Capital* that insisted that these capitalist dynamics were necessarily spatial. Capitalism sought 'a spatial fix' – it made and remade space (through mobilising labour and the means of production, including technology), relentlessly pursuing profit, endlessly looking to accumulate.

To Eliot Hurst (1980), however, embracing Marxism would always be insufficient. In a subsequent essay he argued that by circumscribing their discipline, geographers were severely constraining what they could contribute towards understanding the world: their discipline was presented as a technical practice only, which 'is merely descriptive of self-selected phenomena and is scientifically bankrupt' (Eliot Hurst, 1985, p. 62). He examined eight separate definitions of geography, all of which were guilty of what he termed spatial fetishism (after Anderson, 1973), whereby space is accorded a separate and substantial reality with its own distinctive powers, and he followed David Slater (1977) in arguing that geography cannot be incorporated within Marxism but has to be transcended and superseded by it; to become Marxist, geographers as a professional group must commit suicide. He concluded that 'geography is irrelevant to contemporary society on two basic grounds' (Eliot Hurst, 1985, p. 85):

1   in terms of substantial content, it has developed 'certain types of ideography, method/ technique-strong, but lacking theory in a strict sense' to provide the sort of material which underpins contemporary capitalism; and
2   in philosophical terms it occupies 'an untheorized point of entry to knowledge' because its dominant 'categories of concern, such as space, spatial differentiation, spatial interaction and uneven development, are mere fetishisms and unscientific'.

Radical reconfiguration is needed in order to promote full understanding of capitalist society, not just of geography but the entire body of 'bourgeois social science'.

Structuralist Marxisms, regulation theory, localities and regions

Structuralisms

The reinvigorated Marxism which influenced radical geographers has sometimes been presented as a variant of the philosophy of structuralism (of which there are many different forms: cf. Johnston, 1983a; Peet, 1998), but which are most indebted to the French Marxist Louis Althusser who published a number of influential reinterpretations of Marx in the 1960s and early 1970s. Structuralism is certainly not the only strand of Marxist geography, for Sheppard and Barnes (1990) have also set out the possibilities of an 'analytical Marxism' in geography and David Harvey charted his own course to an historical geographical materialism. On Harvey, Castree (1996) therefore argues that:

what is distinctive about Harvey's Marxism . . . is how little indebted it is to other readings of Marx: reading *Capital* has for him long been preferred to *Reading Capital* (Althusser and Balibar 1970). . . . This is quite unusual among Anglophone Marxian theorists: few appropriate Marx in an 'unreconstructed' way and far more read him through interpretations provided by earlier heirs of Marx's legacy. . . . This is why Harvey . . . sees both Althusserian and analytical Marxism – perhaps the two leading Marxist paradigms of the last two decades – as radical reworkings of Marx. . . . Harvey, by contrast, very much regards his own work as a more or less direct, 'faithful' continuation and extension of Marx's own practices and ideas: in short, as a peculiarly classical version of Marxism bought forward from the last *fin de siècle* to our own.

(p. 344)

Meanwhile, analytical Marxism seeks to evaluate Marx's work and insights using tools available from contemporary social sciences. Influential within radical branches of economics, for geography it has amounted to:

the use of analytical methods . . . to construct a wide-ranging model of the capitalist space economy incorporating commodity production, natural resources, the built environment, class formation, technical change and industrial organization.

(Sheppard and Barnes, 1990, pp. 1–2)

Describing some of the inspiration to this project, Sheppard and Barnes (1990, p. 1) also note how:

As a model for the kind of account we wish to present, we pay homage to David Harvey's (1982) *Limits to Capital*, from which we drew many insights and ideas. Yet our argument is different in style from his, and indeed from many others whose ideas we share. For at the core of this book is the belief that a clear account of the political economic approach is obtained by interrogating our subject matter with an analytical (mathematical) logic.

Nevertheless, structuralism merits further consideration here, in part because of its indirect impacts as an influence on the regulationist approaches that we will also consider.

Perhaps where structuralism had the most direct impact within anglophone geography was in reactions to the work of the Catalan theorist, Manuel Castells (whose original work was mostly written in French). Peet (1989) notes how:

structural[ist] work on cities and urban planning (Preticeille 1976; Pickvance 1976) began to appear in English in the middle 1970s. . . . Most importantly, a translation of Manuel Castells's (1977) structuralist work on cities, *The Urban Question*, had a profound effect, especially on urban geography.

(p. 112)

Peet, however, goes on to note how:

The move towards structuralism, never complete in geographical thought, represented a search for greater theoretical coherence and rigor. It is also the most misrepresented and underappreciated period in social and geographical thought.

(p. 112)

Structural Marxism was also an influence in Doreen Massey's (1984a) account of *Spatial Divisions of Labour: Social structures and the geography of production*. This soon proved to be an influential book (see p. 212, this book), although its debt to structural Marxism was only explicitly specified in the second (1995) edition. Structural Marxism also came to influence a generation of work on the unevenness, contradictions and limits of capitalist development in the Third World (for example, Blaut, 1975; Cannon, 1975; Slater, 1973, 1977), even if its influence on development geography was soon being criticised for its perceived 'economism'. Thus, Corbridge (1986, p. 64) claimed that 'Althusserians have not escaped the sins of economic determinism. . . . In their curiously changless theories, capitalism is endowed with an endless and ageless capacity to secure its own perpetuation.'

In response, Michael Watts (1988) stressed the continued relevance and richness of Marxist perspectives on geographies of (uneven) development. Althusser's work can thus be read as creating radically new formulations of Marxism, especially his development of the concepts of 'over-determination' and 'condensation'. In this he adapted terms from psychoanalysis ('condensation' is used in Freud's interpretations of dreams) to refer to the way that elements of a complex whole are expressed in each of its parts and vice versa. The potential application of these to human geography was not quickly realised, however. Moreover, the sense that structural Marxism has often been mis-represented or caricatured as unduly rigid and mechanistic is also a point of critique for Castree (1994, 1995a), who reminds us of both the continued diversity of Marxist writings, their consid-erable nuances and differences, and the range of works in geography that they lend themselves to. Moreover, as Peet shows, the impacts of structuralism within and beyond Marxist-influenced geo-graphy have been complex. Perhaps the most significant influence has been a rather indirect one, via the way that structuralist ideas shaped regulation theory, which in turn became an influential strand of geographical work in the 1990s. We will consider this strand in a moment.

First, however, it should be noted that structuralism more widely has had a long-standing concern with the dynamics and rules of systems of meaning and power. In particular, structuralism proceeds through examining the relations and functions of the smallest constituent elements of such systems in relation to each other (see p. 111, this book). The emphasis is on this relation, on the relationship of parts to the whole. The work of Ferdinand de Saussure, undertaken just prior to the First World War in the field of linguistics, has been a prototype and inspiration. Setting the agenda for wider structuralist analysis, Saussure's study was centred less on speech itself, but rather on the fundamental rules and conventions enabling language to operate. Structuralism has since found further application in the fields of linguistics and textual analysis, as well as anthropology and wider studies of culture, prior to coming into fruitful interactions with Marxism in the form of Althussers's work. The fact that Marx's method contained a parallel commitment to study the dynamic relations between the whole and parts of the capitalist system enabled this encounter. Marxism adopts a dialectical approach to its subject matter, since this represents how societies develop. It involves the resolution of opposites, usually represented by the trilogy of:

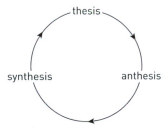

in which each synthesis becomes the next thesis; thus, there is no permanent resolution, each synthesis containing within itself the seeds of its own destruction. Dialectic processes operate at a variety of spatial and temporal resolutions (Castree, 1996; Harvey, 1996b; Ollman, 1993). Indeed, Castree (1996) reminds us that:

Just as Marxism has developed into a complex discursive tradition [and we might add complex political practice too] – that is, a variegated set of distinct but related intellectual positions and practices whose particularity and specific genealogies must be respected – dialectics too has assumed manifold forms and accumulated multiple meanings within [and we might add beyond] this Marxist tradition. Strictly speaking, then, the language of 'Marxism' and 'dialectics', in the singular, must be abandoned for a more discriminating attention to Marxisms and dialectics, in the plural.

(p. 344)

A key structuralist theme is that of relations between levels of analysis:

1   the level of appearances, or the superstructure;
2   the level of processes, or the infrastructure; and
3   the level of imperatives, or the deep structure.

Most forms of structuralism have no deterministic relationship between these levels. A process can result in a considerable number of outcomes, depending on a variety of enabling and constraining factors, most of them within the superstructure. To illustrate this, the French social anthropologist Claude Lévi-Strauss used the analogy of the cam-shaft driving a machine for cutting the outlines of jigsaw puzzles (Figure 6.1). The machine is constrained to certain movements only, but the order in which they come is random, which means that it can produce a large number of unique puzzles. Study of one of these puzzles alone could not reveal the nature of the machine; only study of the machine could do that. Structuralism is the study of the machines but, unlike the analogy, these are not available for investigation. Structuralist researchers must develop a theory of the machine and test whether the contents of the superstructure are consistent with that theory; the contents of the superstructure cannot be predicted, however, because, to continue the cam-shaft analogy, the particular operation of the machine cannot be foreseen.

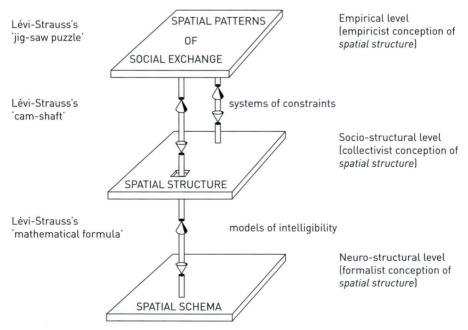

Figure 6.1 A geographical interpretation of Lévi-Strauss's cam-shaft for structuralism

In some forms of structuralism, such as Lévi-Strauss's social anthropology (E. R. Leach, 1974), it is further argued that the processes operating in the infrastructure reflect imperatives located in the deep structure. Theoretical study of the patterns in the superstructure should reveal the nature of the deep structure, therefore (in somewhat similar fashion to the identification of essences in phenomenology: p. 179, this book). Thus, according to Lévi-Strauss, all incest taboos are variants of a basic taboo which is imprinted in everybody's conscience. That basic form occurs in different ways in separate societies and cultures, but careful comparison of all taboos should reveal their common elements, and thus that aspect of human neural structure.

Few geographers have taken structuralism this far, although it might be claimed that the imperatives of deep structures have often appealed to geographers as a domain of analysis (see Golledge, 1981a; Sack, 1983; Sibley, 1998). Moreover, there are those who pointed to the consequences – in their view, problematic ones – of Marxist geography (in the form of David Harvey's work) replicating this appeal to structure at the perceived expense of agency and contingency (Jones, 1999). We return to this theme later in this chapter.

Regulation theory

One way in which structuralism has found productive expression has been via geographers' adoption of regulation theory. This studies the procedures and practices whereby capitalism is reproduced ('regulated') – the sets of checks and balances that keep the system going, albeit from one period of crisis to another. It thus draws on a nuanced understanding of both ideology and power – drawing on the rediscovery and translation of the (pre-Second World War) works of the Italian Marxist Antonio Gramsci (more recently re-presented to geographers by Joel Wainwright, 2010, and Ekers et al., 2013), as well as the concepts of multiple determination (so-called overdetermination) which shape the balances between 'state', 'society', production, consumption, and social and technological forces which are seen to propel and embody socio-spatial transformations (this owes much to Louis Althusser; see Lipietz, 1993). This is much more than a formal interest in government rules and regulations; what is understood by regulation in the common sense of the word. Instead, as Peck and Tickell (1995, p. 16) note:

> Although by no means a completed theoretical project, it has nevertheless matured to the extent that proponents now work from a well-established set of conceptual foundations. The theory holds that capitalist development proceeds through a succession of historically specific phases of stability and crisis, sustained growth and intense restructuring. Emphasis is placed on the role of state forms, social mores, laws and habits; the so-called mode of social regulation (MSR) in underwriting sustained phases of growth. These phases, in which the schema of reproduction (the complex of production, consumption, and distribution systems) is broadly in synchronisation are termed regimes of accumulation.
>
> (p. 16)

Though regulation theory first developed in France outside geography (see Boyer and Saillard, 1995; Macleod, 1997), 'Anglo-American' human geographers and allied social scientists have since added greatly to it through careful attention to the relationships between reconfigurations in systems of production, different geographical scales and their roles in regulatory reconfigurations (e.g. Amin, 1994; Amin and Robins, 1990; Clark, 1992; Dunford, 1990; Gertler, 1992; Goodwin et al., 1993; Tickell and Peck, 1992), including reconfigured rural geographies (Cloke and Goodwin, 1992). In turn, this has reinvigorated geographical literature on the intersection and production of different scales of social and political life (Herod and Wright, 2002, review these debates). Related notions of the 'geographies of governance' have noted how they are today less clearly organised along state lines (and national boundaries – hence, such work has consequences for political geography) within the context of rescalings (e.g. Brenner, 2004; Brenner et al., 2003; G. Clark, 1992; Herod et al., 1998; Herod and Wright, 2002) and the shifting regulation and scales of state policy (see the reviews by MacLeod and Holden, 2009; Painter, 2000; Peck, 2000).

At the same time as regulation theory offered fruitful pathways to renew radical geography's concern with the structural dynamics of capitalism, other work had returned to the points and relations of production (though not just industrial production), charting labour geographies (see the review in Wills, 2002) and changing geographies of class relations (e.g. Herod, 2001; Peck, 1996), including those of universities in which academic geographers and many relatively low-paid support and technical staff work (Roberts, 2000). (For an extended discussion of changes within economic geography over the second half of the twentieth century, see Scott, 2000.)

Spatial divisions of labour and localities

The localities research programme originated in part from Massey's (1984a) work on the changing geography of economic activity in the UK. (A measure of the significance of her book on *Spatial Divisions of Labour* is the collection of ten commentaries by Australian-based authors in *Environment and Planning A*: Fahey et al., 1989.) Her central argument was that understanding changing locational patterns of industries requires appreciation of the links between economic and social change. Local social structures vary in how the labour process is organised, and from comparing two case study regions she concluded that:

> although both . . . are now being drawn into a similar place in an emerging wider division of
> labour, their roles in previous spatial divisions of labour have been very different; they have
> different histories. They bring with them different class structures and social characteristics,
> and, as a result, the changes which they undergo . . . are also different.
>
> (p. 194)

Thus, changing industrial geography is linked to a changing social geography as new 'layers of investment' (and disinvestment) are superimposed on those of earlier eras (see also J. Anderson et al., 1983).

Those processes generate uneven development; a regional mosaic of unique areas reflecting the interpretations of local contexts by the actors involved in creating the UK's changing geographies of production. Massey (1984b) generalised this in her essay 'Geography matters', in terms of each of its three traditional concerns – space, environment and place. Regarding space, for example, she argued that 'aspects of "the spatial" are important in the construction, functioning, reproduction and change of societies as a whole and of elements of society. Distance and separation are regularly used by companies to establish degrees of monopoly control' (p. 5). Regarding the environment, she argued that conceptions of the 'natural', and thus interpretations of local environmental potentials, are socially produced, and could vary among areas as a reflection of their separate social systems, whereas with regard to place she contended that general processes can have particular outcomes in unique areas – hence her introduction of the concept of the 'global sense of place' (Featherstone and Painter, 2013). Thus, the study of geography involves unravelling the unique and the general: 'the fact of uneven development and of interdependent systems of dominance and subordination between regions on the one hand, and the specificity of place on the other' (p. 9) is thus a central concern for the discipline.

Massey illustrated her general contention by a number of case studies in *Spatial Divisions of Labour* and provided others in an essay on women in the labour force (McDowell and Massey, 1984; see also Damaris Rose, 1987). Her work was also instrumental in the establishment of a major research programme (financed by the UK's Economic and Social Research Council), into the Changing Urban and Regional System (CURS); this was introduced by its coordinator (Cooke, 1986) and summarised in a volume that reported on the programme's seven case studies (Cooke, 1989a). The substantive focus was the spatially varying processes of economic restructuring taking place in the UK during the 1980s. Thus,

> The overall objectives of the programme were to explore the impact of economic restructuring at national and local levels, and to assess the role of central and local government policies in enabling or constraining localities, through their various social and political organizations, to deal with processes of restructuring.
>
> (Cooke, 1989a, p. ix)

In addition:

> an important dimension of the research involved seeking to establish the conceptual status of the idea of 'locality' by taking account of a wide range of social scientific theory and research.

The concept of locality was chosen after both the traditional terms 'community' and 'locale' were rejected:

> There is a gap in the social science literature when it comes to a concept dealing with the sphere of social activity that is focused upon place, that is not only reactive or inward-looking with regard to place, and that is not limited in its scope by a primary stress on stability and continuity.
>
> (Cooke, 1989a, p. 10)

Locale was rejected because its spatial scope is vague; it suggests a passive rather than an active context for action and lacks any specific social meaning. (Giddens, 1984, defined a locale as a 'setting for interaction', which implies a spatial component but no more; a virtual site could be described as a locale with that definition.) A locality, on the other hand, is 'the space within which the larger part of most citizens' daily working and consuming lives is lived' (Cooke, 1989a, p. 12) and in which their citizenship rights are defined. Citizenship, according to recent debates, involves individuals having obligations to their community (via the state) as well as entitlements from it – or duties as well as rights (Smith, 1989; see also Bennett, 1989b). The contest over rights and duties is fought out in places, at a variety of scales; their nature involves 'the construction of an identity complete with a package of known rights and obligations related, which posits residence in a definable place or (commonly quite sizeable) territory as the basis for the nurturing and preservation of this identity' (Painter and Philo, 1995, p. 111). We become what we are because of where we are and the categories used in the structuring of civil society there (and we are implicated in those local struggles over our rights, obligations and identities).

At the end of his summary volume on the localities project, Philip Cooke argued that the seven case studies sustained his earlier contention regarding the crucial role of localities in the restructuring process and in the creation and recreation of uneven development; they illustrate

> the argument that the relationship between the different scales is not simply a one-way street with localities the mere recipients of fortune or fate from above. Rather localities are actively involved in their own transformation, though not necessarily as masters of their own destiny. Localities are not simply places or even communities: they are the sum of social energy and agency resulting from the clustering of diverse individuals, groups and social interests in space. They are not passive or residual but, in varying ways and degrees, centres of collective consciousness. They are bases for intervention in the internal workings of not only individual and collective daily lives but also events on a broader canvas as affecting local interests.
>
> (Cooke, 1989a, p. 296)

This conclusion, and the path towards it, however, became the subject of considerable debate, much of it unfavourable on conceptual grounds because what is necessarily local, and what is contingently so, has not been defined (Cox and Mair, 1989).

One of the initial critiques was Neil Smith's (1987b) argument that the CURS programme as formulated was likely to be submerged in a morass of statistical information; it contained within itself the potential for producing no more than those earlier empiricist studies, 'which deliberately examined individual places for their own sake, and [did] not attempt to draw out theoretical or historical conclusions' (p. 62; see also Jonas, 1988). He was also concerned about the vagueness of the spatial scale in defining localities, but welcomed the attempt to blend theoretical analysis with local understanding. (As discussed in the next chapter, the nature of scale, as a social construction, has been extensively explored by geographers: see Cox, 2014, chapter 5, for example.) Philip Cooke (1987) refuted the charge of empiricism, arguing that the objective of the CURS initiative was 'theorised interrogation' (p. 75) of available data. His general position was supported by Urry, who initially argued that 'there are some significant locality-specific processes' (Urry, 1986, p. 239), which he followed with a list of ten different ways in which social scientists have addressed them (Urry, 1987; see also Urry, 1985).

Cochrane (1987, p. 355) wondered whether CURS was 'just a cover for structural Marxism with a human face, or . . . the cover for a return to empiricism with a theoretically sophisticated face', and concluded that the programme contained the danger that as a guide to political action it might suggest that local struggle could suffice (what he termed 'micro-structuralism') rather than the realisation that parts cannot readily be isolated from wholes. Gregson (1987a) was perhaps even less sanguine, arguing that the theoretical rationale for the seven case studies was far from clear, thus making the likelihood of falling into the empiricist trap high; without a properly articulated theoretical core 'CURS simply replicates the mistakes of previous local studies; with such a core it could be so much more' (p. 370). Beauregard (1988) claimed that the programme lacked any clear directions for practice, for using the radical theory to achieve social change.

Simon Duncan's (1989b) full critique accepted that the concept refers to something important – spatial variability and specificity – but he concluded that 'the locality concept is misleading and unsupported' (p. 247; he concluded elsewhere – Duncan and Savage, 1989, pp. 202–4 – that it is 'confused, unsatisfactory, and largely redundant . . . a mystification'). He accepted that 'space makes a difference' in three ways:

1   social processes are constituted in places, which may differ because of previous 'layers of investment' (to use Massey's term);
2   actions take place locally and so can vary spatially; and
3   spatially varying actions can create spatially varying contexts.

But the concept of locality implies 'social autonomy and spatial determinism' (p. 247), both of which he rejects. He is not convinced that local differences are very important in creating uneven development, relative to more general processes: 'Locality is . . . only important if and when locality effects are part of the causal group explaining any event. And locality may well not be important', which implies a verdict of 'not proven' (p. 248).

Philip Cooke's (1989b) response to these critiques (and to the Duncan and Savage paper in particular) was trenchant:

> Local social processes are clearly an abiding feature of contemporary social life. Duncan and Savage's injunction to ignore them and settle on the structural level, supra-local, supra-national or whatever, in order to describe spatial variation in terms which deny agency to the social groups comprising localities, is both dated and redundant.
>
> (p. 272)

And he concluded that: '"Locality" can be seen to be a fascinating, complex concept of considerable value to geographical theory and empirical research.'

Griffiths and Johnston (1991) outlined a clearer framework for studying how localities differ, using three components of local economic, social and political structures drawn from their study of responses to the 1984–5 National Union of Mineworkers' strike in the UK. Johnston (1991) later applied this framework to a range of studies of uneven development at a number of scales, including the southern USA.

The debate over 'localities' was part of a much wider range of work which addressed the role of place in the creation of social scientific understanding. Agnew and Duncan (1989), for example, opened their introduction to *The Power of Place* with the statement that the book is:

> an attempt to make the case for the intellectual importance of geographical place in the practice of social science and history . . . [through] bringing together what can be called the geographical and sociological 'imaginations'.
>
> (p. 1)

Agnew (1989) analysed why place had been devalued in 'orthodox social science' in recent decades, focusing on the confusion of place with community (itself a term with many definitions: Colin Bell and Howard Newby, 1976). The orthodox treatment identifies a decline in community with 'modernization' and its replacement by a 'placeless' society, a transition seen as 'natural, lawful, and universal' (Agnew, 1989, p. 16), with nationalism growing as a 'place-transcending ideology'. Similarly, the alternative view derived from Marxism devalued place with its emphasis on 'freeing people from places' (p. 22; Entrikin, 1989, provides a parallel, elegant argument). The goal of *The Power of Place* is to correct those tendencies, and, through both theoretical argument and empirical illustration, 'argue against the prevalent tendency in history and social science to overvalue the sociological imagination at the expense of the geographical' (Agnew, 1989, p. 7; see also Agnew, 1990). He adduced five reasons for a place perspective (Agnew, 1987a):

1 it allows abstract categories – such as class – to be analysed in the context of everyday situations;
2 it avoids the search for 'universal' laws;
3 it allows a focus on structure–agency interrelations;
4 it shows that societies in different places change at different rates and along different trajectories, thereby removing any tendency to identify 'historical stages' to which all should conform; and
5 it illustrates the operation of cultural agents acting outside economic determinism.

A 'new regional geography'?

The debate over locality research became part of a wider literature on the importance of studying the specific characteristics of places. The numerous works using an implicit contextual approach include Agnew's (1984, 1987a) argument that voting behaviour is strongly influenced by the electorate's local context (see also Johnston, 1986d; Johnston and Pattie, 2006) and a detailed analysis of variations between local states in their provision of public housing (Dickens *et al.*, 1985). Some translated this general concern into a plea for a revived, but restructured, regional geography. Massey (1984b, p. 10), for example, ended such a plea with the argument that it is:

> necessary to reassert the existence, the explicability, and the significance, of the particular. What we [must do] . . . is take up again the challenge of the old regional geography, reject the answers it gave while recognizing the importance of the problem it set.
>
> (p. 10)

This echoes Gregory's (1978a) contention that:

> Ever since regional geography was declared to be dead . . . geographers, to their credit, have kept
> trying to revive it in one form or another . . . This is a vital task . . . we need to know about the consti-
> tution of regional social formations, of regional articulations and regional structures . . . [producing]
> a doubly human geography: human in the sense that it recognizes that its concepts are specifically
> human constructions, rooted in specific social formations, and capable of – demanding of – continual
> examination and criticism; and human in the sense that it restores human beings to their own worlds
> and enables them to take part in the collective transformation of their own human geographies.
>
> (pp. 171–2)

This call was not immediately taken up, although Fleming (1973), Steel (1982) and Hart (1982a) all called for a revival of traditional regional geography, which to Hart meant 'producing good regional geography – evocative descriptions that facilitate an understanding and an appreciation of places, areas and regions' (p. 2; as an example of such work – what might best be termed regional landscape history – see Haggett, 2012). Some have followed Gregory's lead (see also Gregory, 1985b), however, and argued – as do Roger Lee (1984, 1985) and Johnston (1984d, 1985c, 1985d) – for a reconstituted regional geography which recognises (Lee, 1985) that:

1 social processes operate in historically and geographically specific circumstances, so their understanding requires a sensitivity to geographical variations (regional mosaics);
2 society is not a fixed phenomenon but something that is constantly being recreated by human actions. Since those actions occur in historically and geographically specific contexts, then societal recreation is similarly historically and geographically variable;
3 those local transformations occur in the context of wider social relationships; and
4 the regions that emerge are not fixed divisions of territory but changing social constructions.

A major goal of geography should be to uncover the nature of those regions, as illustrated by some of the essays in Johnston et al. (1990).

The emphasis on change in these arguments implies the absence of a distinct niche for historical geography within the discipline's overall programme. The claim that a major route to understanding the present lies in study of the past is not new, but adoption of approaches based on structuration and contextual theory clarifies the importance of an historical perspective. Not surprisingly, therefore, historical geographers have made major contributions to the debates (see Baker and Gregory, 1984), drawing on such sources as the writings of the French *Annales* school of history (Baker, 1984; Pred, 1984c, draws on Braudel, for example). Thus, interpretations of the long-term evolution of a society (e.g. Dunford and Perrons, 1983) and its regional components (Langton, 1984), of basic transformations in a region, whether industrial (e.g. Gregory, 1982a) or agricultural (Pred, 1985, 1986), and of the constitution of particular places (as in Harvey's, 1985c detailed analysis of nineteenth-century Paris as an example of how consciousness is created in a particular context) all illustrate how changes take place as the result of general tendencies being played out in specific milieux by particular human agents. (Not all historical geographers agreed, however: see Meinig, 1978, especially p. 1215, and Dodgshon, 1998.)

The differences between the 'new' and the 'traditional' approaches to regional geography are clarified by Pudup (1988, p. 374), who characterised the latter as empiricist:

> Theoretically neutral observations are the basis for areal description. A reconstructed regional
> geography has foundations that rest in a clarified status of regions as objects of study – put
> simply, why geographers bother to study regions in the first place.
>
> (p. 379)

The answers to that 'Why?' question are provided in works such as Pred's (1984c) on southern Sweden and Gregory's (1982a) on West Yorkshire: regions are (not necessarily formally bounded) territorial entities, produced, reproduced and transformed through human agency; regions are the places in which people learn a culture, and contribute to its continuation (what Thrift, 1983, after Giddens, calls 'settings for interaction'). The nature of those processes is appreciated through a theorized approach, with the appreciation being provided through a narrative that draws on a defined vocabulary and permits 'theory to speak through subsequent empirical accounts' (Pudup, 1988, p. 383; see also Sayer, 1989a, 1989b). What those empirical accounts should focus on is still debated, and the need for a subdiscipline called regional geography has been contested. Warf (1988, p. 57) contended that with the replacement of traditional regional geography by positivism, 'a geography of "regions without theories" quickly became a geography of "theories without regions"', but Johnston (1990a, p. 139) argued that whereas 'we do not need regional geography . . . we do need regions in geography'.

Pudup's characterisation of the 'new' regional geography was extended by Anne Gilbert (1988), who identified three separate concerns with regional specificity in recent writings:

1   a concern with regions as local responses to capitalist processes, which she identified as probably the most prominent among English-speaking writers, who set the study of local variations within a political economy (usually Marxist) framework;
2   a concern with the region as a focus of identification (or 'sense of place'), which is especially strong among French writers concerned with the analysis of culture. To them, appropriation of a place (or region) is part of the creation and recreation of cultural identity; and
3   a concern with the region as a medium for social interaction, playing 'a basic role in the production and reproduction of social relations' (p. 212). The work on 'localities' fits within this concern.

All three represent a break from 'traditional' regional geography, Gilbert claimed, through their recognition that the persistence of regional diversity in the face of the homogenising tendencies within capitalism provides regional geography with a practical significance, in the mobilisation of resistance to those tendencies (see also Peet, 1989). The 'new' work depends on structuralist theory (as illustrated in Agnew's, 1987b, book on the USA within the capitalist world economy); on the recognition that regional processes rely on dialectical rather than naturalistic theories; and on the importance of human agency in the creation, recreation and transformation of regions. Together, these suggest a mode of study committed to understanding and achieving social change, which provides the challenge of making 'geography a science useful for society' (Gilbert, 1988, p. 223).

Subsequently, as already noted, regulationist work has also informed accounts of regions and regional configurations of governance and accumulation (e.g. MacLeod, 1998; MacLeod and Goodwin, 1999), as well as making what Kevin Ward (2001, p. 130) terms 'a substantial contribution to the study of urban politics'. Both strands of work, on regions and on urban localities, have been informed by (and blended with) social, cultural and political geographies, and wider social and political theory (e.g. Murphy, 1991; Paasi, 1991; Thrift, 1990, 1991, 1993, 1994). Thus, according to MacLeod and Jones (2001):

> geographers were encouraged to embrace elements of social theory, sociology, anthropology, and political economy in order to transcend the purported limitations of an earlier regional geography, not least its emphasis on the natural environment as the key factor in establishing the regional shape of society . . . for the new regional geographers, amid a range of rapid economic and political restructuring, regions were being presented as arenas through which to view the movements and dramas of individuals and groups in a whole series of economic, political and cultural contexts.
>
> (p. 671)

Noting how regional autonomy and devolution movements in many countries (especially in the European Union) had fostered much discussion of regions and regionalism since the 1970s, MacKinnon (2009, p. 233) also charts geographers' reworking of 'their regionalist heritage' drawing on the insights of political economy and subsequent strands of thought (including those considered in Chapter 7) 'has fostered a new concern with "the relational region" which reimagines regions as open and discontinuous spaces, defined by the wider social relations in which they are situated'. MacKinnon (2009, p. 233) cites work on southeast England (Allen *et al.*, 1998):

> the emblematic growth region of neoliberal Britain in the 1980s and 1990s, defined by its close con-
> nections to the global economy through the City of London in particular. Within the South East, areas
> of economic decline and deprivation exist, complicating and confounding the overarching image of
> growth and prosperity, while the boundaries of the region can be seen as open and porous (see
> Figure 6.2).

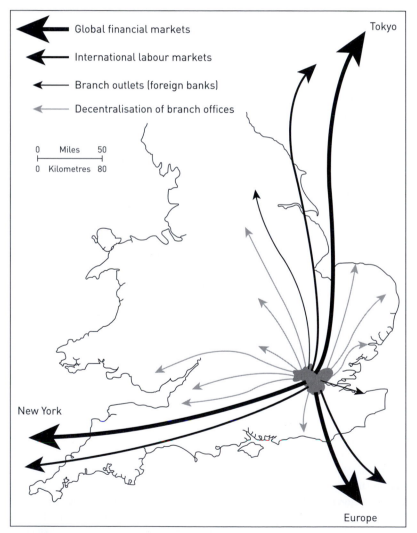

Figure 6.2 International financial links of South East England

Source: Allen *et al.* (1998) *Rethinking the Region*, p. 49, reprinted in MacKinnon (2009, p. 233).

The relationships between territorial, place bound and relational, networked conceptions of regions continue to offer fruitful grounds for investigation (Jonas, 2012).

Critical realism and structuration theory

Marxism has long been interested in exploring different levels of analysis and relations; between economy and culture, for example, or between production and reproduction as well as between different scales and places – classically the contrast between the 'country and the city', but also between 'core' and 'periphery', and the spatial and temporal dynamics of growth and decline. While Marxism has a materialist base, arguing that the basic underpinning of society comprises a set of economic processes and behaviours, these may only be reproduced through social, cultural and political processes, and in turn those also become aspects of what Marxists characterise as the means and forces of production. Debates about the relations between these and their 'relative autonomy' have long characterised Marxist thought. Later – in the 1990s and since – regulation theory has promised geographers a path through these.

In the 1980s, however, human geographers also explored two approaches which claimed to offer advances in the way that structure and agency (and the role of space) could be conceptualised. The first of these, the philosophy of realism, is based on the separation of three domains (Bhaskar, 1978):

1   the domain of the empirical is concerned solely with experiences, with the world as it is perceived;
2   the domain of the actual is concerned with events as well as experiences, accepting that an event (a particular item of human behaviour, for example) may be interpreted in different ways by individuals (by the actor, perhaps, and by another person experiencing it); and
3   the domain of the real is concerned with structures that cannot be apprehended directly, but which contain the mechanisms that lead to the events and their empirical perception.

Realists argue that all science incorporates these three domains, but the natural and the social sciences differ in the type of mechanism with which they deal. In particular, natural science deals with 'closed systems' in which the same interrelationships can be observed on innumerable occasions, allowing the development of universal laws based on replicable results. Social science rarely if ever deals with such closed systems, however, both because of the difficulties of abstracting independent parts from the whole (most attempts to do this produce what Sayer, 1985a, terms 'chaotic conceptions') and because major components of the system – people – learn from their experiences, so that experiments can never be replicated.

The goal of realism, like that of positivism, is to explain events, to discover their causes by 'finding out what produces change, what makes things happen, what allows or forces change' (Sayer, 1985a, p. 163). It differs from positivism because it contends that 'what causes something to happen has nothing to do with the matter of the number of times it has happened or been observed to happen and hence with whether it causes a regularity' (p. 162). Indeed, 'regularities' (i.e. laws and lawlike generalisations) are very unlikely to occur in the social sciences. Andrew Sayer (1984, 1992a) argues that two conditions are necessary if regularities are to occur:

1   the mechanisms must be invariant; and
2   the relationships between the mechanisms and the conditions in which they occur must be constant.

If both hold, then scientists will be studying closed systems, with identical conditions leading to identical events (either naturally or in laboratory conditions). If one or both is absent, however, the object of study is an open system, and regularities will then not occur. The social sciences study open systems:

> we can interpret the same material conditions and statements in different ways and hence learn new ways of responding, so that effectively we become different kinds of people. Human actions characteristically modify the configuration of systems, thereby violating the . . . [first] condition . . . while our capacity for learning and self-change violates the . . . [second] condition.
>
> (Sayer, 1992a, p. 123)

The mechanisms underpinning the operations of human societies are interpreted in different ways, therefore, with the outcomes of those interpretations creating new sets of conditions within which future interpretations will be constrained. The mechanisms are abstract concepts which are actualised in contingently related conditions, as for example with:

> the law of value, which concerns mechanisms which are possessed necessarily by capital by virtue of its structure (as consisting of competing and independently directed capitals, each producing for profit and being reliant on the production of surplus value, etc.), produces effects which are mediated by such things as the particular kinds of technology available, the relative power of capital and labour and state intervention. (p. 140). [.] Now no theory of society could be expected to know the nature and form of these contingencies in advance, purely on the basis of theoretical claims.
>
> (p. 142)

A major issue for those adopting a realist approach concerned the role of space is: does it, as raised by the title of Sayer's (1985b) essay, 'make a difference'? In realist terms that question can be pitched at the level of either the necessary or the contingent – is space important in theoretical work, or only in concrete realisations of the outcome of general processes? Regarding the former, he noted that 'Abstract social theory need only consider space insofar as necessary properties of objects are involved, and this does not amount to very much' (p. 54). All matter has a spatial location, so all events occur in space but, while recognising this, abstract theory need say very little about actual spatial forms. 'Hence, while it is important for abstract theory to be aware of the existence of space, the claims that can be made about it are inevitably rather indifferent ones' and refer to concrete realisations only. He concludes that to claim that space is contingent in its operation does not mean that it is unimportant: 'The difference that space makes in concrete situations can never be ignored, and therefore in that sense "geography matters"' (p. 109).

The study of causation in a realist context involves what Sayer and Morgan (1985) term intensive research programmes, whose questions take the form 'What did the agents actually do?'; answering them involves examining causal processes in a particular case or group of cases. The answers cannot be generalised beyond those cases, in contrast to those produced in extensive research programmes, commonly adopted in empiricist/positivist human geography, which are 'mainly concerned with discovering some of the common properties and general patterns in a population as a whole' (p. 150). It is evident that realism enabled and inspired some rich empirical research in human geography. It is not hard to discern why many were attracted to the mixture of critical potential and rigour that it claimed to offer. Reviewing this, Pratt (2009, p. 381) notes how:

> Critical realism has had a significant impact in the discipline of geography, and in particular it offered a resolution of the disenchantment with positivism, especially – although not exclusively – within human geography, heralding an approach that, it is claimed, avoids the perceived shortcomings of subjectivism (as in the humanistic or phenomenological geography . . .).
>
> (p. 381)

A second path, which some geographers advocated in the 1980s as an overall framework to research and as a means to explore the constitutive role of space in social life, was structuration – a theoretical approach first developed by Giddens (1984) to account for the ways in which people learn about and transform social structures. This is a place-bound (and time-bound) process; one's context (including one's language: see p. 376–7, this book) and a major influence on one's individual development. For exegeses and applications by geographers, see Moos and Dear (1986); Dear and Moos (1986); Phipps (2001).

Structuration, as an example of a contextual approach, clarifies the falseness of any distinction between social relations and spatial structures; as Gregory and Urry (1985, p. 3) express it, 'spatial structure is now seen not merely as an arena in which social life unfolds, but rather as a medium through which social relations are produced and reproduced'. Together they produce the process of spatiality (or socio-spatial dialectic), used by Soja (1980, 1985) to characterise the conjoint social production of space and the spatial construction of society. Giddens and others identify much common ground between his conception of structuration and Hägerstrand's of time geography (though see Gregson, 1986, 1987b). Pred used the two in a number of essays to illustrate their relevance to understanding both the unfolding of his own career (Pred, 1979, 1984a) and the development of particular places (Pred, 1984b, 1984c). Thus:

> The meanings of interpretations I have imposed upon the past, the there and then in Berkeley and Sweden, are a result of the knowledge, attitudes and values I hold now, the ever fading here and now in Berkeley. Yet, the knowledge, attitudes and values I hold here and now are rooted in my past path, my past participation in projects and the institutional and societal context which generated those projects.
>
> (Pred, 1984a, p. 101)

Also, 'Biographies are formed through the becoming of places, and places become through the formation of biographies' (Pred, 1984b, p. 258).

In the end, relatively few geographers adopted structuration as a research methodology, although, as we have noted, it did appear promising. Derek Gregory (1994, p. 113) judged that 'its philosophical and theoretical rigor dovetailed with the sensibilities that had been put in place by spatial science, by a number of post-positivist geographies, and by the development of historico-geographical materialism'. Hence, Cloke et al.'s (1991) text contained an entire chapter on structuration theory, juxtaposed to the one on realism, and Phipps (2001) subsequently explored structuration theory's empirical applications. All adopted the basic argument – derived from Marx – that people make and remake themselves in the places that they make and remake.

The production of 'nature'

The holistic conception of relations and dynamics in Marxist thought has also lent itself to work on one of the oldest of concerns in geography – human–environment relations. Marxist work on uneven development thus includes work on the relationships between society and nature. Liberal concerns over environmental issues had increased substantially during the late 1960s and early 1970s. Harvey (1974d) introduced a Marxist perspective to this issue, arguing that contemporary statements on resource–population ratios are ideologically based, an argument sustained through examination of the works of Malthus (Mayhew, 2014), Marx and Ricardo. Resources are not defined outside the context of societal appraisals of nature, and those appraisals reflect the current mode of production. Thus, predictions such as those of the *Limits to Growth* study (p. 328, this book) represent a status quo theoretical perspective (p. 316, this book), whereas employment of other perspectives may lead to different conclusions:

> let us consider a simple sentence: 'Overpopulation arises because of the scarcity of resources available for meeting the subsistence needs of the mass of the population.' If we substitute our definitions [of subsistence, resources and scarcity] into this sentence we get: There are too many people in the world because the particular ends we have in view (together with the form of social organisation we have) and the materials available in nature, that we have the will and the way to use, are not sufficient to provide us with those things to which we are accustomed.
>
> (Harvey, 1974d, p. 272)

With the first version of the sentence, the policy option seems clear – population reduction. The second allows Harvey to identify three possibilities, however: (1) changing both the desired end and the social organization; (2) changing technical and cultural appraisals of nature; and (3) changing views about the capitalist system and its scarcity basis. The Ehrlich–Commoner arguments regarding the origins of current environmental problems accept the first version of the sentence, however, and Harvey claimed that they have been grasped by the elite of the capitalist world to provide an underpinning for a popular ideology (the need for birth control). Similarly, Keith Buchanan (1973) argued that the arguments for birth control in the Third World are part of the 'white north's' imperialist policy of ensuring its continued access to the resources of the international periphery (even to the extent of stimulating environmental destruction and famine: Bradley, 1986; Blaikie, 1986), and Johnston *et al.* (1987) have explored the geography of what Buchanan (1973, p. 9), along with other peace scientists, terms structural poverty: 'The poverty which is regarded as symptomatic of reckless population growth is rather a structural poverty caused by the irresponsible squandering of world resources by a small handful of nations'.

Under capitalism, the environment (or nature) is treated as a commodity, to be bought, exploited and sold. Neil Smith (1984) points out that nature is presented in capitalism's ideology as external to society, as the 'antithesis of human productivity . . . the realm of use-values rather than exchange values' (p. 32). But a 'second nature' is produced by capitalism:

> In its uncontrolled drive for universality, capitalism creates new barriers to its own future, it creates a society of needed resources, impoverishes the quality of those resources not yet devoured, breeds new diseases, develops a nuclear technology that threatens the future of all humanity, pollutes the entire environment that we must consume in order to reproduce, and in the daily work process it threatens the very existence of those who produce the vital social wealth.
>
> (p. 59)

Nature, like spatial organization and reorganization, becomes a product of the continuing processes of capitalist accumulation (Fitzsimmons, 1989); as an abstract concept it is of no value to society and it is of interest only as a sphere of human activity (Pepper, 1984, 1996). This representation of nature as a social construction is not new: Zimmerman's (1972) classic work on natural resources defined a resource as only existing when it was perceived as such, and Glacken's (1967) classic survey shows how cultural conceptions of nature have varied over centuries. In his adaptation, however, David Pepper (1984) separates the 'historicity of nature' (how the concept of nature in a particular time and place is related to the human activities there) from the 'naturalness of history' (the relationship between natural conditions and human activity – a relationship crudely expressed in theories of environmental determinism, p. 43, this book). Part of the creation of a society involves its historicity of nature:

> Labour is the means whereby man converts nature into forms useful to him. In the main nature does not offer ready-made subsistence to man, neither does man take direct possession of nature's resources. He has to transform them. This process of transformation is a social one – it

is done with other people who are organized in a particular way. . . . Thus through shaping nature, men shape their own society and their relations with their fellows.

(p. 162)

Thus, just as the creation of social relations (the means of exploiting surplus value from labour by capital) is part of the making of a society, so too is the creation of society–nature interrelations; and just as those social relations contain within them the seeds of crisis for capital accumulation (the class conflict) so too do society–nature interrelations (environmental problems). Johnston (1989, 1996d) has carried this argument forward in an account of society–nature relationships under various modes of production and of the role of the state in their regulation (see also Taylor, 1994); the essays in Blaikie and Brookfield (1987) illustrate the general theme.

Harvey's (1974d) early Marxist work on ideology, social and economic life, and their complex interrelationships with nature, has since been extended by others into a holistic conception of the social mediation of nature and the inevitability and necessity of (geographically) uneven development under capitalism (e.g. Braun and Castree, 1998; Castree, 1995b, 2003; Castree and Braun, 2001; Peet and Watts, 1996; Smith, 1984) and the possibility of 'liberation ecologies' (Peet and Watts, 2004) and political ecology's wider connections with geopolitics, power and domination (Peet et al., 2011). Recognising the wealth of work on 'nature', Andrew Jones's (2009, p. 474) review of the trajectory of Marxist geography notes that while it also influenced development geography (Peet, 1991), political geography (Taylor, 1982) and urban geography (Smith, 1996):

> The fourth and final strand of a revised Marxian human geography that has generated an identifiable literature can be loosely described as a revised eco-Marxist human geography. . . . Much of this thought has drawn on a burgeoning eco-Marxist literature beyond human geography, although it has also developed the Marxian perspectives on the production of nature developed by Neil Smith and others in the 1980s. More recent contributions represent a form of revised Marxist geography insofar as they have increasingly sought to engage with postmodern epistemological arguments concerning society/nature and new forms of politics. Such eco-Marxist human geography has been concerned, for example, with debates about environmental political movements, the rights of indigenous groups in the developing world, and environmental values.

In turn, similar relationships between nature, production, ideology and power were also problematised in Marxist-influenced work on 'landscapes'. Stephen Daniels (1989) thus stressed how conceptions of landscape were rooted in reactions to Western capitalist society (see the review by Cosgrove, 2003). Building on this, but in a more direct line to Marx, Don Mitchell (2003, p. 238) notes that 'the landscape can be understood to be a product of human labor, of people going to work on the land to make some thing out of it'. Thus:

> To understand landscape, and to understand the 'culture' within which it exists, requires an examination of human practices – of forms of labor. Through labor the landscape is both made and made known. And in the process of working on and in landscape so too are people changed, whether that change is understood at the level of culture . . . or at the level of the individual human body . . .

(p. 239)

Radicals in debate

The 'radical camp' was never a united body of scholars. Although Marxism remained a focus, it was attacked by others, themselves associated with 'radical' stances. In 1986, for example, Peter Saunders and Peter Williams equated the recent literature in urban studies with an unchallenged 'taken for granted orthodoxy ... [that] thrives on an unspoken and largely unexamined political and theoretical consensus' (p. 393). They claimed that this has opposition to positivism as its linchpin, and arguments within it occupy 'a very narrow spectrum embracing left Weberianism and the different varieties of Marxism, but excluding almost everything else'. They concluded that, as a consequence, urban studies currently:

> makes little pretence of being 'value-free' or 'ethically neutral' and . . . is sheltered from the possibility of empirical disconfirmation. This in turn means that approaches (such as the philosophy of the so-called 'New Right') which cannot be subsumed under the orthodoxy can be dismissed almost a priori. Such work is rarely read, still less seriously considered on its own terms. Rather a pejorative label is attached to it (for example, 'Thatcherism', 'Reaganism', or 'authoritarian populism') which enables us to pigeonhole it within our existing conceptual apparatus . . . without ever having to engage with its intellectual content. In this way alternative ideas are dismissed but never discussed, explained away but never critically evaluated.
>
> (pp. 393–4)

Elements of both Marxism and realism were then criticised, with the latter being characterised as a 'justification for subordinating history to theory' (p. 394) and the source of 'causes which can only be identified theoretically and which are guaranteed immunity from falsification even where there is no manifest evidence for their existence' (p. 395); works which are explicitly anti-positivist are therefore castigated for not being subjected to positivist and critical rationalist procedures.

According to this critique, theories of structures have primacy in realist and Marxist approaches and as such are outdated; some major changes (such a changing gender relations and reconfigurations of culture, 'race' and identity) in British society over the last century are ignored, it is claimed – 'we still employ an essentially outdated class theory in our analyses' (p. 397) – and these approaches are also criticised as:

> having done little or nothing to change the essential features of capitalism. As social change takes place under our noses, so we risk a situation where our methods and theories ensure that we pay it little heed. The orthodoxy is safeguarded and reproduced as the society changes.

One of the strongest continuing adherents to a Marxist position, and a major focus of the Saunders–Williams critique, is Harvey, who characterised the goal of 'my academic concerns these last two decades . . . [as] to unravel the role of urbanisation in social change, in particular under conditions of capitalist social relations and accumulation' (Harvey, 1989b, p. 3). He responded to Saunders and Williams by noting 'a marked strategic withdrawal from Marxian theory within the field of urban analysis and a broadening reluctance to make explicit use of Marx's conceptual apparatus in articulating arguments' (Harvey, 1987, p. 367), and launched an attack against those he identified as abandoning the 'tough rigour of dialectical theorizing and historical materialist analysis', because:

> The case for retiring Marx's *Capital* to the shelves of some antiquarian bookstore . . . is not yet there. Indeed, in many respects the time has never been more appropriate for the application of Marx's conceptual apparatus to understanding processes of capitalist development and transformation. Furthermore, I believe the claim of Marxian analysis to provide the surest guide to the construction of radical theory and radicalizing practices still stands.

His critics had caved in too readily to right-wing pressures, he argued.

Harvey (1987, p. 368) agreed with Saunders and Williams that realism and the concept-
ualisation of agency and structure 'are nothing more than weak disguises, soft versions of a
traditional left orthodoxy ranging from left Weberianism to "different varieties of Marxism"'. But
he disagreed with their identified way forward. He claimed that three 'myths' have pervaded the
critique of Marxist scholarship. The first is economism, as illustrated by James Duncan and David
Ley's critique (1982). The second is the claim that 'The abstractions of Marxian theory cannot
explain the specificities of history and the particularities of geography' (p. 370), to which he took
great exception. Harvey believes that 'it is in principle possible to apply theoretical laws to understand individual instances, unique events' (p. 371), particularly since Marx's most interesting
lawlike statements were about capitalist processes, not events. He illustrated this with the chapter in
Capital on 'The working day', arguing that such a detailed empirical description of how humans
respond to their conditions on a daily basis allow 'categories like money, profit, daily wage, labour
time, the working day, and ultimately value and surplus value [to] arise through an examination of
historical materials' (p. 372), thereby illustrating how theory is both derived from and developed
through the unravelling of particular situations.

Harvey saw the second myth being advanced from 'within the ranks of the left itself' (p. 373);
Massey and Sayer's 'very deep and serious concerns for the particularities of places, events and
processes' were noted. Sayer's realist approach was presented as involving a combination of
'wide-ranging contingency with an understanding of general processes', however:

> The problem with this superficially attractive method is that there is nothing within it, apart from
> the judgement of individual researchers, as to what constitutes a special instance to which special
> processes inhere or as to what contingencies (out of a potentially infinite number) ought to be
> taken seriously. There is nothing, in short, to guard against the collapse of scientific understandings
> into a mass of contingencies exhibiting relations and processes special to each unique event.
>
> (Harvey, 1987, p. 373)

This, Harvey feared, is a path to 'simple empiricism' (p. 374), which he believed Saunders and
Williams (1986) were also promoting; he characterised their agenda as 'nothing short of an
abrogation of scientific responsibility and a caving in of political will'. Against that, Harvey insisted
on 'the viability of the Marxist project', with its focus on 'universalising statements and abstractions'
(p. 375) and its ability to guide political practice.

Finally, the third myth (the 'never-never land of nontotalising discourses into which
Marxists cannot enter because they insist on talking about totalities', p. 374) was presented as an
attack on 'totalizing discourses', meta-theories and meta-narratives. While rejecting the notion that
Marxism is 'inherently totalizing', Harvey nevertheless argued that 'nothing appears more totalizing
to me than the penetration of capitalist social relations and the commodity calculus into every niche
and cranny of contemporary life'. If social theory can look only at the parts, then the whole can
never be apprehended and challenged (part of Harvey's argument against postmodernism: see
Chapter 7). To Harvey (1987):

> every single local study I read . . . points to how locality is caught up in universal processes of
> financial flows, international divisions of labour, and the operations of global financial markets.
> Claims to the contrary are dangerous, for they involve avoiding confrontations with 'the realities
> of political economy and the circumstances of global power'.
>
> (p. 375)

As Harvey (2002, pp. 184–5) later recorded, he rarely responded to critiques – 'because it had
always seemed to me that the general stance of opposition to capitalism was best served by making
creative contributions rather than engaging in destructive slanging matches'. But he 'had been

somewhat taken by surprise' by the feminist critiques (see Chapter 8) of *The Condition of Postmodernity* (Harvey, 1989a: cf. Deutsch, 1991; Massey, 1991) and felt he had little option but to reply; this was published in *Antipode* (Harvey, 1992) – 'Nobody took much notice'.

Harvey's stance was supported by Neil Smith (1987c), who defended Marxism as providing both a broad analytical framework and a 'quintessentially political discourse'; he found that the 'realist project . . . has become the theoretical justification for the belief that there can be no general theory at all concerning questions of geographical space, and that any attempt to devise such a theory is fundamentally misconceived' (p. 379). Others, however, were less supportive; Ball (1987, p. 393), for example, took issue with Harvey's:

> total dismissal of anyone who does not repeatedly declare their Marxist label, who does not believe that everything Marx said is unambiguous and correct, and who fails directly to apply the most abstract propositions of Marxist theory to the empirical situations they are investigating.

Sayer (1987, p. 395) also resented Harvey's impugning of motives, and responded to both Saunders and Williams, and Harvey that 'even if it means breaking off from what I had hoped and still hope for is a broad but common project, the search for a social science with an emancipatory potential'. Sayer agreed very much with Harvey that empirical research enables a clarification of theoretical understanding, it being 'partly responsible for making me revise my abstract ideas about the nature of capital, competition, class and the division of labour' (1987, p. 397; see also Sayer, 1995, 2013). He also defended realism against the charges of both reductionism and theoreticism (exemplified by Saunders and Williams) on one hand, and of empiricism (Harvey) on the other.

Saunders and Williams (1987, pp. 427–8) began their response with the following categorisation:

> Harvey eschews fraternization with the enemy, and he adopts a quasi-religious, almost messianic tone in delivering his epistle. He tells us . . . of his unswerving belief that Marxism provides the surest guide to radical salvation . . . Harvey's statement provides a good example of precisely that tendency in contemporary urban studies which we suggested could stifle fresh initiatives and hamper intellectual debate. If you believe, as Harvey apparently does, that the eternal verities have largely been established by Marx's Capital, then you have effectively closed off the possibility of open debate, and even more the possibility of learning from others who disagree.

Gould (1988) similarly criticised Marxist writers in general and Harvey in particular for what he termed their 'claim to exclusiveness'. Saunders and Williams then attacked Harvey for sustaining a 'totalizing' form of theory, which embraces the whole of society and thus has a privileged starting-point (or set of initial assumptions) which is not open to empirical refutation/ confirmation:

> It is difficult to see how you can get to the whole by studying the parts and building up from there. . . . Totalising discourses are thus always unalterably committed to a priori assumptions – Marx's theories of exploitation, class struggle, and historical evolution came, not from studying people in Manchester, but from ideas about society 'as a whole', and these ideas were then mapped onto existing empirical observations.
>
> (Saunders and Williams, 1987, pp. 428–9)

Thus, Harvey's Marxism is not susceptible to effective falsification; nor, they assert, despite Sayer's arguments, is realism. Both are presented as 'closed' approaches, and although Saunders and Williams share the same goal as Harvey and Sayer, they believe that 'you will not develop an emancipatory social science before social science itself is opened up' (p. 430). Hence the debate, which the editor argued was necessary because capitalist society is changing; the nature of politics is altering and 'social theory has to take account of these cumulative changes' (Dear, 1987, p. 363).

Liberals, radicals and post-marxists in debate

Alongside debates within the 'radical' approach there have been others which, in the published contributions at least, have been much more heated than that generated by the 'quantitative revolution'. They developed in more depth and detail through the 1970s and 1980s, using forums, such as the journal *Area*, which were not available 20 years earlier, when publication outlets for 'views and opinions' were few. Whereas the behaviourists caused little real concern within the discipline, and their views were soon co-opted within the corpus of acceptable approaches, the radicals had much more impact, because they attacked not only the basis of most prior geographical work but also, in the clear interdisciplinarity of their approach, both the bureaucratic structure of the discipline and the existence of geography itself (Johnston, 1978a); some, such as Eliot Hurst (1980, 1985), argued explicitly for the abolition of geography as a separate discipline (see also Taylor, 1996). A clear early example of the liberal–radical polarisation was a debate on the geography of crime initiated by Peet (1975b), who argued that in attempting to make their work relevant geographers avoided asking 'relevant to whom?' (on which, see Harvey, 1974c); the political consequences of their work were ignored. The studies of crime reviewed (e.g. Harries, 1974) referred only to the surface manifestations of a social problem and could not provide solutions, only ways of ameliorating the problems: 'So it is that "useful" geography comes to be of use only in preserving the existing order of things by diverting attention away from the deepest causes of social problems and towards the details of effect' (Peet, 1975b, p. 277). Furthermore, geographers study only the crimes for which statistics are collected, thereby accepting the definition of crime by the elite; their maps, which are useful to police patrols, can therefore be employed to help maintain the status quo of power relations within society. The implicit position of the geographers involved is one of protecting the 'monopoly-capitalist state'.

Harries (1975) responded by attacking the simplistic nature of Peet's arguments, claiming that geographers would have no influence at all if they argued merely that crime is a consequence of monopoly capitalism. He contended that it is best to work within the system, to make the administration of justice more humane and equitable, to protect the potential victims of crime and to provide employment opportunities for graduate geographers. Approaches based on crime control are likely to be more influential than polemics relating its cause to the mode of production: as Yuk Lee (1975, p. 285) also pointed out, Peet 'failed to provide us with any clues as to how he or other radical geographers would study crime' (p. 285).

Peet (1976) responded to Yuk Lee's challenge, outlining a radical theory which would 'contribute directly, through persuasion, to the movement for social revolution' (p. 97). Capitalism harnesses human competitive emotions and generates inequalities in material and power rewards. Aggression is an acceptable part of being competitive and is often released on the lower classes, who are encouraged to consume but are provided with insufficient purchasing power. As the contradictions of this paradox increase, so does the pressure to turn to crime. Thus, crime occurs where the lower classes live, and at their spatial interface with the middle class. Harries (1976) replied that being a radical was a luxury few academics could afford, because working at a publicly financed university demanded a pragmatic rather than a revolutionary approach. To him, Peet's theory is overly economic and deterministic; it fails to account for cultural elements, such as the disproportionate criminal involvement of blacks, the subculture of violence in the southern USA and other areas, and the fact that all economic systems produce minorities disadvantaged in terms of what they want and what they can get by socially legitimate means. He could offer no alternative theory, however: 'I do not carry in my head a theory of crime causation, and I am quite incapable of synthesizing and attaching value judgements to existing theoretical formulations within a couple of pages of typescript' (p. 102). He encouraged Peet to come off the fence, and to get involved in the production of change within the present system. Laurence Wolf (1976), on the other hand, claimed that by concentrating on the traditional concerns of Marxism, Peet was not radical enough.

Two geographers who were involved in a considerable, often virulent, debate in the 1970s were Brian Berry and David Harvey. Berry (1972a) initiated the exchanges with comments on Harvey's (1972) paper on revolutionary theory and the ghetto. He wondered whether Harvey's rational arguments on the need for a revolution would be accepted, 'because of "commitment", the opposition will quietly drift into corners, the world will welcome the new Messiah, and social change will somehow, magically, transpire' (p. 32). The power to achieve change, he contended, needs more than logical argument – 'nothing less than cudgels has been effective' (p. 32) – and Harvey's belief in logical rationalism will be to no avail. He also argued that Harvey was wrong about the ghetto, for liberal policies were succeeding and the inequalities between blacks and whites were being reduced (Berry, 1974a). Harvey (1974a) responded that scarcity must continue in a capitalist economy, which will leave some people – probably those in the inner city – relatively disadvantaged. (One of the consequences of recent neoliberal economic policies and the 'rolling back of the welfare state' has been the rapid increase of inequality within most countries of the 'developed world' over the last three decades, as illustrated in several atlases and other geographical works: Glasmeier, 2005; Dorling, 2010.)

Berry's (1974b) review of *Social Justice and the City* (Harvey, 1973) criticised Harvey's dependence on economic explanations. Based on Daniel Bell's (1973) characterisation of post-industrial society, Berry (1974b) claimed that the economic function is now subordinate to the political:

> the autonomy of the economic order (and the power of the men who run it) is coming to an end, and new and varied, but different, control systems are emerging. In sum, the control of society is no longer primarily economic but political.
>
> (p. 144)

Harvey (1974b) responded that Marxism could not be considered as passé while the selling of labour power and the collusion between the economically and the politically powerful continue, and that the state has to be considered within a marxist framework too (Harvey, 1976; which he brought up-to-date in his writings on neoliberalism and imperialism – e.g. Harvey, 2005b, 2005c – and capitalism's current crises, Harvey, 2014). Berry (1974b) retorted:

> I believe that change can be produced within 'the system'. Harvey believes that it will come from sources external to that system, and then only if enough noise is produced at the wailing wall. . . . The choice, after all, is not that hard: between pragmatic pursuit of what is attainable and revolutionary romanticism, between realism and the heady perfumes of flower power.
>
> (p. 148)

Harvey's (1975c) response came in a review of Berry's (1973b) *The Human Consequences of Urbanization* – a study of urbanisation processes at various times and places, and of the planning responses they stimulated – which concluded that the book is 'all fanfare and no substance' (p. 99), revealing that Anglo-American urban theory is 'substantively bankrupt' and that 'it is scholarship of the Brian Berry sort which typically produces such messes' (p. 99):

> It is doubtful if it makes any sense even to consider urbanisation as something isolated from processes of capital formation, foreign and domestic trade, international money flows, and the like, for in a fundamental sense urbanisation is economic growth and capital accumulation – and the latter processes are clearly global in their compass.
>
> (p. 102)

To Harvey, Berry has nothing to say of any substance, but he recognised that Berry 'is influential and important . . . his influence is potentially devastating' (p. 103 – whether he meant this influence as within geography only, or beyond it too, is not clear). Berry's only response at the time was a general comment on the Union of Socialist Geographers (Halvorson and Stave, 1978):

> there's no more amusing thing than goading a series of malcontents and kooks and freaks and dropouts and so on, which is after all what that group mainly consists of. There are very few scholars in the group.
>
> (p. 233)

Nearly thirty years later, Berry (2002c, p. 443) revisited that debate, claiming that Harvey's 'approach was that of the armchair intellectual contrarian . . . [he] came to exemplify the tenured radical who lives comfortably on the rewards provided by the society that is the objective of his disdain'. Harvey (2002, p. 173), on the other hand, recorded that 'I had a relatively secure position in the field. I could use my intellectual resources and powers to a political end and was fiercely determined . . . to do so to the hilt, much as I had as an undergraduate when confronted by aristocratic privilege' at Cambridge.

Apart from these very polarised exchanges, several others indicated that whereas some 'liberals' were prepared to consider the radical case seriously, others tended to avoid the issues. Michael Chisholm, for example (1975a, p. 175), claimed that:

> while I am fully sympathetic to the view that the 'scientific' paradigm is not adequate to all our needs, and must be supplemented by other approaches, I am not persuaded that it should be replaced. . . . Harvey wants us to embrace the marxist method of 'dialectic'. This 'method' passes my understanding; so far as it has a value, it seems to be as a metaphysical belief system and not – as its protagonists proclaim – a mode of rational argument.
>
> (p. 175)

More frequently, reviewers accepted that the radical view added to their appreciation, but argued that it was not, for them at least, tenable in its entirety. Thus, Morrill (1974) wrote of *Social Justice and the City* that 'I am pulled most of the way by this revolutionary analysis but I cannot make the final leap that our task is no longer to find truth, but to create and accept a particular truth' (p. 477). Leslie King (1976) also sought a middle course:

> An economic and urban geography that will be concerned explicitly with social change and policy. . . . Such a middle course will not find favour with the ideologues, who will see it either as another obfuscation favouring only 'status quo' and 'counter-revolutionary' theory, or as a distraction from the immediate task of building elegant quantitative-theoretic structures, but some paths are being cut through the thicket of competing epistemologies, rambling lines of empirical analysis, and gnarled branches of applied studies that now cover the middle ground.
>
> (pp. 294–5)

He accepted that much quantitative-cum-theoretical geography had sought mathematical elegance as an end in itself; he believed that social science must feed into social policy and generate social change; and he accepted the 'intellectual power' of Marxist analysis but believed that prescriptions based on it are acceptable only if the ideological framework is also. His conclusion suggested the need for more quantification, which was operationally useful rather than mathematically elegant (see also Bennett, 1985a, 1985b), and later noted that 'space . . . should be seen as an element in the political process, an object of competition and conflict between interest groups and different classes' (King and Clark, 1978, p. 12). Finally, David Smith (1977, p. 368) concluded that:

> Marx may have been able to dissect the operation of a capitalist economy with particular clarity, and see the essential unity of economy, polity and society that we so often miss today. But Marx does not hold the key to every modern problem in complex, pluralistic society.

The debates continued, with many geographers unable/unwilling to accept both the Marxist analysis of society and the Marxist programme for action. In part, this reflected a partial reading of Marxism – in particular, a concentration on strict structural interpretations which neither allow for the activities of knowing individuals nor accept or acknowledge the considerable heterogeneity of Marxism. Ley (1980, p. 12), for example, saw in Marxism not only 'some hidden transcendental phenomenon . . . directing the course of human society', which commits an epistemological error by denying or at least suppressing the subjective, but also a theoretical error that devalues the power of human action to redirect the course of events, and a moral error, which makes humans into puppets and threatens basic freedoms of speech, assembly and worship. Muir (1978) attacked interpretations of Marxism and similarly implied that it threatens individual freedom of the academic 'to pick and choose from among the . . . literature' (p. 325). Respondents pointed to the lack of such 'intellectual orthodoxy' and 'sterilized geography' (Manion and Whitelegg, 1979; Duncan, 1979), but Muir (1979, p. 127) remained convinced of the threat implicit in 'the commands from such little men as Marx, Lenin, Trotsky, Stalin and Mao concerning the primacy of activism, the obligations of party membership and the necessity to subordinate individual judgement to the will of the party'. He claimed that radical geography is contributing to understanding, but to call it Marxist is to give it a certain programmatic base. Walmsley and Sorensen (1980), on the other hand, presented Marxism as just irrelevant, deflecting attention from 'reformism and relevance'.

James Duncan and David Ley (1982) published an extensive critique of marxism in geography, containing four major themes.

1   Marxist analysis is a form of holism, in which the whole – variously termed capital, the economic structure, economic processes, etc. – is given a life of its own: an abstraction is assumed to exist. This reification (see also Gould, 1988) offended their belief in individuals as conscious, free agents (see J. Duncan, 1980, on holism in cultural geography, and Agnew and Duncan, 1981) – although they did not proclaim an idealist alternative and accepted that 'individual action cannot be fully explained without reference to the contexts under which individuals act' (p. 32).
2   Individuals are represented as agents of the whole – the means of implementing its goals – not as free decision-makers in their own right.
3   The materialist infrastructure of Marxism is a form of economism – it presents economic processes as the ultimate cause of all behaviour, which excludes many other influences.
4   'The attempt to cast explanation continually and everywhere in terms of economic imperatives, [leads] . . . to a crisis in empirical exposition' (p. 47).

The last comprises what is essentially a positivist critique: 'the form of the explanations is both tautological and empirically untestable. The result is a mystification in explanation of how real processes operate' (p. 55). They concluded that structural Marxism in human geography presents a passive view of the individual, offering explanations in terms of abstract wholes which are obfuscatory and not verifiable. Later work by James Duncan (1985) showed that he was attracted to the structuration approach, however, which allows for human agency operating within structural constraints, as indeed do most interpretations of Marxism.

Others who criticised structural Marxism disagreed with its treatment of the individual, and wanted more convincing models of the interrelationships between infrastructure and superstructure in the process of societal change. Gregory (1981) presented four such models (Figure 6.3):

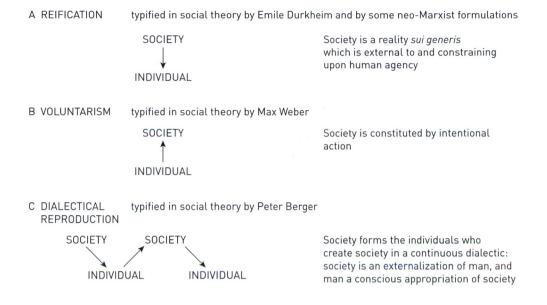

A REIFICATION   typified in social theory by Emile Durkheim and by some neo-Marxist formulations

SOCIETY

INDIVIDUAL

Society is a reality *sui generis* which is external to and constraining upon human agency

B VOLUNTARISM   typified in social theory by Max Weber

SOCIETY

INDIVIDUAL

Society is constituted by intentional action

C DIALECTICAL   typified in social theory by Peter Berger
   REPRODUCTION

SOCIETY   SOCIETY

INDIVIDUAL   INDIVIDUAL

Society forms the individuals who create society in a continuous dialectic: society is an externalization of man, and man a conscious appropriation of society

D STRUCTURATION   typified in social theory by Jürgen Habermas and Anthony Giddens

SOCIETY ———————— SOCIETY

INDIVIDUAL ———————— INDIVIDUAL

Society systems are both the medium and the outcome of the practices that constitute them: the two are recursively separated and recombined

Figure 6.3 Models of the relationships between society and individual

Source: Gregory, 1981, p. 11.

1. reification, in which the individual's actions are determined by the whole, as in a structural Marxism that limits human agency;
2. voluntarism, in which society has no separate identity but is constructed from the actions of free individuals;
3. dialectical reproduction, in which the whole creates the individuals, whose actions then influence the whole, which in turn creates the next generation of individuals; and
4. structuration, which starts with individuals rather than with the whole and portrays them in a continuous dialectic with society from then on.

These different approaches are based on a variety of conceptions of the nature of the human agent (see Barnes, 1988; Claval, 1983; Harrison and Livingston, 1982; and van der Laan and Piersma, 1982). Gregory opted for the fourth, because it integrated humanist and materialist (or structuralist) perspectives, in a way that critiques such as Duncan and Ley's did not (see also Thrift, 1983). Such an integration would recognise the materialist base to society (the infrastructure) while accepting the important role of human agency in the superstructure; humanistic geography should not compete with a scientific approach, but should acknowledge 'the recurrent and recursive relations between the individual and society as being fundamentally implicated in the production and reproduction of both social life and social structure' (p. 15).

Alongside debates on the relevance of Marxist ideas, other geographers were defending quantitative spatial science against critiques such as Gregory's (1978a, 1980). Bennett (1981c, p. 24), for example, argued that much of the critique of quantitative geography as 'positivism' had been misplaced and accepted uncritically the representation of science as positivism given by Harvey in 1969. This has had a pernicious and destructive effect on the subject in three main ways, he claimed: first, it has suggested that scientific and empirical enquiry is largely socially worthless; second, it has often rejected the links between physical and human geography; and third, it has often rejected the existence of geography as a discipline at all. (Elsewhere, he characterised the critique as 'at best a mis-representative irrelevance, and at worst a fatuous distraction': Bennett and Wrigley, 1981, p. 10.) Quantification need not be allied with positivism, according to Bennett, who identified three messages from Harvey's *Explanation in Geography*: geography is primarily inductive; geography is an objective science; and geography seeks universal laws. But 'each is only a partial representation of the literature and ideas it seeks to describe' (Bennett, 1981c, p. 13). Radical and quantitative approaches must be integrated 'by empirical analyses, from the integration of environmental, social, political, and economic aspects with space, historical stimuli, and specific modes of thought and their spatial-political manifestations . . . by the re-establishment of the geographical subject matter' (p. 24). A separate discipline of geography (human plus physical), in which quantitative geography – 'never . . . truly positive' (Bennett and Wrigley, 1981, p. 10) – occupies a central place, would answer 'the fundamental questions of social norms, social distribution, policy impacts and humanistic concerns which the critics rightly emphasize' (p. 10). Many 'radicals' would categorise this as a status quo approach (p. 316, this book).

In a later provocative essay, Bennett (1989b) argued that much 'radical geography' was out of touch with what he termed the 'spirit of the times'. He claimed that the welfare state, with its emphasis on rights, had led geographers to concentrate not only on inequalities but also to adopt a welfarist view that 'morally they should be overcome or at least ameliorated' (p. 279). This leads to the argument that all differences imply relative deprivation and automatic entitlements to state action which would remedy them. But the newly dominant social theories (associated in the UK with 'Thatcherism') stress not 'the negative aspects of capitalism's capacity to create new wants and hence new "relative deprivation" but rather the liberating potential of markets'. According to his argument, geographers should reorient their work to focus on the proper role of the state in such a society, identifying when it should interfere with market forces to ensure basic human rights, rather than promoting a welfare-corporate state that must eventually fail (see Johnston, 1992, 1993d).

Thrall (1985) reported on a conference on 'scientific geography' which reached the consensus that 'Research in the unified areas of theory and modelling, data measurement and simulation, estimation and verification is central to the discipline of geography' (p. 254); such a perspective, he contended, should be used to promote the image of geography among the sciences and the 'acceptance of scientific approaches within geography'. His equation of science with mathematics, statistics and computer literacy was challenged by Driver and Philo (1986), who argued that 'science is more than a matter of technique' (p. 161) and that a clearly technocratic approach 'effectively marginalizes all other modes of interpretation and explanation' (p. 162). Thrall's (1986) response identified three major geographical traditions: humanistic, scientific and pure theoretical. Scientific geography, he claimed, is a 'philosophy of research . . . clearly distinct from the earlier quantitative geography movement' (p. 162); he advocated research involving hypothesis-testing leading to theory creation and verification, rather than either the empiricism of regional geographies or the sterile output of pure theoretical research.

This presentation of geography as an applied, quantitative discipline faced continuing criticism. David Smith (1984), for example, concluded an autobiographical essay with the statement that:

> Who else but geographers would dignify their puerile pursuit of statistics, models and paradigms as a 'revolution' as though it mattered to anyone but themselves? . . . And who else would want to read such trivia? How can we take it all so seriously, when it contributes so little to the

improvement of the human condition? Most geography is inconsequential claptrap, and never more so than during the 'quantitative revolution'.

(p. 132)

Mercer (1984, p. 194) argued in the same vein, though noting that:

the last few years have witnessed a small – though perhaps temporary – retreat from the more lunatic excesses of flat earth quantitative geography towards a growing recognition that reality is not in fact beautifully ordered but that it is characterized much more by contradictions, tension and disharmony. The daunting . . . task for the critical geographer – whether 'Marxist' or humanist – [includes] . . . the fight against the hegemony of naive, blinkered, technocratic thinking.

He wanted to unmask 'technocratic geography' and divert attention away from topics such as 'Where are the regions of healthcare need based on access to hospitals?' to more fundamental questions like 'What leads to ill health?'. His basic theme was that selection of a particular research style, with the connotations of how its output may be used, involves an (albeit possibly unconscious) ideological choice: different conceptions of science are based on different views of both its utility and the social order within which that utility is to be employed. David Seamon (1984) was more forthright:

the humanistic, experiential and phenomenological thrusts swelling in the human *and* natural sciences today have much relation to the human rights movement proceeding in larger society. The manipulative, explanatory, predictive style of positivist science closely parallels the masculine, centralized, materialist power structures that not only subjugate minorities and even entire national populations but also manipulate and exploit the natural environment, bringing on ecological damage and collapse.

(p. 217; see too Chapter 8, this book)

Other geographers, however, sought to integrate elements of the positivist, humanistic and structuralist approaches to human geography (e.g. Christensen, 1982; for a fuller review, see Johnston, 1986c). Some earlier attempts at fusion or dialogue have been short-lived, such as an essay on the links between catastrophe theory and the discontinuities central to Marxist economic theory (Day and Tivers, 1979, pp. 54–8; Alexander, 1979, pp. 228–30). Hay (1979a,) argued for an empirical, analytical geography that is:

at the same time a nomological geography which seeks, for example, to understand the workings of urban rent theory as positivistically observed, a hermeneutic geography which seeks to identify the meaning of the urban rent system for those who are participants (active or passive) within it, and a critical geography which points to the extent to which present urban rent systems are themselves transformations of the capitalist system, but which admits that some of its features may indeed be 'invariant regularities'.

(p. 22)

On a parallel track, Livingstone and Harrison (1981, p. 370) had presented a case for 'a humanistic geography which is, at the same time, critical, in questioning rather than bracketing our presuppositions, hermeneutic, in interpreting the meanings behind action, and empirical, in examining the subjectively interpreted objective world'.

Thirty years later, Wyly (2009) called for a 'strategic positivism', challenging what he sees as 'an assumption that there was something inherent and essential in the links between positivist epistemology, mathematical/statistical/quantitative methodology, and elite conservative politics' (p. 314). He argues that a 'blurred collective memory' about how positivism came to be criticised and avowedly rejected by many radical geographers has emerged. Presumably this has been enabled by textbooks recounting (as *Geography and Geographers* has) the ways that 'prominent figures' (Wyly, p. 314, mentions 'struggles between Berry and Harvey and other prominent figures') argued out the case in the 1970s 'in vivid memorable terms'. He argues that this is not helpful, since it obscures the ways that corporations and states had adopted some of the conceptual categories and language that critical social sciences thought was their own, and misses the fact that many geographers associated with positivism and the quantitative revolution also had critical political orientations. Moreover, he celebrates strands of positivism and quantitative method that combine 'analytical rigor, scholarly accountability, and progressive strategic relevance' (p. 316).

Many of the critics of positivism had argued that such attempts at integration were not feasible, however, because the approaches are incompatible: Gregory (1978a, p. 169) terms it 'inchoate eclecticism'. (See also Gregory, 1982b; Eyles and Lee, 1982; Hudson, 1983. Positivist and structuralist approaches sit unhappily together in Rhind and Hudson, 1981.) The biggest problem was held to be with positivism – not quantification (Walker, 1981b; though see Sayer, 1984, 1992a, and the response in Johnston, 1986a, 1993e) – because of its nomological orientation and belief that human geographers can discover 'invariant regularities'.

Some of the debates were not directly concerned with different philosophies of geography, but rather with their utility. As a consequence of the demands on education and research from governments of the 'New Right', considerable pressure to advance the study and practice of applied geography built up. (Taylor, 1985a sets such pressures in historical context, themes returned to in Chapter 9.) Some, such as Bennett (1985a) and Openshaw (1989), saw geographers' technical competence being applied to situations in the public and private sectors without any necessary reference to philosophical issues; applied geographers are thus technicians (as in Gatrell, 1985), who implicitly accept the context within which their skills are used. The philosophical issues remain important to others, however; Golledge *et al.* (1982) countered Hart's (1982a, p. 5) case for regional geography that 'We cannot allow ourselves to be intimidated by those who flaunt the banner of science' with:

> We equally cannot allow ourselves to be intimidated by those who flaunt the banner of anti-science, those who would reject all that is scientific about the discipline, and those who would urge a return to the descriptive morass from which we have recently emerged.
>
> (p. 558)

Most arguments for applied geography (again, see Chapter 9 for more details) implied that it should be based in empiricist/positivist philosophies, however; the counter-arguments that humanistic science can be applied to improve self- and mutual awareness and that realistic science can be applied to advance social transformation were rarely presented (Johnston, 1986a).

Others believed that an eclectic pluralism is possible (see the discussion in Johnston, 1986c); some conducted empiricist work (often in the behavioural geography mould) unconcerned with any philosophical ramifications of that practice (see Flowerdew, 1986), however, whereas others generally accepted the realist case and situated their activities within that context. And years later, others argued that radical (or its later variant, critical) geography need not be shy of statistics (Wyly, 2009, 2011). Marxism certainly influenced many geographers (and the work of radical geographers, such as Harvey and Massey, made impacts in other disciplines). Gould (1985a), for example, contended that:

> There is no question in my mind that the appearance of Marxist concern in geography, and its concomitant shaping of the lens through which the world is seen, have greatly enriched our

methodological approach. There is an insistence that the things at the surface are not always what they seem, and that it is crucial to dig down underneath the superficial appearances to get at the deep structures. I think this is quite right.

(p. 671)

However, he criticised Marxists for their claims to truth, their condescending attitudes to 'non-believers', their over-concentration on economic forces, and the 'messianic claims that seem to lead so readily and so often to the sacrifice of human beings today for some promise tomorrow' (p. 300). Nearly two decades on, one of a set of papers re-evaluating David Harvey's (1982) landmark *The Limits to Capital*, notes how:

It is written by someone with an acutely sensitive historical-geographical imagination, but barely contains any conventional history or geography. It is relentless in its precision and chiselled in its logic, but eschews a mathematical vocabulary that is the embodiment of precision and logic. It is a political-theoretical tract in favour of the working class, but they barely appear in its pages.

(Barnes, 2004d, p. 412)

For another geographer: 'Reading Harvey, he says, is an exercise in being convinced and then engaging in the hard task of working out why you shouldn't be so convinced' (Gregory, 2006, p. 24, quoting Don Mitchell). Meanwhile, while Harvey and the contributions of other Marxist geographers continued to be reassessed, and as the initial political moment that gave birth to radical geography faded into history, the wider global geographies of power and production had been shifting, accompanying many other global political and social transformations.

What's left

Just as wider political pressure for radical transformations around the world provided the initial backdrop to the emergence of radical geography at the end of the 1960s, so the relative eclipse of socialism in the Third World and its collapse in what was hitherto the Second World of the Soviet Union and Eastern Europe produced new contexts. The political events of the late 1980s and early 1990s in Eastern Europe thus demanded critical reflection. As Sayer (1995) notes:

For the Right they were a vindication: communism was defeated, capitalism had won.... Strangely, it was assumed by many that it also meant that Marxism was defeated, as if Western Marxists ought to have recanted as soon as the statues of Lenin and Marx started being toppled. This ignored the fact that the vast majority of Marx's work and most Western Marxism is concerned with capitalism and that most Western Marxists were overwhelmingly critical of the state socialist regimes.

(p. 13)

This, Sayer argued, diverted attention to capitalism's continued failings and to the failure of what he termed radical political economy to suggest viable alternatives. Hence, his detailed analysis of work in that genre, which he located within 'Marxism's lack of a sufficiently materialist understanding of the social division of labour and its associated division and dispersion of knowledge in advanced economies' (p. 12). Instead of attacking this problem, researchers (in the context of the 'cultural turn' – see the next chapter) have switched their attention away from issues of power, domination and subordination linked to class and capital (p. 13):

This seems an entirely reasonable response, as analysis of these matters was long overdue. The uncharitable explanation is that faced with the challenge of the New Right and the weakening of Marxism, radicals shifted their attention to new concerns which did not require them to make any painful concessions to the Right. Socialism as an alternative political system was increasingly difficult to articulate and defend, but in any case there were other important issues to turn to which provided a convenient escape. While understandable, this escapist response is surely less defensible. In my view, both of these explanations of the shift from economy to culture – charitable and uncharitable – are right. Although post-Marxism has been generally a progressive development in terms of broadening radical interests it has neglected political-economic theory. There is therefore much unfinished business.

(p. 13)

Sayer's approach to that unfinished business concentrates on the complexity of the social division of labour within contemporary capitalism (see also Sayer and Walker, 1992). Social divisions occur between enterprises involved in the creation of a vast range of products and services, whereas technical divisions (on which Marx and his followers concentrated) occur within enterprises, and are concerned with their ownership and structure, and the nature of labour contracts (see Rose and Pevalin, 2003). Individual enterprises can be planned in terms of how much will be produced and marketed, by how many people, and at what cost. Whole conglomerations of enterprises – Sayer follows Hayek in terming them 'catallaxies' – cannot be so planned because of their internal variety and complexity, however; their activities must be coordinated in other ways, which at present means the market economy celebrated by the New Right's identification of the 'end of history' (Fukuyama, 1992; Peet, 1993; Johnston, 1994d). But the inequalities that follow pose major challenges (Sayer, 1995):

Political economy ... has to address issues of division of knowledge and catallaxy, horizontal control, coordination, allocation and economizing, which generally elude democratic control ... To ignore these issues is simply to duck the most serious problems of political economy.

(p. 210)

So that:

Radical political economy cannot continue to follow Marxism in standing apart from the debates of normative political theory, nor can it embrace a postmodernist celebration of fragmentation and rejection of the search for a better social framework. It is now more clear than ever that struggles which are directed against domination and oppression but which lack any normative direction in terms of alternative frameworks are unlikely to be successful. Though it has recently suffered from neglect we need a radical political economy more than ever before.

(p. 252)

The events of 1989 and the early 1990s in the former Soviet Union and its client regimes in Eastern Europe (and their variants in the Third World) thus posed both difficulties and an opportunity for geographers of a Marxist persuasion, as many of their opponents argued that the collapse of the communist states indicated the fallibility of Marxist political programmes. (Neil Smith, 1991, p. 406, noted that 'it has become a common argument in left circles that the crumbling of Communist Party control throughout Eastern Europe and the Soviet Union provides definite proof that planning and state control of the economy don't work and can't work'.) Folke and Sayer (1991) argued that to claim that 'they [the ex-communist states] weren't really socialist' anyhow is an unsatisfactory response, although they recognise that the events mean that 'genuine socialist visions and utopian dreams will have a hard time in Eastern Europe in the coming years' (p. 248); new blueprints of how to manage complex societies in the interests of all are needed (see Sayer, 1995), and they must recognise, as Neil Smith (1991, p. 416) argued, that '1989 teaches us the profound depth of human resistance to all forms of economic, social and political repression': successful revolutions must be broadly based and sustained. Sayer (1992b, p. 217) responded that:

> The problems that the left now faces are not just a consequence of the rise of the right, or of mistakes made in the 1920s; they are also a result of the vacuum on the left concerning socialist alternatives which are feasible as well as desirable. I think we would do better to confront that fatal weakness.

Subsequent work in a political economy tradition has charted the contradictions and unevenness of the capitalist transition in post-communist societies (Pickles and Smith, 1998). Such works claim the continued relevance of Marxist and allied political economy approaches in an avowedly post-Marxist world. Notable too was an online and activist 'People's Geography Project':

> The major goal of the People's Geography Project is to popularize and make even more relevant and useful to ordinary people the important, critical ways of understanding the complex geographies of everyday life that geographers have and continue to develop. Our contention is that such knowledge is an important tool not just in learning to cope with constantly developing and transforming relations of power that are deeply geographical, but in learning how to actively transform those relations in the name of social and economic justice.
>
> (www.peoplesgeographyproject.org/ last accessed 1 May 2015)

Mitchell (2009, 116) describes people's geography as having 'many roots'. These include similar work in history, Bunge's (1971, and Bunge and Bordessa, 1975) earlier commitment to radical 'geographical expeditions' that would map and document injustice, poverty, racism and exclusion in American cities (on Bunge's project, see Merrifield, 1995, and a more recent set of commentaries in *Progress in Human Geography*, pp. 35, 712–20). On another scale, the role of 'global cities' that are the 'control centres' of the international (or 'globalizing') corporate economy (Beaverstock *et al.*, 2000; Knox and Taylor, 1995; Derudder *et al.*, 2012) has become a significant theme of contemporary research (see the online Globalization and World Cities Study Group and Network at www.lboro.ac.uk/gawc/), whose origins can partly be traced to radical geography's attention to power and inequality and its understanding of urbanisation and uneven development as integral to capitalist dynamics.

Others brought Marxist influences to bear within a critique of the globally hegemonic ideas and practices of 'neoliberalism' (see the special issue of *Antipode* – volume 34, 2002, especially Peck and Tickell, 2002), including David Harvey (as noted at p. 206, this book). Thus, Jane Wills (2002) felt able to claim that:

> Political-economic geographers are seeking to develop new theory and analysis to understand capitalism and political change. Encouragingly, I detect a move back to the big picture, to unpacking capitalism, while also developing a more sophisticated theorization of political agency, which incorporates economy and culture, class and identity. . . . There is a huge canvas on which a new generation of geographers could get to work and it is critically important that new students are attracted into political-economic research within the discipline. Indeed this may happen through an initial engagement with cultural and social geography as much as through economic or political geography.
>
> (96–7)

And while, as Wills points out, Marxism also continued as an important influence in reconstituted cultural geography (see the debates covered in Chapters 5 and 7), on the part of others there was continued caution and sober reassessment about the limits to Marxism (e.g. Jones, 1999; Storper, 2001) and/or the value of supplementing it with other approaches and qualifications. Debates in *Antipode* about the future of radical geography have periodically reviewed its status and roles, as well as reconsidering the commitment of geographers to progressive social change, and the policies and politics that might be most productive in such a project, and the role of 'theory' therein (see Hague, 2002; Castree, 2002; Castree *et al.*, 2010; Larner, 2011). For example, Marvin Waterstone (2002, p. 666) claimed that: 'What we have is, once again, a radical critique from the margins of a truly staid discipline that moves along mostly untouched by what happens in the pages of this journal.' To which Jane Wills and Jamie Peck (2002, p. 669) responded that 'the fact that radical geography can no longer be dismissed as a fringe concern represents a remarkable victory for the project, even if in the process it has taken on some 'mainstream' connotations'.

A few years later, Neil Smith (2005a), in exchanges with Amin and Thrift (2005), forcefully argued that radical geography had been tamed, co-opted and was increasingly disciplined by the forces of audit, and the dominant neoliberal agenda of universities and government (especially in the UK, where he compared radical geography's fate to the right-ward course of the Labour Party under Tony Blair's leadership). Amin and Thrift had argued that the Left was in a healthy state, with its continuing political orientation, belief that there must be better ways of organising the world, its critique of power structures and its internal reflexivity, but they doubted that Marxism's privileged position underpinning those values could be maintained; they welcomed the proliferation of alternative theoretical positions as foundations for charting ways forward: 'it is time that the Left in geography came to accept this vibrant pluralism, instead of insisting that certain perspectives necessarily hold a privileged insight into the ways of the world' (p. 238; see also Amin and Thrift, 2013).

Arguably, however, periodic conferences of the International Critical Geography Group (established in 1997, see Desbiens and Smith, 1999; Bachmann and Belina, 2012) as well as new journals (such as www.acme-journal.org) published online and/or collectively owned (www.hugeog.com) still embody something of the original radical spirit. Equally, campaigns about the RGS receiving sponsorship from a multinational oil company with a controversial reputation regarding environmental and human rights issues in its zones of extraction (Chatterton and Maxey, 2009; Gilbert, 1999, 2009) and calls to boycott (by not publishing in or reviewing articles submitted to) the journals owned by a large commercial academic publisher who also arranged events where sales of lethal military equipment are promoted (Chatterton and Featherstone, 2007) are inspired by radical geography. Attention also to the geography of a fresh generation of protest (on issues such as anti-globalisation or more mundane ones like the commodification of public services, including education: see Sparke, 2013) rework radical – or as it has more often been termed today – critical geography. Throughout virtually all human geography, the scale and force of the radical contribution and legacy means that it has proved hard to ignore. Meanwhile, *Antipode* (Chatterton *et al.*, 2011) itself finessed its radical vision, setting out new and long-standing areas of interest (geographies of poverty, marginality and power; urban transformations in the Global South; food, agriculture and environment; counter-imperialism; capitalism and neo-populisms).

It may long have been the case that, as Peter Taylor (2003) states:

> Soon after its emergence, radical geography, whether of Peet, of Harvey, of Blaut, or of Bunge, became explicitly Marxist in nature . . . to be a radical scholar no longer inevitably means taking a Marxist, 'neo' or otherwise, approach to understanding social change. In the contemporary academy there is an array of schools of thought that aspire 'to turn the world upside down'.
>
> (pp. 55–6)

However, it is a measure of the ongoing presence of radical geography and of Marxism within it that avowedly 'post-Marxist' geographers such as Richard Smith and Marcus Doel (2001, p. 137) allowed that 'it is clear that many postmodernists and poststructuralists have extended or reacted to Marxism in a number of useful and interesting ways which do not necessarily mean the jettisoning of a radical left-wing politics'. Such possibilities are also among the subjects of our next chapters here.

Conclusions

This chapter has illustrated the depth and breadth of discussions that occupied much of the 1970s and 1980s. They continued in attenuated forms in the 1990s and offer continuing sources of inspiration. Peet (1998, p. 294) feels that 'A discipline can support only a few true radicals; the rest have functional utility, oblique though this often may seem.'

Yet in the turbulent decades since the late 1960s, a growing proportion of geographers wished to be involved in reshaping societies, either through ameliorative correction of current problems and trends or by designing desirable spatial organisations (Berry, 1973b); some saw this as necessary if human geography was to retain its institutional position. Their motives ranged from 'pure altruism' to 'devoted self-seeking'; their methods varied from those who accept the present mode of production and see humanitarian goals as achievable within its constraints, to those who subscribed to the phenomenological view (Buttimer, 1974) that:

> the social scientist's role is neither to choose or decide for people, nor even to formulate the alternatives for choice but rather, through the models of his discipline, to enlarge their horizons of consciousness to the point where both the articulation of alternatives and the choice of direction could be theirs, to those who believed that a revolution is necessary to remove the causes of society's myriad problems and replace them by an equitable social structure.
>
> (p. 29)

Perhaps paradoxically, many 'revolutionaries' in the above classification were not 'activists' in the sense of being deeply involved in contemporary issues. The 'liberals' advanced the strongest arguments for geographical contributions to the solution of societal problems, particularly those which involve public sector intervention (e.g. Bennett, 1983), and pressed for academic engagement in policy-making; many of the 'radicals', on the other hand, argued that their longer-term goals were best served through educational programmes (e.g. Huckle, 1985), although some were involved in policy-making activities for institutions (such as local governments) that promoted alternative (socialist) strategies, especially in the support for employment initiatives (e.g. Duncan and Goodwin, 1985; see, however, Harvey, 1995 and Massey, 2001, on the tensions this can create). The two groups were members of the same discipline and practised in the same academic environments, yet their goals and methods seemed incommensurate. As Chapter 9 illustrates, key issues around which these arguments pivoted concerned the applications, relevance and prospective impacts of human geography.

At this point, however, we should note that perhaps the key to the influence of radical geography is that Marxist work was able to set many of human geography's leading intellectual agendas. Thus, Neil Smith (2001) comments that:

> Within 'establishment geography', the dramatic rise of Marxism in the early 1970s was first ignored, later despised, and eventually resented, even if it was eventually if begrudgingly accommodated. Yet there was little that establishment geographers, who completely controlled the discipline in North America, Britain, and Australia, could do to prevent it. Responses to Marxist work in this period invariably lacked the intellectual sophistication that Marxist theory

brought, and criticisms often came across as shallow and contradictory. . . . A medical metaphor captures the situation. The allergy of English-speaking geography (as a discipline) to social theory throughout most of the twentieth century now meant that the discipline had no immune system against Marxism. Marxism in the 1970s and 1980s ... offered powerful insights concerning the political questions of the day, and offered a breathtaking global vista of of the geographies of exploitation, oppression and injustice – and their causes. . . . Just as important, it provided the most sophisticated social theory that many geographers had come in contact with, and its opponents had few if any social theoretical resources for counteracting its influence. Unlike anthropology or sociology where various major figures had been socialists, geography as a discipline lacked all immunity against Marxism. Intellectually if not institutionally, the old guard was defeated almost before the struggle began.

(p. 9)

Yet by the 1980s, anglophone geography as a whole was still characterised by a plurality of approaches (Gould and Olsson, 1982), which can perhaps be classified into three intertwined philosophical strands (or sets of strands), each with a distinctive epistemology, or theory of knowledge.

1   First was the stress on science and putative break with the geography inherited from before the Second World War. This was associated with positivism and quantification, a belief in the objectivity of scientific description (empiricism) and analysis of the world, its goal of formulating laws about that world, and its assumption that explanation (causal laws) can be derived by studying the outcomes of the laws; the laws of spatial organisation and behaviour can be revealed by analysing spatial patterns. At least some of those who promoted the empiricist foundation to this philosophy denied that it necessarily leads to the full positivist commitment. They also disagreed over how to use scientific procedures for the evaluation of hypotheses, however, as in the critical rationalism usually associated with Popper (Hay, 1985a; Marshall, 1985; Bird, 1989). Others later stressed the need to study perceptions and cognition; the move to behavioural geography that seemed to offer a way forward. Going beyond this, however:

2   Second were the humanistic philosophies based on a belief that people live in subjective worlds that need to be interpreted and cannot be modelled in any straightforward way. Their actions cannot be explained (predicted) as examples of general laws of behaviour, but only understood, or appreciated, through methodologies that appreciate their subjectivity.

3   Finally, there were various radical moves (often allied with variants of Marxism) which argued that explanations for observed patterns cannot be discovered through analysis of the patterns themselves, but only by the development of theories of the underlying processes that generate the conditions within which human agents can create those patterns. Foremost within this group was Marxism, which argues that the processes are themselves changing – and can be altered by concerted political action – so that no eternal laws of spatial organisation are possible; at least not outside an understanding of the social dynamics of particular societies and modes of production. When he advocated 'a revolution in geographical thought', David Harvey (1973) had argued that:

the most fruitful strategy at this juncture is to explore that area of understanding in which certain aspects of positivism, materialism and phenomenology overlap to provide adequate interpretations of the social reality in which we find ourselves. This overlap is most clearly explored in Marxist thought.

(p. 129)

However, the trajectory of Marxism through geography has been complex, and contained detours such as realism and structuration theory (see Cox, 2013). Moreover, these trajectories (as in the case of structuralist Marxisms' indirect influence, via the work of Massey and in regulation theory) have not always been clear to all who contributed to or contested a variety of Marxist geographies. Weaving across the entire path was also a nascent feminist critique (the focus of Chapter 8) and reworking of the human–environment tradition in geography via Marxist literatures on ecology, landscapes and nature.

All this plurality provided a focus of debate, and even a source of confusion, among human geographers. For some, it presented a polarisation that can only be solved when one approach (paradigm) is proved triumphant. Others accepted that this cannot be done, since there are no common criteria for comparing the approaches. For some, the plurality offered potential for developing a newer, more robust human geography (e.g. Wilson, 1989b), though the nature of that development remained unclear to many. (The summaries of the statements made by six geographers – Peter Hall, Peter Jackson, Doreen Massey, Brian Robson, Nigel Thrift and Alan Wilson (1987) – to a review by the UK Economic and Social Research Council on Horizons and Opportunities in Social Science clearly illustrated the pluralism.) However complex and uncertain this may have looked to many, geography had undoubtedly been transformed, and although it was still a relatively small discipline (especially in the USA), its relative status among the social sciences was greatly enhanced. As both Peet (1998) and Cox (2014) argued, spatial science initiated three major changes in human geography – the importance of explicitly addressing and developing theory; the need for methodological rigour; and the crucial role of space in human affairs. But it addressed the wrong type of theory, used inappropriate methods and had a limited conception of space. The 'radical (and other) revolutions' that followed accepted those foundational needs and built new analytical structures accordingly.

Looking back and celebrating this, Don Mitchell (2009) notes how:

> If the so-called quantitative revolution had shaken geography's old guard but nonetheless seemed to be aligned with the political status quo (it was largely a technocratic development), then radical geography hoped to upend both the discipline and the world of which it was part. Radical geography was thus enthused with experimentation . . . it sought to be directly relevant in ordinary and oppressed people's lives in ways that academic knowledge production rarely has.
>
> (p. 117)

As this chapter indicates, how far radical geography succeeded in those aims remains a subject of debate. Moreover, the world in which geographers work and which they study, and the political and philosophical influences upon them, continued to change. Without any definitive resolution of the debates outlined in this chapter, by the 1990s, geographers were being presented with further intellectual and practical challenges. Writing in 2006, Castree could note how: 'Those Leftists who passed through bachelors programmes and graduate school from the early nineties were inculcated into ways of thinking which, in the main, defined themselves as post- or non-Marxist' (2006b, p. 248). Thus, further developments outside the discipline influenced it by the mid-/late 1980s, and within a decade these were the source of further significant reorientations in what many human geographers did and why. They are the focus of the next two chapters.

Chapter 7

Postmodernism, poststructuralism and postcolonialism

The impact of postmodernism in the field of geography has been particularly intense and far-reaching, perhaps more so than in any other social science . . . postmodern and related poststructuralist critiques of traditional modes of interpreting and explaining geographical phenomena have influenced scholarly research and teaching in nearly all branches of the discipline.

(Edward Soja, 2001a, p. 11860)

We begin by outlining the essence of postmodern ideas. This is no easy task.

(Phil Hubbard *et al.*, 2002, p. 73)

[W]e will only unlock the power of poststructuralist geography to the extent that we embrace nothing but relations and co-relations, their folding and unfolding.

(Marcus Doel, 1999. p. 147)

In certain moments, in certain milieus, particular cities generate a number of issues that incite attention. The work produced by the Los Angeles School of postmodern urbanism (hereafter LA School or LA scholars) is a diary of one such occasion . . . [describing] the onset of what has been called the 'postmodern condition' – that is, a changing perception and experience of space and time and an intensified, sometimes celebratory, consciousness of the 'new'.

(Barbara Hooper, 2009, p. 293)

Perhaps the best – and last – bit of advice that can be given to readers of this book is to not accept anyone else's definition of postmodern geographical praxis.

(Edward Soja, 2001b, p. 294)

Posts- and turns

According to Claudio Minca (2009):

any attempt at defining the state of the art of postmodern geography is doomed to fail; if only for the very fact that the introduction of postmodernism to geography – as to other fields of enquiry – has implied the breaking down of any paradigmatic logic and scepticism toward any linear reconstruction of the history of the discipline.

(p. 363)

Minca dedicated an earlier collection on *Postmodern Geography: Theory and praxis* 'to all those who have had the courage to try to free themselves of the strangleholds of modernity . . . this book can represent a sign that you are not alone' (Minca, 2001, p. xiii). Like others who propose new approaches (though postmodernists would scarcely see themselves as part of a new 'paradigm'), individual postmodern geographers may have faced isolation or resistance at first. Avowedly poststructuralist geography, that soon followed postmodernism, often met 'scornful derision at one end [of the scale of reactions] to bemused indifference at the other' (Dixon and Jones, 1996, p. 767). But their collective presence in the discipline soon made them familiar to many geographers and a significant presence in classrooms, texts and journals. Thus, David Clarke (2006, p. 107) reflects on the way:

> The term postmodern burst unceremoniously onto the geographical scene in the mid to late 1980s. . . . Against the sober promises of realism to rethink science for human geography and the wise counsel structuration theory offered in attempting to resolve the long-standing family feud between structuralists and humanists . . . the reckless, dizzying antics of postmodernists seemed to throw reason itself into doubt . . .
>
> (p. 107)

He also claims that:

> The very first thing to say about the post-modern wor(l)d is that it's *inherently confusing*. (And postmodernists are wont to perform cheap tricks like that, so we now no longer know if we're talking about *words* or *worlds*, and are left feeling thoroughly disorientated regarding their relationship – which we once took for granted as a straightforward matter of representation.)
>
> (p. 107)

It might therefore be rather unpostmodern of us to give our readers the clear structure of this chapter in advance. We also might rearrange things, like Soja's (1989) text on *Postmodern Geographies* which begins on page one with a combined Preface and Postscript – 'to shake up the normal flow of the linear text to allow other, more "lateral" connections to be made'. Alternatively, like Michael Dear and Steven Flusty's (2002, p. xiii) *The Spaces of Postmodernity*, we might 'forego a "Conclusion" for an "Inconclusion", refusing even to contemplate a closure that, like so many previous (anti)climaxes, would function only to suppress different ways of knowing'. But like many geographers, a structured and linear narrative might be what you have come to expect from diagrams, texts, tables, maps and published papers in the discipline, so allow us some pointers here.

This chapter is concerned with a number of influences and transformations which informed human geography in the 1980s and 1990s, and that led to considerable reworking of human geography in the twenty-first century. These include postmodernism, poststructuralism and what was termed a 'cultural turn'. The latter saw a reinvigoration not only of cultural geography, but also a bringing to bear of cultural themes on a wider range of human geographies (including economic geography: such as in Lee and Wills, 1997; Thrift and Olds, 1996). Feminist geography also had a substantial and growing impact since the 1980s; we will consider this in the next chapter. Here, however, we also consider the impact of postcolonialism, in particular a reconsideration of geography's complicity in colonialism and tentative moves to decolonise geography. The former – geography's colonial history – is something that radical geographers and positivists had both sought, in different ways, to put behind them. Positivists had advocated a new scientific geography which would have little or nothing to do with geography's prior trajectory of regional description and narration of difference (geography as an inventory of regions and their peoples). For their part, radicals had often proclaimed their anti-imperialism. Postcolonialism builds on this sense of critique.

Beyond the discipline, the 1980s and 1990s saw extensive changes in the economic, social/technological and political *contexts* within which academic geography was set and in the *practices* employed in other social science and humanities disciplines. The former comprised the globalisation and flexible accumulation trends within capitalism, and associated shifts in the means and spaces of production/consumption (including digitisation and the birth of the internet) described earlier, along with major changes in the nature and role of the state apparatus (and the eclipse of communism – at least that form of it which had been known through most of the twentieth century), which had profound effects on the funding and expectations of research in higher education. The latter involved attempts to come to terms intellectually with those changes in the context of widespread disillusion with current academic practices. On both sides of the North Atlantic and in the other anglophone contexts that are the focus of this book, the private and public sector structures in place since the 1940s were being partly dismantled and replaced by others in which 'markets' (and models from business) played a greater role. In the private sector, for example, many of the large manufacturing industries were very significantly 'downsized', often moving to cheaper sites of production in what was called the 'Third World', and their replacements were often in service industries, usually offering less job security and fewer fringe benefits, and with weaker trade union organisation. In the public sphere, social democratic and 'left' political parties appeared to have lost some of their core support and had to reinvent themselves (in the UK case, the Labour Party branded itself 'New Labour' and in so doing accepted many of the 'market' shifts that Conservative governments had instigated in the 1980s and early 1990s). Change was rapid and was everywhere; understanding why it was happening and what was replacing it provided human geographers with major challenges (as suggested by the essays in Johnston, 1993f.), but the changes were also affecting the institutions in which they worked, often increasing pressures associated with such rapid changes, combined with increased audit and so-called 'efficiency' gains (see Chapters 1 and 10). At the same time, however, many of the demographic and social changes and pressures of the 1960s were also working through the system, with attendant generational shifts and a sense of new possibilities. The radical geography which began then had also come of age and continued to be a point of departure and reference – it had opened up new ways of thinking and new engagements for human geography – and would be re-evaluated by new generations of geographers. Hence, in Neil Smith's (2001) words:

> The rapid deepening of Marxist research in that decade [of the 1980s] was therefore matched by an even more rapid broadening as Marxist work explored and eventually fused with all kinds of social theory.
>
> (p. 10)

For Hubbard *et al.* (2002), however:

> much contemporary critical human geography (though not all) is fundamentally different from the more homogeneous radical geography practised in the late 1960s and 1970s. . . . As such, structural and materialist accounts have been complemented (and in many instances replaced) by positions more sensitive to human agency and questions of culture.
>
> (p. 63)

Cultural turns?

Although it might also partly be summarised by reference to the growth of qualitative methods (Crang, 2002), a more frequent way of summarising new trends in human geography during the later part of the 1980s and the 1990s has been to adopt the label of 'cultural turn'. Chris Philo, who had earlier (1991) identified a nascent cultural turn in human geography, had the following to say of that earlier moment (and of the 1991 collection of essays *New Words, New Worlds: Reconceptualising social and cultural geography* which he had compiled and which contributed to the sense of new departures):

> the main achievement was arguably to heighten our senses of how all things cultural might be raised to a much more prominent position in studies throughout the corpus of human geography (and not just in one or two neatly parceled-off subdisciplines). It was to take much more seriously than hitherto all manner of things that might be construed as constituting the cultural 'stuff' of human life, not just phenomena routinely designated as cultural . . . but also the complete panorama of meaning systems both collective (e.g. religions and nationalisms) and more individual (built up in personal psychic economies). . . . This was to anticipate the amazingly rich arc of the cultural turn within human geography as it has now arisen: a turn which obviously had great ramifications for both social and cultural geography, but one too which has undoubtedly sent shockwaves throughout the length and breadth of human geography, leading to debates (more or less explicit, more or less heated) about the merits of a cultural turn within (say) economic geography, political geography, population geography, environmental geography and elsewhere.
>
> (Philo, 2000, p. 28)

For Philo, the cultural turn had been many years in the making. Thus, what he terms the 'valuable maneuvers' (p. 32) and 'twin routes out of spatial science after c.1970 signposted by, respectively, radical/Marxist geography (with its focus on social structure) and behavioural/ humanistic geography (with its focus on human agency)' (p. 31) together:

> paved the way for the cultural turn of more recent vintage . . . the wider horizons of late 1980s human geography which gave birth to the cultural turn included a variety of developments which extended, and in many cases recast, the Marxist and humanistic revolutions of the previous decade. . . . The reworkings of Marxist geography – the spatializing of Marxist concepts; the restructuring, spatial divisions of labour and locality studies debates; the advent of regulation and regime theories; the critical realist encounter; the agitations around post-Marxism . . . have all created a receptiveness to the ways in which . . . cultural processes become implicated in political-economic spaces . . . [and] the reworkings of humanistic geography – the reappraisal of the landscape tradition; the growing attention to intersubjective meaning systems; the fragmenting of the singular figure of 'Man' in recognizing the sheer diversity of peoples occupying this planet; the new mapping of the human subject(s); the alertness to the psycho-dynamics of gender and sexuality – have all fostered a sensitivity to the many dimensions of immaterial culture which enter into the making of virtually all human spaces imaginable.

Nayak and Jeffrey (2011, p. 115) point out that 'the cultural turn was never a solitary exercise . . . geography was only one of a number of social science disciplines that was influenced by this movement'. Instead, they describe a 'coming together of different strands and activities' encompassing literary criticism, a reinvigorated cultural studies and 'new insights on power and subjectivity'. In human geography however, Barnett (2009) argued that the cultural turn signified three related trends:

1 the emergence of a 'new' cultural geography;
2 an increased attention to culture in a variety of subfields, including economic and historical geography;
3 a claim that culture had become more central in wider social, political and economic processes (as in 'cultural industries' or 'cultural clashes').

Moreover, for Barnett:

> Theoretically, the cultural turn has promoted a greater degree of pluralism in human geography, drawing on concepts from other disciplines and focusing attention on multiple dimensions of difference . . . the cultural turn has underwritten a commitment to investigate the contingent and constructed qualities of phenomena.

(p. 134)

The notion of a 'new' cultural geography (as in the first trend identified by Barnett, 2009) suggests a link with the subdiscipline of cultural geography, in particular a contrast with how it was traditionally practised – what was widely known as the Berkeley school, associated with the work of Carl Sauer (see p. 49, this book). Marie Price and Martin Lewis (1993a) thus suggested that beyond the use of the adjective 'cultural', the 'old' and the 'new' 'share precious little' (p. 2). They argued that the 'new cultural geographers' (they cited Cosgrove, Duncan and Jackson as 'standard bearers') had 'reinvented' the nature of 'old' cultural geography, in order to criticise it, with the Berkeley school as 'its currently most misinterpreted form' (p. 3). In return, Cosgrove (1993) accused them of inventing a conspiracy, claimed that he had not wholeheartedly embraced the term 'new cultural geography', and welcomed the growing accommodation between the 'theory-free' Berkeley school and the theoretically grounded approaches. Jackson (1993b) was also accommodating, while pointing to a wide range of other traditions. More stridently, James Duncan (1993b, p. 518) responded that 'I cannot accept that it is possible to employ the term culture in one's work and yet employ no theory of culture', to which Price and Lewis (1993b) responded by restressing their case that Duncan (1980) misrepresented the Berkeley canon, both in its orientation and in its implicit acceptance of 'the tiresome superorganic debate' (Price and Lewis, 1993b, p. 521). Since those exchanges, debates about the dynamics and merits (or otherwise) of the cultural turn(s) continued (notably in Don Mitchell's, 1995, work as detailed in Chapter 5; see too, Barnett 1998a, 1998b, 2004, and Sayer, 2000). In Nayak and Jeffrey's (2011, p. 117) view, a key to the consolidation of the cultural turn was a 'new generation . . . [of] PhD students who had broken away from Sauerian traditions to explore how culture is linked to power, ideology and the broader field of representation'. In what follows, we return to aspects of these debates and the paths they opened.

Meanwhile, the cultural turn runs through most themes that follow, but these are not contained by it (or them, for, as Cook et al., 2000, p. xii, pointed out, 'there has been less a cultural turn than a series of cultural insights, turns, multiple circuits'). So this chapter will be somewhat more disjointed than its predecessors, as we work through a series of interwoven, postmodern, poststructuralist and postcolonial strands of contemporary human geography.

Postmodernism

Whose definition?

Postmodernism became increasingly popular in the social sciences and humanities in the 1980s and early 1990s attracting considerable attention from geographers. Its rise, according to Soja (1989), represented an attack on the predominance of historicism in modern thought, with its emphasis on biography (both individual and collective) and time and the consequent neglect of spatiality and geography. He defined historicism as:

> an overdeveloped historical contextualization of social life and social theory that actively submerges and peripheralizes the geographical or spatial imagination . . . [and which produces]

an implicit subordination of space to time that obscures geographical interpretations of the changeability of the social world.

(p. 15)

Because of this, he claimed that geographers had failed to attain their 'rightful position' within the social sciences for much of the twentieth century. A 'superficially similar historical rhythm' (p. 33) was assumed to occur across all places, rather than a realisation – as stressed in the localities project and new regional geography projects discussed earlier (p. 215, this book) – that social processes are differently constituted in different places, so that the historical flow is not the same everywhere; not surprisingly, some sought to link the localities project to postmodernism (see Cooke, 1990). Thus, for example, postmodern novels have either a non-linear or an apparently chaotic structure when they try to represent different things happening simultaneously in different places, and postmodern architecture does not rely on the kinds of clear functional structure that characterised most modern architecture (Knox, 1987).

This problem of synchronicity has long been recognised by geographers (as well as poets, novelists and film-makers), as Massey (1992) observed with reference to Darby's (1962) paper on 'The problem of geographic description'. Darby pointed out that:

> A series of geographical facts is much more difficult to present than a sequence of historical facts. Events follow one another in time in an inherently dramatic fashion that makes juxtaposition in time easier to convey through the written word than juxtaposition in space. Geographical description is inevitably more difficult to achieve than is historical narrative.

(p. 2)

But histories have geographies, and relating concurrent changes over both time and space simultaneously provides particular problems, especially since all mapped distributions are simply static snapshots of continuing processes (Blaut, 1962). In a sense then, in his concern for the combinations of place and space Soja was returning to an old theme in the geographical tradition. He did so, however, in the context not of a rural or isolated region but of what he described (adopting a slogan from the *Los Angeles Times*) as:

> the place where 'it all comes together' . . . the sprawling urban region defined by the sixty mile (100 kilometer) circle around the centre of the City of Los Angeles a prototopos, a paradigmatic place; or, pushing inventiveness still further, a mesocosm, an ordered world in which the micro and the macro, the idiographic and the nomothetic, the concrete and the abstract, can be seen simultaneously in an articulated and interactive combination.

(p. 191)

So Soja writes a 'free-wheeling essay on Los Angeles . . . a decidedly postmodern landscape, a search for revealing "other spaces" and hidden geographical texts' (p. 2). He captures the putative postmodernism of these spaces by drawing upon the Argentinean writer Jorge Luis Borges. Thus, Soja (p. 2) quotes from Borges's short story (originally published in Spanish in 1945) of:

> 'The Aleph' – the only place on earth where all places are, a limitless space of simultaneity and paradox, impossible to describe in less than extraordinary language. Borges' observations crystallize some of the dilemmas confronting the interpretation of postmodern geographies:
>
> *Then I saw the Aleph. . . . And here begins my despair as a writer. All language is a set of symbols whose use among its speakers assumes a shared past. How, then, can I translate into words the limitless Aleph, which my floundering mind can scarcely encompass? . . . Really what I want to do is impossible, for any listing of an endless series is doomed to be infinitesimal. In that single*

gigantic instant I saw millions of acts both delightful and awful; not one of them amazed me more than the fact that all of them occupied the same point in space, without overlapping or transparency. What my eyes beheld was simultaneous, but what I shall write down will be successive, because language is successive. Nonetheless, I will try to recollect what I can.

While for Soja it is found in Los Angeles, whose 'spatiality and historicity are archetypes of vividness, simultaneity, and interconnection' (p. 248), postmodernism has been attributed a wide range of explicit and implicit meanings, and its core is hard to identify; Cloke *et al.* (1991, p. 19) write of it as 'infuriatingly difficult to define'.

According to Michael Dear (1994a, p. 3):

Postmodernity is everywhere, from literature, design and philosophy, to MTV [Music Television], ice cream and underwear. This seeming ubiquity only aggravates the problem in grasping its meaning. Postmodern discourse seems capable of instant adaptation in response to context and choice of interlocutors.

(p.3)

Nevertheless, he believes 'we can cut to the heart of the matter by identifying three principal constructs in postmodernism: style, epoch and method', with the first having provided the source of the initial explosion of interest.

1   Postmodernism as style originated in literature and literary criticism, and spread to other artistic fields such as design, film, art, photography and architecture; the general trend involved the promotion of difference and the lack of conformity to overriding structural imperatives. Dear found trends within architecture especially revealing:

the search for the new was associated with a revolt against the formalism and austerity of the modern style epitomised by the unadorned office tower. . . . The burgeoning postmodern architecture was disturbingly divorced from any broad philosophical underpinnings, taking the form of an apparently-random cannibalizing of existing architectural archetypes, and combining them into an ironic collage of (or pastiche) of previous styles.

2   Postmodernism as epoch portrays current developments within society as a major radical break with the past – hence use of the term 'postmodernity' to contrast it with the modernity of the previous epoch. These 'new times' are characterised by difference, so that study of the postmodern epoch involves grappling with the fundamental problem of theorising contemporaneity, i.e. the task of making sense out of an infinity of concurrent social realities. Any landscape is simultaneously composed of obsolete, current and emergent artefacts; but how do we begin to codify and understand this variety?

3   Postmodernism as method is, according to Dear, likely to be the most enduring of the three main trends. It eschews the notions of universal truths and meta-theories which can account for 'the Meaning of Everything'. No portrayal can claim dominance over another; separate theories are incommensurable and so their evaluation is always relative and contingent: 'even the attempt to reconcile or resolve the tensions among competing theories should a priori be resisted'.

Epoch, style and method are undoubtedly tangled up and can be differentiated in a variety of ways. Such tangles are what led Cloke *et al.* (1991) to propose a twofold approach to postmodernism – as an approach or method and as object of analysis – while Demeritt (cited in Hoggart *et al.*, 2002, p. 3) located three strands of postmodernism:

1   a concern for new and non-universalist models of human subjectivity and rationality (e.g. Pile
     1991, 1993; Sibley, 2000) that emphasise complexity and diversity;
2   an epistemological project critiquing concepts of truth and claims of universal knowledge;
     and
3   an emphasis on 'the radical phenomenological and social constructivist implications of
     postmodernism in suggesting that our knowledge of nature, and, in some sense, the nature of
     reality itself (and of nature), are culturally relative' (see Demeritt, 1998).

Either way, postmodernism's emphasis on 'heterogeneity, particularity and uniqueness' (Gregory,
1989a, p. 70) undoubtedly attracted some human geographers to it. Human geographers under the
sway of modernism had emphasised order in their promotion of spatial science, when their
empirical observations could really only identify disorder, which suggested the absence of generally
applicable theories and universal truths (Barnes, 1996). Postmodernism gave them a philosophical
hanger, recognising (Gregory, 1989a):

> [that] there is more disorder in the world than appears at first sight is not discovered until that
> order is looked for . . . we need, in part, to go back to the question of areal differentiation: but
> armed with a new theoretical sensitivity towards the world in which we live and to ways in which
> we represent it.

>   (pp. 91–2)

With regard to theory, Michael Dear (1988) observed that postmodernism suggests that
geographers should be wary of aspirations to 'grand theory' (p. 272). Yet recognition that all
knowledge is not only time–space specific but also expressed in language which reflects what such
specificity (see below) meant, to him, that geography can 'claim its place alongside history as
one of two key disciplines concerned with the time–space reconstruction of human knowledge'
(p. 272).

Understanding the attractions of this postmodern stance to geographers requires prior
appreciation of why modernist spatial science came to be so influential. According to Gregory
(1989b), throughout much of the nineteenth and twentieth centuries geographical practice has
been influenced by particular aspects of, first, anthropology, then sociology, and finally economics.
Two features dominated the adopted modernist paradigm:

1   Its firm base in naturalism, which was likely to have 'special significance in a discipline like
     ours, where human geography is yoked to physical geography' (p. 352) and the consequent
     reliance on the aims and procedures of the natural sciences as relevant to the study of humans
     and their societies.
2   A totalising conception of space, which involves the search for a 'systematic order whose
     internal logic imposes a fundamental coherence on the chaos of our immediate impressions'
     – hence, the dominance of spatial science within that project.

Postmodernism presented a substantial critique to the approaches that dominated much of
geography in the period from the 1950s through the 1980s, with their emphases on order and
'grand theory'. As Cloke et al. (1991, p. 200) noted:

> human geographers will increasingly come to recognise the gravity of the challenge that
> postmodernism as attitude poses to the most conventional theorisations of the human world,
> and will begin to appreciate that a sensitivity to the geography of this world – to its fragmentation
> across multiple spaces, places, environments and landscapes – is itself very much bound up
> with (and an impetus for) a postmodern suspicion of modernist 'grand theories' and

'metanarratives'. . . . These manoeuvres will doubtless cause much unease and controversy, however, given that they cast doubt on the stability of the foundations from which most human geography had proceeded over the last thirty years or so.

Soja's (1989) book represented mostly an engagement with what it identified as postmodernism as a place (or assembly of places). As already noted, the Los Angeles metropolitan area is, he argued, difficult to represent in a narrative because its many images seem 'to stretch laterally instead of unfolding sequentially' (p. 222), and because 'it too seems limitless and constantly in motion, never still enough to encompass, [it seems] too filled with "other spaces" to be informatively described'. All he says he can offer is 'a succession of fragmentary glimpses' into a 'particularly restless geographical landscape' (p. 223) – what he terms an 'interjacent medley' (p. 247). Its environment is too multilayered, created by too many authors to be identified: there are too many 'discordant symbols drawing out the underlying themes'.

Soja's work was subjected to an engaged critique by Gregory (1990; see also Gregory, 1994, chapter 4). He argued that modernist theories of the capitalist city also stressed 'geographically uneven development via simultaneous tendencies towards homogenization, fragmentation and hierachization' (Soja, 1989, p. 50). He accepted that the Los Angeles landscape contains 'an economic order, an instrumental nodal structure, an essentially exploitative spatial division of labor', but claimed that these cannot be summarised into a 'totalising description'; all that can be offered is a 'series of fragmentary glimpses' (p. 246). And yet, Gregory stressed, Soja advised those wishing to explore the metropolis that 'it must be reduced to a more familiar and localized geometry to be seen' (Soja, 1989, p. 224) and its emphases are (Gregory, 1994):

> Towers and freeways, sites and districts, zones and areas, enclaves and pockets, gradients and wedges: a landscape without figures . . . Soja's essay becomes a morphology of landscape that, like Sauer's original, is rarely disturbed by the human form . . . Soja's essay [is] so astonishingly univocal. We never hear the multiple voices of those who live in Los Angeles – other than Soja himself– and who presumably learn rather different things from it.
>
> (pp. 300–1)

To Gregory, Soja excluded much of the difference which is supposedly at the heart of a postmodern approach; he ignored the various social struggles that underpin the making and remaking of the Los Angeles landscape, plus the distinctive urban cultures of ordinary people's everyday lives. Thus, whereas Gregory applauded Soja's thesis that 'it is impossible to recover human geographies from a contemplation of their abstract geometries' (p. 304), nevertheless he concluded that Soja 'renders the landscape of Los Angeles as a still life'. Soja had demonstrated that 'postmodernism can have an insistently critical edge' (Gregory, 1990, p. 312), but there is a clear hermeneutic problem in his work: 'his master-narrative is sometimes so authoritarian that it drowns out the voices of other people engaged in making their own human geographies'. Indeed, according to Nancy Duncan (1996, p. 443) Soja's 'goal of reconstructing geography along postmodernist lines . . . belies the fact that Soja is not really postmodernist!'.

Dear's work has also attracted considerable criticism, notably a paper on 'Postmodern urbanism' (Dear and Flusty, 1998), which also uses the case of Los Angeles – and the work of a 'Los Angeles School' of urbanists (which they contrast with the 'Chicago School' of the 1920s–1940s) – to argue that a new form of urbanism is emerging. They contrasted five features of 'modern' townscapes (those analysed by the Chicago school) – mega-structural bigness; straight-space; rational order and flexibility; hardness and opacity; discontinuous serial vision – with five characteristics of their postmodern successors: quaintspace (or cuteness); textured façades; stylishness; reconnection with the local; and pedestrian–automobile split. They then identify ten features of 'Southern Californian urbanisms' which are synthesised into a 'theory' of protopostmodern urbanism: globalisation and restructuring are generating both political–economic polarisation and cultures of heteropolis (i.e. increased cultural variety stimulated by immigration). Together these are leading to the development of interdictory spaces – or segregation – and a greater variety of spatial forms, such as edge cities and fortified enclaves, as well as a new politics of nature. The result is that:

> Conventional city form, Chicago-style, is sacrificed in favor of a noncontiguous collage of parcelized, consumption-oriented landscapes devoid of conventional centers yet wired into electronic propinquity and nominally unified by the mythologies of the disinformation superhighway. Los Angeles may be a mature form of this postmodern metropolis; Las Vegas comes to mind as a youthful example.
>
> (p. 67)

Nijman (2000, p. 135) offered Miami as an alternative 'paradigmatic city' illustrating the 'fundamental traits and trends' of the emerging urban system.

Critiques of Dear and Flusty included arguments that they had not identified a clear modern–postmodern break: 'The theory of postmodern urbanism is a mixture of the old and the new, of breaks and continuities. It is indebted to and rooted in modernism, just as Los Angeles is less of the sheer novelty than its devotees proclaim' (Beauregard, 1999, p. 398). Jackson (1999) contended that the contrast with the Chicago school was poorly drawn, given the lack of ethnographic studies of life in these paradigmatic cities (on which, see Amin and Thrift, 2002). Sui (1999) went further, arguing that postmodernism was likely to lead geographers into a new 'Dark Age of intellectual inquiry' because of its 'assumed or implied ontological relativism, epistemological nihilism, and methodological neologism' (p. 408; see also Symanski, 1994). Dear and Flusty's (1999) response not only defended their position against Beauregard and Jackson's criticisms but also argued – contra Sui – that postmodernism will not go away (p. 415): 'To encompass the full richness of urban life requires much more work and that we invent new ways of writing, because conventional academic writing is singularly ill-equipped for this task.'

More generally, Michael Dear (1994b, p. 299) has argued that postmodernism has encountered such hostility from those geographers (as it has within other disciplines too) who

> perceived their intellectual authority being threatened; [plus] incomprehension on the part of those who (for whatever reason) failed to negotiate its arcane jargon; and the indifference of a majority who ignored what was presumably perceived as the latest fad ... [yet] despite the combined armies of antipathy and inertia, postmodernism has flourished.

In some cases, it seems, the debate has been conducted in highly personal terms (Dear, 2001). To others, it is just a reflection of changes in not only theoretical positions but also substantive concerns (see, for example, Peach's, 2002 comparisons of the 'old' and the 'new' cultural geography). For Nicholls (2011) the 'LA School' had not only been an exponent of 'postmodern' urbanism, but had integrated Marxist and poststructuralist theories (another 'post' trend in human geography) to create a distinctive analytical framework. Whatever the language and concerns, however, by the end of the twentieth century, new textbooks (in the classic Kuhnian sense) came to see postmodernism as an integral part of 'new theories, new geographies' (Hubbard et al., 2002).

Time–space compression and landscapes of postmodernity

Other geographical work considering postmodernism and putatively postmodern landscapes explored a diversity of postmodernisms, refusing to draw clear boundaries around who or what is or is not 'postmodern' (Amin and Thrift, 2002; Dear 2000; Dear and Flusty, 2002). Leaving aside who and what is and is not 'really' modern or postmodern (for it should be clear that these terms are contested and defined with reference to each other), the most influential geographical treatment of postmodernism turned out to be a wide-ranging critique of it launched by Harvey (1989a) in *The Condition of Postmodernity* (on which, see Harvey, 2002). This soon became an academic bestseller: Neil Smith (2001, p. 10) notes that it 'went on to sell almost 100,000 copies in various languages, was widely influential throughout the social sciences and humanities, and was voted one of the best 100 books of the second half of the twentieth century by the *New Statesman/New Society*'.

Harvey drew on his prior Marxist work (see Chapter 6) on the dynamics of capitalism to argue that the epoch and phenomenon of postmodernism was less a break with the modernism that had supposedly gone before and instead might be interpreted as a variant on a old theme. Harvey's basic thesis on postmodernism is clearly expressed in his brief summary page, entitled 'The argument', which is reproduced here in full:

> There has been a sea-change in cultural as well as in political-economic practices since around 1972.
>
> This sea-change is bound up with the emergence of new dominant ways in which we experience space and time.
>
> While simultaneity in the shifting dimensions of time and space is no proof of necessary or causal connection, strong a priori grounds can be adduced for the proposition that there is some kind of necessary relation between the rise of postmodernist cultural forms, the emergence of more flexible modes of capital accumulation, and a new round of 'time–space compression' in the organization of capitalism.
>
> But these changes, when set against the basic rules of capitalistic accumulation, appear more as shifts in surface appearance rather than as signs of the emergence of some entirely new postcapitalist or even postindustrial society.
>
> (p. vii)

The observed changes, especially but not only in modes of economic organisation and much facilitated by technological changes (notably, but not only, in information technology), are the outcomes of the processes of continual restructuring by which capitalism seeks to overcome crisis tendencies (on which see Harvey, 1982; Harvey and Scott, 1989; their term for this restructuring was 'flexible accumulation') and within which cultural forms are produced and reproduced (as set out in Harvey, 1985c). For Harvey, geography matters in understanding the arrival of a postmodern epoch, style or landscape. Indeed, the key is the way that capitalism and its attendant means of production and circulation have compressed space and time in pursuit of 'spatial fixes' to the ever-present capitalist tendency to crisis and over-accumulation (and hence declining profits). Thus, next to a figure indicating a sense of a shrinking world (see Figure 7.1), Harvey (1989a) notes:

> In what follows I shall make frequent use of the concept of 'time–space compression'. I mean to signal by that term processes that so revolutionize the objective qualities of space and time that we are forced to alter, sometimes in quite radical ways, how we represent the world to ourselves. I use the word 'compression' because a strong case can be made that the history of capitalism has been characterized by speed-up in the pace of life, while so overcoming spatial barriers that the world sometimes seems to collapse inwards upon us. . . . The experience of time-space compression is challenging, exciting, stressful, and sometimes deeply troubling, capable of sparking, therefore, a diversity of social, cultural and political responses.
>
> (p. 240)

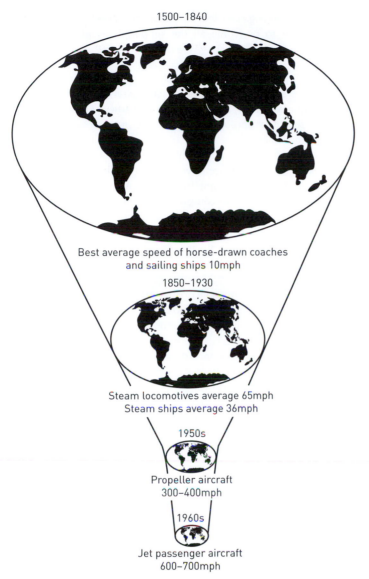

Figure 7.1 The shrinking map of the world through innovations in transport that 'annihilate space through time'

Source: Harvey, 1989a, p. 241.

Harvey's assessment of postmodernism identified some positive elements:

> in its concern for difference, for the difficulties of communication, for the complexity and nuances of interests, cultures, places and the like, it exercises a positive influence. The meta-languages, meta-theories, and meta-narratives of modernism ... did tend to gloss over important differences, and failed to pay attention to important disjunctions and details.
>
> (pp. 113–15)

But he also rejected what he identified as the negative elements, with postmodernism presented as 'a wilful and rather chaotic movement to overcome all the supposed ills of modernism'. Postmodernism, to him, reflected:

> a particular kind of crisis within . . . [modernism], one that emphasizes the fragmentary, the ephemeral, and the chaotic side . . . (that side which Marx so admirably dissects as integral to the capitalist mode of production) while expressing a deep scepticism as to any particular prescriptions as to how the eternal and immutable should be conceived of, represented, and expressed.
>
> (p. 116)

And so it gets the balance wrong, because its adherents are unwilling to grapple with the need for a theory which promotes appreciation of the nature of capitalism and its manifest contradictions, especially, for geographers, those relating to time–space compression through changed transport and other technologies and the political strategies which promote the fortunes of people in different places. The shifts identified in the first sentence of his summary require a dynamic theory of historical materialism, within which he identified four main agenda items:

1 The treatment of difference and 'otherness' (see p. 273, this book) within the dialectics of social change, thereby 'recuperating such aspects of social organisation as race, gender, religion, within the overall frame of historical materialist enquiry (with its emphasis on the power of money and capital circulation) and class politics (with its emphasis upon the unity of the emancipatory struggle)'.
2 Recognition of the importance of cultural practices, including the production of images and discourses, on the reproduction of the social order.
3 Recognition of the importance of space and time in the geopolitics of capitalism, now that 'Historical materialism is finally beginning to take its geography seriously.'
4 Acceptance that historical–geographical materialism is an open-ended and dialectical mode of enquiry rather than a closed and fixed body of understandings. Meta-theory is not a statement of total truth but an attempt to come to terms with the historical and geographical truths that characterise capitalism in general as well as in its present phase.

(p. 355)

Harvey, in turn, was critiqued. Deutsch (1991) criticised his 'totalizing vision' of society, with aspects of difference merely appended to Marxist theory, and Massey (1991) similarly pointed out that the book fails to consider the possibility of a feminist interpretation (also see Chapter 8, this book) of postmodernism (and of the images that Harvey uses to illustrate modernism and postmodernism in The Condition of Postmodernity). Massey points out that 'gender is a determining factor in cultural production' (p. 51), but Harvey promotes a form of Marxism within which gender differences can be subsumed. Likewise, Michael Dear (1991b) claimed that Harvey 'seems incapable of tolerating difference' (p. 536) so that 'The catalogue of different voices that are consequently denied relevance in Harvey's discourse is long and depressing' and his goal appears to be to 'dissolve differences' (p. 537).

Much of the postmodern literature is ignored, Dear argued: 'Harvey is a much better political economist than he is cultural critic'. He summarised Harvey's position as follows:

1 all aspects of social processes can be encompassed within historical materialism; so that
2 no alternatives need be addressed;
3 the shift to flexible accumulation accounts for recent social, economic and political changes; and
4 there is no need for any other theory beyond historical materialism.

He then concluded that 'Perhaps it is time that Harvey tried to transform his Marxism, rather than obliging the world to fit into it'. Harvey recognises difference, but refuses to incorporate it to his thinking.

In response, Harvey (1992) noted that he was influenced by some feminist writers, and that incorporation of their insights would have strengthened his book; the work of Young (1990) in particular had considerably shaped his later work on justice (Harvey, 1993a). He accepted the role of situation and position in the production of knowledge (see p. 272, this book) but found some of their applications in feminist work 'rather more vulgar' because their focus on individual biographies is not dialectically placed in an evolving mode of production and because of their denial of the veracity of other accounts. Thus, his book was written from a particular (privileged, white male) position and:

> emphasized the commonality of our condition as users of commodities and money and as partici-
> pants in labor markets and the circulation processes of capital. But emphasizing commonality
> does not deny difference. Properly done, it can enhance the understanding of differences at the
> same time as it provides a critical basis from which to evaluate the work of those who purport to
> write critical theory outside the confines of what capitalism as a social system is all about.
>
> (p. 304)

In this context, he found the critiques 'unnecessarily personalized, hurtful and sometimes abusive' (p. 308), part of a strategy of creating a feminist position which he claims shows that 'they exhibit not the slightest hint of concern to grapple with the deeper problematic' involved with dissolving gender differences. He wants debate to focus on these and finishes by pointing to a major paradox regarding truth and generalization in all postmodern writing:

> postmodernists ... cannot criticise *The Condition* as wrong, misguided, or fundamentally
> misconceived without deploying truth terms of their own which presuppose they have an
> ultimate line on a truth they theoretically claim cannot exist. I am reminded here of William
> Blake's great aphorism: 'to generalise is to be an idiot; to particularise is to achieve the greatest
> distinction of merit', which sounds great . . . until it is recognized as a generalization and thereby
> self-condemned as idiotic.
>
> (p. 322)

Elsewhere, Dear (1991a, p. 549) criticised both Harvey and Soja, claiming that each rejects the pluralism and celebration of difference that characterises postmodernism:

> By insisting on their totalizing and reductionist visions, Soja and Harvey squander the insights
> from different voices and alternative subjectivities. Difference is relegated to the status of an
> obstacle hindering 'our' view of a coherent theoretical and political praxis.

Trevor Barnes (1996), too, concluded that Harvey's commitment to what he terms the Enlightenment project – 'a belief in rational progress, the individual subject, a monolithic order, and universal truth' (p. 3) – runs counter to prevailing trends:

> Although [Harvey] wants to recognize that truths are socially constructed at a given moment in
> time and space, he also wants to claim that his theory is a valid and accurate representation of
> the capitalist world.

Michael Dear's evaluation of postmodernism's promise for human geography was unequivocally positive (1994a):

> Simply stated, we live in an era of postmodern consciousness; there is no choice in this matter, unless we are prepared to declare in favor of ignorance or the status quo. I believe that a revolution of sorts is occurring in geographical thinking.
>
> (p. 9)

He supported this last claim by arguing that in the decade since 1984 we had witnessed: (a) a truly unprecedented increase in quality scholarship devoted to the relationship between space and society; (b) reassertion of the significance and role of space in social theory and social process; (c) an effective reintegration of human geography with mainstream social science and philosophy; (d) the establishment of theory and philosophy as the sine qua non for the discipline's identity and survival; (e) a new appreciation of diversity and difference, and a consequent diversification of theoretical and empirical interests; and (f) a self-conscious questioning of the relationship between geographical knowledge and social action. He further claimed that a large number of publications that appeared in 1989 or later 'reveal a significant postmodern consciousness' (p. 7) in seven areas:

1   cultural landscapes and 'place-making';
2   the economic landscapes of flexible accumulation and post-Fordism;
3   philosophical and theoretical disputes about space and language;
4   problems of representation in geographical writing and image-making;
5   the politics of postmodernity and difference;
6   the construction of the individual; and
7   a realisation of the importance of nature and the environmental question.

This leads him to an optimistic conclusion, provided that geographers are prepared to grapple with the postmodern challenge:

> Postmodernism places the construction of meaning at the core of geography's problematic. The key issue here is authority; and postmodernism has served notice on all those who seek to assert or preserve their authority in the academic or everyday world. And yet I understand that geographers, like everyone else, cling tenaciously to their beliefs. Knowledge is, after all, power, and we are all loathe to relinquish the basis for our claims to legitimacy.
>
> (p. 9)

Thus:

> To ignore the postmodern challenge is to risk disengaging geography once more from the mainstream. To accept it is to encourage new ways of seeing, to relish participating at the cutting edge of social and philosophical inquiry, to convince our peers of the significance of space in contemporary social thought and social process, and to help forge a new politics for the twenty-first century.

Many human geographers remained concerned as to how these could be achieved, as Dear (1995) recognised in responding to an evaluation of developments since his 1988 paper. Elspeth Graham (1995, p. 175) accepted that in evaluating Dear's (1994b) claim that postmodernism has flourished:

> If judged by the volume of literature which mentions postmodernism then this is certainly true. Postmodernism has become the lingua franca of intellectual discourse (Ley, 1993). As yet, however, few are fluent and communication easily breaks down, inevitably producing a disturbingly disorganised discussion (a concern also of Warf, 1993).

Such debates continued (see Minca, 2001) and became tangled up with others; about consumption and its roles in the (post)modern economy and society (e.g. Gregson *et al.*, 2002; Glennie and Thrift, 1992; Gregson, 1995; May, 1996a; Thrift and Olds, 1996), for example, and concerning the merits, possibilities (or otherwise), pitfalls and problematics of poststructuralism and postcolonialism. But there is possibility here, too, in the startling recognition that all geographical (and other) interpretations:

> are local and contingent, and that any conclusions represent no more than a temporary hegemony of a favoured belief. Such studied relativism makes it a condition of knowing and practice that we make explicit the conditions and criteria under which knowing is to occur and decisions to be deliberated.
>
> (Dear, 1995, pp. 179–80)

Such a notion of knowing one's position has been problematised in poststructuralist (and many feminist, see Chapter 8) conceptions of the individual a complex and sometimes contradictory or divided subject. It is to such reconceptualisations that we now turn.

Poststructuralism: power, representation and performance

Poststructuralism's geography

We have already noted how for some geographers the fragmented city of Los Angeles becomes a place to investigate postmodernity. What has come to be known as 'poststructuralism' also has its own geography. It is hard to say where postmodernism and poststructuralism divide – for some, the latter is viewed as variant of the former. And while Richard Smith (2003, p. 71) warns us that 'the term was not widely used in France' and that the label poststructuralism 'obliterates the differences between quite different authors', Peet (1998), however, begins his account of 'Poststructuralism, postmodernism and postmodern geographies' across the Atlantic, in a city that has at times been presented as an archetype of modernity:

> Poststructuralism and postmodernism in their theoretical guises are very much the direct products of events in France. In May 1968, a Parisian alliance of student agitators and working-class militants rose in spontaneous revolutionary upheaval. It seemed for a short while that the dreams of radical modernism were about to be realized in their ultimately radical, social-anarchistic form (worker self-management, participatory democracy . . .). But the eventual, eternal return of bourgeois rational normalcy to Paris began a period of contemplative reaction on the left. This turned against structuralism and Marxism immediately, hence a series of poststructural and post-Marxist ideas. But thinkers began also to react once more against the aspirations of modernism as a whole, hence the postmodern aspect to poststructural philosophy.
>
> (194–5)

A key concern of 'poststructuralist' work has been the issue of struggles over representation: the complex of linguistic, cultural and symbolic processes which poststructuralists argue is intimately tied up with (express and constitute) power. Although humanistic geography (see Chapter 5) and some of the debates about regional geography did raise issues about how language (and other modes of geographic description) embody, enable and constrain power, poststructuralism made these a more central concern:

> Pieces of the world . . . do not come with their own labels, and thus representing 'out there' to an
> audience must involve more than just lining up pieces of language in the right order. Instead it is
> humans that decide how to represent things, and not the things themselves.
>
> (Barnes and Duncan, 1992, p. 2)

Thus, writing is integral to the complex and contested social construction of knowledge and meaning. If the written word is not a straightforward mirror of the world it seeks to represent, but is itself understood as a complex social creation, several consequences follow (Gregory and Walford, 1989).

1   Severing the assumed mirror link between 'reality' and 'text' means that reality cannot be apprehended outside the language used to describe it: the world is always narrated. The texts that we produce draw on other texts, which are the sources of the images that we are trying to convey, such as the metaphors that we use to describe the 'unknown' (Barnes, 1996). This is termed intertextuality: meanings are created in a continual transition process from text to text (Barnes and Duncan, 1992, p. 3):

> new worlds are made out of old texts, and old worlds are the basis of new texts. In this world
> of one text careening off another, we cannot appeal to any epistemological bedrocks in
> privileging one text over another. For what is true is made inside texts, not outside them.

2   Writing reveals as much about the writers, and their position, as it does about what is being written about. We write, Barnes and Duncan argue, from our own 'local' setting; thus the worlds we represent are inevitably stamped with our own particular set of interests. To understand critically our own representations, and also those of others, we must therefore know the kinds of contexts bearing upon an author that makes an account come out the way that it does.
3   All writing involves the use of literary apparatus, such as metaphors and other rhetorical devices, whose use we must appreciate since they are central to the construction and circulation of meanings.

These are illustrated by Barnes and Duncan, and in the essays in their edited volume, with reference to landscape descriptions. Landscapes, following Cosgrove and Daniels (1988, p. 1), may be 'represented in a variety of materials and on many surfaces – in paint on canvas, in writing on paper, in earth, stone, water, and vegetation on the ground. A landscape park is more palpable but no more real, nor less imaginary, than a landscape painting or poem.' Thus the 'texts' to be appreciated may be produced through the medium of words, drawings, paintings or other media, or may be inscribed in the landscape itself.

Trevor Barnes and James Duncan (1992) identified three major concepts as central to the study of representation: text, discourse and metaphor. Text embraces a wide range of cultural products (p. 263, this book) that involve the author rewriting what has been 'read' in a hermeneutic exercise (p. 189, this book) and is open to a range of interpretations:

> the meaning of a text is unstable, dependent upon the wide range of interpretations brought to
> bear upon it by various different readers. Similarly, social productions and institutions also
> address a wide range of possible interpreters. But those interpreters are not free to make of the
> text what they like, but are subject to discursive practices of specific textual communities. Thus
> both how we produce a text and how we interpret one depends upon our textual community – on
> the language (even the particular form of a language) that we use, reflecting our individual
> compositional and contextual positions.
>
> (p. 6)

Discourses are the larger structures from within which texts are constructed and within which others are read. They comprise (p. 8):

> frameworks that embrace particular combinations of narratives, concepts, ideologies and signifying practices, each relevant to a particular realm of social action. Between discourses words may have different connotations, causing people who ostensibly speak the same language to talk past one another, often without realizing it.

The relationship between a component of the text, such as a word (the signifier), and that to which it refers (the signified) is a complex social construction, undertaken within the specific textual community. Thus:

> discourses are practices of signification, thereby providing a framework for understanding the world. As such, discourses are both enabling as well as constraining: they determine answers to questions, as well as the questions that can be asked. More generally, a discourse constitutes the limits within which ideas and practices are considered to be natural. . . . These limits are by no means fixed, however. This is because discourses are not unified, but are subject to negotiation, challenge and transformation.
>
> (p. 8)

Metaphors are major devices for representing meanings. The world is apprehended and known through study, which requires a language, or some other form of textual representation, for the transmission of meanings. Metaphors are extremely valuable, since they provide a means of describing the unknown using the vocabulary of the known; the unfamiliar is illuminated by comparing it to the familiar. In much science, for example, the use of metaphor steers the study of the unknown, providing a framework for its investigation – as was the case with applications of the gravity model in spatial science (described in Chapter 6 of this book; Barnes, 1996). Understanding a discourse therefore involves appreciating its metaphors.

If metaphors dominate textual discourse, however, they also ensure instability in the ongoing transmission of meanings, or the reproduction of knowledge. As Trevor Barnes (1996) argued, drawing on the work of Derrida:

> the meaning of words and concepts (signifiers) can never be directly tied to particular things (signified). For meaning is derived from a signifier's position with respect to all other signifiers in the system [i.e. a metaphor can only be understood in the context in which it was developed]. According to Harland (1987, p. 135), 'In Derrida's conception, one signifier points away to another signifier, which in turn points away to another signifier, which in turn points away to another signifier, and so on ad infinitum'. There is no anchor of some final presence or some ultimate origin point of meaning. Meaning, rather, is always produced through displacement and deferral, shaped as much by what is absent as by what is present.
>
> (p. 166)

If meaning can only be conveyed through 'an orchestration of signifiers' (metaphors), therefore, then we are led 'to deny identification of an unimpeachable presence. For, if there is no ultimate signified and only a shifting system of signifiers . . . there can only ever be the flux of meaning and no constant presence' (Barnes, 1996, p. 166).

Hubbard *et al.* (2002) similarly registered the developing impacts of Derrida in geography and how 'deconstruction' (which also refuses to recognise any clear distinction between the metaphorical and the literal) opens up the meanings of geography:

> For Derrida, when we describe something as having a certain characteristic ('x'), we inevitably recognize that it lacks another characteristic (i.e. is not 'y') and we are able to speak the difference between these (so that one becomes deemed as lacking in relation to the other: 'x' > 'y') . . . language is a play of signifiers that cannot articulate absolute difference, only lack or excess. The implication of Derrida's thinking is that the meaning of texts can never be [definitively] pinned down; language is seen to defer, rather than yield, truth and meaning.
>
> (p. 87)

They illustrated this through looking up the word 'geography' in the *Oxford English Dictionary*. They note that it contains three definitions (referring to places, area and arrangement). In turn, they look up the words (such as climate) that crop up in one of these definitions, only to find that each also has several meanings:

> Tracking through the dictionary highlights how meaning is always intertextually deferred, always reliant on words that are themselves reliant on others and so on. Only when located in discourse and grounded in context do words take on meanings.
>
> (Hubbard *et al.*, 2002, p. 88)

In turn, however, the limits to such contexts are hard to specify. In other words, any geographical categories that construct boundaries are susceptible to deconstruction, whereby the boundary is shown to be to profoundly arbitrary and to rest on excluding what is held to belong (usually difference and diversity) only on the 'other' side of the boundary.

Metaphor is not the only device used in geographical writing, however. According to Jonathan Smith (1996), geographers use four different modes in their story-telling: Romance – as in biographical narratives of individual and group struggle, especially where it leads to radical change; Tragedy – as in representations of deterministic systems, many of which involve prognostications of doom; Comedy – which represents harmony and reconciled conflict; and Irony – in which the detached observer occupies a superior position. These represent release, resignation, reconciliation and removal respectively, and in them the author employs one or more tropes, or figures of speech:

1   Metaphor involves the use of comparisons to introduce concepts, describing 'the remote in terms of the immediate, the exotic in terms of the domestic, the abstract in terms of the concrete, and the complex in terms of the simple' (p. 12).
2   Metonymy uses technical terms to provide (accurate) descriptions for specialised audiences.
3   Synecdoche promotes understanding through synthesising the general and the particular to impart meanings.
4   Irony suggests that representation and understanding are futile, and that general apprehensions can never be produced.

James Duncan and David Ley (1993, pp. 2–3) identified four major modes of representation within human geography:

1   Description of observations obtained through fieldwork, which dominated cultural geography until relatively recently; its underpinning assumption was that 'trained observation transcribed into clear prose and unencumbered by abstract theorising produces an accurate understanding of the world'.
2   Mimesis, whereby the world is reflected in media other than words, as with the mathematical modelling of spatial science that had little impact upon cultural geography.

3   Postmodernism, which 'distrusts and interrogates all meta-narratives including those of the researcher' and, as indicated earlier, rejects the search for 'universal truths'.

4   Hermeneutic interpretation, which acknowledges the role of the interpreter and therefore rules out mimesis – 'reality' cannot be faithfully reproduced.

Yet, however many discrete styles are identified and whatever way they are classified, the rhetorical styles produced by geographers reflect both the general approach to their subject matter and the audiences they consider themselves to be writing for – for example, irony is characteristic of many postmodernists; metonymy (and perhaps tragedy) of many analysts of spatial systems; and metaphor and romance of the educator seeking to open eyes to the world. Above all, language is crucial in the construction and transmission of meanings, for, as Barnes (1996) explained:

> we can never have direct access to things in and of themselves. This is because in order to understand those terms, they must already be expressed in language. But if they are part of language, then their meaningfulness only becomes about through a play of difference among signifiers, which . . . exclude any fundamental signifieds or presences.
>
> (p. 167)

Appreciation of those meanings involves understanding the metaphors employed (as Olsson, 1980, did when exploring the gravity model metaphor which underpinned his early spatial science). This involves appreciating the translation process, represented by Olsson (1992, p. 86, and 1991) as distortion, and therefore an exercise in the use of power (see the sympathetic review by Philo, 1994). Doel (1993), for example, uses irony in his characterisation of the futility of much geographical writing's failure to represent the world the author claims to portray, whereas Pred (1989, 1990) explored a variety of linguistic repertoires ('words, variable meanings, pronunciation, grammar, sign-tactical arrangements, rules of interpretation and expressive bodily gestures at one's command' (1990, p. 33)) to show how their production and reproduction are inherent to local struggles. Further, like Olsson, he has employed various writing strategies in order to 'subvert the taken-for-granted (and thereby ideology-riddled and power-laden) nature of the academic printed word' (Pred, 1990). This:

> seeks to make the taken-for-granted format of representation appear strange and yet comprehensible, that seeks to make the reader understand and mentally see what she otherwise might not understand or mentally see, that seeks, somehow, to push through the filter of preconceptions and interpretative predispositions deeply inset in the reader's social, biographical and disciplinary past. Thus, I occasionally resort to chameleon like (mis)spellings, hyphenations and word-couplings that are de- or re-signed either so as to trigger previously unmade associations, or so as to convey the ambiguity, the shifting subtleties, the multiplicity of meanings characteristic of on-the-ground practices and social relations in any place.
>
> (p. 48)

Thus, as Trevor Barnes and Derek Gregory (1997) note, attention to the poetics of geographical writing is also attention to their politics:

> It might seem odd that we foreground 'poetics'. Some geographers read and write poetry during their off-hours, and occasionally use it in their books and articles, but there is nothing conventionally poetic about most contemporary geographical writing. Some critics would say that post-positivist geographers are among the least poetic of all: that their prose is all too often stilted, leaden, inflated, and opaque. . . . By poetics we have in mind the interpretation provided by the anthropologist and cultural critic James Clifford. . . . For Clifford, all ethnographic

accounts – and, we would argue, geographical ones too – are rhetorical constructions, textual artefacts that seek to persuade us of their claims through an amalgam of 'academic' and 'literary' genres. But such accounts are not just texts; they also reach out to wider contexts of power and resistance, institutional constraint and innovation.... Words, as poets and contributors to this volume know very well, are extraordinarily powerful. We need to use them with care, with sensitivity, and above all with a critical passion: a poetics of inquiry is thus also a politics of inquiry.

(pp. 3–4)

Deconstructing maps and critical geopolitics

Of course, geography is not only writing – words-(and sometimes numbers and formulae)-in-order-on-a-page. Geography also involves mapping. Although it may be particularly evident in the case of propaganda maps (see Monmonier, 1996; Pickles, 1992), all maps (and Geographical Information Systems, see Curry, 1998; Pickles, 1995a; Flowerdew, 1998, among the other debates prompted by the arrival of GIS that were considered in Chapter 4) are interpretations, a point that was stressed in a series of papers by a cartographic historian, Brian Harley. His starting-point was that (Harley, 1989):

we still accept uncritically the broad consensus, with relatively few dissenting voices, of what cartographers tell us maps are supposed to be. In particular, we often tend to work from the premise that mappers engage in an unquestionably 'scientific' or 'objective' form of knowledge creation. Of course, cartographers believe that they have to say this to remain credible but historians do not have that obligation. It is better for us to begin from the premise that cartography is seldom what cartographers say it is.

(p. 1)

Cartographers' 'scientistic rhetoric' was becoming more strident with the development of computer-assisted map-making, he claimed, but he used deconstruction procedures 'to break the assumed link between reality and representation which has dominated cartographic thinking' (p. 2); by 'reading between the lines' of maps, he sought to identify their 'silences and contradictions', their metaphors and rhetorical flourishes.

Cartographers have traditionally tended to accept that the objects they wish to represent on their maps are 'real and objective' pregivens. Their goal is to display these accurately, hence their search for ever greater scientific precision; they can come closer to an exact mirror in their representation and can reject maps (especially old ones) that fail to conform to their canons of conforming to the rules. But, Harley argued, the production of maps is governed by cultural as well as scientific rules. Thus, for example, most societies are ethnocentric in producing maps which have their territories at the centre of their world, thereby helping to promote geopolitical world views. In selecting what to show, and what prominence to give it, cartographers frequently employ a 'hierarchicalization of space':

it is taken for granted in a society that the place of the king is more important than the place of a lesser baron, that a castle is more important than a peasant's house, that the town of an archbishop is more important than that of a minor prelate, or that the estate of a landed gentleman is more worthy of emphasis than that of a plain farmer. Cartography deploys its vocabulary accordingly so that it embodies a systematic social inequality. The distinctions of class and power are engineered, reified and legitimated in the map by means of cartographic signs.

(p. 7)

Thus, partial representations are undertaken 'behind a mask of a seemingly neutral science', yet the 'rules of society will surface. They have ensured that maps are at least as much an image of the social order as they are a measurement of the phenomenal world of objects.'

If we accept maps as cultural texts, Harley contends, we can interrogate them and come to learn of their functions within the society for whom they were created. Map-making involves a series of steps, he argues – selection, omission, simplification, classification, the creation of hierarchies, and 'symbolization' – all of which are rhetorical devices. Cartographic rhetoric is involved in the production of all maps, and is implicated in the exercise of power:

> Power is exerted on cartography. Behind most cartographers there is a patron; in innumerable instances the makers of cartographic texts were responding to external needs. Monarchs, ministers, state institutions, the Church, have all initiated programs of mapping for their own ends. In modern Western society maps quickly became crucial to the maintenance of state power – to its boundaries, to its commerce, to its internal administration, to control of populations, and to its military strength.

Thus, mapping became a state business, and the publication of maps became subject to laws regarding state security.

The map becomes 'juridical territory'; it facilitates surveillance and control. Maps are still used to control our lives in innumerable ways. A mapless society would now be politically unimaginable. All this is power with the help of maps. It is an external power, often centralised and exercised bureaucratically, imposed from above and manifests in particular acts or phases of deliberate policy. As a consequence, cartography and cartographers are not just one element in a power structure (p. 13): 'Cartographers manufacture power: they create a spatial panopticon. It is a power embedded in the map text.' Their power is not exercised over people directly, but rather over the knowledge made available to them; 'maps, by articulating the world in mass-produced and stereotyped images, express an embedded social vision' (p. 14; Harley, 1992, exemplified this in his essay on the role of maps in the Columbian encounter with the 'new world'; see also Schulten, 2001).

Deconstructing maps serves three functions:

1   It challenges the myth that technological improvements 'always produce better delineations of reality'.
2   It allows appreciation of the role of maps in the historical processes of creating a socially constructed order to the world.
3   It promotes the understanding of other texts as the meaning of maps is discovered.

(Harley, 1989, p. 15)

Harley later extended this argument, asserting that 'As a discourse created and received by human agents, maps represent the world through a veil of ideology, are fraught with internal tensions, provide classic examples of power-knowledge, and are always caught up in wider political contexts' (Harley, 1990, p. 1). This is illustrated by a range of examples including, for example, cartographic complicity in racial stereotyping through place-name labelling, and the exclusion of 'places to avoid' (such as the shanty towns that are absent from the maps of most South American cities). Cartographers' interests, he claimed, are dominated by mechanical issues related to 'efficiencies' associated with new technologies, so that the ethical issues of what is and is not depicted go largely unconsidered.

Harley concluded his 1990 paper by noting that 'The challenge to and continual crisis of representation is universal and not peculiar to cartography' (p. 18). Earlier, he had referred to GIS as extending the crisis of representation to the crisis of the machine. GIS are new ways of presenting the world, new texts to be deconstructed to illustrate meanings and power relationships. This argument has been taken up by Pickles (1995a, p. 3), who identified as a central characteristic of all GIS that they involve 'the production of electronic spatial representations' of data. Those media not only facilitate data capture, entry and reproduction and speed up operations on the data but also allow 'new forms of representation'.

Thus, like maps and other texts, the use of GIS relies on signs and representations (signifiers and signified) which call for deconstruction: 'We are . . . entering a potential new phase of ways of worldmaking for which we desperately need new ways of wordmaking' (Pickles, 1995a, p. 5) because GIS are much more than counting machines with greatly increased efficiency; they are enabling new ways of representing the world, and hence new 'realities', and their use by the technically skilled involves new power relationships.

For David Gilbert (1996), wider developments of 'hypermedia' (within which GIS could be situated) would allow exactly the sorts of explorations of textual and other media that postmodernists and others promote:

> In a computer hypertext the reader can use a mouse to point and click on a word, and be instantly taken to related ideas elsewhere in the text. The freedom created for the reader to jump from place to place, to compare, contrast, or simply juxtapose different elements radically changes the nature of writing (or 'authoring' . . .) and of reading . . . the hypertext form marks a move away from a modernist concern for 'objects, positions, order and stability' towards a postmodern emphasis on 'processes, relations, chaos, and instability'.
>
> (p. 7)

Images (still and moving), diagrams, sounds and other media can be incorporated, allowing a 'polyphonic' (Crang, 1992) approach to geographical writing and study similar to the postmodern novel: the 'reader' becomes an 'active co-author'. For Pickles (2004), the entire world is 'geo-coded', so the boundaries between maps, meanings and the world have become hard to discern. Building on this (and noting the importance of Harley's work and the debates about GIS), Kitchin and Dodge (2007) insist that maps *emerge* through practices. In other words:

> Rather than cartography being narrowly understood as the scientific pursuit of how best to represent the spaces of the world (focused on issues such as form and accuracy), cartography becomes understood as the pursuit of representational solutions (not necessarily pictorial) to solve relational, spatial problems.
>
> (p. 343)

The powers and limits of maps, and of what Pickles terms a 'critique of cartographic reason' also form one of the themes in Olsson (2007). Like other works by Olsson (see p. 189, this book) this book defies easy summary. According to one sympathetic reviewer: 'the writing exudes a captivating playfulness and pleasure . . . the productive questioning of geometric propositions, not their fixing and binding of terms' (Pickles, 2007, pp. 396 and 397).

Others have also set out the consequences and possibilities of deconstructive approaches in geography, in particular through more careful engagement with the works of the French philosopher Jacques Derrida (Barnett, 1999; see also p. 259, this book) and such approaches (in fertile combination with Foucault's work and the legacies of radical geography) have also enabled a critical reengagement with geopolitical discourses (Dodds and Atkinson, 2000; Ó Tuathail, 1996; Dalby and Ó Tuathail, 1998) that examines the imagination, narration and scripting of global and strategic spaces. As was noted in Chapter 2, a 'political trend' associated with the birth of geopolitics was among the six foundational strands of human geography described by Freeman (1961). Geopolitics rapidly acquired its own trajectory: 'an enigmatic, shadowy, contested and sometimes shameful category' (Atkinson and Dodds, 2000, p. 1). One response however, inspired by poststructuralism, has been to interrogate such traditions through a 'critical geopolitics' (initiated by Taylor's, 1990c, pioneering book). Thus, according to Ó Tuathail and Dalby (1998):

> Critical geopolitics bears witness to the irredeemable plurality of space and the multiplicity of possible constructions of space. Thus . . . it pays particular attention to the boundary-drawing practices and performances that characterize the everyday life of states. In contrast to conventional geography and geopolitics, both the material borders at the edge of the state and the conceptual borders designating this as a boundary between a secure inside and an anarchic outside are objects of investigation. Critical geopolitics is not about the 'the outside' of the state but about the very construction of boundaries of 'inside' and 'outside', 'here' and 'there', the 'domestic' and the 'foreign'.
>
> (p. 3)

Such work has examined both elite (government and state) and popular (e.g. media or cinematic) geopolitical narratives and the connections between them (Dittmer, 2010; Sharp, 2000).

A similar critical sense of the taken-for-granted spaces of (geo)politics and cartography informed Martin Lewis and Karen Wigen's (1997) critique of conventional notions of continents (Europe distinct from Asia, for example, or the notion of Australia as a continent apart from Asia and the Pacific); the 'common-sense' and naturalised view of the world, which they show to be historically contingent, rather than eternal and fixed. And while some have wondered if any rendition of geopolitics ('critical' or otherwise) is compatible with a cosmopolitan, multicultural society (Heffernan, 2000), Ó Tuathail (writing under both the anglicised and Irish spellings of his name) responded that: '"critical geopolitics": a project that involves commitments, subjectivities and positions which can be described . . . as a positional geo-politics which "one cannot not want"' (Toal/Ó Tuathail, 2000, p. 386).

Poststructuralist geographies

While 'critical geopolitics' has offered a rich vein of ideas and poststructuralist departures and others have revisited complex and wider entanglements of power (see the survey by Allen, 2003), Doel's (1999) *Poststructuralist Geographies: The diabolical art of spatial science* retraced human geography's trajectory through an avowedly poststructuralist lens:

> Letting space take place: that is the ambition of geography . . . geography is simply an inclination towards the event of spacing. This is not spacing in the paranoiac sense of dissociating one position from another, of forcing distanciated identities onto space – a tendency that I have dubbed 'pointillism'. Rather than a poststructuralism that could get taken up by geography and geographers, I want to demonstrate that poststructuralism is always already spatial: that it attends from the off to the 'difference that space makes'.
>
> (p. 10)

Doel makes clear that his explorations of poststructuralism and geography are operating in the spaces opened by Marxist geography (he thus engages creatively with Harvey's work), yet the result is not reducible to a reworking (or even a post-) Marxist geography. For example, he posits a 'Deleuzian' notion of origami rather than socio-spatial dialectics:

> the world can be (un)folded in countless ways, with innumerable folds over folds, and folds within folds, but such a disfiguration never permits one (or more) of those folds to become redundant, nor for one (or more) of them to seize power as a master-fold. Every fold plays its part in lending consistency to the thing that is folded and since every fold participates in the

> lending of consistency to 'something = x' without ever belonging to it ... folds cannot be distinguished in terms of the essential and the inessential, the necessary and the contingent, or the structural and the ornamental. Every fold plays its part: every fold spays 'it' apart. The event of origami is in the (un)folding, just as the gift is in the wrapping: not as content, but as process.
>
> (p. 18)

The book is subsequently divided into three sections: (a) the space of poststructuralism: a review of the spatiality of selected writings by Baudrilliard, Deleuze, Foucault, Irigaray and Lyotard; (b) a schizoanalysis of the geographical tradition – exploring spatial science and the drawing of lines; and (c) a move to a 'Poststructuralist geography'. Doel, however, explains that:

> Accordingly what I want to achieve in this book is neither a secure identity for poststructuralist geography, which one or two readers may attest to being the 'Real Thing', nor a justification for reducing the geographical idiom in all of its heterogeneity to the theoretical practices of figures such as Deleuze, Derrida, Lyotard, Irigary, Baudrillard and Olsson. I hope that my actual aim is much more modest. I want to participate in generating a different way of feeling about events, about the world, about others, and about theoretical practice.
>
> (p. 18)

Doel's book, and similar works by him and his collaborators elsewhere (Doel and Clarke, 1999), elicited a variety of reactions from hostility to fascination and bemusement. Thus, Peet (1998, p. 241) labelled it an 'Amoral geography': 'an extreme form of the postmodern position in the discipline of geography, which does manage to lose virtually all connection with any project of emancipatory politics'.

Responding to an earlier paper (Clarke and Doel, 1995), in this case on 'transpolitical geography', however, Philo (1994) begins with an observation that:

> It goes without saying that papers by David Clarke and Marcus Doel are tricky to read, and that many readers – whether they would call themselves political geographers, cultural geographers, urban geographers or whatever – are wary about the effort required to see if anything valuable can be recovered from them. Questions immediately arise as to whether readers can fully understand the papers, whether they are indeed supposed to be able to, whether the onus is deliberately being shifted from the author(s) to the reader(s), and about whether the latter can ever achieve more than one creative 'misreading' after another.
>
> (p. 525)

Philo concludes with a mixture of reservations and possibility:

> I ... would want to be less cavalier than Clarke and Doel in consigning large portions of existing political geography to the dustbin of outdated or untenable geographical inquiries, but I nonetheless have sympathy for a paper which is so consistently unnerving and unswerving in its commitment to finding the 'holes' in an established geography ...
>
> (p. 531)

Philo (1992) had earlier sought to explore the geographies in the influential works of the French (poststructuralist) thinker Michel Foucault (cf. Matless, 1992). Philo presents this as: 'Not a matter of "creating yet another intellectual base to defend" ... but of stirring another voice into the richness of recent "geographical" debates ...' (p. 138).

Foucault is interpreted as suspicious and subversive of supposed historical (and geographical) certainties. Foucault's heterodox notions of order and power (the former as highly contingent to different times, societies and places and the latter as diffuse; not something which is simply held, so much as enacted and exercised) provide us with:

> a blueprint for . . . a postmodern geography in which details and difference, fragmentation and chaos, substance and heterogeneity, humility and respectfulness feature at every turn, and an account of social life which necessarily brings with it a sustained concern for the geography of things rather than a recall for the formal geometries of spatial science.
>
> (p. 159)

Moreover:

> Foucault's geography emerges directly from his own suspicion of the certainties (the order, coherence, truth, reason) supposed by most historians and social scientists to lie at the heart of social life, and as such I think that it can be adjudged a 'truly' postmodern human geography in a manner that, say, Edward Soja's postmodern geographies cannot. We might not like this Foucauldian version of a postmodern human geography, but I think that there is much we can learn from it, even if we then chose to retain our faith in a more obviously modernist conceptual, practical, and political geographical project.
>
> (p. 137)

Twenty years later, Philo (2012, p. 496) noted how 'There are [now] many works in anglophone human geography deriving nourishment from Foucault conceptually, methodologically and substantively, including one compendium of essays.' He argued that the appearance of new material (both French transcripts of some of Foucault's lectures and new material in translation) and cross-fertilisations with some of the more recent currents of anglophone human geography signalled how:

> Foucault not only has many of the answers, he still has many of the as-yet little-asked questions. To me, academic geography is not yet ready to move beyond Foucault; or, if we do move, then he needs to come along with us as a peculiarly interesting travelling companion. If we recognise that Foucault is still publishing new work, more so than many scholars still alive, then this assertion acquires even more force.
>
> (p. 496)

Foucault and other francophone philosophical-historical thinkers (such as Deleuze and Guattari) were interpreted in another book-length text on *Post-structuralist Geography*. In this, Murdoch (2006) reviewed the impacts and implications of poststructuralist theory for human geography, claiming that:

> In the wake of post-structuralism's incursion, geographers arguably investigate a broader range of socio-spatial phenomena than was the case previously and do so using innovative research methods. . . . Writing styles have also changed, with less attention now paid to the scientific rigour and rather more emphasis placed on the aesthetic and inventive character of geographical discourses and texts . . .
>
> (p. 1)

Although Murdoch considered both Foucault's approach to power and variety of 'relational' approaches to space, place and nature (building on Massey, 2005), his book also contained a chapter on 'Spaces of heterogeneous association' dedicated to new ways of under-standing networks (as 'actor-networks', comprising complex assemblages of human and non-human components) and referred to limits of what geographers can know with reference to an emerging 'non-representational theory'. Both of these have developed as significant strands of poststructuralist influence enlivening human geography.

Actor-network theory (ANT), affect and non-representational theories (NRTs)

Actor-network theory was first developed in (very broadly poststructuralist inspired) work on the histories and applications of science and technology (see Law and Hassard, 1999, for a summary). According to Murdoch (1998) it was:

> a useful way of thinking about how spatial relations come to be wrapped up into complex networks. Moreover, the theory is also believed to provide a means of navigating those dualisms, such as nature/society, action/structure and local/global, that have afflicted so much geographical work.
>
> (p. 357)

The study of networks has long engaged human geographers – as in spatial science and spatial analysis (see Chapter 4). 'Relational' spaces had also been explored earlier in humanistic and radical geographies (Chapters 5 and 6). Actor-network theory (ANT) however, interprets 'actors' (or actants as they are termed) as complex assemblages of humans and artefacts:

> What ANT [Actor-Network Theory] adds to the more commonplace understandings of relational spaces is a concern with networks. While the term network is commonly utilised in social science to describe technological relations, economic forms, political structures and social processes, ANT uses the term in a way which is quite distinct from such applications. Or rather, it might be argued that ANT bundles all these network applications together for it concerns itself with the heterogeneity of networks; that is, ANT seeks to a analyse how social and material processes (subjects, objects and relations) become seamlessly entwined within complex sets of association.
>
> (Murdoch, 1998, p. 359)

ANT was soon adopted by geographers interested in cyberspaces and more mundane technologies (Bingham, 1996; Hinchliffe, 1996). ANT also proved a fruitful means to explore human-nature interactions (Whatmore, 1999, 2002). Indeed, for Thrift (2000):

> all the usual boundaries from which and with which western knowledge is constituted – between humans and things, NATURE and CULTURE, tradition and MODERNITY, inside and outside – must be put aside. These divides have made it impossible to see the world for what it really is: a collection of heterogeneous activities which are constantly in formation.
>
> (p. 5)

Whatmore (2002) draws on ANT to investigate 'hybrid geographies', whereby humans are caught in networks with animals/nature and technologies. In a subsequent symposuim about this work, Whatmore (2005) notes that her aim was:

> To experiment with ways of writing more-than-human geographies that were as imaginative as they were materialist, at odds with the analytical terms set by a rivalry between political economy and cultural studies that prevailed during my formative years as a geographer.
>
> (p. 843)

Such oppositions – like others such as nature/human – have long haunted human geography. For this reason the 'more-than-human' geographies in Whatmore's work are celebrated by some readers. Philo (2005, p. 826), for example, celebrates how in Whatmore's analysis: 'everything just gets so entangled, so translated into everything else – the delicate networks of nature and culture, economics and politics, text and context, discourse, documents and devices'.

ANT was subsequently used in an increasing range of work, from scholarship on the planning of Sydney's urban fringe (Ruming, 2009), to the study of waste in Hungary (Gille, 2010), or the reconceptualisation of landscape (Allen, 2011). Reviewing the expansion of ANT-approaches, Bosco (2006) claimed that:

> as exemplified by the number of books, articles and conference presentations, the conceptualization of networks provided by ANT has surpassed other more traditional ideas and theories about networks that have been around in geography for more than four decades . . .
>
> (p. 140)

While Bosco judges that 'ANT is an excellent framework to describe the complex and mutable composition of networks of heterogeneous actors' (p. 143), it is not always seen in these terms. For one veteran Marxist geographer, earlier radicals (he cites Jim Blaut as an example: see Mathewson and Wisner, 2005, and Harvey, 2005) 'said more about space than a thousand actor network theorists ever could . . . ANT has to be the biggest fraud ever visited on social theory' (Peet, 2005, p. 166). ANT did rapidly become influential, however (see the review by Jóhannesson and Bærenholdt, 2009). Moreover, it has influenced other recent geographical departures, especially those using terms such as affect, performativity and non-representational theory (NRT).

Hence, soon after ANT became influential, other strands of broadly poststructuralist work in human geography emerged. One questioned the validity of notions of (nested or interacting) scales that had emerged from Marxist geography since the 1970s (see Chapter 6), urging geographers instead to attend to consider how they might develop a 'Human geography without scale' (Marston *et al.*, 2005). They argue that the literature on scale has been characterised by binaries (place–space, local–global) and by presuppositions that one side of the binary has more causal force than the other. As a consequence, certain processes/scales are relegated to the status of a mere 'local' case study:

> This is why, we believe, localities researchers more often looked 'up' to 'broader restructurings' than 'sideways' to those proximate or even distant localities from which those events arguably emerged . . . thereby eviscerating agency at one end of the hierarchy in favour of such terms as 'global capitalism' . . . 'larger scale forces' . . . while reserving for lower rungs examples meant to illustrate . . . these processes in terms of local outcomes and actions. . . . What is ignored in these associations is the everydayness of even the most privileged social actors who, though favourably anointed by class, race and gender, and while typically more efficacious in spatial reach, are no less situated than the workers they seek to command.
>
> (p. 421)

Poststructuralists also renewed debates about the difficulties, potentials and pitfalls of geographical representation and contexts (old themes, but given new twists by poststructuralism, in the 1980s and 1990s, as already noted). In revisiting these, Richard Smith (2003) drew on Baudrillard, Paul Harrison (2002) on Wittgenstein, and John Wylie (2005) on Merleau-Ponty; all stressed the (drawing/framing of) limits to (geographical) explanation/representation of practices. Summing up such departures, Dewsbury *et al.* (2002) provided the following indications:

in the performances that make us, the world comes about. This is about giving space to the event of the world, to make primary its emergent nature, and to the active role we too play in actualizing that which happens – we are thinking here ephemerally felt; the desire that lights up a room, the turning you didn't take (but which still haunts you), the anxiety of completing the next task. . . . These are not arcane concerns; they speak directly to our practices as social scientists, to the way our techniques are attentive to aspects of the world's unfolding. We are thinking of an expanded socio-logic, of mobilising other sources of expression (literature, art, performance), and above all of rearticulating what counts.

(p. 439)

Thrift and Dewsbury (2000) illustrated this via a musical score for a jazz performance (Figure 7.2), drawing attention to the 'dots' in the diagram: 'those concentrations, those moments of intensity, events if you like – that allow dead geographies to come alive as they are performed . . . through the full range of senses – an art of evocation' (p. 427).

In related terms, work on 'affect' raised both the social-spatial relations of emotion – and spaces of human thoughts/actions embodied via expressions, movement and visceral feelings (Anderson, 2006; Anderson and Harrison, 2006; Mohammad and Sidaway, 2012; Pile, 2010; Thien, 2005). This work evolved against the backdrop of what Kingsbury and Pile (2014, p. 6) describe as a growing 'breadth, depth and maturity of psychoanalytically inspired approaches to geography'. In McCormack's (2003, p. 488) terms, affect means 'the ways in which the world is emergent from a range of spatial processes whose power is not dependent upon their crossing a threshold of

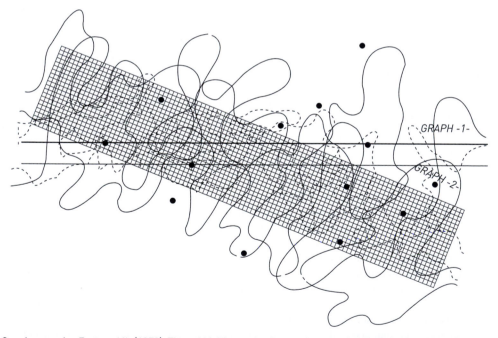

Cage's score for *Fontana Mix* (1958). The grid laid over the dots and serpentine lines guides the performer.

Figure 7.2 Cage's score for *Fontanta Mix* (1958)

Source: reprinted in Thrift and Dewsbury, 2000, p. 427.
Fonatana Mix by John Cage
© Copyright by Henmar Press, Inc., New York
Reproduced by kind permission of Peters Edition Limited, London.

contemplative cognition'. In other words, human geography must attend to social forces, human feelings, actions and movements that are not always explicitly thought through. Considering affect's interface with American geopolitics after '9/11', Ó Tuathail (2003) describes how:

> the United States is not the only place where the politics of affect loom large and the desire for revenge and symbolic empowerment abound. Across the globe, from Casablanca to Chechnya, Bali, Riyadh, and Tel Aviv, suicide bombers are blowing their bodies apart to become temporarily potent heroic martyrs, unleashing the affect within as deadly violence all around.
>
> (pp. 868–9)

To some extent, these literatures (ANT, perfomativity, affect) have coalesced around 'non-representational theories' (Thrift, 2008), yielding another abbreviated term: NRT. While Barnett (2008, p. 186) notes how NRT draws on 'work on the phenomenology of everyday urban life', hence suggesting that there might be links with the phenomenological approaches developed more than twenty years before (related to humanistic geography, see Chapter 5), NRT arguably owes more to the work of French 'poststructuralist' authors Gilles Deleuze and Félix Guattari (whose work has also entered geography through other routes, see Bonta, 2007) and a diverse range of work on performance and philosophy. However, the introductory chapter to a collection-taking stock of NRT and geography explains that:

> non-representational styles of thinking can by no means be characterised as anti-representation per se. Rather what pass for representations are apprehended as performative presentations, not reflections of some a priori order waiting to be unveiled, decoded, or revealed.
>
> (Anderson and Harrison, 2010, p. 19)

One widely cited account of the rise of NRT approaches prefers the term 'more than representational' (Lorimer, 2005), signifying all that exceeds and cannot be straightforwardly captured in geographical representations. However, a more recent 'interested sceptic' claims to:

> identify the heart of NRT through its insistence on the following: on the practical and processural fluidity of things (rather than the finished and fixed); on the production of meaning in action (rather than through pre-established systems and structures) . . . the possibilities of things emerging surprisingly (rather than being predetermined); on a wide definition of Life as humans/ with/plus (rather than humanistic) . . .
>
> (Cresswell, 2012, p. 97)

Nayak and Jeffrey (2011, p. 291) claim that NRT 'is still something of a cult pursuit rather than mainstream practice' while acknowledging how 'the fullness of experience can never be reduced to mere words and sparse representational codes' (p. 301). They interpret the onset of NRT in geography as, in part, a reaction to the focus on contests over meaning/representation that was central to the cultural turn as it developed in the 1980s and 1990s. In their view, much NRT is thereby 'consciously written against the representational modes familiar in earlier [work]' (p. 302). Situating NRT, however, Nayak and Jeffrey 'suggest that this work might do better to place itself within, rather than against, the "cultural turn"' (p. 302). For Barnett (2008), the recent growth of work on emotions, affect and NRT might be overstating the emotional/unconscious in an ironic reversal of earlier models in human geography that privileged the conscious and rational, whereas for Tolia-Kelly (2006) much of the consideration of affect in human geography is marked by ethnocentrism, mistaking white/Western experiences of these as universal and hence valid for everyone. Such critical reflection on 'race', eurocentricity and imperialism (past and present), have been advanced by postcolonial geographies.

Postcolonialism: decolonising human geography?

As Chapter 2 charted, geography has long had a close relationship to colonialism and racism. Jane Jacobs (2001, p. 11838) describes 'a special relationship . . . between geography and the making of empires . . . surveying, mapping and description were the frontier technologies', and Mayhew (2011b, p. 36) notes 'a genesis drenched in blood' for the 'set of willed appropriations we label "geography"'. Some radical geographers had drawn attention to this legacy in the late 1960s–1970s (see Chapter 6), part of the initial demands for a new, more critical geography. Since then, calls to 'decolonise' geography have multiplied.

These demands yielded more explicit awareness of the class, 'racial' and gendered backdrop to geographical work (also linked to the impacts of feminism, which will be considered in Chapter 8); an enhanced and politicised sense of the 'positionality' of the author. Thus, according to Jackson (1993a):

> Contemporary society is characterized by multiple forms of exploitation and oppression, which suggests that our politics should also be increasingly positional. . . . Those of us who wish to change the discipline and have ambitions to change the world should start from modest beginnings, recognizing our own positionality with respect to the fundamental inequalities of gender, 'race' and class. For if those dimensions are a source of power to us, they are as surely a source of oppression to those around us. If our geographical imagination is to develop in ways that are genuinely and constructively oppositional, we should begin by changing ourselves.
>
> (pp. 210–11)

Gregory (1994, p. ix) claims that he recognised this need when he moved from the University of Cambridge to the University of British Columbia (Vancouver, Canada) while writing *Geographical Imaginations*. He had to come to terms with what he calls his own 'situatedness', clarified to him by three features of his new home:

1   Canada had a double colonial legacy, linked to both England and France, plus close ties to the USA.
2   Canada is an 'avowedly multicultural society' and many of his students had cultural roots different from his own.
3   The University of British Columbia was taking issues of gender and sexuality much more seriously than he had been accustomed to.

As a consequence, he abandoned his first draft of the text and began again, although Guelke (1995) argued that:

> in the text there is scarcely a hint of Canadian culture or intellectual life. This work could have been written anywhere but Canada . . . Gregory has a desk in Canada, but he has ignored its artists and intellectuals as thoroughly as any British colonial official might have kept his focus on London in the heart of Africa. Canada for Gregory is an invisible country and its peoples have evidently produced little of importance for him.
>
> (p. 185)

However, Gregory claims to have been awakened to issues of positionality in North America, such as colonial status and equal rights, which altered his stance on the issues and literature that he addressed without requiring a particular 'Canadian position'.

For Gregory, a major source for geographers' appreciation of their own positionality and hence partial representation of the world (in every sense of the term partial) has been the writing of

Edward Said, especially his classic *Orientalism* (Said, 1978), which argued that 'Westerners' created an 'imaginative geography' of the Orient as part of the imperialist project. Orientals were presented by 'Westerners' as 'others', with different (implicitly if not explicitly inferior) cultures that contrasted with themselves (for examples and elaborations, see Kasbarian, 1996, and Taylor, 1993). The Orient did not exist as a straightforward pre-given object. It was imagined by the British and French in the nineteenth century (an activity promoted by the various geographical societies which transmitted this view of the Oriental 'other' to home audiences: see M. Bell *et al.*, 1995), for example, in part to promote its colonial dispossession and exploitation (Gregory, 1994, p. 171): 'Orientalism was an active process of othering, the exhibiting of "the" Oriental in a profoundly worldly set of texts which, in a quite fundamental sense, made colonization and dispossession possible' (see also Gregory, 1995b).

The 'othering' of the Orient (and the wider non-West) also provided its residents with a position from which they could later challenge the imperial and colonial hegemonies, not only politically but also culturally and intellectually. This includes, for example, the field of 'subaltern studies' whose goal is to enable the voice and recognise the experience of hitherto marginalised colonial subjects, thereby challenging the intellectual hegemony of 'Western' narratives. That 'Western' hegemony is based on partial views of the rest of the world, often promoted for political and other reasons (see Sharp, 1993). Emancipation and exchange requires that such partiality be countered. Gregory (1994) argued that:

> To assume that we are entitled to speak only of what we know by virtue of our own experience is not only to reinstate an empiricism: it is to institutionalize parochialism. Many of us have not been very good at listening to others and learning from them, but the present challenge is surely to find ways of comprehending those other worlds – including our relations with them and our responsibilities toward them – without being invasive, colonizing and violent. If we are to free ourselves from universalizing our own parochialisms, we need to learn how to reach beyond particularities, to speak to the larger questions without diminishing the significance of the places and the people to which they are accountable.
>
> (p. 205)

Thus, the call for postcolonial perspectives in geography that:

> highlight the importance of representing people and places across different cultures, traditions and contexts but also point to the difficulties of such endeavours. At the same time, postcolonial critiques stress the need to destablise what might be taken for granted and assumed in our own cultures, traditions and contexts. So for example, postcolonial studies challenge the production of knowledges that are exclusively western and ethnocentric by not only focusing on the world beyond 'the West' but also by destabilising what is understood and taken for granted about 'the West'.
>
> (Blunt and Wills, 2000, p. 168)

Others, such as Daniel Clayton (2001, 2002, 2011), Jonathan Crush (1994), Eric Pawson (1999), Jennifer Robinson (2003a, 2003b), Sarah Radcliffe (2005) and James D. Sidaway (2000a) similarly characterised the possibilities and difficulties of a postcolonial geography, from a variety of subdisciplinary vantage points. For example, Crush (1994) also specifies what the aims of a postcolonial geography might be:

> the unveiling of geographical complicity in colonial dominion over space; the character of geographical representation in colonial discourse, the de-linking of local geographical enterprise from metropolitan theory and its totalizing systems of representation; and the

recovery of those hidden spaces occupied, and invested with their own meaning, by the colonial underclass.

(pp. 336–7)

Blunt and McEwan (2002, p. 3) detect a possibility for geography here, in so far as 'an emphasis on the spatiality of postcolonial thought can help us move beyond the impasse of thinking [about postcolonialism] primarily in temporal terms'. While geography has traditionally been associated with Western knowledge and colonialism, they stress 'the diversity of postcolonial perspectives in analyzing space, power, identity and resistance' (p. 6). Moreover:

> Although postcolonialism might not have had much impact on the power imbalances between North and South, the diverse body of approaches identified as postcolonial are a significant advancement and offer a great deal to the possibilities of a meaningfully decolonized geography.

Daniel Clayton (2001, pp. 749–50) suggested the following departures for 'postcolonial-geographical lines of enquiry that (variously and among other things)':

1   seek to document and theorise multiple, contradictory and fragmented spaces of identity;
2   work with postcolonial theory in grounded ways;
3   be attuned to the discordant postcolonial politics of places studied;
4   acknowledge that all studies of colonialism are 'situated knowledges' that stem from and feed into particular sites of study, learning and memory;
5   work at a variety of scales and seek to build 'middle range' narratives;
6   treat colonial contact as a two-way and contradictory process;
7   take archives and the past seriously, but use them to question or transform understandings of the present; and
8   have an anti-colonial effect.

However, for Clayton, this list:

> does not amount to some sort of postcolonial template for geography. Rather, we have here the rudiments of a postcolonial-geographical way of thinking, and one that I think needs to be explored and debated in much greater depth, not least because there is considerable disagreement about the appropriate aims and methods of postcolonial studies and the status of colonialism as a spatial problematic.

(p. 750)

Jazeel (2014, p. 101) wants 'a postcolonial geographical research imagination to more effectively, and far less imperially, engage radical alterity'. Offering five pathways for postcolonial geography, Sidaway et al. (2014) suggest:

1   Narrating the planetary: instead of references to the global, with assumptions of universality, they focus on 'grounded genealogies of the uneven co-production of categories, sites and landscapes' (p. 6).
2   Acknowledging other postcolonialisms: they suggest the value of work in postcolonial theology for example, as well the importance of postcolonial scholarship from outside the Anglo-American mainstreams and in other languages.
3   Planetary indigeneity: calling for further thoughtful engagement with categories, movements and epistemologies associated with indigenous peoples.

4   Seeing like an empire: examining the presence and reproduction of imperial ways of thought and action not only in the West but also with reference to other powers, such as China's narratives about its place in the world system.

5   Problematising translations: asking how the translation of geographical terms and categories invariably involve intersections 'on imperial ground' (p. 16).

The arrival of an undergraduate textbook on *Geographies of Postcolonialism* (Sharp, 2009) might signify that postcolonial issues had already become part of anglophone human geography's mainstream (Sharp examines Western representations of the 'Other' and how these shaped European self-images, the continuing legacies of colonialism and postcolonial theory). But for Joel Wainwright (2013), the critical moves required to decolonise geography have scarcely begun, and risk being outmanouevred by renewed links between applied geographical technologies (such as GIS, remote sensing and surveillance of geodata) and the American military. Moreover, writing from two territories (Canada and Ireland) that were once parts of the British Empire, Gilmartin and Berg (2007, p. 120) argued that 'much of what passes for postcolonial theory in British geography reinforces new forms of colonial epistemologies and colonial hierarchies, while destabilising their older forms'. In particular, they complain that not only do geographers in imperial 'core' countries continue to dominate theory production in human geography (including postcolonial theory) at the expense of those in the formerly colonised 'peripheries', but that:

> postcolonial geography is dominated by critiques of the discursive construction of historical colonialism, rather than focusing on the aftermath or continuation of colonialism. This effective hijacking of the postcolonial within geography means that it loses theoretical force, becomes one of a long list of 'posts' that often alienate geographers who might like to, or already do, engage with the challenges posed by postcolonialism to the ways in which we construct knowledge.
>
> (p. 123)

In developing this argument, they draw on Derek Gregory's (2004) book on *The Colonial Present*. In this Gregory returns to Said's (1978) account of Orientalism, since 'Orientalism is abroad again, revivified and hideously emboldened' (p. 18). He argues that this 'revivification' became most evident via the 'war on terror' after 9/11, culminating in Western occupations of Afghanistan and Iraq as well as the ongoing colonisation of Palestine through Israeli settlement. However, according to Gregory:

> – for us to cease turning on the treadmill of the colonial present – it will be necessary to explore other spatializations and other topologies, and to turn our imaginative geographies into geographical imaginations that can enlarge and enhance our sense of the world and enable us to situate ourselves within it with care, concern, and humility.
>
> (p. 262)

'Race', racisms and ethnocentrism

Jacobs and Jackson (1996) argued that postcolonialism demands sustained argument with geographies of 'race' and racism. This raises both the complicity of the discipline in the racist categorisations of colonialism (and all the attendant debates about environmental determinism considered in Chapter 2) as well as the contemporary whiteness of the discipline (see Bonnett, 1997; McGuiness, 1999) and therefore, in Peake and Kobayashi's (2002, p. 50) terms: 'the various (and often hidden) racist practices and discourses that permeate the epistemological foundations of geography and the institutional structures and practises that shape our work and environment'. For them, an anti-racist geography embraces:

1   research on the geographies of racism;
2   extending anti-racist principles throughout the 'institutional practices' of the discipline, particularly in the classroom;
3   increasing the participation and presence of non-white geographers in the discipline; and
4   linking all of these to activism beyond (but building on insights offered by) the discipline, academic life and the university.

For these geographers (and others contributing to a 2002, special issue of *The Professional Geographer* on 'Race, racism and geography') the key would be a less white discipline as well as a discipline that is more self-critically *aware* of its historic whiteness (and anti-racist in practice: Kobayashi and Peake, 2000; Nayak and Jeffrey, 2011, pp. 192–8; Mahtani, 2006; Sanders, 2006). Echoing the ways that the increased presence, power and visibility of women in the discipline since the 1970s has played a role in the making of feminist geographies (see Chapter 8), Pulido (2002) argues that:

> We need sufficient scholars to generate an intellectual synergy around race. A similar process can be seen in feminist geography. Only after women gained access to academia did the study of gender flourish. Men could have taken the lead, but they did not. Because patriarchy was a problem for female geographers, they studied it seriously.
>
> (pp. 45–6)

She insisted, therefore, that while white geographers have produced incisive and critical work on race and that there is no inherent reason why 'scholars of color are more likely to have more penetrating insights and analyses' (p. 45), 'more people of color could create a "critical mass", which currently does not exist'. Thus, a call to change the culture of a discipline that institutionally 'is nearly as white an enterprise as country and Western music, professional golf, or the supreme court of the United States' (Delaney, 2002, p. 12).

And whiteness itself should be seen as a marker of difference, rather than the often taken-for-granted norm against which supposed minorities (in global terms, arguably majorities) are defined. Thus, Bonnett and Nayak (2003) argued that:

> it is only by understanding such normative terms as 'white' and 'western'– the ones against which others are defined as exotic – that wider systems of racial privilege can be bought into view. By making it clear that categories such as whiteness are also the products of racialization, that they too have a history and a geography and, hence, are changeable, we can help transform the critique of race and ethnicity from a 'subfield' into an essential theme running throughout a rigorous geographical education.
>
> (p. 309)

Similarly, the ethnocentrism of geography curricula in British schools (Winter, 1997) had attracted fresh attention – and historical scrutiny (Ploszajska, 1999). More pertinent to much of the other material being considered in this chapter, however, have been critiques of the ethnographic assumptions of much of the research on affect, which tend to assume they are universally valid, across space and time. Challenging this leads Tolia-Kelly (2006, p. 216) to demand 'distinctly anti-racist theorizations of emotional economies of "love" and "hate"' along with others examining the 'racialization' of affect (Crang and Tolia-Kelly, 2010; Nayak, 2010; Saldanha, 2010; Swanton, 2010).

Almost a decade on from the special issue of *The Professional Geographer* on 'Race, racism and geography' that had foregrounded the issues, two critical reflexive papers returned directly to the problematics of race and power in the discipline. In one of these papers, Patricia Price (2010,

p. 166) asks: 'what does it mean to be anti-racist if there is no outside to racialized geographies? How does one go about really *doing* anti-racist work in a wholly racialized world?' For Price, it is both troubling and easy to document 'the remarkably persistent whiteness of geography's practitioners' (p. 156) such that the growing popularity of work on whiteness in the discipline might have become 'a comfort zone rather than a space of truly critical engagement with racism' that 'may in fact not simply reflect but also unwittingly act to reinforce white dominance in geography'.

Also reiterating the point about whiteness of geography (writing from a location in western Canada), Berg (2012, p. 509) remarks how often white geographers 'fail to recognize (let alone contest) the actually existing forms of white supremacy that operate in contemporary colonial settler and imperial societies in places like Europe, North and South America and Australasia'. For Berg, this is about more than white hegemony; it is part and parcel of white *supremacy*. Those white folk in these societies who disavow racism or seek to disaffiliate themselves from racism also need to recognise that they are 'benefitting in very real material ways from ongoing white supremacy' and that 'All geographers are invited into white supremacy . . . but not under the same conditions nor with the same penalties for refusal to enter' (p. 515).

However, charting the shifts from mapping 'race' and ethnic segregation towards geographies of the social construction of such categories, Claire Dwyer and Caroline Bressey (2008, p. 7) described how moves 'to engage more critically with categories of identity and place testify to the depth and dynamism of current work on geographies of race and racism'. Meanwhile, 'ideas about "race(s)" (and their relationship to territorially bounded nations) remain among the most powerful sources of human identity and division within the contemporary world', in the terms of Jackson and Penrose (1993, p. 1) written early on in the growth of such work. But although they are often taken for granted, they are social constructions, 'the product of specific historical and geographical forces, rather than biologically given ideas whose meaning is dictated by nature'. Thus, those who speak from a 'racial' or national position are doing so within categories defined by imperial and colonial histories and geographies – in many cases (notably with regard to racism) from positions of (often unacknowledged) privilege or, in Berg's (2012, p. 8) view, 'from ongoing white supremacy'. It is still frequently the case that 'resistance is often couched in terms which do not challenge such dominant modes of representation' (Penrose and Jackson, 1993, p. 203). Only when the social construction of identity is recognised does a politics of identity emerge, whereby individuals and groups 'attach significance to certain dimensions and contest the relevance of other designations'.

The politics of identity, like all politics, is a struggle for power and, as Penrose and Jackson (1993) argued, 'takes place within a hegemonic system of social relations':

> at any given place and time positions of hegemony are being employed to exercise and preserve power [which] . . . includes the capacity to set the parameters for negotiation within any given society. It also gives the freedom to define 'difference' and to enforce this vision through hegemonic institutions of government, law and education. In a system where 'sameness' and conformity are rewarded, the power to define 'difference' becomes the power to disadvantage and disempower.
>
> (p. 207)

These and other complex geographies of power, domination and resistance (Pile and Keith, 1997), or 'entanglements of power' as one collection (Sharp *et al.*, 2000) termed them, became rich domains of geographical enquiry since the 1980s. Moreover, the practice of geography, the circulation of its texts and journals, and the structure and sociology of the academic discipline came to be interpreted:

> as much a product of specific social, cultural and geographical constructions as the taken-for-granted geographies they [geographers] wish to contest and transform and this affects their analyses (and their disagreements with each other). Such geographies are always constructed within relations of power in which some spaces (and ways of understanding them) are effaced by dominant and hegemonic geographies.
>
> (Berg, 2003, p. 308)

Related concerns, of course, were also evident in the debates initiated by radical geography, as Chapter 6 documented. And we will return to them in Chapters 8, 9 and 10.

Conclusions

Postmodernism, poststructuralism, postcolonialism and the cultural turn intersected human geography in complex and intertwined ways and have been subject to a wide variety of definitions, understandings and interpretations. However, Dixon and Jones (2005) noted how, during the 1990s:

> work that labelled itself as 'post-structuralist' or 'post-modern', as well as a host of concepts and methods under the heading of 'social theory' and 'literary theory', were increasingly deployed as a means of critiquing the ontological presumptions and claims to scientific rigour of what were then considered to be the dominant 'paradigms' within the discipline, namely spatial science, critical realist/Marxist, and humanist geographies.
>
> (p. 243)

Some geographers voiced concern over what these would mean for the identity and social relevance of the discipline or drew attention to the degree of fashionability associated with geography's 'cultural turn' (Barnett and Low, 1996; Barnett, 1998a, 1998b; Castree, 1999; Mohan, 1994), and others condemned the 'excesses' of the cultural turn (and associated encounters) for its neglect of urgent social questions about inequality and social justice or for its lack of application (e.g. Sayer, 2000) and its 'disciplinary idealism' (Peet, 1996b). While personally sympathetic to many of its departures, McDowell (2000) also noted that there was a considerable unease and angry reaction from some regarding the 'cultural turn'. Thus, Hamnett (2003) was concerned:

> that the rise of a 'post-modern' human geography, with its stress on textuality and texts, deconstruction, critique, 'reading' and interpretation, has led human geography into a theoretical playground where its practitioners stimulate or entertain themselves and a handful of readers, but have in the process become increasingly detached from contemporary social issues and concerns. The risk is that much of human geography will cease to be taken seriously in the world beyond the narrow confines of academe. It will be seen simply as a corner of the post-modern theoretical playground, possibly entertaining to study for a while, but something which can be safely ignored while the grownups get on with the business of changing the world, often for the worse.
>
> (p. 1)

Others, notably Don Mitchell (1995), criticised what they see as the reification of culture in geography's putative cultural turn: 'Like "race", "culture" in itself possesses no explanatory value. . . . Our goal, therefore, should be one of figuring out how the idea of culture becomes socially solidified as a thing, ream, attribute or domain' (p. 113).

Some others might sympathise with Mikesell's (1999, p. 441) claim that 'Most of what has been written by self-styled radical, new, or social-cum-cultural geographers is superficial or worse'; similar sentiments are voiced by Gould (1999). Yet many welcomed and took pride in a human geography which is more engaged with (and in some ways contributing to) wider social theory. Gregory (1994, p. 4) thus claimed that:

> Many of those working in other humanities and social sciences have also become interested in questions of place, space and landscape. . . . Their studies have multiple origins, but they treat the production of social space, of human spatiality, in new and immensely productive ways.

Gregory went on to state – reflecting a human geography more deeply informed by and situated within a wider social theory, revealing both a sense of optimism and of limits about what human geography has achieved – that:

> The production of space is not an incidental by-product of social life but a moment intrinsic to its conduct and constitution, and for geography to make a difference – politically and intellectually – it must be attentive to difference.
>
> (p. 414)

In complementary terms, Amin and Thrift (2000) argued that economic geography was currently 'narrowly specialised' (p. 4) and drew on only a thin range of work within the contemporary discipline of economics – what they saw as its 'certain type of rigour; its emphasis on well-defined models of limited domains' (p. 5). Geographers should look beyond those traditional areas of economic scholarship to the 'new areas of economic study that are currently both flourishing and provided a genuine ground for the kind of contributions we can make' (p. 8). This stimulated a number of responses, several of which – like Ron Martin and Peter Sunley's (2001) – warned against 'throwing out the baby with the bathwater' by abandoning contacts with 'traditional' economic geography: one of the headings in their response – 'Beware of vague theory and thin empirics' – was repeated by others in a variety of guises. Rodríguez-Pose (2001) warned of the possibility of 'killing economic geography' with a 'cultural turn overdose', for example, and Plummer and Sheppard (2001) asked 'Must emancipatory economic geography be qualitative?'. Elsewhere, Plummer (2001, p. 761) argued that increasingly geographers' 'visions of the economic landscape are being constructed on the basis of, at best, vaguely formulated and strictly imposed theoretical constructs' which are only evaluated by either 'casual empiricism or self-selected case studies' – leading him to align himself with Markusen's (1999) query – 'How do I know it when I see it?'.

We have perhaps not been able to do proper justice in this chapter to all aspects of the developments of the last few decades, for we are also mindful of Chris Philo's (2000) point that:

> It is important, I feel, to resist the impression of sitting in some kind of satellite circling human geography's cultural turn, claiming the 'scopic power' to see clearly all that is taking place but which others closer by cannot themselves comprehend.
>
> (p. 27)

Nearly a decade later, Philo (2009, p. 466) criticised the 'conflation' of the cultural turn, postmodernism and poststructuralism in textbooks. For example, of the sixth (2004) edition of *Geography and Geographers'* attempt to negotiate these in a single chapter (as in this seventh edition), Philo argues that:

> It would surely be preferable *not* to run together these theoretical stances, but rather to inspect them carefully, critically, but sympathetically on their own (really quite different and distinct) terms, and even more importantly they should not be homogenized as somehow constitutive, indicative, or derivative of a more overarching cultural turn. To do so risks losing conceptual

precision, and thereby evacuating the cultural turn of whatever focus, specificity, and coherence *can* be detected . . .

(p. 446)

However, part of the cultural turn (and perhaps postmodernism and poststructuralism) was the reconfiguration of many subdisciplinary boundaries (epitomised in the scope and heterodoxy of edited 'companions' to and handbooks for economic, cultural and political geography; Anderson *et al.*, 2003; Agnew *et al.*, 2003; Sheppard and Barnes, 2000) as well as some inter- and transdisciplinary trends. Hubbard *et al.* (2004, p. 61–2) provided a useful précis of these intricate turns – and their limits:

> the importance of culture has been seized upon widely across the discipline, with social, political and economic geographers seeking to undertake culturally sensitive analyses [they cite Lee and Wills, 1997, as an example of such engagements by economic geographers for instance] . . . while the combination of the 'cultural turn' in geography and a wider 'spatial turn' across the social sciences and humanities might create the semblance of a more fluid, post-disciplinary landscape, disciplinary boundaries do continue to exist as administrative units within institutions and within the popular (academic) imagination.

Similar thoughts led Mike Crang and Nigel Thrift (2000) to reflect that:

> Within the discipline there has been a burgeoning interest in social thought that has both extended and pluralized the influences drawn upon by geographers. Beyond the discipline social thought appeared to be increasingly smitten with a geographical idiom of margins, spaces and borders. However, this spatial turn was not a cause for disciplinary triumphalism that others were turning to geography since much of it seemed resolutely ignorant of geography and geographers as a discipline. Indeed, it seemed at various times to show both deliberate ignorance of geography while – lest anyone might become chauvinistic or proprietary over the claims of the discipline – also displaying how limited much geographical thought had been.

(p. xi)

Indeed, their book illustrated the wide range of philosophical and social-theoretical sources on which geographers and others came to draw in their considerations of space. They identified six 'spaces' within this corpus of work: spaces of language; spaces of the self and other; metonymic spaces; agitated spaces; spaces of experience; and spaces of writing. Furthermore, 'exotic new hybrids continue to be produced' (pp. 24–5), and the interactions among geographers, social scientists and scholars in the humanities over issues of space and spatiality are likely to increase (see Merriman *et al.*, 2012), even though many contemporary treatments by the last group are undertaken in apparent ignorance of geographical interest in such topics (as in Miller and Hashmi, 2001). Some have since claimed that poststructuralism has been *distinctive* in demanding:

> rigorous interrogation of those core concepts – such as objectivity and subjectivity, center and margin, materialism and idealism, truth and fiction – that underpin much of modern-day academia, including the majority of geographic thought and practice.

(Woodward *et al.*, 2009, p. 396)

Yet these more recent developments within human geography *share* something with those charted earlier in this book. They were all based on dissatisfaction with what has been, and is being, done in the name of the discipline, and a wish to change it, at least by adding new perspectives if not by also expunging some of their predecessors.

Chapter 8

Feminist geography

Geography, could you be my world?

. . .

By looking within and without, upside down and inside out,
Come alive geography, come alive!

(Clare Madge, 1997, pp. 32–3)

In 1984, when feminist geography was just getting going, and when the field was characterised by a relatively small number of individuals, the WGSG [Women and Geography Study Group, established in 1980 within the professional Institute of British Geographers] produced the text *Geography and Gender*. Aimed at first year undergraduates, this ground-breaking book was also important for the collective way in which it was written. Rather than being written by one or two named individuals, the book was written by several people working and writing together, as 'a collective', and has as its 'author' the WGSG. Taking their cue from wider feminist politics, the feminist geographers who wrote *Geography and Gender* wished both to challenge the accepted conventions of academic writing (conventions which celebrate and reward the individual, apparently working in glorious isolation) and to acknowledge the genuinely collaborative and supportive ways in which feminist geography was emerging in Britain.

(Gillian Rose *et al.*, 1997, pp. 1–2)

To the pioneers of the 1970s, a journal devoted to feminist geography must have seemed a distant dream. Although the women's liberation movement motivated many to take action in the streets, few dared to move that action into the halls of the academy. Yet feminist voices persisted and slowly gained momentum. And from these modest yet courageous beginnings, feminist geography has now permeated most, if not all, corners of the discipline and has become a force that cannot be ignored in accounts of contemporary geographical knowledge. Our hope is that *Gender, Place and Culture* – a journal devoted to feminist geography – will serve to celebrate and consolidate that presence.

(Liz Bondi and Mona Domosh, 1994, p. 3)

Presumably every history of geography textbook written today and each seminar taught on this subject must contend with the history of feminist geographies, of which the founding of the journal *Gender, Place and Culture* comprises a significant event.

(Mona Domosh and Liz Bondi, 2014, p. 1068)

Since the 1970s, overlapping the flowering of humanistic geography, the growth of radical geography and subsequent postcolonial/modern/structuralist geographies, the issue of gender has been increasingly foregrounded. Much of this has been led by critical encounters between geography and feminism; raising issues about the ways that space, place and landscape are experienced differently by men and women and exploring the power relations related to these variations. Sexuality and space also become foci of investigation and debate. Thus, human geographers began to study how and *where* people experience themselves as sexual beings, in particular their relationship to categories (such as 'lesbian' or 'heterosexual') commonly attributed to different expressions of sexuality. The advent of geographical research on gender and sexuality reflects wider social changes accompanying the women's movement in new waves of feminism, the birth of the 'women's' and subsequently 'gay' liberation movements and changing social and sexual mores in Western society. The increased participation of women in the paid labour force (albeit often receiving lower salaries than men, notwithstanding equal opportunities legislation that makes formal discrimination illegal) also produced new gendered divisions of labour. The sociospatial roles of men and women shifted. These, and the attendant struggles, also shaped what human geographers researched as the gender division of labour in the discipline itself provided considerable initial impetus (and has remained a key issue) for feminist geography.

Although for many years and in most places at least half of geography undergraduate students have been women, more of the research publication and much of the teaching (both human and physical geography) have tended to be done by men. This is still true in a number of university geography departments. Long taken for granted, this male dominance (and the wider 'masculinity' of geographical scholarship) became a theme of extensive debate and focus of demands for progressive transformation since the 1970s. For example, in the UK of the early 1980s – a period when many of the changes associated with the radical and humanistic critiques of positivism made for a lively human geography – only 25 per cent of graduate students in geography (the data were for both physical and human geography) were women (Johnston and Brack, 1983). Johnston and Brack went on to quantify some other facts and trends relating to the relative status of women in the discipline:

> of the 1106 [UK-based] academic geographers identified in this survey, only 126 (11.4 per cent) were female. Furthermore, the percentage of females obtaining posts has declined. In the 1930s, 16.8 per cent of geographers obtaining their first posts (plus those in post in 1933) were female. In the 1940s, the figure was 30.4 per cent, but it fell from then on, to 11.6 per cent in the 1950s, 8.5 per cent in the 1960s and 7.9 per cent in the 1970s. . . . Of the 142 geographers who have held readerships [between 1933 and 1983] only 6.3 per cent were females (the chances of a male becoming reader were 1 in 7.4, whereas for a female they were 1 in 14). Of the 142 who have occupied chairs, only 6 (4.3 per cent) have been women.
>
> (p. 110)

Even in those few departments where larger numbers of women entered postgraduate training and research in geography, they frequently found limited prospects for employment as academic geographers. Monk (1998), claims that Clark University in Massachusetts – where in the 1920s Ellen Churchill Semple (see p. 44, this book) spent the later part of her influential career – was an 'original hearthland' of women geographers from the 1920s to the 1970s), but nevertheless:

> Even though Clark was 'welcoming', when the women graduated they found restricted opportunities. Occupational segregation was quite evident. Realistically, women geographers, even those with doctorates, could aspire to positions in teacher education institutions, the few women's colleges, high schools, or outside academia. . . . University positions were almost exclusively open to men.
>
> (p. 28)

In many departments there remains a numerical prevalence of men, particularly in more senior positions. Introducing a focus section of *The Professional Geographer* on 'Women in geography in the 21st century', Karen Falconer Al-Hindi (2000, p. 701) notes how:

Many women geographers are the only one, or perhaps one of two, in their geography departments. Some are even the first woman ever hired by their department. I have long felt indebted
to the two women geographers who preceded me at my institution. Although neither was here
more than a couple of years, each had a significant impact on the department and helped make
it a place where I could remain and achieve tenure. While I cannot repay my debt to these women
directly, I can – as can each of us – through strategic participation in the structures of the
academy help to make it more welcoming (and someday perhaps even equally open) to the wide
variety of women and men who are geographers.

(p. 710)

Feminist geography has become a rich field. In addition to frequent papers in all the main
disciplinary journals and the prominent, specialist, journal *Gender Space and Culture*, there are many
collections and texts. For example, a 1990s text on *Methods and Techniques in Human Geography* (Robinson,
1998) contains a chapter on 'Feminist geographies' among others on 'multivariate analysis',
'generalised linear models' and 'categorical data analysis'. Other collections such as *A Feminist Glossary
of Human Geography* (McDowell and Sharp, 1999), *Space, Gender and Knowledge: Feminist readings* (McDowell
and Sharp, 1997), *Full Circles: Geographies of women over the life course* (Katz and Monk, 1993), *Thresholds in
Feminist geography* (Jones et al., 1997), and *A Companion to Feminist Geography* (Nelson and Seager, 2005)
bear witness to the wealth of feminist work in geography. One of the more recent of these
collections, *Feminisms in Geography: Rethinking space, place and knowledges* (Moss and Falconer Al-Hindi,
2008), was keen to stress the range and diversity of scholarship in feminist geography, while noting
that 'Along its path to the present, one of feminist geography's arcs has been an active political
commitment, a praxis' (p. 4). Specialist feminist geography research groups and organisations also
exist in many countries and at the international level (Morin, 2009). Most new textbooks (such as
the ones referenced in the Preface here) have chapters on feminist geography and all the companions
and encyclopedias that we also mention there contain extensive references to the field. At the same
time, references to 'man and the environment' that were taken for granted as somehow including
the half of humanity who are female are much less frequent. Papers now seldom say 'he' when
referring to an unnamed geographer, but usually write 'he or she' or sometimes 's/he'. Many
journals require submitted papers to be written in non-sexist language. It was not always so.

The other half

The first paper raising the issue of the relative status of women in the discipline appeared in 1973
(Zelinksky 1973a) – and was penned by a man – although around this time the first papers on
feminist geography appeared in *Antipode* (Burnett, 1973; Hayford, 1974). Subsequent papers tended
to focus on the facts and consequences of the exclusion and isolation both of women geographers
and of the distinctive geographies of women (for example, differing experiences and uses of urban,
paid work and 'home' spaces). The titles of subsequent papers that used terms such as 'sex
discrimination in geography' (Berman, 1974), 'the other half', 'geographical study of women'
(Tivers, 1978), and 'on not excluding half of the human in human geography' (Monk and Hanson,
1982) reveal the concern to produce a more inclusive human geography. Looking back at this
formative period, Bowlby and Tivers (2009, p. 59) recognise the earlier presence of some women
as geography academics, many of whom 'found important roles within the geographical community,
and did not question the male bias of its subject content'. In the 1970s, however, a new generation
of women entered geography who had been influenced by a fresh wave of feminism that reflected
wider political demands and social shifts since the 1960s. Bowlby and Tivers (2009) argue, however:

that the development of early feminist geography was not the heroic creation of a few pioneers.
Rather, in many places during the 1970s, women and some men in geography began to explore

the geographies of women, and in time, this lead to the examination of a range of explanations for the subordination of women and to the development of more sophisticated feminist geographies.

(p. 63)

Looking back and celebrating the 'amplification in the range of questions asked – and change in the nature of some of the questions asked' by geographers, Susan Hanson (2004, p. 719), pointed out that:

Certain questions simply were not on geography's agenda until women began raising them. Examples are:

- the geography of everyday life;
- the links between unpaid work of caring and work in the paid labor force;
- the impacts of international monetary policy on the lives of women and children;
- the relationship of international migration to child care, domestic work, and the sex trade; and
- women's role in changing the face of the earth.

Changes in who 'we' are has not just enlarged the range of questions asked; it has also helped to expand the approaches we use in collecting and analyzing data and to alter the nature of the theories that guide our views of the world.

Early feminist analyses in human geography stressed that the different experiences that structure women's views of the world are created in contexts that have a clear power gradient between the oppressors (men) and the oppressed (women): the collective experiences of men and women are distinctive and unequal. Thus, the feminist project is necessarily a political one, seeking to remove the power gradient through emancipatory and other processes. Just as Marx argued that his project involved the emancipation of the working classes, providing them with the means to control their own destinies, so the feminist project is the basis for 'identity politics' –'an emancipatory politics of opposition . . . resisting and challenging the fraudulent claims of dominant groups' (Bondi, 1993, pp. 86–7). This calls for 'consciousness raising' and transformation, out of which identities can be constructed and reconstructed (in places) as bases for action. While it came late to geography, feminism has a long history as an intellectual and political movement and there are diverse strands of feminist thought and politics. Nayak and Jeffrey (2011) hence preface the chapter on 'Feminist geographies' in their textbook on *Geographical Thought* with an account of the longer intellectual (and political) contexts in which feminism emerged. These comprise a series of 'waves': the first was a late nineteenth- and early twentieth-century struggle for basic rights of citizenship – to vote; the second a wider struggle for equality and against stereotypes and discrimination in the 1960s and 1970s; and the third, more recent wave, focused on diversity, identity and difference. The impact on academic geography came in the latter part of this second wave and into the transition to the third. According to Nayak and Jeffrey (2011):

Second wave feminism constituted both a political movement aimed at countering women's discrimination in a patriarchal society and an intellectual movement directed towards the gendered ways in which knowledge is produced. This combination of political and intellectual objectives is reflected in the emergence of a feminist geography as a distinct field of the discipline.

(p. 136)

By the 1980s, therefore, an engaged and diverse feminist geography had emerged and was establishing itself as a vibrant domain of research and teaching. According to Linda McDowell

(1986a, p. 151), such a feminist geography project 'emphasizes questions of gender inequality and the oppression of women in virtually all spheres of life', and its goal includes uncovering and countering such inequality and discrimination within the geographical profession itself (on which, see that early essay by Zelinsky, 1973b, plus Zelinsky *et al.*, 1982; Jackson *et al.*, 1988; Johnson, 1989). That is part of a much larger task (McDowell, 1989):

> to demonstrate that women do matter in geography, and to argue that the failure to take gender differences into account impoverishes both geographical teaching and scholarship ... [in addition] it is not enough just to add women in as an additional category. Feminist geography, as opposed to a geography or geographies of women, entails a new look at our discipline. It poses several awkward questions about how we currently divide the subject matter into convenient academic parcels and also challenges current practice in teaching and research.
>
> (p. 137)

However, McDowell (1993b) also noted how:

> As in feminist scholarship more generally, that part of our discipline subsumed under the heading 'feminist geography' has become such a diverse and pluralistic enterprise that it is increasingly inaccurate to define feminism as a single perspective.
>
> (p. 305)

Refining this a few years later, McDowell (1999) identified 'the specific aim' of feminist geography as:

> to investigate, make visible and challenge relations between gender divisions and spatial divisions, to uncover their mutual constitution and problematise their apparent naturalness. Thus the purpose is to examine the extent to which women and men experience spaces and places differently and to show how these differences themselves are part of the social construction of gender as well as that of place.
>
> (p. 91)

Citing her work too, Nayak and Jeffrey (2011) note how:

> One of the initial tasks for feminist geographers seems, at first sight, relatively straightforward: to make women, and issues relating to women, visible within the discipline ... this was a critical moment ... staking a claim within a discipline that had either ignored women or constructed them as the 'other'.
>
> (p. 137)

While these concerns remained important, others soon supplemented and sometimes displaced them. Nayak and Jeffrey (2011, p. 145) judge that this was 'under the philosophical turn of post-modernism'. Certainly, postmodern feminism, along with other encounters, theoretical departures and politics, has produced diversity in feminist geography.

Diverse strands

In the 1990s new encounters between feminism and geography soon led to broader critiques and concerns. Documenting these for the fourth edition of *The Dictionary of Human Geography*, Geraldine Pratt (2000, p. 261) thus noted how 'Since the late 1980s, many feminist geographers have moved away from an exclusive focus on gender and class systems. This new phase can be identified as feminist geographies of difference.' She identified three characteristics of this new phase:

1   The contestation of gender categories of man and woman. Feminist geography has refused to
     take such categories for granted, instead exploring the ways that humans learn to perform the
     accustomed roles of being a man or a woman and how these vary across time and space. Such
     concern has led feminist geography to be 'increasingly attentive to the differences in the
     construction of gender relations across races, ethnicities, ages, religions, sexualities [the
     differing experiences and meanings of gay, lesbian bisexual and heterosexual roles for
     example], and nationalities, and to exploitative relations among women who are positioned in
     varying ways along these multiple axes of difference' (p. 261). In other words, the commonality
     of women's position and experiences – something that was often taken for granted in earlier
     renditions of feminist geography – is replaced with a more nuanced understanding of the
     diversity of what it means to be a man or a woman in different places, among different classes
     and ethnicities, and at different times.

2   The resort to a broader array of social and (reflecting 'the cultural turn') cultural theory. This
     includes both poststructuralist thought and psychoanalysis, seeking a deeper appreciation of
     how gender relations and identities (including sexuality) are formed and shaped. While this
     has produced a flow of works (and some attention to feminist geography in wider feminist
     studies), another consequence of these developments is that 'theoretical differences
     among feminist geographers are more obvious than in the past' (p. 261). In other words, it is
     perhaps no longer appropriate (if it ever was) to speak of a singular feminist geography.
     Rather, feminist geographies have come to embody the wider debates and polemics across the
     more or less Marxist, poststructuralist and postcolonial positions that earlier chapters have
     examined. So, while feminist geography has often had to struggle to establish itself against
     sexist and sometimes misogynist dismissal on the part of other geographers, ideological
     differences among feminist geographers frequently surfaced (see Valentine, 1998; Johnson,
     1994; McDowell, 1990; and Peake, 1994, on different 'personal' aspects of all this), indicating
     continuing debate.

3   A move towards the concept of 'situated (feminist) knowledges'. This has meant exploring the
     basis of different geographical claims and the diversity of vantage points from which they are
     made. While this shares much with a wider postmodernism and poststructuralism in geography
     and social sciences, in feminist geographies, it has also led to a stress on what is a stake in
     creating feminist alliances across differences of class and 'race' (Gibson-Graham, 1994; Jacobs,
     1994b).

Pratt illustrated these (and prior) developments through tabulating three strands of feminist
geography that have evolved very broadly in sequence. She described the first phase as a concern
to bring the effects of gender inequality into geography – hence, 'the geography of women'.
A second strand, influenced by Marxism, sought to theorise these inequalities in terms of their
relationships to the capitalist economy – hence, 'socialist feminist geography'. Finally, there
had been a wider stress on construction of gendered differences, both of men and women,
straight, gay and bisexual, from different places, classes and 'races', hence 'feminist geographies of
difference'. In the equivalent table published nine years later in a subsequent edition of The Dictionary
of Human Geography (see Figure 8.1) Pratt (2009, p. 247) added a fourth strand that 'could be called
"transversal" feminist geography' in which 'Connections are being drawn in many directions' and
'The lessons from the debates about difference have been learnt' around 'an ethics of care without
erasing the differences' (pp. 247–8).

The table in Pratt (2009) was entitled 'Interwoven strands of feminist geography'. However,
these putative strands do not always divide clearly or neatly at all. For example, Gibson-Graham's
influential work The End of Capitalism (As We Knew It): A feminist critique of political economy (1996, reissued
2006) was simultaneously Marxist, feminist and poststructuralist. Throughout the many connections
made by feminist geography, the original concern with the sociology of the discipline long endured

GEOGRAPHY OF WOMEN

Topical focus	*Theoretical influences*	*Geographical focus*
Description of the effects of gender inequality	Welfare geography, liberal feminism	Constraints of distance and spatial separation

SOCIALIST FEMINIST GEOGRAPHY

Topical focus	*Theoretical influences*	*Geographical focus*
Explanation of inequality, and relations between capitalism and patriarchy	Marxism, socialist feminism	Spatial separation, sedimentation of gender relations in place

FEMINIST GEOGRAPHIES OF DIFFERENCE

Topical focus	*Theoretical influences*	*Geographical focus*
The construction of gendered heter(sexed) identities; differences among women; gender and constructions of nature; hetropatriarchy and geopolitics	Cultural, post-structural, post-colonial, psychoanalytic, queer, critical race theories	Micro-geographies of the body; mobile identities; distance, separation and place; imaginative geographies; colonialisms and post-colonialisms; environment/nature

FEMINIST TRANSVERSAL GEOGRAPHIES

Topical focus	*Theoretical influences*	*Geographical focus*
Citizenship; migration; nationalism; transnationalism; ethnographies of the state; development; political ecology; geopolitics; state violence; relations between global North and global South; material objects; progressive possibilities for mapping and GIS; affect and emotions	Theories of transnationalism, globalization and transversal networks and circuitries; non-representational theory; political ecology; Agamben; political economy; theories of affect	Global networks and circuits; multi-scalar and multi-site focus on connections, relations and processes; constructions and disruptions of scale; space of exception; borders and border breakdowns; embodiment and connectivity; dispossession

Figure 8.1 Interwoven strands of feminist geography

Source: Pratt, 2009b, p. 248.

(and the concern with the hitherto relatively marginal position of women within its power structures). Blunt and Wills (2000) thus noted how:

> While feminist geographies are diverse, they share several common themes. . . . One important site of resistance is against sexism within the discipline, departments and other institutions of geography. Resisting the status of detached and disembodied analyses, feminist geographies often share a commitment to situating the production of knowledge in more embodied and contextualised ways. Finally, just as feminist theory and politics transcend disciplinary boundaries, they also transcend the subdisciplinary boundaries within geography, inspiring a wide range of connections within and between economic, social, cultural, historical, political and urban fields of geographical interest.
>
> (p. 91)

Some of the early militancy of feminist geography may have been tempered by time, however. Although her account is of the specific trajectory of feminist geography in Australia, Louise Johnson (2012) describes how it arrived there in the 1980s, quickly:

> Producing critical evaluations of male-dominated geography departments, curriculum, and journals, feminist geographers proceeded to stake claims in each of these spheres while also substantially revising the content of geographical research. . . . However, as the new century dawned, the agenda changed and the anger and urgency dissipated as the broader and university contexts altered. It was a period of consolidation, as feminist insights and approaches were focused on key subject areas – such as the home, identity and sexuality – and became more mainstream.
>
> (p. 345)

She wonders, however, if this work and presence of women in the academy might be 'an indication of success or of co-option?'

In the meantime, work on the geography of masculinity displayed how feminist ambitions to study the gendering of space and the social construction of gender could not be confined to the scrutiny of women or women's experiences of patriarchy, since gender is no more solely an attribute of women than ethnicity is of black people. Feminist geography (and allied work) has therefore stimulated more attention to masculinity, *both* within geography as a discipline (the starting point of much critique, as already noted) and within wider society. Geographies of performing manhood thus came under scrutiny (e.g. McDowell, 2002; McDowell and Court, 1994; Nayak, 2003; Peck and Tickell, 1996; Smith and Winchester, 1998; Jones, 1998; Jackson, 1991b) as well as their relationship to nationalism and the geopolitical (Dalby, 1994; Hyndman, 2001; Radcliffe, 1996; Radcliffe and Westwood, 1996; Sparke, 1994b) plus the 'public' or political sphere of citizenship (e.g. Staeheli and Cope, 1994), together with work on the diversity of (or 'new') femininities (Laurie *et al.*, 1999). Surveying emerging research on geography and masculinity, Berg and Longhurst (2003, p. 351) pointed to masculinity's 'temporal and geographical contingency'. This contingency has since been illustrated by Hopkins and Noble (2009) and via an edited collection assembled by Van Hoven and Hörschelmann (2005) on *Spaces of Masculinities*, although a review of the latter pointed to the still 'sometimes tentative state of the art of geographical research on masculinities' (Phillips, 2006, p. 552), and Blunt and Dowling (2006, p. 112) argued that, 'there is a paucity of research on masculinity and home' (though see the review by Gorman-Murray, 2008).

Reassessing histories of geography

The aspiration of feminist geographers is not simply to add a further fragment to the academic discipline, with the potential for its ghettoisation, therefore, but rather to ensure that a feminist perspective informs all work within human geography. Feminist geography, according to Johnson (1989), involved recognising women's common experience of, and resistance to, oppression by men, and a commitment to end it 'so that women can define and control themselves' (p. 85). Evaluation of geographical practice will demonstrate that it is 'sexist, patriarchal and phallocentric' and will open the way to emancipation, by providing a guide to political practice (Bowlby *et al.*, 1989). One of the fullest such evaluations was undertaken by Gillian Rose (1995). In a critique of histories of the discipline, for example, she argued that:

> The writing of certain kinds of pasts is legitimated by, and legitimates, only certain kinds of presents. . . . Traditions are constructed: written, spoken, visualized, taught, lived. . . . Certain people or kinds of people are included as relevant to the tradition under construction and others are deemed as irrelevant.
>
> (p. 414)

This is a procedure which, she claimed, allowed Stoddart (1991) to dismiss the writings of Victorian women travellers (Domosh, 1991a) as outside what he interpreted as the geographical tradition and thereby irrelevant to any history he might write. Women have been written out of geography's history, and not only by Stoddart, so that:

> even if we can no longer be certain exactly what geography was in the past, in virtually all histories of geographical knowledges one apparently incontrovertible fact remains: geography, whatever it was, was almost always done by men.
>
> (Rose, 1995, p. 414)

Stoddart (1991) argued against the perceived need for a 'feminist historiography' of geography expressed in Domosh's (1991a) critique of the treatment of women in his *On Geography* (Stoddart, 1986). He argued that such an enterprise is unnecessary for an understanding of the work of the small number of women who were influential on nineteenth-century geography: 'No feminist historiography is required to analyse their contributions: they looked after themselves, their careers and their scholarship perfectly well without such assistance' (p. 485), and to do other is to impose 'the explanatory whims of an evanescent present on an increasingly distant past' (p. 486). Domosh (1991b) responded that women should not be 'tacked on to the list of heroes'; their particular viewpoints and experiences should be explored through a feminist approach.

Domosh (1991b) and Rose (1995) thus force us to question the delimitation of geographical tradition – how it is that which is classified as geography became defined in such a way that its tradition is so male dominated, a 'patriarchy'. Citing this work and others leads McEwan (1998) to comment how:

> Stoddart's delimitation of the discursive boundaries of geography produces a definition that is both epistemologically narrow and contestable. As Blunt [1994] points out, institutions outside the academy were [also] important in shaping views of women and their travels, and produced a wider constituency for geographical knowledge. Rose's suggestion that a strategy is required to unsettle the traditional boundaries of geography is both radical and holds out the potential of producing more inclusive histories of the discipline. The real problem, it seems, in (re)writing histories of geography lies with the apparent need to position women (and subaltern groups) and their 'knowledges' at various historical junctures within or beyond the 'tradition', or within an alternative 'tradition'. The fact remains that women travellers, for example, were not geographers, nor did they define themselves as such.
>
> (p. 373)

Out of these tangles, McEwan (1998) proposes that:

> A particular problematic in writing feminist histories of geography has been the apparent need to claim women travellers in a disciplinary sense, in order that they can be located in the 'geographical tradition(s)'. A critical feminist approach need not concern itself with such myopic considerations.
>
>   [for]
>
>   Despite a recognition of their exclusion from the traditions and practice of scientific geography, however, the experiences of nineteenth-century women [travellers] are nonetheless of relevance to the histories of geography and imperialism.
>
> (p. 374)

She notes that women were both actively involved in imperialism – in a variety of roles – and influenced imperial policies as well as making notable contributions (through travel writing) to imperial discourse. Thus, 'Their status as writers, rather than their place in "heroic" histories of geography, lends significance to the works of women travellers.'

Moreover, influenced by postcolonial arguments (see Chapter 7 here), McEwan argues that the very notion of a Western patriarchal geographical tradition demands deconstruction, which would show it to have drawn extensively (and often without admitting it) on many others ('native' and women's labour and knowledges, for example). The question of the role and relative significance of women's geographical work in anglophone geography has returned in subsequent debates.

Focusing on its more recent history (especially the period since 1945 that is also our main concern here), Gillian Rose's (1993a) major book-length contribution to the debate on the role of women, masculinity and men within geography (entitled *Feminism and Geography*) involved the argument that:

1 'the academic discipline of geography has historically been dominated by men' (p. 1);
2 within the profession, women have often been patronised, sometimes harassed and frequently marginalised;
3 feminism remains 'outside the project' of geography (p. 3);
4 'the domination of the discipline by men has serious consequences both for what counts as legitimate geographical knowledge and who can produce such knowledge. [men] . . . have insisted that geography holds a series of unstated assumptions about what men and women do, and that the discipline concentrates on spaces, places and landscapes that it sees as men's' (p. 2). Moreover, 'to think geography—to think within the parameters of the discipline in order to create geographical knowledge acceptable to the discipline—is to occupy a masculine subject position' (p. 4).

This leads her to conclude that the social construction of the discipline of geography is such that it is now necessarily 'masculinist' – it concerns itself primarily with the issues classically of interest to men (p. 4; see also Pile, 1994). This is not presented as a straightforward conscious strategy, even if at times it may have been such. Rather, geography has adopted a particular set of positions and practices which attract both men and women who subscribe, often either subconsciously or by default, to masculinist positions. Most of her book details the four points of her main argument – via a series of careful and creative engagements with many of the aspects of and 'paradigms' within human geography that have also concerned us here (hence Rose examines work on landscape, for example, on spatial science as well as political economy, structuration, locality, and so on). The final chapter sets out the basis for a feminist approach to geography via thinking about the complexity and richness of what Gillian Rose terms 'paradoxical space', and considers alternatives on which resistance to the masculinist hegemony can be based.

In part inspired by Rose's critique, another book-length study (Maddrell, 2009) sought 'To offer a new perspective on the history of British geography by focusing on the geographical work of women from 1850 to 1970' (p. 1). In *Complex Locations: Women's geographical work in the UK 1850–1970*, Maddrell provides evidence that 'Women incontrovertibly were "doers" of geography before 1970, which leads on to ask two key questions: what were they doing? And why does so little of it feature in our disciplinary histories?' (p. 6). In answering these, she reconsiders the question of the place of women travellers in the history of geographical thought, arguing that their work was often more connected with 'the interests of state' (referring to the imperial state, and nineteenth-century British geography's close involvement with imperial exploration, discovery and survey) than has been acknowledged. However, the evidence assembled in her book and argument is a broader one about the institutionalisation of academic geography in the UK. Maddrell (2009, p. 186) points out that:

> Women students played an important role in the demand for geography courses in the 1900s. It
> is also significant that the first women lecturers in geography in higher education can be traced

to the foundation of the first geography departments and first formal qualifications in geography. This can be seen as representing something of a strategic alliance between women gaining access to higher education (including teaching posts) and the needs of an emerging discipline within the university sector.

The sheer numbers of women involved in the early history of British geography as a university discipline have been underestimated and their contributions often marginalised. She argues that histories of geography need to take more account of these figures, some significant authors in their own right (especially figures such as Eva Taylor and Hilda Ormsby, who produced historical and regional geographies: Crowther *et al.*, 1967; Maddrell, 2006), but others of whom many were university teachers, editors or research assistants whose published record was modest and so has tended to be forgotten or who are underrepresented in subsequent disciplinary histories – even those who have excavated geographers' wartime roles (Balchin, 1987), when British women played an important part in intelligence work among other key, but often behind-the-scenes roles (Maddrell, 2008). According to Maddrell (2008, p. 142), such 'ordinary stories', 'provide the warp and weft of contextual history and there is no justification for these stories to be hidden histories because they do not fit the heroic traditions found in some histories of geography'. More widely, however, Maddrell (2009) argues that:

> Gender as an analytical concept needs to be more fully incorporated into the historiography of geography, but this should not depend only on women undertaking a gendered division of labour: anyone engaging with the history of the discipline should be sensitive to constructions of femininities and masculinities within geographical discourses and practices, as well as other axes of difference. Such an approach can only enhance our appreciation of the intricacies of the detailed tapestry – to use a feminine metaphor – of the history of geography and geographical ideas.
>
> (p. 339)

On reading *Complex Locations*, Norcup (2010) in Ward *et al.* (2010) declares:

> Let the book be not only illuminative to the reader about the diversity of geographical endeavour, but also a reminder to contemporary geographers of how the legacies of the subject being made today – in an era of departmental closures and short-term contracts – might be written up or written out in the future. It is about being careful and mindful of all those lives making the subject.
>
> (p. 396)

Sixteen years before, Gillian Rose (1993b) had pointed to the ways in which feminist work continued to face marginalisation. She thus pointed out how 'two recent books discussing – and thus implicitly constituting – geography as an academic discipline' either disregard or incorporate feminist work:

> The first of these, Livingstone's (1992) account of *The Geographical Tradition*, is a roll call of great men which completely ignores over a decade of work on gender issues; both the women who write on gender, and their work, are simply erased as if they never happened. The second adopts a different strategy towards feminist work; [Tim] Unwin's (1992) *The Place of Geography* devotes one paragraph to feminist work in the discipline, which has the effect of rendering feminism as just one more, rather small, example of the rich diversity of, well, the geographical tradition: any possibility that feminism might offer a fundamental critique of the discipline is thus lost.
>
> (p. 531)

Similarly, Mayer (1989) felt that the representation of women in basic human geography textbooks was still characterised by 'consensus and invisibility'. While subsequent books describing the changing structure of human geography (Hubbard *et al.*, 2002; Holloway and Hubbard, 2001; Peet, 1998) do accord feminist critiques more attention, Cloke *et al.*'s (1991) earlier and influential *Approaching Human Geography: An introduction to contemporary theoretical debates* does not. Instead, they self-consciously note both the exclusion of geographical debates conducted in languages other than English (a theme we return to in Chapter 10) and the limited attention that they pay to feminist geography, 'and to the very real difference feminist arguments are undoubtedly making to the theory and practice of human geography more generally' (p. x). They note that feminist arguments are at least as influential in human geography – and in some ways more so – than the series of debates (Marxist, humanist, postmodernist etc.) that their book concentrates on. However, they feel that feminist geography 'is itself influenced by and to some degree fragmented into Marxist, humanist, structurationist, realist and postmodernist approaches' (p. x–xi). Their wish is not to 'ghettoise' feminist geography through according it a separate chapter, on grounds both of its importance throughout human geography (they say that 'feminism in geography should surely be seen as more than just "another" approach', p. xi) and, because they feel that while men can meaningfully discuss feminism in geography, 'at present such discussion is most ably pursued by women within the discipline who are developing new ways of theorising and practising a human geography sensitised to gender issues' (p. xi).

Rose (1993a, p. 3) claims that Cloke *et al.*'s (1991) position only 'demonstrates and sustains' the continued exclusion of feminism, by refusing to address its impact on geography, and thus relates 'to the very nature of hegemonic geographical knowledge itself'. Nearly a decade on, Cloke *et al.*'s tactic was less sustainable and Peet's (1998) book on *Modern Geographical Thought* culminated with a chapter on feminist theory and the geography of gender. Peet notes the initial close association of feminist geography with radical geography although, as he goes on to chart the range of debates that has characterised feminist geography since the 1970s, he recognises that the latter soon developed its own set of debates and sense of purpose and directions. Although Peet recognises that by the 1980s there were differences 'between feminist and masculinist radical geography' (as well as 'between perspectives in feminist geography': p. 271), he reminds us that while some radical geography may have been as sexist in its language and assumptions as the mainstream, feminist and radical arguments were 'often made by the same people, simultaneously radical and feminist' (p. 270). It is thus no accident that the earliest papers formulating what became feminist geography appeared in *Antipode: A Journal of Radical Geography* that today, according to its guidelines for authors (appearing on every inside back cover and online), 'publishes papers which offer a radical (Marxist/socialist/anarchist/anti-racist/feminist/queer/green) analysis of geographical issues and whose intent is to engender the development of a new and better society'. There too, in past copies of *Antipode*, much of the extensive debate about the relationship of capitalism to patriarchy and the spaces through which this is mediated can be found.

From the geography of women to socialist feminisms

By the mid-1980s, a rich debate about the relationship between class and gender relations had developed (see, for example, Foord and Gregson, 1986; Lauria and Knopp, 1985; and the reviews of the debates in Rose, 1993a, and Peet, 1998). These debates embodied a range of approaches and positions. Walby (1986) identified five approaches to the study of gender inequality:

1   the demonstration that it is either theoretically insignificant or non-existent;
2   indications that it is derivative of capitalist relations, and has no independent status;

3   suggestions that it results from an autonomous **patriarchal system** that is the primary form of
     social inequality;
4   demonstrations that it results from patriarchal **relations that are** so intertwined with capitalist
     social relations that they comprise a single system of capitalist patriarchy; and
5   arguments that it results from the interaction of **autonomous** systems of patriarchy and
     capitalism.

Bowlby *et al.* (1989) showed that the second and fourth of these have been more commonly
adopted among geographers, thus focusing on ways in which patriarchy is structured by
capitalism's imperatives. Foord and Gregson (1986) adopted the fifth strategy, however, arguing
that although patriarchy and capitalism are empirically linked, they are theoretically separate.
Just as capitalism can be presented as a particular example of a necessary condition (a mode
of production within which economic and social life is structured), with unique individual
instances at separate times and places, so, they argued, patriarchy can be represented as a particular
example of a general condition (gender relations) which also has its time- and space-specific
realisations.

Foord and Gregson (1986, p. 199) argued that the necessary interrelationships between
humans and nature call for a form of social organisation to ensure human survival and social
reproduction, so that 'all social relations must involve gender, and gender ... relations will be
embedded in all forms of social relations'. They could not conceive of a society lacking gender
relations, therefore, so that the latter are independent of any mode of production, let alone any
particular instance of it. Patriarchy is a particular form of gender relations, in which men dominate
the processes of species reproduction; empirical work focuses on the nature of that domination 'in
particular periods and places' (p. 206): 'Just as other relations vary and combine differently over
time and space, so too must the practices of which comprise these relations' hence the professed
importance of studying them within the context of the localities programme, as illustrated by
McDowell and Massey (1984; see also p. 213, this book). The assumption that politics and social
relations could be read-off the socioeconomic characteristics of the locality was, however,
problematised by Rose (1989), who takes into account the roles of unwaged women's labour and
thus disturbs the assumption that a locality is reducible to the structure of its paid employment
alone (cf. Jackson's, 1991, wider cultural critique of locality studies).

Similarly, because they regard gender and social relations as semi-autonomous spheres of
activity, Foord and Gregson contended that Marxist and feminist theories could not easily be integrated. Against this view, McDowell (1986b) advanced a class analysis and rejected the notion of a
universal female experience. Biological reproduction is part of the process of capitalist reproduction, and 'Patriarchal social relations are further strengthened by the political and ideological functions of the state that has a vested interest in supporting the domination of individual women in
the exploited class by individual men in that class' (p. 313). The contradiction underpinning the
need to create surplus value within capitalism is thus the source of women's oppression, not any
necessary gender relations: 'the social construction of male sexuality and the dominance of family
forms based on sexuality and kinship networks to class societies are historical resolutions of the
contradiction' (p. 317). Subsequently, McDowell insisted on supplementing regulationist approaches
(see p. 21, this book) with an understanding of the importance of unwaged work (overwhelmingly
performed by women) in the home and the significance of gendered subjects in the (paid) workplace. In terms of the overall relationships between gender, space and society, Gier and Walton
(1987) took a slightly different approach, disagreeing with Foord and Gregson's contention that
gender relations are necessary to all social relations:

> Evidence from anthropology and history as well as other disciplines indicates that gender has
> not always been used to identify male and female sexual difference and its attendant physical

and psychological archetypes . . . . [so that] the very identification of the concept is the product
of human consciousness and human society.

(pp. 56–7)

According to this view, significant gender differences need not always be present (see also Knopp
and Lauria, 1987). Gregson and Foord (1987, pp. 373–4) responded by defending their position
that mode of production and gender relations are 'distinct and separate objects of analysis which
interlock as particular forms (capitalism and patriarchal gender relations) but not as conceptual
categories', but were unprepared to admit that this was a preface to the creation of a universal
theory of women's oppression.

Although some of the earlier work presented as feminist geography involved demonstrating
that women are subordinated and thereby have distinctive spatialities in contemporary societies
(e.g. Women and Geography Study Group, 1984; Little *et al.*, 1988), wider issues were soon
introduced. Pratt and Hanson (1994), for example, moved from demonstrating differences between
men's and women's commuting patterns to analyses of how those patterns were exploited by
capital to reproduce segmented labour markets (see Hanson and Pratt, 1995). Indeed, the issue of
mobility and gendered differences in its range and mode continued, albeit redefined and expanded,
by incorporating broader social and cultural geographies of mobility (associated with the 'cultural
turn': Law, 1999), including work on the gendered geography of urban fear and danger, and its
consequences for women's experience and use of space (Pain, 1991).

The variety of perspectives adopted in feminist geography as it broadened is illustrated in
McDowell's (1993a, 1993b) review articles. During the 1980s and early 1990s feminist geography,
while addressing the discipline's three main concepts of space, place and nature, shifted from
analyses of gender differences to concerns over the social creation of gendered beings in particular
places, which brings feminist geography (as distinct from feminist geography) closer to the wider
feminist project – 'the study of the lives, experiences and behaviour of women' (1993a, p. 161).
Three main themes are identified in the early work:

1   Spatial differences in women's status – demonstrating 'man's inhumanity to women' (p. 163)
    – a largely empirical task that emphasised Western experience and was increasingly criticised
    for its ethnocentrism.
2   Gender and place: women and the urban environment, which stressed that most women were
    excluded from analyses of urban areas that focused on public rather than private activities, at
    scales which were larger than the individual household, and so ignored the gender relations
    that underpinned the home, childbirth and the unpaid labour involved in social reproduction
    (pp. 165–6):

> in common with the other social sciences, geography takes for granted the Enlightenment
> distinction between the public and the private, and, implicitly, the gendered associations of
> those spheres . . . [which involves the patriarchal assumptions that] women are the angels of
> the hearth, to be confined to domesticity preferably in sylvan and suburban surroundings,
> while men join the fray of public life in the bustling city centres, returning home for emotional
> and sexual solace, and hot meals to fit them for their continuing labours.

3   Patriarchal power, which illustrated the 'blindness' of (urban and other) geographers to the
    'embodiment of conventional gender divisions' (p. 167) in the built environment on both large
    (the structuring of urban land-use patterns) and small (the design and layout of buildings) scales.

Thus, gender inequalities were added to the others identified by those involved in portraying
'unfairly structured cities' (e.g. Badcock, 1984) as major elements in the reproduction processes of

'patriarchal capitalism'. This 'feminist empiricism' (McDowell, 1993a, p. 174) paralleled concerns of other geographers for social justice, and stimulated a 'common focus on excluded or oppressed groups – be they women, the working class or ethnic minorities – [that] united "radical" geographers'. These concerns needed to be linked to wider theoretical considerations, however. McDowell (like Sheppard, 1995) saw a bifurcation within human geography (McDowell, 1993a):

> It is perhaps not too much of a caricature to argue that throughout the 1970s and early 1980s human geography consisted of two phalanxes going off in sharply different directions. In the early part of the period, the opposing forces consisted of those who held on to mathematically modelled, rational choice, context-free notions facing the grand theoreticians of structuralism whose forward march in a single direction admitted women only in the tail position. Despite their relegation, it was in this camp that feminists found common interests, not least because of its intellectual and political attraction . . . [nevertheless] Despite an enormous shift over the last 15 years or so within economic and urban geography from a structuralist Marxism influenced by Althusser towards a geohistorical materialism alive to specificity, complexity and local struggles . . . gender issues remained marginal to the concerns of 'left' geographers.
>
> (p. 174)

Yet in 1989 Susan Christopherson was moved to publish a critical reflection on the ways in which feminism and feminists still remained outside what she saw as the geographical project. Feminist empiricism was a foundation for a challenge to the masculinist domination, even epistemological foundations, of human geography (p. 175):

> this diverse set of theoretical and empirical books and papers published throughout the 1980s succeeded in placing gender divisions on the geographical agenda. It made women and men visible as both academic authors and subjects, thus challenging the implicit masculinity of the discipline. It also succeeded as geography, documenting some of the range of variation in women's social position and circumstances in different parts of the world and raising questions about gender identity and place.

It also laid the groundwork for advance on other fields:

> this work has enabled more complex theoretical questions about the extent of variety in the spatial constitution of gender and the specific ways in which the characteristics of masculinity and femininity vary between spaces, classes and ethnicities to begin to be raised. Thus by the end of the decade the particular geographic interest in variety coincided with a more widespread interest in the deconstructionism in social theory to enlarge the scope of the agenda for feminist geography in the forthcoming years . . .

– in which it was joined by those interested in other sources of difference, such as sexual orientation (see p. 300, this book) and race.

Notably, feminist work on the intersections of race, class and gender explored how these are registered in the lives of migrant women workers (e.g. England and Stiell, 1997; Kofman and England, 1997; Pratt, 1999; Radcliffe, 1990; Stiell and England, 1997), extending socialist-feminist influenced works on geographies of social class, patriarchy, maternity and childcare (e.g. Dyck, 1990; England, 1996; Gregson and Lowe, 1994) and the wider reworkings of gender, class and space (Pratt and Hanson, 1988). In tandem, work on the geographies of children (Holloway and Valentine, 2000) and on different forms and geographies of illness, impairment and disability (Butler and Parr, 1999; Imrie, 2000), while not simply reducible to feminist geography, were certainly influenced by its concerns and methods, and are a measure of its broad impacts.

Meanwhile, feminist geography itself registered a growing concern with 'the body' (e.g. Longhurst, 1995, 2000), both as a subject in space – produced through a variety of discourses (think of the role of advertising and media representations): 'in order to understand questions around sexual specificity, the differences between bodies, women's social subordination to men, and the mutually constitutive relationships that exist between bodies and spaces' (Longhurst, 1997, p. 486) – and about how geography (in tandem with other Western social sciences) has assumed a masculinist separation between mind and body:

> Cartesian dualism underlines our thinking in a myriad of ways, not least in the divergence of the social sciences from the natural sciences, and in a geography which is based on the separation of people from their environments. Thus while geography is unusual in its spanning of the natural and social sciences and in focusing on the interrelations between people and their environments, it is still assumed that the two are distinct and one acts on the other. . . . Geography, like all of the social sciences, has been built upon a particular conception of the mind and body which sees them as separate, apart and acting on each other.
>
> (Louise Johnson, 1989, cited in Longhurst, 1997, p. 492)

Thus, too, feminist work sought to transform approaches to the study of landscape by relating it to the way that it is represented ('appreciated'), in ways that are analogous to the heterosexual male gaze directed towards the female body (Nash, 1996). And subsequent geographical work on embodiment owes some of its inspiration to feminism. According to Dyck (2011, p. 357), the body is framed by the interactions of biology, technology, environment and science with 'categories of difference, such as those based on perceived ethnicity or gender, which frame our understanding of appropriate bodily practices in particular places as well as notions of belonging'. Moreover, concerns about the body as a contested site and for the Cartesian distinction between mind and body have been challenged in postmodern and poststructuralist feminist geographies.

Feminist geographies of difference

Although explorations of postmodernism and poststructuralism in geography have sometimes proceeded without due attention to gendered and sexual differences, the postmodern impulse to accommodate and give voice to difference has been seen as a possibility for feminist perspectives and feminist geographies. Building on that foundation involved two other feminist perspectives emerging alongside the 'rationalist or empiricist feminism' (McDowell, 1993b):

1   Anti-rational, or feminist standpoint theory celebrates gender differences and, rather than present them as unfair or unjust, promotes them and has as its goal the elimination of the traditional allocation of superiority to everything associated with masculinity. Knowledge is socially created, in context, so that women's experiences of, for example, menstruation, child-birth and lactation lead to the construction of different self-identities from those involved in masculine experience, which also vary between places, because of their separately constructed gender relations. Such different standpoints provide a basis not only for understanding but also for practice, hence the development of various forms of radical feminism, including ecofeminism.

2   Postrational or postmodern feminism argues that treatment of women as a single category involves linking together very different groups with separate experiences and needs. Many feminists initially identified this as a threat to the project, but McDowell argues that the

encounter with postmodernism stimulated a debate about 'situated knowledge' which has led to the situation (McDowell, 1993b) whereby:

> geographers sympathetic to the postmodern deconstructive project are . . . reluctant to abandon gender as a difference that makes a difference, if no longer the difference. The current aim within feminist geography is a move towards . . . 'partial' or 'situated knowledges' that recognize that the positionings of white British women in the academy, to take but one example, are not the same as those of other women, women from different ethnic or class backgrounds, and that this makes a difference to knowledge construction.
>
> (p. 310)

For geographers, this emphasis on difference within the 'oppressed' group raises important issues regarding the role of place in identity-creation. Knowledge is both local and gendered – and also linked to other socially constructed categories. Feminists argue that 'some differences are more significant than others' (McDowell, 1993b):

> their work demands a theoretical analysis of differences, building up an understanding of the mutual interrelations of gender. . . . of sexuality, household and family structures and the political economy of domestic and workplace relations within and between places. It is a feminism located in a theoretical understanding of differences between women, rejecting both the transhistorical and cross-class search for the origins of patriarchy that concerned us at an earlier moment . . . and notions of a cultural feminist essentialism that denies the structure of power relations between women. It is a feminism that recognizes the existence of a material world, or women living in different social formations, engaged in struggles in which their interests may converge or diverge.
>
> (p. 315)

This presents a major challenge to all geographers, who, as this and the preceding chapter have illustrated, are having to come to terms with a wide range of situated knowledges. Bondi and Domosh (1992, p. 210) have argued, however, that although postmodernism attacks universal truth claims and so should sustain feminist arguments, its application has failed to do so because 'there is a major impulse within postmodernism that continues a tradition of masculine discourse in which the stereotyping of women is intrinsic to its operation'. However, both these writers and other feminist geographers have welcomed aspects of postmodernism. Thus, Nancy Duncan (1999), writing on postmodernism/postmodernity in *A Feminist Glossary of Human Geography* (itself a marker of the vitality of the field) notes how:

> Many feminist geographers are attracted to postmodernism's rejection of REASON as being gendered, historical and ethnocentric rather than disembodied and neutral, as is usually claimed. Others agree with postmodernism's refusal of ESSENTIALISM and the replacement of POSITIVIST and empiricist theories of representation and truth with a diversity of viewpoints.
>
> (p. 212, author's capitals)

Similarly, the emergence of work in a psychoanalytic tradition in geography (see Chapter 7) was enabled by feminist geography even if it is not necessarily confined to it. The subsequent evolution of this work, which has intersected with both emotional geography and geographical debates about 'affect' (Bondi, 2005; Curti *et al.*, 2011; Dawney, 2011; Pile, 2010, 2011) have complex relationships to feminism. But they would be hard to imagine without feminism's influence on the discipline. More recently, too, a geographical encounter between the non-representational theory (NRT) that was described in Chapter 7 (p. 271, this book) and feminism has commenced. While Bondi (2005):

> would suggest that feminist geographers find research informed by non-representational theory too abstract, too little touched by how people make sense of their lives, and therefore too 'inhuman', ungrounded, distancing, detached and, ironically, disembodied. Conversely, those informed by nonrepresentational theory find feminist work too reliant on cognitive ways of knowing (including especially individual accounts of experience), and insufficiently 'transhuman'.
>
> (p. 438)

Colls (2012) similarly registered this uneasy relationship, in which feminist geography and NRT have tended to regard each other as outside their remit. But she advocates:

> positivity as a way of affirming what feminist geography is rather than what it *is not* in relation to other ways of knowing and doing . . . of what feminist geographies of sexual difference *could be* if, through a critical alignment of sexual difference theory and non-representational geographies, we begin to question 'how might non-representational geographies allow us to think feminist geography differently and to think differently as feminist geographers?'.
>
> (p. 442)

Therefore, while the relationships between feminist geography and the variety of 'post' (-modern, -colonial and -structuralist) geographies that were mapped in Chapter 7 have sometimes been fraught, equally often there are creative combinations and exchanges. Another useful way to begin to take stock of some these is to focus on feminist methodologies in human geography.

Feminist methods?

According to Thien (2009, p. 74), feminist geographers 'have incorporated both quantitative and qualitative methods, with preferences which have shifted over time'. Thus, by the early 1990s, most feminist work in geography employed qualitative methods. Subsequently, however, this has been reappraised:

> Quantitative methods are neither absent nor rejected in feminist methodologies. . . . As the special issue of *The Professional Geographer*, 'Should women count?' [Mattingly and Falconer Al-Hindi, 1995] exemplifies, quantitative methods are not, in and of themselves, antifeminist; rather it has been the underlying politics and the epistemological positionings of such methods within masculinist claims to objective truth which have resulted in feminist critiques.
>
> (p. 75)

Feminist geography's concern with embodiment encouraged qualitative methodologies and renewed debates about geographical fieldwork (a debate that was usefully summarised in Peet's, 1998, chapter on 'Feminist theory and the geography of gender' in his Modern Geographical Thought), including assumptions about gender and sexuality in geographical field-work (Cupples, 2002). These related to debates about 'positionality' – the vantage points (class, gendered, national and 'racial' formations) from which geographers write and teach (see England, 1994; the debates between Sidaway, 1992, and Madge, 1993; and McDowell, 1992, and Schoenberger, 1992, plus Herod, 1999, and other papers in that 1999 special issue of Geoforum). While such debates were not confined specifically to feminist geography, they were appreciably enhanced by feminist critiques (see Rose, 1997, for a critical review).

Moves to revive ethnography (participant observation in which the researcher observes and interacts closely with the society under study; see the articles in a special issue of *The Professional Geographer* on critical feminist perspectives: Nast, 1994) as a method in human geography also partly rested on feminist groundings (Jackson, 1985; Herbert, 2000). At times, there seemed no end to the potential for feminist geographies to make a difference to the assumptions and norms of human geography. Thus, quantitative geographies (Falconer Al-Hindi, 2001; Mattingly and Falconer Al-Hindi, 1995; Rocheleau, 1995) as well as issues such as remote sensing and GIS methodologies came under distinctively feminist scrutiny. Summarising a set of papers (published in *Gender, Place and Culture*) on feminist geography and GIS, Hanson (2002) noted how:

> denying one's connections to, embeddedness in, and complicity with an intellectual tradition facilitates classifying that tradition as 'the other'; such denial legitimizes intellectual boundary maintenance and dismissing the other tradition as irrelevant. Years ago, the idea that feminism and quantification were incompatible had to be dispelled by feminist geographers carefully demonstrating how quantitative approaches could support a feminist analysis. Similarly, this collection of articles effectively refutes the notion that feminism and GIS have nothing to offer each other.
>
> (p. 301)

At the same time, Mei-Po Kwan (2002, p. 645) sought 'to reimagine GIS as a method in feminist geography and describe feminist visualisation as a possible critical practice in feminist research'.

Feminist geography (again sometimes in affinity with postmodernism and poststructuralism) also encouraged attention to 'writing practices' (Bondi, 1997) and the development of alternative narrative styles, genres and methods (England, 1994; and see also the reading group engagement in Bondi *et al.*, 2002). Thus, Gregson and Rose (1997) noted how:

> The pre-eminence of the dispassionate, distant and disembodied voice is something which has been challenged repeatedly in human geography over the past 20 years or so, for example by those who initiated the first critiques of positivism. Humanistic geographers, for example, argued that such a writing voice simply could not evoke the creative and emotional senses of place.... Marxist geographers suggested that the supposed neutrality of that voice served merely to hide its lack of critique of capitalism; it was a voice whose apparent objectivity masked a conservative and reactionary set of values. More recently, feminist geographers have also criticized that voice as expressive of certain kinds of masculinities. Particularly in the academy, where cool and calm rationality is the desired norm of both behavior and debate (which is not to suggest that this is what actually happens!), to be masculine often means not to be emotional or passionate, not to be explicit about your values, your background, your own felt experiences.
>
> (p. 23)

As represented in one of the epigraphs opening this chapter, therefore, the conventional process of academic writing (which foregrounds the names and therefore authority of individual or joint authors) was challenged by more collective modes held to be in the spirit of feminist politics. Thus, two key textbooks of feminist geography were authored by a collective, the Women and Geography Study Group of the IBG (WGSG, 1984, 1997). Yet Gregson and Rose (1997) detected a tension for feminist geography here. On one hand, the possibility that to write in alternative, embodied or personalised positions risks having feminist work which does this being dismissed by the mainstream. But, on the other hand, the potential and power of a human geography that explicitly draws on personal experiences and feelings is evident.

Sexuality and space

The intersections of sexuality with gender and class and space started to be mapped in the in the late 1970s. As the human body became an explicit domain of geographical analysis and influenced by feminism, by the 1990s a plethora of works considered sexuality and spaces, including a landmark edited book, *Mapping Desire: Geographies of sexualities* (Bell and Valentine, 1995). Reflecting on the preceding decade of such works on sexuality and space, Hubbard (2000, p. 191) thought that:

> Although traditionally regarded as 'squeamish' about sexual matters [see too Symanski, 2002] . . . recent research by geographers has begun to demonstrate that space is inevitably sexed in a variety of complex ways, placing issues of sex and sexuality firmly on the geographical agenda. As such, there is now a substantial body of critical geographic scholarship which has indicated that space is fundamentally shaped by the dynamics of human sexuality, reflecting the ways in which sex is represented, perceived and understood.

In similar terms, England (1999, p. 95) recorded that:

> The 1990s have seen a rapid expansion of published research on sexualities and sexual identities. Most major English language geography journals have published articles related to sexualities (although there are notable holdouts). There is certainly no longer a shortage of materials that can be used to teach the geographies of sexualities although there is a shortage of materials looking explicitly at heterosexualities.

Where there may once have been 'squeamishness' there is now a wealth of material. Gavin Brown *et al.* (2011, p. 295) note how:

> A quarter of a century ago, early sexual geographers taunted their colleagues for their 'squeamishness' about the spatial aspects of sexuality. . . . Today many introductory human geography textbooks include some reference to 'gay space' and the geographies of sexual difference, and journal articles abound . . .
>
> (p. 265)

Elsewhere, in the introduction to an edited collection, taking stock of *Geographies of Sexualities*, they note how 'the relatively young field of geographies of sexualities has blossomed over the past decade or so' (G. Brown *et al.*, 2009, p. 1).

The first papers on sexuality and space appeared at the end of the 1970s and in the early 1980s (Weightman, 1981). These, and much of the subsequent work on sexuality and space, focused at first on the spaces and places associated with expressions and geographical consequences of lesbian and gay male sexuality. Such research and associated manifestations of lesbian, gay and bisexuality within geography did not establish themselves without resistance. Citing Carter's (1977) objections, Michael Brown and Lawrence Knopp (2003, p. 314) note that in the mid-1970s: 'The simple act of arranging meetings of gay and lesbian geographers at Association of American Geographers' meetings precipitated extraordinary nasty public (and published) denouncements from established and secure figures in the discipline.'

Subsequent cases of harassment (Valentine, 1998) revealed continued discord among geographers over the status and positioning of work on sexuality and those who have fostered it. However, a critical mass of scholarship and degree of wider acknowledgement of sexuality and space as a challenging field of study became evident by the mid-1990s. A mark of this was the publication of that influential collection on *Mapping Desire: Geographies of sexuality* (Bell and Valentine, 1995). In this and other work that accompanied or preceded it (e.g. Adler and Brenner, 1992; Brown, 1995; Forest, 1995; Valentine, 1993a, 1993b), the parameters of work on sexuality and space both work with the early concerns about gay and lesbian spaces (and their experiences in mainstream 'heterosexual space') as well as complementing this with a wider sense of the paradoxical spaces of sexuality (Bell *et al.*, 1994) and bodies, sexuality and the sociospatial dynamics of capitalism (Duncan, 1996a; Knopp, 1992). Much of the earlier work was on urban quarters or zones associated with the gay and lesbian 'scene', including the role of gay men as agents in the wider gentrification of urban neighbourhoods, supplementing established Marxist work on gentrification (Knopp, 1990). Subsequently, work in rural geographies and on migration was enriched by attention to the roles of sexualities. Summarising all this, Binnie and Valentine (1999) conclude that:

> geographies of sexualities have come a long way in the last decade. Progress in the study of lives of dissident sexualities has been most striking, most notably in three key areas: urban geography, rural lesbian and gay geographies, and the geography of sexual citizenship. In each of these areas, there have been significant developments in both geographical understandings of lesbian, gay and bisexual lives and signs that sexual dissidents are beginning to have an impact on the heteronormative nature of the geographical knowledge.
>
> (p. 182)

The latter proved to be an enduring challenge, to which we will return in a moment. However, at this point we should add that Binnie and Valentine also noted the unevenness of geographical work on sexuality (something which has also been evident in terms of the development and impacts of feminism in geography). In terms of sexuality and space, they note two dimensions with respect to this. First, 'the ethnocentricity of the literature on sexuality and space remains largely unchallenged' (p. 183); in other words the sexualities and spaces studied had mostly tended to be Western and white. Second, they note how some subdisciplines have been much more receptive than have others:

> Indeed, there are many areas within the discipline of geography where discussion of sexuality has been notable for its absence, for example, transport geography and population geography. The different philosophical approaches that dominate different subdisciplinary areas may explain the uneven impact of work on dissident sexualities within geography. Geography remains a highly contested enterprise. While social and cultural geography have been very receptive towards contemporary developments in social theory (particularly with respect to the postmodern emphasis on 'difference') and therefore towards dissident sexualities, many other fields of geographical enquiry remain wedded to their positivist tradition. However this does not explain why the hegemonic sexual identity, heterosexuality, has thus far received so little attention from geographers.

They note partial exceptions, such as Philip Crang (1995) and McDowell (1995) – both, however, perhaps better known as social-cultural and economic geographers respectively – amid a relative lack of attention to 'mainstream' heterosexuality. Subsequently, a few geographers mapped and theorised heterosexuality, notably Hubbard (2000). The whiteness and Western focus of the first wave of work on sexuality and space remained an issue that concerned some (Nast, 2002), although that work has increasingly been supplemented by case studies and theorisations of sexuality and space based on the experiences of non-whites (early examples include Peake, 1993; Mohammad, 1999; the papers introduced by Puar *et al.*; Haritaworn, 2009; and Rouhani, 2007).

Queering geography

Much as feminist geography concerned itself first with the status of women, before turning to consider wider constructions of femininity and masculinity (and thus men), so the work attentive to sexuality and space tended first to focus on the relative invisibility of the geographies associated with male homosexuality, lesbianism and bisexuality in Western society ('insignificant others' to mainstream human geography, according to David Bell, 1991) before turning either to examine the construction and problems associated with such categories or to the wider geographies of sexuality – 'straight', gay, or whatever. One aspect of this is 'queer theory' (representing a subversion of a hitherto mostly pejorative term) which begins with, but then goes beyond, the issue of sexuality. Thus, as Elder (1999), a South African-born, white male geographer who worked in the USA and represented his sexuality as 'gay', noted:

> I see queerness as a theoretical and personal insistence to study the relationships between social boundaries (like 'race', gender, sexuality and class). Queerness is not only a device for thinking through issues related to sexuality. It follows that for me the attraction to queer theory is an obvious one; the language of geography as well as the queer geography that shapes my daily life. The queer geography that I refer to is learning to perform on boundaries (coming out) and realizing the consequences and effects of those performances . . . I am not only referring to the sexual aspects of my identity: I am perceived by students to be too white to be 'African' and too foreign to be "American". As a condition, queerness captures a delightful sense of unbounded chaos and uncertainty and it helps me to think about identity.
>
> (p. 88)

For Kitchin and Lysaght (2003, p. 490): 'Queer theory, a theoretical position founded on poststructuralist ideas, posits that sexual identity is never fixed, but is always in the process of becoming . . . it critiques the notion of being able to draw coherent boundaries around sexual identities.' What role, then, possibly for a queer geography?

> [–] an extension of queer theory suggests that, given . . . that all space is sexed, all space is inherently queered and it is only ever temporarily fixed as heterosexual (or homosexual for that matter). That is, the sexing of space is in a constant process of 'becoming', created through the (re)production of (sexual) discourse and practice.
>
> (p. 491)

In turn, however, for others too sexuality (be it 'queer' or not) is, like the gender relations of which it forms a part, seen to be caught up with wider economic, social, cultural and political geographies. For Hubbard (2000), this was:

> beginning (belatedly) to clarify the importance of heterosexuality in maintaining and reproducing geographical order, particularly in their attempts to elucidate the importance of domestic space in maintaining the 'family values' which lie at the heart of heteronormality.
>
> (p. 211)

Geography as a discipline has also been seen as embodying similar heteronormality, an often taken-for-granted assumption that the spaces and lives we study and the outlook of the geographers who do the studies are 'straight'. This led Binnie (1997) to ask the following questions:

> How have systems of (geographical) thought from positivist social science to poststructuralism treated sexual dissidence? Are they inherently, essentially heterosexist? Are some systems of

thought and theoretical frameworks more damaging or better at promoting the interests of an antihomophobic 'project' than others?

(p. 223)

It does not take Binnie long to conclude that spatial science was problematic, by virtue of its indirect association with a scientific discourse that assigned deviance to homosexuality ('the medicalization and pathologicalization of homosexuality' in his words (p. 224)). Binnie recognised more potential in postmodern, feminist and poststructuralist geographies. Yet he remained wary of the 'cultural turn', declaring that:

> I remain deeply suspicious of the 'cultural turn' in geography and social theory more generally, as the post-modern emphasis on difference (although opening up a space for work on sexuality) has also been guilty of skirting around the edges of sex and sexuality, preferring a more abstract engagement with, for example, Lacanian psychoanalysis.

(p. 225)

Binnie also took issue with much feminist geography. For example, he was concerned that otherwise 'invaluable, insightful critique of the masculinism of the discipline [here he refers to Rose, 1993a, 1995] . . . does have a tendency to rant about "men and their theories"'. The use of the term 'rant' here resorts to a pejorative language that has often been applied by sexist men to feminism and feminists. But perhaps it is indicative of how strongly Binnie perceived a relative failure of some feminist geographers to recognise diversity among men and masculinities. For Binnie, there seems to be a world of difference within what maleness and especially 'male' sexuality means (queer, bi, straight, camp, sadomasochist, for example). He feels therefore that those feminist geographies that, for example, posit landscape as a domain of sexualised masculine vision foreclose other ways that men and women might look and be seen, amid 'the celebration of queer pleasure, sexual diversity, and exuberance' (p. 227).

A few years later, Howell (2009) reappraised the geography of sexuality in the light of these and other contributions, such as a book-length study of *Closet Space* (Brown, 2000) which had blended poststructuralist, queer and other social theories to provide an account of the interactions of hidden and 'out' gay male lives, sites and spaces. According to Howell (2009), therefore, 'Geographers of sexuality' have:

> been directed by the career of queer theory to a strikingly expansive and engaging new topography of sexuality. We recognize that all space is sexed, in that it encodes and endorses societal norms of sexual behavior and identity, but also that, because of the unstable and contradictory nature of these sexual categories, all space is actually or potentially queer.

(pp. 123–4)

It might be expected too that more will ask broader questions about a queer epistemology in geography, like Michael Brown and Lawrence Knopp (2003, p. 313) who argue that:

> Serious engagements with sexuality, then, necessitate a careful reconsideration of some fundamental ontological, epistemological and methodological issues. These include the relationship between nature, society and human agency; the nature of identity; problems of naming and counting; of drawing inferences and conclusions; of the roles of qualitative and quantitative methods in social science (how can we understand the social consequences of sexualities without understanding them as lived experiences?); objectivity and subjectivity (can sexualities ever be understood as strictly objective or subjective phenomena?); and more.

These issues have been heightened in the case of transgendered people, whose identity either does not conform to the biological sex that they were assigned at birth, and/or who may experience being bi-gendered, a-gendered or seek to actively transform their gender. As Browne *et al.* (2010, p. 573) noted: 'Gender geographies have focused on normatively gendered men and women, neglecting the ways in which gender binaries can be contested and troubled.' They list a few exceptions (such as Doan, 2007, 2010) and more appear in the special section of *Gender, Place and Culture* that their paper introduces, but judge that generally:

> geographical enquiry has yet to explore the lives and experiences of people, including trans people, that trouble and call into question these hegemonic, normative binaries. Such omissions mean that assumptions predicated on a straightforward gender mapping onto biological sex organs and gender roles and relations grounded in male/female and man/woman separations are often uncritically reproduced. To date, geographical scholarship has not fully engaged with the challenges posed by gender diversity particularly as it has been taken up in trans and queer scholarship.
>
> (p. 573)

The challenges raised by transgendered geographies that do not readily fit into a male/female binary are distinctive. Others, however, had sometimes felt themselves to be excluded from 'the project' (a reference to Christopherson's, 1989, paper designating how feminist work was poised to play a defining role in 'the project' of reconstituting human geography), as in the case of the concerns and insights of disabled geographers (Chouinard and Grant, 1995), who complained of 'not being anywhere near the project' (see too the paper by Butler and Bowlby, 1997, and others in that special issue of *Environment and Planning D: Society and Space* on Geography and Disability). Arguably this form of marginality continues. Notwithstanding a few pioneering works, Wilton and Evans (2009) judge that:

> disability – as a dimension of oppression and a subject for inquiry – does not have the status of race, class, gender, and sexuality in the broader subfields of social, cultural, urban, economic, or political geography. This is despite the fact that disabled persons constitute a significant – and growing – minority of the population in Western countries and they continue to confront and struggle against manifold barriers to full participation in, and access to, housing, paid employment, political participation, public space, social lives, and cultural representation.
>
> (p. 209)

Conclusions

Things have changed since the end of the 1980s, when Jackson (1989) remarked that:

> gender and sexuality are still too rarely regarded as part of the central agenda of human geography . . . regarded as peripheral, private, and personal issues, not suitable for academic debate or public discussion. . . . What could possibly be geographical, critics imply, about such intimate personal subjects as gender and sexuality?
>
> (p. 104)

Feminist geography has undoubtedly made a difference. There have been, however, some significant boundaries to this.

First, the challenge of feminist geographies working across different cultures and languages has been hard to negotiate, an issue raised in Radcliffe (1994) and Staeheli and Nagar (2002). Subsequently, Garcia-Ramon *et al.* (2006, p. 3), writing from Catalonia, Denmark and Greece, raised the question of anglophone hegemony in the journal that had become a key outlet for feminist geography (*Gender, Place and Culture*). They reconsider, for example:

> the very concept of gender, including the distinction between sex and gender. Many non-English languages, our own among them, have only one word and are therefore linguistically unable to act out the sex/gender distinction. . . . We would however argue that the lack of a distinction should not necessarily be seen as a disadvantage. The sex/gender distinction is a historical construct, developed for the specific political purpose of formulating a defence against sexism and biologism. Even though it actually did the job, it is not a necessary condition for developing the argument. Many non-English-speaking feminists have managed to develop non-essentialist arguments without access to these particular terms. It would be possible to argue that they, in that specific case, avoid thinking in the binary oppositions that Angloamerican poststructuralist feminists so painfully struggle to deconstruct.
>
> (p. 3)

Yet while one of the key insights of feminist geography has been to bring home (so to speak) the social and cultural contingency of gender relations, it was challenging to negotiate the global range of these (given different expectations and norms of gender roles – and of gendered spaces – in different sites around the world), notwithstanding important feminist work on geographies of globalisation (Nagar *et al.*, 2002) or collections such as Falah and Nagel (2005) on *Geographies of Muslim Women*. Subsequently, however, Louise Johnson (2009, p. 52) claims that feminist geography had gone 'well beyond the dualism of women and men to admit race, ethnicity, age, disability, nationality, religion, and sexuality into their works' to document 'how multiple axes of difference are realized in and through space'.

Second, the impact of feminism in human geography invariably developed unevenly; it took longer for political geography to register feminist themes (see Drake and Horton, 1983; Hyndman, 2001; Kofman and Peake, 1990) than social and cultural, urban and economic geography, for example. And other subdisciplines hardly seemed to notice it at all. Much human geography continued as usual, outside (but a subject of criticism from) the influences of feminist traditions. There were also resistances, tensions and 'backlashes'. For example, a senior male geographer complained that in feminist geography, especially work focused on language, he claimed to have encountered 'not just anger, which is perfectly understandable, but something close to hate' (Gould, 1994). This was countered by Peake (1994, p. 205), who characterised Gould's position as:

> woe betide any feminist who adopts a mode of expression that offends his notions of clearness, effectiveness, and beauty . . . this situation is fine only if we accept the male prerogative to legislate language . . . a proprietary attitude that could be espoused only by someone who has appointed himself as the arbiter of language. . . . He is refusing women the right to make linguistic changes because, in his dominant subject position, he cannot admit to change, to the recognition of difference.
>
> (p. 197)

Gould (1999) responded that although:

> the various genres of writing emerging out of an excitingly refurbished social and cultural geography are frequently enlightening, thought-provoking, and intellectually satisfying, in that they help us to see something in a different way, from a perspective or viewpoint we had not taken before . . . some throw up (the image is deliberate) writing that is so totally biased, shrill and predictable that it eventually becomes boring.
>
> (p. 82)

Writing in the early 1990s, Susan Hanson (1992, p. 569) noted that in geography, feminists 'have been likened to invaders from outer space – alien, ugly, women warriors, come to destroy the cozy tranquility and predictability of an established order of earthly life'. But she turned this science fiction metaphor around, by claiming that feminism could expose certain claims of the irrelevance of gender 'as science fictions'. She advocated a creative 'collision' of feminism and geography that would transform both. Geography would not be the same after feminism, but also:

> Geographers can transform feminism by pulling it down to earth, by grounding it, by showing how gender is and continues to be shaped by real geographies. With appropriate infusions of the geographical imagination, feminists can come to see the importance of the real geography behind the geographic metaphors that have suffused feminist writing,
>
> (p. 583)

By the start of the new century, feminist concerns and influences had become visible across many areas of human geography. The relative success (partial victories?) of feminist geography in transforming human geography had also served as inspiration for others seeking other transformations (as we documented in Chapter 7 with respect to the 'whiteness' of the discipline). Some of the impetus to social geographies of childhood (Aitken, 2001; Holloway and Valentine, 2000), as well as non-human/more-than-human geographies of other sentient life in relation to humans (e.g. Philo and Wilbert, 2000; Whatmore, 2002; Wilbert, 2009) may also have been encouraged by the example of how feminism *broadened* the range and understanding of what constitutes 'human' geography.

Strategic tensions remain, as between the idea that feminist geography should (or has) become mainstream and its sense of being on the margins and therefore iconoclastic and more critical. There are questions regarding the importance of differences between the genders vis-à-vis acknowledging differences among women and the stability of the terms 'male' and 'female'. These questions were most recently rearticulated within the Women and Geography Study Group (WGSG) of the RGS (with the Institute of British Geographers). The WGSG had been an important vehicle for the development of feminist geography in the UK; sponsoring conference sessions and meetings and commissioning textbooks – as noted in one of the opening quotes for this chapter. However, following debates and a poll of its members, it adopted the title of Gender and Feminist Geography Research Group in 2013. Writing in favour of the change, Brickell *et al.* (2013) felt they:

> have to ask could the title of the WGSG be read as essentialised and symbolically exclusive? Would the WGSG lobby on behalf of men where they were excluded because of their gender? Are we in danger of embodying that very exclusion? Are there other gender related concerns and advocacy issues for feminist scholars that don't come under the heading of 'women'? For those of us writing here, as feminists, it is important that the WGSG is seen to 'practice what it preaches' in terms of gender inclusion and this includes the group's – symbolic – name. Hence we think the name should be changed to a more inclusive form.
>
> (p. 11)

This did not proceed without controversy, however, with Browne *et al.* (2013) arguing that:

> the proposed removal of 'women' from WGSG would be a detrimental step. It negates the understanding that inequalities still exist between men and women (as well as the ways in which those who transgress this category are policed).
>
> (p. 7)

The WGSG had always been open to men. Though very few men joined (and two who had joined argued also that the name change was not needed (Hopkins and Jackson, 2013)), many more were influenced by feminist geography in different ways, from those who found intellectual engagement and inspiration to those (the majority?) who became more aware of (and hence ceased or moderated) sexist language in their writing, teaching and at academic meetings.

The feminist movement within geography from the 1970s on was a major, unsettling challenge to the discipline and its then male hegemony. By the power of its arguments against then current disciplinary practices and silences – as well as the structuring of its academic communities – it rapidly achieved major changes, though not, of course, without resistance. But its impact was much wider than the unequal treatment of women, and later of sexual orientations. It awakened human geographers to the importance of positionality (as illustrated in the discussions of postcolonialism in the previous chapter), thereby widening the general appreciation of the important role of context (including geographical context) in the structuration of societies in general and geography in particular, especially its political implications.

Yet while differences may have become more evident than in the formative days of feminist geography, it would be wrong to emphasise these at the expense of missing the broad changes that feminist geography has wrought. Feminist geography also tests notions of straightforward disciplinary progress and a neat sequence of paradigms in human geography. Did the scientific and technical revolution of quantification constitute 'progress' on what went before from a feminist vantage point? Feminist geography must operate within and negotiate wider academic structures (those described in Chapter 1), but is also critical of disciplinary and academic norms and has questioned the boundaries of geographic thought and asked who makes and remakes these. More than four decades on from its arrival, feminist geography is integral to the networks, history and structures of the discipline. Arguably, therefore, it remains the case that, as Gregson *et al.* (1997, pp. 199–200) postulated:

> to position feminist geography entirely on the margins of contemporary human geography is misplaced. Rather for this collective [here WGSG, 1997], feminist geographies occupy the centre and margins simultaneously: another example of paradoxical space.

Chapter 9

Applied geography and the relevance debates

If the past fifty years have been spent in developing geographical methods of survey and analysis, surely the time has now come to apply those methods towards the understanding and interpretation of some of the features of the world today. Further the time has surely come when those same methods of survey and analysis can be used in helping towards the solution of some of the great world problems.

(L. Dudley Stamp, 1960, p. 10)

Geography is by its nature *applied*. Geographers' systems of interest are of such direct concern to people, businesses, government, other organisations, and NGOs that any understanding achieved within the discipline is likely to be applicable.

(Robert Bennett and Alan Wilson, 2003, p. 463)

There was a time when American geographers ... graced the covers of national news magazines and engaged in policy at the highest levels of political decision making. Recall that Woodrow Wilson's preparations for the post-World War I peace efforts were headquartered at the American Geographical Society, calling on the expertise of seven past and future presidents of the Association of American Geographers ...

(Billie Lee Turner II, 2005, p. 462)

The history of geography is one of a subject and an academic discipline with a strong applied value − or relevance, as it is often termed. Indeed, the subject matter from which the discipline emerged comprised useful knowledge − maps, information about environmental conditions, material on trade potential etc. And as that information was collated, so it was put to wider uses in the service of the state apparatus − to Mackinder, one of the discipline's founders in the UK, geography was an 'aid to statecraft' (Parker, 1982; Kearns, 2009; Morin, 2011). As Bennett and Wilson (2003) point out, applied geography was being promoted there very early in the twentieth century (Keltie, 1908), with another book having the same title appearing soon after the First World War (Stevens, 1921). Nevertheless, and despite much 'applied work' by individuals (both academic geographers and those with geographical training working in other professions), notably though not only in subdisciplines such as medical and health geography that remain avenues for applied work (Kwan, 2014; J. Mohan, 2000; Cliff and Haggett, 2003), more explicit concerns about geography's applicability and relevance did not take centre stage within academic debates until well after the Second World War.

The period since the late 1960s has contained many testing years in the countries being studied here. The underlying problem has been economic uncertainty. After two immediate postwar decades of relatively high rates of economic growth and a sense of increasing prosperity, the

American and British economies began to experience serious difficulties, along with many of their neighbours. At times, each of the post-1970s decades appeared to offer a return to growth, only for the booms to be punctuated – in some areas at least – by sudden reversals, notably in 2007–2008. Further, it became increasingly clear that the prosperity was not being shared by all, even within the richest countries let alone at a global scale. This was highlighted in the USA by the growing tempo of the civil rights movement and in the UK by the turmoil in Northern Ireland, which began as a civil rights movement (and violent state reactions to it), and a series of 'inner-city riots' which were in part responses to racism and deprivation/discrimination. Student and youth protest erupted in both countries (as well as elsewhere) during 1968, related both to protest against the American war in Vietnam (that extended to Cambodia and Laos) and a wider upsurge in militancy, plus increasing concern about human-induced degradation of the physical environment. Discrimination against women was also brought to the forefront of attention (including the geographical profession: Zelinsky, 1973b), whereas since the 1980s growing economic and associated social and political inequalities, both within most Western countries as well as internationally, attracted increasing attention (Glasmeier, 2005; Dorling et al., 2010).

Much of the protest of the late 1960s centred on particular issues and was relatively ephemeral. For many of those involved, the aim was to win reforms within society, in the classical liberal/social democrat manner, while leaving its major structure untouched. For some, however, disillusion stimulated what Peet (1977) termed a 'breaking-off' from liberalism and a move to more radical political stances:

> The starting point was the liberal political social scientific paradigm, based on the belief that societal problems can be solved, or at least significantly ameliorated, within the context of a modified capitalism. A corollary of this belief is the advocacy of pragmatism – better to be involved in partial solutions than in futile efforts at revolution. Radicalization in the political arena involved, as its first step, rejecting the point of view that one more policy change, one more "new face", would make any difference.
>
> (p. 242)

The substantive concerns of those termed 'radicals' from the late 1960s on were treated in detail in Chapter 6. Although there are significant overlaps with that, the current chapter is concerned with general issues of applied geography and its role within the discipline. As noted in Chapter 1, the pressures for more applied work built up from the 1970s on in response to economic, social and political conditions. In addition, government policies towards higher education, especially in the UK, placed an increased emphasis on applied work, both in terms of providing vocational and professional training for students in fields where employment prospects were good and in the pressure to cover institutional costs through 'earnings' from research activity, whereas in the USA there was an increasing emphasis on geographic education providing students with technical skills (notably though not solely in GIS) to enhance their employment and career potentials. Thus, the debate over applied geography concerned pragmatic issues of disciplinary survival as well as concerns with regard to relevance.

Tracking the development of applied geography is not as easy as study of other aspects of the discipline, because relatively little of its output is published in academic journals. This is partly because of the nature of the research contracts and the associated funding for researchers. Some contracts make publication in academic journals difficult, because of commercial sensitivity of the research results, and even where this is not a particular problem the time pressures on those involved – to complete one project and then go on to the next, if for no other reason than to secure a continuing salary for the contract researchers – militate against preparation of academic papers and mature reflection on the theoretical and other implications of particular (usually empirical) research findings. As universities have been pressed to do more of this type of work, however, and as geographers have identified commercial niches for themselves, so the practice of applied geography has somewhat disappeared from general view. Nevertheless, its nature and validity have been debated, and that forms the focus of the present discussion.

Disenchantment and disillusion in academic geography

A forthcoming revolution in human geography, against the innate conservatism of behavioural studies and spatial science, was foreseen by Kasperson (1971):

> The shift in the objects of study in geography from supermarkets and highways to poverty and racism has already begun, and we can expect it to continue, for the goals of geography are changing. The new men [sic] see the objective of geography as the same as that for medicine – to postpone death and reduce suffering.
>
> (p.13)

Through the 1970s, an important aspect of debates about the application and relevance of geography concerned the status of 'radical geography'. This can be illustrated by one senior geographer's writings then. Wilbur Zelinsky was President of the Association of American Geographers (AAG) in 1973; his views may not be entirely typical, and are certainly more firmly stated than those of most of his peers (but see Symanski, 2002, 2007), but they reflect both growing disillusion within geographical circles with past achievements (see also Cooke and Robson, 1976) and a concern about future directions. Zelinsky's (1970, p. 499) first statement began:

> This is a tract. . . . The reader is asked to consider what I have come to regard as the most timely and momentous item on the agenda of the human geographer: the study of the implications of a continuing growth in human numbers in the advanced countries, acceleration in their production and consumption of commodities, the misapplication of old and new technologies, and of the feasible responses to the resultant difficulties.
>
> (p. 499)

He developed three basic arguments:

1 that people are inducing for themselves a state of acute frustration and a crisis of survival;
2 that these conditions originate, and can only be solved, in the 'advanced' nations; and
3 that the current 'growth syndrome' has profound geographical implications.

Material accumulation can no longer be considered progress, he argued, because it is unsustainable; effort is currently misallocated on a massive scale, and there is a major geographical task involved in its sensible reallocation.

Zelinsky identified five typical academic reactions towards the problems of the 'growth syndrome':

1 ignore them;
2 accept that major consequences will occur, but only eventually;
3 admit that problems exist, but argue that they are easily solved by the free market, perhaps with state guidance;
4 claim that nothing can be done, but that in any case we will survive; and
5 realise the potential for immediate, unprecedented trouble.

His own reaction clearly fell into the fifth type, but many others suggested that it was (and remains) an overreaction, as illustrated by Beckerman's (1995) critique of the 'sustainable development' thesis. Zelinsky identified three roles which geographers could play in facing up to perceived oncoming disasters. The first – which involves a minimal political commitment and 'should not offend even the most rock-ribbed conservative scholar' (p. 518) – is the geographer as diagnostician, applying 'the geographic stethoscope to a stressful demography' (p. 519) and mapping what he calls geodemographic load, environmental contamination, crowding and stress. The second involves the geographer as prophet, projecting and forecasting likely futures. Finally, there is the geographer as architect of utopia, educating with regard to problems and possible solutions and providing support for the unknown leaders who have the political will to guide society through the coming 'Great Transition'.

Ending his 'declaration of conscience' on a pessimistic note, Zelinsky argued regarding geographers:

> how woefully deficient we are in terms of practitioners, in terms of both quantity and quality, how we are still lacking in relevant techniques, but most of all that we are totally at sea in terms of ideology, theory and proper institutional arrangements.
>
> (p. 529)

Those criticisms were not confined to geographers, however; he applied them to scientists en masse in his AAG presidential address (Zelinsky, 1975). Science, he contended, is the twentieth-century religion, and has failed to avert the oncoming crisis (see also Harvey, 1973, 1989a, 1994). Its disciplinary specialisms and separatism 'fog perception of larger social realities' (p. 128) while 'fresher, keener insights, along with much better prose' (p. 129) are produced by the brighter contemporary journalists.

Zelinsky identified five crucial axioms as the foundations of science:

1   that the principle of causality is valid for studying all phenomena;
2   that all problems are soluble (see Johnston's, 1990b, response to Pacione, 1990a, regarding the difference between the solution and resolution of a problem);
3   that there is a final state of perfect knowledge;
4   that findings have universal validity; and
5   that total scientific objectivity is possible.

The social sciences have failed to live up to these, he claims, for several reasons: their immaturity; their use as a refuge for mediocre personnel; the difficulty of their subject matter concerning interpersonal relationships; their problems of observation and experimentation; and the political and other problems involved in applying their proposed solutions. Furthermore, and the major cause of their failure, the natural science model is irrelevant to the study of society:

> If we are in pursuit of nothing more than information or knowledge, then there is some value in copying the standard formula of a research paper in the so called hard sciences. . . . But if we are in pursuit of something more difficult and precious than just knowledge, namely understanding, then this simple didactic pattern has very limited value.
>
> (p. 141)

The natural science approach adopted by positivist spatial-scientists helps to describe the world, he claims, but not to understand it. Berry, on the other hand, argued to the contrary that all other approaches are flawed: science, as he defines it, 'is the basis for rational action' (Berry, 2002d, p. 502). If geographers practise it, then they may have substantial influence on 'how environmental influences modify thought and behavior by modifying brain structure and functioning . . . provided . . . we resist the linguistic excesses and the seductive but ultimately self-defeating siren song of self-congratulatory postmodernism' (Berry, 1999, p. 590). For him, following Edward O. Wilson (1998), science is differentiated from pseudo-science on five criteria: the repeatability of its findings by independent analysis; the economy of its explanations; its unambiguous statements; its heuristic power for future work; and its consilience – its appreciation that 'all living phenomena are ultimately obedient to the laws of physics and chemistry . . . and that all biological phenomena are products of evolution by natural selection' (Berry, 1998, p. 95). Geography needs to meet all these criteria if it is to contribute to constructing a better world. In a later piece (Berry, 2002e, p. 1), he presents a 'toolkit' for detecting whether a claim is valid as a series of four questions – 'Can it be tested?'; 'Is it supported by evidence?'; 'Has it been confirmed?'; and 'Is it logical?' – and wonders 'how much of what we see published in our flagship journals meets the test of my Cheshire cat's smile, and of that which does not, how much is pseudo-science, anti-science, or pure baloney' (p. 2).

There were two major components to Zelinsky's case: geographical research should be relevant to the solution of major societal problems; and the positivist-based spatial-science methodology may be inappropriate for such a task. Several were quick to point out that neither was particularly new, especially the first. House (1973), for example, reviewed the tradition of involvement in public policy by British geographers, and Stoddart (1975b, p. 190) identified the late nineteenth-century views of Reclus and Kropotkin as 'the origins of a socially relevant geography' – the latter was later rediscovered by the 'radicals' (Peet, 1977). Nor was the more 'revolutionary' approach of those who had 'broken off' from liberalism particularly novel: Santos (1974) reminded English-speaking geographers of the Marxist-inspired works of Jean Dresch on capital flows in Africa and Jean Tricart on class conflict and human ecology, both prior to the Second World War (see Clout, 2014), for example, and Keith Buchanan's presidential address to the New Zealand Geographical Society on the need for studying 'the absolute geographical primacy of the state, especially in the non-Western world' (Buchanan, 1962), produced an acid response from Spate (1963; on that episode more widely and related debates within New Zealand geography, see Moran, 2000).

Zelinsky's views were echoed in another AAG presidential address, with Ginsburg (1973) writing that:

> Much so called theory in geography . . . is so abstracted from reality that we hardly recognize reality when we see it. . . . The increasing demand for rigor to cast light on trivia has come to plague all of the social sciences . . . the most important questions tend not to be asked because they are the most difficult to answer.
>
> (p. 2)

At about that time, the Association established a Standing Committee on Society and Public Policy (Ginsburg, 1972), which White (1973, p. 103) hoped would 'be alert to distinguishing the fatuous problems and the activities that are pedestrian fire-fighting or flabby reform'. This statement was made at a session on geography and public policy at the Association's 1970 conference; a similar theme was chosen for the 1974 conference of the Institute of British Geographers (IBG). Not all geographers accepted White's aim – using learned societies to influence public policy – while not denying the value of geographic method in social engineering. Trewartha (1973) stated that:

> I must demur when he proposes that it should be a corporate responsibility of our professional society to become an instrument for social change. . . . From the beginning, the unique purpose of the Association of American Geographers has been to advance the cause of geography and

geographers; it was never intended to be a social-action organization. . . . All kinds of research, pure as well as applied, should be equally approved and supported by the AAG.

(p. 73)

Relevance to what and for whom?

Claims that geographical work should be more relevant to major societal problems raised queries about the nature of that relevance, and it soon became apparent that there was no consensus on what should be done, and why. The ensuing debate is illustrated by a number of contributions to the British journal *Area* during the early 1970s.

Michael Chisholm (1971b) opened it, identifying differences between governments, with their interests in cost-effective research and their primacy in evidence-based decision-making, and academics, some of whom are concerned to protect their academic freedom and their right to be the sole judges of what they study and publish. Traditionally, geographers had advised governments in the roles of information gatherers and 'masterful synthesisers' and they had not been involved in the final stages of policy-making; they had been delvers and dovetailers, but not deciders. On the latter role, unfortunately:

> The magic of quantification is apt to seem rather less exciting when the specifications of the goods it can deliver are inspected at close quarters (p. 66). . . . The danger with empirical science is the absence of guidance at the normative level as to which of various options one should take.
>
> (pp. 67–8)

The challenge to human geography, according to Chisholm, was to define such norms. He himself later worked for several UK governments: as a member of independent commissions (the Local Government Boundary Commission in the 1970s and the Local Government Commission in the 1990s) he was involved in advising governments on the best way to redraw important components of England's administrative map (Chisholm, 1975b, 1995) – although he has been critical of many of the procedures with which he was involved, and was apparently not reappointed to the Local Government Commission because of this (Chisholm, 2000; Chisholm and Leach, 2008). Eyles (1971) responded that the focus of relevant research should be 'some of the social and spatial inequities in society' (p. 242), and the first challenge is to study the distribution of power in society, thereby identifying the mechanisms for allocating scarce resources. Research could then isolate the disadvantages of relative powerlessness and provide the basis for policy aimed at redistributing resources.

British geographers were introduced to the ongoing debates in American geography in two reports on the 1971 AAG conference. (This was not the first at which major social issues had been raised. The 1969 meeting should have been in Chicago but was transferred to Ann Arbor as a protest over police violence directed at anti-war protesters at the 1968 Democratic Party Convention in Chicago. The radical journal *Antipode* was launched there in 1969, and those attending were confronted by the problems of the inhabitants of Detroit's black ghetto, small groups of whom 'invaded' several of the sessions to present their concerns at the academics' 'unworldliness'. Bunge led an 'unofficial' field trip to Detroit which highlighted the massive racialised disparities across Detroit's suburbia.) By 1971, 'many geographers were deeply frustrated by a sense of failure' to deal with major social issuegs (Prince, 1971b, p. 152), but although some members were taking notice of 'the sufferings of the outside world' (p. 152) other scholars remained 'locked in private debates, preoccupied with trivia, mending and qualifying accepted ideas' (p. 153).

David Smith's (1971) report on that AAG conference suggested that American geography was about to undergo another revolution, to counter a situation in which 'geography is overpreoccupied with the study of the production of goods and the exploitation of natural resources, while ignoring important conditions of human welfare and social justice' (p. 154). This forthcoming revolution would involve a fundamental re-evaluation of research, teaching activities and basic social philosophies, and was represented at the conference by activities such as the foundation of SERGE (the Socially and Ecologically Responsible Geographer) by Zelinsky and others and by a motion at the Annual General Meeting condemning USA involvement in Vietnam. Smith (1971) was unsure whether this revolution, with its emphasis on social as against economic concerns, would spread to the UK, however:

> The conditions which have helped to spawn radical geography in the United States include the existence of large oppressed racial minorities, inequalities between rich and poor with respect to social justice, a power structure and value system largely unresponsive to the needs of the underprivileged, and an unpopular war which is sapping national economic and moral strength. These conditions do not exist in Britain or exist in a less severe form, and the stimulus for social activism in geography is thus considerably less than in America.
>
> (pp. 156–7)

By the mid-1990s, however, he was writing of the need 'to place social justice at the heart of human geography' (Smith, 1994a/b, p. 1). Dickenson and Clarke (1972) responded to his 1971 claim by arguing that British geographers had long been concerned with 'relevant' issues, with particular respect to the Third World.

Another commentary on AAG's 1971 conference came from Berry (1972b), who felt that what he had observed was just a new fad involving 'new entrants to the field seeking their "turf"'. He could identify no real commitment:

> The majority of the new revolutionaries, it seems, are essentially white liberals – quick to lament the supposed ills of society and to wear their bleeding hearts like emblems or old school ties – and quicker to avoid the hard work that diagnosis and action demand. A smaller group of hard-line Marxists keeps bubbling the potage of liberal laments. *In neither group is there any profound commitment to producing constructive change by democratic means. . . . If either of these will be the 'new geography' of the 1970s, count me out.*
>
> (pp. 77–8, original italics; see also Berry, 2002c)

To Berry, academic geographers should provide a knowledge base on which policy can be built, which implies close involvement with the education of future generations of policy-makers (on which, see Yeates, 2001). To Blowers (1972), however, 'The issue is not how we can cooperate with policy-makers, but whether and in what sense we should do so. It is a question of values' (p. 291); he argued that the sorts of activities proposed by Berry would be strongly supportive of the status quo, and unlikely to produce fundamental social reform.

David Smith's (1973a, p. 1) response to Berry was that 'bleeding hearts sometimes help to draw attention to important issues, and Marxists can make valuable contributions in the search for alternatives to existing institutions and policies', and he pointed out that the current 'fad' was no more pronounced than that of the quantifiers a decade earlier. The results of their 'revolution' offered little for the solution of social problems, however, and Smith doubted the value of the large projects established by the AAG as part of its geography and public policy drive. (One of those projects produced a major series of books on metropolitan America, though their applied relevance was not clear (Adams, 1976).) Research should highlight particular problems, and teaching should place emphases on 'a man in harmony with nature rather than master of it, on social health rather than economic health, on equity rather than efficiency, and on the quality of life rather than the quantity of goods' (p. 3). Michael Chisholm (1973) advocated caution in the corridors of power, because geographers had done insufficient substantive research to back up a 'hard sell' (a point made three decades earlier by Ackerman: see p. 76, this book); Eyles (1973, p. 155) argued that 'any entry to those corridors assumes that the structure underlying policy alternatives is basically sound'; and Blowers (1974) wrote that in the corridors one can only influence, not decide, and that for the latter task geographers must develop their political convictions and act accordingly. (Blowers was active in English local government, where political parties dominate, for some decades, focusing on environmental issues, on which he also published major academic works (Blowers, 1984). On another geographer's role as an active local politician, see Pawson, 2011.)

This debate on if, and how, geographers should contribute to the solution of societal problems was a major issue at the 1974 IBG annual conference. Coppock's (1974) presidential address presented the challenges, opportunities and implications of geographical involvement in public policy, an involvement which he felt the current generation of students welcomed. He argued that policy-makers were largely ignorant of potential geographical contributions, while at the same time geographers seemed unaware that 'there is virtually no aspect of contemporary geography which is not affected to some degree by public policy' (p. 5). Coppock wanted to change this, to have geographers identify the contributions that they could make, to encourage research relevant to those contributions, and to enter a dialogue with those who advise on and implement public policy. (At Edinburgh in the 1960s he had established a Tourism and Recreation Research Unit within the Department of Geography which undertook consultancy work for both the public and private sectors as well as providing the material for research publications (Coppock and Duffield, 1975; Clout, 2002).)

Other conference contributors were neither as optimistic nor as committed as Coppock. Hare (1974), himself an adviser to the Canadian government, reacted to cries that geographers were not consulted enough with the reply 'Thank goodness'; geography, as a discipline, is irrelevant to the separate domain of public policy-making, although geographers, as individuals, because of the breadth of their training, could offer much that was valuable, so his conclusion was 'Geography no, geographers yes' (p. 26). His was a different response to mounting social concerns from Steel, who told the British Geographical Association (Steel, 1974):

> As geographers we often get hot under the collar over the number of theoretical economists who are called on to advise the governments of developing countries. We comment on how much better World Bank surveys of countries would be if they were prepared, at least in part, by geographers. . . . We wonder why university departments of geography are not engaged on a consultancy basis more often than they are, and we marvel that the Overseas Development Administration in London has only a handful of geographers on its staff where, we feel, an army would be more appropriate.
>
> (p. 200)

Hare (1977) subsequently argued that a major reason for a lack of geographical contributions to public policy may be the poverty of their academic training; in recent years we have 'swept geography departments into the social-science divisions of faculties of arts and sciences where, from playing second fiddle to geologists or literary critics, we learned to play second fiddle to economists and sociologists' (p. 263). Geographers would have to rebuild their discipline based on the centrality of society–environment interactions, he argued, with a new brand of physical geography that leans heavily on biological ideas and sources: 'We must reassert the old, essential truth that geography is the study of the earth as the habitat of man, and not some small sub-set of that gigantic theme' (p. 266; see also Steel, 1982; Hart, 1982a). Hare's views on the current irrelevance of geography to public policy were supported by Peter Hall (1974, p. 49), who argued that:

> geography, most clearly of all the social sciences, has neither an explicit nor an implicit normative base . . . spatial efficiency . . . is rather a description of what men seek to do in actuality . . . not . . . an objective to be achieved or objective function to be maximized.

Policy-makers must seek their norms elsewhere; geographers, meanwhile, must develop a new political geography which will aid them in understanding the crucial role of political decisions in structuring spatial systems (Johnston, 1978c).

Two other papers given at the 1974 IBG conference argued against Coppock's programme. Bridget Leach (1974), for example, claimed that governments, as paymasters, already constrained what geographers could do research on, and as a result geographers were being used; their only alternative was political action. Harvey's (1974c) contribution was entitled 'What kind of geography for what kind of public policy?'. Individuals wishing to become involved in policy-making were, he argued, stimulated by motives such as personal ambition, disciplinary imperialism, social necessity and moral obligation; at the level of the whole discipline, on the other hand, geography had been co-opted, through the universities, by the growing corporate state, and geographers had been given some illusion of power within a decision-making process designed to maintain the status quo. He portrayed the corporate state as 'proto-fascist' (p. 23), a transitional step on the path to the barbarism depicted in Orwell's *Nineteen Eighty-Four.* The function of academics, he claimed, was to counter such trends, to expunge the racism, ethnocentrism and condescending paternalism from within their own discipline and to build a radical subject, thereby assisting all human beings 'to control and enhance the conditions of our own existence' (p. 24).

Harvey (1973) had previously argued that the current mode of analysis in geography offered little for the solution of pressing societal concerns:

> There is an ecological problem, an urban problem, an international trade problem, and yet we seem incapable of saying anything of depth or profundity about any of them. When we do say anything, it appears trite and rather ludicrous. . . . It is the emerging objective social conditions and our patent inability to cope with them which essentially explain the necessity for a revolution in geographic thought.
>
> (p. 129)

He recognised three types of theory:

1   Status quo, which represents reality accurately but only in terms of static patterns, and therefore cannot make predictions which will lead to fundamental social change.
2   Counter-revolutionary, which also represents reality, but obfuscates the real issues because it ignores (either deliberately or accidentally) the important causative factors and so can be used to promote changes that will not bring about significant alterations to the operation of those factors. It is 'a perfect device for non-decision making, for it diverts attention from fundamental issues to superficial or non-existent issues' (p. 151).
3   Revolutionary, which is grounded in the reality it seeks to represent, and is formulated so as to encompass the contradictions and conflicts which produce social change.

Harvey wished to write revolutionary theory, thereby overthrowing the current paradigm; his blueprint for geography:

> does not entail yet another empirical investigation. . . . In fact, mapping even more evidence of man's patent inhumanity to man is counter-revolutionary in the sense that it allows the bleeding-heart liberal in us to pretend we are contributing to a solution when in fact we are not. This kind of empiricism is irrelevant. There is already enough information. . . . Our task does not lie here. Nor does it lie in what can only be termed 'moral masturbation' of the sort which accompanies the masochistic assemblage of some huge dossier of the daily injustices. . . . This, too, is counter-revolutionary for it merely serves to expiate guilt without our ever being forced to face the fundamental issues, let alone do anything about them. Nor is it a solution to indulge in that

emotional tourism which attracts us to live and work with the poor for a while. . . . These . . . paths . . . merely serve to divert us from the essential task at hand.

This immediate task is nothing more nor less than the self-conscious and aware construction of a new paradigm for social geographic thought through a deep and profound critique of our existing analytical constructs. This is what we are best equipped to do. We are academics, after all, working with the tools of the academic trade . . . our task is to mobilize our power of thought, which we can apply to the task of bringing about a humanizing social change.

(pp. 144–5)

To Harvey, then, relevant geography involves building new geographic theory on a Marxist base, and its dissemination will achieve social reform through the education process. (This gradualist view, promoting reform through education, is only implicit in Harvey's work (Johnston, 1974, p. 189). Harvey later discussed the problems of creating a new society, and outlined a scenario of how that might come about, and with what effects (Harvey, 2000).) Blaut (1979) contended that Marxist theory offered benefits because it can handle two crucial issues which positivist theory cannot: increased injustice and heightened economic and social instability.

Harvey (1984) presented a forceful elaboration of his views in a paper subtitled 'an historical materialist manifesto'. Geography, he contended, not only records, analyses and stores information about society but also 'promotes conscious awareness of how such conditions are subject to continuous transformation through human action' (p. 1). The nature of the knowledge that it produces and propagates reflects the social context of place and time, hence the role of geography in the 'Bourgeois era' as an 'active vehicle for the transmission of doctrines of racial, cultural, sexual, or national superiority' (p. 3). The positivist movement had sought to establish a universal science of spatial relations, which was countered by both humanistic and Marxist critiques. Together the latter suggested the development of a revitalised geography, but they lacked:

a clear context, a theoretical frame of reference, a language which can simultaneously capture global processes restructuring social, economic and political life in the contemporary era and the specifics of what is happening to individuals, groups, classes, and communities at particular places at certain times.

(p. 6)

Historical materialism provides that framework, he argues (though see Eliot Hurst's, 1985, argument that if Harvey accepted his own (1972) contention that disciplinary boundaries are counter-revolutionary, then it is hard to understand how he can later promote a disciplinary-based manifesto).

Harvey argued that geographers cannot be neutral; their work must be of value to some special interest group within society. For him, that group should not be 'generals, politicians, and corporate chiefs' (p. 7) but the disenfranchised. A people's geography 'threaded into the fabric of daily life with deep taproots into the well-springs of popular consciousness . . . must also open channels of communication, undermine parochialist worldviews, and confront or subvert the power of the dominant classes or the state. It must penetrate the barriers to common understanding by identifying the material base to common interests' (p. 7). It would not only reveal to the disenfranchised how societies are structured and restructured so that 'centers exploit peripheries, the first world subjugates the third, and capitalist powers compete for domination of protected space (markets, labor power, raw materials). People in one place exploit and struggle against those in another place' (p. 9). It would also help them to:

Define a political project that sees the transition from capitalism to socialism in historico-geographical terms . . . we must define, also, a radical guiding vision: one that explores the realms of freedom beyond material necessity, that opens the way to the creation of new forms of society in which common people have the power to create their own geography and history in the image of liberty and mutual respect of opposed interests. The only other course . . . is to sustain a present geography founded on class oppression, state domination, unnecessary material deprivation, war, and human denial.

(p. 10)

Harvey returned to this theme in 1989, in a retrospective volume reporting on a conference held to mark the twentieth anniversary of the publication of *Models in Geography* (see p. 93, this book). He set out a brief critique of the scientific approach adopted in the earlier book, claiming that the type of modelling adopted could not be used to tackle what he identified as the major geographical questions (such as the 'evolutionary path of capitalism itself' (Harvey, 1989b, p. 212); in this context, he portrays 'all geography as historical geography'). Indeed, 'when put in the context of these grander questions, the modelling effort appeared both puny and not particularly revealing'. Many geographers retained that modelling approach, however, by fragmenting the discipline and focusing on narrow questions. They had restricted the nature of the questions asked and as a consequence:

we can now model spatial behaviour like journey-to-work, retail activity, the spread of measles epidemics, the atmospheric dispersion of pollutants, and the like, with much greater security and precision than once was the case. And I accept that this represents no mean achievement. But what can we say about the sudden explosion of third world debt in the 1970s, the remarkable push into new and seemingly quite different modes of flexible accumulation, the rise of geo-political tensions, even the definition of key ecological problems? What more do we know about major historical-geographical transitions (the rise of capitalism, world wars, socialist revolutions and the like)? Furthermore, pursuit of knowledge by the positivist route did not necessarily generate usable configurations of concepts and theories. There must be thousands of hypotheses proven correct at some appropriate significance level in the geographical literature by now, and I am left with the impression that in toto this adds up to little more than the proverbial hill of beans.

(pp. 212–13)

Harvey found it hard to understand why Marxism had had such little influence on geography until the 1970s, and noted that, when it did, 'geographers who turned to Marx were swept up in that, and many were so submerged in it that they entirely forgot their own disciplinary identity' (p. 215). The submersion was necessary, but so was the identity; the production of knowledge is a political project, and the history and future of the discipline have to be appreciated in that context. Thus:

any project to 'remodel' contemporary geography must take the achievements of the Marxist thrust thoroughly into account at the same time as it recognizes the limits of positivism and the restricted domain of the modelling endeavours that derived therefrom. This is not to rule all forms of mathematical representation, data analysis and experimental design out of order, but to insist that those batteries of techniques and scientific languages be deployed within a much more powerful framework of historical-materialist analysis.

This, he accepts, must be ideological 'because it is necessarily political and built upon some conception of our collective agency in history', as illustrated in his book on *Rebel Cities* (Harvey, 2012).

Liberal contributions

Liberalism is defined here as combining 'a belief in democratic capitalism with a strong com-mitment to executive and legislative action in order to alleviate social ills' (Bullock, 1977, p. 347). Associated in the USA with the left of the Democratic party, and in Australia, Canada, New Zealand, the UK and parts of Europe with social/liberal democratic, socialist and Labour parties, liberals are concerned that all members of society do not fall below certain minimum levels of well-being (variously defined), and are prepared for state action within the capitalist structure in order that this can be achieved. There are diverse strands within liberalism, some more radical/left and others more centred on market liberalism and so connected with centre-right politics and neoliberal policies. Within geography however, much of the early work conducted in liberal ethos focused on description rather than theory-construction; Chisholm's (1971a) investi-gation of the potential of welfare economics as a basis for normative theory which does not involve profit-maximisation goals was one of the few exceptions (see also Wilson, 1976b).

There is a long tradition of liberal contributions by 'applied geographers', both in their research and associated activities, and in their teaching (see House, 1973; Hall, 1981b, 2003, reviews UK geographers' work in the fields of urban planning, for example, and Hudson, 2003, similarly outlines their work on issues of regional development and planning). In the UK, applied geography as information gathering and synthesising has a substantial record, starting with Stamp's Land Utilization Survey in the 1930s, and his subsequent involvement in the preparations for postwar land-use planning (Stamp, 1946b, 1949). Other aspects of land-use planning were of interest to geographers in the 1930s and 1940s, and were discussed several times at the RGS (Freeman, 1980b); similar geographical involvement occurred in the USA (Kollmorgen, 1979). During the Second World War geographers made many contributions as information synthesisers and gatherers, the latter including the development of air-photo interpretation (on British contributions, see Balchin, 1987, and Clout and Gosme, 2003; on USA activity, see several of the essays in E. W. Miller, 1993).

After the Second World War, land-use and transport planning were established on a large scale, and trained geographers provided a substantial proportion of its personnel. The relationship between land-use planning (often termed town and country planning in the UK and spatial planning elsewhere − terminology increasingly adopted in the UK too) and geography led Phelps and Tewdwr-Jones (2008) to ponder if planning is geography's disciplinary alter ego. Moreover:

> Further similarities between the disciplines of geography and planning emerge when we consider not only critical reflection over what each of the disciplines is, but also critical reflection on the evolution of ideas in both. Again, it is worth rehearsing for a geographical readership the degree of intellectual ferment that has characterised the planning discipline over the past four decades and more. Broadly speaking, academic planning has undergone the same evolution of ideas and thought as human geography with key texts embracing positivism, behavioural and humanistic approaches, structuralism and post-modern approaches, as well as specific theories associated with these approaches.
>
> (p. 571)

In addition to planning, after 1945, many academic geographers became active in applied work in other ways. Technical developments in cartography and data handling saw them involved in redistricting for Congressional elections (Morrill, 1981), a wide range of mapping activities (Rhind and Adams, 1980; Rhind later became the first academic to head the British national mapping agency − the Ordnance Survey: Rhind, 2003), and developing regional bases for the presentation of census statistics (Coombes et al., 1982; Berry et al., 1969), for example, while developments in GIS during the 1980s and 1990s stimulated very substantial increases in mapping and enumeration (Charlton *et al.*, 1995; Martin, 2002). Alongside this, various policies have been evaluated, such as those aimed at changing the distribution of industrial activity, and much of the entropy-maximising systems modelling (p. 118, this book) was intended to provide procedures for planning the joint activities of land-use and transport planning (Wilson, 1974; Bennett and Wilson, 2003; on whether it did, see Batty, 1989). Elsewhere, Hägerstrand, who made many original and highly influential contributions to the development of spatial analysis (p. 132, this book) was deeply involved in a wide range of spatial planning projects in Sweden, in part through his pioneering work in computer cartography (Öberg, 2005). Developments in what has been termed 'participatory GIS' (see too Chapters 4 and 6) have promoted open access to the hardware and software, thereby overcoming the 'expert-customer' relationship (Elwood, 2006) and the associated creation of 'neogeography' provides online access to map-making tools for non-professionals (Goodchild, 2009; Rana and Joliveau, 2012; see also the New Mappings Collaboratory at the University of Kentucky: http://newmaps.as.uky.edu/online-public-engagement), though scepticism about neogeography as 'an attempt to reduce geography to an exercise in instrumental rationality' (Leszczynski, 2014, p. 75) reiterates some of the prior debate about for whom and what kinds of (applied) geography is being proposed (and enacted) for what kind of policies

In a different context, after the events of 9/11 in 2001, the Association of American Geographers published a book illustrating its editors' conception of geography's role – especially in its deployment of GIS and associated technologies – in tackling such terrorist activities and reducing the probability of their recurrence (Cutter *et al.*, 2003). This was substantially criticised in reviews (several were published, for example, in the December 2004 issue of the *Annals of the Association of American Geographers*). The editors defended their chosen perspective by claiming that 'through our discussions with hundreds of geographers, it was clear that geography had a great deal more to offer on this issue than political commentary' (Cutter *et al.*, 2004, p. 1001), with their collection reflecting a discipline ready to respond to major national events with material about 'how to reduce risks of terrorism in the country and the world at large', thereby demonstrating 'disciplinary citizenship' (p. 1002).

Most of the work presented as exemplifying applied geography used an empiricist and (usually implicit) positivist framework. Geographers are perceived as having valuable skills in the collection and ordering of data, as in land-use surveys, for example, and the presentation of such data frequently assumed the existence, and desirability of maintaining, certain causal relationships; planning agricultural land use, for example, often assumed a clear causal link between the physical environment and agricultural productivity and that of industrial location assumed the need for efficiency via the minimisation of total travel costs. (Interestingly, some critics pointed out that the greatest use of such optimising models occurred not in the 'capitalist West' but rather in the 'socialist East'.) As positivist work on the allocation of land uses and traffic flows increased, so the potential for geographic inputs to spatial planning was promoted. Most of this was pragmatic application of technical skills, though there were also evaluations of policy impacts (e.g. Hall *et al.*, 1973) and attempts to develop decision-making theories in this context (Hall, 1981b, 1982, 1998).

Geographers' empiricist role continued to be advanced in response to economic crises and the perceived need for valid data. In the UK, for example, the government introduced a three-year programme in 1983 to provide 'new blood' for university research activities; 792 lectureships were available, in open competition between universities, with a further 146 in the fields of information technology. Geography departments received 11.5 per cent of these posts (1.1 per cent of the total; the staff of geography departments comprised 2 per cent of all university posts in 1982–3). David Smith (1985) argued not only that geography lost out relatively in that contest (the 'winners' were engineering, technology, and the physical and biological sciences), but also that within geography the posts allocated were selectively focused on certain aspects of the discipline: five were for research in remote sensing/digital mapping, for example, and three more were for various aspects of mathematical modelling (two in physical geography); only one reflected 'place-specific' issues, a post in historical–cultural geography. Smith interprets this as follows:

> The predominance of remote sensing (and associated digital mapping) reflects a view of geography as a technologically sophisticated means of gathering and displaying information, in the tradition of the geographer as map-maker linked to the contemporary preoccupation with information technology . . . [that] appears to conflate the needs of the discipline with a conception of the needs of society in which the emphasis is much more economic than social. It is hard to see more than one or two of the posts contributing much to the solution of social problems. The predominant impression is of an a-social view of the world, in which social relations, class structure and political power seem strangely absent.
>
> (p. 2)

Clayton (1985a) responded that this was unlikely to be the only example of more direct state influence on the direction of academic research (see also his analysis of how geographers should react to that: Clayton, 1985b). In 1986 the University Grants Committee produced its first rating of the research record of every university department – on criteria that included externally earned research income as well as publications, information on which was subjectively examined by 'expert panels' which graded every department; for the latest exercise (2014) 20 per cent of the final grading was based on a series of 'impact case studies'. (The Research Assessment Exercises – RAEs – were replaced by the Research Excellence Framework – REF – for the 2014 exercise.) Using the results of the first exercise, it began a selective allocation of funds to universities that reflected those ratings (see D. M. Smith, 1986). Similar rating exercises were undertaken in 1989, 1992, 1996, 2001 and 2006, and 2014 (each used peer review and was more transparent than the first: Thorne, 1993, though see Collini, 2012, and Sayer, 2015). Virtually all the research money distributed to the universities by the funding councils (some £20 million annually in 2003 across geography departments) is now allocated by a formula linked to those ratings, with most money going to the largest departments whose research records and plans win the highest ratings (on the funding implications, see Johnston, 1993a, 1995b; on the ethics of such a distribution, see Smith, 1995, 1996; Rhind, 1996; Curran, 1996; and Curran, 2000, 2001a modelled the distribution of 'success' in these exercises). In the early twenty-first century, this allocation procedure was made even more selective, with only a minority of the country's geography departments getting substantial allocations because only work graded 3* or 4* (the highest two grades) qualified for funding, with the implication that the remainder were largely being funded for their teaching activities only. (Similar exercises were adopted by the Australian and New Zealand governments.)

Alongside those who argued for a greater commitment to applied geography in the empiricist/ positivist mould, and therefore for an implicit acceptance of a particular ideology (Johnston, 1981a), others challenged this as the best way to respond to pressing societal problems. They argued for a reappraisal of how geographers could assist in understanding the genesis of those problems rather than in the suggestion of solutions which rarely tackled the root causes. The major contributions of the two groups are the subject of the next sections.

Mapping welfare

A range of quantitative research was published in the 1960s and 1970s under the general title of factorial ecologies, applications of multivariate statistical procedures to large data matrices as a means of representing spatial variations in population characteristics. These, according to David Smith (1973b, p. 43), were over-reliant on certain types of census data and therefore provided little information on social conditions. Earlier attempts had been made to structure analyses of such data towards particular ends, as in the work of rural sociologists on farmers' levels of living

(Hagood, 1943) – a concept introduced to the geographical literature by G. Malcolm Lewis (1968) and in John Thompson *et al.*'s (1962) investigation of variations in levels of economic health among different parts of New York State.

Factorial ecology procedures were adapted to the task of portraying spatial variations in social welfare in the 1970s, led by two workers. Knox (1975) promoted the mapping of social and spatial variations in the quality of life as a fundamental objective for geography, to provide both an input to planning procedures and a means of monitoring policies aimed at improving welfare. He divided the concept of 'level-of-living' into three sets of variables – physical needs (nutrition, shelter and health); cultural needs (education, leisure and recreation, and security); and higher needs (to be purchased with surplus income) – and used statistical procedures to portray their spatial variations. With the resultant maps, geographers must then decide whether they are playing a sufficient role in awakening human awareness of the extent of the disparities or are 'under an obligation to help society improve the situation' (p. 53).

Similar work by Smith (1973b) was set in the context of the American social indicators movement and the growing belief there that GNP and national income 'are not necessarily direct measures of the quality of life in its broadest sense' (p. 1). He initiated the collation and dissemination of territorial social indicators in order to illustrate the extent of discrimination by place of residence which occurs in the USA; multivariate statistical procedures generated the maps, on interstate, intercity and intra-urban scales. (Cox, 1979 extended the treatment to the international scale.) Smith's (1979, p. 11) goal was to present 'the basis for a better understanding of the origins of inequality as a geographical condition and of the difficulties in the way of plans to promote greater equality in human life chances'.

In Chisholm's terms (p. 313, this book), the works by Cox, Knox and Smith represented the geographer as delver and dovetailer, producing information on which more equitable social planning could be based, but with little indication as to how. Other studies performed similar roles, but also suggested spatial policies which could lead to improvements. Harries (1974), for example, analysed spatial variations in crime rates and the administration of justice, and argued that predictive models of criminal patterns could assist the organisation of police work (see also Haining, 2009); Shannon and Dever (1974) and Joseph and Phillips (1984) investigated variations in the provision of healthcare facilities and argued for spatial planning that would improve the services offered to the sick (which is different from a geography of prophylaxis: Fuller, 1971); and Morrill and Wohlenberg (1971) studied the geography of poverty in the USA, proposing both social policies – higher minimum wages, guaranteed incomes, guaranteed jobs, and stronger anti-discrimination laws – and spatial policies (such as an extensive programme of economic decentralisation to a network of regional growth centres) that would alleviate this major social problem. More recently, research in social geography has been undertaken with the twin goal of both uncovering and accounting for geographies of 'social problems' and contributing towards their amelioration/ eradication (e.g. on alcoholism: Jayne *et al.*, 2011; on children's use of online resources: Holloway and Valentine, 2000). An alternative, highly personal, programme of mapping variations in human welfare was advanced by Bunge (1971), who prepared a 'geobiography' of his home area in Detroit's black ghetto. His deeply humanitarian concern for the future was interpreted as a need to ensure a healthy existence for children; he proposed a 'dictatorship of the children' (Bunge, 1973b, p. 329) with regions –'may the world be full of happy regions' (p. 331) – designed for them. This requires a reduction in the worship of machines, which are inimical to children's health (Bunge, 1973c), and a mapping of the sorts of variables never collected by external agencies and therefore requiring the development of geographical expeditions within the world's large cities. These maps would include roach regions, parkless spaces, toyless regions and rat-bitten-children regions (Bunge, 1973d), and some were prepared for both Detroit and Toronto (Bunge and Bordessa, 1975). This work was later described by Merrifield (1995, p. 57) as involving excursions 'beyond the cloisters of the academy' and 'a redefinition of the research problematic and intellectual commitment of the researcher away from a smug campus career to one incorporating a dedicated community perspective'. The task of empathising with and situating oneself authentically within an impoverished community is difficult, and the researcher's own biography is likely to influence the process of learning about and becoming committed to such a community. This raises the issues of voice that have increasingly concerned those studying various forms of representation. By becoming involved, the geographer will become 'a person of action, a radical problem-raiser, [and] a responsible critical analyst participating with the oppressed' (Merrifield, 1995), but:

> there is always an immanent hazard that the voice heard in the supposed symbiosis between academic geographer and folk geographer is skewed towards the overzealous – though well-meaning – academic geographer. As the voice of the oppressed is muted, the expedition program degenerates into a paternalism reminiscent of 19th century Western missionaries and settlement houses.
>
> (p. 63)

More mapping has been undertaken since the pioneering efforts – as in Seager's atlases (Seager and Olson, 1986; Seager et al., 1995; Seager and Enloe, 2011), but increasingly, in line with Harvey's critique of such descriptive analyses, efforts have been redirected towards an appreciation of the underlying processes generating such geographies, not just describing 'who gets what, where' but also analysing 'who gets what, where . . . and how' (Johnston, 1975). Indeed, the ease of producing maps by computer has resulted in a large number being published at all scales, with many being welcomed for the spatial insights they provide into the geographies of, for example, inequality (e.g. Dorling et al., 2010; Dorling and Thomas, 2011), even though such volumes are much stronger on description than analysis and explication, let alone policy prescriptions; they stimulate questions but have no answers (in the 'geographer as information gatherer' mould identified by Chisholm, 1971b), and some are basically political polemic/rhetoric (compare Dorling, 2014, and Sayer, 2015).

Attempts at understanding

The mapping investigations just discussed were – and their successors largely still are – very largely descriptive, and any prescriptions offered based on limited theoretical foundations. Attempts were made to develop the necessary theoretical understanding, however. Kevin Cox (1973), for example, looked at the urban crisis in the USA – the racial tensions and riots, municipal bankruptcies, and the role of the government in the urban economy – presenting his analysis in terms of conflict over access to sources of power (see also Adams, 1972, on the role of spatial constraints as a generating influence on ethnic unrest in American cities in the late 1960s). This was intended as part of an educational exercise, for:

> It would be utopian to think that we can propose solutions on the basis of our analysis. The locational problems and locational consequences of policies weave too intricate a web for that to be possible. All we can hope to do is inform. To be aware of the problems and of their complexity may induce some sensitivity in a citizenry which has shown as yet precious little tolerance for the other point of view.
>
> (K. Cox, 1973, p. xii)

Nevertheless, his final chapter, entitled 'Policy implications', discussed two imperatives towards greater equity in the provision of public services – moral and efficiency (the latter applies to the total level of welfare in society as well). The policies presented involved spatial reorganisation to achieve the desired equity (including metropolitan government integration), community control, population redistribution and transport improvements. Massam's (1976) review of geographical contributions to social administration also focused on spatial reorganisation; he evaluated service provision using the spatial variables of distance and accessibility, with major chapters on the size and shape of administrative districts and on the efficient allocation of facilities within such districts (see also Hodgart, 1978).

Cox's work heralded an increased geographical interest in a much neglected field, the role of the state in capitalist society (see also Cox et al., 1974; Dear and Clark, 1978; Johnston, 1982b; Cox, 1979). Political geography had traditionally been mostly concerned with the state at the macro-scale only, dealing with (geo-)political regions, with national and intra-federal boundaries, and with the operations of the international political system (e.g. Muir, 1975). The similarly underdeveloped field of electoral geography had highlighted spatial variations in voting, but there had been little work on either the geographical inputs to voting or the geographical consequences of the translation of votes into political power (Gudgin and Taylor, 1979; Taylor, 1978, 1985b; Taylor and Johnston, 1979). The state is involved in many aspects of economic and social geography, however, as both Buchanan (1962) and Coppock (1974) had stressed, but few geographers had investigated this involvement in any detail, or the electoral base on which it is founded (Brunn, 1974; Johnston, 1978c).

A framework for understanding spatial variations in well-being was presented by Coates et al. (1977). They defined the components of well-being, whose variations were mapped at three scales (international, intra-national and intra-urban), and argued that those variations reflected three sets of causes: the spatial division of labour; accessibility to goods and facilities; and the political manipulation of territories (see also Cox, 1979). Spatial policies aimed at the reduction of spatial inequalities were evaluated, such as various forms of positive discrimination by areas, leading to the conclusion that the division of labour is the primary determinant of levels of social well-being. Creation of this division is a social and not a spatial process, though it has clear spatial consequences, so that:

> The root causes of spatial inequalities cannot be tackled by spatial policies alone, therefore. Inequalities are products of social and economic structures, of which capitalism in its many guises is the predominant example. Certainly inequalities can be alleviated by spatial policies . . . but alleviation is not cure: whilst capitalism reigns, however, remedial social action may be the best that is possible . . . the solution of inequalities must be sought in the restructuring of societies.
>
> (pp. 256–7)

Smith (1977; see also Smith, 1979) essayed a more ambitious attempt at explaining such spatial variations, arguing that 'the well-being of society as a spatially variable condition should be the focal point of geographical enquiry . . . if human beings are the object of our curiosity in human geography, then the quality of their lives is of paramount interest' (pp. 362–3). This was the foundation of his case for 'a restructuring of human geography around the theme of welfare . . . to provide both positive knowledge and guidance in the normative realm of evaluation and policy formulation' (p. ix). His book proceeded from theory through measurement to application. The theoretical section is an amalgam of normative welfare economics with Marxian perspectives on the creation of value, plus the political conflict for power:

> The analysis will inevitably reveal certain fundamental weaknesses of the contemporary capitalist-competitive-materialistic society, but the temptation to offer a more radical critique of existing structures has been resisted, in favour of an approach that builds on the discipline's established intellectual tradition.
>
> (p. xi)

He concluded that:

> As geographers we have a special role – a truly creative and revolutionary one – that of helping to reveal the spatial malfunctionings and injustices, and contributing to the design of a spatial form of society in which people can be really free to fulfil themselves. This, surely, would be progress in geography.
>
> (p. 373)

Inequality, justice and ethics

David Smith's (1994a) *Geography and Social Justice* took forward his concerns with inequality and welfare, reflecting changing circumstances (p. xiii). He had been:

> involved [in] a variety of engagements with geographical aspects of inequality and human welfare. Reflecting the dominant mood of the times, explicit concern for social justice remained muted, for the most part. But the theme was still there, and increasingly required reassertion as social change on the world stage – east, west and south – began to resurrect some basic questions concerning the distribution of benefits and burdens under alternative economic and political arrangements.

Smith determined to explore those aspects of moral philosophy related to the issue of distributive justice, to 'see what can be made of this in the geographical context'. This involved a return to normative thinking, 'with how we conceive of what is right or wrong, better or worse, in human affairs lived out in geographical space'. He concluded that 'social justice should not be left to market forces' because 'to commend observed market outcomes on the grounds that they are the results of a just process is not credible' (p. 279), and he argued instead for a foundation in egalitarianism which, despite postmodern critiques, should embrace certain universal needs (see also p. 246, this book). Harvey (1993a, 1993b, 1996a) also addressed issues of justice, focusing on the interplay between universal conceptions and local circumstances.

Smith took this agenda forward in a series of essays on 'moral geographies', at a time when ethical issues came to the fore in debates within UK geography, as illustrated by the controversies related to Shell's sponsorship of the RGS (see pp. 334, this book; similar issues had previously been debated at American geography conferences (see p. 313) but much less so in the UK). He identified this 'contemporary geography and ethics movement' (Smith, 1997a, p. 584) as involving a greater interrogation of the literature of philosophy than previously, with particular reference to the role of space and place therein – what he terms 'the contextual thickening of moral concepts in the particular (local) circumstances of differentiated human being' (p. 587), an area that philosophers tend to ignore. (See Miller and Hashmi, 2001, who don't ignore it, but do ignore the work of geographers.) The study of ethics thus required appreciation of the place(s) in which people were morally socialised. Space too was important, as illustrated by essays on the spatial scope of care and beneficence: is there a distance-decay effect to moral action (Smith, 1994b, 1998)? As with others (e.g. Sayer and Storper, 1997), Smith also found stimuli for this turn to ethical and normative concerns in the writings of environmentalists and feminists as well as philosophers (see also Whatmore, 1997, and Lawson, 2007, on the ethics of care).

For Smith (1997b), an 'ethical turn' in geography was necessary because of the pressing need for public debate on moral issues, 'Otherwise, humankind risks losing any residual grip on the meaning of the good life, any capacity to recognise and challenge evil' (Smith, 2000, p. vii). For him, concepts central to geographic enquiry – 'landscape, location and place, locality, proximity and distance, space and territory, development and nature' (p. viii: these are the subject matter of his book's main chapters) – are significant to debates about 'questions of justice and the good life'. Difference is highly significant in ethics, not so much difference at an individual level (relevant though this is to fair treatment of all) but rather difference between groups, much of which is rooted in places, hence the importance of geography in what Smith terms 'geographically [or context-]sensitive ethics'. He sees the choice between universal ethical principles, universally applied, and local particularity, lacking any transcendent values, as a false dualism, with either:

> it would be hard to see how anything resembling moral truth might be discovered, or created, except universals so vacuous as to be impotent in the face of our sorry reality, or beliefs so restricted in scope as to be part of the problem rather than part of the solution of human conflict.
>
> (p. 202)

His final message advances:

> the imperative of developing more caring relations with others, especially those most vulnerable, whoever and wherever they are, within a more egalitarian and environmentally sustainable way of life in which some of the traditional strengths of community can be realised and spatially extended.
>
> (p. 208)

Cloke (2002) made a similar argument for 'living ethically and acting politically', though his search for the geographies that might underpin and secure this is based on spiritual beliefs and his focus is on the recognition of what he terms 'ordered evil'.

Others joined Smith in his effort to develop a 'moral (or ethical) imagination' within geography (Proctor, 1998a), which includes calls for reconsidering the human/nature binary division that underpins so much of contemporary geography (Proctor, 1998b; Whatmore, 2002). Apart from David Smith's (2000) book, several other contributions addressed a broad range of issues in and exemplars of geographically sensitive ethics (see Proctor and Smith, 1999, which included essays on space, place, nature and knowledge; Bondi et al., 2002; Valentine, 2003). For Sack (1997, p. 1), for example, the fact that as humans we are geographical beings, transforming the earth and then ourselves being transformed by the altered environments that we inhabit, requires a moral focus to our work; the nature and force of geographical agency calls for moral agency, for an 'understanding of the consequences of our actions on nature and culture, locally and globally' (p. 2). The impact of these arguments was almost immediate: in 1998 a journal on *Ethics, Place and Environment* was launched to carry material that illustrated and advanced the case for a more theoretically developed and explicit moral commitment within geography. (In 2010, it was merged with the formerly separate journal *Philosophy and Geography*.)

Environmentalism

The late 1960s saw a rapid increase in concern about environmental problems; in the USA (Mikesell, 1974):

> Towards the end of the 1960s the American public was overwhelmed with declarations of an impending environmental crisis. . . . Since that time, crisis rhetoric and a yearning for simple answers to complicated questions have given way to a more sophisticated and deliberate search for environmental understanding. Ecology has been institutionalized.
>
> (p. 1)

Two of the leaders of the public debate disagreed over the cause of the problems (O'Riordan, 1976, pp. 65–80). Ehrlich argued for the primacy of population increase, and popularised the concept of zero population growth; Commoner claimed that technological advances and the consequent rapid depletion of resources plus deposition of pollutants created the major problems. Both arguments have clear geographical components, and it was stressed that geographers have a strong record of activity in resource conservation: George Perkins Marsh had written on the topic in 1864 (Lowenthal, 1965, 2001) and the climatologist Warren Thornthwaite had been closely involved in the 1930s' soil conservation movement established as a consequence of the Dust Bowl phenomenon (Mather and Sanderson, 1996). The interest in landscape modification was advanced by Sauer and his followers, and reflected in the 1955 symposium Man's Role in Changing the Face of the Earth. Similar interest elsewhere was exemplified by Cumberland's (1947) pioneering classic on soil erosion in New Zealand. Nevertheless, Mikesell argued that 'developments in geography have been such that the several phases of national preoccupation with environmental problems have not produced a general awareness of our interests and skills' (p. 2). Similarly, Eden (2003, p. 213; see also Castree, 2004) argued that because of 'a detour to other topics even as other disciplines discovered the environment as a topic of interest', British geographers were not prominent in the early years of this movement, leaving them 'playing "catch-up" in the late twentieth century, as the discipline sought to reoccupy the ground previously abandoned'. (Turner, 2002, on the other hand, argued that human–environment interrelationships now form a viable core area of geographical scholarship in the USA, hence his welcome for the creation of a Human Environmental Sciences Section in a recent reorganisation of the National Academy of Sciences with which most of the geographers who are members of the Academy are affiliated.)

As part of the AAG's increased commitment to public affairs in the early 1970s, its Commission on College Geography established a Panel on Environmental Education and sponsored a Task Force on Environmental Quality. The latter reported (Lowenthal et al., 1973) that geographers would make excellent leaders for the educational tasks in hand, because of:

1   the breadth of their training and their ability to handle and synthesise material from a range of sources;
2   their acceptance of the complexity of causation;
3   the range of information which they are trained to tap;
4   their interest in distributions; and
5   their long tradition of study in this area (as illustrated by Lowenthal's 1958, 2001, biographical studies of George Perkins Marsh).

All had fostered expertise in work on environmental perception, on vegetation succession, and on relationships between land use and soil erosion, which could be used as the bases for environmental impact statements, the elaboration of environmental choices and international research collaboration.

Two types of work characterised geographers' activities on society–environment interrelationships at this time. The first involved the traditional geographical concern with description and analysis. Review volumes such as Perspectives on Environment (Manners and Mikesell, 1974) were prepared, and a particular interest in problems of the physical environment of urban areas was generated (Detwyler and Marcus, 1972; Berry and Horton, 1974; Berry (with others) 1974; a later, British, addition was Douglas, 1983, 2013). The second type focused on issues of environmental management (O'Riordan, 1971a, 1971b), with particular emphasis on its economic aspects and on societal response to environmental hazards (Hewitt, 1983); as Kates (1972, p. 519) pointed out, economics provided the theories and prescriptions of the 1960s (and later, see Judith Rees, 1985). A topic of special interest was leisure, particularly the growing demand for recreational facilities, and the impact of recreational and tourism activities on the environment (Patmore, 1970, 1983; see the critique of much of that work in Owens, 1984); a separate journal on Tourism Geographies was launched in 1999.

Despite such activity, Mikesell concluded that geographical contributions to environmentalism had not been great up to 1974. He commented regarding the prognostications of *The Limits to Growth* (Meadows *et al.*, 1972), for example, that, 'the debate on this most relevant of all issues has attracted remarkably little attention from geographers' (Mikesell, 1974, p. 19, though see Eyre, 1978). He concluded more generally that 'one must add hastily that many of the environmental problems exposed in recent years and also many of the social and philosophical issues debated during the environmental crusade have not been given adequate attention by geographers'. This conclusion was supported by analysis of the contents of recent geographical journals and O'Riordan's (1976) lengthy bibliography, and sustained by Goudie (1993) nearly two decades later. Billie Lee Turner (2002) suggested that this had been rectified, however, with research on the human–environment interface forming a major component of (American) geography. The volume of work published in journals such as *Applied Geography* supports this conclusion (although not all of the papers published are based on applied work per se), as does the rapid growth of work in subfields identified as cultural ecology and political ecology (Robbins, 2004), for which the *Journal of Political Ecology* was launched in 1994. (For evaluations of work in political ecology and its applications, see Blaikie 2008, 2012; Dwyer and Baird, 2014.)

A powerfully argued case not only for more work on the society–environment interface but also for its centrality to the whole of geographical activity was presented by Stoddart (1987; see also Stoddart, 2001), who contended that instead of celebrating the achievements of a century of professional geography, many of his colleagues were 'despondent, morose, disillusioned, almost literally devoid of hope, not only about Geography as it is today but as it might be in the future' (p. 328). This, he believed, was because so many of them 'have either abandoned or failed ever to recognize what I take to be our subject's central intent and indeed self-evident role in the community of knowledge' (p. 329). For him, geography had become diffuse, lacking a central focus which should be 'Earth's diversity, its resources, man's survival on the planet' (p. 331). This requires a unified discipline, human and physical, in which 'The task is to identify geographical problems, issues of man and environment within regions – problems not of geomorphology or history or economics or sociology, but geographical problems: and to use our skills to work to alleviate them, perhaps to solve them.' Focusing on 'the big questions, about man, land, resources, human potential' would involve geographers reclaiming 'the high ground' (p. 334; examples include Williams, 1989, 2003) and abandoning much that is currently done:

> Quite frankly I have little patience with so-called geographers who ignore these challenges. I cannot take seriously those who promote as topics worthy of research subjects like geographical influences in the Canadian cinema, or the distribution of fast-food outlets in Tel Aviv. Nor have I a great deal more time for what I can only call the chauvinist self-indulgence of our contemporary obsession with the minutiae of our own affluent and urbanized society. . . . We cannot afford the luxury of putting so much energy into peripheral things. Fiddle if you will, but at least be aware that Rome is burning all the while.

James Bird (1989, p. 212) pointed out, however, that although Stoddart called for a geography that is 'real, united and committed . . . we are not told exactly what it is, though for Stoddart such a geography obviously exists'.

Similar, though less strident, calls have been made by others (e.g. Douglas, 1986; Goudie, 1986b; see also Cosgrove and Daniels, 1989); in response, the relevance of much that Stoddart would disregard in contemporary social science has been promoted as necessary to an appreciation of society–environment relations (Blaikie, 1985; Blaikie and Brookfield, 1987; Johnston, 1989 – the latter was written to counter the relative naiveté of much writing by physical geographers regarding the potential role of the state in the resolution of environmental problems: Trudgill, 1990; Pitman, 2005; see also Johnston, 1996d, 2006). The nature of risk was closely scrutinised by

Adams (1995), for example, as were varying perceptions of nature (Simmons, 1993) and political approaches to environmental problems (Pepper, 1996); a successor to the 1954 Wenner-Gren Symposium (The Earth as Transformed by Human Action: B. L. Turner et al., 1990) explored human exploitation of the environment in considerable depth, as have others on specific issues such as desertification (Granger, 1990; Middleton and Thomas, 1994, discuss the extent to which it was a political construction).

Stoddart's argument was clearly set in the cultural ecology mould. A similar case was made by Kates (1987), who regretted the dominance of spatial science within geography and argued that when environmental issues became important on the public and political agenda in the early 1970s:

> No discipline was better situated than was geography to provide intellectual and scientific leadership. The natural science for the environmental revolution should have been the science of the human environment. Instead, intellectual leadership was split among biology, economics, and engineering, each of which transferred onto the human environmental realm their own theories of nature, of economy, or of technology, but none of these offered a truly integrated view.... The theory of the human environment, then, was the theory of plant or animal ecosystems, or of pervasive externalities, or of technological and managerial fixes.
>
> (p. 526)

For geographers, a perceived inability to respond to the demand for environmental scientists (producing 'geographers who can sit astride the natural and social science boundary to provide analysis, integration and leadership') was an opportunity lost, a 'road not taken'. But the Malthusian dilemma remains (Mayhew, 2014), posing great questions for society to which geographers can bring their special disciplinary advantages in the search for answers:

> We possess more than passing knowledge of both the natural and social sciences.... We have some useful tools for organizing data and information.... We possess a strong tradition of empirical field research.... And perhaps most important we have and we teach a respect for other peoples' theories. Our answer, then, as to why geography ... is that we are needed and that we are useful. When they go forth, our students understand the nature of the great questions, have more than a passing knowledge of natural and social science, have been in the field, have collected and organized new data, and have placed these data into a theoretical perspective.
>
> (Kates, 1987, p. 532)

To be sure that they will be called upon to perform in this way, he argued, geographers must put their house in order and some university departments should rebuild centres of expertise in the human–environment tradition. The University of Oxford established an Environmental Change Centre alongside its School of Geography, for example, and several of its directors have had an established international reputation in the study of the interactions between climatic change and human activity (Parry and Duncan, 1995; Liverman, 2009). Other universities have similarly created 'environmental institutes' within which geographers play important roles; in some countries, such as Australia, geography has somewhat disappeared as a separate academic discipline, instead only being present in universities as part of some institutional structure focused on environmental science/studies (Holmes, 2002).

One of the major concepts to emerge from the increased concern over environmental problems in the late 1980s was sustainable development, a considerably ambiguous term that is generally taken to imply continual increases in material living standards without any diminution in environmental capacity to meet the needs of future generations (Turner, 1993). This was a focal concern of the 1992 Earth Summit at Rio de Janeiro, but the difficulties of implementing global environmental policies have been frequently stressed (see the essays in Johnston et al., 1995, 2002), although others believe that the problems have been overstated and can readily be addressed, within a satisfactory time scale, using instruments of economic policy that promote continued development (Beckerman, 1995; see also Kates, 1995).

While these broader concerns were being aired, geographers were continuing to be active in the tradition established by White (see p. 126, this book). In 1937, for example, his memorandum to President Roosevelt helped to convince him, against Congressional advice, not to make the Secretary of War responsible for flood control projects, thereby bypassing local and regional planning bodies, and in 1965 he prepared a paper for the Bureau of the Budget on national policy options for dealing with floods and similar hazards. The result was the National Flood Insurance Program, after which White held annual seminars in Boulder to bring public-sector officials and academics together to discuss disaster response and mitigation policies (Platt, 1986). In this, he and the many others who have studied environmental policy and contributed to its implementation have adopted what Feldman (1986) termed the citizen-scholar model:

> The scholar chooses his or her research concerns on the basis of perceived social need but attempts to conduct the inquiry free of influence from outside the inquiry itself. The scholar is committed to social and political action as follow-up to the research, based on the findings and quality of the inquiry. But political activity is, during the course of the scholarship, kept separate . . . [In addition] the citizen-scholar teaches public affairs both within a curriculum and by example, letting her or his own activities demonstrate good citizenship and good scholarship without being didactic.
>
> (p. 189)

Earlier, White (1972, p. 322) had asked, 'What shall it profit a profession if it fabricate a nifty discipline about the world while that world and the human spirit are degraded?', and in seeking to avoid that problem he developed an approach which Wescoat (1992) compares to that of the philosopher of pragmatism, John Dewey. It had four components:

1   recognition of the precariousness of existence;
2   a pragmatic conception of the nature of inquiry – problems arise and are tackled within situations, for there are no absolute truths; hence
3   a tradition of learning from experience; and
4   a belief in public discourse and democracy.

All these fitted White's deep involvement with the work of the Society of Friends. (For a fuller discussion of his career, see White, 2002; Hinshaw, 2006.)

The breadth of study in 'modern' environmentalism was illustrated by O'Riordan's (1981) work, much of it set in the liberal humanitarian tradition already illustrated here and from which he drew four conclusions:

1   modern environmentalism challenges many aspects of Western capitalism;
2   it points out paradoxes rather than clear solutions;
3   it involves a conviction that better modes of existence are possible; and
4   it is a politicising and reformist movement, based on a realisation of the need for action in the face of impending scarcity and a lack of faith in the western democracies (pp. 300–1).

A new social, environmental order is required and O'Riordan identified three possibilities: centralised, authoritarian and anarchist. He chose a middle-of-the-road, liberal option:

> we must individually and collectively seize the opportunities of the present situation to end the era of exploitation and enter a new age of humanitarian concern and cooperative endeavour

with a driving desire to re-establish the old values of comfortable frugality and cheerful sharing.

(p. 310)

Such a new era, involving a new political order based on a combination of local self-determination and supra-nationalism, can be achieved through education, he claimed, so environmental education will form a preparation for citizenship. Others were more doubtful: Pepper (1996, pp. 324–5), for example, was sceptical about the liberal arguments that 'a basic unselfishness and communalism which is human nature would come to the fore in ecotopia's unalienated society, as it has never been allowed to do in industrial capitalism' (arguments that he terms 'simply hard to believe') and promoted instead the view 'that by making humanist, egalitarian and socialist aspirations a prerequisite for an ecological society, rather than something that is supposed automatically to follow, we can avoid making ecological society a repressive dystopia'. Social change must precede the resolution of environmental problems.

Some writers have argued that the revival of interest in environmental issues – mainly through the study of resources and their management – provides a contemporary linking of human and physical geography. Relatively little of the research and textbook writing indicates any integration of the two, however (Johnston, 1983b, 1989), because the focus is almost invariably on the processes studied in one of the subdisciplines only. (This is illustrated by most of the papers in the journal *Applied Geography*, which focuses on such issues and by those in *The Geographical Journal*, which the RGS decided would also concentrate on such topics from 2001 on; geographers also contribute to specialist journals in this field such as *Global Environmental Change – Human and Policy Dimensions*.) Physical geographers portray trends such as demographic growth, technological sophistication, urbanisation and demands for resources as catalysts for environmental processes and changes, but they take those trends for granted and seldom address what processes generate them. Similarly, human geographers largely take the resources of the physical environment as given and do not enquire about the physical and chemical laws underpinning their genesis (though see Hudson, 2012, on the transformation of those materials in changing industrial processes, and Gregson and Crang, 2014, on the ecology of recycling). For human geographers their links with other social scientists are very much stronger than those with environmental scientists (as illustrated from the outset, for example, by the lack of references to the physical geography literature in O'Riordan, 1976, and Pepper, 1996); physical geographers, in turn, direct much of their scientific activity away from geographical journals towards work by their peers in other disciplines such as geology (especially of the Holocene), glaciology and hydrology (Johnston, 2003a). Alongside more theoretical work by human geographers stressing the social construction of nature and the false nature–society dichotomy (Whatmore, 2002; Castree, 2005), however, recent years have seen growing collaboration between human and physical geographers in work on environmental issues (e.g. Landström et al., 2011; Pappenberger et al., 2015).

Geographers have also turned their attention to other 'great questions' within contemporary society, such as nuclear weapons and nuclear power. They have criticised attitudes to civil defence policies and likely deaths from nuclear blast and fall-out (Openshaw et al., 1983), for example, with others setting that concern within a developing geographical contribution to peace studies (Pepper and Jenkins, 1985), and they have addressed issues relating to the siting of nuclear power stations (Openshaw, 1986) and the transport of nuclear waste. (Openshaw's work on these topics involved innovative prototypes of GIS technology and was used to identify clusters of cancers that by their location appeared to be linked to the presence of nearby nuclear installations.) Arguments for geographical contributions to other aspects of peace studies, such as international relations (van der Wusten and O'Loughlin, 1986; Agnew, 2005, 2009; Flint, 2005), were advanced (Pepper and Jenkins, 1983), none more forcibly than Gilbert White's (1985, p. 14), which saw the human family 'tortured and driven by its newfound capacity to throw the whole set of processes out of kilter by more violent action'. The two editions of *A World in Crisis?* (Johnston and Taylor, 1986, 1989; see also the two volumes on *Geographies of Global Change*; Johnston *et al.*, 1995, 2002) focused on many of these issues, ranging from the macro-scale in the study of geopolitics to the investigation of individual human rights. And at the beginning of the twenty-first century, the attacks on New York's World Trade Center and the Pentagon led American geographers – under the auspices of the AAG – to address issues of geography's role in the promotion of 'homeland security' (Cutter *et al.*, 2003; substantial critiques of this book and the approach it promoted were published in the AAG's flagship journal *Annals*, volume 94 (4), 2004).

Geographers and policy

Many of the studies referred to here have been concerned with identifying problems and suggesting solutions. Underlying their varied approaches has been the basic thesis that geographers should be much more involved in the creation and monitoring of policies. But what sort of policies, and what sort of involvement?

Berry (1973b) categorised planning policies into four types:

1 Ameliorative problem-solving involves identifying problems and proposing immediate solutions, as with the removal of a traffic bottleneck. Such solutions are likely to stimulate further problems in the future, since it is only the proximate cause that is tackled (the features of the bottleneck) rather than the real cause (the growth of traffic).
2 Allocative trend-modifying: planning towards the future involves identifying trends, evaluating what is likely to be the best outcome of the several which they imply, and then allocating resources to steer the system being planned towards that end.
3 Exploitative opportunity-seeking: planning with the future identifies trends and then seeks to gain the maximum benefit from them, irrespective of the possible long-term consequences. Compared with the previous category, this one has a dominantly short-term focus.
4 Normative goal-oriented planning for the future begins with a statement of goals, a vision of the future, and then prepares a strategy which will ensure that they are achieved.

Relatively little policy-making is of the fourth type; most involves elements of the other three, with general statements about goals but no clear strategy regarding a foreseeable future. Some critics (like Harvey, 1973, 1974c) interpreted this as meaning that geographers who become involved in policy-making and evaluation are likely to accept the dominant forces in society uncritically, which leads to arguments that their claimed scientific objectivity and neutrality are an (often unrealised) cloak for ideological political judgements about the nature of society. Whatever their individual motives, such geographers are acting on behalf of interest groups (private and public) whose sustenance depends on maintaining an unjust and unequal structure to society.

Gordon Clark (1982, p. 43) accepted that 'the academic community is not independent: there are no objective standards between competing explanations and thus policy advice'. Nevertheless, he agreed 'with the principle of policy analysis and the involvement of academics in policy-making' (p. 48). That involvement cannot be presented as neutral and objective, however, since all social science 'explanations' are incomplete and compete with others to provide plausible accounts and prescriptions. He suggested that academic contributions to policy analysis should be guided by four propositions (pp. 55–9):

1 academics must acknowledge their own values and beliefs in presenting policy alternatives and impact assessments;

2   policy analysts must be advocates for particular causes rather than supposedly independent and objective adjudicators of knowledge;

3   policy science should be critical of the status quo; and

4   sponsoring institutions must encourage advocate briefs and make those briefs accessible to the public.

In accepting these, academics who become involved in policy analysis should be considered 'part of the political process' (p. 57) whose role would be to ensure that 'Choices would be brought squarely into the open and be dependent upon the political, as opposed to expert, process' (p. 59).

The present situation idolises experts, he claimed, and promotes a myth of social science as objective, neutral knowledge, although most governments carefully choose those experts who fit their own ideological presuppositions and whose advice is therefore very likely to be consistent with the desired direction of 'evidence-based policy', rather than policy-based evidence (Johnston et al., 2015); there are always elements that distrust 'experts', however, especially if it can be shown that their research (as in climate change) is supported by commercial and other vested interests. Clark argues that social scientists cannot be neutral experts, because their 'values, interests and normative views of the world' mean that their presentations, although scholarly and rigorous, are necessarily partial. Sayer (1981) made a similar case in a slightly different context, arguing that any rigorous social scientific inquiry must be based on rationally defended value judgements; objectivity does not require neutrality. (Clark, 1991, illustrates this from his experience as an expert witness in court.)

In an essay on the role of urban geographers in applied work, Pacione (1990a) argued that practitioners have paid insufficient attention to conceptual issues underlying what they do. He derived the following 'principles or guidelines' to remedy those failings:

1   the notion of value-free research is an illusion;

2   towns, as examples of places, are meaningful entities on which to work;

3   a spatial perspective is of substantial value;

4   the main emphasis of applied urban geography is on problem-solving;

5   a realist position provides the context for such work (see p. 219, this book);

6   analysis must integrate various spatial scales;

7   a wide methodological tool-kit of quantitative and qualitative procedures must be employed; and

8   geography must integrate the findings of many disciplines.

Johnston (1990b) responded by posing six questions to Pacione: What is a problem? Are problems always soluble? What is science? What is a geographic perspective? Who solves? What sort of society? He concluded that some of the principles/guidelines are trivial and/or irrelevant, some are unsupported, and some are wrong, and that unless Pacione is prepared to address the fundamental issues of what problems are and how they can be tackled, he is unlikely to help those who want both an end to currently perceived problems and a society which would no longer produce such problems. Furthermore, few problems are soluble in the true sense of that term, which points to the need for resolution between opposing points of view, none of which have any straightforward claim to absolute truth (as earlier argued by Wolpert, 1970). This point was made in another way in debates between Palestinian and Israeli geographers over the geography of Israel/Palestine (Waterman, 1985; Soffer and Minghi, 1986; Newman and Portugali, 1987; Falah, 1989; Waterman and Kliot, 1990; Falah and Newman, 1995; Newman, 1996).

Undertaking research characterised as 'relevant' – pertinent to tackling societal problems – raises major issues for the individuals concerned. Bruce Mitchell and Dianne Draper (1982, p. 2) argued that 'when functioning as an advocate or consultant, the geographer must consciously decide how to resolve a conflict which may arise regarding the promotion of one perspective versus critical assessment and balanced judgement about all viewpoints'. There are also issues of ethics: 'when functioning as a pure researcher, the geographer must balance a concern for obtaining necessary information against a concern for respecting the dignity and integrity of those people or things being studied' (p. 3), which also applies to 'relevant' research. They claimed that geographers have largely ignored these ethical issues, and their professional bodies, unlike those of other disciplines, have promulgated no codes of conduct. Conflict is frequently likely between striving to 'discover truth' and respecting the rights of those being studied, however, and they advocated individual, institutional and external controls; most research institutions now have ethics committees seeking to ensure that those rights are respected, and journals require authors to declare both any funding that has supported their research and whether they have any conflicts of interest.

An example of the problem was raised in 1996 within the newly merged RGS-IBG. The Shell oil company was one of the RGS's commercial sponsors prior to the merger, donating £40,000 per annum towards the costs of its Expeditions Advisory Service; this was presented as an example of the sort of commercial sponsorship for geographical work which academic geographers could benefit from within the merged society. In late 1995, however, several members of the Ogoni people in the Niger Delta region of Nigeria, including a well-known author Ken Saro-Wiwa, were sentenced to death by a military tribunal (in a trial which Western observers claimed violated their basic human rights) and subsequently executed, despite political pressure on the Nigerian government, which culminated in its exclusion from the Commonwealth Heads of Government conference held in New Zealand in November 1995. Their 'crimes' were related to protests over Shell's treatment of the Ogoniland environment, where it was involved in exploiting major oil and gas fields. A number of academic geographers argued that Shell's corporate patronage of the RGS-IBG should be ended because of its environmental record there and its alleged complicity in the trial of the Ogoni leaders. A motion to that effect was passed by a very substantial majority at the Research Section (formerly the IBG) conference in January 1996 (an event that attracted considerable media attention worldwide) and the RGS-IBG as a consequence set up a working party to consider all aspects of corporate sponsorship, which proposed a code of practice and ethical guidelines but had little apparent long-term impact (on the debate, see Gilbert, 1999; see also Watts, 2000).

Most of the critiques of the arguments for greater involvement by geographers in policy analysis are entirely sympathetic. They make a case for sensitive geographical involvement; others question the grounds for such involvement and instead focus on the development of revolutionary theory (see p. 316, this book). The former focus almost entirely on applied work in the tradition of positivist, empirical science, in which the goal is perceived to be to improve well-being through contributing to one or more of:

1   the preparation of public (i.e. state) policies;
2   the development of commercial (i.e. profit-making) strategies; and
3   the attack on environmental problems.

Virtually none of this work is concerned with the applied 'arms' of the two other types of science identified in Chapter 1, leading to either greater self- and mutual awareness (see Chapter 5) or emancipation (Chapter 6). To that extent, most applied geography in the English-speaking world accepts, and seeks to serve, the ruling ideas – and so also the '(new) ruling class' Harvey (2003).

The nature of applied geography has been restricted by some to work at the 'society–environment interface'. When launching the journal *Applied Geography* in 1981, for example, its founding editor followed Stamp (1960, p. 10), who defined the discipline's applied role as addressing 'some of the great world problems – the increasing pressure of population on space, the development of underdeveloped areas, or the attempt to improve living conditions'. Two decades later, according to Briggs (1981, p. 2), 'These problems remain. Indeed, they are growing; environmental problems such as pollution, damage to wildlife, destruction of habitats, soil erosion and resource depletion; the problems of human deprivation and inequality.' He set out the fundamental basis of applied geography as use of resources:

> The exploitation of scarce resources represents a dominating theme to human existence. It is from the pursuit of these resources and from the attempt to decide between alternative policies of exploitation, that not only environmental damage, but also the greater part of political, social and economic problems emerge; they can be seen as expressions of man's inability to organize himself and his world to his best, long-term advantage.

To overcome that inability, the applied geographer must be brave (p. 6): 'He needs to commit himself before he knows all the answers. He needs to be able to make public mistakes. But he must also be prepared to learn from them.'

Much applied work undertaken by academic geographers has been commissioned by one or more arms of the public sector and many (perhaps most) geographers who have applied their skills in non-academic careers have probably entered the public sector too. Given the importance of the welfare state in the first four post-Second World War decades, this has not been surprising. Nevertheless, work has always been done for the private sector too – for example, geographers such as Applebaum (1954) were involved in work for supermarket and similar companies in the USA during the 1950s and geographers were central to the establishment of an Institute for Retail Studies at the University of Stirling in the mid-1980s (Dawson et al., 2006; see also Wrigley, 2002). Others, building on their expertise in spatial data handling, have moved into information management, including some who work in prestigious university business schools.

The pressure on university academics to obtain more external financial support for research activities since the 1980s stimulated much more work for the private sector (Breheny, 1989) – a trend accentuated by the attempts to reduce the size of the public sector and to increase competitiveness and efficiency in the provision of public goods (including, of course, higher education). Some of this work has involved spatial analysis in its various forms – as by the GMap company established at the School of Geography at the University of Leeds, which has applied and extended Alan Wilson's pioneering work on location-allocation models to a wide range of problems in, for example, operations research for healthcare providers and the location of franchises for major car firms (Birkin et al., 1990, 1996; Birkin et al., 1995; Birkin et al., 2002, 2010); the company was later sold by the university into the private sector. Other work has used geographers' skills in spatial data handling (what Openshaw and others refer to as 'adding value' to geocoded data) in the growing field of geodemographics, whereby marketing and other campaigns are spatially targetted to people living in areas where demand for particular goods and services is most likely to be generated (Batey and Brown, 1995; Birkin, 1995; Harris et al., 2005; Longley, 2012). More widely, a number of UK university geography departments have followed the early lead from Edinburgh (p. 315, this book) and Leeds, and established units involving their academic staff and others appointed for their research skills, only to undertake applied work (much of it interdisciplinary) in a wide range of fields; examples include the Flood Hazard Research Centre at the University of Middlesex (Penning-Rowsell and Pardoe, 2012), the International Boundaries Research Unit at the University of Durham (which publishes two series of books – the International Boundary Studies Series and the World Boundaries Series – as well as regular briefings: www.dur.ac.uk/ibru/) and the Personal Finance Research Centre at the University of Bristol (www.bristol.ac.uk/geography/research/pfrc/about/).

In many cases, because of commercial sensitivity such work – especially its applied components – is not fully reported in academic journals and thus the discipline as a whole may not be aware of it. An example is Alice Coleman's (1985) research on the effects of building design on behaviour, based on Oscar Newman's (1972) concept of defensible space; alongside her academic publications she had a considerable impact on contemporary debates about neighbourhood design and UK £50 million was invested in experiments to test her ideas (Jacobs and Lees, 2013). Where the goal is to influence (usually public) policy in general rather than a specific project, publication in academic journals alongside the private advice is more feasible, as in work on the operation of housing markets (e.g. Smith, 2008, 2015; Smith et al., 2006) led to policy proposals (Parkinson et al., 2009; Smith and Searle, 2010; Smith et al., 2008).

A geographer who perhaps made more contributions to applied geography was Peter Hall. Initially a historical geographer, he published many detailed analyses of urbanisation and the cultural context of urban growth, including his *magnum opus Cities in Civilization* (Hall, 1998). He also researched the nature of the UK planning system in comparative context, including a major two-volume critique of its impact (Hall et al., 1973). His research led him not only to make many contributions to debates over future urban form and its planning – not only as critic and commentator but also as consultant and adviser to several governments – but also to paint scenarios of those futures, as in London 2001 (Hall, 1989). His was a humanistic approach, as indicated by the title of one of his last books – *Good Cities: Better lives* (Hall, 2014: for a full appreciation of his career and work, including his own *apologia per vita sua*, see Tewdwr-Jones et al., 2014).

Changing contexts and applied geography

The demands for greater involvement in what is generally termed applied geography have grown in recent decades, largely as a response to changes in the societal contexts within which Anglo-American human geographers work. This is not a surprising trend, according to Peter Taylor's (1985a) analysis of the history of geography. In periods of economic recession, cutbacks in public funding for higher education and research can be expected, and to counter the loss of support academic researchers are forced to seek financial backing elsewhere (including arms of the state which contract for research); that backing is only likely to come if individuals within the discipline, and even organisations representing it, can convince potential sponsors of the value (i.e. potential 'profitability') of investing in geography and geographers. From this observation, and following Gräno (1981), Taylor (1985a, p. 100) identified two external influences on disciplinary developments: 'Within academia geographers had to be given an intellectual foundation to satisfy intellectual peers, and outside in the wider world geography had to be justified as a useful activity on which to spend public money.' These produce 'pure' and 'applied' geography respectively. Both are necessary to a discipline's future, but the relative emphasis placed on each will vary over time and space:

> Outside pressures will be particularly acute in periods of economic recession when all public expenditure has to prove its worth. All disciplines will tend to emphasize their problem-solving capacity and we can expect applied geography to be in the ascendancy. . . . In contrast in a period of expanding economies and social optimism outside pressures will diminish and academia can be expected to be under less external pressure. Geographers will thus be able to contemplate their discipline and feel much less guilty about this activity. We can expect bursts of pure geography to occur in these periods.

Taylor identified cycles of pure and applied geography corresponding to cycles in economic prosperity, and Hägerstrand (1977, p. 329) reached a similar conclusion: 'When the world is stable and/or unhampered liberalism prevails, then there is probably not much to do for geographers except surviving in academic departments trying to keep up competence and train schoolteachers in how wisely arranged the world is.' (Applied geography is also an increasing focus in many degree curricula which focus on transferable skills – notably in GIS – that students can deploy in a variety of occupations: Schlemper et al., 2014; Singleton and Spielman, 2014.)

The period since the late 1970s should demonstrate applied geography in the ascendancy, according to Taylor's analysis, an expectation generally supported by experience of those years. Pressure to justify the discipline in terms of utility to economic goals was strong, and the search for research contracts of all types became much more determined (so that learned journals and newsletters began to list new grants and contracts won by geographers, and institutions such as the IBG prepared documents to promote the discipline that emphasised its utilitarian aspects, including one which graduates could show to potential employers). That pressure for applied geography continued into the neoliberal environment of the 1980s and following decades, with academic departments continually encouraged to canvass for research funds from outside sponsors (which they were apparently very successful at: Johnston, 1995b). In the 2013 UK Research Excellence Framework , one-fifth of the evaluation was based on assessments of the impact of their research on outside users, with each department having to produce case studies showing how the work of individual academics or groups have directly implicated either public policy or commercial activities (Collini, 2012; Pain et al., 2011). Increasingly, too, many of them promote their degree programmes to potential students in terms of the skills that they will learn which will be of benefit to them in the labour market (with considerable emphasis in recent years on GIS and related technologies, as discussed on p. 157, this book). Such non-partisan academic advice based on research evidence informing public debate and legislation is now frequently published by learned societies such as the British Academy (e.g. Hix et al., 2010; Balinski et al., 2010).

The consequences of 'The Impact Agenda' for geography became the subject of heated debates, led by academics based in the UK, where the Research Excellence Framework institutionalised audits and claims about impact (Pain et al., 2011; Slater, 2012). As the introduction to a set of essays on impact and human geography noted:

> Academics across diverse disciplines within the UK are therefore grappling with the changing expectations and pressures placed on research, its practices and relationships. The impact agenda requires that impact must be directly traceable back to a piece of published research, even though in reality that relationship may not be direct or linear.
>
> (Rogers et al., 2014, p. 4)

More generally, geographers – allied with other social scientists – have increasingly 'sold' their relevance to government and other potential 'stakeholders' – the UK Academy of Social Sciences, for example, published a volume on *Why the Social Sciences Matter* (Michie and Cooper, 2015), a booklet on *The Business of People* (Campaign for the Social Sciences, 2015), and a volume researching *The Impact of the Social Sciences* (Bastow et al., 2014).

The place of applied geography

While in the UK the impact agenda has led to a further phase of discussion in the recent years, the desirability of applied geography has long been debated . Some clearly accepted and argued for an applied ethos, but their main concern was either with geography using its perceived particular skills in what Harvey (1973; see p. 316, this book) termed the application of status quo theory (as in Pacione, 1990a, 1990b) or with exploring how geographers could make a committed contribution towards the achievement of change (Clark, 1982).

Bennett (1989b) based a case for a reorientation of what geographers do on a belief that there has been a major shift in the 'culture of the times', away from 'welfarism' – a consensus ideology aimed at 'improving the quality of life and provision of needs through collective and governmental intervention' (p. 273) – towards what he terms 'post-welfarism', which emphasises individual rather than collective decision-making, and the role of markets rather than states:

> the emergent 'culture of the times' has been happier to see the market as both the creator and the provider of new wants. Rather than markets being seen as an inhumane and exploitative system, socialism and even corporatist social democracy have come to be associated with the odious and paternalistic treatment of individuals.
>
> (p. 286)

This led him to criticise geographers who work with 'social theory' and accept the welfarist ethos, as being necessarily of the 'political left', and he challenged their 'core concept of relative deprivation and its consequential focus on relative intervention', which inevitably leads to a situation of 'total state intervention in everything', Bennett argued that:

> Where the Thatcher era has heralded consumer choice and economic change, social theory and socialist politics have sought to defend the mode of production and to trap people in labour-intensive work practices and unattractive jobs vulnerable to technological change. The spirit of market freedom of individuals has heralded a consumer and service economy which has offered the release from the least attractive toils and labours, and has seemed to offer the potential to satisfy many of people's most avaricious dreams.
>
> (p. 286)

This market-oriented society poses an important challenge to geographers' applied role. It was embraced by a number of geographers, including some who work with GIS. Longley (1995, p. 127), for example, argued that spatial analysis using GIS technology can enhance business management systems, enabling users 'to gain competitive advantage in sophisticated consumer-led markets'. He criticised the 'deskilling of geography and planning' associated with arguments such as Harvey's (1989b) that quantitative research has produced little relevant output. Not only does the development and teaching of GIS equip students with 'a range of flexible skills for use in continually restructuring labour markets' (p. 129; see also NAS-NRC, 1997; NRC, 2011), but in addition the range of applications:

> provides clear testimony that quantitative spatial analysis is most certainly not preoccupied with techniques that do not work to analyse problems that do not matter. Social science that does not show interest in real world issues of popular concern is doomed to remain on the sidelines of academic respectability and perceived social relevance, and reinvigorated spatial analysis is central to the measurement and modelling of economic and social aspects of human behaviour.

Bennett portrayed the geographical research he criticised as focused on aiding government intervention in economy and society, without questioning its validity. He believed such intervention to be no longer viable and also rejected its underpinning critical stance, based on a critique of capitalism. He identified the intellectual challenge for geographers to involve dismissing notions of a welfare state founded on a social theory which emphasises rights and relative needs, because policies based on such theory 'cannot be proved to work. . . . Even social democracy offers no easy solutions' (p. 287). But markets do fail, so:

> The two key questions for a post-welfarist society are: first how can support be improved by practical policies that can be demonstrated to work and are reasonably cost-effective; and second at what point does governmental action end. . . . Hence what is needed is a better definition of what is 'socially' possible through collective action and what is not. Social theory and

social democracy have, perhaps, promised too much and hence led to disillusionment in its own promises.

(pp. 287–8)

Answering those questions is central to the role he cast for geographers: 'The grand objective of the discipline should be to contribute to the debate around these issues. But it must be a contribution to practice.' This implies a discipline in which practical concern with means is more important than theoretical debate about ends. In a post-welfarist society:

> The welfare state, and its associated public decision-making, no longer have the privileged status that it can be justified by mere statements of belief in public or governmental goods: public goods and the policy that provides them have to be demonstrated to be effective in meeting social needs; policies have to work and be more effective than alternatives.

(pp. 288–9)

Bennett provided no detail regarding what such a post-welfarist geography would contain, though he referred approvingly to Openshaw's (1989) arguments (see p. 156, this book). He offered two brief suggestions, however. The first is a criticism of Marxist and related work:

> The key aspect . . . is a better understanding of the structure of economic incentives and rights, rather than class. . . . We need to identify a new language not of class but of 'rights' or nature. By which I mean choice conceptions of rights which promote autonomy, freedom, self-determination and human development, and not 'interest' conceptions of rights which make people passive beneficiaries of the services of others.

(p. 289)

The other is a recommendation that:

> For the academic discipline of geography this means adaptation of its frameworks of teaching and research. I would argue that one aspect of this requires more intensive training in analytical methods including model-based approaches, information systems and elementary analytical skills.

Thus, geographers are called upon to participate in a re-evaluation of the welfare state, focusing attention on the limits to both individual choice and collective action, and to develop the analytical skills which will advance policy appraisal and so contribute 'useful knowledge to the research process, to policy debate and to practice' (p. 290). In his commentary at the end of the book in which Bennett's chapter appears, Macmillan (1989b, p. 306) comments that 'the idea that the welfarist tradition is in terminal decline seems highly debatable', though developments in subsequent decades indicated that many governments are reconsidering the nature of their welfare provision in more market-oriented terms and the consequences of the post-2008 debt and fiscal crises in many countries saw the implementation of widespread reconfiguration of and cuts in welfare eligibility and expenditure, in the context of (contested) political rhetoric that austerity policies are needed to counter the crises.

The 1990s and subsequent decades did not see a return to the 'traditional' welfare state, therefore, but rather a continuation of the trends set by Thatcherism in the UK and Reaganism in the USA. The hegemonic ideology is associated with 'neo-liberalism' (Brenner and Theodore, 2002a, p. 350), underpinned by a 'belief that open, competitive and unregulated markets, liberated from all forms of state interference, represent the optimal mechanism for economic development'. As they pointed out, however, this belief is generally associated with new sets of policies of state interference in non-economic spheres – usually conceived as associated with promotion of economic success (Brenner and Theodore, 2002b) – as illustrated by the rise of the 'workfare state' (Peck, 2001; Painter, 2002) and increased levels of prison incarceration in the USA (Gilmore, 2002).

Continuing debates: grey, public and participatory geographies

Geographers have undertaken a great deal of applied work over recent decades, in both public and private sectors and across a wide range of subject matter. In addition to the activities of those working within academia, many trained geographers are putting their expertise into practice, not least in the deployment of GIS (Wright, 2012), in which geographers play central roles in the training of students and practitioners and the development of software for a wide range of applications. This latter activity is probably more visible in the USA, where the AAG has a large applied geography specialty group and holds regular conferences as well as sessions at the Association's annual meetings; these occur alongside specialist GIS conferences such as GISRUK and ESRI's international user conferences.

Despite all of this, however, and in part reacting to contemporary changes in the discipline, a number of geographers bemoan their discipline's apparent relative failure to address major contemporary issues within society – i.e. to be sufficiently applied and relevant, and therefore influential. Peck (1999), for example, argued in an editorial entitled 'Grey geography?' that 'human geographers have on the whole been conspicuous by their absence from substantive policy debates' (p. 131). Academic practice increasingly privileges research outputs in the form of journal papers (which enhance career prospects and a department's ratings) over 'practical and policy-oriented research':

> Policy research seems to have become the grey 'other' of academic research. While academic research appeals, with its cerebral and 'pure' processes of library-based learning and thoughtful contemplation, to the privileged scientific canon, policy research is often tainted by its association with cash, clients, contracts and reports-in-cardboard-covers. An appropriate analogy here might be with the distinction between manual and mental labour, where policy research is associated with 'getting one's hands dirty'.

The division between 'academic' and 'practical' research is potentially damaging to geography, he claims, with the latter being significantly neglected and undervalued. Such research need not be constructive – using the term in the sense of sustaining the status quo distribution of wealth and power, but it should be engaged and effective – 'Active participation in the formal policy process, and in the wider political domain in which this is embedded, is surely something that geographers would dispense with at their peril' (p. 133). He cautions, however, that some policy-research is 'shallow and simple' (p. 134), aimed at quick fixes, as compared with that which is 'deep and complex'; unfortunately, the former is more likely to have an immediate impact within policy communities, so that geographers wishing to influence policy have to find ways of being influential without compromising their ('deep and complex') academic strengths.

Massey (2000) expanded this argument. She noted that there had long been an antipathy towards the social sciences within political and policy communities, especially towards the 'soft' social sciences (i.e. all but economics), but that the UK's New Labour government (1997–2010) sought to change that in its drive towards 'evidence-led policy'. However, one of the problems with such a stance is that the politicians and policy-makers remain those who pose the questions and call for the evidence which, in Peck's terms, may call for 'shallow and simple' responses. She argues that social scientists (including geographers) should be involved in both reformulating the questions and raising others that are rarely, if ever, asked. To do that requires a 'more in-depth and sustained relationship with policymakers' (p. 132) than has been the norm. Massey exemplified her case regarding geographers' lack of impact with regard to regional uneven development in the UK, suggesting that public debates about this in 2000–1 were being conducted as if no research had been done on the issue. A government adviser had said in her presence that academics were 'a waste of time' and 'never came up with anything of real use to those involved in "real politics"' (Massey, 2001, p. 11). To some extent this confirmed her view that much of what we write as academics for academics (and perhaps their students) has no wider impact, but it also implied that 'the only way to be politically or socially relevant was to come up with advice or answers on government policy' (p. 12). There is a need for a much wider engagement, deploying geographic theories and evidence to promote policy initiatives rather than accepting the agenda of others, but to do that geographers needed to tackle the negative public image of their discipline as 'boring, as being a joke subject, as being all about capes and bays'. Geographers have to tackle their intellectual image if they are to be heard and respected in the policy-making corridors. (See also the essays, including her own, in her *festschrift* volume: Featherstone and Painter, 2013.)

A similar theme was developed by Leyshon (1995) and reiterated by Ron Martin (2001a; Martin and Sunley, 2011; Wrigley, 2013), who were concerned that geographers were not making a greater impact in debates about major contemporary issues, such as growing inequality (though see Philo, 1995). To Ron Martin (2001a, p. 267), important but rarely asked questions include 'What are we doing geography for?' and 'For whom are we doing it?'. In this context:

> Part of our endeavour must surely be to expose and explain the inequalities and injustices that our socio-economic and [sic] system produces and reproduces. And following from this, we have an obligation, indeed a duty, to assess and debate policy responses to those inequalities and injustices, with a view to exposing the limitations of existing approaches and helping to reshape political and public opinion as to possible alternatives. In short, there is a key role for the geographer as social critic, and a strong case for making geography a more activist endeavour.

However, it seems that we aren't, so that there is a missing agenda with regard to geography and public policy, and associated with this is a failure to meet what he sees as a 'moral duty'. That duty involves an agenda with three main items: exposing and explaining inequalities (as in Dorling, 2010); interrogating and evaluating existing policies and practices; and 'seeking to exert a direct influence on policy-making processes, at all scales, with the aim of producing more appropriate and more effective forms of policy intervention' (Martin, 2001b, p. 190). (The implication is that this should be an organised collective response, but later (p. 194) he acknowledges that there is a case that such policy activity should be a separate response by individual geographers operating within their own academic and political agenda – a restatement of Hare's earlier argument; p. 315, this book). The main reason why we currently lack influence, Martin argues, is not because we publicise our work insufficiently or that we lack confidence in arenas beyond our immediate profession but, rather, because 'much of what is done under the banner of human geography is unlikely to be seen by policy-makers as being remotely germane to policy issues. The fundamental problem is that there is no readily discernible policy research agenda in the discipline' (p. 191).

A major cause for geography's policy irrelevance, Ron Martin argued (2001b, p. 194), was 'the postmodern/textualist/discursive and cultural "turns" that have had such a pervasive impact across the subject in the last few years'. These have enriched the discipline in a number of ways but nevertheless resulted in its further retreat from policy research and modes of enquiry: 'it is difficult to envisage how the vague abstractions and epistemic and ontological relativism of much of human geography research . . . can form the basis of a critical public policy analysis' (p. 196). Hamnett (2003) agreed, claiming that geographers are not addressing major issues of global and national significance to anything like the level they should. For him, too, 'much recent cultural geography . . . has been a retreat from substantive political engagement and social analysis in favour of superficial academic radicalism' (p. 2):

> parts of geography are the new intellectual dilettanti: relevance has been replaced by irrelevance, reality by representation and social criticism by theoretical critique. For some, contemporary human geography has become an arena for theoretical play, little more. Regrettably, such frothy theoretical constructions are likely to be viewed in the outside world as . . . nice to look at, nice to taste but insubstantial and not to be taken seriously.
>
> (p. 3)

Their prognoses differed, however. For Martin, human geography is suffering from a lack of empirical and explanatory rigour, as well as an intellectual bias against policy issues and a lack of political commitment. Insufficient numbers of geographers lack one or more of these. As he notes, to many empirical rigour implies 'positivistic, quantitative methods and formal (statistical and mathematical) analysis, and what is widely regarded as a misplaced search for general principles' (p. 197). It may, but it need not: rigour calls for a 'sustained attempt to interrogate . . . evidence critically, to contextualize it thoroughly, to test propositions or to assess its wider relevance'. For Hamnett, on the other hand, 'Quantitative techniques and aggregate social research have been largely abandoned. . . . Analysis of large data sets has become totally passé, the object of suspicion or even derision as "empiricist"' (p. 2; but see Johnston et al., 2003; Johnston et al., 2014a, 2014b).

In responding to these critiques, several authors have presented partial defences of the contemporary situation, while at the same time agreeing with the force of much that has been written. Several, too, highlight the constraints on academic work in this area, given other career pressures. Thus, for example, both Banks and MacKian (2000) and Pollard et al. (2000) argue that Peck focused too narrowly on a particular genre of policy research – emphasising macro-economic issues to the neglect of others in social, environmental and cultural policy realms on which much is being done, especially at local scales: 'geographers are not abandoning policy – it is simply that we are approaching it from many more angles, guises and positions than Peck gives us credit for' (Banks and MacKian, 2000, p. 253). Peck (2000) responded positively, while at the same time stressing his preference for engaged policy research, 'which gets behind the backs – and, if necessary, under the feet or up the noses – of policy makers and elite actors' (p. 255). This involves creating a different image for geographers and their fundamental concerns; in contemporary society, neoclassical economic thinking is generally accepted as 'the commonsense of our age' (p. 256), whereas critical geographical thinking is not.

The response from Dorling and Shaw (2002) called for geographers to participate more in debates beyond their academic discipline and spend less time on 'internal discussions'. In order to get their arguments across to such wider audiences, they claim, geographers need concrete (i.e. empirical, quantitative) not abstract arguments, presented without any deployment of 'elitist jargon'; the former are more likely to be reported in the press and eventually to infiltrate the policy-making arena. Geographers also need, according to their prescription, to interact more with other social scientists. At present, they do not value policy-related work, and are not very good at it when they try; they are also too inclined to jump on new bandwagons – and perhaps geography 'is an intellectual safety net, an academic refugee camp – a place where academics can work on whatever they wish to work on and not be disturbed by the need to conform to the traditions of other disciplines . . . a home for intellectual anarchists' (p. 638). Nevertheless, on a positive note they also conclude that:

> Academic research does influence policy. . . . It also helps to form the values of both individuals and institutions. Most current policies are informed in some way by findings which were derived

from university-based research at some point in the past. Most individual views are now very much informed by what children learn in schools from teachers, almost all of whom are to some degree influenced by what they learned at university. Similarly, academics in universities may perhaps have their greatest impact through their teaching rather than publication.

(pp. 637–8)

So, as long as enough geography is taught in universities, eventually it will trickle through the educational system and influence later generations of policy-makers. Ron Martin (2002) largely agreed with them, pointing to a malaise within geography – its 'inferior standing and profile in the wider academic, educational and public domains' (p. 643; see also Martin Powell and George Boyne, 2001, and the response by John Mohan, 2003). For Massey (2002), however, Dorling and Shaw have a narrow conception of the 'political': one can become embedded in civil society in a wide variety of ways and arenas, gaining influence indirectly as a consequence. A discussion of British urban policy illustrates one such route (Amin et al., 2000), whereas Susan Smith's (1986) classic monograph on *Crime, Space and Society* also illustrates the wider applicability of geographic research. Meanwhile, Bonnett (2003) called on university geographers to capitalise on public interest in their subject matter and connect with this in critical and engaging ways. In a related move, Kevin Ward (2006) joined those exploring what might be meant by 'public geographies' and how these could be fostered. He claimed that these public geographies were about more than policy per se, and 'more about the public or publics' (p. 496) with whom geographers might usefully and productively engage:

Let's acknowledge that both [*policy* and *public* geographies] are legitimate means of involving geographers and geography in the lives of people, in a world in which geography structures the very social relations that hold together the human (and the non-human world). And let us also reflect on how the public engagement of today might become the policy reform of tomorrow.

(p. 501)

This notion of public geography broadly coincided with the development of what are known as participatory geographies, a form of action-orientated research (Pain, 2003) involving engagement with, rather than just study of, individuals, groups and communities; the research itself is collaborative and involves not just enhancing appreciation of situations but assisting the collaborators in campaigning for change in their (usually local) circumstances (Kindon et al., 2007; mrs kinpaisby, 2008). A Participatory Geographies Research Group was established within the RGS–IBG in 2007, taking forward ideas and approaches first launched by Bunge in the late 1960s – which themselves were partly inspired by the anarchist writings of Kropotkin and Reclus (Springer, 2012). Emancipatory in its aims, this work is largely local in its orientation, although Harvey (2012) has linked grassroots movements with the larger goal of replacing capitalism, and Mason et al. (2013) have suggested a broader agenda aimed at 'seeking out the workings of power and resulting injustices and thence striving to transform such actions' (p. 255).

Other work that has a broadly emancipatory goal includes critical work in geopolitics, broadly defined. Books like Gregory's (2004) *The Colonial Present*, Harvey's (2003) *The New Imperialism* and Neil Smith's (2005b) *The Endgame of Globalization* are all aimed at much wider audiences than their academic peers, and have been reviewed in the non-academic media. Furthermore, their authors have become public intellectuals (Castree, 2006a), not only through their writing but also their public performances and commentaries on contemporary events (as illustrated by Gregory's website: http://geographicalimaginations.com/).

Conclusions

Many of the arguments for applied geography have located it within the spatial-science approach to the discipline; that which is applicable is that which is based on empiricism, as illustrated in Pacione's (1999) substantial overview of 'useful knowledge' in geography. But each of the three types of science identified in Chapter 1 has its 'applied arm'; hermeneutic sciences have as their goal the development of self- and mutual understanding, a project shared with the critical sciences, whose emancipatory aspirations seek similar understanding within the wider context of appreciation of the economic, social, cultural and political contexts within which individuals and groups are embedded. Whereas radical geography sought emancipation largely in the context of class interests (e.g. Harvey, 2012), therefore, that imbued with the cultural turn promoted a wide range of identity politics, within which class is only of the foundations (May, 1996b), and which the gender and sexual politics that feminist and allied work in geography brought to the fore within the discipline. Such emancipatory intentions may extend further, both opening people's eyes to the nature of the world and leading them towards alternative scenarios but, as the collection of papers introduced by Michael Woods and Graham Gardiner (2011) illustrates, applied policy research can create dilemmas for those whose roots are within the critical geography, what they identify as the trade-offs between principles and pragmatism. Thus, Blunt and Wills (2000, p. x) introduced their book on dissident geographies by classifying them as sharing 'a political commitment to overturning prevailing relations of power and oppression'. Their stance was further illustrated by the title and subtitle of a book, and many of the essays therein, published to mark the fortieth anniversary of the first edition of the journal *Antipode* (Castree *et al.*, 2010). Only a few scholars, however (e.g. Harvey, 2000), have laid out in detail what such an alternative (applied) geography might look like (see Gibbons, 2001), or explored how it might be brought about (though see Harvey, 2012). Yet the development of wider debates reconsidering the nature and place of applied geography, the arrival of public geography and a renewal of moves to enact participatory geographies indicate that Harvey's (1974c) earlier arguments remain pertinent to fresh generations of geographers, whereby the crucial questions to be asked about any work claiming to be 'relevant' continue to be 'relevant to whom?' and 'for what?' Three decades later, a survey and review of 'The Complex Politics of Relevance in Geography' indicated, however, that these questions could be answered in a diversity of ways. Therefore:

> relevance is not easily measured, and may not be directly observable. While this perspective may be at odds with the performance – or productivity-based outcomes that increasingly dominate evaluations of research, we argue that our approach recognizes the ineluctably political nature of relevance and the diverse goals that we, as a community of scholars, promote.
>
> (Staeheli and Mitchell, 2005, pp. 357–8)

Whereas the debates about the impact agenda in the UK shaped the way that relevance and applied geography have been conceptualised in recent years, the fact that responses from UK-based geographers include discussions of 'creativity' (Phillips, 2010) and 'engaging' (Wills, 2014) is testimony to ongoing contest about what/who is geography for. These contests transcend national frameworks, however, taking their place in a wider disciplinary nexus of anglophone geography whose past, present and futures we reappraise in the next chapter.

Chapter 10

A changing discipline?

I was only fifteen when I chose geography as my field. Twelve years later, in 1957, I received my doctoral degree. So ended my long period of formal training. Ever since, I have not only taught and done research in geography but I have breathed and lived it. How was (is) that possible? How can geography, a rather down-to-earth-subject, have such a hold on me, offer me 'salvation' when, from time to time, my personal life seemed to be the pits? I couldn't have answered properly as a teenager or even as a newly minted PhD. I can give a well-rounded answer only late in life – in retrospect, for the meaning of geography has expanded over a lifetime. It grew as I grew.

(Yi-Fu Tuan, 1999, p. 93)

[Geographers] appear to live in an intellectual world characterized by groups of people plowing their own theoretical furrows, with little outright objection to others doing their own thing. . . . There is very little engagement between geographers today and geography over 20 years ago.

(Tim Cresswell, 2013, pp. 196, 15)

I wondered if the institutional position of the discipline had strengthened with its recently enhanced intellectual position, and what the geography of geography's institutional presence and strength was. For sure, anecdotal stories circulate regarding the opening or closure of a geography department, a renaming, a split and so forth . . .

(Lily Kong, 2007, p. 13)

Shifts of major importance [in geographical practices] do occur, but they seldom encompass the whole scientific community – old ideas and concepts remain with us to a large extent; new discoveries may sometimes have the character or mutations – and usually they look more like the rephrasing of old truths.

(Arild Holt-Jensen, 2009, p. 124)

The preceding chapters have outlined the major debates among anglophone human geographers since 1945 regarding their discipline's practices – what it studies, how and why. The temporal (post-1945) and spatial (UK and anglophone North America, though with some English-language material from elsewhere) foci have been placed in broader historical and geographical contexts. The chapter titles themselves indicate that several very different approaches to the discipline have been advocated; their contents suggest a large number of academics promoting new arrangements but being resisted by defenders of the old order – to a greater or lesser extent. The purpose of this final chapter is not to assess progress within those approaches (Lowe and Short, 1990), let alone to establish whether they have contributed towards the attainment of 'higher levels of intellectual, social and physical well-being for [our] fellow men [and women]' (Wise, 1977, p. 10); indeed, according to some interpretations, progress in the Enlightenment sense of that term is not feasible

(T. Barnes, 1996; Johnston and Sidaway, 2015). For, as Bassett (1999, p. 28) sets out in detail, one's determination of whether there has been progress depends on how that concept is defined. Geography may have made one or more of: institutional progress – becoming more firmly established and influential within academia; empirical progress – in that it can better predict some aspects of the 'reality' it studies; explanatory progress – being better able to explain through concepts and theories that which it studies; conceptual progress – with its theories having wider scope and greater internal consistency; progress in intersubjective understanding – we can now appreciate better how we and others see the world; pragmatic progress – our understanding allows better public policy; and emancipatory progress – we are freed from illusions about how the world works. (Tambolo, 2015, focuses on just three theories of progress: two derive from Popper – increased explanatory power and closer approximations to the truth; the other to Feyerabend – a steady increase of competing alternatives.)

Over the period discussed in this book, progress along several, if not all, of those dimensions may have been achieved. Assessing that is not the basic concern here, however. Rather, the evaluation returns to the issues raised in Chapter 1, where several models of the development of scientific disciplines were presented. No formal testing of those models is presented, for no methodology has been outlined that would allow such a task. Instead, their general relevance is assessed against the material outlined in the earlier chapters.

Human geographers and models of disciplinary progress

Along with members of almost every other scientific and social science discipline, human geographers have been attracted to the ideas and language of Kuhn's paradigm model (M. Harvey and Holly, 1981):

> the use of the word paradigm has become fashionable in geography as well as having become a pivotal concept for courses in geographic thought on both sides of the Atlantic. Thomas Kuhn has become as familiar to students of geography as Hartshorne or Humboldt.
>
> (p. 11)

Kuhnian ideas were applied in the 1960s and 1970s with relatively little reference to the major debates they had stimulated throughout Anglo-American academia, however. Most human geographers relied, it seems, on the first (1962) edition of *The Structure of Scientific Revolutions*, apparently unaware either that 'The loose use of "paradigm" in his book has made [it] amenable to a wide variety of incompatible interpretations' (Suppe, 1977a, p. 137) or that 'Kuhn's views have undergone a sharply declining influence on contemporary philosophy of science' (Suppe, 1977c, p. 647). Furthermore, Kuhn (1977) substantially revised his ideas later (see also the exegesis of his work in B. Barnes, 1982) and argued that it would be rare for a paradigm as he understood it (puzzle-solving within a normal science framework) to appear in the human sciences because their goal is 'new and deeper interpretations' (Kuhn, 1991, p. 23). Geographers at that time displayed a general tendency to adopt ideas from other disciplines rather uncritically. Agnew and Duncan (1981), for example, argued that:

> Recent reviews and programmatic statements concerning trends in Anglo-American human geography leave the impression that little attention has been given by geographers to the philosophical compatibility of borrowed ideas . . . that the political implications of different ideas have largely been ignored . . . and that controversy on source disciplines or literatures has not excited much interest.
>
> (p. 42)

Their examples do not include the import of Kuhnian ideas into geography, but their conclusions certainly hold in this case. (For a critical discussion of geographers' importing Kuhnian ideas, see Mair, 1986. It could have been used as one of the examples in Symanski and Agnew's (1981, p. 2) treatment of 'order and skepticism in geography', in which they argue that 'geographers have given too much attention to the search for order and not enough to skepticism'.)

Kuhnian concepts were first used in the geographical literature by Haggett and Chorley (1967), as part of a normative argument for a revolution in geographical method – and, perhaps surprisingly, in a book that they produced from a conference designed to introduce school-teachers to their 'new geography'. *Models in Geography* was reprinted in three separate paperbacks and widely used as an undergraduate text (Haggett, 2015). Nevertheless, the context of its production echoes Goodson's (1988) claim that geography is one of the few academic disciplines for which the pressure for its establishment in universities came largely from school-teachers (on which, see Chapter 2). Haggett and Chorley argued that the then dominant paradigm, as they defined it, could not handle either the explosion of relevant data for geographical research or the increasing fragmentation and compartmentalisation of the sciences. They proposed a new 'model-based' paradigm:

> able to rise above this flood-tide of information and push out confidently and rapidly into new data-territories. It must possess the scientific habit of seeking for relevant pattern and order in information, and the related ability to rapidly discard irrelevant information.
>
> (p. 38)

A similar situation some fifty years later has been suggested with the rise of 'big data' (Kitchin, 2013; Graham and Shelton, 2013).

That new paradigm had been launched more than a decade previously (see Chapter 3); Haggett and Chorley's goal was to spread the new ideas, and to win British converts to a new orientation of work within geography. In the same book, Stoddart (1967b, p. 512) used the concept to promote an 'organic paradigm' – as a 'general conceptual model'. When that essay was reprinted in 1986, he added as a footnote 'that I would now take a less enthusiastic view of Kuhn's analysis' (Stoddart, 1986, p. 231).

Haggett and Chorley were attracted by the concept of scientific revolutions. A similar argument was presented in a paper that appeared at almost the same time as Kuhn's book. Burton (1963, p. 113) introduced the term 'quantitative and theoretical revolution' and argued that:

> An intellectual revolution is over when accepted ideas have been overthrown or have been modified to include new ideas. An intellectual revolution is over when the revolutionary ideas themselves become part of the conventional wisdom. When Ackerman, Hartshorne and Spate are in substantial agreement about something, then we are talking about the conventional wisdom. Hence, my belief that the quantitative revolution is over and has been for some time. Further evidence may be found in the rate at which schools of geography in North America are adding courses in quantitative methods to their requirements for graduate degrees.

Wayne Davies (1972b) also used Kuhnian terminology in the title of his book, *The Conceptual Revolution in Geography*. He identified a change from contemplation of the unique to adoption of 'the more rational scientific methodology' (p. 9) as a revolution; none of the contributions reprinted in his collection refers to Kuhn, but their context is clearly influenced by Kuhnian ideas. A year later, Harvey (1973) used the term, like Haggett and Chorley, in a normative sense in his search for a new world view; the goal was to promote a revolution in geographical practice rather than – as Kuhn portrayed it – describing a revolution emerging from the anomalies thrown up by failures in normal-science puzzle-solving.

The authors cited so far used the paradigm concept at either the macro-scale of a world view or the meso-scale of a disciplinary matrix (see p. 19, this book). Others employed it either as a general descriptive tool (e.g. Buttimer, 1978b, 1981; Holt-Jensen, 1981, 1988; see also Asheim, 1990; paradigm had the largest number of entries in the index to Preston James and Geoffrey Martin, 1981, and remained one of the most commonly cited terms in the next edition – Martin and James, 1993 – but had just four index entries in the fourth, Martin, 2005), or as a framework for summarising subdisciplinary changes (Herbert and Johnston, 1978). Kuhn's micro-scale concept of a paradigm as an exemplar has also been deployed: Taylor (1976), for example, identified seven separate revolutions during the preceding decades and Webber (1977) proposed an entropy-based paradigm (see p. 115, this book).

Milton Harvey and Brian Holly (1981, p. 31) identified five paradigms within geography during the last century, associating each with an individual scholar: 'we can tentatively assign paradigmatic status to . . . Ratzel with the paradigm of determinism, Vidal with that of possibilism, Sauer with the landscape paradigm, Hartshorne with the chronological paradigm and Schaefer with the spatial organization paradigm'. They were focusing on 'schools of thought' associated with particular scholars, some of which ran concurrently rather than consecutively. They claimed that a single paradigm dominated the 1960s. Whether it was stimulated by Schaefer as they claim, despite the views of Bunge (1979) and others, is open to question. Cox (1995, p. 306) claimed that 'Schaefer arguably was rescued from intellectual oblivion only in order to provide some philosophical justification for what was happening in human geography at that time', though Getis (1993) reported that Schaefer's paper was widely read at the University of Washington in the late 1950s (see also Berry, 1993). Some of James Wheeler's (2002a) data cast doubts on Harvey and Holly's claim regarding dominance. The 1970s were rather characterised by a 'diversity of viewpoints' (Harvey and Holly, 1981, p. 37), within which spatial organisation remained important. Zelinsky (1978, p. 8) called that decade one:

> of confused calm, or rather of pluralistic stalemate, as geographers explore a multiplicity of philosophical avenues and research strategies . . . without that single firm conviction as to destination that guided most of us in the past.

The discipline was technically more capable, substantively more catholic, philosophically more mature, internationally more merged, socially more relevant, and academically more linked to other disciplines, he claimed, and these would probably lead to a continued plurality of approaches: 'I happen to believe that, more than anything else, this philosophical coming of age, this rising above the superficiality and tunnel vision that blemished so much geographical work earlier in this century, justifies' (p. 10) the title of his edited collection – *Human Geography: Coming of age*.

Although some were uncertain about the relevance of Kuhn's concepts to changes in the 1970s and 1980s, in comparison with the 1950s and 1960s, others were less equivocal. In his introduction to a collection of essays on *The Nature of Change in Geographical Ideas*, Berry (1978a, pp. vii, ix) claimed that 'The changes in geographical ideas that we have discussed are distinctly Kuhnian.' The other contributors' essays provided little supporting evidence for this claim, however, and Berry's own contribution suggested that pluralism and interparadigm conflict were much more common than periods of normal science. Writing of geographical theories of social change, he noted (Berry, 1978b, pp. 19, 22, citing Mikesell):

> a diversity arising from the mosaic quality of modern geography . . . [geographers] have . . . moved from one paradigm to another, and in the last decade they have been extremely dynamic. . . . With multiple ideas and multiple origins, modern geography could rightly be characterized as a mosaic within a mosaic.
>
> (Mikesell, 1969)

Gauthier and Taaffe (2003) used the term 'revolutions' to describe what they identified as three major changes in American geography during the twentieth century according to six criteria: type of change; pace of change; intensity of accompanying debate; operational characteristics (or core concepts); impact of the change; and context for the change. Paradigms were also the core organising framework of a discussion of changes in geomorphology (Orme, 2002).

Human geographers' critiques of Kuhnian applications

Kuhnian concepts and terminology were widely used by human geographers for several decades, therefore, although some treatments of disciplinary history entirely ignored this literature (Freeman, 1980a, 1980b) and others made little use of it. Buttimer (1993, p. 70) has a single, brief, reference to Kuhn, for example (as does Cox, 2014). Livingstone (1992), after noting that 'In the wake of Kuhn's treatise, a batch of historians working in various disciplines set out on a paradigm hunt, looking for paradigms, paradigm shifts, and what not' and that 'Geographers were no exception' (p. 14), concluded that some, like Berry, Harvey, and Haggett and Chorley, used the concept of a paradigm as 'little more than a flag for rallying the troops' (p. 24) to a new cause. Gould (1994, p. 196) argued that 'They challenge, sometimes with deliberate overstatement, using polemic to grab the ear of an established coterie, and rhetoric to persuade it of its folly.' Berry (1993) claimed that this was necessary; in the context of Hartshorne's responses to Schaefer (p. 61, this book), those promoting the new approach in the late 1950s were engaged in a debate with the 'discipline's luminaries':

> Mainline geographers were suspicious, threatened, antagonistic; and we reciprocated. We felt we had to fight and fight we did, earning reputations for brashness and abrasiveness. So be it: if we had not been aggressive, geography would have rolled over us. Instead, we tried to roll geography.
>
> (p. 438)

In that fight they sought support from the emerging field of regional science (see p. 83, this book), and as a consequence of their victory, according to Morrill (1993):

> I believe we saved geography from extinction as a serious university discipline, by attracting and training good students, by writing articles and books that developed theory and method, by gaining a foothold in science at large, and by applying these methods and theories to contemporary social problems.
>
> (p. 442)

Nevertheless, he still found geography's position 'weak, fragile and almost invisible . . . we survived but did not succeed in the goal of propelling geography into the mainstream' because 'much of geographic work . . . was not good enough' and some of the best spatial scientists 'defected' (p. 443).

Livingstone's (1992) attitude to discussions of Kuhn and the revolutionary fervour of some spatial scientists was that:

> The details of these (and other similar cases) need not be reviewed here. Suffice to say that their revolutionary gung ho spirit of triumphalism was scarcely what Kuhn had in mind as he portrayed the mega-level Gestalt-shifts in the history of science.
>
> (p. 24)

While the reception of Kuhn's work in the half-century since its publication has been complex and part of a wider opening-up of the history of sciences (Gordon, 2012), many of the presentations and uses of Kuhn, however, were superficial (see Graves, 1981), both cavalier and uncritical in basing their descriptions on Kuhnian foundations. Mair (1986, p. 357) includes earlier editions of this book in that category, claiming that it 'rarely reaches beyond a superficial comparison of the Kuhnian model and the recent history of human geography'. Nevertheless, he did recognise that it, like other evaluations, concludes 'either that Kuhn's ideas are irrelevant to the historiography of geography, or that they "fit" only when stretched almost beyond recognition' (p. 345). Hubbard *et al.* (2002, p. 26) were also critical of the use of the paradigm concept in geography and its treatment in earlier editions of this book (although we should clarify here that the fifth edition of *Geography and Geographers* had also concluded that 'Kuhn's model does not fit the experience of human geography since 1945 therefore', p. 389). They note that:

> In particular, the idea that geography has moved through unified (and generational) paradigms glosses over the ideas and practices associated with those who did not conform to the dominant or fashionable way of doing things. The consensus among geographers at any one time that there is a best way of doing things has seldom been complete or stable, and to pretend that it has been so is to obliterate the voices of many researchers. In relation to the recent history of Anglo-American human geography, we therefore need to be mindful of the fact that it is often white, English-speaking, middle-class, heterosexual, able-bodied male academics who seek to define the Zeitgeist and identify which ideas are most useful to progress. The net result of this is that dissenting voices – and alternative traditions within geography – are often marginalised or obliterated in the pages of geographical history . . . you should be wary that most histories of geography can serve to legitimate the careers of an academic elite while obliterating the views of others regarded as insignificant.
>
> (p. 26)

Livingstone (1992, p. 25) found that 'it became plain that much of this writing [about the applicability of the paradigm concept] amounted to misdirected effort', but agreed with Mair that 'even among [the] . . . critics the flavour of Kuhn's work still lingers. The Kuhnian ghost, it seems, is proving rather hard to exorcize from the history of geography.' Nevertheless, he welcomed the sociological approach to disciplinary history which the adaptation of Kuhn's ideas heralded; it countered the belief in 'conceptual cumulation, disciplinary progress, and internal chronology', and encouraged contextual readings in its stead.

Other geographers have been very sceptical of the value of a Kuhnian interpretation, with two claiming that it has 'distorted even perverted the development of geography' (Haines-Young and Petch, 1978, p. 1). Just over a quarter of a century later, Kwan (2004) described 'the influence of Kuhn's (1962) model of scientific revolutions' as a problem:

> Since Kuhn was taken to conceive of disciplinary change as a succession of perspectives, each eclipsing the other, and to advocate a clean break with existing practices and the dominance of a singular vision . . . difference and diversity in perspectives have no role in this interpretation of his model . . . [which] tends to intensify antagonism within geography because it suggests that the normal state of human geography involves one perspective being victorious over another and that there is something wrong with the persistence of an incompatible viewpoint. It is therefore important to recognize that the Kuhnian model (as we used it) is not suitable for a discipline like geography, in which a variety of perspectives and methodologies coexist at the same time.
>
> (p. 759)

The concepts of revolutions and normal science have also been criticised as providing poor descriptions of geography in recent decades in another textbook (Holt-Jensen, 1988, 2009).

Individual geographers may have experienced personal revolutions and rapid shifts from one paradigm to another – at the meso- if not the macro-scale. (Harvey, 1973, 2002, indicates this for himself; see also the essays in Billinge *et al.*, 1984: the editors (p. 11) quote Cox, Gould, Olsson, Scott and David Smith as additional examples.) Sheppard (1995, p. 287), employing citation data, contrasted a drastic shift in the discipline's 'master weavers' – the term is drawn from Bodman's, 1991, 1992, depiction of the most-cited human geographers – from spatial science to social theory during the 1980s and also argued that:

> It is of interest to note that 8 of the 12 social theorists [all of the geographers discussed are classified by him as either spatial scientists or social theorists], as well as a large number of other influential individuals in the social-theory group, began their careers within the research traditions of spatial analysis.

Among them, Sheppard identified three cohorts:

1 Those who wrote classic spatial analysis papers of the 1960s (he cites Bunge, Cox, Harvey, Johnston, Olsson, Pred, Scott, Soja and M. J. Webber).
2 Those 'whose early work was solidly within spatial analysis, and recognised as such, but for whom this work did not represent a major period in research careers that shifted rapidly from spatial analysis to Marxism and social theory' (he lists Dear, Massey, Peet, P. Taylor and Thrift).
3 Those whose careers began in the late 1970s with contributions to spatial analysis when it was already under heavy criticism, 'only to re-identify themselves prominently with the concerns of social theory' (the names given, and trajectories discussed, are Gordon Clark, John Paul Jones – and himself!).

For Sheppard, therefore, there have been only two major competing world views – the spatial-scientists' versus the social theorists'; most of the debates discussed in this book can presumably either be encapsulated within that macro-scale competition or involve controversy within a world view (over disciplinary matrices and/or exemplars, but not basic orientation, which Barnes, 1996, portrays as for or against the Enlightenment project with its belief in progress and universal truths). Indeed, much of the debate, especially in the latter twentieth-century decades, was about methods, specifically 'quantitative vs. qualitative' (with many textbooks focusing only on one of the pair: Johnston, 2006).

To claim that such individual experiences can be amalgamated into disciplinary revolutions strikes some observers as both inapt and inconsistent with the evidence (Bird, 1977, p. 105): 'Perhaps so many revolutions in so short a time indicate in themselves either a continuously rolling programme, or something basically wrong with the overturning metaphor'. Bird (1978, p. 134) suggested that mono-paradigm dominance of a discipline is inconsistent with 'the fact that society itself is organised around more than one major principle'. The focus of Bird's criticism is not clear, however; it is much more convincing on the world-view scale than at that of the disciplinary matrix, for example, and even less so at the scale of the exemplar.

Stoddart's criticism was even more pointed; he initially saw some value in the paradigm model (Stoddart, 1967b) but later argued (Stoddart, 1977) that:

> the concept sheds no light on the processes of scientific change, and readily becomes caricature. I suggest that as more is understood of the complexities of change in geography over the last hundred years, and especially of the subtle interrelationships of geographers themselves, the less appropriate the concept of the paradigm becomes.
>
> (p. 1)

Stoddart (1977, p. 2) also argued that 'There is scope for sociological enquiry into the extent to which the concept has been used in recent years as a slogan in interactions between different age groups, schools of thought, and centres of learning.' In 1986 he reiterated that 'the adoption of Kuhn's terminology, far from clarifying history, actively distorts it, largely by reducing the participants to caricature figures' (p. 13).

His analysis showed both the absence of consensus (normal science) and the slow pace of change (which is more readily represented in Lakatos's schema (p. 19, this book)). In his view the paradigm concept became part of the 'boosterism' image with which geographers conducted debates (Stoddart, 1981a):

> the paradigm terminology has been used to illuminate either the establishment of views of which a commentator approved, or to advocate the rejection of those he did not (p. 72) . . . the concept of revolution bolsters the heroic self-image of those who see themselves as innovators and who use the term paradigm in a polemical manner . . . those who propound the Kuhnian interpretation have done so in ways which tend to make it self-fulfilling.
>
> (p. 78)

Continuing the argument, Stoddart (1986) argued that it was such wish-fulfilment and strategic use by those who:

> propound Kuhnian interpretation . . . rather than its value as a framework for studying historical change, that makes the paradigm idea of interest to the historian of science: as itself an object of study, rather than as means of understanding the complexities of change.
>
> (pp. 17–18)

Billinge et al. (1984, p. 6) claimed that Chorley and Haggett's initial use of Kuhn's terminology was 'In some measure . . . only gestural' because although they clearly distinguished between normal science and extraordinary research 'the "anomalies" within the traditional paradigm which were central to Kuhn's thesis were never identified in any detail'. Chorley and Haggett were pressing for a revolution in the nature of geography, but it was not a revolution generated by the failure of the previous paradigm (using failure in a Kuhnian context). Billinge et al. pointed out that the term 'paradigm' was being used more for propaganda than for historiography – thereby potentially gaining prestige by locating their would-be revolution alongside those in physics described by Kuhn (Taylor, 1976). They accepted that the Kuhnian model should be rejected, but argued that since Kuhn himself did not expect the model to fit the social sciences, such a conclusion is hardly surprising. Nevertheless, they did identify insights in Kuhn's work that may be of value in the study of a social science like human geography, such as his concept of the paradigm as what constitutes, or 'gives form to' (Billinge et al., 1984, p. 16) scientific activity by providing the 'maps' with which scientists are socialised into a discipline or subdiscipline. Science takes place in 'heterogeneous settings' with their own 'rules and resources' (p. 17); it is a 'localised' activity which occurs within (scientific) communities and hence its study is necessarily a study of community sociology.

Mair (1986, p. 359) took the criticism further, contending that, with the single exception of Billinge et al.:

> geographers have entirely misinterpreted Kuhn's contributions. They have been wrong on Kuhn in the simple sense that the skeletal Kuhnian model is misrepresented as Kuhn's major contribution. More fundamentally, however, they have been wrong on Kuhn in misconceiving his entire project.

For him, the main values of Kuhn's work were:

1   the concept of the exemplar as an analogy to be used in the creative process;
2   the notion of incommensurability in the comparison on competing paradigms; and
3   the clear need to study the sociology of scientific communities.

Geertz (1983) also stressed the last of these points. For others, the important issue is the absence of clear 'revolutions'. Michael Chisholm (1975a), for example, concluded that:

> by using prose as the medium of exposition and by concentrating on the essential underlying ideas, it is manifest that continuity of thought dominates over the apparent discontinuity which some have called the 'quantitative revolution'.
>
> (p. 89)

Later in the same book, however, he admitted that 'in respect of analytical techniques geography has indeed experienced a revolution. However, in terms of the subject-matter studied the position has changed less rapidly' (p. 173; there is a confusion of scales implicit here, between exemplars and disciplinary matrices, if not world views) – and it was only later that the problems addressed by geographers 'caught up' with their analytic ability to handle them (see also Gregory, 1978a, pp. 52–3; 1994, p. 55). Although they use the term 'revolutions', Gauthier and Taaffe (2003) are close to Chisholm in their characterisation, writing of a:

> continuity model with surges superimposed to represent the three 'revolutions'. During the surges the weaknesses of the previous paradigm are usually stressed and caricatured. . . . Later, as the surge subsides, we begin hearing more about hapless conceptual babies being thrown out with the bathwater of the old paradigm. The merits of some aspects of the previous paradigm are discussed and recognized, and a certain amount of continuity sets in – perhaps only leading to a wave of enthusiasm about a new paradigm.
>
> (p. 523)

Despite such criticisms (which according to Livingstone, 1992, p. 25, 'began to come thick and fast' once 'the first flush of paradigm enthusiasm died down and a more measured consideration of Kuhn's relevance to geographical history was undertaken'), relatively little was done until the 1980s and 1990s to suggest alternative sociologies based on other models of the history of science, including some developed specifically to represent the situation in human geography. Livingstone's (1992) work (first noted in the Preface here), drew on contextual histories of science. He notes how, 'In the wake of Kuhn's treatise, a batch of historians working in various disciplines set out on a paradigm hunt, looking for paradigms, paradigm-shifts and what not. Geographers were no exception' (p. 14). He did not wish to repeat the exercise, while recognising that:

> Whatever the internal conceptual irresolutions within the corpus of Kuhn's writings, and the conceptual sloppiness of the manner in which it was imported into geographical history, there can be no doubting the benefit that a broadly sociological rendering of both science and geography has wrought.
>
> (p. 25)

Reflecting wider debates on the history and sociology of knowledge since Kuhn, Livingstone favoured a broader 'approach to geography's history that will do justice to the intellectual and social context within which geographical knowledge was produced' (p. 23). Livingstone then wants to suggest:

> that it might be helpful if we were to think of geography as a tradition that evolves like a species over time. As I say, this is a risky analogy, for my colleagues are sure to sniff all the problems of earlier organic analogies or to suspect some underlying evolutionary epistemology . . . I judge the risk worth taking because I think the image helps us see that ideas in general, and geographical ideas in particular, are historical entities that change, transform, evolve over time in different cognitive and social environments. As I see it, geography is a tradition that, like a species, has undergone historical transformation.
>
> (p. 30)

But tradition was seen by some to be no less problematic than paradigm as a way of narrating disciplinary history. Drawing on a set of critical reactions to Livingstone's arguments (Driver, 1995; Matless, 1995; Rose, 1995), Anne Godlewska (1999) argued that:

> Taking a history of ideas approach to the concept of 'tradition' as does Livingstone – following a particular line of descent – de-emphasizes the exclusions, the contestations, the lulls, the gaps, and the collapses in the history of geography and understates the power of discursive formations to limit and form research and practice. I welcome Gillian Rose's contention that 'tradition' implies exclusion. I think that discursive formations do also and I agree that those exclusions, whether gender-based or otherwise, should be a particular focus of attention.
>
> (p. 315)

Godlewska's (1999) critique of Livingstone appears in an endnote to a book-length study 'about the nature of French geography about two hundred years ago. It is not a disciplinary history' (p. 1). Inspired by Foucault, she seeks:

> to get at changes that alter discursive patterns: change that alters the objects of the study of particular discursive formations, the operations they may engage in, the apparent relevance and importance of their concepts, and the theoretical options open to thinkers; change that alters the boundaries of a given discursive field and the means by which it reinvents itself; and change that overturns hierarchies in the intellectual division of labor and consequently the orientation of research.
>
> (p. 1)

Her focus on the development of geographic language – both textual and graphic (modes of mapping and visualising spatial and natural relations) – has inspired others (Johnston and Sidaway, 2015) considering the relationships between geographical concepts, and disciplinary change.

A decade before Livingstone's (1992) account of *The Geographical Tradition*, Wheeler (1982) found Lakatos's work attractive, arguing that several separate programme cores could be identified in contemporary human geography (he cites areal differentiation, spatial science, cognitive-behavioural approaches and marxist structuralism), in each of which the operation of positive heuristics (see p. 19, this book) can be identified. These research programmes are in continuing competition, whose nature will change over time (Wheeler, 1982):

> Given the deficiencies in Kuhn's scheme and the rather ill-defined nature of geography, it seems unlikely that the future of the discipline will be characterized by sequences of revolutionary change interspersed with efficient problem solving. Moreover expectations of revolutionary

progress appear to be unjustified. Instead it seems that a variety of approaches, which will undoubtedly wax and wane in popularity, will continue to be employed.

(p. 4)

Giles Mohan (1994), drawing on the notion of a product cycle, suggested that a rapid turnover of theories had recently occurred within geography as a consequence of pressures within the academic profession. He identified 'four "big" approaches over the preceding decade hailed as capable of explaining social reality' alone (p. 387; the four were critical realism, structuration, postmodernism and postcolonialism). He associates this rapid turnover with the growing 'commodification of knowledge' and the competition for status within academia, so that:

> these ideas were not paradigmatic and their diffusion into geography uneven. Likewise some of these intellectuals [those promoting the 'big' approaches] have persisted with their theoretical frameworks and have not continually experimented with new ones . . . [nevertheless] Many academics fight for currency and exploit new markets in knowledge. This increased competition has, in the spirit of flexible specialisation, resulted in the 'niching' of academic thought despite the recent emphasis on trans-disciplinary pursuits.
>
> (p. 387)

Some of the forces impelling these trends are general within academia, but one at least is specific to geographers (see also Crampton and Elden, 2007):

> As a trans-disciplinary subject geography has often lacked kudos in wider social theory. The result is that geography has tended to borrow and incorporate any social theory that appears spatial or contains spatial metaphors. The increased commodification of knowledge has served to heighten this tendency. For example, we have had Giddens' locales, Foucault's disciplinary spaces and Mohanty's contested cartographies. It seems that few have time to actually apply any of these potentially useful ideas before the next theoretical innovation superseded them.
>
> (p. 389)

Disciplinary matrices and exemplars come and go at an alarming rate according to this view, with geography equated to a supermarket in its concern for 'turnover time' and short shelf lives as its practitioners seek status through the novelty of their approaches.

Geography and its environment

Some geographers have argued (in similar terms to ours here) that the major influence on their discipline's content and approaches comes from a combination of its wider environment, particularly its economic, social and political milieux, and changes in other subjects. In general, it is believed that geography is more likely to change because of developments in other disciplines than vice versa. Stoddart (1981b, p. 1), for example, introduced a book of essays as demonstrating 'that both the ideas and the structure of the subject have developed in response to complex social, economic, ideological and intellectual stimuli'. Evidence for this was presented, as a second theme of the essays, in the reciprocal relationship between geographers and their milieux: 'throughout its recent history geographers have been not only concerned with narrowly academic issues, but have also been deeply involved with matters of social concern'. This conclusion had less purchase in subsequent decades, however, with the growth of what became known as a 'spatial turn' in the humanities and other social sciences (Warf and Arias, 2009; Morton, 2011; Porter, 2011) and an increasing awareness of the importance of incorporating space in general, and distance in particular, in economic analyses. (Many of those who 'discovered' geography did so very partially, however, often either selecting those geographical writings and practices which fitted their predilections or largely ignoring much contemporary geographical scholarship, as in Kaplan, 2012.)

Hubbard *et al.* (2002) argued that:

> contemporary theoretical approaches to human geography need to be understood contextually, in relation to at least three things . . . :
>
> 1 The *history* of geography – how geography has actually developed in terms of what geographers have studied (and how).
> 2 The *sociology* of geography – how institutions, social networks, journals and educational structures (particularly universities) have shaped the development of geography.
> 3 The *psychology* of geography – how individual geographers have adopted ways of thinking about and interpreting the world, whether conformist or confrontational.
>
> (pp. 25–6, original italics)

One of the most detailed deployments of the contextual approach is Livingstone's (1992) *The Geographical Tradition*, in which he refers to the history of geography as 'situated messiness' with the absence of any 'essential nature' (p. 28). In the discussion of any particular theory within geography, therefore, this means that:

> it will never be wrong to ask of any theory: Why was it put forward? Whose interests did it advance or retard? In what kind of milieu was it conceived and communicated? How adapted was it to its cognitive and social environment?
>
> (p. 29)

Livingstone's chapters on geography in the late nineteenth century ('A sternly practical pursuit; geography, race and empire'), the first half of the twentieth ('The regionalizing ritual: geography, place and particularity') and the 1950s–1960s ('Statistics don't bleed: quantification and its detractors') illustrate his theme that:

> geography changes as society changes, and that the best way to understand the tradition to which geographers belong is to get a handle on the different social and intellectual environments within which geography has been practised.
>
> (p. 347)

Within those environments, the character of geography was contested, though not necessarily by all of its practitioners, reflecting the particularities of any one time and place: 'Sometimes the conversations have admitted a range of geographers, from time to time only a select group were equipped, or permitted, to take part' (p. 358). But as we get closer to the present, the latter condition became rarer, with fewer 'partisan apologists . . . [striving] to monopolize the conversation in order to serve their own sectarian interests'. Likewise, Gregory (1994, p. x) writes that the concept of a 'discipline' such as geography implies 'a set of sovereign concepts . . . [and] a rigorous policing operation' when in effect it involves individuals and groups traversing 'nomadic tracks . . . [and] new lines of fight' as they explore alternative ways of producing and reproducing knowledge – as fully illustrated in the various chapters of *Geographical Imaginations*. Elsewhere, Livingstone has argued for the importance of studying both 'life geographies' (the career trajectories of individual geographers, including their spatial movements) and 'intellectual spaces' – the places in which ideas are crystallised and developed (Johnston, 2004e; Livingstone, 2000a).

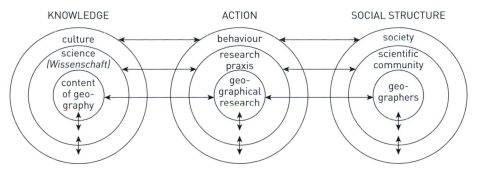

Figure 10.1 Geography's context

Source: Gräno, 1981, p.19.

Many geographers who employ a contextual approach stress that any changes in a discipline's contents (revolutionary or not) are associated with significant events in its milieux. Thus, Berdoulay (1981, p. 10) wrote of the role of the *Zeitgeist* (spirit of the age) influencing what geographers do. Gräno (1981) took this much further (Figure 10.1); geographers are a subset within the community of scientists, which is itself a subset of wider society, whose culture includes a scientific component that strongly influences the content of geography. The community of geographers is an 'institutionalizing social group' (p. 26), providing the context within which individual geographers are socialised and defining their disciplinary goals within the constraining and enabling characteristics of the external structures. A major goal when geography was established as an academic discipline was to create an identity, to 'establish an object of study that could be regarded as geography's own and that differed from that of other disciplines' (p. 30). This, he claimed, was a 'passive education role' (p. 32) disseminating knowledge of society–environment interrelationships, but it was replaced in the 1970s when 'geography began to contribute actively to a transformed and replanned world. Applied geography was created and geography became a profession outside the small world of the university' (p. 32).

Geographers made and continue to remake geography: 'It was the external goals of society that brought the establishment of geography as an academic discipline. This took place without any noticeable contribution from any other scientists' (Gräno, 1981, p. 30; see, however, Stoddart, 1986). The initial period of institutionalisation involved geography operating largely as a pedagogic subject, meeting the needs for training teachers whose activities would promote the interests of the expanding 'nation-states' (Capel, 1981; Taylor, 1985a; Goodson, 1981, 1988; Fuller, 2000). Capel (1981, p. 36) referred to the creation of geography as a discipline resulting from:

> the presence of geography in primary and secondary education at the time when the European countries began the rapid process of diffusion of elementary education; the necessity to train geography teachers for primary and middle schools was the essential factor which led to the institutionalization of geography in the university and the appearance of the scientific community of geographers.
>
> (p. 36)

(See also Freeman 1961, 1980a, and, for a more critical account, Rieser, 1973.)

As the context changed, however, so did geography for (Capel, 1981):

> The established community employs strategies tending to reproduce and amplify itself. Never will it opt for self-liquidation; the community will defend its survival, even if other communities of scientists investigate similar problems with like methods, or if the logical incoherence of the

conceptions that they defend is revealed . . . Everything will be sacrificed for the reproduction and growth of the community, including the coherence of the very conception of the discipline: different conceptions can defend themselves in distinct moments or even simultaneously, without putting into doubt the continuity of the science practised.

(p. 66)

The promotion of national interests, especially national commercial interests, by nineteenth-century geographical societies provided the context for the creation of geography as an academic discipline (see the essays in Bell *et al.*, 1995, and also Morin, 2011), and subsequent changes in economic, social and political structures and needs generated new demands, to which geographers responded. (See, for example, Taylor's, 1993, evaluation of different geographical perspectives on 'the global' during the twentieth century.)

Scott's (1982) answer to the question 'Why do geographers, regional scientists, urban economists and others study the spatial patterning of social events?' (p. 141) provides an example of disciplinary reconstruction in the context of changed circumstances. Writing in North America in 1982 (a decade later, the context was very different), he concentrated on the ways in which late capitalism is organised, not through market relations, but rather involving bureaucratic intervention by an all-embracing state. Because 'the geography of late capitalist society is shot through with problems and predicaments' (p. 145), state action is required in the 'more subtle forms of social, cultural, and psychological management' (p. 146) on which science thrives (Scott here uses 'geography' not in the professional but rather the vernacular sense: Johnston, 1986b.) By participating in those managerial tasks, however, scientists contribute to the creation of countervailing forces, with which they may also become involved:

the endemic crisis of economic production and growth in late capitalist society creates the need for specific problematics and policy discourses out of which technical control may be accomplished. But technical control creates an advanced set of social conditions in which a countervailing set of human predicaments makes its appearance – alienation, the destruction of affective human relations, the repoliticization of human and regional planning, and so on.

(p. 152)

Hence, the radical and humanistic responses to spatial science.

The link between the external environment and the practice of economic geography was examined further in a later essay, which Scott (2000, p. 18) began by claiming that:

Over time, economic geography has behaved in a manner quite different from that which might be expected of a rationally ordered discipline pursuing some pre-ordained epistemological mission. Its historical course has been notably responsive to changes in external economic conditions, to the unfolding of political events, and to the play of professional ambitions and rivalries. . . . These imply that any answer to the question 'what are the central problems of economic geography?' is likely to be historically contingent, even if the inertia of tradition and its institutionalization in a system of higher education means that important continuities can be deciphered over time.

(p. 18)

In this context, 'one of the few general principles that seems to be reasonably evident in this connection is that the concrete questions and problems that society faces at any given historical moment tend also, in one way or another, to become burning questions and problems for practising social scientists' (p. 19), though it is not clear whether this simply reflects social scientists responding to the perceived world or rather social scientists being required to reorient their concerns by external 'paymasters'. (In other cases, however, changes in the external environment and their appreciation stimulate scholars to reorient their intellectual interests, as in Derek Gregory's, 2004, *The Colonial Present*, writing of which was stimulated by his response to the 2001 and 2003 American-led invasions of Afghanistan and Iraq, and the ongoing Israeli occupation of Palestine.) Thus, economic geography can be characterised by a series of episodes – spatial analysis and regional science; behavioural geography (a 'brief interlude'); political economy; localities (another 'brief interlude'); and a world of regions within globalisation (see also Clark *et al.* 2000). Scott eschews Kuhnian terminology – neither paradigms nor revolutions are mentioned – in part because none of his episodes or interludes has entirely ended:

> I have tended to treat the main episodes in the postwar history of economic geography as occurring in relatively discrete blocks of historical time. To be sure, each of these episodes can be described in terms of a moment of emergence, a subsequent period of rapid efflorescence, and a period of decline, where the latter period is often extended over a number of decades. As a consequence, almost all of the episodes identified here continue to leave traces on the work of practicing economic geographers today, just as adherents of this or that tendency whose heyday is long past continue to produce and publish scholarly work right down to the present. The intellectual landscape of economic geography at any given moment in time is therefore best represented as a sort of intellectual palimpsest rather than as a unified front.
>
> (pp. 32–3)

Scott does not assess the value of work published long after the episode's heyday, however. Further, he implies – in part by his criticism of Trevor Barnes's (1996) work – that the response to economic changes by economic geographers is largely determined by those changes, leaving relatively little room for debate and decision-making (with just a few economic geographers left practising an obsolescent craft long after others have moved on). For Barnes (2000), context is both constraining and enabling:

> The discipline arose because of a set of contextual factors, and as those factors changed so did economic geography. Further, those factors . . . entered into economic geography's very constitution, shaping the questions it asked, the methods and tools it used, and the range of answers it deemed acceptable. For sure, the discipline had its own geniuses, people who were brilliant in thinking up new ideas, but their creativity was always . . . informed and tempered by a wider context.
>
> (p. 12)

Thus, for him, economic geographers 'invented' and 're-invented' economic geography through the accepted practices of the subdisciplinary community(ies), as exemplified by Peck and Tickell's (2012) revisiting of their theorization of neoliberalism twenty years earlier (Peck and Tickell, 1994; Tickell and Peck, 1992). Scott (2000, p. 34), too, accepts that changes in disciplinary practices are invariably subject to 'a welter of critique and counter-critique' and may reflect 'the changing fortunes of different professional groups as those whose human capital is sunk in waning problematics engage with those pressing forward with new approaches in the endless struggle to come out on top in the stock-market of ideas', but seems less prepared than Barnes to identify particular individuals having a significant role in this struggle.

For Thrift (2000, p. 692), the cultural turn – Scott's final episode – 'saved economic geography from what might otherwise have been a musty oblivion. By the 1980s, economic geography was in a pretty moribund state, at risk of boring its audience to death.' Subsequently, however, work in economic geography – and other social science disciplines too – no longer took place on economists' terms: the nature of the economic, and the assumptions and principles of the discipline which studies it, is open to debate (see also Amin and Thrift, 2000).

There is no necessity for geography, however, no 'specific necessities in scientific knowledge' (Gräno, 1981, p. 65). Rather, according to Peter Taylor (1985a, p. 93):

> Geography is a social institution. Like all such institutions its value to society varies over time and place. The creation of any social institution is a result of a group of people who identify a particular need and are able to find the resources to meet that need. As needs change the institution has to adopt to survive.
>
> (p. 93)

It may fail; the forms of scientific discourse created and practised may not command sponsorship and resources, and may lose their positions within educational curricula and institutions. Scott (1982, p. 151) argued strongly that success requires identification of clear social needs: 'only discourses that are posited upon existing problems of social life and practice, and upon existing political interests, stand any likelihood of commanding a significant consensus of scholars and scientists'.

There is nothing deterministic in these arguments, for they depend on geographers identifying what is and is not sustainable in a particular context and successfully promoting themselves as able to meet the perceived needs. How they do this will reflect what Berdoulay (1981) calls their individual 'circles of affinity': social networks extending outside their disciplinary base. To appreciate those one must appreciate biographies, as Buttimer (1981) argued and several volumes of autobiographical recollections illustrate (Buttimer, 1983; Billinge et al., 1984; P. Moss, 2001; Gould and Pitts, 2002; L. King, 2008; see also the fuller presentations in Haggett, 1990, and Gould, 1999). Buttimer (1983, p. 3) defends such an approach because 'each person's life echoes the drama of his or her times and milieu; in all, to varying degrees, the propensity to submit or rebel. Through our own biographies we reach toward understanding, being and becoming.' Autobiography provides what she terms 'choreographic awareness', the 'moral, aesthetic, and emotional commitments which are related to lived experience and which underpin a scholar's eventual choice of epistemological presentation and style of practice' (p. 12). It can indicate aspects of disciplinary history that are absent from the written record of research papers and student texts, isolating the influence of 'pioneers of geographic thought whose inspiration flowed through their teaching and field experiences, through their counselling and listening' (Buttimer, 1981, p. 88; for critical approaches to autobiography, see Roche, 2011, and Shaw, 2013). Biographies offer an alternative source, but there are few on geographers who have contributed significantly to the contemporary period (Paterson, 1985); and the few biographical dictionaries available (most of which, like the serial Geographers: Bibliographies, deal only with dead geographers and so have little resonance to contemporary geography) are limited in their treatment of geographers' intellectual contributions relative to their career paths (Larkin and Peters, 1993).

Geographers and their networks

The autobiographical theme was taken up explicitly by Pred (1979, 1984a), using the language of time geography (p. 131, this book). For him (1979):

> under contemporary circumstances, the 'life content' and the spatial and temporal attributes of an individual's life path are apt to be highly influenced by the sequence of independently existing specialized roles – created by institutions and technology – that a person is able to occupy. . . . For the academic whose writings are rooted complexly in the past, it is evident that the content of those works would be different if the individual's life path had been linked differently with a variety of specific, independently existing limited-access roles of both an academic and a non-academic character.
>
> (p. 175)

Thus, he was able to trace the evolution of his academic interests from his interactions with people who were with him in particular places at certain times – their 'path convergences' – as well as items that he read and appreciated.

A particular feature of a scientist's autobiography, however, is that the encounters which influence career development and change need not be interpersonal. James Bird (1975, 1985) used Popper's concept of World Three (the world of recorded knowledge) to show how geographers can be strongly influenced by reading both ancient and contemporary sources, so that our milieux are not as bounded (in time and space) as are those of people who rely much more on interpersonal transmission of information (see Johnston, 1984e, 2004a) – though, of course, we are reliant on what materials are, or can be, made available to us locally, which is increasing very rapidly through developments in information technology (Adams, 1995). As noted earlier with regard to actor-network theory (p. 268, this book), non-human artefacts – notably though far from solely published items – are crucial elements in the circulation of knowledge and as influences on individuals' academic developments; those artefacts are not themselves immutable, however, since (as Livingstone, 2005, and others have argued) interpretations of texts often vary according to the context in which they are read. Technological developments have speeded up this circulation process, with major impacts on changes in knowledge construction, although, as always, uneven accessibility to relevant items can have a significant influence: items that are widely circulated are more likely to be influential, as was the case with the mimeographed discussion papers circulated by graduate students at the heart of the 'quantitative revolution' in Seattle in the 1950s (Johnston, 2004a). Nevertheless, whether such 'knowledge communities', many of which are located in a particular place (Pinch, 1998), eventually significantly influence the direction that their discipline takes depends on their relative power to change others' research practices.

Geographers have extended their work on contextual influences on their discipline's content and direction by drawing on the ideas of sociologists of science regarding 'situated, or local, knowledge'. In presenting an 'alternative history' of economic geography, for example, Trevor Barnes (1996, p. 105) argued that understanding geographers' practices requires knowing 'something about the local context and not overarching principles of rationality'. Thus:

> Economic geographers will continue to do economic geography in the way that they have always done, that is, by creatively drawing upon existing ideas and beliefs and responding – sometimes with great originality and imagination, other times with less of each – to the concerns of their own local context . . . this is all there is to practice. There is no foolproof method of directing inquiry.
>
> (p. 124)

The nature of the 'local' is not addressed in depth, however, so that the interaction of 'place as context' with 'World Three as context' remains unexplored. (For Barnes himself, for example, the key local context was the Department of Geography at the University of Minnesota in the late 1970s where he was a graduate student, working with Fred Lukermann and Eric Sheppard and alongside Michael Curry. A decade later, he was working in Vancouver and still writing with Curry and Sheppard, who were in Los Angeles and Minneapolis respectively.) But Barnes's later work illustrates his theme through the examples of economic geography and locational analysis: rather than linear sequences of individuals building on their predecessors' work, he emphasises the 'halting, local, contingent factors that shaped it' (Barnes, 2003a, p. 91). While there may be progress in the technologies deployed – as with computing power and communications – there is no 'relentlessly progressive' trajectory of thought but rather a series of 'enormously bright, imaginative, intellectually vibrant and creative' individuals carving out their own histories 'reflecting the particular circumstances of their origins'. Like Livingstone (2000b), Barnes argues that all knowledge is locally produced, geographically and historically bounded in its generation – though as communication without travel becomes easier, one wonders for how long the local can be defined in terms of places rather than as networks – the 'locales' (or settings for interaction) of knowledge-production within geography may increasingly be a punctiform set of sites (themselves far from permanent) among which there are frequent flows, rather than a relatively small, bounded space such as the Department of Geography at the University of California, Berkeley, in the 1930s. (Keighren's, 2010, example of how the dissemination of the arguments in Semple's 1911 classic *Influences of Geographical Environment* was aided by both her lecture tours and visiting appointments at various North American universities illustrates how those locales were linked in actor networks; part of Keighren's exploration was based on a survey of the book's impact reported by John Wright, 1962, and their uncovering of how the book was interpreted and deployed further exemplifies the argument for a 'geography of reading'.)

The case for understanding the changing nature of geography contextually closely parallels that promoted by structurationists for appreciating all human activity; the operation of human agency must be analysed within the constraining and enabling conditions provided by its environment. Thus, Johnston (1983c, p. 4) enlisted structuration theory as a framework for analysing the changes in human geography that are the concern of this book, arguing that 'the content of a discipline at any one time and place reflects the response of the individuals involved to external circumstances and influences, within the context of their intellectual socialization'. Mikesell (1981) took a similar position, though without the language of structuration, describing the history of geography in the USA as a sequence of 'temporary enthusiasms or episodes' (p. 9), which he claimed were responses to contemporary stimuli. Hence, the popularity of urban studies then, the relative decline in foreign-area studies for some years after the American retreat from Vietnam and the impact of national concern about the physical environment in the 1970s. He also suggested that physical milieux influenced the culture of individual geographers:

> The first generation of American geographers grew up in a country that was still strongly influenced by the mores of small towns. Most of the students now attracted to geography are products not only of an urban but increasingly of a suburban environment . . . the geographical profession has changed and is changing as a consequence of the suburban, middle-class origin of most of its current membership.
>
> (p. 12)

Philip Porter (1978) advanced similar arguments, identifying two types of American geography: a midwest version that 'was a characteristically optimistic, action-oriented, "can-do" kind' (p. 17); and a Californian geography, based on Sauer that was 'historical, uncompromisingly academic, speculative, suspicious of government, keenly interested in cultures other than the dominant Anglo culture of the United States' (p. 18).

This environmental influence thesis should not become environmental determinism, however. Mikesell noted that urban geography flowered in the 1950s and 1960s not only in Chicago but also in Iowa. But it stresses the importance of studying human geography and human geographers in context; as Capel (1981) noted, the disciplinary community (or major elements within it) will seek to maintain its identity by bending to perceived shifts in their milieux (as clearly demonstrated for anthropologists by Patterson, 1986). Mikesell (1981, p. 13; see also Johnston, 1996e) also argued, following Harvey (1973) and others, that reactions to environmental shifts involve individual scholars seeking not only to defend and promote their chosen discipline but also to defend and promote their own status and careers within it:

> innovation will continue to be regarded as a virtue. Much of the development that has already taken place in American geography is a consequence of the attempt of individual scholars to stake a claim for themselves, to be or at least to seem to be different from their rivals. The fact that most academics see virtue in innovation means that there is reward for innovation.

Mikesell did not discuss the scale of the innovation, however; does it have to involve developments within the positive heuristic of a research programme, to use Lakatos's terms, or does it require the launch of a new research programme? Can one bring potentially greater rewards, and disasters, than the other?

One little explored aspect of this sociological approach is that it focuses on the winners rather than the losers, on those whose ideas have a major impact on at least parts of their discipline. Many ideas are not taken up at the time they are initially published, but are later revived, whereas others are never followed through because the published versions languish largely unread. Why some individuals and their ideas are much more influential than others is an intriguing aspect of a discipline's history, as illustrated by the belated recognition of Hägerstrand's seminal work (Duncan, 1974b: Morrill, 1984, p. 62, notes that his visit to Seattle in 1959 'had an electrifying impact'); the slight impact of Wreford Watson's ideas regarding the social geography of cities, which predated other now better known studies by several years but were almost ignored (Robinson, 1991; Johnston, 1993b); the focus on only one aspect of Gottmann's voluminous writings (Johnston, 1996c); and the lack of influence that Dickinson's pioneering work on functional regions and urban areas had on 1950s–1960s urban geography (Johnston, 2000c, 2002a). In general, it seems that the most influential are those who are well-networked with other influential scholars, especially those who have been trained in large graduate schools with a clear agenda for the discipline. Morrill recognised this with regard to the developments he was involved in during the 1950s and 1960s, at both the University of Washington (see p. 69, this book; also Berry, 1993, on the depth of the resources available there, outwith as well as inside the geography department) and Northwestern University (Morrill, 1993):

> This happened when and where it did because at a few universities there was a fortunate combination of an emphasis on interdisciplinary study, existing development of theory in these related fields, notably regional economics and sociology, some students ready for change, and some faculty ready to encourage them.
>
> (p. 442)

Following this, leading proponents of the new approach were able to promote the discipline more widely within the social sciences, thus achieving what Leslie King (1993, p. 545) refers to as 'the professionalization of geography as a social science'.

Context is not deterministic, of course: it provides opportunities. Thus, Hanson's (1993) reflections on graduate school at Northwestern University, while stressing the positive aspects of the education provided (such as the concern for theory and explanation, the importance of space, and the emphasis on standards of evidence), also noted that:

> Coexisting with these considerable strengths, counterbalancing and – at the time – sometimes overwhelming them, were blind spots and inflexibilities bred in revolutionary fervor and in earlier bruising conflicts with the reigning paradigm. Our mentors exuded a certain intolerance, a certain hubris, a certain arrogance, a certain certainty that there was only one way to the Truth, and they knew it. There was a caricaturing and a trivializing of the idiographic/regional approach, much as today critical social theorists caricature and demean spatial analysis and scientific geography.
>
> (p. 553)

Hanson and some of her contemporaries were involved in the emergence of behavioural geography as a counter to some of the 'certainties' being advanced by the 'space cadets' (her term for the Seattle-trained spatial-scientists then on the staff at Northwestern, p. 534): 'By dethroning economic man and replacing him with a variety of decision makers, behavioral geography ignited the first glimmer of the notion that each viewpoint is partial, incomplete and dependent on the subject's [i.e. the decision-maker's] location.' She was led, by her questioning of the assumptions which underpinned the then popular theories, to the realization that (pp. 555–6) 'people's decision making and behavior are not immune to context but are conditioned, inter alia, by time, place, class and gender', an argument which applies as much to geography itself as to that which it studies – indeed, the recent growth of GIS reflects geographers (among others) capitalising on what they have seen as a major opportunity for their discipline.

Although context is undoubtedly important at some stages of any scholar's career – where they are and who they interact with there, influencing what they studied and how, it need not always be important. Some major contributions to the expansion of knowledge are made by 'lone scholars' who work almost entirely independently, as indicated by the citations to their research publications; good examples of this are provided by recent biographical studies of both Michael Williams (Clout, 2011b) and, especially, Robin Donkin (Baker, 2011).

A broader way of placing geographers in networks has been to develop insights from actor-network theory (ANT) (Barnes, 2001a). Debate about the merits of ANT as an approach for geography became the subject of disciplinary debate in the 1990s and have continued (see p. 268 in Chapter 7). When applied to the evolution of the discipline itself, ANT takes the notion of a community of scholars linked together in networks, but takes it further by extending the nodes on that network to inanimate objects as well – such as publications – many of which may be mobile rather than fixed, and immutable rather than transient. A discipline, in this context, is thus an interlocked network of entities, 'material and non-material', 'social' and 'natural' (Barnes, 2001a, p. 528) and understanding its development involves studying how strong network links are developed: the powerful entities (probably though not necessarily individuals) in those networks are those that have greatest influence on disciplinary change, as Barnes (2001a, 2001b) illustrates for two different periods in economic geography.

A structure for the discipline?

The structure of geography can be represented as a network of nodes and flows – very much like Haggett's representation of the spatial economy (Figure 4.1, p. 102). The nodes are individual geographers and the flows are the ideas and research findings they share with others in formal situations such as teaching and conference sessions, during informal conversations, and in their publications. Some of those flows are situated – in classrooms, conference halls, and the informal settings for conversations; others are widely available in many places, such as the libraries of geography books and journals; an increasing proportion is even more generally available, to anybody with access to electronic libraries. Individuals move between settings (in some cases while remaining in the same physical place), and as they do so may be either or both influenced by others or influential on them. They also refer to the flow of material, again hoping to be influenced by what they access, and submit further material to the network, hoping to influence others indirectly through what they write.

Within those networks, some individuals and some flow materials are more important than others – i.e. the network nodes can be structured into a hierarchy of influence. In part, the degree to which either an individual or a flow material is influential reflects settings. With regard to direct interpersonal contact, for example, it is often suggested that dense concentrations of workers with linked interests are more likely to stimulate important developments through interaction: there are intellectual economies of scale and scope. This is the basis for the growth of major graduate schools, and of focused research groups of staff in many university geography departments (perceived as necessary in the UK to meet the criteria for the post-1980 Research Assessment Exercises); they provide the critical mass whereby ideas are generated, refined and disseminated, both within the school and outwards, as graduates get posts elsewhere and convey the message to new settings. Examples of this can be found in the twentieth-century history of human geography, as with the

'Berkeley school' under Sauer (Spencer, 1979), the historical geographers trained by Darby (Prince, 2000), the 'Washington school' under Garrison (see p. 69, this book), the 'Chicago school' of urbanism (Yeates, 2001; Berry, 2002c), and its Los Angeles successor (Dear, 2002; Dear and Dahmann, 2008; plus a number of more specialised groupings: see, e.g., Peach, 2003). To some extent the influence may be serendipitous: a student may enrol in a department not expecting to be supervised, or even influenced, by a particular staff member (or even other students) there. Unanticipated conversations at a conference may have a similar impact, and so forth.

The degree to which individuals may influence each other's mode of working is likely to vary, however, even if other possible factors are held constant. Some individuals may not be appointed to a large graduate school, for example, and so lack the access to a potential band of disciples. Others are, but for some reason or another do not capitalise on that access; they may be 'lone scholars', difficult to work with, or studying topics that are currently relatively unpopular or, as in the examples of Donkin and Michael Williams cited earlier, working independently without any strong links to established networks (and, usually in such cases, without their own graduate students following their example). Chance, charisma and a number of other factors can result in some individuals having much more influence over the development of their discipline than others do.

The same argument can be made about the impact of various publications: some are widely read and influential; others may attract few readers and be largely ignored (whatever their potential merits). To some extent this may also reflect the settings in which they are placed: a paper in a relative obscure publication that has a small circulation will be relatively invisible and may pass largely unnoticed, compared to one in a widely circulated and referred to journal. Similarly, some publishers are better able to get books widely purchased and adopted as texts and some publications are identified by reviewers and other commentators as important exemplars of particular types of work (Johnston, 2009a). But there is more to the distribution of influence than such material conditions. There is little doubt that some authors are considered more innovative and stimulating than others – especially, perhaps, those who challenge the conventional wisdoms and whose writings cannot be categorised, in a general sense, as simply contributing to normal science activity within the context of a given exemplar. They may move work on in a small way along established paths, but not stimulate wider debates about means and ends. Thus, some authors' writings are more likely to be read and referred to (i.e. through formal media – class reading lists and book reviews – and informal media) than others', as is suggested by the large variations in the numbers of citations to particular authors and works. Given the pressure on time, and the great explosion of publications in recent decades, selectivity is necessary: people choose to read that which looks most relevant to their current concerns, whether that is a course that has to be taught or a research project advanced. Their selectivity is based on a number of criteria, such as whether the potential reader knows (of) the author and rates her/his work highly, or advice that a piece is worth consulting, or just chance while browsing along a bookshelf (or through a bookseller's website), through the pages of a journal (including the book reviews) or, increasingly, by both Internet searches and customised information sources delivered electronically.

What we have, then, in the continuing operation of an academic discipline is a massive network of potential contacts only some of which are actualised. Knowing which have been, when and where, is not easy to discover. Some clues are given in published material Media – many authors acknowledge individuals who have helped them in their work, for example; references in a paper indicate the sources considered worthy of mention (although not necessarily all those accessed); and known biographical details (where a person was in graduate school, and when, for example) are suggestive. Autobiographical and biographical essays may fill in gaps – though these are relatively few – as also might obituaries and memoirs. But in most cases we have only circumstantial evidence of who influenced whom, who read what and to what effect, and so forth, and such evidence is the stuff of books like this. (The items we have referred to in this book indicate our own selectivity in deciding what was of value to read and use in compiling our account.) Almost certainly, detailed studies of disciplinary networks would find them divided up into subcommunities – Geertz's (1983) 'intellectual peasant villages'.

All of these contacts – direct and indirect, positive and negative – are not only the basic evidence for books like this, therefore; they are also the foundation of disciplinary change. Almost all intellectual endeavour involves and results from interpersonal interaction of some form, usually mediated by the written word (World Three in Popper's terminology). What is of interest, however, even if difficult to determine, is how those micro-level contact patterns have any significant impact on the directions that a discipline takes at the macro-level. Or, in other words, do some individuals and groups (or individual written pieces) have a disproportionate impact, so that if they had not been on the networks the discipline might have taken other directions? There seems little doubt that in the early days of geography as an academic discipline this was indeed the case: among the small number of practitioners a few powerful professors, by virtue of a combination of and ability to disseminate ideas, were highly influential on geography's foundations (on which, see Lichtenberger, 1984). Since then, as the number of geographers has grown and the size of the network expanded exponentially, influence has been much more widely dispersed. Nevertheless, it may well be that certain paths would not have been followed – at least when and where they were, and with what effect – if particular events had not occurred. Almost certainly geography would have taken a quantitative turn at some stage in the 1960s, but would it have taken the particular form that it did if, for example, Ullman, Hudson and Garrison had not taken up posts at the University of Washington in the 1950s, if a group of talented graduate students had not clustered there at the same time (some of them attracted there by Ullman's presence rather than Garrison's), if they had been unable to spread their ideas through discussion papers (Johnston, 2004a), and so forth? And would this book have taken its particular form – even been written at all! – if Chorley and Haggett had not been influenced by Kuhn's ideas in the mid-1960s (and a publisher had not perceived a potential market for it)?

Such questions are, of course, ultimately unanswerable counterfactuals, although by posing them it is possible to illuminate what did happen. They also illustrate the all-important contingency of academic life. A discipline like human geography is not a disembodied phenomenon with a life and trajectory of its own; it is shaped and continually reshaped by geographers reacting to circumstances in particular milieux, the temporal and spatial situations in which they are placed and place themselves. We can present the history of geography as a changing melange of ideas and approaches, but that melange is a (far from planned) human creation; it is the sum of a large number of biographies.

Human geography's response to challenges

The impact of external circumstances on geographical practice has been illustrated repeatedly in this book. The initial establishment of geography as a university discipline was associated with arguments about its strategic (military and imperial) and educational (training geography teachers who could then instruct children about the place of their nation) value. Yet doubts about the rigour and value of the discipline (especially, but not only in the USA) have cropped up repeatedly, sometimes resulting in the closure of departments. Such challenges arguably fostered some of the disciplinary revolutions and debates that *Geography and Geographers* documents, as human geographers sought to make their subject more rigorous, relevant or critical and/or to reorient it vis-à-vis other social or physical sciences and humanities. Reflecting on the post-1945 challenges to geography in the USA, Murphy (2007) argued that:

> The shadow cast by Harvard's [1948] decision to close its department worked against the establishment of geography departments at smaller public institutions, private universities and

liberal-arts colleges, as the argument could be (and was) made that a great university did not have to have geography. . . . The situation was exacerbated when three other leading universities – the University of Pennsylvania, Stanford and Yale – followed Harvard's lead and dropped their geography departments in 1963, 1964 and 1967, respectively. Hence, when the higher education juggernaut slowed in the mid-1970s, the academic discipline of geography was vulnerable.

(p. 124)

Murphy goes on to note how:

The situation deteriorated further in the 1980s when formerly prestigious departments were closed at the University of Michigan (1982), Columbia University (1986), Northwestern University (1986), and the University of Chicago (1987). As with the earlier closures at Harvard, Penn and Yale, each of these programme losses could be seen as the result of the actions of unsympathetic administrators or of internal problems (lack of attention to general undergraduate education, feuds among faculty, etc.). But focusing on such explanations directs attention away from the larger issue: many in the US academic elite had come to view geography as a dispensable subject in institutions of higher learning.

(p. 124)

Hence, further challenges have been noticeable since the 1970s as a consequence of economic recessions, cut-backs in higher education funding, and attacks on social-science research. In 1996 the House of Representatives voted to end funding for the social science divisions of the National Science Foundation – although this was not enacted then, there have been several attempts since, the latest in 2014, to restrict funding for social science research. A decade earlier the UK Secretary of State for Education insisted on the term 'social science' being removed from the title of the Social Science Research Council, and sought to make its successor – the Economic and Social Research Council (ESRC) – more responsive to government and commercial agenda, and the policy prescriptions of the 'New Right' designed to sustain what has been termed 'flexible accumulation' (Hudson, 1988) and 'neoliberalism' (Peck and Tickell, 2002) through ideas about a 'free economy and a strong [though restructured and pro-enterprise] state' (Gamble, 1988; Johnston and Pattie, 1990). These pressures have continued. Thus, since 2012, when inviting reviewers to comment on grant applications, the ESRC asked not only about the project's originality, theoretical and methodological foundations, potential contributions to knowledge, and value for money, but also its 'potential for policy and practitioner impact': had the applicants provided 'a clear impact summary'?; had they 'fully considered the potential of their research for policy and/or practitioner impact . . . [and] set out [feasible] appropriate plans for achieving impact'?; and had they 'identified appropriate stakeholders . . . [and] how they will benefit from the research'?

Many reacted to this by promoting human geography as an 'applied discipline', offering relevant skills for the attack on contemporary problems. In the December 1981 issue of the AAG Newsletter an article entitled 'A survival package for geography and other endangered disciplines' indicated that 'On 19 June, 1981, the Board of Regents of the University of Michigan decided, by unanimous vote, to terminate the Department of Geography at the end of the 1981–82 academic year' (Kish and Ward, 1981, p. 8). Their experience of this decision led George Kish and Robert Ward to suggest how other departments of geography could counter similar attacks. They emphasised the teaching role, recognising the need to attract students in a competitive market and present 'our wares in a stimulating and excellent way' involving 'the virtues of applied geography. As students become more conscious of careers, there may be a corresponding need to increase a skill-oriented curriculum. This could demonstrate the relevancy of geography and enhance its appeal to students' (p. 14). Kish and Ward – 'To appease some traditionalists' – did not advocate reorganising the entire corpus of geography 'but only [an] attempt to broaden the appeal of geography to the student population'. (See also Ford, 1982; Powell, 1981, and Wright and Koch, 2009; on a more recent closure, see Chan, 2011.) Their prescriptions were written at a time in North America when, according to Haigh (1992, p. 185), geographers' peers saw their discipline as 'small, marginal and perhaps immature': the late 1970s had seen a net loss of 32 university departments, including one-sixth of those in 'private and denominationally funded institutions'. Avoiding the discipline's demise was presented as a crucial task (Wilbanks and Libbee, 1979), a problem stimulated by the absence of geography in most American high schools and the very small number of undergraduate students intending to take geography courses (Murphy, 2007). Despite the vitality of geographical research, therefore, 'Geographical education is still regarded as a marginal activity by American schools and colleges' (Haigh, 1992, p. 189); the discipline there was not only 'small, often beleaguered' (Koelsch, 2002, p. 274) but:

> it would be difficult to identify any other university study that in the course of the 20th century had achieved full departmental independence and programmatic recognition at the doctoral and research level in so many leading American universities, only to lose both by the end of that same century. At the pre-college level . . . geography remains largely a minor component of social studies or of global education, taught most often by teachers with little or no background in the discipline.

Koelsch puts this parlous state down to the 'fracturing of any consensual view of what the discipline and its practitioners should be about', although Murphy (2007) judged that:

> the closing of the departments at Columbia, Northwestern and Chicago [in th 1980s] appears to signal the end of an era of decline. Geography has not experienced a single major programme loss since that time. In a few places geography has been folded into a larger unit with some loss of influence or standing, and a handful of mid-size programmes have closed at regional colleges (e.g. West Georgia College). Nonetheless, the last 15 years have generally seen a steady expansion of geography's institutional base. New programmes have been founded, many existing programmes have expanded, and student enrolment has generally risen (although not without a temporary downturn in undergraduate enrolments in the late 1990s – perhaps due to the rapid expansion of interdisciplinary environmental studies programmes).
>
> (p. 125)

There was some rejoicing within the geographical discipline when, as Tomlinson (2009) puts it, there was a 'reintroduction of geography to Harvard'. But it was a narrowly conceived reintroduction, involving a Center for Geographical Analysis which the university's president promoted because 'Geographic information systems will let us change the nature of questions that are asked in a wide diversity of sciences and humanities.' The Center's website presents itself as administering infrastructure, collecting and disseminating spatial datasets and providing training and consultation in the use of geospatial technologies.

Yet elsewhere, challenges to the position of the discipline persist. In Australia since the 1990s almost all the separate university departments of geography have been merged into larger units, thereby losing a separate identity for the discipline in some cases (Holmes, 2002). This was part of a larger shift in Australian universities whose traditional medium for teaching and research:

> the academic disciplines, was to some extent destabilised and displaced amid the mergers and rapid growth, the rationalistation of degreees and programs, the sudden changes in mission and decision structures, the explosive growth in international education. Unless they could gain autonomous resources by selling themselves, individual disciplines began to lose purchase on their own agendas.
>
> (Marginson, 2004, p. 220)

In the UK, too, mergers and closures characterised the first years of the millennium with more than one-tenth of the country's geography departments 'disappearing' – as separate entities, if not totally – causing concern among disciplinary leaders (Cooke, 2002), not least because the trends appeared to be linked with a reduction in the number of secondary (high) school students taking geography, which for more than a century was the foundation for successful university departments (see p. 39, this book). More recently, Tim Hall *et al.* (2015) note that:

> The number of autonomous, single subject Geography departments in the UK has fallen from 47 in 1995 to 30 in 2013. The sharpest decline, from 41 to 33, occurred between 2003 and 2005. Correspondingly the numbers of multi-disciplinary departments of various kinds within which Geography is located rose from 29 in 1995 to 47 in 2013.
>
> (p. 58)

Especially in the USA, there was frequently a commitment to secure geography through promotion of it as an applied discipline (part of the wider debates considered in Chapter 9) in research as well as teaching. The announcement of a new editorial policy for *The Professional Geographer* in the AAG *Newsletter* for April 1992 reported that (p. 1) '[The new editor] hopes to emphasize work in applied geography and to include information on corporation activities, state and local government projects, Federal government activities, activities by United Nations and other international agencies, and research projects.' The need for more applied work was widely accepted among American human geographers as a necessary means for their discipline's vitality. Mikesell (1981) had suggested that:

> geographers fortunate enough to have secure teaching positions will worry about what they should do. The best response to this concern could be a decision to do what they have been doing, but with a keener appreciation of context and a greater willingness to be influenced by environment.
>
> (p. 14)

The response took a variety of forms, with many arguing for a greater concentration on cartographic, remote sensing and geographical information systems skills.

This response was illustrated by a group of papers published by the AAG in 1995. In 1993, the National Academy of Sciences/National Research Council undertook its first major review of geography for nearly three decades (see p. 87, this book). It established a committee of 16 charged with 'Rediscovering geography: new relevance for the new century' and given five objectives (Wilbanks, 1995):

1   identify critical issues and constraints for the discipline of geography;
2   clarify priorities for teaching and research;
3   link developments in geography as a science with national needs for geography education;
4   increase the appreciation of geography within the scientific community; and
5   communicate with the international scientific community about future directions of the discipline in the USA.

Within this context, a report was commissioned from the AAG on the supply of and demand for geographers. Introducing papers based on that report, Wilbanks (1995) noted some challenges and opportunities:

> many geography programs in universities across the United States are already experiencing unprecedented growth in student demand. . . . A major concern for geography should be that,

because of supply limitations (e.g. a lack of additional faculty positions in geography departments)
the demand will be artificially and arbitrarily truncated by limitations on course enrollments.

(p. 316)

This is described as a 'very welcome but very difficult disciplinary transition from relative penury
to relative abundance'.

The committee's report on the supply side of the equation (Gober et al., 1995a) referred to an
earlier study (Goodchild and Janelle, 1988) on the internal structure of the discipline which found
that 'Technical expertise and interest in geographical information systems (GIS) were burgeoning,
especially among young geographers, while regionally oriented specialities were shrinking' (Gober
et al., 1995a, p. 317). (Turner and Varlyguin's, 1995, report on foreign-area dissertations, in the
same issue of The Professional Geographer, shows that the latter trend was relative not absolute.) To identify which skills and interests university departments of geography were focusing on, department
chairs were asked about specialisations and the occupations students were being prepared for. Of
the 212 respondents, 139 identified at least one specialisation, and the full list was:

led by programs in environmental/resource management, techniques (GIS, cartography, and
remote sensing), and urban planning. These tracks appear to be designed to prepare students
for the occupations in which geographers traditionally have found work rather than to develop
their interests in regional geography or the systematic specialties like urban, economic or
physical geography that have traditionally formed the core of the academic discipline.

More specifically:

We also asked chairs to indicate the specific occupations for which students were being prepared.
The occupations indicated by the highest numbers of departments were (1) GIS/remote sensing
specialist, (2) secondary school teacher, (3) cartographer, (4) environmental manager/technician,
and (5) urban/regional planner. . . . GIS was by far the most popular occupational trajectory
(involving almost 11% of students enrolled in programs offering GIS training).

The committee also surveyed the labour market experiences of graduates (Gober et al., 1995b, p.
331), finding that 'Among those who were employed and listed occupations closely related to
geography, respondents clustered into five predictable occupations: teacher (15.6%), environmental
manager/technician (12.9%), GIS/remote sensing specialist (10.5%), cartographer (8.2%), and
planner (6.7%).' They also studied the employers and job-seekers using the placement service at the
AAG's 1994 annual meeting: 45 per cent of the employers were looking for candidates with technical
qualifications, and they comprised 80 per cent of all non-academic employers using the service.

Finally, the committee surveyed the future labour market by asking department chairs the areas
of expertise of staff who would be retiring over the next decade, and also of those which they
would be recruiting (Gober et al., 1995c). Technical areas of the discipline predominated in the
latter category, with GIS specified in 17 per cent of the cases – nearly twelve times more frequently
than it was among the retirements. As the discipline reproduced itself over the next decade,
according to these figures, the relative number of technical specialists would increase substantially,
whereas traditional areas such as agricultural, historical, political, cultural and economic geography
would decline substantially. The committee's conclusions, based on interviews with the AAG's
corporate sponsors, implied a major shift in the nature of the discipline as taught to undergraduates
(see also Miyares and McGlade, 1994):

The debate over geography as a broad-based liberal arts discipline or as a technical, semiprofessional
field ignores the realities of the current labor market. Sponsors told us they want employees who

can combine technical skills with a broad-based background. Geography's comparative advantage over other social sciences lies in its ability to combine technical skills with a more traditional liberal arts perspective. Successful geography programs will be those that are able to find the appropriate balance of field-based technical skills like GIS, cartography and air-photo interpretation with competence in literacy, numeracy, decision making, problem solving, and critical thinking.

(p. 346)

In this context, the review's conclusions – indeed, its whole tenor – focused on technical skills; its first recommended response was that the discipline should 'Improve geographic analysis in a new era of data and analytical tool availability, related to broader needs of science' (NAS-NRC, 1997, p. 162). Its student base could be enhanced if it were to:

Implement programs to support and assure the quality of the science content in kindergarten through grade 12 geography education . . .

[and]

Foster conceptually sound general education courses in geography as part of a liberal arts college education . . .

[while at the same time it should]

Develop programs that bring geographic perspectives to bear more effectively on business, government and other organizations at national to community levels.

(p. 164)

Little of this has occurred, however, and although geography was recognised in the mid-1990s as a 'core discipline' for contemporary education, no funding to develop school geography programmes had been made available (Richardson, 2003); in the 1920s, according to Schulten (2001), geography academics in the USA distanced themselves from their discipline's status in the country's schools, and this still appears very largely to be the case. A lesser distancing has occurred more recently in the UK. Until the 1970s the community of academic geographers was closely linked to that of teachers of the subject in the country's (mainly secondary/high) schools; they participated in developing school syllabuses and curricula, and the setting and marking of the external examinations which were used as indicators of students' potential for undertaking degree courses in the discipline, for example, and a number were active in the Geographical Association, the learned society for geography teachers (Balchin, 1993). Such involvement is now rare, in part because of other pressures on scholars' time (such as higher student–staff ratios and a greater emphasis on research for career advancement in universities). A large number of students who have studied geography at high school and wish to continue with the discipline at university remain the strongest foundation to the discipline's strength in the latter institutions, but academic geographers now do relatively little to foster that key interaction (Sidaway and Johnston, 2007).

External pressures became just as great in the UK during the 1980s, with expression of the need for applied geography and the development of 'transferable skills' (problem-solving, group-working, IT awareness, etc.) within undergraduate curricula (see Matthews and Livingstone, 1996; the *Journal of Geography in Higher Education* has published widely on this general theme since its inauguration in 1977). Bennett (1982), for example, found at the 1982 conference of the Institute of British Geographers that it was:

possible to discern a strong and growing set of foci which, if they do not yet demark a new core, at least show an emerging commonality of interest. For this writer these foci were a widespread assertion of 'relevant' research, the reassertion of quantitative and analytical methods, and the rejection of recent anti-empirical movements.

(p. 69)

He welcomed these, noting that 'hot [i.e. relevant] issues are not ones which British geographers are particularly noted for tackling en masse – and, as a result, the discipline has suffered a lack of public exposure, and a marked inability to influence public and private decisions'. Furthermore:

> At a time when higher education as a whole is under considerable challenge, and when geography as a discipline may suffer particularly severe pressure in some institutions it is heartening to see the emergence of concern with the hot issues expressed at this meeting.
>
> (p. 71)

Beaumont (1987) echoed this a few years later:

> the issues raised for the next twenty years are a practical and developmental, rather than a research, orientation. The future is unknown, but it could be exciting, if geographers are prepared to become involved (probably with new partners) in doing geography.
>
> (p. 172)

Richards and Wrigley (1996) presented the early 1990s as a period of even greater and faster change, arguing that in contrast earlier periods:

> seem in retrospect almost to have been years of relative stability and calm, prefacing the maelstrom of change which swept through the whole British education system in the early to mid-1990s. . . . This period has been one of system-wide expansion, of curriculum and quality control through external review . . . and of institutional change. . . . These have combined to alter, quite fundamentally, the size, nature and structure of British geography.
>
> (p. 41)

The assessment of teaching in university departments of geography was set in the context of a national agenda concerned not only with the quality of the education provided (which some believe conflicts with an overemphasis on research: Gibbs, 1995; Johnston, 1996f.) but also with the development of 'transferable skills' beyond the traditional focus on subject-based knowledge and a critical intellect (Johnston, 1996a). The assessment of research produced not only a grading of all departments but also substantial differentials in funding (Johnston, 1993a), while government directives on research funding made it (Richards and Wrigley, 1996):

> likely that future research will be constrained to ask specific questions, will be increasingly 'applied', and will be manipulated by government, business and industry for ends which are unlikely to assist the less material aspects of the quality of civilized life; and critical (social or environmental) science will be marginalized.
>
> (p. 47)

To a considerable extent, this was ameliorated because geographers in UK universities have been able to rely on a continuing stream of potential undergraduates wanting to read for degrees in geography. At the end of the twentieth century, for example, there were some 7,000 or so places available each year on geography honours degrees (in which the majority of the student's time would be spent on geography courses, especially in the final years), compared to only 4,000 or so who major in geography from all USA universities. This large stream of UK would-be geographers results from the discipline's strength in the country's secondary schools and place in national curricula – itself a product of successful political lobby by bodies such as the Geographical Association (Goodson, 1988; Balchin, 1993; Walford, 2001; Johnston, 2003b). Almost all the 7,000 students arriving at UK universities each year to study geography have a strong foundation in the subject, having studied it for the previous ten years – a situation very different from that in the USA where most students arriving at university have not encountered geography as an academic discipline at high school, and their understanding of the discipline rarely extends beyond basic geographical knowledge of where places are. Geography has not been perceived as a 'problem discipline' in many UK universities, therefore; (most of) its departments can recruit excellent students and their graduates can get good jobs. There is relatively little need to focus on courses that will attract students by the 'employability value added' they might provide – though there is attention to this. The breadth of material in cultural geography and related subject matter taught in many of the departments currently reflects this: academic geographers in the UK are freer to teach a wide range of material at undergraduate level than is the case for their North American counterparts.

Despite this important difference between the two sides of the North Atlantic, reflecting long-established differences in the role of geography in the respective school systems, in the UK, as elsewhere, there is a need for continual vigilance in order to protect and promote the discipline's interests. In his 2002 Presidential Address to the RGS, Ron Cooke (2002) identified several challenges facing the discipline in the UK. Key among these was the potential reduction of time given to geography in the school curriculum: the Society was working to sustain geography's central role, because it is 'unquestionably fundamental to public understanding of such enduring issues as social equity, globalization and the relations between environment and society' (p. 260; see also Thrift, 2002; Johnston, 2002b).

The beginning of the new century's second decade saw a new threat to geography – and other social science and humanities disciplines – in the UK, with the removal of any direct public subsidy for its teaching in universities. From 2012 on, undergraduate students were required to repay loans covering the full cost of their courses (set at most universities as the maximum allowed – £9,000 per annum), after graduation and achievement of a threshold annual income. This stimulated an immediate substantial decline in the number of applicants, and concerns among geographers as to the attractiveness of their degrees, relative to those with clearer career prospects, to students who would face debts (covering both their tuition and living expenses) of £50,000 or more. In addition, with such debts there was concern whether many students would want subsequently to continue into graduate schools, many of which are sustained by recruiting large numbers of overseas students who pay premium fees. The number of independent departments of geography had declined in the previous decade, especially in the more recently created institutions where relatively small units were integrated with larger portmanteau, multidisciplinary creations, and this major change in funding (albeit to one much closer to the long-established American model) accentuated concerns for the medium-term health of the discipline.

A generational model?

The importance of context as an influence on the nature of the practice of human geography suggests that the paradigm model might be rephrased as a generational model (Johnston, 1978a, 1979; Taylor, 1976). Changes in the external environment provide necessary, but not sufficient, stimuli to changes within the discipline which may be interpreted as attempts to develop new research programmes, if not to launch a revolution and create a new normal science (see the discussion of Scott's work on p. 359, this book).

Associated with the external changes must be a set of conditions within the discipline itself which is sympathetic to the new demands of the milieux. In most cases, these conditions are best met by younger members of the discipline. Stegmuller (1976, p. 148) argues that 'it is mostly young people who bring new paradigms into the world. And it is young people who are most inclined to champion new causes with religious fervour, to thump the propaganda drums.' To win influence, however, especially in times of resource shortage within higher education, younger workers need the patronage of some established members of the discipline (see Chapter 1). According to Lemaine *et al.* (1976), for example:

> Mendel's work [on what is now termed genetics], and that of his successors, was a response to scientific problems. But the scientific implications of their results were not pursued until there existed a strong group of scientists who, owing to their academic background and their position in the research community, were willing to abandon established conceptions.
>
> (p. 5)

Such responses may involve attempts to create a new paradigm or new research programme, or they may only require new branches of an existing paradigm/programme. Whichever it is, even when establishment support has been obtained, success is more likely when certain criteria are met, including (van den Daele and Weingart, 1976):

1 an autonomous system of evaluation and reputation;
2 an autonomous communication system;
3 acknowledgement of the new ability to solve puzzles within the confines of the disciplinary matrix;
4 a formal organization providing training programmes which allow reproduction and expansion of the new group's membership;
5 an informal structure with leaders; and
6 resources for research.

These rarely create problems if resources are available; Capel (1981), Mikesell (1981) and Taylor (1976) all indicated that innovation is encouraged if it brings status, charisma and resources to the discipline. During periods of stagnation and retrenchment, however, conditions are less favourable because of the lack of resources with which to develop innovations and major shifts – if any – are more likely to be achieved by revolutions among existing members of the discipline. The generational model suggests that the latter is rare. With regard to the research record, John Law (1976) counselled that:

> it may well be the case that scientists do lay special emphasis on the accounts in scientific papers, but my hunch is that there is immense (and non-trivial) variation between scientists on this count. For some, science is something you do in the laboratory, something you talk about, and something you get excited about. For others, science is what they write and what they read in the journals. I would even hypothesize (in conformity with the invisible college notion) that those who are generally felt to be of higher status locate science less in their journals than in their own and other people's heads.
>
> (p. 228)

Many academics are less active as researchers, and some get 'left behind' as changes proceed. The 'normal science' that they continue to teach is probably based on the world view, disciplinary matrix, and exemplars into which they were socialised. But their colleagues, socialised later and influenced by different environmental conditions, operate in different ways. The academic career cycle, in combination with a changing milieu, can produce a multiparadigm teaching discipline.

The key elements in the generational model (modified after Johnston, 1978a) are:

1 the external environment is a major influence on a discipline's contents, especially in the social sciences that are closely linked with that environment;

2   at times, the nature of this environment may change significantly, provoking a reaction among a minority of members of the discipline who try to stimulate change in disciplinary practice by its established members and to generate interest in that change among the youngest generation of research workers – the latter is usually much more successful than the former;

3   together, this grouping presents a new 'school of thought', although in some cases opposing new schools may be stimulated;

4   the new school is co-opted into the disciplinary career structure;

5   the publications of the new school come to dominate the disciplinary research output, as the productivity of the earlier generation declines;

6   students face two or more separate generational schools in a department's teaching syllabus; and

7   over time, members of the new school attain seniority and political influence within the discipline (and write the textbooks that define their perspectives on the current 'normal science': Johnston, 2006).

Several consequences may follow. If a discipline fails to react to a changing milieu it could stagnate, and so innovators are encouraged. Another is that some potential innovators may be unable to influence their discipline, either because conditions are not conducive, the 'establishment' does not react positively to their suggestions, or they cannot obtain permanent positions within the academic career structure. Simon Duncan (1974b, p. 109) illustrated these 'processes of resistance' using the example of Hägerstrand's (1968) ideas on spatial diffusion that were originally published in 1953; citation analysis shows that widespread recognition was much delayed, compared to the average for all publications in geography (Stoddart, 1967a). This was not, Duncan claims, because of either language or Hägerstrand's relative isolation in Sweden (though see Getis, 1993, p. 519, on the 'excitement' generated by Hägerstrand's visit to Seattle in 1960, which led to Morrill doing his Ph.D. work in Sweden), but the apparent irrelevance of his work to those steeped in another paradigm. Only when spatial science had been established was Hägerstrand's contribution recognised:

> Hägerstrand's own attempts to disseminate his work met rejection from adherents to orthodoxy, but later enthusiasm from those pioneering spatial science. The eventual relay of information to this community owed more to dogged personal effort than to the formal communication system of normal science, and general recognition was not achieved until professional allegiances were reorganized.
>
> (p. 130; see also Johnston, 1993c, 1996c)

Even when a new idea has been recognised, however, several virtually independent groups may be involved in its development (Gatrell, 1982).

In all these debates about changing communities within geography and orientations to the discipline, the threefold hierarchy introduced by Kuhn should be remembered (see p. 19, this book). Changes, whether revolutionary or not, are much more likely to occur at the lowest level of the exemplar than at the level of the disciplinary matrix, and even less likely with regard to the world review. Charting an evolving discipline involves separating the changes that occur regularly within communities, almost as part of normal science, from those which involve much more radical rethinks of what is done, and why.

Geographies of geography?

Given the importance of context being stressed by observers of geography's recent history, it should not be surprising to find that there are significant differences in how the discipline has been (and is being) practised, at a variety of scales and across a range of cleavages. It has been suggested, for example, that there are four main cleavage lines within contemporary geography (Johnston, 1997b, pp. 30–1):

1   the substantive divisions according to subject matter studied;
2   the epistemological divisions according to views on the nature of knowledge and its production;
3   divisions in rationale, reflecting the reasons for undertaking work; and
4   community divisions which can be found at a variety of scales – language used (macro-scales); national schools (meso-scales); and intellectual schools (local communities at the micro-scale).

To a considerable extent these all overlap, creating a myriad academic communities and networks within the discipline with points of contact and interaction, but also considerable difference and distance.

One of the clearest axes of contemporary differentiation in geographical practices is that provided by different meso-scale linguistic communities. In the nineteenth century, when geography was being launched as an academic discipline, there was considerable international contact and developments in several linguistic realms – not least the English-speaking world – were much influenced by pioneering ideas emanating from Germany and France (see Chapter 2). These were enhanced by the personal contacts developed during periods of study by individual geographers at European universities (as with Dickinson – Johnston, 2000c, 2002b; and Ogilvie – Clout and Stevenson, 2004) as well as at international congresses and occasional events such as the Transcontinental Excursion of 1912 when some sixty European geographers participated in a 21,560km journey across the USA with their American hosts (Chisholm et al., 1913; Clout, 2004; Maclean, 2011). These contacts waned, however; they were fairly weak by the time that the First World War further ruptured international intellectual communities, and were not easily re-established in the subsequent decades.

During the second-half of the twentieth century, geography as practised in countries where English was either the main language or widely used in universities (including Sweden and the Netherlands) came to dominate: 'International' journals almost invariably published in English only, which created difficulties for 'outsiders', not only in reading the material but, even more so, in getting their research findings and ideas before wide audiences (Aalbers, 2004; Gutiérrez and López-Nieva, 2001; Banski and Ferenc, 2013). For some commentators this has amounted to the formation of an Anglo-American disciplinary hegemony – a debate that was touched upon in the Preface to this book. Minca (2000, p. 285), for example, reported that 'the past fifteen years or so of the postmodern debate in geography have been an almost exclusively Anglo-American domain', marginalising those who 'navigate only on the edges of the Anglo-American academic empire – but are firmly placed within other geographical traditions', with a joint conference of Italian and Anglo-American geographers providing:

> further confirmation of the essentially undisputed dominion of Anglo-American geography . . . today the boundaries as well as the rules/coordinates of what passes for 'international' debate within our discipline are determined from within the Anglo-American universe.
>
> (p. 287)

For many (especially younger) European geographers this created problems of co-location in their home academic traditions and the 'languages and methodologies of Anglo-American geography', with the (so-called) 'international' journals published in English offering a 'formidable instrument of internal regulation and legitimation for the Anglo-American community [but] . . . not a propitious terrain for encounter and debate among diverse geographical traditions'. The result was a claimed 'hegemonic closure': all debates in 'international journals' were 'essentially internal to the Anglo-American geographical community' (p. 288) from which others were effectively excluded:

> It is therefore deceiving, if not intellectually dishonest, to identify these leading journals as the locus of international debates – unless, of course, one retains the belief that the entirety of relevant contemporary theoretical contributions in geography can be limited to the Anglo-American world and to English-language publications.

Samers and Sidaway (2000) took up this debate, citing a wider literature and pointing out the relative marginalisation of some scholars within the English-language realm; they related the Anglo-American (i.e. North Atlantic) apparent intellectual hegemony to the wider post-1945 American economic and geopolitical influence. The period was marked by an upsurge of influence of British geography (Whitehand and Edmondson, 1977). Nevertheless, they argued that work in other traditions had influenced that undertaken in the Anglo-American realm (citing regulation theory as an example; see p. 211, this book), and that all geographical traditions were inherently hybrid (or 'transnational'). Garcia-Ramon (2003, p. 1) was unconvinced by this argument, however, and restated the case regarding Anglo-American hegemony. For Simonsen (2002, p. 391) there was a paradox that growing cultural globalisation was associated with Anglo-American predominance within human geography, in which 'a presumed universality is constructed that tends to conceal a particularity based, to a large extent, on specific experiences of the USA and United Kingdom'.

There are also significant differences within the Anglo-American realm, however, not least between the UK and the USA. To some extent these reflect the different inheritances – the intellectual palimpsests identified by Scott (p. 358, this book) – as with the cultural geography developed by Sauer and the Berkeley school for several decades from the 1920s on and the parallel school of historical geographers, with its emphasis on cross-sections, established by Darby in the UK at about the same time. They also reflect the different educational contexts. In the UK (as noted on p. 372), geography has long been a strong discipline in the schools, because of successful lobbying campaigns by institutions such as the RGS and the Geographical Association; indeed, according to Steve Fuller (2000, p. 228), who based his argument on Goodson (1988), 'geography . . . is perhaps the only major academic discipline to have been built from the grassroots of sub-university-level teachers', although his claim that it 'did not have a clear research agenda until elementary- and secondary-school teachers lobbied Parliament to get the subject established in the universities' is historically inaccurate.

This strength in the country's schools long ensured a continued flow of students to the UK universities who wish to take a degree in the subject and have a firm foundation in it. This not only allows more specialist teaching from an early stage than is the case where many students come to geography new when they reach university but also, because the discipline has little difficulty recruiting undergraduates, means less direct pressure on its departments to offer 'relevant' curricula. In the USA, on the other hand, geography has typically been weak in the country's high schools, and most students have had little exposure to the discipline beyond some basic learning of locational facts in primary classes. Geography departments are therefore present in only a small minority of the country's universities, and they need to attract students once they get to the campus by offering courses that are attractive, not least because they suggest career paths and benefits that a geographical training might bring. Thus, introductory courses are usually pitched at a very general level, and much emphasis is currently put (as discussed on p. 370, this book) on providing courses on 'saleable, transferable skills', notably those associated with GIS. Graduate programmes also recruit widely – whereas most academic geographers in the UK have first degrees in the subject, many in the USA do not (as shown by the autobiographical accounts in volumes such as Gould and Pitts, 2002) – and many American graduate programmes have particular orientations, not least because most of them are small, in terms of the numbers of staff (faculty), when compared with the major UK departments. Furthermore, in the USA geography is competing for status within the social sciences with a number of much larger disciplines – notably, economics, political science and sociology – which considerably constrains its promotion and growth. In the UK, on the other hand, the creation of university geography departments largely predated that of sociology and politics departments (if not economics), giving it a comparative advantage, alongside that provided by its school presence, in promoting itself to potential students.

There are, then, complex geographies of geography, some of whose parameters we have started to map here. Few departments (even the largest in the UK) cover the full range of the current subdisciplines in their teaching programmes, for example, let alone in their research concentrations. Most specialise, reflecting the interests of senior staff at particular times in their development and institutional resourcing decisions. Thus, the practice of geography as an academic discipline itself reflects its own fundamental precepts. There are general features that apply to most geography programmes but also particularities that reflect local characteristics and individual decision-making. In geography, as in so much else, place matters.

Recapping: human geography as paradigms, streams or traditions?

Several models of scientific progress have been applied to the task of interpreting changes within human geography over the last six decades, therefore. This final discussion reconsiders the general relevance of Kuhn's concept of a paradigm (at all three scales of definition), without necessarily suggesting either substantial periods of mono-paradigm dominance and normal science for the discipline as a whole or major revolutionary events that involved large numbers of geographers switching from one paradigm to another. (In some ways, paradigms as disciplinary matrices can be equated with research programmes in Lakatos's terminology; the Kuhnian term is preferred here.) In seeking to understand the relative popularity of various paradigms at different times, however, it is necessary to employ the contextual approach ignored by some advocates of a Kuhnian approach within geography. In this section we are also mindful of the other ways of narrating disciplinary change that have been flagged in this and the introductory chapter.

The 1950s and 1960s

Regionalism, with its empiricist and implicit exceptionalist philosophy, dominated the discipline until the 1940s, though in several versions. In the USA, for example, areal differentiation was stressed by geographers emanating from the midwest, notably Chicago and Madison, whereas those from Berkeley focused on the evolving cultural landscape, with a third distinct school based in Clark University (Bushong, 1981; Prunty, 1979; Koelsch, 1988; Martin, 1998; Monk, 1998). The regional theme dominated in the UK, too, though with a greater emphasis on physical geography and less evidence of distinct 'schools' (Freeman, 1979; Johnston and Gregory, 1984).

Systematic studies increasingly replaced regional descriptions through the 1950s; their stated (though often unrealised) aims were to increase the content of regional descriptions, advancing the understanding of particular places through knowledge of the general processes (especially physical) which interact to produce unique local characteristics. Nevertheless, regional synthesis remained the central focus of many (senior) geographers' conception of their discipline, and regional courses predominated in many undergraduate degree programmes.

Regionalism was a disciplinary matrix with several competing exemplars, therefore. In the 1950s and 1960s, there was what Entrikin (1981, p. 1) terms 'transition between reigning orthodoxies . . . in which the spatial theme superseded the regional theme'. This was marked by the further growth of systematic studies, the distancing of many geographers from the core belief in regional synthesis, and an increasing emphasis on finding laws of spatial organisation, involving distance as a basic influence on human behaviour. As Guelke (1977a, 1977b, 1978) and Entrikin (1981) have indicated, the distancing was not rapid – in part no doubt because of the political need to sustain the unity and identity of the discipline – and some of those promoting the new paradigm only argued that quantitative analysis and spatial science offered better and more rigorous procedures for identifying and understanding regions (as in Berry, 1964a). This was seen by some as a methodological shift only (Chisholm, 1975a), which involved new exemplars but not a new disciplinary matrix, let alone a new world view.

The methodological shift alone was increasingly portrayed as insufficient, however, and a philosophical shift slowly emerged (elsewhere termed a 'quiet revolution': Johnston, 1978a, 1981b, 1979). Regional synthesis as the disciplinary core was unattractive to many younger geographers (e.g. Gould, 1979), for whom contact with other social scientists (as in the Second World War – see p. 59, this book) introduced the excitements of systematic specialisms practised according to the positivist model of science, which implied the search for laws and applicable research findings. The links with the regional core were severed and a new disciplinary matrix of spatial science established, although Batty (1989) argued that by the time its models were well developed and applicable the demand for them had declined: 'It is an irony of history that such good models finally exist which could well have produced excellent advice in their day had they been available. But that day has passed' (p. 156). The term 'region' took on a very different meaning (Johnston, 1984a). The shared values of the new disciplinary matrix differed from those of regionalism, and to the extent that the adherents of spatial science, most of them from the new generation of geographers who were part of the educational boom of those decades, came to dominate the discipline (see Mikesell, 1984, on whether they did) so a revolution can be said to have occurred. Whether it was a revolution in Kuhnian terms is doubtful, however, since, as noted earlier, the shift hardly fits into the 'response to anomalies' component of Kuhn's model. That would involve not only a shift in disciplinary matrix but also a shift in world view, in the conception of the nature of science, but as Hartshorne made clear (see p. 61), traditional regional geography as he conceived it did not deny the possible relevance of generalisations about processes for the understanding of places. James Bird (1989) identified a 'one and only' revolution, however, dating it to June 1966 with the publication of Bunge's (1966b) note which criticised those who argued that the uniqueness of location meant that positivism was inapplicable in human geography.

The new disciplinary matrix of spatial science was firmly established in Anglo-American human geography by the end of the 1960s, and has been sustained since. Over nearly 50 years it has seen many shifts in exemplars, a lot of them linked to methodological – especially technical – developments in data collection, collation, analysis and display. (Compare, for example, Haggett, 1965c, with Haggett et al., 1977, for shifts in the general orientation; Cliff and Ord, 1973, with Cliff and Ord, 1981, for a single methodological issue; and Hägerstrand, 1968, with Cliff et al., 1987, and Haggett, 2000, for a particular substantive topic.) Such paradigm shifts at the level of exemplar are closer to Kuhn's original presentation (and later clarified intention), with new ways of doing research accepted as superior to those previously used. They brought greater substantive success too: Haggett (1978, p. 161), writing on 'The spatial economy', claimed that it:

> is more carefully defined than before, we know a little more about its organization, the ways it responds to shocks, and the way some regional sections are tied to others. There now exist theoretical bridges, albeit incomplete and shaky, which span from pure, spaceless economics through to a more spatially disaggregated reality.

In addition to the shifts among methodological exemplars there were also changes within the spatial science disciplinary matrix which had a wider import. The first was the reorientation away from normative modelling, which involved testing observed spatial organisation against a priori models, and towards behavioural studies; as Kevin Cox (1981, 2014) and Hanson (1993) indicate, the disciplinary matrix was not queried but there were major shifts in emphasis and style (see also the discussion of Openshaw's views on p. 156, this book). The second was the advocacy for 'welfare geography' (p. 337), described by Eyles and Smith (1978) as a response to social conditions and a desire to make contemporary human geography more relevant to them.

The 1950s and 1960s saw the establishment of a new disciplinary matrix for human geographers, therefore, with twin concentrations on the spatial organisation of society and human spatial behaviour. During those decades, and much more so since, particular exemplars have waxed and waned. They have introduced new methodological procedures and new substantive foci, in part as reactions to anomalies thrown up within the discipline (the failure of certain normative models, for example, and the shortcomings of some quantitative procedures) and in part as responses to trends in society, as illustrated by the many branches and sub-branches of the disciplinary matrix, reflecting the substantive interests of geographers (urban social, agricultural, etc.) and the technical arsenal that they deploy. Reviews of American geography have been almost entirely structured within those systematic specialisms (Gaile and Willmott, 1989, 2004).

The 1970s and 1980s

The spatial-science disciplinary matrix expanded rapidly in the 1960s and early 1970s and in many departments it was relatively unchallenged until the mid-1970s (Taylor, 1976). Responses to the output from that disciplinary matrix and to events and issues elsewhere in society then stimulated two major challenges, contesting not just the exemplars or even the disciplinary matrix, but the world view implicit in the disciplinary matrix of spatial science.

The first of these challenges – termed humanistic here – was a response both to the nature of spatial science and to the ideology of society that it reflected. Spatial science is technocratic in its orientation and application; it tends to reduce people to terms in equations; it thereby ignores their individuality and freedom of action (Ley, 1981) and ignores the immense variety among places in favour of a universalistic view of how people think and act. As Trevor Barnes (1996, p. 6ff.) describes it, spatial-scientists adhered (implicitly in most cases) to the 'Enlightenment project' launched in seventeenth-century Europe (Livingstone and Withers, 1999; Withers, 2007), which emphasised:

1   The notion of progress in the development of scientific understanding, through the application of the power of reason and the rationality of scientific methods.
2   A belief in autonomous, sovereign, self-consciously directed individuals, so that all are in essence the same, able to make the same moral and other judgements.
3   A belief that 'the world had order and humans could find it' through application of their rational methods, notably those based on logic and mathematics.
4   An acceptance that there are universal truths, which hold at all times and in all places, and whose discovery is the goal of all scientific endeavour.

These tenets are questioned by humanistic geographers, whose reading in other social sciences and philosophy led them to argue for a focus on subjectivity, which was clearly incommensurable with positivistic spatial science, so that the choice of which to practise was ideological (Johnston, 1986a). According to Barnes (1996, pp. 8–10), they had their own core beliefs (shared with other counter-Enlightenment scholars) which:

1   Rejected 'both the epic of progress and the power of rationality and reason' (p. 8).
2   Contended that individuals are 'shaped from the outside rather than from the inside' (p. 9; i.e. by their context).

3   Rejected any notions of a 'monolithic order'.
4   Argued that science advances not through the use of universal constructs but rather through 'a set of local social practices of a given time and place'.

Cloke *et al.* (1991) sustained this case, and criticised spatial science on the grounds that it can involve treating individuals as members of categories, which they saw as denigrating – at best, it is only 'a partial treatment of people' and at worst it involves their 'complete neglect' (p. 67). It treats them as 'little more than dots on a map, statistics on a graph or numbers in an equation . . . exercises . . . that effectively convert . . . human beings into "dehumanised" entities drained of the very "stuff" (the meanings, values and so on) that made humans into humans as opposed to other things living or non-living' (p. 69).

Humanistic geography was introduced as an alternative science of geography, therefore, and not as a reorientation of the existing way, though some proponents claimed roots in earlier geographical practices such as those of the French geographer Vidal de la Blache (Buttimer, 1978a). It had its own disciplinary matrix, and its own variety of exemplars, in terms of both their philosophy and their subject matter, as set out in Chapter 5. Its introduction did not stimulate a major switch in the contents of the discipline as a whole, however. Rather, it offered an alternative conception of human geography and competed (ultimately unsuccessfully) with the others on that basis.

The second challenge – from radical geography – was also a response to the contents of the spatial-science disciplinary matrix, but was much more influenced by the wider political conjuncture than was the humanistic. As detailed in Chapter 6 earlier, those who launched what was initially known as radical geography were concerned with the failure of positivist spatial science to tackle and solve pressing societal problems. They advocated applied geography, but defined relevance in a very different way (Harvey, 1974c; Johnston, 1981a). As with humanistic geography, this involved promoting a scientific revolution at the level of the world view; the realist science that they advanced was incommensurable with both positivist and humanistic stances.

Once it was reasonably well established, radical geography also became a kind of disciplinary matrix, with a variety of exemplars. The core of its programme is the desire to uncover the mechanisms that drive society, providing accounts of how people act and how the empirical world is organised. How such understanding can be achieved, and how the knowledge should be used, has been the subject of much debate – both between Marxists and non-Marxists and among various types of Marxist – and exemplars have waxed and waned rapidly in recent years as a consequence. The debates within the realist conception of science are about how to achieve agreed scientific goals; the debates with positivist and humanistic geographers are about the nature of science. The former produce internal revolutions; the latter seek to promote major disciplinary revolutions. (Note, however, Trevor Barnes's (1996, p. 23) critique of Harvey, the leading advocate of marxism within geography, and Andrew Sayer, the leading advocate for realism, as both implicit part-adherents to the Enlightenment project, and hence of the same world view as the spatial-scientists: 'my argument is that as with Harvey, Sayer totters between Enlightenment and anti-Enlightenment views'.)

These decades were turbulent for human geography, which Ley (1981) found both exciting and confusing, in part because of the rapidity with which human geographers explored new ideas. He believed that such exploration was often superficial, with ideas discarded almost as soon as they had been adopted: 'In true North American fashion, obsolescence is setting in more and more speedily' (p. 209). Others argued that ideas were often divorced from their original context, and that geographers who imported them may have been unaware of the controversies surrounding their use (Agnew and Duncan, 1981; Duncan, 1980; Mohan, 1994). Some sought to reconcile the various world views (Harrison and Livingstone, 1982; Hay, 1979a; Johnston, 1980b, 1982c; Livingstone and Harrison, 1981); some argued that is impossible (Eyles and Lee, 1982); and others still contended that each of the various world views is informed by the others (e.g. Thrift's, 1987, p. 401, claim that Marxist political economy now 'forms a vital subtext to most theorising'), so that geographers should use their synthesising powers to integrate the various perspectives (Brookfield, 1989, p. 314).

Whatever the detailed interpretations and attempts to reconcile the competing world views, the 1980s were characterised by a pluralism that not only challenged the would-be hegemony of the approaches launched in the 1960s but also presented problems for geographers in presenting their discipline to others. The intense conflict that these debates engendered is exemplified in a retrospective essay by Berry (2002c, p. 441). His hope was that:

> Urban geography, with a rapidly developing reputation within the social sciences and a penchant for transdisciplinary inquiry, was to be directed towards and receive its inspiration from a commitment to help meeting national needs. It was my hope that the challenges of use-oriented research would stimulate new rounds of theoretical development and leverage the field to new levels of understanding. I took seriously the notion that geography is the science of which planning is the art.

But:

> By 1980 the excitement had waned, however. The paradigm appeared lost. There was a new fragmentation as young geographers did what so many have done in earlier epochs of the discipline's history – they changed the channel to a different program. The common interest that seemed to be at hand succumbed to bickering and feuds.

Urban geography – until then in the disciplinary vanguard according to Berry's view – diverged along a series of 'incommensurate paths' (p. 443) as a result of both internal and external forces. The latter reflected the sentiments associated with the anti-Vietnam war protests, with many graduate students mobilised, according to Berry, by 'left-wing activists' who turned their anti-war sentiment into an 'anti-American anticapitalist radicalism'. Instead of 'use-inspired basic research', these radicals took the stance of the 'armchair intellectual contrarian', as exemplified for him by David Harvey:

> There could be no meeting of the minds, no common ground. David disavowed our fumbling attempts to build a social science, positioning his agenda squarely within the realm of politics and imprinting in the minds of many young geographers the idea that science is merely one political viewpoint among many.
>
> (p. 443)

The new generation that followed the radical and humanistic geographers 'rejected the Enlightenment ideal of the perfectibility of the human mind – the ideal of an ever-progressing rationality – devaluing the speaking subject in favor of a highly theoretical linguistic and cultural determinism' (p. 444). As a result, Berry shifted away from geography as an institutional base, while remaining active within his chosen orientation. Twenty years later, he was identifying with 'a political economy that we are trying to enrich with an emergent geographic information science that offers the prospect for new rounds of use-oriented investigation supported by new and more powerful forms of spatial analysis'. The potential is exciting for him, and 'once again the social and environmental sciences are looking to us', while, however:

> There is great promise in the idea of spatially integrated social science, but geography's traditional threat, the congenital inability to pull together, remains with us. We should not fritter

away yet another opportunity because we lack the fortitude to withstand the contrarian responses
of our professional dialecticians.

(p. 444)

For him, at least, the debates of the 1970s and 1980s were not past history: the world view that he
and others pioneered some 50 years previously remains central to their view of the discipline's
future. For others, the debate seems to have been lost. Looking back at the 1980s, Ford (2002,
p. 438) identified a polarisation of urban geography between 'social theory and software' with no
central ground between them: 'most of the [current] applicants for urban positions are interested
in the city only tangentially, and instead present themselves as experts in aspects of either theory or
technology'.

The 1990s and the turn of the century

The 1990s witnessed yet more turbulence, again reflecting both dissatisfaction with the various
world views and a rapidly changing external environment – economically, socially, politically,
technically and culturally. Within it, the challenge of humanistic geography very substantially
withered, in that very few followed the postulates of idealism, phenomenology and the other
approaches discussed in Chapter 5, although, as stressed in Chapter 6, rejection of the Enlightenment
model of the individual was at the core of the various critiques of Marxism. The 'new generation'
emphasised difference, although – in part via the 'cultural turn' and the influences of feminism and
poststructuralism – in a rather different way from that promulgated by Hartshorne some half a
century earlier (Campbell, 1994). Spatial science did not wither, let alone disappear, however. Its
world view remained strong, although more marginal to the discipline, many of whose members
(in part on political-ideological grounds) rejected not only the implicit positivism but also the
emphasis on technical rigour in data analysis. Some commentators claim otherwise: according to
Hamnett (2003, p. 2), 'Quantitative techniques and aggregate social research have been largely
abandoned. . . . Analysis of large data sets has become totally passé, the object of suspicion or even
derision as "empiricist".' (Similar arguments appear in, inter alia, Cloke et al., 1991, and Peet, 1998;
for responses see Chapter 4, this book; Johnston et al., 2003, 2014a, 2014b; Unwin, 2005.)

By the 1990s, the discipline was substantially fragmented, not only in its world view and
disciplinary matrices but also in its substantive concerns. It comprised a variety of approaches held
together as much by the realpolitik of university politics and funding as by any adherence to core
beliefs, with relatively few of its practitioners interacting across subdisciplinary frontiers (Johnston,
1991, 1996b). This view was challenged by Gould (1994, p. 194): 'I find the idea of fragmentation
absurd, and too frequently raised by people who long for others to conform to their monolithic
ideological positions.' He argued that geographers with different specialist interests (as reflected,
for example, by their affiliations to institutional study groups) nevertheless are following his advice
of 'don't specialize . . . read eclectically' and:

> Despite some contrary opinion, [there] is no evidence that they are disconnecting from the rich
> body of ideas that constitute the geographic 'way of looking' today. For their research, they may
> be using remotely sensed images, geographic information systems, structural-Marxist concepts,
> hydrological science, socially constructed ideas of the environment, Michel Foucault and the
> power of surveillance, deconstructions of official government reports, goal programming, a
> heightened awareness of gender issues . . . I mean, you name it! . . . to illuminate the spatio-
> temporal human condition at a place . . . embedded in a larger regional and national space. With
> all that conceptual and methodological integration, where, literally on earth, is the fragmentation?

Others were much more concerned about fragmentation, however. When the AAG Council
decided to initiate a follow-up volume to that edited by James and Jones (1954) thirty years earlier,

it rejected a proposal from Gaile and Willmott (subsequently published independently: Gaile and Willmott, 1989) because it drew largely on the AAG's specialty groups and so reflected a fragmented view of the discipline rather than focus on 'cross-cutting and unifying concepts and methods' (Abler *et al.*, 1992a, p. xvii). The book which they edited for the AAG (*Geography's Inner Worlds: Pervasive themes in contemporary American geography*) had four main sections:

1   What geography is about – geography's worlds, places and regions, representations of the world.
2   What geographers do – observation, analysis, modelling, communication.
3   How geographers think – location, place, region and space, movements, cycles and systems, the local–global continuum, scale in space and time.
4   Why geographers think that way – paradigms for enquiry?, humanism and science in geography, applications of geographic concepts and methods, the peopling of American geography.

This was done to counter the perceived fragmentation (Abler *et al.*, 1992b, p. 2):

> Specialization and specialty groups foster better communication in a multi-faceted discipline. They permit scholars and practitioners with common interests to achieve identity without forming independent associations . . . [but they also] foster intellectual isolation by retarding the cross-fertilization that occurs when geographers encounter unexpected ideas. Despite the fact that the fission evident within geography is common in the physical and social sciences, specialization may have gone as far as it can or should go in American geography.

Hence their book was structured to bring geographers together:

> to highlight the common elements within a discipline whose practitioners are in danger of forgetting their shared heritage and ideals. In their preoccupation with the great diversity of geographic problems with which they deal on a day to day basis, geographers are prone to overlook how much they share emotionally and intellectually with other geographers that they do not share with colleagues in other disciplines.

A few of the chapters are multi-authored, presumably as a way of bringing individuals with diverse interests together to identify common elements, and the outcome led the editors to conclude that the discipline is shown to be 'simultaneously less fragmented than we had feared it would be and less coherent than we had hoped' (Abler *et al.*, 1992c, p. 391), but nevertheless 'wonderfully diverse in substance, method and philosophy':

> Compared with other disciplines, geography has always been tolerant of variety, and often enthusiastic about multiplicity. With the possible exception of the period before 1920 when environmental determinism achieved a measure of intellectual hegemony, American geography has never been a normal, paradigmatic science of the kind postulated by Thomas Kuhn.
>
> (p. 395)

Thus, although concerned about fragmentation they also promote it as a 'geographical good thing':

> American geography was postmodern long before the term was invented. It has historically been eclectic and self-contradictory in many respects. It has often playfully delighted in juxtaposing dissonant substantive interests and intellectual traditions. American geography will prosper in

the 1990s not so much because geography will change radically, but rather because the discipline's social and intellectual environments have now evolved to where geography has long been. Geographers are at last professing and practicing a postmodern discipline in a postmodern age.

Nevertheless, if geography is to prosper, the basis for coherence (necessary in a post-modern discipline?) has to be realised, which they argued involves: overcoming the indifference to physical geography among some human geographers; finding 'the language and narrative forms that will engage the attention of American society' (p. 398); and ensuring mutual recognition between the two main 'camps' within human geography – termed spatial scientists and social theorists elsewhere in this chapter ('The critical matter is that all parties avoid the whole-sale rejection of each others' viewpoints that occurred in geography after 1920 and again after 1955'; p. 400).

A few years later, members of the Association returned to the general issue in a forum held at its Centennial Conference on 'Where we have come from and where we are going'. In his introduction, the then president noted that for all the apparent successes of the preceding hundred years (Murphy, 2004):

The challenge of expanding geographic understanding is as pressing as ever, geographical research remains marginal to the larger research enterprise, and the study of geography is still largely absent from many schools, colleges and universities across the United States.

(p. 701)

The collected essays cover a wide range of topics and agendas for geography's future practices, but with the general case, as expressed by Hanson (2004, p. 715) that 'the research questions we ask should serve society, a society that is richly diverse'. Reflecting contemporary societal concerns, therefore, much of the discussion focused on environmental challenges and research that bridges the human and physical sciences (a characteristic also of the NAS-NRC, 2001, review a few years later). For Kwan (2004, p. 759) this required a 'post-social-theory, post-spatial analysis future' which finds 'ways to make geography a respectable discipline and to enhance its status without erasing difference within the discipline'. Like many before her (surveyed in Matthews and Herbert, 2004), she sought a more unified discipline (especially in its public face), when all the evidence of debates within it over the preceding decades suggested the near-impossibility of its achievement. Several of the authors, too, sought a future in which geography/geographers made an impact but Don Mitchell (2004, p. 766, following Eliot Hurst, 1985) argued that geographers should not collaborate 'with the manifest injustice that makes our wealth, our relative peace, and our relative good health possible . . . a world and geography in which others have the chance to have what we have'. Instead:

I say we resist. I say we turn the age of extremes into the age of emancipation, that we turn the geography of violence and oppression into the geography of justice . . . [using our] limited power . . . to develop and promote a liberal and liberating education (for all).

As suggested in the previous chapter, it is hard to merge such differing views of what an 'applied-applicable' geography might be.

A second feature of the contemporary world, which is particular to academic life in many of its aspects yet at the same time linked to changes in the wider milieux, relates to the contexts within which universities operate. Since the 1980s, as illustrated earlier with regard to the pressures for applied work and skills-based training (see Chapter 9), universities have come under increasing pressure to adopt a range of business-like practices and to be accountable for their activities, including research. This has stimulated what has been termed an 'audit and accountability culture' and a 'quality industry' in the UK (Johnston, 1994b), and many have claimed that it has redirected research (some would say skewed it) in particular directions.

Luke (2000, p. 216) referred to many aspects of this restructuring of the academic milieu in the USA as a 'discipline of reputational development'. Within academic networks power is exercised to ensure conformity to both general academic and specific disciplinary norms, and access to positions of power and influence is in general restricted to those who conform and win reputation points as a consequence. (It may also, it is suggested, require conformity in the way that academics write: Dewsbury, 2014; Badley, 2015.). Thus:

> Academic life is an existence pegged to perpetual examinations: seminar discussions, research papers, dissertation defenses, conference papers, journal submissions, book contracts, teaching evaluations, committee assignments, tenure hearings, academic promotions, annual reports ... these facts about the routines of continual judgment in the disciplinary operations of the discipline [are] ... those necessities that 'constitute a recurrent aspect of academic life'.
>
> (p. 216)

As we noted in Chapter 1, academic life is a chosen career, and those involved want to get their share of its rewards, which involves facing up to the continual judgements posed by the 'perpetual examinations' – and to a considerable extent this means conformity, since the allocation of those rewards is very largely in the hands of others.

Such conformity operates at two levels. At the first it operates within the discipline – or at least a significant section of it. As Luke expresses it, professional training and acculturation from graduate school onwards inculcates norms of conformity: each discipline is a 'powerful regime of normalizing thoughts and behaviors' which is 'intent on creating the correctly disciplined disciple of its knowledge via personal surveillance and methodological instruction' (p. 217). This, among other things, involves a great deal of ranking, of both individuals and departments, according to their reputations:

> The professionally correct gauge the value and position of individual persons and departments, willingly and openly, as 'names'. One is 'a big name', 'a name', or 'a no-name'. . . . Reputation and its development in rankings, therefore, boil down, first, to building one's meganymic profile, and, second, avoiding anymic obscurity or micronymic ineffectuality.

For Luke, this building a reputation for oneself – and the building of a collective departmental reputation based on those of its individual members – is not a function of what one studies (of disciplinary substance) but simply of successful performance on the key reputational indicators, which may be grant/contract income obtained, for example, or publications in certain journals, or citations to one's published work. To some extent, even, an individual's scholarly reputation is a function of the university and/or department that he/she works in:

> Who gets what, where, when, and why in university departments, professional associations and grant competitions depends on these ranking systems of invidiously ranked distinction. While the intrinsic worth of any widely sought after rewards – pay raises, tenure, promotion, corner offices, or administrative powers – are declining, the battle for them continues at full tilt. And, the strategies for winning success or evading failure totally are denominated in these codes of professional correctness: the normalizing judgment of superiors, peers, and inferiors iterate their evaluations in matrices or performative success and failure.
>
> (p. 218)

This, in turn, influences the content of the discipline, at least in the short term, since the works of those who win in the ranking contests are presumably considered as providing the best exemplars to follow by those still seeking reputational preferment (unless they perceive that greater renown as likely to result from successfully challenging the conventional wisdom and power bases). Thus, a discipline is divided into a range of subcommunities, each of which has its 'meganyms' whose ideas and opinions have a substantial impact on not only the allocation of rewards and resources (leading to claims of overweening power in some instances (Short, 2002)), but also what is considered good and bad practice within that community.

Many of Luke's claims relate to the internal operations of disciplines and subdisciplines but increasingly, as we have already suggested, those operations are subject to external audit and constraint. This is especially marked in the UK, where reputational rankings of departments (and therefore, at least implicitly, of their members) is the basis for the allocation of large sums of public money, especially with regard to research; teaching quality is also evaluated but the results are not used in selective disbursements. Although these judgements are made by peer review procedures involving academics (mainly 'meganyms' (Short, 2002)) evaluating each other, they are constructed to meet political agenda regarding accountability and self-reliance (through the attraction of external support for research work, including contracts for applied work for public and private sector contractors). Thus they are informed, at least, by reputational metrics such as income earned and citations to papers published, and the quality of research publications is to some extent conditioned by the standing of the journals in which they appear.

The first decades of a new century

The most recent decade in human geography's intellectual history has been one of considerable contrasts between the earlier and later years. Until 2007, most anglophone universities in general and research in particular blossomed in a relatively benign economic context, albeit one where neoliberalism was increasingly hegemonic. With some exceptions, university geography departments were able to expand – both their staff and their research students – and grants were available to take a wide range of work forward. By the decade's end, however, financial crisis and political shifts saw major changes in higher education and research expenditure, such as the much higher fees that were to be charged by English universities from 2012 and the likelihood of students graduating with both large debts and encountering very tight labour markets. For those finishing postgraduate studies and seeking a first university appointment as an academic geographer, prospects were often bleak, especially for those whose skills and training did not give them technical (read potentially applied) expertise (Gillen, 2015). A decade before, a UK-based survey of 'The future for new geographers' (Shelton, 2005) had been more upbeat, notwithstanding a concern about the proliferation of fixed-term academic jobs.

In times of relative abundance the contest for resources within universities and their departments is less, and the potential for conflict smaller, than when opportunities are few. If expansion is feasible, then it is possible for all types of geographical practices to gain new resources and for the discipline's breadth to flourish. If the various practices are acceptable (Johnston, 2006), it illustrates how during this decade some textbook writers/editors, like their 1960s predecessors, excluded certain practices using the strategies identified on p. 24, this book). Given the imperative of the previous decade for university departments, especially the larger ones (more characteristic of the UK than the USA), to specialise in particular research themes in order to enhance their 'research ratings', potential conflict over which types of geography should be preferred in which places was less than might otherwise have been the case; departments became more specialised and few offered a full range of research opportunities within the discipline (or even a full range of advanced courses in their undergraduate, let alone graduate programmes). The landscape of geography had become much more varied, and the first years of the twenty-first century fully illustrated that.

As a consequence, although major differences at the scale of both the world view and the disciplinary matrix were clearly present, there was little overt conflict between adherents of – as some simplified the situation (Sheppard, 1995, Kwan, 2004) – spatial analysis and social theory approaches. Revolution was not in the air and relative peaceful co-existence was generally possible (There were exceptions, such as the brief and relatively mild 'spat' initiated by Hamnett, 2003.) Thus, human geography throughout the decade was characterised by a wide range of communities and networks of scholars – some much more closely knit than others – each of which took forward its own research practices. Those communities had their own journals, conferences (and conferences within the big conferences) and interaction between them was relatively slight. There were debates within many of them – at the level of the exemplar (such as that within one of the 'spatial analysis' communities on the measurement of segregation: Johnston et al., 2010) – but these were relatively low-key and had little impact beyond those involved. And they sometimes change what they study as well as how. In the 1950s–1970s, for example, spatial analysts were much influenced by central place theory and spent a great deal of energy seeking a particular form of (hierarchical) spatial order. That model's underlying assumptions have much less contemporary purchase, so there are few studies of shopping centres and their organisational matrices (though see Ó hUallacháin and Leslie, 2013); order is sought in other contexts, as in work that incorporates ideas taken (at least implicitly) from structuration theory which relate how people behave to their local contexts, and in turn how their behaviour changes those contexts and how others then behave.

Despite this generally optimistic situation there were concerns, many of them linked to the interactions between the discipline and wider society, and especially the latter's expectations of academics. In the USA, for example, the continued concern about geography's place in its educational and research institutions and its wider status within academia stimulated debates regarding disciplinary futures (as at the AAG's centennial conference: Murphy, 2004) and the commissioning of a further report promoting the role of the spatial sciences generally and geography in particular (NRC, 2011; see also Cutter et al., 2003). In the UK, the continued stress on research assessment (and the associated league tables – not an exercise confined to the UK – Castree et al., 2006; Cupples and Pawson, 2012; Liu and Zhan, 2012) raised concerns that certain types of work would be privileged over others, especially when the terms of the next national exercise (the 2014 Research Excellence Framework) was to involve assessments of not only the quality of published work and the local intellectual environment, but also the relative importance to be placed on evaluations of the demonstrated impact of research on society, especially on the economy in general and 'wealth creation' in particular. This was seen to privilege particular types of research practice, a fear exacerbated by public money being directed towards some (notably 'spatial analysis'; the ESRC launched a number of programmes to enhance training and research in quantitative analysis, for example, and it was a condition of the British Academy's public sector settlement in 2011 that it devote significant parts of that resource to numeracy).

One of the significant features of these years – and to a slightly lesser extent of the 1990s too – is that there has been much less debate in the discipline's journals regarding its nature. Contemporary human geography is well described by Cresswell's representation of it (quoted at the start of this chapter) as a series of linked intellectual worlds between which there is relatively little contact, with mutual tolerance but also misunderstanding and sometimes misrepresentation – to some extent paralleling Rusu's (2012) characterisation of contemporary sociology and misrepresentations (on which, see Johnston et al., 2014a, 2014b). The reason for this change – which is a major contrast to the debates of the 1950s–1980s – perhaps reflects the volumetric expansion of geography since the Second World War. A relatively small discipline could readily be dominated by a small number of gatekeepers, so those who wanted change had to displace them in order to promote their alternative view of the discipline – which is certainly what was attempted in the so-called 'theoretical and quantitative revolutions'. That is much less feasible in a much larger discipline, and though there is competition for resources within it, attempts to reconfigure the entire discipline are unlikely. Cases are made for changes in practices – as described in the later chapters of this book – but, save at the exemplar level in Kuhn's hierarchy of paradigms, these rarely involve attacks on other types of work. This fragmentation of a large discipline into many relatively isolated intellectual villages – a number of them co-existing with the same university departments – probably also links to Cresswell's other representation of contemporary geography – a lack of contact with disciplinary history; so much contemporary geography has few, if any, roots in geography as practised no more than a few decades ago. This is undoubtedly linked to the absence of a 'geographical canon', key writings to which practitioners return and students are directed as core to the discipline's development (on which see the forum essays convened by Keighren *et al.*, 2012, and a second set in the *Journal of Historical Geography* (Powell, 2015)); human geography is perhaps held together not by its key texts but rather by its key concepts (Johnston and Sidaway, 2015; Agnew, 2014). Geography was a vibrant discipline for several decades; it is now a congeries of vibrant, many of them sustainable, subdisciplines. The attempted 'revolution' of the early post-Second World War decades, as both Peet (1998) and Kevin Cox (2014) stated: introduced the importance of explicitly addressing and developing theory; stressed the need for methodological rigour; and focused geographers' attention on space – but to many who succeeded the 'quantifiers', they introduced the wrong theory, the wrong methods and a narrow conception of space. With the core concepts – identified by some as place, space and environment – these imperatives are at the core of the wide diversity of geographical practices and scholarship.

An abundance of turbulence

Evaluating these turbulent years, when the turbulence shows little sign of abating, is not easy. Kuhn's model, sometimes presented to geographers as periods of normal scientific progress punctuated by major revolutions, is not sufficient to account for what has happened within their discipline since 1945. But each of the basic Kuhnian definitions of a paradigm (world view, disciplinary matrix and exemplar) is relevant to appreciating what has occurred and is occurring within geography, as Mair (1986) argued for the concept of the paradigm as a scientific community (or disciplinary matrix) with shared values.

At the lowest definitional level of a paradigm, the exemplar, human geographers socialised within any of the available disciplinary matrices have shifted the orientation of their work as 'better' ways have been suggested to them and different topics have come to the foreground; minor revolutions have occurred – frequently. Such shifts have gone in a variety of directions, presenting an apparent anarchy – not chaos, which is the vernacular use of that term, but 'free and voluntary cooperation of individuals and groups' (Labedz, 1977, p. 22). But the core values hold.

There is also some evidence of the relevance of the largest scale definition of a paradigm, the world view, to changes in human geography. Very different and incommensurable conceptions of the nature of science have competed for geographers' attention. They differ in their scientific and their societal goals, and demand choice. Some made a very clear choice, switching from one conception of science to another because, as Harvey (1973, p. 17) expressed it with regard to Marxism: 'I can find no other way of accomplishing what I set out to do or of understanding what has to be understood': thirty years later, he noted: 'now, when the [Marxist] text is so pertinent, scarcely anyone cares to consider it' (Harvey, 2000, p. 8).

Some argued that the choice is really between two world views only (spatial science and social theory (Sheppard, 1995)); others identified three and, according to James Duncan and Trevor Barnes (1993, p. 248), one of them triumphed:

> R. J. Johnston's *Geography and Geographers Since 1945* [*sic*] . . . could have been subtitled 'Modernism and its Discontents'. For its narrative structure is compellingly organized around the contest for intellectual supremacy among the triad of approaches: empiricist, positivist and

modernist social theory. Although from Johnston's perspectives the social theorists win the day, and the others are relegated to the dustbin of geographical history, it is only modernism that really triumphed, because it is the only game in town.

Each of those positions, they contend, seeks to provide the 'one best method to explain geographical phenomena' and 'None has seriously entertained the notion that there is no best method' (p. 249). The differences among the three approaches are, they claim, 'relatively minor when compared to their shared assumptions', and they fear that the most radical aspects of postmodern epistemology have been rejected by geographers who 'wish to incorporate certain of its important insights into a reconstructed modernist [i.e. Enlightenment] project'. If the postmodern project was fully embraced, they argue, the discipline would become much more varied since it would no longer be characterised by the views of 'trained academics':

> The problem lies in our conceptualization of difference. Difference for geographers is other spaces, other places, other regions or other landscapes. To embrace difference is in Gregory's ... words to embrace 'areal differentiation'. It is instructive, however, to compare the postmodernist geographers' conception of difference to that of postmodernist ethnographers. For the latter, difference is other people and, to be more precise, other peoples' voices. Their plea is for an end to monovocality and authorial authority. Let us hear direct, they say, from those from whom we as academics have for so long spoken.
>
> (p. 253)

Such a shift in the nature of geography was politically threatening because of its challenge to the academics' authority. Furthermore:

> Multivocality (taking others' voices seriously) is not something that can be easily accomplished in geography, for we have no tradition of ethnography. Unless we begin to focus more of our energies on developing our techniques for listening to others our calls for difference are highly suspect, for we will continue systematically to silence difference. Those outside the academy have no voice in our work other than the one we choose to give them ... to speak for another is not a politically innocent act. We have appropriated their voice – colonized their perspective.

Buttimer (1993, p. 23), drawing on Stephen C. Pepper (1942), identified four 'root metaphors' or world views underpinning the practice of geography, each of which 'projects a distinct interpretation of reality':

1   the world as a mosaic of patterns and forms, which she sees as the 'central root metaphor for geography's chorological tradition, the most practiced of all four' (p. 22);
2   the world as a mechanism of causally integrated interacting systems;
3   the world as an organism, as a whole which comprises unity in diversity; and
4   the world as arena, as the context within which 'spontaneous and possibly unique events may occur'.

[To which could be added a fifth:]

5   the world as text, as the landscape is a means of understanding its creators' intentions and cultures (an activity also associated with both archaeology and, in different ways, some forms of theology).

Buttimer did not present her four as exclusive, nor are they treated as such: 'Most creative scholars ... avail themselves of more than one root metaphor in the course of a career. Only the rare dogmatist clings to one throughout.' Nevertheless, she sees them as 'vectors of distinct thought styles whose appeal has varied through different moments in Western social history' (p. 24). Which styles are dominant when depends on the interaction of people and context:

> Acceptance or rejection of a particular paradigm, model, or method within the discipline of geography has as much to do with the aesthetic, emotional or moral connotations of a root metaphor as it does with purely epistemological reasoning. The succession of metaphors within any tradition raises questions about the interplay of internal and external circumstances. Career stories of geographers reveal important clues about their succession and relative appeal. At any moment of disciplinary history, all four root metaphors . . . may be simultaneously co-present, although one or more may appear dominant within particular periods.

Her descriptions of the metaphors, based to a considerable extent on autobiographical data, do not lead to a detailed appreciation of why one or more has been more visible in some periods than others. Instead, she identifies a major tension between, on the one hand, 'the integrated approaches of organism and mechanism [which] have invited research of wider scale and have apparently enhanced the status of geography within the academy' which is desirable given 'the political expediency of the times' and the 'political economy of research grantsmanship', and, on the other, 'The dispersed approaches of mosaic and arena [to which we can add text] have yielded more sensitive accounts of life and landscape at local and regional scales of inquiry' (p. 212). Her own preference is clearly for the latter pairing, for 'the humanist's emancipatory hope' and 'communication and understanding' (p. 219). Humanism must be 'the leaven in the dough and not a separate loaf in the smorgasbord of geographic endeavor' (p. 220); but she does not seek to impose (p. 212):

> The integrity of disciplinary practice . . . demands a flexibility to changing educational needs and to new substantive research challenges. Ideally each individual, department or research team, given its resources, aims and context, should assume responsibility for designing and adapting its agenda to such changing demands.

Each world view's disciplinary matrices identify the framework within which research is conducted. The shift from regionalism to spatial science involved a change of disciplinary matrix, from mosaic to machine, for example; while the world view remained constant – that of positivist science with its foundation in empiricism (although it was hidden until relatively late in the shift) – the shared goals altered very considerably. The relative popularity of Marxism, realism, structuration and postmodernism at different times reflects changes in the disciplinary matrix of the 'geography as social theory' world view.

Peet (1998) also mapped some of these trends – and earlier 'schools of geographical thought'– as a set of movements, flows and channels in 'disciplinary space' (see Figure 10.2). Though Peet indicates some self-consciousness at such a bold and clear sketch of geography:

> Theoretical organizational devices are political tools of power formation. So the map may give a 'bird's-eye' view but the bird is an eagle, symbol of the state, and soaring on high, rather than a working sparrow pecking up the crumbs of its localized existence.
>
> (p. 10)

A decade later, Chris Philo (2008, p. xxi) commented how Peet's map 'has a nicely 'organic' feel suggestive of a braiding river with meanders and eddies devoid of any one main channel'. Some might prefer to visualise disciplinary nodes and networks, but the appeal of such geo-fluvial metaphors is strong. Gerike (2012, p. 11) talks about geographical scholarship as a 'braided' stream that today 'reflects changing, braided channels and the merger of rivers of theory and

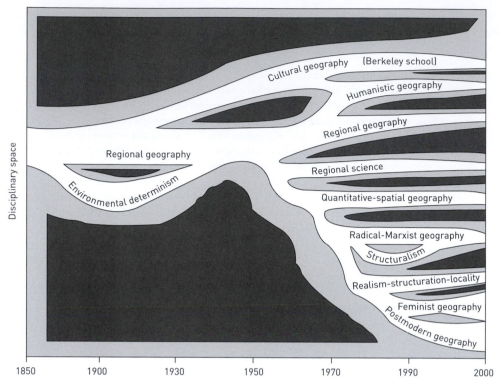

Figure 10.2 Schools of modern and postmodern thought

Source: Peet, 1998, p. 10.

perspective from other ontological and epistemological river systems'. And Janice Monk (2015, p. 272) says when:

> Searching for a geographic metaphor that would capture over four decades of research, writing, and professional engagement in geography, I settled on 'braided streams.' This fluvial form is characterized by divergent and convergent channels, mostly occurring 'where there are almost no lateral confining banks'.

Philo (2008), however, draws another map (see Figure 10.3), while recognising that any such map would really need to be three-dimensional:

> Permitting us to see that, even after their eras of greatest popularity, most approaches have continued as a presence 'behind the scenes', adding to the general intellectual background out of which subsequently more prominent approaches have emerged, whether as outgrowth or opponent, and occasionally returning to positions of enhanced recognition. Three dimensionality would also allow us to discern how some approaches presently sited some distance apart on the diagram have in actuality ended up 'folding' together; and I am particularly aware of how feminist geography, while emerging on the 'structure' side of the map, has reached over in various ways to parts of the humanistic, cultural and psychoanalytic features drawn on the 'agency' side of the map.

(p. xii)

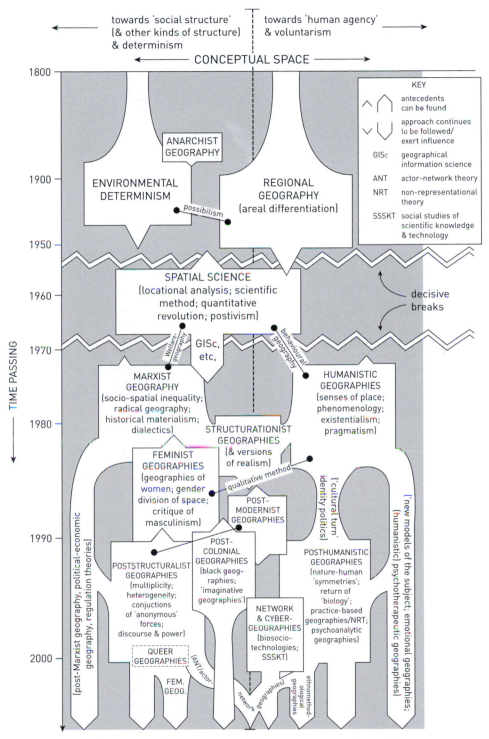

towards 'social structure'
(& other kinds of structure)
& determinism

towards 'human agency'
& voluntarism

CONCEPTUAL SPACE

Figure 10.3 Disciplinary space: the 'map' of changing approaches in human geography

Source: Philo, 2008, p. xxi.

He recognises too that extensive reading and toil are needed to make sense of 'the chaos and fuss integral to being "down there" working with the depicted approaches on a daily basis' (p. xxiii).

A relatively rare attempt to (re)narrate the history of geography from such other 'down there' viewpoints – drawing upon a departmental geographical magazine produced by students at the University of Glasgow – was also produced by Philo (1998). He notes that:

> little appears to have been said about how academic visions of geography have been received by the countless students who filter through 'our' lecture theatres, tutorial rooms and libraries. It seems to me that this is an omission of some magnitude, and that therefore more might be done to ascertain how all manner of shifts in academic geography, as fostered by small numbers of 'visible' professional geographers translate into the thinking, writing, fieldworking and murmurings of the many 'invisible' student geographers whose responses are almost never afforded sympathetic scrutiny.
>
> (p. 345)

Lorimer and Spedding (2002) also stress the wealth, complexity, and value of departmental histories, reconstructing something of the contextual histories of geography – as practised by staff and students – at another Scottish department (Aberdeen: see also Findlay and Werritty, 2010, on Dundee), and Withers (2002, p. 306) has pursued a similar task at Edinburgh:

> kindled by hearing stories from older colleagues . . . [and] motivated by the 'discovery' at one time and another of large amounts of textual material relating to an earlier department: students' work, personal papers of former staff, photographs, examination papers and old glass slides. I say 'discovery' for a reason: colleagues in my department knew of its existence, but not until an interest was expressed in 'that old stuff' was any value attached to it as a nascent department archive.

Either way, human geography has long been characterised by a multiparadigm situation at the world-view level, by competition between disciplinary matrices within at least two of those world views, and by a wealth of exemplars on which research is based in all three (all features that appear consistent with Feyerabend's notion of progress). It is because of the last element, and because many human geographers have not been deeply schooled into any one disciplinary matrix, let alone the use of any one exemplar, that the discipline appears replete with fickle allegiances, as individuals explore various ways of practising geography (Johnston, 1981b, pp. 313–14):

> Much geographical work is exploratory, and is conducted by individuals who operate independently. Indeed, many, although influenced by what they read, are in no sense socialized into a particular matrix or set of exemplars which might be associated with a 'research school' and its leader. The charisma of certain individuals and their published works may occasionally produce the prophet and disciple situation. Much more usual, however, is a situation of fickle allegiances.

Such fickleness suggests anarchy, in the sense used by Feyerabend (1975; Johnston, 1976c, 1978a). Some human geographers shift frequently among exemplars, sometimes between disciplinary matrices, and just occasionally between world views. And some, as part of a refutation of Kuhn, have proposed explicitly hybrid geographies that value differences and diversity within the discipline, calling for:

the proliferation of hybrids, or geographies and geographers of the third kind; those that cut across the divides between the social-cultural and the spatial-analytical, the qualitative and the quantitative, the critical and the technical, and the social-scientific and the arts-and-humanities. It is a future not of 'either/or' but of 'both-and' . . .

(Kwan, 2004, p. 760)

The last six decades have seen individual human geographers occasionally promoting a particular paradigm (world view, disciplinary matrix or exemplar) in opposition to prevailing practice. That has rarely been done in isolation; contacts with others or reading (or both) suggest the argument, and support is then sought within the discipline, and perhaps resources and sponsorship from within society too. Why are some people better able to promote paradigm change than others? Little work has been done on such aspects of the sociology of geography as a discipline, especially with regard to power over the acceptance/rejection of proposed paradigm shifts and the ability to convince others of the 'rightness' of any approach. Studies have also been undertaken on influential figures in the discipline's early development (e.g. in Blouet, 1981, and Smith, 2003) and, as we noted in the Preface, there is a wealth of information and reflection on these themes recorded in the 552 'Geographers on film' interviews assembled by Maynard Western Dow, although the process of digitising and making these readily available is incomplete (Martin, 2013). In more conventional written format, a 'biobibliography' series was initiated in 1977 (Armstrong and Martin, 2000). This has now published over 400 biobibliographical essays, forming, in the words of the editors of a recent volume:

an ongoing critical enquiry into the lives and works of persons who have contributed to what, at various times and places, has been taken to be geographical knowledge . . . a scholarly site in which to raise questions about what it means to do geography, then and now, and to think about it biographically.

(Lorimer and Withers, 2007, p. 4)

The nature of academic geography in the UK has also been explored in sociological and historical studies (Johnston and Brack, 1983; Sidaway, 1997), but aside from the survey by Tim Hall (2014) these dynamics have not been the subject of further detailed analysis in more recent years (notwithstanding reflections on impact agendas, and audits of research 'excellence' in the UK referenced in earlier chapters). Nor, interestingly, has the impact of the 'commercialisation of knowledge' by publishing houses been much scrutinised (though the issues were flagged in Barnett and Low, 1996). The increased stress on research publications in the UK from the mid-1980s at first led to problems for textbook publishers (Davey et al., 1995). However, the more recent proliferation of geographical textbooks, readers, companions and dictionaries that we noted in the Preface indicates that publishers soon found ways to renew their role and influence. Most scholarly journals are either owned or produced by commercial publishers too (although their editors are invariably academics), leading to a growing debate about alternatives to commercial, for-profit publishing in human geography and other disciplines (Moss et al., 2002).

Meanwhile, within the discipline, analyses of citation patters within such journals were used:

1   to identify frequently referenced works that can be categorised as exemplars (Whitehand, 1985; Wrigley and Matthews, 1986; Bodman, 1991, 1992; Yeung, 2002);
2   to isolate research communities, groups of scholars who refer to each other's writings on a particular topic and so occupy a disciplinary matrix focused on particular exemplars (Gatrell, 1984a, 1984b);
3   to rate journals and chart their interdependence (Whitehand, 1984; Gatrell and Smith, 1984);

4   to rank departments in the USA and the UK in terms of publications produced, publications
     cited and peer evaluation, for example (Morrill, 1980; L. Jones, Lindsey and Coggeshall, 1982;
     B. L. Turner and Meyer, 1985, on the USA; on the UK, see Bentham, 1987; D. Smith, 1988b;
     Johnston, 2003a);
5   to conduct exploratory studies of particular subdisciplines and the main exemplars which have
     stimulated how individuals practise (M. Phillips and Unwin, 1985); and
6   to identify citation patterns across cultural/linguistic divides (Schuermans et al., 2010).

All describe aspects of the discipline's structure but tell us little of its processes; like spatial science
they provide valuable descriptive data and indicate connections and relations, but not necessarily
much understanding. Whitehand (2002, p. 518–19) admits that:

> Citations provide an entrée to a communication network whose understanding is an aspect of the
> solution of questions about the geography, history, psychology, demographics, sociology, eco-
> nomics, and politics of knowledge – but to understand that network we need to know more about
> the environment in which it functions, including the different kinds of intellectual traffic using
> the links in the network and the academic life courses of those that generate the traffic.

For that understanding, it is necessary not only to appreciate the contexts in which people
worked but also to realise that those contexts are not straightforwardly determinate. A change in a
country's economic, social and/or political trajectories will not necessarily bring forth a certain
reaction from geographers; these will be uneven, individuals may respond to their interpretations
of that change, and their responses may stimulate others to follow them. This is the most likely
reason for shifts at the level of the world view and, probably, the disciplinary matrix too, as
exemplified by the introduction of a radical/structuralist world view and by the shift within the
empiricist world view from regionalism to spatial science. But such shifts can stimulate counter-
reactions, as with the advancement of humanistic geography as a response to spatial science; those
promoting the latter cause were influenced both by currents of thinking outwith geography and by
their explorations of World Three. At the level of the exemplar, shifts are most likely to occur as a
response to events within the discipline, as Kuhn suggests; this occurred within the spatial-science
disciplinary matrix with the movement from normative to behavioural analyses, but the later move
towards welfare geography was strongly influenced by external factors.

No simple model can be applied to the changes within human geography to provide an
explanation of why the discipline has altered in the ways described here. Kuhn's work provides a
valuable vocabulary and organising framework within which to describe the communities, but
what has occurred reflects the perceptions and actions of individual human geographers. Like all
other aspects of society, geography is a discipline created by and for geographers, and is continually
fought over and recreated by them, at a variety of scales and in social, economic and political
contexts that vary over space and through time. The last point is crucial, for – as the structuration
approach stresses – people are socialised in particular contexts and they then create part of the
milieux within which others are socialised; these processes are presented by Bourdieu (1988,
1990) as the construction and evolution of 'fields' (see the exegeses by Bassett, 1996, and Painter,
2000) and by Purcell (2003) as 'islands of practice'. Those places (fields and islands) are not
isolated, and changes in one can influence changes in others. But, as illustrated here, there have
been substantial differences from one place to another within Anglo-American human geography
regarding how the discipline is practised, reflecting both the nature of those places and the
people in them; on the broader international scale, the differences are even greater (Johnston and
Claval, 1984).

What of Kuhn's model in the context of other frameworks of interpretation that emerged
subsequently? As the number of symposia and other events that occurred to celebrate the fiftieth

anniversary of the first publication of *The Structure of Scientific Revolutions* (1962) testifies, Kuhn's ideas still attract a great deal of attention (Sismondo, 2012) as do other evaluations of his impact (Machamer and Osbeck, 2003). Few students of the history of science and social science cleave to his full model of periods of normal science punctuated by revolutions but many identify considerable value in the sociological approach that he promoted and its emphasis on scientific communities (Dear, 2012), on paradigms as exemplars that 'structure the problems, the activities and the material environments of normal science ... rather than through rules that guide or dictate science' (Sismondo, 2012, p. 416; see also Fuller, 2000; Bird, 2003; Delanty, 2003). Such a sociological approach also characterises Latour's (1987) *Science in Action*, published 25 years after Kuhn, which used a different and broader definition of a community – as an actor network – that Michael Lynch (2012, pp. 452–3) compares with Kuhn's disciplinary matrices:

> It is not a giant step from Kuhn's disciplinary matrix to Latour's actor-network. Both are heterogeneous, and both fold social, semiotic, and material elements into an overall nexus that is at once technical and communal ... both stress that the linkages that integrate and hold together disciplinary matrices and actor-networks are not limited to semiotic links between words or citations. The outlines of research 'communities' may be indicated by common symbolic generalizations and citation networks, but the networks are integrated by material practices and infrastructures. Universities and research institutes provide sites and facilities, but as Latour so strongly emphasizes, a particular laboratory (or a field station, or some other concentration of people, instruments, techniques, specimens and archives) is the indispensable 'passage point' for specific research ventures.

What Kuhn's and especially Latour's sociological approach to scientific communities/networks makes clear, as Mialet (2012, p. 457) expresses it, is that scientists' work does not conform to universal norms or share common presuppositions and paradigms: science is 'something much messier, because its practitioners [are] human beings who argue, disagree, misunderstand, ignore, betray, talk past one another, and sometimes even agree, but who always, and in every eventuality, talk and write (a lot)'. This messiness has been illustrated by the varying human geographical practices over the last six decades. Attempts to define those communities and networks precisely (as in the use of citation analysis (Bodman, 2010)) find it difficult to specify their boundaries. Although there may be cores defined by their membership – for short periods at least – many researchers move between them, interact less intensively in some networks at certain periods than others and may spend considerable periods working largely independently of others (on Les Hepple as an excellent exemplar of this, see Johnston *et al.*, 2010). Occasionally, a new group may cohere and not only interact intensively but have a wider impact – drawing others into their orbit; and just occasionally a group make seek to define the discipline more generally (as illustrated by two groups of textbooks used – largely by what they exclude – to circumscribe the discipline (Johnston, 2006). But for most of the time individuals, perhaps with a small number of (and not always the same) collaborators, take their own ideas forward, publishing them as illustrations of how they are increasing knowledge in a particular way and context without either explicit preaching of 'this is what human geography should be' or seeking to denigrate other practices. Human geography for most of the post-1945 period has comprised a fluid structure of 'micro-worlds' in which individuals, in many cases temporarily but in a few permanently, find 'islands of relative stability' within which to undertake their own very particular (if not peculiar) practices (Pickering, 2012, p. 468); only occasionally do iconoclasts and change agents come along who try and establish 'island empires', knitting together various communities and networks into some larger, hopefully coherent whole – often for a political rather than a scientific purpose. For most of the time, geographers – like most other academics in the social sciences and humanities – practice in situations of 'situated messiness'; few regard themselves as permanently located within a well-defined 'community' but operate, much more actively at some times than others, within (sometimes minimally) overlapping networks – what Badley (2011a, 2011b, 2015) terms nomadic fabricators and weavers of stories and texts (see also Hall, 2014).

And the future?

> human geography is a profoundly trendy subject, in part because it still tends to import most of its ideas from outside the discipline. New generations of postgraduates therefore tend, rather more than in other and larger disciplines, to follow new trends. The discipline is therefore characterized by generational cycles which leave little room for much work outside the norm.
>
> (Thrift, 2000, p. 692)

In the early 1970s, several leading geographers were asked their views of the discipline's future (Chorley, 1973). Little of what they suggested is reflected in what happened; the contents of a similar book produced twelve years later (Johnston, 1985a) are very different. For J. P. Jones and Dixon (2004, p. 381) however (in an introduction to a set of essays on 'what next . . . [for] geographical theorizing in light of some thirty-plus years of scholarly encounters with critical social theory'), the answer is:

> 'Nothing. We are now repeating ourselves.' We have come to this position in spite of the fact that binaries, as socially and relationally constructed, would seem to open up new and dynamic fields of inquiry. Our view, however, is that the relations among oppositions have long since been stabilized by the gravitational force field of research practices that hold them together as systems, and that the form of binary analysis employed in most social theory (that is, construction 'deconstruction' reconstruction) has long since worked its way across the limited number of possibilities. It is difficult, in the words of the business world, 'to think outside the box'.

More recently, however, Gregory and Castree (2012, p. xxv–vi) judged that:

> the field of human geography is remarkably fecund. . . . In large part this positive judgement rests on a rethinking of what we mean by a discipline. An older meaning – inculcation into a canonical body of knowledge, a sort of academic holy writ, upon which one slowly builds to become a 'disciple' – has given way to a newer meaning: exposure to a variety of knowledges that share a family resemblance and which provide the means for critical, creative inventions not mere supplements to the existing stocks of knowledge.

Debates about how geography should be practised and taught and the extent and nature of any canon have recently resurfaced (Keighren *et al.*, 2012), reflecting not only internal divisions but also trends in the wider academic milieu and the societies to which human geographers belong. In the context however of turbulence, university restructurings and uncertainty over funding and priorities, Richard Powell (2011, p. 2) argues that:

> I do not think anyone should go so far as to identify a canon, but should students perhaps be required to read Mackinder, Isaiah Bowman or Ellen Semple or other geographical thinkers . . .? Without a sense of these tests, geographers become increasingly divorced from any sense of a community of practice or, indeed of a shared, contested enterprise.

In similar terms, John Agnew and David Livingstone (2011, p. 1) note that:

> Particularly before the creation of university departments or degree programs in geography, the label 'geographer' or 'geographical writer' was not a self-evident one for many who we might judge today as central figures in the 'geographical canon' (e.g. Immanuel Kant or Alexander von Humboldt). Geography as a discipline is therefore, very largely a retrospectively constituted tradition. Though there are good grounds for feeling uneasy and self-conscious about the invention of tradition, we cannot do without a tradition if we are to engage in common dialogue, avoid historical superficiality, think critically and creatively about the nature of the discipline, prepare the next generation of students, and ground commitments to our fields of study in rapidly changing institutional settings. Traditions are inescapable. . . . The issue is to ensure that they remain vital conversations between past and present . . .

The future contents and contexts of those conversations and other new directions cannot be readily predicted; one could suggest with some certainty, however, that human geography will not be characterised by mono-paradigm dominance in the next few decades. The intersections of generations, ideas, approaches from past and present and the recombinations that have long been present in the discipline should also make us cautious about predictions. David O'Sullivan (2014, p. 41) recently argued that 'It makes sense to think of the quantitative revolution in geography as an unfinished revolution.' But we might add that human geography comprises a series of intersecting partial revolutions that do not readily lend themselves to straightforward explanatory narrative. Three decades earlier, in the introduction to an autobiographical essay, David Smith (1984, p. 118) wrote that:

> my own professional activities seem to have been a continual struggle to come to terms with (or keep up with) the rapidly shifting focus of human geography. The struggle arises in large measure from the difficulty of breaking free from one's own intellectual heritage. . . . If my own struggles represent anything more than one half-life experience, it may well be the theme of the geographer or social scientist as creature of his or her times. . . . If anything is to be learned from the instant replay of such recent events . . . it is that 'scientific' advance is not conducted in a social vacuum but as an integral part of human history, within which the element of chance arising from individual personality and creativity plays an important part. So let us proceed with the recollection of one of the random variables.

We are all random variables, it seems. Robson (1984, p. 104) used a similar turn of phrase: 'it is clear how small a part in my own development seems to have been played by clearly guided aims and how much has been contributed by the collage of rather random influences and serendipitous events to which I both responded and contributed'. Hence our 'individual' and collective projects and life-paths (our 'geographies') can be appreciated (perhaps yielding 'joy' as Kearn et al., 2014, see it) and set in context, but no more. We are therefore often uncertain where we, as individuals, are going, let alone exactly where geography as a set of linked yet frequently anarchic communities is heading. Yet we are making the future of geography as we practise it, just as we have been remaking its past by writing this book.

References

Aalbers, M. B. 2004. Creative destruction through the Anglo-American hegemony: A non-Anglo-American view on publications, referees and language. *Area* 36, 319–22.

Abler, R. F. 1971. Distance, intercommunications, and geography. *Proceedings, Association of American Geographers* 3, 1–4.

Abler, R. F. 1993a. Everything in its place: GPS, GIS, and geography in the 1990s. *Professional Geographer* 45, 131–9.

Abler, R. F. 1993b. Personal perspectives on the fiftieth anniversary of the founding of the ASPG. In E. W. Miller (ed.) *The American Society for Professional Geographers: Papers presented on the occasion of the fiftieth anniversary of its founding.* Washington, DC: Occasional Publication of the Association of American Geographers.

Abler, R. F., Adams, J. S. and Gould, P. R. 1971. *Spatial Organization: The geographer's view of the world.* Englewood Cliffs, NJ: Prentice-Hall.

Abler, R. F., Marcus, M. G. and Olson, J. M. 1992a. Preface. In R. F. Abler, M. G. Marcus and J. M. Olson (eds) *Geography's Inner Worlds: Pervasive themes in contemporary American geography.* New Brunswick, NJ: Rutgers University Press, xv–xx.

Abler, R. F., Marcus, M. G. and Olson, J. M. 1992b. Contemporary American geography. In R. F. Abler, M. G. Marcus and J. M. Olson (eds) *Geography's Inner Worlds: Pervasive themes in contemporary American geography.* New Brunswick, NJ: Rutgers University Press, 1–8.

Abler, R. F., Marcus, M. G. and Olson, J. M. 1992c. Afterword. In R. F. Abler, M. G. Marcus and J. M. Olson (eds) *Geography's Inner Worlds: Pervasive themes in contemporary American geography.* New Brunswick, NJ: Rutgers University Press, 391–402.

Abrahamsson, C. and Gren, M. (eds) 2012. *GO: On the geographies of Gunnar Olsson.* Farnham: Ashgate.

Abrahamson, H. S. 2010. *National Geographic: Behind America's lens on the world*, 2nd edn. Bloomington, IN: iUniverse.

Ackerman, E. A. 1945. Geographic training, wartime research, and immediate professional objectives. *Annals of the Association of American Geographers* 35, 121–43.

Ackerman, E. A. 1958. *Geography as a Fundamental Research Discipline.* Chicago: University of Chicago, Department of Geography Research Paper 53.

Ackerman, E. A. 1963. Where is a research frontier? *Annals of the Association of American Geographers* 53, 429–40.

Adams, J. S. 1969. Directional bias in intra-urban migration. *Economic Geography* 45, 302–23.

Adams, J. S. 1972. The geography of riots and civil disorder in the 1960s. *Economic Geography* 48, 24–42.

Adams, J. S. (ed.) 1976. *Urban Policymaking and Metropolitan Dynamics: A comparative geographical analysis.* Cambridge, MA: Ballinger.

Adams, J. S. 1995. *Risk.* London: Routledge.

Adams, J. S. 2002. The quantitative revolution in urban geography. *Urban Geography* 22, 530–9.

Adams, P. C. 1995. A reconsideration of personal boundaries in space-time. *Annals of the Association of American Geographers* 85, 267–85.

Adams, P. C., Hoelscher, S. and Till, K. E. 2001. Place in context: Rethinking humanistic geographies. In P. C. Adams, S. Hoelscher and K. E. Till (eds) *Textures of Place: Exploring humanist geographies.* Minneapolis, MN: University of Minnesota Press, xiii–xxiii.

Adler, S. and Brenner, J. 1992. Gender and space: Lesbians and gay men in the city. *International Journal of Urban and Regional Research* 16, 24–34.

Agnew, J. A. 1984. Place and political behaviour: The geography of Scottish nationalism. *Political Geography Quarterly* 3, 191–202.

Agnew, J. A. 1987a. *The United States in the World-Economy: A regional geography*. Cambridge: Cambridge University Press.

Agnew, J. A. 1987b. *Place and Politics: The geographical mediation of state and society*. Boston, MA: Allen & Unwin.

Agnew, J. A. 1989. The devaluation of place in social science. In J. A. Agnew and J. S. Duncan (eds) *The Power of Place*. Boston, MA: Unwin Hyman, 9–29.

Agnew, J. A. 1990. Sameness and difference: Hartshorne's *The Nature of Geography* and geography as areal variation. In J. N. Entrikin and S. D. Brunn (eds) *Reflections on Richard Hartshorne's* The Nature of Geography. Washington, DC: Occasional Publication of the *Association of American Geographers*, 121–40.

Agnew, J. A. 1993. Representing space: Space, scale and culture in social science. In J. S. Duncan and D. Ley (eds) *Place/Culture/Representation*. London: Routledge, 251–71.

Agnew, J. A. 1994. The territorial trap: The geographical assumption of international relations theory. Review of International Political Economy 1, 53–80.

Agnew, J. A. 1997. Commemoration and criticism: Fifty years after the publication of Harris and Ullman's 'The nature of cities'. *Urban Geography* 18, 4–6.

Agnew, J. A. 2005. *Hegemony: The new shape of global power*. Philadelphia, PA: Temple University Press.

Agnew, J. A. 2009. *Globalization and Sovereignty*. Langham, MD: Rowman & Littlefield.

Agnew, J. A. 2014. By words alone shall we know: Is the history of ideas enough to understand the world to which our concepts refer? *Dialogues in Human Geography* 4, 311–19.

Agnew, J. A. and Duncan, J. S. 1981. The transfer of ideas into Anglo-American human geography. *Progress in Human Geography* 5, 42–57.

Agnew, J. A. and Duncan, J. S. 1989. Introduction. In J. A. Agnew and J. S. Duncan (eds) *The Power of Place*. Boston, MA: Unwin Hyman, 1–8.

Agnew, J. A. and Livingstone, D. N. 2011. Introduction. In J. A. Agnew and D. N. Livingstone (eds) *The Sage Handbook of Geographical Knowledge*. London, Thousand Oaks, New Delhi and Singapore: Sage, 1–17.

Agnew, J. A., Mitchell, K. and Toal, G. (eds) 2003. *A Companion to Political Geography*. Oxford: Blackwell.

Aitken, S. C. 1991. A transactional geography of the image-event: The films of Scottish director, Bill Forsyth. *Transactions of the Institute of British Geographers* NS16, 105–18.

Aitken, S. C. 2001. *Geographies of Young People: The morally contested spaces of identity*. London and New York: Routledge.

Aitken, S. C. 2013. Book review: GO: On the geographies of Gunnar Olsson. Edited by Christian Abrahamsson and Martin Gren. *Geographical Review* 103, 580–3.

Aitken, S. C., Cutter, S. L., Foote, K. E. and Sell, J. S. 1989. Environmental perception and behavioral geography. In G. L. Gaile and C. J. Willmott (eds) *Geography in America*. Columbus, OH: Merrill, 218–38.

Akatiff, C. 1974. The march on the Pentagon. *Annals of the Association of American Geographers* 64, 26–33.

Alexander, D. 1979. Catastrophic misconception? *Area* 11, 228–30.

Alexander, J. W. and Zahorchak, G. A. 1943. Population-density maps of the United States: Techniques and patterns. *Geographical Review* 33, 457–66.

Alker, H. R. 1969. A typology of ecological fallacies. In M. Dogan and S. Rokkan (eds) *Quantitative Ecological Analysis in the Social Sciences*. Cambridge, MA: MIT Press, 69–86.

Allen, C. D. 2011. On actor-network theory and landscape. *Area* 43, 274–80.

Allen, J. 2003. *Lost Geographies of Power*. Oxford: Blackwell.

Allen, J., Massey, D. and Cochrane, A. 1998. *Rethinking the Region*. London: Routledge.

Althusser, L. and Balibar, E. 1970. *Reading 'Capital'*. London: New Left Books.

Amedeo, D. and Golledge, R. G. 1975. *An Introduction to Scientific Reasoning in Geography*. New York: John Wiley.

American Association of Geographers (AAG) 1946. Lessons from the wartime experience for improving graduate training for geographic research. *Annals of the Association of American Geographers* 36, 195–214.

Amin, A. (ed.) 1994. *Postfordism: A reader*. Oxford: Blackwell.

Amin, A. and Robbins, K. 1990. The reemergence of regional economies: The mythical geography of flexible accumulation. *Environment and Planning D: Society and Space* 8, 7–34.

Amin, A. and Thrift, N. J. 2000. What kind of economic theory for what kind of economic geography? *Antipode: A Radical Journal of Geography* 32, 4–9.

Amin, A. and Thrift, N. J. 2002. *Cities: Reimagining the urban*. Cambridge: Polity Press.

Amin, A. and Thrift, N. J. 2005. Just the future. *Antipode: A Radical Journal of Geography* 37, 220–38.

Amin, A. and Thrift, N. J. 2013. *Arts of the Political: New openings for the Left*. Durham, NC: Duke University Press.

Amin, A., Massey, D. and Thrift, N. J. 2000. *Cities for the Many Not the Few*. Bristol: Policy Press.

Anderson, B. 2006. Becoming and being hopeful: Towards a theory of affect. *Environment and Planning D: Society and Space* 24, 733–52.

Anderson, B. and Harrison, P. 2006. Questioning affect and emotion. *Area* 38, 333–5.

Anderson, B. and Harrison, P. 2010. The promise of non-representational theories. In B. Anderson and P. Harrison (eds) *Taking Place: Non-representational theories and geographies*. Farnham and Burlington, VT: Ashgate, 1–34.

Anderson, J. 1973. Ideology in geography: An introduction. *Antipode: A Radical Journal of Geography* 5(3), 1–6.

Anderson, J., Duncan, S. and Hudson, R. 1983. *Redundant Spaces in Cities and Regions?: Studies in industrial decline and social change* (special publication series, Institute of British Geographers No. 15). London: Academic Press.

Anderson, K. 1995. Culture and nature at the Adelaide Zoo: At the frontiers of 'human' geography. *Transactions of the Institute of British Geographers* NS20, 275–94.

Anderson, K., Domosh, M. Pile, S. and Thrift, N. (eds) 2003. *The Handbook of Cultural Geography*. London, Thousand Oaks, CA and New Delhi: Sage.

Anon., 1968. A preliminary contribution to the geographical analysis of a Pooh-scape. *IBG Newsletter* 6, 54–63.

Anselin, L. 1988. *Spatial Econometrics: Methods and models*. Dordrecht: Kluwer.

Anselin, L. 1995. Local indicators of spatial association – LISA. *Geographical Analysis* 27, 93–115.

Anselin, L. 1998. Exploratory spatial data analysis in a geocomputational environment. In P. A. Longley, S. M. Brooks, R. MacDonnell and B. Macmillan (eds) *Geocomputation: A primer*. Chichester: John Wiley, 77–94.

Anselin, L. 2012. From SpaceStat to CyberGIS: Twenty years of spatial data analysis software. *International Regional Science Review* 35, 131–57.

Anselin, L. and Rey, S. (eds) 2010. *Perspectives on Spatial Data Analysis*. Berlin: Springer-Verlag.

Anselin, L. and Rey, S. J. 2012. Spatial econometrics in an age of CyberGIScience. *International Journal of Geographical Information Science* 26, 2211–26.

Anselin, L., Florax, R. and Rey, S. (eds) 2004. *Advances in Spatial Econometrics: Methodology, tools and applications*. Berlin: Springer-Verlag.

Applebaum, W. 1954. Marketing geography. In P. E. James and C. F. Jones (eds) *American Geography: Inventory and prospect*. Syracuse: Syracuse University Press, 245–51.

Appleton, J. 1975. *The Experience of Landscape*. London: John Wiley.

Appleton, J. 1994. *How I Made the World: Shaping a view of landscape*. Hull: University of Hull Press.

Archer, J. C. and Taylor. P. J. 1981. *Section and Party*. Chichester: John Wiley.

Armstrong, M. P. 1993. On automated geography. *Professional Geographer* 45, 440–2.

Armstrong, M. P. 2000. Geography and computational science. *Annals of the Association of American Geographers* 90, 146–56.

Armstrong, P. 1999. Charles Darwin's image of the world: The influence of Alexander von Humboldt on the Victorian naturalist. In A. Buttimer, S. D. Brunn and U. Wardenga (eds) *Text and Image: Social construction of regional knowledges*. Leipzig: Leibniz Institute for Regional Geography, *Beiträge zur regionalen Geographie* 49, 46–53.

Armstrong, P. H. and Martin, G. J. (eds) 2000. *Geographers: Biobibliographical studies*. *Volume 20*. London: Continuum.

Ash, J. and Simpson, P. 2014. Geography and post-phenomenology. *Progress in Human Geography*. doi: 10.1177/0309132514544806

Asheim, B. T. 1990. How to confuse rather than guide students: A review of Holt-Jensen's *Geography – History and Concepts*. *Progress in Human Geography* 14, 281–92.

Atkinson, D. and Dodds, K. 2000. Introduction to geopolitical traditions: A century of geopolitical thought. In K. Dodds and D. Atkinson (eds) *Geopolitical Traditions: A century of geopolitical thought*. London and New York: Routledge, 1–24.

Attoh, K. 2014. Imagining a 'cultural turn' in transportation geography. *Journal of Cultural Geography* 31, 141–51.

Bachmann, V. and Belina, B. 2012. Crisis, critique and the 6th International Conference of Critical Geography. *Antipode: A Radical Journal of Geography* 44, 555–9.

Backhaus, G. 2009. Phenomenology/phenomenological geography. In R. Kitchin and N. Thrift (eds) *International Encyclopedia of Human Geography*. Amsterdam: Elsevier, 137–44.

Badcock, B. A. 1970. Central-place evolution and network development in south Auckland, 1840–1968: A systems analytic approach. *New Zealand Geographer* 26, 109–35.

Badcock, B. A. 1984. *Unfairly Structured Cities*. Oxford: Basil Blackwell.

Badley, G. F. 2011a: Academic scribbling: A frivolous approach? *Teaching in Higher Education* 16, 255–66.

Badley, G. F. 2011b: Case notes for the impeachment of an ability traitor: A textor's inquiry. *Qualitative Inquiry* 17, 483–92.

Badley, G. F. 2015. Playful and serious adventures in academic writing. *Qualitative Inquiry* 21. doi: 10.1177/1077800415569785

Bahrenberg, G., Fischer, M. M. and Nijkamp, P. (eds) 1984. *Recent Developments in Spatial Data Analysis: Methodology, measurement, models.* Aldershot: Gower Press.

Bailey, T. C. and Gattrell, A. C. 1995. *Interactive Spatial Data Analysis.* London: Longman.

Baker, A. R. H. 1972. Rethinking historical geography. In A. R. H. Baker (ed.) *Progress in Historical Geography.* Newton Abbott: David & Charles, 11–28.

Baker, A. R. H. 1979. Historical geography: A new beginning? *Progress in Human Geography* 3, 560–70.

Baker, A. R. H. 1981. An historico-geographical perspective on time and space and on period and place. *Progress in Human Geography* 5, 439–43.

Baker, A. R. H. 1984. Reflections on the relations of historical geography and the *Annales* school of history. In A. R. H. Baker and D. Gregory (eds) *Explorations in Historical Geography: Interpretive essays.* Cambridge: Cambridge University Press, 1–27.

Baker, A. R. H. 2003. *Geography and History: Bridging the Divide.* Cambridge: Cambridge University Press.

Baker, A. R. H. 2011. Robert Arthur Donkin, 1928–2006. *Proceedings of the British Academy* 172, 115–39.

Baker, A. R. H. and Gregory, D. 1984. Some *terrae incognitae* in historical geography: An exploratory discussion. In A. R. H. Baker and D. Gregory (eds) *Explorations in Historical Geography: Interpretive essays.* Cambridge: Cambridge University Press, 180–94.

Balchin, W. G. V. 1987. United Kingdom geographers in the Second World War. *Geographical Journal* 153, 159–180.

Balchin, W. G. V. 1993. *The Geographical Association: The first hundred years, 1893–1993.* Sheffield: Geographical Association.

Balinski, M., Johston, R. J., McLean, I. and Young, H. P. 2010. *Drawing a New Constituency Map for the United Kingdom: The Parliamentary Voting System and Constituencies Bill 2010.* London: The British Academy.

Ball, M. 1987. Harvey's Marxism. *Environment and Planning D: Society and Space* 5, 393–4.

Ballabon, M. B. 1957. Putting the economic into economic geography. *Economic Geography* 33, 217–23.

Ballas, D., Clarke, G. P. and Turton, I. 2002. A spatial microsimulation model for social policy micro-spatial analysis. In B. Boots, A. Okabe and R. Thomas (eds) *Modelling Geographical Systems: Statistical and computational applications.* Dordrecht: Kluwer, 143–68.

Banks, M. and MacKian, S. 2000. Jump in! The water's warm: A comment on Peck's 'grey geography'. *Transactions of the Institute of British Geographers* NS25, 249–54.

Banski, J. and Ferenc, M. 2013. 'International' or 'Anglo-American' journals of geography? *Geoforum* 45, 285–95.

Barnes, B. 1974. *Scientific Knowledge and Sociological Theory.* London: Routledge & Kegan Paul.

Barnes, B. 1982. *T. S. Kuhn and Social Science.* London: Macmillan.

Barnes, T. J. 1985. Theories of international trade and theories of value. *Environment and Planning A* 17, 729–46.

Barnes, T. J. 1988. Rationality and relativism in economic geography: An interpretive review of the homo economicus assumption. *Progress in Human Geography* 12, 473–96.

Barnes, T. J. 1989a. Place, space and theories of economic value: Contextualism and essentialism in economic geography. *Transactions of the Institute of British Geographers* NS14, 299–316.

Barnes, T. J. 1989b. Structure and agency in economic geography and theories of economic value. In A. Kobayashi and S. Mackenzie (eds) *Remaking Human Geography.* Boston, MA: Unwin Hyman, 134–48.

Barnes, T. J. 1996. *Logics of Dislocation: Models, metaphors and meaning of economic space.* New York: Guilford Press.

Barnes, T. J. 2000. Inventing Anglo-American economic geography, 1889–1960. In E. Sheppard and T. J. Barnes (eds) *A Companion to Economic Geography.* Oxford: Blackwell, 11–26.

Barnes, T. J. 2001a. 'In the beginning was economic geography' – a science studies approach to disciplinary history. *Progress in Human Geography* 25, 521–544.

Barnes, T. J. 2001b. Lives lived and lives told: Biographies of geography's quantitative revolution. *Environment and Planning D: Society and Space* 19, 409–430.

Barnes, T. J. 2001c Retheorizing economic geography: From the quantitative revolution to the 'cultural turn'. *Annals of the Association of American Geographers* 91, 546–65.

Barnes, T. J. 2002. Critical notes on economic geography from an aging radical. Or radical notes on economic geography from a critical age. *ACME: An International E-Journal for Critical Geographies* 1, 8–14.

Barnes, T. J. 2003a. What's wrong with American regional science? A view from science studies. *Canadian Journal of Regional Science* 26, 3–26.

Barnes, T. J. 2003b. The place of locational analysis: A selective and interpretive history. *Progress in Human Geography* 27, 69–95.

Barnes, T. J. 2004a. The rise (and decline) of American regional science: Lessons for the new economic geography? *Journal of Economic Geography* 4, 107–29.

Barnes, T. J. 2004b. Placing ideas: Genius loci, heterotopias and geography's quantitative revolution. *Progress in Human Geography* 28, 565–95.

Barnes, T. J. 2004c. A paper related to everything but more related to local things. *Annals of the Association of American Geographers* 94, 278–83.

Barnes, T. J. 2004d. 'The background of our lives': David Harvey's The limits to capital. *Antipode: A Radical Journal of Geography* 36, 407–13.

Barnes, T. J. 2006a. Geographical intelligence: American geographers and research and analysis in the Office of Strategic Services 1941–1945. *Journal of Historical Geography* 32, 149–68.

Barnes, T. J. 2006b. Between deduction and dialectics: David Harvey on knowledge. In N. Castree and D. Gregory (eds) *David Harvey: A critical reader*. Oxford: Blackwell, 26–46.

Barnes, T. J. 2008a. Stuck in a mess (again): A response to Johnston, Fairbrother, Hayes, Hoare and Jones. *Geoforum* 30, 1807–10.

Barnes, T. J. 2008b. Geography's underworld: The military-industrial complex, mathematical modelling and the quantitative revolution. *Geoforum* 39, 3–16.

Barnes, T. J. 2014. Geo-historiographies. In R. Lee, N. Castree, R. Kitchin et al. (eds) *The SAGE Handbook of Human Geography*. London, Thousand Oaks, CA, New Delhi, Singapore: Sage, 202–28.

Barnes, T. J. and Abrahamsson, C. 2015. Tangled complicities and moral struggles: The Haushofers, father and son, and the spaces of Nazi geopolitics. *Journal of Historical Geography* 47, 64–73.

Barnes, T. J. and Duncan, J. S. 1992. Introduction: Writing worlds. In T. J. Barnes and J. S. Duncan (eds) *Writing Worlds: Discourse, text and metaphor in the representation of landscape*. London: Routledge, 1–17.

Barnes, T. J. and Farish, M. 2006. Between regions: Science, militarism, and American geography from world war to cold war. *Annals of the Association of American Geographers* 96, 807–26.

Barnes, T. J. and Gregory, D. (eds) 1997. *Reading Human Geography: The poetics and politics of enquiry*. London: Arnold.

Barnes, T. J. and Minca, C. 2013. Nazi spatial theory: The dark geographies of Carl Schmitt and Walter Christaller. *Annals of the Association of American Geographers* 103, 669–87.

Barnett, C. 1995. Awakening the dead: Who needs the history of geography? *Transactions of the Institute of British Geographers* NS20, 417–19.

Barnett, C. 1998a. The cultural turn: Fashion or progress in human geography. *Antipode: A Radical Journal of Geography* 30, 379–94.

Barnett, C. 1998b. Cultural twist and turns. *Environment and Planning D: Society and Space* 16, 631–4.

Barnett, C. 1999. Deconstructing context: Exposing Derrida. *Transactions of the Institute of British Geographers* NS24, 277–93.

Barnett, C. 2004. A critique of the cultural turn. In J. S. Duncan, N. C. Johnson and R. H. Schein (eds) *A Companion to Cultural Geography*. Malden, MA and Oxford: Blackwell, 38–48.

Barnett, C. 2008. Political affects in public space: Normative blind-spots in nonrepresentational ontologies. *Transactions of the Institute of British Geographers* NS33, 186–200.

Barnett, C. 2009. Cultural turn. In D. Gregory, R. Johnston, G. Pratt, M. J. Watts and S. Whatmore (eds) *The Dictionary of Human Geography*, 5th edn. Malden, MA and Oxford: Wiley-Blackwell, 134–5.

Barnett, C. and Low, M. 1996. Speculating on theory: Towards a political economy of academic publishing. *Area* 28, 13–24.

Barrows, H. H. 1923. Geography as human ecology. *Annals of the Association of American Geographers* 13, 1–14.

Barton, R. 2003. 'Men of science': Language, identity and professionalization in the mid-Victorian scientific community. *History of Science* 41, 73–119.

Bassett, K. 1996. Postmodernism and the crisis of the intellectual: Reflections on reflexivity, universities and the scientific field. *Environment and Planning D: Society and Space* 14, 507–27.

Bassett, K. 1999. Is there progress in human geography? The problem of progress in the light of recent work in the philosophy and sociology of science. *Progress in Human Geography* 23, 27–47.

Bassett, K. and Short, J. R. 1980. *Housing and Residential structure: Alternative approaches*. London: Routledge & Kegan Paul.

Bastow, S., Dunleavy, P. and Tinkler, J. 2014. *The Impact of the Social Sciences: How academics and their research make a difference*. London: Sage.

Batey, P. and Brown, P. 1995. From human ecology to customer targeting: The evolution of geodemographics. In P. Longley and G. Clarke (eds) *GIS for Business and Service Planning*. Cambridge: GeoInformation International, 73–103.

Batty, M. 1976. *Urban Modelling: Algorithms, calibrations, predictions*. London: Cambridge University Press.

Batty, M. 1978. Urban models in the planning process. In D. T. Herbert and R. J. Johnston (eds) *Geography and the Urban Environment*, Vol. 1. Chichester: John Wiley, 63–134.

Batty, M. 1989. Urban modelling and planning: Reflections, retrodictions and prescriptions. In B. Macmillan (ed.) *Remodelling Geography*. Oxford: Basil Blackwell, 147–69.

Batty, M. 2007. *Cities and Complexity: Understanding cities with cellular automata, agent-based models, and fractals*. Cambridge, MA: MIT Press.

Batty, M. 2013. *The New Science of Cities*. Cambridge, MA: MIT Press.

Batty, M. and Longley, P. 1994. *Fractal Cities: A geometry of form and function*. London: Academic Press.

Beaumont, J. R. 1987. Quantitative methods in the real world: A consultant's view of practice. *Environment and Planning A* 19, 1441–8.

Beaumont, J. R. and Gatrell, A. C. 1982. An introduction to Q-analysis. CATMOG 34, Norwich: Geo Books.

Beauregard, R. A. 1988. In the absence of practice: The locality research debate. *Antipode: A Radical Journal of Geography* 20, 52–9.

Beauregard, R. A. 1999. Break dancing on Santa Monica Boulevard. *Urban Geography* 20, 396–399.

Beaver, S.H. 1962. The Le Play Society and fieldwork. *Geography* 47, 225–40.

Beaverstock, J., Smith, R. G. and Taylor, P. J. 2000. World city network: A new metageography of the future? *Annals of the Association of American Geographers* 90, 121–34.

Becher, T. and Trowler, P. R. 2001. *Academic Tribes and Territories*, 2nd edn. Buckingham: Open University Press.

Beckerman, W. 1995. *Small is Stupid: Blowing the whistle on the greens*. London: Duckworth.

Beckinsale, R. P. 1976. The international influence of William Morris Davis. *Geographical Review* 66, 448–66.

Beckinsale, R. P. 1997. Richard J. Chorley: A reformer with a cause. In D. R. Stoddart (ed.) *Process and Form in Geomorphology*. London: Routledge, 3–14.

Bell, C. and Newby, H. 1976. Community, communion, class and community action: The social sources of new urban politics. In D. T. Herbert and R. J. Johnston (eds) *Social Areas in Cities. Volume 2: Spatial Perspectives on Problems and Policies*. Chichester: John Wiley, 189–208.

Bell, D. 1973. *The Coming of Post-Industrial Society*. New York: Basic Books.

Bell, D. J. 1991. Insignificant others: Lesbian and gay geographies. *Area* 23, 323–29.

Bell, D. J. and Valentine, G. (eds) 1995. *Mapping Desire: Geographies of sexualities*. London and New York: Routledge.

Bell, D. J., Binne, J., Cream, J. and Valentine, G. 1994. All hyped up and no place to go. *Gender, Place and Culture* 1, 31–47.

Bell, M. and McEwan, C. 1996. The admission of women fellows to the Royal Geographical Society, 1892–1914: The controversy and the outcome. *Geographical Journal* 162, 295–312.

Bell, M., Butlin, R. J. and Heffernan, M. (eds) 1995. *Geography and Imperialism: 1820–1940*. Manchester: Manchester University Press.

Bennett, R. J. 1974. Process identification for time-series modelling in urban and regional planning. *Regional Studies* 8, 157–74.

Bennett, R. J. 1975. Dynamic systems modelling of the Northwest region: 1. Spatio-temporal representation and identification. 2. Estimation of the spatio-temporal policy model. 3. Adaptive parameter policy model. 4. Adaptive spatio-temporal forecasts. *Environment and Planning A* 7, 525–38, 539–66, 617–36, 887–98.

Bennett, R. J. 1978a. Forecasting in urban and regional planning closed loops: The examples of road and air traffic forecasts. *Environment and Planning A* 10, 145–62.

Bennett, R. J. 1978b. *Spatial Time Series: Analysis, forecasting and control*. London: Pion.

Bennett, R. J. 1979. Space-time models and urban geographical research. In D. T. Herbert and R. J. Johnston (eds) *Geography and the Urban Environment*, Vol. 2. London: John Wiley, 27–58.

Bennett, R. J. 1981a. A hierarchical control solution to allocation of the British Rate Support Grant. *Geographical Analysis* 13, 300–14.

Bennett, R. J. (ed.) 1981b. *European Progress in Spatial Analysis*. London: Pion.

Bennett, R. J. 1981c. Quantitative and theoretical geography in Western Europe. In R. J. Bennett (ed.) *European Progress in Spatial Analysis*. London: Pion, 1–32.

Bennett, R. J. 1981a. Quantitative geography and public policy. In N. Wrigley and R. J. Bennett (eds) *Quantitative Geography*. London: Routledge & Kegan Paul, 387–96.

Bennett, R. J. 1982. Geography, relevance and the role of the Institute. *Area* 14, 69–71.

Bennett, R. J. 1983. Individual and territorial equity. *Geographical Analysis* 15, 50–87.

Bennett, R. J. 1985a. Quantification and relevance. In R. J. Johnston (ed.) *The Future of Geography*. London: Methuen, 211–24.

Bennett, R. J. 1985b. A reappraisal of the role of spatial science and statistical inference in geography in Britain. *L'Espace Géographique* 14, 23–8.

Bennett, R. J. 1989a. Demography and budgetary influence on the geography of the poll tax: Alarm or false alarm? *Transactions of the Institute of British Geographers* NS14, 400–17.

Bennett, R. J. 1989b. Whither models and geography in a post-welfarist world? In B. Macmillan (ed.) *Remodelling Geography*. Oxford: Basil Blackwell, 273–90.

Bennett, R. J. and Chorley, R. J. 1978. *Environmental Systems: Philosophy, analysis and control*. London: Methuen.

Bennett, R. J. and Haining, R. P. 1985. Spatial structure and spatial interaction: Modelling approaches to the statistical analysis of geographical data. *Journal of the Royal Statistical Society A* 148, 1–36.

Bennett, R. J. and Thornes, J. B. 1988. Geography in the United Kingdom, 1984–1988. *Geographical Journal* 154, 23–48.

Bennett, R. J. and Wilson, A. G. 2003. Geography applied. In R. J. Johnston and M. Williams (eds) *A Century of British Geography*. Oxford: Oxford University Press for the British Academy, 463–501.

Bennett, R. J. and Wrigley, N. 1981. Introduction. In N. Wrigley and R. J. Bennett (eds) *Quantitative Geography*. London: Routledge & Kegan Paul, 3–11.

Bennett, R. J., Haining, R. P. and Wilson, A. G. 1985. Spatial structure, spatial interaction and their integration: A review of alternative models. *Environment and Planning A* 17, 625–46.

Bentham, G. 1987. An evaluation of the UGC's rating of the research of British university geography departments. *Area* 19, 147–54.

Berdoulay, V. 1981. The contextual approach. In D. R. Stoddart (ed.) *Geography, Ideology and Social Concern*. Oxford: Blackwell, 8–16.

Berdoulay, V. 2011. German precursors and French challengers. In J. A. Agnew and J. S. Duncan (eds) *The Wiley-Blackwell Companion to Human Geography*. Chichester: Wiley-Blackwell, 73–88.

Berg, L. D. 1993. Between modernism and postmodernism. *Progress in Human Geography* 17, 490–507.

Berg, L. D. 2003. (Some) spaces of critical geography. In A. Rogers and H. R. Viles (eds) *The Students' Companion to Geography*, 2nd edn. Oxford: Blackwell, 305–11.

Berg, L. D. 2012. Geographies of identity I: geography – (neo)liberalism – white supremacy. *Progress in Human Geography* 36, 508–17.

Berg, L. D. and Kearns, R. A. 1998. America Unlimited. *Environment and Planning D: Society and Space* 16, 128–32.

Berg, L. D. and Longhurst, R. 2003. Placing masculinities and geography. *Gender, Place and Culture* 10, 351–60.

Berman, M. 1974. Sex discrimination in geography: The case of Ellen Churchill Semple. *Professional Geographer* 26, 8–11.

Berry, B. J. L. 1958. A critique of contemporary planning for business centers. *Land Economics* 25, 306–12.

Berry, B. J. L. 1959a. Ribbon developments in the urban business pattern. *Annals of the Association of American Geographers* 49, 145–55.

Berry, B. J. L. 1959b. Further comments 'geographic' and 'economic' economic geography. *Professional Geographer* 11(1), 11–12.

Berry, B. J. L. 1964a. Approaches to regional analysis: A synthesis. *Annals of the Association of American Geographers* 54, 2–11.

Berry, B. J. L. 1964b. Cities as systems within systems of cities. *Papers, Regional Science Association* 13, 147–63.

Berry, B. J. L. 1965. Research frontiers in urban geography. In P. M. Hauser and L. F. Schnore (eds) *The Study of Urbanization*. New York: John Wiley, 403–30.

Berry, B. J. L. 1966. *Essays on Commodity Flows and the Spatial Structure of the Indian Economy*. Chicago: University of Chicago, Department of Geography, Research Paper 111.

Berry, B. J. L. 1967. *The Geography of Market Centers and Retail Distribution*. Englewood Cliffs, NJ: Prentice-Hall.

Berry, B. J. L. 1968. A synthesis of formal and functional regions using a general field theory of spatial behavior. In B. J. L. Berry and D. F. Marble (eds) *Spatial Analysis*. Englewood Cliffs, NJ: Prentice-Hall, 419–28.

Berry, B. J. L. 1969. Review of B. M. Russett *International Regions and the International System*. *Geographical Review* 59, 450–1.

Berry, B. J. L. (ed.) 1971. Comparative factorial ecology. *Economic Geography* 47, 209–367.

Berry, B. J. L (with others) 1974. *Land Use, Urban Form and Environmental Quality*. Chicago: Department of Geography, Research Paper 155, University of Chicago.

Berry, B. J. L. 1972a. 'Revolutionary and counter-revolutionary theory in geography' – a ghetto commentary. *Antipode: A Radical Journal of Geography* 4(2), 31–3.

Berry, B. J. L. 1972b. More on relevance and policy analysis. *Area* 4, 77–80.

Berry, B. J. L. 1973a. A paradigm for modern geography. In R. J. Chorley (ed.) *Directions in Geography*. London: Methuen, 3–22.

Berry, B. J. L. 1973b. *The Human Consequences of Urbanization*. London: Macmillan.

Berry, B. J. L. 1974a. Review of H. M. Rose (ed.) *Perspectives in Geography 2: Geography of the ghetto, perceptions, problems and alternatives. Annals of the Association of American Geographers* 64, 342–5.

Berry, B. J. L. 1974b. Review of David Harvey, *Social Justice and the City. Antipode: A Radical Journal of Geography* 6(2), 142–9.

Berry, B. J. L. 1978a. Introduction: A Kuhnian perspective. In B. J. L. Berry (ed.) *The Nature of Change in Geographical Ideas*. DeKalb, IL: Northern Illinois University Press, vii–x.

Berry, B. J. L. 1978b. Geographical theories of social change. In B. J. L. Berry (ed.) *The Nature of Change in Geographical Ideas*. DeKalb, IL: Northern Illinois University Press, 17–36.

Berry, B. J. L. 1993. Geography's quantitative revolution: Initial conditions, 1954–1960: A personal memoir. *Urban Geography* 14, 434–41.

Berry, B. J. L. 1995. Whither regional science? *International Regional Science Review* 17, 297–306.

Berry, B. J. L. 1998. On consilience. *Urban Geography* 19, 95–97.

Berry, B. J. L. 1999. Beyond postmodernism. *Urban Geography* 20, 589–90.

Berry, B. J. L. 2002a. The Chicago School in retrospect. *Urban Geography* 22, 559–61.

Berry, B. J. L. 2002b. Clara voce cognito. In P. Gould and F. R. Pitts (eds) *Geographical Voices: Fourteen autobiographical essays*. Syracuse: Syracuse University Press, 1–26.

Berry, B. J. L. 2002c. Paradigm lost. *Urban Geography* 23, 441–5.

Berry, B. J. L. 2002d. Big tents or firm foundations? *Urban Geography* 23, 501–2.

Berry, B. J. L. 2002e. My Cheshire cat's smile. *Urban Geography* 23, 1–2.

Berry, B. J. L. and Baker, A. M. 1968. Geographic sampling. In B. J. L. Berry and D. F. Marble (eds) *Spatial Analysis*. Englewood Cliffs, NJ: Prentice-Hall, 91–100.

Berry, B. J. L. and Garrison, W. L. 1958a. The functional bases of the central place hierarchy. *Economic Geography* 34, 145–54.

Berry, B. J. L. and Garrison, W. L. 1958b. Recent developments in central place theory. *Papers and Proceedings, Regional Science Association* 4, 107–20.

Berry, B. J. L. and Horton, F. E. (eds) 1974. *Urban Environmental Management: Planning for Pollution Control*. Englewood Cliffs, NJ: Prentice-Hall.

Berry, B. J. L., Goheen, P. G. and Goldstein, H. 1969. *Metropolitan Area Definition: A re-evaluation of the concept and statistical practice*. Washington, DC: US Bureau of the Census.

Best, U. 2009. Critical geography. In R. Kitchin and N. Thrift (eds) *International Encyclopedia of Human Geography*. Amsterdam: Elsevier, 345–57.

Best, U. 2014. Competitive internationalisation or grassroots practises of internationalism? The changing international practises of German-language critical geography. *Social and Cultural Geography*. doi: 10.1080/14649365.2014.979862

Bettencourt, L. and West, G. 2010. A unified theory of urban living. *Nature* 467, 912–13.

Bhaskar, R. 1978. *A Realist Theory of Science*. Brighton: Harvester Press.

Billinge, M. 1977. In search of negativism: Phenomenology and historical geography. *Journal of Historical Geography* 3, 55–68.

Billinge, M. 1983. The mandarin dialect: An essay on style in contemporary geographical writing. *Transactions of the Institute of British Geographers* NS8, 400–20.

Billinge, M., Gregory, D. and Martin, R. L. (eds) 1984. *Recollections of a Revolution: Geography as spatial science*. London: Macmillan.

Bingham, N. 1996. Object-ions: From technological determinism towards geographies of relations. *Environment and Planning D: Society and Space* 14, 635–7.

Binnie, J. 1997. Coming out of geography: Towards a queer epistemology? *Environment and Planning D: Society and Space* 15, 223–37.

Binnie, J. and Valentine, G. 1999. Geographies of sexuality – a review of progress. *Progress in Human Geography* 23, 175–87.

Binnie, J., Madge, C., Pain, R., Raghuram, P. and Rose, G. 2005. Working a fraction and making a fraction work: A rough guide for geographers in the academy. *Area* 37, 251–9.

Bird, A. J. 2000. *Thomas Kuhn*. Chesham: Acumen Books.

Bird, A. J. 2003. Three conservative Kuhns. *Social Epistemology* 17, 127–33.

Bird, J. H. 1975. Methodological implications for geography from the philosophy of K. R. Popper. *Scottish Geographical Magazine* 91, 153–63.

Bird, J. H. 1977. Methodology and philosophy. *Progress in Human Geography* 1, 104–10.

Bird, J. H. 1978. Methodology and philosophy. *Progress in Human Geography* 2, 133–40.

Bird, J. H. 1985. Geography in three worlds: How Popper's system can help elucidate dichotomies and changes in the discipline. *Professional Geographer* 37, 403–9.

Bird, J. H. 1993. *The Changing Worlds of Geography: A critical guide to concepts and methods*, 2nd edn. Oxford: Clarendon Press.

Birkin, M. 1995. Customer targeting, geodemographics and lifestyle approaches. In P. Longley and G. Clarke (eds) *GIS for Business and Service Planning*. Cambridge: GeoInformation International, 104–49.

Birkin, M. and Clarks, M. 1988. SYNTHESIS: A synthetic spatial information system: Methods and examples. *Environment and Planning A* 20, 645–71.

Birkin, M. and Clarke, M. 1989. The generation of individual and household incomes at the small area level using synthesis. *Regional Studies* 23, 535–48.

Birkin, M. and Wilson, A. G. 1986. Industrial location models 1. A review and integrating framework and 2. Weber, Palander, Hotelling and extensions within a new framework. *Environment and Planning A* 18, 175–206 and 293–306.

Birkin, M., Clarke, G. and Clarke, M. 2002. *Retail Geography and Intelligent Network Planning*. Chichester: John Wiley.

Birkin, M., Clarke, G. and Clarke, M. 2010: Refining and operationalizing entropy-maximizing models for business applications. *Geographical Analysis*, 42, 422–45.

Birkin, M., Clarke, M. and George, F. 1995. The use of parallel computers to solve nonlinear spatial optimisation problems: An application to network planning. *Environment and Planning A* 27, 1049–68.

Birkin, M., Clarke, G., Clarke, M. and Wilson, A. G. 1990. Elements of a model-based GIS for evaluation of urban policy. In L. Worrall (ed.) *Geographic Information Systems: Development and applications*. London: Belhaven Press, 131–62.

Birkin, M., Clarke, G., Clarke, M. and Wilson, A. G. 1996. *Intelligent GIS: Location decisions and strategic planning*. Cambridge: Geoinformation International.

Blaikie, P. M. 1978. The theory of the spatial diffusion of innovations: A spacious cul-de-sac. *Progress in Human Geography* 2, 268–95.

Blaikie, P. M. 1985. *The Political Economy of Soil Erosion*. London: Longman.

Blaikie, P. M. 1986. Natural resource use in developing countries. In R. J. Johnston and P. J. Taylor (eds) *A World in Crisis? Geographical perspectives*. Oxford: Basil Blackwell, 107–26.

Blaikie, P. M. 2008. Epilogue: Towards a future for political ecology that works. *Geoforum* 39, 765–72.

Blaikie, P. M. 2012. Should some political ecology be useful? *Geoforum* 43, 231–9.

Blaikie, P. M. and Brookfield, H. C. (eds) 1987. *Land Degradation and Society*. London: Methuen.

Blalock, H. M. 1960. *Social Statistics*. New York: McGraw Hill.

Blaug, M. 1975. Kuhn versus Lakatos, or paradigms versus research programmes in the history of economics. *History of Political Economy* 7, 399–419.

Blaut, J. M. 1962. Object and relationship. *Professional Geographer* 14, 1–6.

Blaut, J. M. 1970. Geographic models of imperialism. *Antipode: A Radical Journal of Geography* 2, 65–85.

Blaut, J. M. 1975. Imperialism: The Marxist theory and its evolution. *Antipode: A Radical Journal of Geography* 7, 1–19.

Blaut, J. M. 1979. The dissenting tradition. *Annals of the Association of America Geographers* 69, 157–64.

Blaut, J. M. 1987. Diffusionism: A uniformitarian critique. *Annals of the Association of American Geographers* 77, 30–47.

Blouet, B. W. (ed.) 1981. *The Origins of Academic Geography in the United States*. Hamden, CT: Archon Books.

Blouet, B. W. 1987. *Halford Mackinder: A biography*. College Station, TX: Texas A&M University Press.

Blowers, A. T. 1972. Bleeding hearts and open values. *Area* 4, 290–2.

Blowers, A. T. 1974. Relevance, research and the political process. *Area* 6, 32–6.

Blowers, A. T. 1984. *Something in the Air: Corporate power and the environment*. London: Harper & Row.

Blumenstock, D. I. 1953. The reliability factor in the drawing of isarithms. *Annals of the Association of American Geographers* 43, 289–304.

Blunt, A. 1994. *Travel, Gender and Imperialism: Mary Kingsley and West Africa.* New York: Guilford.

Blunt, A. and McEwan, C. (eds) 2002. *Postcolonial Geographies.* New York and London: Continuum.

Blunt, A. and Wills, J. 2000. *Dissident Geographies: An introduction to radical ideas and practice.* Harlow: Pearson.

Blunt, A. and Dowling, R. 2006. *Home.* London and New York: Routledge.

Boal, F. W. and Livingston, D. N. 1989. The behavioural environment: Worlds of meaning in a world of facts. In F. W. Boal and D. N. Livingstone (eds) *The Behavioural Environment.* London: Routledge, 3–17.

Boddy, M. J. 1976. The structure of mortgage finance: Building societies and the British social formation. *Transactions of the Institute of British Geographers* NS1, 58–71.

Bodman, A. R. 1991. Weavers of influence: The structure of contemporary geographic research. *Transactions of the Institute of British Geographers* NS16, 21–7.

Bodman, A. R. 1992. Holes in the fabric: More on the master weavers in human geography. *Transactions of the Institute of British Geographers* NS17, 108–9.

Bodman, A. R. 2010. Measuring the influentialness of economic geographers during the 'great half century': An approach using the h-index. *Journal of Economic Geography* 10, 141–56.

Bond, D. W. 2014. Hegel's geographical thought. *Environment and Planning D: Society and Space* 32, 179–98.

Bondi, L. 1993. Locating identity politics. In M. Keith and S. Pile (eds) *Place and the Politics Identity.* London: Routledge, 84–101.

Bondi, L. 1997. In whose words? On gender identities, knowledge and writing practices. *Transactions of the Institute of British Geographers* NS22, 245–58.

Bondi, L. 2005. Making connections and thinking through emotions; between geography and psychotherapy. *Transactions of the Institute of British Geographers* NS 30, 433–48.

Bondi, L. and Domosh, M. 1992. Other figures in other places: On feminism, postmodernism and geography. *Environment and Planning D: Society and Space* 10, 199–213.

Bondi, L. and Domosh, M. 1994. Editorial. *Gender, Place and Culture* 1, 3–4.

Bondi, L., Avis, H., Bingley, A., Davidson, J., Duffy, R., Einagel, V. I., Green, A.-M., Johnston, L., Lilley, S., Listerborn, C. Marshy, M., McEwan, S., O'Connor, N., Rose, G., Vivat, B. and Wood, N. 2002. *Subjectivities, Knowledges and Feminist Geographies: The subjects and ethics of social research.* Lanham, MD: Rowman & Littlefield.

Bonnett, A. 1993. Contours of crisis: Anti-racism and reflexivity. In P. Jackson and J. Penrose (eds) *Constructions of Race, Place and Nation.* London: UCL Press, 163–80.

Bonnett, A. 1997. Geography, 'race', and whiteness: Invisible traditions and current challenges. *Area* 29, 193–99.

Bonnett, A. 2003. Geography as the world discipline: Connecting popular and academic geographical imaginations. *Area* 35, 55–63.

Bonnett, A. 2003. Response to Stannard. *Area* 35, 323–4.

Bonnett, A. and Knoop, A. 2003. Cultural geographies of racialization – the territory of race. In K. Anderson, M. Domosh, S. Pile and N. Thrift (eds) *The Handbook of Cultural Geography.* London, Thousand Oaks, CA and New Delhi: Sage, 300–12.

Bonnett, A. and Nayak, A. 2003. Racialization – the territory of race. In K. Anderson, M. Domosh, S. Pile and N. Thrift (eds) *The Handbook of Cultural Geography.* London, Thousand Oaks, CA and New Delhi: Sage, 300–12.

Bonta, M. 2007. Review of Deleuze and space by Ian Buchanan and Gregg Lambert (eds) *Annals of the Association of American Geographers* 97, 811–13.

Boots, B. N. and Getis, A. 1978. *Models of Spatial Processes.* Cambridge: Cambridge University Press.

Bosco, F. J. 2006. Actor-network theory, networks, and relational approaches in human geography. In S. Aitken and G. Valentine (eds) *Approaches to Human Geography.* London, Thousand Oaks, CA and New Delhi: Sage, 136–46.

Bosker, M., Brakman, S., Garretsen, H. and Schramm, M. 2010. Adding geography to the new economic geography: Bridging the gap between theory and empirics. *Journal of Economic Geography* 10, 793–823.

Bosman, J. 2009. The changing position of society journals in geography. *Tijdschrift voor economische en sociale geografie* 100, 20–32.

Bourdieu, P. 1988. *Homo Academicus.* Cambridge: Polity.

Bourdieu, P. 1990. *The Logic of Practice.* Cambridge: Polity.

Bowd, G. P. and Clayton, D. W. 2013. Geographical warfare in the tropics: Yves Lacoste and the Vietnam War. *Annals of the Association of American Geographers* 103, 627–46.

Bowen, M. 1970. Mind and nature: The physical geography of Alexander von Humboldt. *Scottish Geographical Magazine* 86, 222–33.

Bowen, M. 1981. *Empiricism and Geographical Thought from Francis Bacon to Alexander von Humboldt.* Cambridge: Cambridge University Press.

Bowlby, S. R. and Tivers, J. 2009. Feminist geography, prehistory of. In R. Kitchen and N. Thrift (eds) *International Encyclopedia of Human Geography*. Amsterdam: Elsevier, 59–63.

Bowlby, S. R., Foord, J. and McDowell, L. 1986. The place of gender in locality studies. *Area* 18, 327–31.

Bowlby, S. R., Lewis, J., McDowell, L. and Foord, J. 1989. The geography of gender. In R. Peet and N. J. Thrift (eds) *New Models in Geography*, Vol. 2. London: Unwin Hyman, 157–75.

Bowman, I. 1921. *The New World: Problems in political geography*. New York: Harrap.

Boyce, R. R. (ed.) 1980. *Geography as Spatial Interaction by Edward L. Ullman*. Seattle, WA: University of Washington Press.

Boyer, R. and Saillard, Y. (eds) 1995. *Regulation Theory: The state of the art*. London and New York: Routledge.

Boyle, M. 2005. Biographical approaches in the teaching of history and philosophy of human geography: Introduction to review essays on *Key Thinkers on Space and Place*. *Environment and Planning A* 37, 161–4.

Boyle, M. J. and Robinson, M. E. 1979. Cognitive mapping and understanding. In D. T. Herbert and R. J. Johnston (eds) *Geography and the Urban Environment*, Vol. 2. Chichester: John Wiley, 59–82.

Boyle, P. J., Cullis, A., Feng, Z., Flowerdew, R. and Gayle, V. 2004. *Adding Geographical Variables and Identifiers to Longitudinal Datasets*. A report to the National Longitudinal Strategy Committee and ESRC.

Bracken, I., Higgs, G., Martin, D. and Webster, C. 1990. A classification of geographical information systems literature and applications. *CATMOG* 52, Norwich: Environmental Publications.

Bradley, P. N. 1986. Food production and distribution – and hunger. In R. J. Johnston and P. J. Taylor (eds) *A World in Crisis? Geographical perspectives*. Oxford: Basil Blackwell, 89–106.

Brakman, S., Garretsen, H. and Vav Marrewijk, C. 2009. *The New Introduction to Geographical Economics*. Cambridge: Cambridge University Press.

Braun, B. and Castree, N. (eds) 1998. *Remaking Reality: Nature at the millennium*. London and New York: Routledge.

Breheny, M. 1989. Chalkface to coalface: A review of the academic–practice interface. *Environment and Planning B: Planning and Design* 16, 451–68.

Breitbart, M. M. 1981. Peter Kropotkin, the anarchist geographer. In D. R. Stoddart (ed.) *Geography, Ideology and Social Concern*. Oxford: Blackwell, 134–53.

Brenner, N. 2004. *New State Spaces: Urban governance and the rescaling of statehood*. Oxford: Oxford University Press.

Brenner, N. and Elden, S. 2009. State, space, world: Henri Lefebvre and the survival of capitalism. In N. Brenner and S. Elden (eds) *Henri Lefebvre, State, Space, World: Selected essays*. Minneapolis, MN, and London: Minnesota University Press, 1–48.

Brenner, N. and Theodore, N. 2002a. From the 'new localism' to the spaces of neoliberalis. *Antipode: A Radical Journal of Geography* 34, 341–7.

Brenner, N. and Theodore, N. 2002b. Cities and the geographies of 'actually existing neoliberalism'. *Antipode: A Radical Journal of Geography* 34, 349–79.

Brenner, N., Jessop, B., Jones, M. and MacLeod, G. (eds) 2003. *State/Space: A reader*. Oxford: Blackwell.

Brickell, K., Maddrell, L. A., Martin, A. and Price, L. 2013. By any other name? The Women and Geography Study Group. *Area* 45, 11–12.

Briggs, D. J. 1981. The principles and practice of applied geography. *Applied Geography* 1, 1–8.

Brittan, S. 1977. Economic liberalism. In A. Bullock and O. Stallybrass (eds) *The Fontana Dictionary of Modern Thought*. London: William Collins, 188–9.

Brookfield, H. C. 1962. Local study and comparative method: An example from Central New Guinea. *Annals of the Association of American Geographers* 52, 242–54.

Brookfield, H. C. 1964. Questions on the human frontiers of geography. *Economic Geography* 40, 283–303.

Brookfield, H. C. 1969. On the environment as perceived. In C. Board, R. J. Chorley, P. Haggett and D. R. Stoddart (eds) *Progress in Geography*, Vol. 1. London: Edward Arnold, 51–80.

Brookfield, H. C. 1973. On one geography and a Third World. *Transactions of the Institute of British Geographers* 58, 1–20.

Brookfield, H. C. 1975. *Interdependent Development*. London: Methuen.

Brookfield, H. C. 1989. The behavioural environment: How, what for, and whose? In F. W. Boal and D. N. Livingstone (eds) *The Behavioural Environment*. London: Routledge, 311.

Browett, J. 1984. On the necessity and inevitability of uneven spatial development under capitalism. *International Journal of Urban and Regional Research* 8, 155–76.

Brown, G., Browne, K. and Lim, J. 2009. Introduction. In K. Browne, J. Lim and G. Brown (eds) *Geographies of Sexualities: Theory, practices and politics*. Farnham and Burlington: Ashgate, 1–18.

Brown, G., Browne, K. and Lim, J. 2011. Sexual life. In V. Del Casino Jr, M. Thomas, P. Cloke and R. Panelli (eds) *A Companion to Social Geography*. Chichester and Malden: Blackwell, 293–308.

Brown, L. A. 1968. *Diffusion Processes and Location: A conceptual framework and bibliography*. Regional Science Research Institute, Bibliography Series 3, Philadelphia, PA.

Brown, L. A. 1975. The market and infrastructure context of adoption: A spatial perspective on the diffusion of innovation. *Economic Geography* 51, 185–216.

Brown, L. A. 1981. *Innovation Diffusion: A new perspective*. London: Methuen.

Brown, L. A. and Moore, E. G. 1970. The intra-urban migration process: A perspective. *Geografiska Annaler* 52B, 1–13.

Brown, M. P. 1995. Ironies of distance: An ongoing critique of geographies of AIDS. *Environment and Planning D: Society and Space* 13, 159–83.

Brown, M. P. 2000. *Closet Space*. London and New York: Routledge.

Brown, M. P. and Knopp, L. 2003. Queer cultural geographies – We're here! We're queer! We're over there, too! In K. Anderson, M. Domosh, S. Pile and N. Thrift (eds) *The Handbook of Cultural Geography*. London, Thousand Oaks, CA and New Delhi: Sage, 313–29.

Brown, R. H. 1943. *Mirror for Americans: Likenesses of the eastern seaboard 1810*. New York: American Geographical Society.

Brown, S. E. 1978. Guy-Harold Smith, 1895–1976. *Annals of the Association of American Geographers* 68, 115–18.

Browne, K., Nash, C. J. and Hines, S. 2010. Introduction: Towards trans geographies. *Gender, Place and Culture* 17, 573–77.

Browne, K., Norcup, J., Robson, E. and Sharp, J. 2013. What's in a name? Removing women from the Women and Geography Study Group. *Area* 45, 7–8.

Browning, C. E. 1982. *Conversations with Geographers: Career pathways and research styles*. Studies in Geography 16, Department of Geography, University of North Carolina, Chapel Hill, NC.

Brunn, S. D. 1974. *Geography and Politics in America*. New York: Harper & Row.

Brunsdon, C. 2014. Spatial science – looking outward. *Dialogues in Human Geography* 4, 45–9.

Brush, J. E. 1953. The hierarchy of central places in southwestern Wisconsin. *Geographical Review* 43, 380–402.

Buchanan, K. M. 1958a. Review of R. J. Harrison Church, *West Africa: A study of the environment and man's use of it*. *Economic Geography* 34, 277–8.

Buchanan, K. M. 1958b. Review of R. W. Steel and C. A. Fisher (eds) *Geographical Essays on British Tropical Lands*. 1956. *Geographical Review*, 48, 447–9.

Buchanan, K. M. 1962. West wind, east wind. *Geography* 47, 333–46.

Buchanan, K. M. 1966. *The Chinese People and the Chinese Earth*. London: Bell.

Buchanan, K. M. 1973. The white north and the population explosion. *Antipode: A Radical Journal of Geography* 5(3), 7–15.

Buchanan, K. M. and Pugh, J. C. 1958. *Land and People in Nigeria: The human geography of Nigeria and its environmental basis*. London: University of London Press.

Buchanan, R. O. 1968. The man and his work. In C. Embleton and J. T. Coppock (eds) *Land Use and Resources: Studies in applied geography. A memorial volume to Sir Dudley Stamp*. London: Institute of British Geographers, Special Publication 1, 1–12.

Bullock, A. 1977. Liberalism. In A. Bullock and O. Stallybrass (eds) *The Fontana Dictionary of Modern Thought*. London: William Collins, 347.

Bulmer, M. 1984. *The Chicago School of Sociology*. Chicago: University of Chicago Press.

Bunge, W. 1962. *Theoretical Geography*, 1st edn. Lund Studies in Geography, Series C: General and Mathematical Geography. Lund: Gleerup.

Bunge, W. 1966a. *Theoretical Geography*, 2nd edn. Lund Studies in Geography, Series C: General and Mathematical Geography 1. Lund: Gleerup.

Bunge, W. 1966b. Locations are not unique. *Annals of the Association of America Geographers* 56, 375–6.

Bunge, W. 1968. *Fred K. Schaefer and the Science of Geography*. Harvard Papers in Theoretical Geography, Special Papers Series, Paper A, Laboratory for Computer Graphics and Spatial Analysis, Harvard University, Cambridge, MA.

Bunge, W. 1971. *Fitzgerald: Geography of a revolution*. Cambridge, MA: Schenkman.

Bunge, W. 1973a. Spatial prediction. *Annals of the Association of American Geographers* 63, 566–8.

Bunge, W. 1973b. Ethics and logic in geography. In R. J. Chorley (ed.) *Directions in Geography*. London: Methuen, 317–31.

Bunge, W. 1973c. The geography of human survival. *Annals of the Association of American Geographers* 63, 275–95.

Bunge, W. 1973d. The geography. *Professional Geographer* 25, 331–7.

Bunge, W. 1979. Fred K. Schaefer and the science of geography. *Annals of the Association of American Geographers* 69, 128–33.

Bunge, W. and Bordessa, R. 1975. *Canadian Alternative: Survival, expeditions and urban change.* Geographical Monograph No. 2, Department of Geography, Atkinson College, York University, Toronto, Ontario.

Bunting, T. E. and Guelke, L. 1979. Behavioral and perception geography: A critical appraisal. *Annals of the Association of American Geographers* 69, 448–62, 471–4.

Burnett, K. P. (ed.) 1981. Studies in choice, constraints, and human spatial behavior. Special issue, *Economic Geography* 57, 291–383.

Burnett, K. P. 1973. Social change, the status of women and models of city form and development. *Antipode: A Radical Journal of Geography* 5(3), 57–62.

Burrows, R. 2012. Living with the h-index? Metric assemblages in the contemporary academy. *Sociological Review* 60, 355–72.

Burton, I. 1963. The quantitative revolution and theoretical geography. *Canadian Geographer* 7, 151–62.

Burton, I., Kates, R. W. and White, G. F. 1978. *The Environment as Hazard.* New York: Oxford University Press.

Bushong, A. D. 1981. Geographers and their mentors: A genealogical view of American academic geography. In B. W. Blouet (ed.) *The Origins of Academic Geography in the United States.* Hamden, CT: Archon Books, 193–220.

Butler, R. and Bowlby, S. 1997. Bodies and spaces: An exploration of disabled people's experiences of public space. *Environment and Planning D: Society and Space* 15, 411–33.

Butler, R. and Parr, H. (eds) 1999. *Mind and Body Spaces: Geographies of illness, impairment and disability.* London: Routledge.

Butlin, R. A. 1982. *The Transformation of Rural England c. 1580–1800.* Oxford: Oxford University Press.

Butlin, R. A. 2009. *Geographies of Empire: European empires and colonies c.1880–1960.* Cambridge and New York: Cambridge University Press.

Buttimer, A. 1971. *Society and Milieu in the French Geographical Tradition.* Chicago: Rand McNally.

Buttimer, A. 1974. *Values in Geography.* Commission on College Geography, Resource Paper 24, Association of American Geographers, Washington, DC.

Buttimer, A. 1976. Grasping the dynamism of lifeworld. *Annals of the Association of American Geographers* 66, 277–92.

Buttimer, A. 1978a. Charism and context: The challenge of *la geographie humaine.* In D. Ley and M. S. Samuels (eds) *Humanistic Geography: Prospects and problems.* Chicago: Maaroufa Press, 58–76.

Buttimer, A. 1978b. On people, paradigms and progress in geography. Institutionen for Kulturgeografi oeh Eeonomisk Geografi vid Lunds Universitet, *Rapporter och Notiser* 47.

Buttimer, A. 1979. Erewhon or nowhere land. In S. Gale and G. Olsson (eds) *Philosophy in Geography.* Dordrecht: D. Reidel, 9–38.

Buttimer, A. 1981. On people, paradigms and progress in geography. In D. R. Stoddart (ed.) *Geography, Ideology and Social Concern.* Oxford: Blackwell, 70–80.

Buttimer, A. 1983. *The Practice of Geography.* London: Longman.

Buttimer, A. 1993. *Geography and the Human Spirit.* Baltimore, MD: Johns Hopkins University Press.

Buttimer, A. 1995. Book review – *Vidal de La Blache: 1845–1918. Un Génie de la Géographie. Annals of the Association of American Geographers,* 85(2), 406–7.

Buttimer, A. 2007. Torsten Hägerstrand (1916–2004). In H. Lorimer and C. W. J. Withers (eds) *Geographers: Biobibliographical studies,* Vol. 27. London and New York: Continuum, 119–57.

Buttimer, A. and Fahy, G. 1999. Imagining Ireland through geography texts. In A. Buttimer, S. D. Brunn and U. Wardenga (eds) *Text and Image: Social construction of regional knowledges.* Leipzig: Leibniz Institute for Regional Geography, Beiträge zur regionalen Geographie 49, 179–91.

Buttimer, A. and Hägerstrand, T. 1980. *Invitation to Dialogue: A progress report.* DIA Paper 1, Lund: University of Lund.

Buttimer, A. and Mels, T. 2006. *By Northern Lights: On the making of geography in Sweden.* Aldershot: Ashgate.

Butzer, K. W. 1989. Cultural ecology. In G. L. Gaile and C. J. Willmot (eds) *Geography in America.* Columbus, OH: Merrill, 192–208.

Butzer, K. W. 1990. Hartshorne, Hettner, and *The Nature of Geography.* In J. N. Entrikin and S. D. Brunn (eds) *Reflections on Richard Hartshorne's The Nature of Geography.* Washington, DC: Association of American Geographers, 35–52.

Cadwallader, M. 1975. A behavioral model of consumer spatial decision making. *Economic Geography* 51, 339–49.

Cadwallader, M. 1986. Structural equation models in human geography. *Progress in Human Geography* 10, 24–47.

Cameron, I. 1980. *To the Farthest Ends of the Earth*. London: Macdonald.

Campaign for Social Sciences 2015. *The Business of People: The significance of social science over the next decade*. London: Academy of the Social Sciences.

Campbell, C. S. 1994. The second nature of geography: Hartshorne as humanist. *Professional Geographer* 46, 411–17.

Campbell, J. A. 1989. The concept of 'the behavioural environment', and its origins, reconsidered. In F. W. Boal and D. N. Livingstone (eds) *The Behavioural Environment*. London: Routledge, 33–76.

Campbell, J. A. and Livingstone, D. N. 1983. Neo-Lamarckism and the development of geography in the United States and Great Britain. *Transactions of the Institute of British Geographers* NS8, 267–94.

Cannon, T. 1975. Geography and underdevelopment. *Area* 7, 212–6.

Capel, H. 1981. *Filosofía y ciencia en la geografía contemporánea: Una introdución a la geografía*. Barcelona: Barcano.

Capel, H. 1981. Institutionalization of geography and strategies of change. In D. R. Stoddart (ed.) *Geography, Ideology and Social Concern*. Oxford: Blackwell, 37–69.

Carey, H. C. 1858. *Principles of Social Science*. Philadelphia, PA: J. Lippincott.

Carlstein, T. 1980. Time, resources, society and ecology. Lund: Department of Geography, University of Lund.

Carlstein, T., Parkes, D. N. and Thrift, N. J. (eds) 1978. *Timing Space and Spacing Time* (3 vols). London: Edward Arnold.

Carr, M. 1983. A contribution to the review and critique of behavioural industrial location theory. *Progress in Human Geography* 7, 386–402.

Carroll, G. R. 1982. National city-size distributions: What do we know after 67 years of research? *Progress in Human Geography* 6, 1–43.

Carrothers, G. A. P. 1956. An historical review of the gravity and potential concepts of human interaction. *Journal, American Institute of Planners* 22, 94–102.

Carter, G. 1977. A geographical society should be a geographical society. *Professional Geographer* 29, 101–2.

Casetti, E. 1999. The evolution of scientific disciplines, mathematical modelling and human geography. *Geographical Analysis* 30, 332–9.

Castells, M. 1977. *The Urban Question*. London: Edward Arnold.

Castree, N. 1994. Teaching history, philosophy and theory: Notes on representing Marxism and 'Marxist geography'. *Journal of Geography in Higher Education* 18, 33–42.

Castree, N. 1995a. Review essay. The lonely hour of the last word: Marx, Althusser, and the critical critics. *Environment and Planning A* 27, 1163–78.

Castree, N. 1995b. The nature of produced nature. *Antipode: A Radical Journal of Geography* 27, 13–48.

Castree, N. 1996. Birds, mice and geography: Marxisms and dialectics. *Transactions of the Institute of British Geographers* NS21, 342–62.

Castree, N. 1999. Editorial: Situating cultural twists and turns. *Environment and Planning D: Society and Space* 17, 257–260.

Castree, N. 2002. The antinomies of Antipode. *Antipode: A Radical Journal of Geography* 34, 672–8.

Castree, N. 2003. Geographies of nature in the making. In K. Anderson, M. Domosh, S. Pile and N. Thrift (eds) *The Handbook of Cultural Geography*. London, Thousand Oaks, CA and New Delhi: Sage, 168–83.

Castree, N. 2004. Environmental issues: signals in the noise? *Progress in Human Geography* 28, 79–90.

Castree, N. 2005. *Nature (Key Ideas in Geography)*. Abingdon and New York: Routledge.

Castree, N. 2006a. Geography's new public intellectuals? *Antipode: A Radical Journal of Geography* 38, 396–412.

Castree, N. 2006b. The detour of critical theory. In N. Castree and D. Gregory (eds) *David Harvey: A critical reader*. Malden, MA and Oxford: Blackwell, 247–69.

Castree, N. 2011. Commentary. The future of geography in English universities. *Geographical Journal* 177, 294–9.

Castree, N. and Braun, B. (eds) 2001. *Social Nature: Theory, practice and politics*. Malden, MA and Oxford: Blackwell.

Castree, N. and Gregory, D. (eds) 2006. *David Harvey: A critical reader*. Malden, MA and Oxford: Blackwell.

Castree, N., Kitchin, R. and Rogers, A. 2013. Anglo-American geography. In N. Castree, R. Kitchin and A. Rogers *A Dictionary of Human Geography*. Oxford: Oxford University Press, 16.

Castree, N., Chatterton, P., Heynen, N., Larner, W. and Wright, M. W. (eds) 2010. *The Point is to Change It: Geographies of hope and survival in an age of crisis*. Chichester: Wiley-Blackwell.

Castree, N., Kitchin, R., Van Weesep, J., Bohle, H.-G,. Kleine, D., Kulke, E., Munton, R., Pawson, E., Aspinall, R., Sheppard, E., Berg, L. D., Hoggart, K. and Powell, J. 2006. Research assessment and the production of geographical knowledge. *Progress in Human Geography* 30, 747–82.

Chan, W. F. 2011. Mourning geography: A punctum, Strathclyde and the death of a subject. *Scottish Geographical Journal* 127, 255–66.

Chapman, G. P. 1977. *Human and Environmental Systems: A geographer's appraisal*. London: Academic Press.

Chapman, G. P. 1982. *The Green Revolution Game*. Cambridge: Marginal Context.

Chappell, J. E., Jr 1975. The ecological dimension: Russian and American views. *Annals of the Association of American Geographers* 65, 144–62.

Chappell, J. E., Jr 1976. Comment in reply. *Annals of the Association of American Geographers* 66, 169–73.

Chappell, J. M. A. and Webber, M. J. 1970. Electrical analogues of spatial diffusion processes. *Regional Studies* 4, 25–39.

Charlton, M., Rao, L. and Carver, S. 1995. GIS and the census. In S. Openshaw (ed.) *Census Users' Handbook*. Cambridge: GeoInformation International.

Chatterton, P. and Featherstone, D. 2007. Intervention: Elsevier, critical geography and the arms trade. *Political Geography* 26, 3–7.

Chatterton, P. and Maxey, L. 2009. Introduction: Whatever happened to ethics and responsibility in geography? *ACME: An International E-Journal for Critical Geographies* 8, 429–39.

Chatterton, P., Gidwani, V., Heynen, N., Kent, A., Larner, W. and Pain, R. 2011. Antipode in an antithetical era. *Antipode: A Radical Journal of Geography* 43, 181–9.

Chisholm, G. G., Beckit, H. O. and Ogilvie, A. G. 1913. The American transcontinental excursion of 1912. *Geographical Journal* 42, 321–60.

Chisholm, G. G. 1895. *Handbook of Commercial Geography*. London: Longman.

Chisholm, G. G. 1899. *The Times Gazetteer of the World*. London: The Times.

Chisholm, M. 1962. *Rural Settlement and Land Use*. London: Hutchinson.

Chisholm, M. 1966. *Geography and Economics*. London: G. Bell & Sons.

Chisholm, M. 1967. General systems theory and geography. *Transactions of the Institute of British Geographers* 42, 45–52.

Chisholm, M. 1971a. In search of a basis for location theory: Micro-economics or welfare economics? In C. Board , R. J. Chorley, P. Haggett and D. R. Stoddart (eds) *Progress in Geography*, Vol. 3. London: Edward Arnold, 111–34.

Chisholm, M. 1971b. Geography and the question of 'relevance'. *Area* 3, 65–8.

Chisholm, M. 1971c. *Research in Human Geography*. London: Heinemann.

Chisholm, M. 1973. The corridors of geography. *Area* 5, 43.

Chisholm, M. 1975a. *Human Geography: Evolution or revolution?* Harmondsworth: Penguin.

Chisholm, M. 1975b. The reformation of local government in England. In R. Peel, M. Chisholm and P. Haggett (eds) *Processes in Physical and Human Geography: Bristol essays*. London: Heinemann, 305–18.

Chisholm, M. 1976. Regional policies in an era of slow population growth and higher unemployment. *Regional Studies* 10, 201–13.

Chisholm, M. 1995. Some lessons from the review of local government in England. *Regional Studies* 29, 563–9.

Chisholm, M. 2000. *Structural Reform of British Local Government: Rhetoric and reality*. Manchester: Manchester University Press.

Chisholm, M. 2001. Human geography joins the Social Science Research Council: Personal recollections. *Area* 33, 428–30.

Chisholm, M. and Leach, S. 2008. *Botched Business 2006–2008: The damaging process of reorganising local government*. Coleford: Douglas Mclean.

Chisholm, M. and Manners, G. (eds) 1973. *Spatial Policy Problems of the British Economy*. London: Cambridge University Press.

Chisholm, M. and O'Sullivan, P. 1973. *Freight Flows and Spatial Aspects of the British Economy*. Cambridge: Cambridge University Press.

Chisholm, M. and Rodgers, B. (eds) 1973. *Studies in Human Geography*. London: Heinemann.

Chisholm, M., Frey, A. E. and Haggett, P. (eds) 1971. *Regional Forecasting*. London: Butterworth.

Chorley, R. J. 1962. *Geomorphology and General Systems Theory*. Professional Paper 500–B, United States Geological Survey, Washington, DC.

Chorley, R. J. 1964. Geography and analogue theory. *Annals of the Association of American Geographers* 54, 127–37.

Chorley, R. J. 1973. Geography as human ecology. In R. J. Chorley (ed.) *Directions in Geography*. London: Methuen, 155–70.

Chorley, R. J. 1995. Haggett's Cambridge: 1957–1966. In A. D. Cliff, P. R. Gould, A. G. Hoare and N. J. Thrift (eds) *Diffusing Geography*. Oxford: Blackwell, 355–74.

Chorley, R. J. and Bennett, R. J. 1981. Optimization: Control models. In N. Wrigley and R. J. Bennett (eds) *Quantitative Geography*. London: Routledge & Kegan Paul, 219–24.

Chorley, R. J. and Haggett, P. 1965a. Trend-surface mapping in geographical research. *Transactions and Papers, Institute of British Geographers* 37, 47–67.

Chorley, R. J. and Haggett, P. (eds) 1965b. *Frontiers in Geographical Teaching*. London: Methuen.

Chorley, R. J. and Haggett, P. (eds) 1967. *Models in Geography*. London: Methuen.

Chorley, R. J. and Kennedy, B. A. 1971. *Physical Geography: A systems approach*. London: Prentice-Hall International.

Chorley, R. J., Beckinsale, R. and Dunn, A. J. 1973. *The History of the Study of Landforms or the Development of Geomorphology. The life and work of William Morris Davis*, Vol. 2. London: Routledge.

Chouinard, V. and Grant, A. 1995. On being not even anywhere near 'the project': Ways of putting ourselves in the picture. *Antipode: A Radical Journal of Geography* 27, 137–66.

Chouinard, V., Fincher, R. and Webber, M. 1984. Empirical research in scientific human geography. *Progress in Human Geography* 8, 347–80.

Chrisman, N. R. 2006. *Charting the Unknown: How computer mapping at Harvard became GIS*. Redlands, CA: ESRI.

Chrisman, N. R., Cowen, D. J., Fisher, P. F., Goodchild, M. F. and Mark, D. M. 1989. Geographic information systems. In G. L. Gaile and C. J. Willmott (eds) *Geography in America*. Columbus, OH: Merrill, 776–96.

Christaller, W. 1966. *Central Places in Southern Germany* (translated by C. W. Baskin). Englewood Cliffs, NJ: Prentice-Hall.

Christensen, K. 1982. Geography as a human science. In P. Gould and G. Olsson (eds) *A Search for Common Ground*. London: Pion, 37–57.

Christopherson, S. 1989. On being outside 'the project'. *Antipode: A Radical Journal of Geography* 21, 83–9.

Cirincione, C., Darling, T. A. and O'Rourke, T. G. 2000. Assessing South Carolina's congressional districting. *Political Geography* 19, 189–212.

Clark, A. H. 1949. *The Invasion of New Zealand by People, Plants and Animals: The South Island*. New Brunswick, NJ: Rutgers University Press.

Clark, A. H. 1954. Historical geography. In P. E. James and C. F. Jones (eds) *American Geography: Inventory and prospects*. Syracuse: Syracuse University Press, 70–105.

Clark, A. H. 1977. The whole is greater than the sum of the parts: A humanistic element in human geography. In D. R. Deskins *et al.* (eds) *Geographic Humanism, Analysis and Social Action: A half century of geography at Michigan*. Ann Arbor, MI: Michigan Geographical Publication No. 17, 3–26.

Clark, G. L. 1982. Instrumental reason and policy analysis. In D. T. Herbert and R. J. Johnston (eds) *Geography and the Urban Environment*, Vol. 5. Chichester: John Wiley, 41–62.

Clark, G. L. 1985. *Judges and the Cities*. Chicago: University of Chicago Press.

Clark, G. L. 1992. Real regulation reconsidered. *Environment and Planning A* 24, 615–27.

Clark, G. L. and Dear, M. J. 1984. *State Apparatus*. Boston, MA: George Allen & Unwin.

Clark, G. L., Feldman, M. P. and Gertler, M. S. 2000. Economic geography: Transition and growth. In G. L. Clark, M. P. Feldmann and M. S. Gertler (eds) *The Oxford Handbook of Economic Geography*. Oxford: Oxford University Press, 3–17.

Clark, K. G. T. 1950. Certain underpinnings of our arguments in human geography. *Transactions and Papers (Institute of British Geographers)* 16, 15–22.

Clark, M. 1991. Developments in human geography: Niches for a Christian contribution. *Area* 23, 339–44.

Clark, W. A. V. 1975. Locational stress and residential mobility in a New Zealand context. *New Zealand Geographer* 31, 67–79.

Clark, W. A. V. 1981. Residential mobility and behavioral geography: Parallelism or interdependence? In K. R. Cox and R. C. Golledge (eds) *Behavioral Problems in Geography Revisited*. London: Methuen, 182–205.

Clark, W. A. V. 1991. Geography in court: Expertise in adversarial settings. *Transactions of the Institute of British Geographers* NS16, 5–20.

Clark, W. A. V. 1993. Applying our understanding: Social science in government and the marketplace. *Environment and Planning A* Anniversary Issue, 38–47.

Clark, W. A. V. 2002. Pacific views of urban geography in the 1960s. *Urban Geography* 22, 540–8.

Clarke, D. B. 2006. Postmodern geographies and the ruins of modernity. In S. Aitken and G. Valentine (eds) *Approaches to Human Geography*. London, Thousand Oaks, CA and New Delhi: Sage, 107–21.

Clarke, D. B. and Doel, M. 1995. Transpolitical geography. *Geoforum* 25, 505–24.

Clarke, D. B., Davies, W. K. D. and Johnston, R. J. 1974. The application of factor analysis in human geography. *Statistician* 23, 259–81.

Clarke, G. P. (ed.) 1996. *Microsimulation for Urban and Regional Policy Analysis*. London: Pion.

Clarke, K. C. 1998. Visualising different geofutures. In P. A. Longley, S. M. Brooks, R. MacDonnell and B. Macmillan (eds) *Geocomputation: A primer*. Chichester: John Wiley, 119–37.

Clarke, M. and Holm, E. 1987. Towards an applicable human geography: Some developments and observations. *Environment and Planning A* 19, 1525–41.

Clarke, M. and Holm, E. 1988. Microsimulation methods in spatial analysis and planning. *Geografiska Annaler* 69B, 145–64.

Clarke, M. and Wilson, A. G. 1985. The dynamics of urban spatial structure: The progress of a research programme. *Transactions of the Institute of British Geographers* NS10, 427–51.

Clarke, M. and Wilson, A. G. 1989. Mathematical models in human geography. In R. Peet and N. J. Thrift (eds) *New Models in Geography*, Vol. 2. London: Unwin Hyman, 30–42.

Clarkson, J. D. 1970. Ecology and spatial analysis. *Annals of the Association of American Geographers* 60, 700–16.

Clauset, A., Shalizi, C. R. and Newman, M. E. J. 2009. Power–law distributions in empirical data. *Siam Review* 51, 661–703.

Claval, P. 1981. Epistemology and the history of geographical thought. In D. R. Stoddart (ed.) *Geography, Ideology and Social Concern*. Oxford: Blackwell, 227–39.

Claval, P. 1983. *Models of Man in Geography*. Syracuse: Department of Geography, Syracuse University, Discussion Paper 79.

Claval, P. 1999. Historians, geography and the general public in France. In A. Buttimer, S. D. Brunn and U. Wardenga (eds) *Text and Image: Social construction of regional knowledges*. Leipzig: Leibniz Institute for Regional Geography, Beiträge zur regionalen Geographie 49, 70–6.

Claval, P. 2009. National schools of geography. In R. Kitchin and N. Thrift et al. (eds) *International Encyclopedia of Human Geography*. Oxford: Elsevier, 236–41.

Clayton, D. 2001. Questions of postcolonial geography. *Antipode: A Radical Journal of Geography* 33, 749–51.

Clayton, D. 2002. Critical imperial and colonial geographies. In K. Anderson, M. Domosh, S. Pile and N. Thrift (eds) *The Handbook of Cultural Geography*. London, Thousand Oaks, CA and New Delhi: Sage, 531–57.

Clayton, D. 2011. Colonizing, settling and the origins of academic geography. In J. A. Agnew and J. S. Duncan (eds) *The Wiley-Blackwell Companion to Human Geography*. Chichester: Wiley-Blackwell, 50–70.

Clayton, D. and Barnes, T. J. 2015. Continental European geographers and World War II. *Journal of Historical Geography* 47, 11–15.

Clayton, K. M. 1985a. New blood by (government) order. *Area* 17, 321–2.

Clayton, K. M. 1985b. The state of geography. *Transactions of the Institute of British Geographers* NS10, 5–16.

Cliff, A. D. and Haggett, P. 1988. *Atlas of Disease Distributions: Analytical approaches to epidemiological data*. Oxford: Blackwell.

Cliff, A. D. and Haggett, P. 1989. Spatial aspects of epidemic control. *Progress in Human Geography* 13, 315–47.

Cliff, A. D. and Haggett, P. 1995. Disease implications of global change. In R. J. Johnston, P. J. Taylor and M. J. Watts (eds) *Geographies of Global Change: Remapping the world in the late twentieth century*. Oxford: Blackwell, 206–22.

Cliff, A. D. and Haggett, P. 1998. On complex geographical space: Computing frameworks for spatial diffusion processes. In P. A. Longley, S. M. Brooks, R. MacDonnell and B. Macmillan (eds) *Geocomputation: A primer*. Chichester: John Wiley, 231–56.

Cliff, A. D. and Haggett, P. 2003. The geography of disease distributions. In R. J. Johnston and M. Williams (eds) *A Century of British Geography*. Oxford: Oxford University Press for the British Academy, 463–501.

Cliff, A. D. and Ord, J. K. 1973. *Spatial Autocorrelation*. London: Pion.

Cliff, A. D. and Ord, J. K. 1981. *Spatial Process*. London: Pion.

Cliff, A. D., Haggett, P. and Ord, J. K. 1987. *Spatial Aspects of Influenza Epidemics*. London: Pion.

Cliff, A. D., Haggett, P. and Smallman-Raynor, M. 1993. *Measles: An historical geography of a major human viral disease from global expansion to local retreat, 1840–1990*. Cambridge: Cambridge University Press.

Cliff, A. D., Hagett, P., Ord, J. K. and Versey, G. 1981. *Spatial Diffusion: An historical geography of epidemics in an island community*. Cambridge: Cambridge University Press.

Cliff, A. D., Haggett, P., Ord, J. K., Bassett, K. and Davies, R. B. 1975. *Elements of Spatial Structure: A quantitative approach*. London: Cambridge University Press.

Cliff, A. D., Smallman-Raynor, M., Haggett, P., Stroup, D. F. and Thacker, S. B. 2009. *Emergence and Re-emergence. Infectious Diseases: A geographical analysis.* Oxford: Oxford University Press.

Clifford, N. 2002. The future of geography: When the whole is less than the sum of the parts. *Geoforum* 33, 421–5.

Cloke, P. 2004. Enumeration. In P. Cloke, P. Crang and M. Goodwin (eds) *Envisioning Human Geographies.* London: Arnold.

Cloke, P. and Goodwin, M. 1992. Conceptualising countryside change: From Post-Fordism to rural structural coherence. *Transactions of the Institute of British Geographers* NS17, 321–36.

Cloke, P. 2002. Deliver us from evil? Prospects for living ethically and acting politically in human geography. *Progress in Human Geography* 26, 587–604.

Cloke, P., Crang, P. and Goodwin, M. (eds) 1999. *Introducing Human Geographies,* 1st edn. London: Arnold.

Cloke, P., Crang, P. and Goodwin, M. (eds) 2005. *Introducing Human Geographies,* 2nd edn. London: Arnold.

Cloke, P., Philo, C. and Sadler, D. 1991. *Approaching Human Geography: An introduction to contemporary theoretical debates.* London: Paul Chapman.

Clout, H. D. 2002. John Terence Coppock, 1921–2000. *Proceedings of the British Academy* 115, 207–24.

Clout, H. D. 2003a. Place description, regional geography and area studies: The chorographic inheritance. In R. J. Johnston and M. Williams (eds) *A Century of British Geography.* Oxford: Oxford University Press for the British Academy, 247–74.

Clout, H. D. 2003b. Albert Demangeon, 1872–1940: Pioneer of la géographie humaine. *Scottish Geographical Journal* 119, 1–24.

Clout, H. D. 2003c. *Geography at University College London: A brief history.* Department of Geography, University College London, London.

Clout, H. D. 2004. Lessons from experience: French geographers and the transcontinental excursion of 1912. *Progress in Human Geography* 28, 597–618.

Clout, H. D. 2009. *Patronage and the Production of Geographical Knowledge in France: The testimony of the first hundred regional monographs, 1905–1966.* Historical Geography Research Series, 41. London: Royal Geographical Society (with The Institute of British Geographers).

Clout, H. D. 2011a. Lionel William Lyde (1863–1947). In C. W. J. Withers and H. Lorimer (eds) *Geographers: Biobibliographical studies,* Vol. 30. London and New York: Continuum, 1–21.

Clout, H. D. 2011b. Professor Michael Williams 1935–2009. *Proceedings of the British Academy* 172, 355–75.

Clout, H. D. 2014. Jean Tricart (1920–2003). In H. Lorimer and C. W. J. Withers (eds) *Geographers: Biobibliographical studies,* Vol. 33. London: Bloomsbury, 11–42.

Clout, H. D. and Gosme, C. 2003. The Naval Intelligence Handbooks: A monument in geographical writing. *Progress in Human Geography* 27, 153–73.

Clout, H. D. and Stevenson, I. 2004. Jules Sion, Alan Grant Ogilvie and the College des Ecossais in Montpellier: A network of geographers. *Scottish Geographical Journal* 120, 181–98.

Coates, B. E., Johnston, R. J. and Knox, P. L. 1977. *Geography and Inequality.* Oxford: Oxford University Press.

Cochrane, A. 1987. What a difference the place makes: The new structuralism of locality. *Antipode: A Radical Journal of Geography* 19, 354–63.

Cockings, S. L., Harfoot, A., Martin, D. and Hornby, D. 2011. Maintaining existing zoning systems using automated zone design techniques for creating the 2011 census geographies for England and Wales. *Environment and Planning A* 43, 2399–2418.

Coffey, W. J. 1981. *Geography: Towards a general spatial systems approach.* London: Methuen.

Cohen, S. B. 1988. Reflections on the elimination of geography at Harvard, 1947–51. *Annals of the Association of American Geographers* 78, 148–51.

Cole, J. P. 1969. Mathematics and geography. *Geography* 54, 152–63.

Cole, J. P. and King C. A. M. 1968. *Quantitative Geography.* London: John Wiley.

Coleman, A. 1985. *Utopia on Trial: Vision and reality in planned housing.* London: Hilary Shipman.

Collini, S. 2012. *What are Universities For?* London: Penguin.

Collins, M. J. 2002. *Cold War Laboratory: RAND, the Air Force and the American state, 1945–1950.* Washington, DC: Smithsonian Institute Press.

Colls, R. 2012. Feminism, bodily difference and non-representational geographies. *Transactions of the Institute of British Geographers* NS 37, 430–45.

Conant, J. B. and Haugeland, J. (eds) 2000. *The Road Since Structure: Philosophical essays 1970–1993, with an autobiographical interview.* Chicago: University of Chicago Press.

Conzen, M. P. 1981. The American urban system in the nineteenth century. In D. T. Herbert and R. J. Johnston (eds) *Geography and the Urban Environment*, Vol. 4. Chichester: John Wiley, 295–348.

Cook, I. and Crang, M. 1995. *Doing Ethnographies (CATMOG 58)*. Norwich: Geo Books.

Cook, I., Crouch, D., Naylor, S. and Ryan, J. R. 2000. Foreword. In I. Cook, D. Crouch, S. Naylor and J. R. Ryan (eds) *Cultural Turns/Geographical Turns: Perspectives on cultural geography*. Harlow: Prentice Hall, xi–xii.

Cooke, P. N. 1986. The changing urban and regional system in the United Kingdom. *Regional Studies* 20, 243–52.

Cooke, P. N. 1987. Clinical inference and geographic theory. *Antipode: A Radical Journal of Geography* 19, 69–78.

Cooke, P. N. (ed.) 1989a. *Localities: The changing face of urban Britain*. London: Unwin Hyman.

Cooke, P. N. 1989b. Locality theory and the poverty of 'spatial variation' (A response to Duncan and Savage). *Antipode: A Radical Journal of Geography* 21, 261–73.

Cooke, P. N. 1990. *Back to the Future: Modernity, postmodernity and locality*. London: Unwin Hyman.

Cooke, R. U. 1992. Common ground, shared inheritance: Research imperatives for environmental geography. *Transactions of the Institute of British Geographers* NS17, 131–51.

Cooke, R. U. 2002. Presidential address. *Geographical Journal* 168, 260–3.

Cooke, R. U. and Harris, R. 1970. Remote sensing of the terrestrial environment: Principles and progress. *Transactions of the Institute of British Geographers* 50, 1–23.

Cooke, R. U. and Reeves, R. W. 1976. *Arroyos and Environmental Change in the American South-West*. Oxford: Clarendon Press.

Cooke, R. U. and Robson, B. T. 1976. Geography in the United Kingdom, 1972–1976. *Geographical Journal* 142, 3–22.

Coombes, M. G., Dixon, J. S., Goddard, J. B., Openshaw, S. and Taylor, P. J. 1982. Functional regions for the population census of Great Britain. In D. T. Herbert and R. J. Johnston (eds) *Geography and the Urban Environment*, Vol. 5. Chichester: John Wiley, 63–111.

Cooper, A. 1992. New directions in the geography of religion. *Area* 24, 123–9.

Cooper, W. 1952. *The Struggles of Albert Woods*. London: Jonathan Cape.

Coppock, J. T. 1974. Geography and public policy: Challenges, opportunities and implications. *Transactions of the Institute of British Geographers* 63, 1–16.

Coppock, J. T. and Duffield, B. S. 1975. *Recreation in the Countryside: A spatial analysis*. London: Macmillan.

Corbridge, S. 1986. *Capitalist World Development*. London: Macmillan.

Corbridge, S. 1988. Deconstructing determinism. *Antipode: A Radical Journal of Geography* 20, 239–69.

Corbridge, S. 1989. Marxism, post-Marxism, and the geography of development. In R. Peet and N. J. Thrift (eds) *New Models in Geography*, Vol. 1. London: Unwin Hyman, 224–54.

Corbridge, S. 1993. Marxists, modernities, and moralities: Development praxis and the claims of distant strangers. *Environment and Planning D: Society and Space* 11, 449–72.

Corbridge, S., Martin, R. L. and Thrift, N. J. (eds) 1994. *Money, Space and Power*. Oxford: Blackwell.

Cormack, L. B. 1997. *Charting an Empire: Geography at English universities, 1580–1620*. Chicago: University of Chicago Press.

Cosgrove, D. 1983. Towards a radical cultural geography: Problems of theory. *Antipode: A Radical Journal of Geography* 15, 1–11.

Cosgrove, D. 1984. *Social Formation and Symbolic Landscape*. London: Croom Helm.

Cosgrove, D. 1989a. Geography is everywhere: Culture and symbolism in human landscapes. In D. Gregory and R. Walford (eds) *Horizons in Human Geography*. London: Macmillan, 118–35.

Cosgrove, D. 1989b. Models, description and imagination in geography. In B. Macmillan (ed.) *Remodelling Geography*. Oxford: Blackwell, 23–44.

Cosgrove, D. 1993. Commentary. *Annals of the Association of American Geographers* 83, 515–17.

Cosgrove, D. 1996. Classics in human geography revisited. *Progress in Human Geography* 20, 197–9.

Cosgrove, D. 2001. *Apollo's Eye: A cartographic genealogy of the earth in western imagination*. Baltimore, MD: Johns Hopkins University Press.

Cosgrove, D. 2003. Landscape of the European sense of sight – eyeing nature. In K. Anderson, M. Domosh, S. Pile and N. Thrift (eds) *The Handbook of Cultural Geography*. London, Thousand Oaks, CA and New Delhi: Sage, 249–68.

Cosgrove, D. and Daniels, S. (eds) 1988. *The Iconography of Landscape*. Cambridge: Cambridge University Press.

Cosgrove, D. and Daniels, S. 1989. Fieldwork as theatre: A week's performance in Venice and its region. *Journal of Geography in Higher Education* 13, 169–82.

Cosgrove, D. and Jackson, P. 1987. New directions in cultural geography. *Area* 19, 95–101.

Couclelis, H. 1986a. Artificial intelligence in geography: Conjectures on the shape of things to come. *Professional Geographer* 38, 1–10.

Couclelis, H. 1986b. A theoretical framework for alternative models of spatial decision and behavior. *Annals of the Association of American Geographers* 76, 95–113.

Couclelis, H. and Golledge, R. G. 1983. Analytic research, positivism, and behavioral geography. *Annals of the Association of American Geographers* 73, 331–9.

Couper, P. 2015. *A Student's Introduction to Geographical Thought: Theories, philosophies, methodologies*. London, Thousand Oaks, CA: Sage.

Court, A. 1972. All statistical populations are estimated from samples. *Professional Geographer* 24, 160–1.

Cowen, D. J. 1983. Automated geography and the DIDS. *Professional Geographer* 35, 339–40.

Cox, K. R. 1969. The voting decision in a spatial context. In C. Board, R. J. Chorley, P. Haggett and D. R. Stoddart (eds) *Progress in Geography*, Vol. 1. London: Edward Arnold, 81–118.

Cox, K. R. 1973. *Conflict, Power and Politics in the City: A geographic view*. New York: McGraw-Hill.

Cox, K. R. 1976. American geography: Social science emergent. *Social Science Quarterly* 57, 182–207.

Cox, K. R. 1979. *Location and Public Problems*. Oxford: Basil Blackwell.

Cox, K. R. 1981. Bourgeois thought and the behavioral geography debate. In K. R. Cox and R. G. Golledge (eds) *Behavioral Problems in Geography Revisited*. London: Methuen, 256–79.

Cox, K. R. 1989. The politics of turf and the question of class. In J. Wolch and M. Dear (eds) *The Power of Geography*. Boston, MA: Unwin Hyman, 61–90.

Cox, K. R. 1995. Concepts of space, understanding in human geography, and spatial analysis. *Urban Geography* 16, 304–26.

Cox, K. R. 2013. Notes on a brief encounter: Critical realism, historical materialism and human geography. *Dialogues in Human Geography* 3, 3–21.

Cox, K. R. 2014. *Making Human Geography*. London and New York: Guildford Press.

Cox, K. R. and Golledge, R. G. (eds) 1969. *Behavioral Problems in Geography: A symposium*. Evanston, IL: Northwestern University Studies in Geography 17.

Cox, K. R. and Golledge R. G. (eds) 1981. *Behavioural Problems in Geography Revisited*. London: Methuen.

Cox, K. R. and Macmillan, B. 2001. W. Bunge, theoretical geography. *Progress in Human Geography* 25, 71–7.

Cox, K. R. and Mair, A. 1989. Levels of abstraction in locality studies. *Antipode: A Radical Journal of Geography* 21, 121–32.

Cox, K. R. and McCarthy, J. J. 1982. Neighbourhood activism as a politics of turf: A critical analysis. In K. R. Cox and R. J. Johnston, (eds) *Conflict, Politics and the Urban Scene*. London: Longman, 196–219.

Cox, K. R., Reynold, D. R. and Rokkan, S. (eds) 1974. *Locational Approaches to Power and Conflict*. New York: Halsted Press.

Cox, N. J. 1989. Modelling, data analysis and Pygmalion's problem. In B. MacMillan (ed.) *Remodelling Geography*. Oxford: Basil Blackwell, 204–10.

Cox, N. J. and Jones, K. 1981. Exploratory data analysis. In N. Wrigley and R. J. Bennett Cox (eds) *Quantitative Geography*. London: Routledge & Kegan Paul, 135–43.

Crampton, J. W. 2009. Cartography: Maps 2.0. *Progress in Human Geography* 33, 91–100.

Crampton, J. W. and Elden, S. (eds) 2007. *Space, Knowledge and Power: Foucault and Geography*. Aldershot: Ashgate.

Crampton, J. W., Dobson, J. E., Smith, N. and Morin, K. M. 2012. Forum: Karen Morin's Civic Discipline. *Geographical Review* 102, 539–62.

Crane, D. 1972. *Invisible Colleges*. Chicago: University of Chicago Press.

Crang, M. 1994. On the heritage trail: Maps of and journeys to olde Englande. *Environment and Planning D: Society and Space* 12, 341–56.

Crang, M. 2002. Qualitative methods: The new orthodoxy? *Progress in Human Geography* 26, 647–55.

Crang, M. and Cook, I. 2007. *Doing Ethnographies*. London, Thousand Oaks, CA and New Delhi and Singapore: Sage.

Crang, M. and Thrift, N. J. (eds) 2000. *Thinking Space*. London: Routledge.

Crang, M. and Tolia-Kelly, D. P. 2010. Nation, race, and affect: Senses and sensibilities at national heritage sites. *Environment and Planning A* 42, 2315–31.

Crang, P. 1992. The politics of polyphony: Reconfigurations in geographical authority. *Environment and Planning D: Society and Space* 10, 527–49.

Crang, P. 1995. It's showtime: On the workplace geographies of display in a restaurant in southeast England. *Environment and Planning D: Society and Space* 12, 675–704.

Cresswell, T. 2010. New cultural geography – an unfinished project? *Cultural Geography* 17, 169–74.

Cresswell, T. 2012. Review essay. Nonrepresentational theory and me: Notes of an interested sceptic. *Environment and Planning D: Society and Space* 30, 96–105.

Cresswell, T. 2013. *Geographic Thought: A critical introduction*. Chichester: Wiley-Blackwell.

Crewe, L. 2000. Geographies of retailing and consumption. *Progress in Human Geography* 24, 275–90.

Cromley, R. G. 1993. Automated geography ten years later. *Professional Geographer* 45, 442–3.

Crowe, P. R. 1936. The rainfall regime of the Western Plains. *Geographical Review* 26, 463–84.

Crowe, P. R. 1938. On progress in geography. *Scottish Geographical Magazine* 54, 1–19.

Crowe, P. R. 1970. Review of *Progress in Geography* Vol 1. *Geography* 55, 346–7.

Crowther, J. G., Cotter, C. H., Wallis, H., Sadler, D. H. and Hague, R. 1967. Eva G. R. Taylor. *Journal of Navigation* 20, 94–101.

Crush, J. 1994. Post-colonialism, decolonization and geography. In A. Godlewska and N. Smith (eds) *Geography and Empire*. Oxford: Blackwell, 333–50.

Cullen, I. G. 1976. Human geography, regional science, and the study of individual behaviour. *Environment and Planning A* 8, 397–411.

Cumberland, K. B. 1947. *Soil Erosion in New Zealand*. Wellington: Whitcombe & Tombs.

Cumbers, A. 2009. Marxism/Marxist geography I. In R. Kitchin and N. Thrift (eds) *International Encyclopedia of Human Geography*. Amsterdam: Elsevier, 461–73.

Cupples, J. 2002. The field as a landscape of desire: Sex and sexuality in geographical fieldwork. *Area* 34, 382–90.

Cupples, J. and Pawson, E. 2012. Giving an account of oneself: The PBRF and the neoliberal university. *New Zealand Geographer* 68, 14–23.

Curran, P. J. 1984. Geographic information systems. *Area* 16, 153–8.

Curran, P. J. 1996. Differential research funding. *Area* 28.

Curran, P. J. 2000. Competition in UK higher education: Competitive advantage and Porter's diamond model. *Higher Education Quarterly* 54, 386–410.

Curran, P. J. 2001a. Competition in UK higher education: Applying Porter's diamond model to geography departments. *Studies in Higher Education* 26, 223–51.

Curran, P. J. 2001b. Remote sensing: Using the spatial domain. *Environmental and Ecological Statistics* 8, 331–44.

Curry, L. 1967. Quantitative geography. *Canadian Geographer* 11, 265–74.

Curry, L. 1972. A spatial analysis of gravity flows. *Regional Studies* 6, 131–47.

Curry, M. 1982a. The idealist dispute in Anglo-American geography. *Canadian Geographer* 26, 37–50.

Curry, M. 1982b. The idealist dispute in Anglo-American geography: A reply. *Canadian Geographer* 26, 57–9.

Curry, M. 1992. Reply. *Annals of the Association of American Geographers* 82, 310–12.

Curry, M. 1994. Image, practice and the hidden impacts of geographic information systems. *Progress in Human Geography* 18, 460–90.

Curry, M. 1995. GIS and the inevitability of ethical inconsistency. In J. Pickles (ed.) *Ground Truth: The social implications of geographical information systems*. New York: Guilford Press, 68–87.

Curry, M. 1996. On space and spatial practice in contemporary geography. In C. Earle, K. Mathewson and M. S. Kenzer (eds) *Concepts in Human Geography*. Lanham, MD: Rowman & Littlefield, 3–32.

Curry, M. 1998. *Digital Places: Living with geographic information technologies*. London and New York: Routledge.

Curti, G. H., Aitken. S. C., Bosco, F. J. and Goesrisch D. D. 2011. For not limiting emotional and affectual geographies: A collective critique of Steve Pile's 'Emotions and affect in recent human geography'. *Transactions of the Institute of British Geographers* NS 36, 590–4.

Cutter, S. L., Golledge, R. and Graf, W. L. 2002. The big questions in geography. *Professional Geographer* 54, 305–17.

Cutter, S. L., Richardson, D. B. and Wilbanks, T. J. (eds) 2003. *The Geographical Dimensions of Terrorism*. London and New York: Routledge.

Cutter, S. L., Richardson, D. B. and Wilbans, T. J. 2004. The geographical dimensions of terrorism: Future directions. *Annals of the Association of American Geographers* 94, 1001–2.

Dacey, M. F. 1962. Analysis of central-place and point patterns by a nearest-neighbor method. In K. Norborg (ed.) *Proceedings of the IGU Symposium in Urban Geography, Lund 1960*. Lund: C. W. K. Gleerup, 55–76.

Dacey, M. F. 1968. A review on measures of contiguity for two and k-color maps. In B. J. L. Berry and D. F. Marble (eds) *Spatial Analysis*. Englewood Cliffs, NJ: Prentice-Hall, 479–95.

Dacey, M. F. 1973. Some questions about spatial distributions. In R. J. Chorley (ed.) *Directions in Geography*. London: Methuen, 127–52.

Dalby, S. 1991. Critical geopolitics: Discourse, difference, and dissent. *Environment and Planning D: Society and Space* 9, 261–83.

Dalby, S. 1994. Gender and critical geopolitics: Reading security discourse in the new world disorder. *Environment and Planning D: Society and Space* 9, 261–83.

Dale, A. 1993. Office of Population Censuses and Surveys longitudinal study. *Environment and Planning A* 25, 83–6.

Daly, M. T. 1972. *Techniques and Concepts in Geography: A review*. Melbourne: Thomas Nelson.

Daniels, S. J. 1989. Marxism, culture and the duplicity of landscape. In R. Peet and N. Thrift (eds) *New Models in Geography*, Vol. 2. London: Unwin Hyman, 196–220.

Daniels, S. J. 1991. The making of Constable country. *Landscape Research* 16, 9–18.

Darby, H. C. (ed.) 1936. *An Historical Geography of England before 1800. Fourteen Studies*. Cambridge: Cambridge University Press.

Darby, H. C. (ed.) 1973. *A New Historical Geography of England*. Cambridge: Cambridge University Press.

Darby, H. C. 1940a. *The Medieval Fenland*. Cambridge: Cambridge University Press.

Darby, H. C. 1940b. *The Draining of the Fens*. Cambridge: Cambridge University Press.

Darby, H. C. 1948. The regional geography of Thomas Hardy's Wessex. *Geographical Review* 38, 426–43.

Darby, H. C. 1951. The changing English landscape. *Geographical Journal* 117, 377–98.

Darby, H. C. 1953. On the relations of geography and history. *Transactions and Papers, Institute of British Geographers* 19, 1–11.

Darby, H. C. 1962. The problem of geographical description. *Transactions and Papers, Institute of British Geographers* 30, 1–14.

Darby, H. C. 1977. *Domesday England*. London: Cambridge University Press.

Darby, H. C. 1983a. Historical geography in Britain, 1920–1980: Continuity and change. *Transactions of the Institute of British Geographers* NS8, 421–8.

Darby, H. C. 1983b. Academic geography in Britain, 1918–1946. *Transactions of the Institute of British Geographers* NS8, 14–26.

Darby, H. C. 2002. *The Relations of History and Geography: Studies in England, France and the United States* (edited by M. Williams, H. Clout, T. Coppock and H. Prince). Exeter: University of Exeter Press.

Davey, J., Jones, R., Lawrence, V., Stevenson, I., Jenkins, A. and Shepherd, I. D. H. 1995. Issues and trends in textbook publishing: The views of geography editors/publishers. *Journal of Geography in Higher Education* 19, 11–28.

Davies, R. B. and Pickles, A. R. 1985. Longitudinal versus cross-sectional methods for behavioural research: A first-round knockout. *Environment and Planning A* 17, 1315–30.

Davies, W. K. D. 1972a. Geography and the methods of modern science. In W. K. D. Davies (ed.) *The Conceptual Revolution in Geography*. London: University of London Press, 131–9.

Davies, W. K. D. 1972b. Introduction: The conceptual revolution in geography. In W. K. D. Davies (ed.) *The Conceptual Revolution in Geography*. London: University of London Press, 9–18.

Davies, W. K. D. 1984. *Factorial Ecology*. Aldershot: Gower.

Davis, W. M. 1906. An inductive study of the content of geography. *Bulletin of the American Geographical Society* 38, 67–84.

Dawney, L. 2011. The motor of being: clarifying and defending the concept of affect: a response to Steve Pile's 'Emotions and affect in recent human geography'. *Transactions of the Institute of British Geographers* NS 36, 599–602.

Dawson, J. and Unwin, D. J. 1976. *Computing for Geographers*. Newton Abbott: David & Charles.

Dawson, J., Larke, R. and Mukoyama, M. (eds) 2006. *Strategic Issues in International Retailing*. London: Routledge.

Day, M. and Tivers, J. 1979. Catastrophe theory and geography: A Marxist critique. *Area* 11, 54–8.

Daysh, G. H. J. (ed.) 1949. *Studies in Regional Planning*. London: Philip & Son.

Dear, M. J. (ed.) 2002. *From Chicago to L.A.: Making sense of urban theory*. London: Sage.

Dear, M. J. 1987. Society, politics and social theory. *Environment and Planning D: Society and Space* 5, 363–6.

Dear, M. J. 1988. The postmodern challenge: Reconstructing human geography. *Transactions of the Institute of British Geographers* NS13, 262–74.

Dear, M. J. 1991a. The premature demise of postmodern urbanism. *Cultural Anthropology* 6, 538–52.

Dear, M. J. 1991b. Review of Harvey, *The condition of postmodernity*. *Annals of the Association of American Geographers* 81, 533–9.

Dear, M. J. 1994a. Postmodern human geography: An assessment. *Erdkunde* 48, 2–13.

Dear, M. J. 1994b. Commentary. Who's afraid of postmodernism? Reflections on Symanski and Cosgrove. *Annals of the Association of American Geographers* 84, 295–300.

Dear, M. J. 1995. Practising postmodern geography. *Scottish Geographical Magazine* 111, 179–81.

Dear, M. J. 2000. *The Condition of Postmodern Urbanism*. Oxford: Blackwell.

Dear, M. J. 2001. The politics of geography: Hate mail, rabid referees, and culture wars. *Political Geography* 20, 1–12.

Dear, M. J. 2011. Historical moments in the rise of the geohumanities. In M. Dear, J. Ketchum, S. Luria and D. Richardson (eds) *GeoHumanities: Art, history, text at the edge of place*. Abingdon and New York: Routledge, 309–14.

Dear, M. J. and Clark, G. L. 1978. The state and geographic process: A critical review. *Environment and Planning A* 10, 173–84.

Dear, M. J. and Dahmann, N. 2008. Urban politics and the Los Angeles school of urbanism. *Urban Affairs Review* 44, 266–79.

Dear, M. J. and Flusty, S. 1998. Postmodern urbanism. *Annals of the Association of American Geographers* 88, 50–72.

Dear, M. J. and Flusty, S. 1999. Engaging postmodern urbanism. *Urban Geography* 20, 412–16.

Dear, M. J. and Flusty, S. 2002. Preface. In M. J. Dear and S. Flusty (eds) *Spaces of Postmodernity: Readings in human geography*. Oxford: Blackwell, ix–xiii.

Dear, M. J. and Moos, A. I. 1986. Structuration theory in urban analysis: 2. Empirical application. *Environment and Planning A* 18, 351–74.

Dear, M. J. and Scott, A. J. (eds) 1981. *Urbanization and Urban Planning in Capitalist Society*. London: Methuen.

Dear, P. 2012. Fifty years of Structure. *Social Studies of Science* 42, 424–28.

Dearden, J. and Wilson, A. G. 2011. Using participatory computer simulation to explore the process of urban simulation. *Transactions in GIS* 15, 273–289.

Delaney, D. 2002. The space that race makes. *Professional Geographer* 54, 6–14.

Delanty, G. 2003. Rethinking Kuhn's legacy without paradigms: Some remarks on Steve Fuller's Thomas Kuhn: A philosophical history for our times. *Social Epistemology* 17, 153–6.

Delyser, D. and Starrs, P. F. 2001. Doing fieldwork: Editors' introduction. *Geographical Review* 91, iv–viii.

Demeritt, D. 1998. Science, social constructivism and nature. In B. Braun and N. Castree (eds) *Remaking Reality: Nature at the millennium*. London: Routledge, 177–97.

Demeritt, D. 2008. Dictionaries, disciplines and the future of geography. *Geoforum* 39, 1811–13.

Dennis, R. J. 1984. *English Industrial Cities in the Nineteenth Century: A social geography*. Cambridge: Cambridge University Press.

Derrida, J. 1976. *Of Grammatology*. Baltimore, MD: Johns Hopkins University Press.

Derudder, B. 2011. Some reflections on the 'problematic' dominance of 'Web of Science' journals in academic human geography. *Area* 43, 110–12.

Derudder, B., Hoyler, M., Taylor, P. J. and Witlox, F. (eds) 2012. *The International Handbook of Globalization and World Cities*. Cheltenham: Edward Elgar.

Desbarats, J. 1983. Spatial choice and constraints on behavior. *Annals of the Association of American Geographers* 73, 340–57.

Desbiens, C. 2002. Speaking in tongues, making geographies. *Environment and Planning D: Society and Space* 20, 1–3.

Desbiens, C. and Smith, N. 1999. The international critical geography group: forbidden optimism? *Environment and Planning D: Society and Space* 18, 379–82.

Detwyler, T. R. and Marcus, M. G. (eds) 1972. *Urbanization and Environment*. Belmont, CA: Duxbury Press.

Deutsch, R. 1991. Boy's town. *Environment and Planning D: Society and Space* 9, 5–30.

DeVivo, M. S. 2015. *Leadership in American Geography: The twentieth century*. Lanham, MD: Lexington Books.

Dewsbury, J.-D. 2014. Inscribing thoughts: the animation of an adventure. *Cultural Geographies in Practice* 21, 147–152.

Dewsbury, J.-D., Harrison, P., Rose, M. and Wylie, J. 2002. Enacting geographies. *Geoforum* 33, 437–40.

Dicken, P. 1986. *Global Shift*. London: Harper & Row.

Dicken, P. 2003. *Global Shift: Transforming the world economy*. London: Sage.

Dicken, P. 2011. *Global Shift: Mapping the changing contours of the world economy*, 6th edn. New York: Guildford Press.

Dickens P., Duncan, S. S., Goodwin, M. and Gray, F. 1985. *Housing, States and Localities*. London: Methuen.

Dickenson, J. P. and Clarks, C. G. 1972. Relevance and the 'newest geography'. *Area* 3, 25–7.

Dickinson, R. E. 1930. The regional functions and zones of influence of Leeds and Bradford. *Geography* 15, 548–57.

Dickinson, R. E. 1933. The distribution and functions of smaller urban settlements of East Anglia. *Geography* 18, 19–31.

Dickinson, R. E. 1938. Landscape and society. *Scottish Geographical Magazine* 55, 1–15.

Dickinson, R. E. 1947. *City Region and Regionalism*. London: Routledge & Kegan Paul.

Dickinson, R. E. 1976. *Regional Concept: The Anglo-American Leaders*. London: Routledge & Kegan Paul.

Dingemans, D. 1979. Redlining and mortgage lending in Sacramento. *Annals of the Association of American Geographers* 69, 225–39.

Dittmer, J. 2010. *Popular Culture, Geopolitics, and Identity*. Plymouth and Lanham, MD: Rowman & Littlefield.

Dixon, D. P. 2014. The way of the flesh: Life, geopolitics and the weight of the future. *Gender, Place & Culture: A Journal of Feminist Geography* 21, 136–51.

Dixon, D. P. and Jones III, J. P. 1996. Editorial: For a supercalifragilisticexpialidocious scientific geography. *Annals of the Association of American Geographers* 86, 767–79.

Dixon, D. P. and Jones III, J. P. 2004. What next? *Environment and Planning A* 36, 381–90.

Dixon, D. P. and Jones III, J. P. 2005. Derridean geographies. *Antipode: A Radical Journal of Geography* 37, 242–5.

Doan, P. L. 2007. Queers in the American city: Transgendered perceptions of urban space. *Gender, Place and Culture* 14, 57–74.

Doan, P. L. 2010. The tyranny of gendered spaces – reflections from beyond the gender dichotomy. *Gender, Place and Culture* 17, 635–54.

Dobson, J. E. 1983a. Automated geography. *Professional Geographer* 35, 135–43.

Dobson, J. E. 1983b. Reply to comments on 'Automated geography'. *Professional Geographer* 35, 349–53.

Dobson, J. E. 1993. The geographic revolution: A retrospective on the age of automated geography. *Professional Geographer* 45, 431–9.

Dodds, K. and Atkinson, D. (eds) 2000. *Geopolitical Traditions: A century of geopolitical thought*. London and New York: Routledge.

Dodgson, R. A. 1998. *Society in Time and Space: A geographical perspective on change*. Cambridge: Cambridge University Press.

Doel, M. A. 1993. Proverbs for paranoids: Writing geography on hollowed ground. *Transactions of the Institute of British Geographers* NS18, 377–94.

Doel, M. A. 1999. *Poststructuralist Geographies: The diabolical art of spatial science*. Edinburgh: University of Edinburgh Press.

Doel, M. A. and Clarke, D. B. 1999. Dark panopticon. Or, attack of the killer tomatoes. *Environment and Planning D: Society and Space* 17, 427–50.

Domosh, M. 1991a. Towards a feminist historiography of geography. *Transactions of the Institute of British Geographers* NS16, 95–104.

Domosh, M. 1991b. Beyond the frontiers of geographical knowledge. *Transactions of the Institute of British Geographers* NS16, 488–90.

Domosh, M. and Bondi, L. 2014. Remembering the making of *Gender, Place and Culture*. *Gender, Place and Culture: A Journal of Feminist Geography* 21, 1063-70.

Donaghy, T. 2014. Walter Isard's evolving sense of the scientific in regional science . . . *International Regional Science Review* 37, 78–95.

Dorling, D. 1995. *A Social Atlas of Britain*. Chichester: John Wiley.

Dorling, D. 1998. Human cartography: When is it good to map? *Environment and Planning A* 30, 277–88.

Dorling, D. 2010. *Injustice: Why social inequality persists*. Bristol: Policy Press.

Dorling, D. 2014. *Inequality and the 1%*. London: Verso Books.

Dorling, D. and Fairbairn, D. 1997. *Mapping: Ways of representing the world*. London: Longman.

Dorling, D. and Shaw, M. 2002. Geographies of the agenda: Public policy, the discipline and its (re)'turns'. *Progress in Human Geography* 26, 629–42.

Dorling, D. and Thomas, B. 2011. *Bankrupt Britain: An atlas of social change*. Bristol: Policy Press.

Dorling, D., Newmman, M. and Barford, A. 2010. *The Atlas of the Real World: Mapping the way we live*. London: Thames & Hudson.

Douglas, I. 1983. *The Urban Environment*. London: Edward Arnold.

Douglas, I. 1986. The unity of geography is obvious. *Transactions of the Institute of British Geographers* NS11, 459–63.

Douglas, I. 2013. *Cities: An environmental history*. London: I. B. Tauris.

Downs, R. M. 1970. Geographic space perception: Past approaches and future prospects. In C. Board, R. J. Chorley, P. Haggett and D. R. Stoddart (eds) *Progress in Geography*, Vol. 2. London: Edward Arnold, 65–108.

Downs, R. M. 1979. Critical appraisal or determined philosophical skepticism? *Annals of the Association of American Geographers* 69, 468–71.

Downs, R. M. and Meyer, J. T. 1978. Geography and the mind. Human geography: Coming of age. *American Behavioral Scientist* 22, 59–78.

Downs, R. M. and Stea, D. (eds) 1973. *Image and Environment*. London: Edward Arnold.

Downs, R. M. and Stea, D. 1977. *Maps in Mind*. New York: Harper & Row.

Drake, C. and Horton, J. 1983. Comment on editorial essay: Sexist bias in political geography. *Political Geography Quarterly* 2, 329–37.

Driver, F, 1995, Sub-merged identities: Familiar and unfamiliar histories. *Annals of the Institute of British Geographers* NS20, 403-4.

Driver, F. 1998. *Geography Militant: Cultures of exploration in an age of empire*. Oxford: Blackwell.

Driver, F. 2003. On geography as a visual discipline. *Antipode* 35, 227–31.

Driver, F. and Philo, C. 1986. Implications of 'scientific' geography. *Area* 18, 161–2.

Drysdale, A. and Watts, M. 1977. Modernization and social protest movements. *Antipode: A Radical Journal of Geography* 9, 40–55.

Duffy, P. 1995. Literary reflections on Irish migration in the nineteenth and twentieth centuries. In R. King, J. Connell and P. White (eds) *Writing Across Worlds: Literature and migration*. London: Routledge, 20–38.

Dunbar, G. S. (ed.) 2002. *Geography: Discipline, profession and subject since 1870*. Dordrecht: Kluwer.

Dunbar, G. S. 1981. Geography in the University of California (Berkeley and Los Angeles) 1868–1941. In G. S. Dunbar, *The History of Geography: Collected essays*. Utica, NY: G. S. Dunbar, 66–75.

Duncan, J. S. 1980. The superorganic in American cultural geography. *Annals of the Association of American Geographers* 70, 181–98.

Duncan, J. S. 1985. Individual action and political power: A structuration perspective. In R. J. Johnston (ed.) *The Future of Geography*. London: Methuen, 174–89.

Duncan, J. S. 1993a. Representing power: The politics and poetics of urban form in the Kandyan Kingdom. In J. S. Duncan and D. Ley (eds) *Place/Culture/Representation*. London: Routledge, 232–50.

Duncan, J. S. 1993b. Commentary. *Annals of the Association of American Geographers* 83, 517–19.

Duncan, J. S. 1994. After the civil war: Reconstructing cultural geography as heterotopia. In K. E. Foote, P. J. Hugill, K. Mathewson and J. M. Smith (eds) *Re-reading Cultural Geography*. Austin: University of Texas Press, 401–8.

Duncan, J. S. and Barns, T. S. 1993. Afterword. In T. J. Barnes and J. S. Duncan (eds) *Writing Worlds: Discourse, text and metaphor in the representation of landscape*. London: Routledge, 248–53.

Duncan, J. S. and Ley, D. 1982. Structural Marxism and human geography: A critical assessment. *Annals of the Association of American Geographers* 72, 30–59.

Duncan, J. S. and Ley, D. 1993. Introduction: Representing the place of culture. In J. S. Duncan and D. Ley (eds) *Place/Culture/Representation*. London: Routledge, 1–21.

Duncan, N. (ed.) 1996a. *Body Space: Destabilizing geographies of gender and sexuality*. London: Routledge.

Duncan, N. 1996b. Postmodernism in human geography. In C. V. Earle, K. Mathewson and M. S. Kenzer (eds) *Concepts in Human Geography*. Lanham, MD: Rowman & Littlefield, 429–58.

Duncan, N. 1999. Postmodernism/postmodernity. In L. McDowell and J. P. Sharp (eds) *A Feminist Glossary of Human Geography*. London: Arnold, 211–13.

Duncan, O. D. 1959. Human ecology and population studies. In P. M. Hauser and O. D. Duncan (eds) *The Study of Population*. Chicago: University of Chicago Press, 678–716.

Duncan, O. D. and Schore, L. F. 1959. Cultural, behavioral and ecological perspectives in the study of social organization. *American Journal of Sociology* 65, 132–46.

Duncan, O. D., Cuzzort, R. P. and Duncan, B. 1961. *Statistical Geography*. New York: The Free Press.

Duncan, S. S. 1974a. Cosmetic planning or social engineering? Improvement grants and improvement areas in Huddersfield. *Area* 6, 259–70.

Duncan, S. S. 1974b. The isolation of scientific discovery: Indifference and resistance to a new idea. *Science Studies* 4, 109–34.

Duncan, S. S. 1975. Research directions in social geography: Housing opportunities and constraints. *Transactions of the Institute of British Geographers* NS1, 10–19.

Duncan, S. S. 1979. Radical geography and Marxism. *Area* 11, 124–6.

Duncan, S. S. 1981. Housing policy, the methodology of levels, and urban research: The case of Castells. *International Journal of Urban and Regional Research* 5, 231–54.

Duncan, S. S. 1989a. Uneven development and the difference that space makes. *Geoforum* 20, 131–9.

Duncan, S. S. 1989b. What is a locality? In R. Peet and N. J. Thrift (eds) *New Models in Geography*, Vol. 2. London: Unwin Hyman, 221–54.

Duncan, S. S. and Goodwin, M. 1985. The local state and local economic policy. *Capital and Class* 27, 14–36.

Duncan, S. S. and Savage, M. 1989. Space, scale and locality. *Antipode: A Radical Journal of Geography* 21, 179–206.

Dunford, M. 1990. Theories of regulation. *Environment and Planning D: Society and Space* 8, 297–322.

Dunford, M. and Perrons, D. 1983. *The Arena of Capital*. London: Macmillan.

Dwyer, C. and Bressey, C. 2008. Introduction: Island geographies: New geographies of race and racism. In C. Dwyer and C. Bressey (eds) *New Geographies of Race and Racism*. Aldershot: Ashgate, 1–13.

Dwyer, M. B. and Baird, I. G. 2014. Principled engagement: Political ecologists and their interactions outside the academy. Introduction to a set of short interventions. *ACME: An International E-Journal for Critical Geographies* 13, 473–7.

Dyck, I. 1990. Space, time and renegotiating motherhood: An exploration of the domestic workplace. *Environment and Planning D: Society and Space* 8, 459–83.

Dyck, I. 2011. Embodied life. In V. Del Casino, M. Thomas, P. Cloke and R. Panelli (eds) *A Companion to Social Geography*. Oxford: Blackwell, 346–61.

Earle, C. V. 1996. Classics in human geography revisited. *Progress in Human Geography* 20, 195–7.

Eden, S. 2003. People and the contemporary environment. In R. J. Johnston and M. Williams (eds) *A Century of British Geography*. Oxford: Oxford University Press for the British Academy, 213–46.

Eilon, S. 1975. Seven faces of research. *Operational Research Quarterly* 26, 359–67.

Ekers, M., Hart, G., Kipfer, S. and Loftus, A. (eds) 2013. *Gramsci: Space, nature, politics*. Oxford: Wiley-Blackwell.

Elden, S. 2013. *The Birth of Territory*. Chicago: University of Chicago Press.

Elden, S. and Mendietta, E. (eds) 2011. *Reading Kant's Geography*. Albany: State University of New York Press.

Elder, G. S. 1999. 'Queering' boundaries in the geography classroom. *Journal of Geography in Higher Education* 23, 86–93.

Eliot Hurst, M. E. 1972. Establishment geography: Or how to be irrelevant in three easy lessons. *Antipode: A Radical Journal of Geography* 5, 40–59.

Eliot Hurst, M. E. 1980. Geography, social science and society: Towards a de-definition. *Australian Geographical Studies* 18, 3–21.

Eliot, M. E. 1985. Geography has neither existence nor future. In R. J. Johnston (ed.) *The Future of Geography*. London: Methuen, 59–91.

Ellegård, K. and Svedin, U. 2012. Torsten Hägerstrand's time – geography as the cradle of the activity approach in transport geography. *Journal of Transport Geography* 23, 17–25.

Elliott, P., Cuzick, J., English, D. and Stern, R. (eds) 1996. *Geographical and Environmental Epidemiology: Methods for small-area studies*. Oxford: Oxford University Press.

Elliott, P., Wakefield, J., Best, N. and Briggs, D. (eds) 2001. *Spatial Epidemiology: Methods and Applications*. Oxford: Oxford University Press.

Elwood, S. 2006. Critical issues in participatory GIS: Deconstructions, reconstructions and new research directions. *Transactions in GIS*, 10, 693–708.

Elwood, S. 2010. Geographic information science: Emerging research on the societal implications of the geospatial web. *Progress in Human Geography* 34, 349–57.

Elwood, S. and Leszczynski, A. 2011. Privacy, reconsidered: New representations, data practices, and the GeoWeb. *Geoforum* 42, 6-15.

England, K. (ed.) 1996. *Who Will Mind the Baby?: Geographies of childcare and working mothers*. London and New York: Routledge.

England, K. 1994. Getting personal: Reflexivity, positionality and feminist research. *Professional Geographer* 46, 80–9.

England, K. 1999. Sexing geography, teaching sexualities [1]. *Journal of Geography in Higher Education* 23, 94–101.

England, K. and Stiell, B. 1997. 'They think you're as stupid as your English is': Constructing foreign domestic workers in Toronto. *Environment and Planning A* 29, 195–215.

Entrikin, J. N. 1976. Contemporary humanism in geography. *Annals of the Association of American Geographers* 66, 615–32.

Entrikin, J. N. 1980. Robert Park's human ecology and human geography. *Annals of the Association of American Geographers* 70, 43–58.

Entrikin, J. N. 1981. Philosophical issues in the scientific study of regions. In D. T. Herbert and R. J. Johnston (eds) *Geography and the Urban Environment*, Vol. 4. Chichester: John Wiley, 1–27.

Entrikin, J. N. 1989. Place, region, and modernity. In J. A. Agnew and J. S. Duncan (eds) *The Power of Place*. Boston, MA: Unwin Hyman, 30–43.

Entrikin, J. N. 1990. Introduction: *The Nature of Geography* in perspective. In J. N. Entrikin and S. D. Brunn (eds) *Reflections on Richard Hartshorne's The Nature of Geography*. Washington: Association of American Geographers, 1–16.

Entrikin, J. N. and Tepple, J. H. 2006. Humanism and democratic place-making. In S. Aitken and G. Valentine (eds) *Approaches to Human Geography*. London: Thousand Oaks, CA and New Delhi: Sage, 30–41.

Economic and Social Research Council (ESRC) 1988. *Horizons and Opportunities in Social Science*. London: ESRC.

Evans, M. 1988. Participant observation: The researcher as research tool. In J. Eyles and D. M. Smith (eds) *Qualitative Methods in Human Geography*. Cambridge: Polity Press, 197–218.

Eyles, J. 1971. Pouring new sentiments into old theories: How else can we look at behavioural patterns? *Area* 3, 242–50.

Eyles, J. 1973. Geography and relevance. *Area* 5, 158–60.

Eyles, J. 1974. Social theory and social geography. In C. Board, R. J. Chorley, P. Haggett and D. R. Stoddart (eds) *Progress in Geography*, Vol. 6. London: Edward Arnold, 27–88.

Eyles, J. 1981. Why geography cannot be Marxist: Towards an understanding of lived experience. *Environment and Planning A* 13, 1371–88.

Eyles, J. 1989. The geography of everyday life. In D. Gregory and R. Walford (eds) *Horizons in Human Geography*. London: Macmillan, 102–17.

Eyles, J. and Lee, R. 1982. Human geography in explanation. *Transactions of the Institute of British Geographers* NS7, 117–21.

Eyles, J. and Smith, D. M. 1978. Social geography. Human geography: Coming of Age. *American Behavioral Scientist* 22, 41–58.

Eyre, J. D. (ed.) 1978. *A Man for All Regions: The contributions of Edward L. Ullman to geography*. Chapel Hill, NC: University of North Carolina at Chapel Hill, Department of Geography, Studies in Geography 11.

Eyre, S. R. 1978. *The Real Wealth of Nations*. London: Edward Arnold.

Fahey, S., Gibson, K., Huxley, M., Johnson, L. C., Mcoughin, J. B., Walker, J., O'Neill, P., Webber, M. J., Fagan, R. H., Massey, D. B. and Fincher, R. 1989. *Spatial Divisions of Labour* in practice. *Environment and Planning A* 21, 655–700.

Falah, G. 1989. Israelization of Palestine human geography. *Progress in Human Geography* 13, 535–50.

Falah, G. 1994. The frontier of political criticism in Israeli geographic practice. *Area* 26, 1–12.

Falah, G. and Nagel, C. (eds) 2005. *Geographies of Muslim Women: Gender, religion, and space*. London and New York: Guildford Press.

Falah, G. and Newman, D. 1995. The spatial manifestation of threat: Israelis and Palestinians seek a 'good boundary'. *Political Geography* 14, 689–706.

Falconer Al-Hindi, K. 2000. Focus: Women in geography in the 21st century. Introductory remarks: Structure, agency, and women geographers in academia at the end of the long twentieth century. *Professional Geographer* 52, 697–702.

Falconer Al-Hindi, K. 2001. Guest editorial. Do you get it? Feminism and quantitative geography. *Environment and Planning D: Society and Space* 19, 505–13.

Fawcett, C. B. 1919. *The Provinces of England*. London: Williams & Norgate.

Featherston, D. and Painter, J. (eds) 2013. *Spatial Politics: Essays for Doreen Massey*. Chichester: Wiley-Blackwell.

Feldman, E. J. 1986. The citizen-scholar: Education and public affairs. In R. W. Kates and I. Burton (eds) *Geography, Resources and Environment*, Vol. 2. *Themes from the work of Gilbert F. White*. Chicago: University of Chicago Press, 188–206.

Fellmann, J. D. 1986. Myth and reality in the origin of American economic geography. *Annals of the Association of American Geographers* 76, 313–30.

Ferguson, R. I. 2003. Publication practices in physical and human geography: A comment on Nigel Thrift's 'The future of geography'. *Geoforum* 34, 9–11.

Feyerabend, P. 1975. *Against Method*. London: New Left Books.

Fieldhouse, E., Pattie, C. J. and Johnston, R. J. 1996. Tactical voting and party constituency campaigning at the 1992 British general election in England. *British Journal of Political Science* 26, 403–18.

Findlay, A. and Werritty, A. 2010. Putting geography in its place. *Scottish Geographical Journal* 126, 215–30.

Fingleton, B. 1984. *Models of Category Counts*. Cambridge: Cambridge University Press.

Fischer, M. M. and Getis, A. (eds) 2009. *Handbook of Applied Spatial Analysis: Software tools, methods and applications*. Berlin: Springer.

Fischer, M. M. and Gopal, S. 1993. Neurocomputing – a new paradigm for geographic information processing. *Environment and Planning A* 25, 757–60.

Fisher, P. F. 1989a. Geographical information system software for teaching. *Journal of Geography in Higher Education* 13, 69–80.

Fisher, P. F. 1989b. Expert system applications in geography. *Area* 21, 279–87.

Fitzsimmons, M. 1989. The matter of nature. *Antipode: A Radical Journal of Geography* 21, 106–21.

Fleming, D. K. 1973. The regionalizing ritual. *Scottish Geographical Magazine* 89, 196–207.

Fleure, H. J. 1919. Human regions. *Scottish Geographical Magazine* 35, 94–105.

Flint, C. (ed.) 2005. *The Geography of War and Peace*. New York: Oxford University Press.

Flowerdew, R. 1986. Three years in British geography. *Area* 18, 263–4.

Flowerdew, R. 1989. Some critical views of modelling in geography. In B. Macmillan (ed.) *Remodelling Geography*. Oxford: Basil Blackwell, 245–54.

Flowerdew, R. 1998. Reacting to ground truth. *Environment and Planning A* 30, 289–302.

Flowerdew, R. 2011. How serious is the modifiable areal unit problem for analysis of English census data? *Population Trends* 145, 106–18.

Folch-Serra, M. 1990. Place, voice and space: Mikhail Bakhtin's dialogical landscape. *Environment and Planning D: Society and Space* 8, 255–74.

Foley, B. and Goldstein, H. 2012. Measuring success: League tables in the public sector. London: The British Academy.

Folke, S. 1972. Why a radical geography must be Marxist. *Antipode: A Radical Journal of Geography* 4(2), 13–18.

Folke, S. 1973. First thoughts on the geography of imperialism. *Antipode: A Radical Journal of Geography* 5(3), 16–20.

Folke, S. and Sayer, A. 1991. What's left to do? Two views from Europe. *Antipode: A Radical Journal of Geography* 23, 240–8.

Foord, J. and Gregson, N. 1986. Patriarchy: Towards a reconceptualisation. *Antipode: A Radical Journal of Geography* 18, 186–211.

Foote, D. G. and Greer-Wootten, B. 1968. An approach to systems analysis in cultural geography. *Professional Geographer* 20, 86–90.

Ford, L. R. 1982. Beware of new geographies. *Professional Geographer* 34, 131–5.

Ford, L. R. 2002. Emerging political paradigms. *Urban Geography* 23, 433–40.

Forer, P. C. 1974. Space through time: A case study with New Zealand airlines. In E. L. Cripps (ed.) *Space-time Concepts in Urban and Regional Models*. London: Pion, 22–45.

Foresman, T. W. (ed.) 1998. *The History of Geographic Information Systems: Perspectives from the pioneers*. Upper Saddle River, NJ: Prentice-Hall.

Forest, B. 1995. West Hollywood as a symbol: The significance of place in the construction of a gay identity. *Environment and Planning D: Society and Space* 13, 133–57.

Forrester, J. W. 1969. *Urban Dynamics*. Cambridge, MA: MIT Press.

Foster, J., Muellerleile, C., Olds, K. and Peck, J. 2007. Circulating economic geographies: Citation patterns and citation behaviour in economic geography, 1982–2006. *Transactions of the Institute of British Geographers* NS32, 295–312.

Fotheringham, A. S. 1981. Spatial structure and distance-decay parameters. *Annals of the Association of American Geographers* 71, 425–36.

Fotheringham, A. S. 1993. On the future of spatial analysis: The role of GIS. *Environment and Planning A Anniversary Issue*, 30–4.

Fotheringham, A. S. 1997. Trends in quantitative methods I: Stressing the local. *Progress in Human Geography* 21, 81–96.

Fotheringham, A. S. and MacKinnon, R. D. 1989. The National Center for Geographic Information and Analysis. *Environment and Planning A* 21, 141–4.

Fotheringham, A. S. and Rogerson, P. A. (eds) 2009. *The SAGE Handbook of Spatial Analysis*. London: Sage.

Fotheringham, A. S., Brunsdon, C. and Charton, M. E. 2000. *Quantitative Geography: Perspectives on spatial data analysis*. London: Sage.

Fotheringham, A. S., Brundson, C. and Charton, M. E. 2002. *Geographically Weighted Regression: The analysis of spatially varying relationships*. Chichester: John Wiley.

Foucault, M. 1972. *The Archaeology of Knowledge*. London: Tavistock Publications.

Franklin, R. S. and Ketchum, J. 2013. Working in a landscape of recession and expansion: Academic jobs in geography in the United States, 1990–2011. *Professional Geographer* 65, 205–20.

Freeman, T. W. 1962. *A Hundred Years of Geography*. Chicago: Aldine.

Freeman, T. W. 1977. *Geographers: Biobibliographical studies*. London: Mansell.

Freeman, T. W. 1979. The British school of geography. *Organon* 14, 205–16.

Freeman, T. W. 1980a. *A History of Modern British Geography*. London: Longman.

Freeman, T. W. 1980b. The Royal Geographical Society and the development of geography. In E. H. Brown (ed.) *Geography, Yesterday and Tomorrow*. Oxford: Oxford University Press, 1–99.

Frickel, S. and Gross, N. 2005. A general theory of scientific/intellectual movements. *American Sociological Review* 70, 204–32.

Fukuyama, F. 1992. *The End of History and the Last Man*. London: Penguin Books.

Fuller, G. A. 1971. The geography of prophylaxis: An example of intuitive schemes and spatial competition in Latin America. *Antipode: A Radical Journal of Geography* 3(1), 21–30.

Fuller, S. 2000. *Thomas Kuhn: A philosophical history for our times*. Chicago: University of Chicago Press.

Gaile, G. L. and Willmott, C. J. (eds) 1984. *Spatial Statistics and Models*. Dordrecht: D. Reidel.

Gaile, G. L. and Willmott, C. J. (eds) 1989. *Geography in America*. Columbus, OH: Merrill.

Gaile, G. L. and Willmott, C. J. (eds) 2004. *Geography in America at the Dawn of the 21st Century*. New York: Oxford University Press.

Gale, N. and Golledge, R. G. 1982. On the subjective partitioning of space. *Annals of the Association of American Geographers* 72, 60–7.

Gamble, A. 1988. *The Free Economy and the Strong State*. London: Macmillan.

Garcia-Ramon, M. D. 2003. Globalization and international geography: The questions of languages and scholarly traditions. *Progress in Human Geography* 27, 1–5.

Garcia-Ramon, M. D., Simonsen, K. and Vaiou, D. 2006. Guest Editorial: Does anglophone hegemony permeate *Gender, Place and Culture*? *Gender, Place and Culture* 13, 1–5.

Garrison, W. L. 1953. Remoteness and the passenger utilization of air transportation. *Annals of the Association of American Geographers* 43, 169.

Garrison, W. L. 1956a. Applicability of statistical inference to geographical research. *Geographical Review* 46, 427–9.

Garrison, W. L. 1956b. Some confusing aspects of common measurements. *Professional Geographer* 8, 4–5.

Garrison, W. L. 1959a. Spatial structure of the economy I. *Annals of the Association of American Geographers* 49, 238–49.

Garrison, W. L. 1959b. Spatial structure of the economy II. *Annals of the Association of American Geographers* 49, 471–82.

Garrison, W. L. 1960a. Spatial structure of the economy III. *Annals of the Association of American Geographers* 50, 357–73.

Garrison, W. L. 1960b. Connectivity of the interstate highway system. *Papers and Proceedings, Regional Science Association* 6, 121–37.

Garrison, W. L. 1962. Simulation models of urban growth and development. In K. Norborg (ed.) *IGU Symposium in Urban Geography*, Lund Studies in Geography B 24. Lund: C. W. K. Gleerup, 91–108.

Garrison, W. L. 1979. Playing with ideas. *Annals of the Association of American Geographers* 69, 118–20.

Garrison, W. L. 1995. Living with and loving a no-win situation. *International Regional Science Review* 17, 327–32.

Garrison, W. L. 2002. Lessons from the design of a life. In P. Gould and F. R. Pitts (eds) *Geographical Voices: Fourteen autobiographical essays*. Syracuse: Syracuse University Press, 99–123.

Garrison, W. L. and Marble, D. F. (eds) 1967a. *Quantitative Geography. Part I: Economic and cultural topics*. Evanston, IL: Northwestern University Studies in Geography, Number 13.

Garrison, W. L. and Marble, D. F. (eds) 1967b. *Quantitative Geography, Part II: Physical and cartographic topics*. Evanston, IL: Northwestern University Studies in Geography, Number 14.

Garrison, W. L. and Marble, D. F. 1957. The spatial structure of agricultural activities. *Annals of the Association of American Geographers* 47, 137–44.

Garrison, W. L., Berry, B. J. L., Marble, D. F., Nystuen, J. D. and Morrill, R. L. 1959. *Studies of Highway Development and Geographic Change*. Seattle, WA: University of Washington Press.

Gatrell, A. C. 1982. *Geometry in Geography and the Geometry of Geography*. Discussion Paper 6, Department of Geography, Salford: University of Salford.

Gattrell, A. C. 1983. *Distance and Space: A geographical perspective*. Oxford: Oxford University Press.

Gatrell, A. C. 1984a. The geometry of a research specialty: Spatial diffusion modelling. *Annals of the Association of American Geographers* 74, 437–53.

Gatrell, A. C. 1984b. Describing the structure of a research literature: Spatial diffusion modelling in geography. *Environment and Planning B: Planning and Design* 11, 29–45.

Gatrell, A. C. 1985. Any space for spatial analysis? In R. J. Johnston (ed.) *The Future of Geography*. London: Methuen, 190–208.

Gatrell, A. C. and Lovett, A. A. 1986. The geography of hazardous waste disposal in England and Wales. *Area* 18, 275–83.

Gatrell, A. C. and Smith, A. 1984. Networks of relations among a set of geographical journals. *Professional Geographer* 36, 300–7.

Gauthier, H. L. 2002. Edward 'Ned' Taaffe (1921–2001). *Annals of the Association of American Geographers* 92, 573–83.

Gauthier, H. L. and Taaffe, E. J. 2003. Three 20th century 'revolutiuons' in American geography. *Urban Geography* 23, 503–27.

Geary, R. C. 1954. The contiguity ratio and statistical mapping. *Incorporated Statistician* 5, 115–41.

Geertz, C. 1983. *Local Knowledge: Further essays in interpretive anthropology.* New York: Basic Books.

Gehlke, C. E. and Biehl, H. 1934. Certain effects of grouping upon the size of the correlation coefficient in census tract material. *Journal of the American Statistical Association,* Supplement 29, 169–70.

Gerike, M. J. 2012. *Explorations in Historiographies of Geographical Knowledges.* PhD thesis. University of Kansas. Available at: http://hdl.handle.net/2097/15043

Gertler, M. 1992. Flexibility revisited: Districts, nation-states, and the forces of production. *Transactions of the Institute of British Geographers* NS17, 259–78.

Getis, A. 1963. The determination of the location of retail activities with the use of a map transformation. *Economic Geography* 39, 1–22.

Getis, A. 1993. Scholarship, leadership and quantitative methods. *Urban Geography* 14, 517–25.

Getis, A. and Boots, B. N. 1978. *Models of Spatial Processes.* Cambridge: Cambridge University Press.

Gibbons, W. 2001. 'Critical of what?': Past and current issues in critical human geography. *History of Intellectual Culture* 1. Available at: www.ucalgary.ca/hic/issues/vol1/4

Gibbs, G. 1995. The relationship between quality in research and quality in teaching. *Quality in Higher Education* 1, 147–57.

Gibson, K. 1991. Considerations on northern Marxist geography: A review from the Antipodes. *Australian Geographer* 22, 75–81.

Gibson, K. and Graham, J. 1992. Rethinking class in industrial geography: Creating a space for alternative politics of class. *Economic Geography* 68, 109–27.

Gibson-Graham, J. K. 1994. Stuffed if I know! Reflections on post-modern feminist social research. *Gender, Place and Culture* 1, 205–24.

Gibson-Graham, J. K. 1996. *The End of Capitalism (As We Knew It): A feminist critique of political economy.* Oxford and Cambridge, MA: Blackwell.

Gibson-Graham, J. K. 2006. *The End of Capitalism (As We Knew It) – With a New Introduction.* Minneapolis, MN: University of Minnesota Press.

Giddens, A. 1976. *New Rules of Sociological Method.* London: Hutchinson.

Giddens, A. 1981. *A Critique of Contemporary Historical Materialism.* London: Macmillan.

Giddens, A. 1984. *The Constitution of Society.* Oxford: Polity Press.

Giddens, A. 1985. *The Nation State and Violence.* Oxford: Polity Press.

Gier, J. and Walton, J. 1987. Some problems with reconceptualising patriarchy. *Antipode: A Radical Journal of Geography* 19, 54–8.

Gilbert, A. 1988. The new regional geography in English- and French-speaking countries. *Progress in Human Geography* 12, 208–28.

Gilbert, D. 1995. Between two cultures: Geography, computing and the humanities. *Ecumene* 2, 1–13.

Gilbert, D. 1996. Between two cultures: Geography, computing and the humanities. *Ecumene* 2, 1–14.

Gilbert, D. 1999. Sponsorship, academic independence and critical engagement: A forum on Shell, the Ogoni dispute and the Royal Geographical Society (with the Institute of British Geographers). *Ethics, Place and Environment* 2, 219–28.

Gilbert, D. 2009. Time to shell out? Reflections on the RGS and corporate sponsorship. *ACME: An International E-Journal for Critical Geographies* 8, 521–9.

Gille, Z. 2010. Reassembling the macrosocial: Modes of production, actor networks and waste regimes. *Environment and Planning A* 42, 1049–64.

Gillen, J. 2015. On the spatiality of the academic job market in critical human geography. *Social and Cultural Geography* 16, 721–9.

Gilmartin, M. and Berg, L. D. 2007. Locating postcolonialism. *Area* 39, 120-4.

Gilmore, R. 2002. Race and globalization. In R. J. Johnston, P. J. Taylor and M. J. Watts (eds) *Geographies of Global Change: Remapping the world.* Oxford: Blackwell, 261–74.

Ginsburg, N. S. 1961. *Atlas of Economic Development*. Chicago: University of Chicago Press.

Ginsburg, N. S. 1972. The mission of a scholarly society. *Professional Geographer* 24, 1–6.

Ginsburg, N. S. 1973. From colonialism to national development: Geographical perspectives on patterns and policies. *Annals of the Association of American Geographers* 63, 1–21.

Glacken, C. J. 1956. Changing ideas of the habitable world. In W. L. Thomas (ed.) *Man's Role in Changing the Face of the Earth*. Chicago: University of Chicago Press, 70–92.

Glacken, C. J. 1967. *Traces on the Rhodian Shore: Nature and culture in Western thought from ancient times to the end of the eighteenth century*. Berkeley, CA: University of California Press.

Glacken, C. J. 1983. A late arrival in academia. In A. Buttimer (ed.) *The Practice of Geography*. London: Longman, 20–34.

Glasmeier, A. K. 2005. *An Atlas of Poverty in America: One nation, pulling apart, 1960–2003*. Abingdon and New York: Routledge.

Gleeson, B. 1996. A geography for disabled people? *Transactions of the Institute of British Geographers* NS21, 63–85.

Glennie, P. and Thrift, N. J. 1996. *Shaping the Day: A history of timekeeping in England and Wales, 1300–1800*. Oxford: Oxford University Press.

Glennie, P. D. and Thrift, N. J. 1992. Modernity, urbanism and modern consumption. *Environment and Planning D: Society and Space* 10, 423–44.

Gober, P., Glasmeier, A. K., Goodman, J. M., Plane, D. A., Stafford, H. A. and Wood, J. S. 1995a. Employment trends in geography, Part 1: Enrollment and degree patterns. *Professional Geographer* 47, 317–28.

Gober, P., Glasmeier, A. K., Goodman, J. M., Plane, D. A., Stafford, H. A. and Wood, J. S. 1995b. Employment trends in geography, Part 2: Current demand conditions. *Professional Geographer* 47, 329–35.

Gober, P., Glasmeier, A. K., Goodman, J. M., Plane, D. A., Stafford, H. A. and Wood, J. S. 1995c. Employment trends in geography, Part 3: Future demand conditions. *Professional Geographer* 47, 336–46.

Goddard, J. B. and Armstrong, P. 1986. The 1986 Domesday project. *Transactions of the Institute of British Geographers* NS11, 279–89.

Godlewska, A. M. C. 1999. *Geography Unbound: French geographic science from Cassini to Humboldt*. Chicago: University of Chicago Press.

Goheen, P. G. 1970. *Victorian Toronto*. Chicago: University of Chicago, Department of Geography, Research Paper 127.

Gold, J. R. 1980. *An Introduction to Behavioural Geography*. Oxford: Oxford University Press.

Gold, J. R. 1992. Image and environment: The decline of cognitive-behaviouralism in human geography and grounds for regeneration. *Geoforum* 23, 239–47.

Gold, J. R. 2010. Reginald Golledge and behavioural geography. *Progress in Human Geography* 34, 683–5.

Golinski, J. 1998. *Making Natural Knowledge: Constructivism and the history of science*. Cambridge: Cambridge University Press.

Golledge, R. G. 1969. The geographical relevance of some learning theories. In K. R. Cox and R. G. Golledge (eds) *Behavioral Problems in Geography: A symposium*. Evanston, IL: Northwestern University Studies in Geography 17, 101–45.

Golledge, R. G. 1970. Some equilibrium models of consumer behavior. *Economic Geography* 46, 417–24.

Golledge, R. G. 1980. A behavioral view of mobility and migration research. *Professional Geographer* 32, 14–21.

Golledge, R. G. 1981a. Misconceptions, misinterpretations, and misrepresentations of behavioral approaches in human geography. *Environment and Planning A* 13, 1325–44.

Golledge, R. G. 1981b. A critical response to Guelke's 'Uncritical rhetoric'. *Professional Geographer* 33, 247–51.

Golledge, R. G. 1983. Models of man, points of view, and theory in social science. *Geographical Analysis* 15, 57–60.

Golledge, R. G. 1993. Geography and the disabled: A survey with special reference to vision impaired and blind populations. *Transactions of the Institute of British Geographers* NS18, 63–85.

Golledge, R. G. 1996. A response to Gleeson and Imrie. *Transactions of the Institute of British Geographers* NS21, 404–11.

Golledge, R. G. 2002. The nature of geographic knowledge. *Annals of the Association of American Geographers* 92, 1–14.

Golledge, R. G. 2007. Building on the down-under experience. In L. J. King (ed.) *North American Explorations: Ten memoirs of geographers from down under*. Victoria: Trafford Publications, 18–35.

Golledge, R. G. and Amedeo, D. 1968. On laws in geography. *Annals of the Association of American Geographers* 58, 760–74.

Golledge, R. G. and Brown, L. A. 1967. Search, learning and the market decision process. *Geografiska Annaler B* 49, 116–24.

Golledge, R. G. and Couclelis, H. 1984. Positivist philosophy and research in human spatial behavior. In T. F. Saarinen, D. Seamon and J. L. Sell (eds) *Environmental Perception and Behavior: An inventory and prospect*. Chicago: Department of Geography, University of Chicago, Research Paper 209, 179–90.

Golledge, R. G. and Rayner, J. N. (eds) 1982. *Proximity and Preference: Problems in the multidimensional analysis of large data sets*. Minneapolis, MN: University of Minnesota Press.

Golledge, R. G. and Rushton, G. 1984. A review of analytic behavioural research in geography. In D. T. Herbert and R. J. Johnston (eds) *Geography and the Urban Environment*, Vol. 6. Chichester: John Wiley, 1–44.

Golledge, R. G. and Stimson R. J. 1997. *Spatial Behavior: A geographic perspective*. New York: Guilford Press.

Golledge, R. G. and Stimson, R. J. 1987. *Analytical Behavioural Geography*. London: Croom Helm.

Golledge, R. G. and Timmermans, H. (eds) 1988. *Behavioural Modelling in Geography and Planning*. London: Croom Helm.

Golledge, R. G. and Timmermans, H. 1990. Applications of behavioural research on spatial problems: I. Cognition. *Progress in Human Geography* 14, 57–100.

Golledge, R. G., Brown, L. A. and Williamson, F. 1972. Behavioral approaches in geography: An overview. *Australian Geographer* 12, 59–79.

Golledge, R. G., Couclelis, H. and Gould, P. R. (eds) 1988. *A Ground for Common Search*. Santa Barbara, CA: The Santa Barbara Geographical Press.

Golledge, R. G., Church, R., Dozier, J., Estes, J. E., Michaelsen, J., Simonett, D. S., Smith, R., Smith, T. R., Strahler, A. H. and Tobler, W. R. and Hart, J. F. 1982. Commentary on 'The highest form of the geographer's art'. *Annals of the Association of American Geographers* 72, 557–9.

Goodchild, M. F. 1992. Geographical information science. *International Journal of Geographical Information Systems* 6, 31–46.

Goodchild, M. F. 1993. The years ahead: Dobson's automated geography in 1993. *Professional Geographer* 45, 444–6.

Goodchild, M. F. 1995a. Future directions for geographic information science. *Annals of GIS* 1, 1–7.

Goodchild, M. F. 1995b. GIS and geographic research. In J. Pickles (ed.) *Ground Truth: The social implications of geographic information systems*. New York: Guilford, 31–50.

Goodchild, M. F. 2009. NeoGeography and the nature of geographic expertise. *Journal of Location Based Services* 3, 82–96.

Goodchild, M. F. 2010. Twenty years of progress: GIScience in 2010. *Journal of Spatial Information Science* 1, 3–20.

Goodchild, M. F. and Janelle, D. G. 1988. Specialization in the structure and organization of geography. *Annals of the Association of American Geographers* 78, 11–28.

Goodchild, M. F. and Mark, D. M. 1987. The fractal nature of geographic phenomena. *Annals of the Association of American Geographers* 87, 265–78.

Goodchild, M. F., Yuan, M. and Cova, T. J. 2007. Towards a general theory of geographic representation in GIS. *International Journal of Geographical Information Science* 21, 239–60.

Goodson, I. 1981. Becoming an academic subject: Patterns of explanation and evolution. *British Journal of Sociology of Education* 2, 163–79.

Goodson, I. 1988. *School Subjects and Curriculum Change: Studies in curriculum history*. Brighton: The Falmer Press.

Goodwin, M., Duncan, S. and Halford, S. 1993. Regulation theory, the local state, and the transition of urban politics. *Environment and Planning D: Society and Space* 11, 67–88.

Gordon, P. E. 2012. Forum: Kuhn's Structure at fifty. Introduction. *Modern Intellectual History* 9, 73–6.

Gorman-Murray, A. 2008. Masculinity and the home: A critical review and conceptual framework. *Australian Geographer* 38, 367–79.

Goss, J. 1995. Marketing the new marketing: The strategic discourse of geodemographic information systems. In J. Pickles (ed.) *Ground Truth: The social implications of geographic information systems*. New York: Guilford, 130–70.

Gottmann, J. 1951. Geography and international relations. *World Politics* 3, 153–73.

Gottmann, J. 1952. The political partitioning of our world: An attempt at analysis. *World Politics* 4, 512–19.

Goudie, A. S. 1986a. *The Human Use of the Environment*. Oxford: Basil Blackwell.

Goudie, A. S. 1986b. The integration of human and physical geography. *Transactions of the Institute of British Geographers* NS11, 454–58.

Goudie, A. S. 1993. Land transformation. In R. J. Johnston (ed.) *The Challenge for Geography: A changing world: A changing discipline*. Oxford: Blackwell, 117–37.

Gould, P. R. 1963. Man against his environment: A game theoretic framework. *Annals of the Association of American Geographers* 53, 290–7.

Gould, P. R. 1965. Wheat on Kilimanjaro: The perception of choice within game and learning theory frameworks. *General Systems Theory* 10, 157–66.

Gould, P. R. 1969. Methodological developments since the fifties. In C. Board, R. J. Chorley, P. Haggett and D. R. Stoddart (eds) *Progress in Geography* 1, London: Edward Arnold, 1–50.

Gould, P. R. 1970a. Is statistix inferens the geographical name for a wild goose? *Economic Geography* 46, 439–48.

Gould, P. R. 1970b. Tanzania 1920–63: The spatial impress of the modernization process. *World Politics* 22, 149–70.

Gould, P. R. 1972. Pedagogic review. *Annals of the Association of American Geographers* 62, 689–700.

Gould, P. R. 1973 [1966]. On mental maps (Michigan Inter-University Community of Mathematical Geographers, Discussion Paper 9, reprint). In R. M. Downs and D. Stea (eds) *Image and Environment: Cognitive mapping and spatial behaviour*. London: Edward Arnold, 182–220.

Gould, P. R. 1975. Mathematics in geography: conceptual revolution or new tool? *International Social Science Journal* 27, 303–27.

Gould, P. R. 1977. What is worth teaching in geography? *Journal of Geography in Higher Education* 1, 20–36.

Gould, P. R. 1978. Concerning a geographic education. In D. A. Lanegran and R. Palm (eds) *An Invitation to Geography*. New York: McGraw-Hill, 202–26.

Gould, P. R. 1979. Geography 1957–1977: The Augean period. *Annals of the Association of American Geographers* 69, 139–51.

Gould, P. R. 1980. Q-analysis, or a language of structure: An introduction for social scientists, geographers and planners. *International Journal of Man-Machine Studies* 12, 169–99.

Gould, P. R. 1981a. Letting the data speak for themselves. *Annals of the Association of American Geographers* 71, 166–76.

Gould, P. R. 1981b. Space and rum: An English note on espacien and rumian meaning. *Geografiska Annaler B* 63, 1–3.

Gould, P. R. 1985a. *The Geographer at Work*. London: Routledge & Kegan Paul.

Gould, P. R. 1985b. Will geographical self-reflection make you blind? In R. J. Johnston (ed.) *The Future of Geography*. London: Methuen, 276–90.

Gould, P. R. 1988. The only perspective: A critique of Marxist claims to exclusiveness in geographical inquiry. In R. G. Golledge, H. Couclelis and P. R. Gould (eds) *A Ground for Common Search*. Santa Barbara, CA: Santa Barbara Geographical Press, 1–10.

Gould, P. R. 1993. Why not? The search for spatio-temporal structure. *Environment and Planning A* Anniversary Issue, 48–55.

Gould, P. R. 1994. Sharing a tradition – geographies from the enlightenment. *Canadian Geographer* 38, 194–202.

Gould, P. R. 1999. *Becoming a Geographer*. Syracuse: Syracuse University Press.

Gould, P. R. and Ola, D. 1970. The perception of residential desirability in the Western Region of Nigeria. *Environment and Planning* 2, 73–8.

Gould, P. R. and Olsson, G. (eds) 1982. *A Search for Common Ground*. London: Pion.

Gould, P. R. and Pitts, F. R. (eds) 2002. *Geographical Voices: Fourteen autobiographical essays*. Syracuse: Syracuse University Press.

Gould, P. R. and White, R. 1974. *Mental Maps*. Harmondsworth: Penguin.

Gould, P. R. and White, R. 1986. *Mental Maps*, 2nd edn. London: George Allen & Unwin.

Graham, E. 1995. Postmodernism and the possibility of a new human geography. *Scottish Geographical Magazine* 111, 175–8.

Graham, M. and Shelton, T. 2013. Geography and the future of big data, big data and the future of geography. *Dialogues in Human Geography* 3, 255–61.

Granger, A. 1990. *The Threatening Desert*. London: Earthscan.

Gräno, O. 1981. External influence and internal change in the development of geography. In D. R. Stoddart (ed.) *Geography, Ideology and Social Concern*. Oxford: Blackwell, 17–36.

Graves, N. J. 1981. Can geographical studies be subsumed under one paradigm or are a plurality of paradigms inevitable? *Terra* 93, 85–90.

Gray, F. 1975. Non-explanation in urban geography. *Area* 7, 228–35.

Gray, F. 1976. Selection and allocation in council housing. *Transactions of the Institute of British Geographers* NS1, 34–46.

Green, N. P., Finch, S. and Wiggins, J. 1985. The 'state of the art' in Geographical Information Systems. *Area* 17, 295–301.

Greenberg, D. 1984. Whodunit? Structure and subjectivity in behavioral geography. In T. F. Saarinen, D. Seamon and J. L. Sell (eds) *Environmental Perception and Behavior: An inventory and prospect*. Chicago: Department of Geography, University of Chicago, Research Paper 209, 191–208.

Greer-Wootten, B. 1972. *The Role of General Systems Theory in Geographic Research*. Department of Geography, Toronto: York University, Discussion Paper No. 3.

Gregory, D. 1976. Rethinking historical geography. *Area* 8, 295–9.

Gregory, D. 1978a. *Ideology, Science and Human Geography*. London: Hutchinson.

Gregory, D. 1978b. The discourse of the past: Phenomenology, structuralism, and historical geography. *Journal of Historical Geography* 4, 161–73.

Gregory, D. 1980. The ideology of control: Systems theory and geography. *Tijdschrift voor Economische en Sociale Geografie* 71, 327–42.

Gregory, D. 1981. Human agency and human geography. *Transactions of the Institute of British Geographers* NS6, 1–18.

Gregory, D. 1982a. *Regional Transformation and Industrial Revolution: A geography of the Yorkshire woollen industry*. London: Macmillan.

Gregory, D. 1982b. Solid geometry: Notes on the recovery of spatial structure. In P. R. Gould and G. Olsson (eds) *A Search for Common Ground*. London: Pion, 187–222.

Gregory, D. 1985a. Suspended animation: The stasis of diffusion theory. In D. Gregory and J. Urry (eds) *Social Relations and Spatial Structures*. London: Macmillan, 296–336.

Gregory, D. 1985b. People, places and practices: The future of human geography. In R. King (ed.) *Geographical Futures*. Sheffield: Geographical Association, 56–76.

Gregory, D. 1989a. Areal differentiation and post-modern human geography. In D. Gregory and R. Walford (eds) *Horizons in Human Geography*. London: Macmillan, 67–96.

Gregory, D. 1989b. The crisis of modernity? Human geography and critical social theory. In R. Peet and N. J. Thrift (eds) *New Models in Geography*, Vol. 2. London: Unwin Hyman, 348–85.

Gregory, D. 1990. Chinatown, Part Three? Soja and the missing spaces of social theory. *Strategies: A Journal of Theory, Culture and Politics* 3, 40–104.

Gregory, D. 1994. *Geographical Imaginations*. Oxford: Blackwell.

Gregory, D. 1995a. Between the book and the lamp: Imaginative geographies of Egypt, 1849–50. *Transactions of the Institute of British Geographers* NS20, 29–57.

Gregory, D. 1995b. Imaginative geographies. *Progress in Human Geography* 19, 447–85.

Gregory, D. 1997. *Explorations in Critical Human Geography*. Heidelberg: Department of Geography, University of Heidelberg.

Gregory, D. 2004. *The Colonial Present: Afghanistan, Palestine, Iraq*. Malden, MA and Oxford: Blackwell.

Gregory, D. 2006. Introduction: Troubling geographies. In N. Castree and D. Gregory (eds) *David Harvey: A critical reader*. Malden, MA and Oxford: Blackwell, 1–25.

Gregory, D. 2009. Geographical imagination. In D. Gregory, R. Johnston, G. Pratt, M. J. Watts and S. Whatmore (eds) *The Dictionary of Human Geography*, 5th edn. Malden, MA and Oxford: Wiley-Blackwell, 282–5.

Gregory, D. and Castree, N. (eds) 2006. *David Harvey: A critical reader*, Malden, MA and Oxford: Blackwell.

Gregory, D. and Castree, N. 2012. Editors' introduction: Human geography. In D. Gregory and N. Castree (eds) *Human Geography* (5 vols). London, Thousand Oaks, CA and New Delhi and Singapore: Sage, xxv–lxxix.

Gregory, D. and Ley, D. 1988. Culture's geographies. *Environment and Planning D: Society and Space* 6, 115–16.

Gregory, D. and Urry, J. 1985. Introduction. In D. Gregory and J. Urry (eds) *Social Relations and Spatial Structures*. London: Macmillan, 1–8.

Gregory, D. and Walford, R. 1989. Introduction: Making geography. In D. Gregory and R. Walford (eds) *Horizons in Human Geography*. London: Macmillan, 1–7.

Gregory, K. J. 2003. Physical geography and geography as an environmental science. In R. J. Johnston and M. Williams (eds) *A Century of British Geography*. Oxford: Oxford University Press for the British Academy, 93–136.

Gregory, S. 1963. *Statistical Methods and the Geographer*. London: Longman.

Gregory, S. 1976. On geographical myths and statistical fables. *Transactions of the Institute of British Geographers* NS1, 385–400.

Gregson, N. 1986. On duality and dualism: The case of structuration and time geography. *Progress in Human Geography* 10, 184–205.

Gregson, N. 1987a. The CURS initiative: Some further comments. *Antipode: A Radical Journal of Geography* 19, 364–70.

Gregson, N. 1987b. Structuration theory: Some thoughts on the possibilities for empirical research. *Environment and Planning D: Society and Space* 5, 73–91.

Gregson, N. 1995. And now it's all consumption. *Progress in Human Geography* 19, 135–41.

Gregson, N. and Crang, M. 2014. Waste, resource recovery and labour: Recycling economies in the EU. In J. Michie and C. Cooper (eds) *Why the Social Sciences Matter*. Basingstoke: Palgrave-Macmillan.

Gregson, N. and Foord, J. 1987. Patriarchy: Comments on critics. *Antipode: A Radical Journal of Geography* 19, 371–5.

Gregson, N. and Lowe, M. 1994. *Servicing the Middle Classes: Class, gender and waged domestic labour in contemporary Britain*. London and New York: Routledge.

Gregon, N. and Rose, G. 1997. Contested and negotiated histories of feminist geographies. In Women and Geography Study Group (eds) *Feminist Geographies: Explorations in diversity and difference*. Harlow: Longman, 13–48.

Gregson, N., Crewe, L. and Brooks, K. 2002. Retailing, space and practice. *Environment and Planning D: Society and Space* 20, 597–617.

Gregson, N., Rose, G., Cream, J. and Laurie, N. 1997. Conclusions. In Women and Geography Study Group (eds) *Feminist Geographies: Explorations in diversity and difference*. Harlow: Longman, 191–200.

Gribbin, J. 2002. *Science: A history 1543–2001*. London: Penguin.

Griffiths, M. J. and Johnston, R. J. 1991. What's in a place? An approach to the concept of place as illustrated by the British National Union of Mineworkers' strike, 1984–1985. *Antipode: A Radical Journal of Geography* 23, 185–213.

Grigg, D. B. 1977. E. G. Ravenstein and the laws of migration. *Journal of Historical Geography* 3, 41–54.

Grossman, L. 1977. Man–environment relationships in anthropology and geography. *Annals of the Association of American Geographers* 67, 126–44.

Gudgin, G. and Taylor, P. J. 1979. *Seats, Votes and the Spatial Organisation of Elections*. London: Pion. (Reprinted 1992 by ECPR Press, Colchester.)

Guelke, L. 1971. Problems of scientific explanation in geography. *Canadian Geographer* 15, 38–53.

Guelke, L. 1974. An idealist alternative in human geography. *Annals of the Association of American Geographers* 14, 193–202.

Guelke, L. 1975. On rethinking historical geography. *Area* 7, 135–8.

Guelke, L. 1976. The philosophy of idealism. *Annals of the Association of American Geographers* 66, 168–9.

Guelke, L. 1977a. The role of laws in human geography. *Progress in Human Geography* 1, 376–86.

Guelke, L. 1977b. Regional geography. *Professional Geographer* 29, 1–7.

Guelke, L. 1978. Geography and logical positivism. In D. T. Herbert and R. J. Johnston (eds) *Geography and the Urban environment: Progress in research and applications*, Vol. 1. London: John Wiley, 35–61.

Guelke, L. 1981a. Uncritical rhetoric: 'A classic disservice'. *Professional Geographer* 33, 246–7.

Guelke, L. 1981b. Idealism. In M. E. Harvey and B. P. Holly (eds) *Themes in Geographic Thought*. London: Croom Helm, 133–47.

Guelke, L. 1982. The idealist dispute in Anglo-American geography: A comment. *Canadian Geographer* 26, 51–7.

Guelke, L. 1995. Review of *Geographical imaginations* by Derek Gregory. *Canadian Geographer* 39, 184–6.

Gutierréz, J. and López-Nieva, P. 2001. Are international journals of human geography really international? *Progress in Human Geography* 25, 55–71.

Gutting, G. 1980. Introduction. In G. Gutting (ed.) *Paradigms and Revolutions*. Notre Dame, IN: University of Notre Dame Press, 1–22.

Habermas, J. 1972. *Knowledge and Human Interests*. London: Heinemann.

Hacking, I. 1983. *Representing and Intervening*. Cambridge: Cambridge University Press.

Hägerstrand, T. 1967. The computer and the geographer. *Transactions and Papers of the Institute of British Geographers* 42, 1–20.

Hägerstrand, T. 1968. *Innovation Diffusion as a Spatial Process*. Chicago: University of Chicago Press.

Hägerstrand, T. 1975. Space, time and human conditions. In A. Karlquist, L. Lundquist and F. Snickars (eds) *Dynamic Allocation of Urban Space*. Farnborough: Saxon House, 3–12.

Hägerstrand, T. 1977. The geographers' contribution to regional policy: The case of Sweden. In D. R. Deskins et al. (eds) *Geographic Humanism, Analysis and Social Action: A half century of geography at Michigan*. Ann Arbor, MI: Michigan Geographical Publications No. 17, 329–46.

Hägerstrand, T. 1982. Diorama, path and project. *Tijdschrift voor Economische en Sociale Geografie* 73, 323–39.

Hägerstrand, T. 1984. Presence and absence: A look at conceptual choices and bodily necessities. *Regional Studies* 18, 373–8.

Haggett, P. 1964. Regional and local components in the distribution of forested areas in southeast Brazil: A multivariate approach. *Geographical Journal* 130, 365–77.

Haggett, P. 1965a. Changing concepts in economic geography. In R. J. Chorley and P. Haggett (eds) *Frontiers in Geographical Teaching*. London: Methuen, 101–17.

Haggett, P. 1965b. Scale components in geographical problems. In R. J. Chorley and P. Haggett (eds) *Frontiers in Geographical Teaching*. London: Methuen, 164–85.

Haggett, P. 1965c. *Locational Analysis in Human Geography*. London: Edward Arnold.

Haggett, P. 1967. Network models in geography. In R. J. Chorley and P. Haggett (eds) *Models in Geography*. London: Methuen, 609–70.

Haggett, P. 1969. On geographical research in a computer environment. *Geographical Journal* 135, 497–507.

Haggett, P. 1972. *Geography: A modern synthesis*. New York: Harper & Row.

Haggett, P. 1973. Forecasting alternative spatial, ecological and regional futures: Problems and possibilities. In R. J. Chorley (ed.) *Directions in Geography*. London: Methuen, 217–36.

Haggett, P. 1978. The spatial economy. Human geography: Coming of age. *American Behavioral Scientist* 22, 151–67.

Haggett, P. 1990. *The Geographers' Art*. Oxford: Blackwell Publishers.

Haggett, P. 1991. Classics in human geography revisited. *Progress in Human Geography* 15, 300–2.

Haggett, P. 1994. Prediction and predictability in geographical systems. *Transactions of the Institute of British Geographers* NS19, 6–20.

Haggett, P. 2000. *The Geographical Structure of Epidemics*. Oxford: The Clarendon Press.

Haggett, P. 2001. *Geography: A global synthesis*. Harlow: Pearson Education.

Haggett, P. 2005. Peter Robin Gould (1932–2000). In P. H. Armstrong and G. J. Martin (eds) *Geographers: Biobibliographical studies*, Vol. 24. London and New York: Continuum, 42–62.

Haggett, P. 2012. *The Quantocks: Biography of an English region*. Chew Magna, Somerset: Point Walter Press.

Haggett, P. 2015. Madingley: Half-century reflections on a geographical experiment. *Geography* 100, 5–11.

Haggett, P. and Chorley, R. J. 1965. Frontier movements and the geographical tradition. In R. J. Chorley and P. Haggett (eds) *Frontiers in Geographical Teaching*. London: Methuen, 358–78.

Haggett, P. and Chorley, R. J. 1967. Models, paradigms, and the new geography. In R. J. Chorley and P. Haggett (eds) *Models in Geography*. London: Methuen, 19–42.

Haggett, P. and Chorley, R. J. 1969. *Network Models in Geography*. London: Edward Arnold.

Haggett, P. and Chorley, R. J. 1989. From Madingley to Oxford. In B. Macmillan (ed.) *Remodelling Geography*. Oxford: Basil Blackwell, xv–xx.

Haggett, P., Clifee, A. D. and Frey, A. E. 1977. *Locational Analysis in Human Geography*, 2nd edn. London: Edward Arnold.

Hagood, M. J. 1943. Development of a 1940 rural farm level of living index for counties. *Rural Sociology* 8, 171–80.

Hague, E. 2002. Intervention roundtable. Antipode, Inc? *Antipode: A Radical Journal of Geography* 34, 655–61.

Haigh, M. J. 1992. The crisis in American geography. *Area* 14, 185–9.

Haines-Young, R. 1989. Modelling geographical knowledge. In B. Macmillan (ed.) *Remodelling Geography*. Oxford: Basil Blackwell, 22–39.

Haines-Young, R. and Petch, J. R. 1978. *The Methodological Limitations of Kuhn's Model of Science*. Salford: University of Salford, Department of Geography, Discussion Paper 8.

Haines-Young, R. and Petch, J. R. 1985. *Physical Geography: Its nature and methods*. London: Harper & Row.

Haining, R. P. 1980. Spatial autocorrelation problems. In D. T. Herbert and R. J. Johnston (eds) *Geography and the Urban Environment*, Vol. 3. Chichester: John Wiley, 1–44.

Haining, R. P. 1981. Analysing univariate maps. *Progress in Human Geography* 5, 58–78.

Haining, R. P. 1989. Geography and spatial statistics: Current positions, future developments. In B. Macmillan (ed.) *Remodelling Geography*. Oxford: Basil Blackwell, 191–203.

Haining, R. P. 1990. *Spatial Data Analysis in the Social and Environmental Sciences*. Cambridge: Cambridge University Press.

Haining, R. P. 2009. Spatial methodologies to support local policing in the UK: Glimpsing the future. *21st Century Society* 4, 161–74.

Haklay, M., Singleton, A. and Parker, C. 2008. Web mapping 2.0. The necrogeography of the GeoWeb. *Geography Compass* 2, 2011–39.

Halas, M. 2014. Searching for the perfect footnote: Friedrich Ratzel and the others at the roots of Lebensraum. *Geopolitics* 19, 1–18.

Hall, P. 1974. The new political geography. *Transactions of the Institute of British Geographers* 63, 48–52.

Hall, P. 1981a. *Great Planning Disasters*. London: Penguin.

Hall, P. 1981b. The geographer and society. *Geographical Journal* 147, 145–52.

Hall, P. 1982. The new political geography: Seven years on. *Political Geography Quarterly* 1, 65–76.

Hall, P. 1989. *London 2001*. London: Unwin Hyman.

Hall, P. 1998. *Cities in Civilization: Culture, innovation and urban order*. London: Weidenfeld & Nicolson.

Hall, P. 2003. Geographers and the urban century. In R. J. Johnston and M. Williams (eds) *A Century of British Geography*. Oxford: Oxford University Press for the British Academy, 545–62.

Hall, P. 2014. *Good Cities, Better Lives: How Europe discovered the lost art of urbanism*. London: Routledge.

Hall, P., Drewett, R., Gracey, H. and Thomas, R. 1973. *The Containment of Urban England*. London: Allen & Unwin.

Hall, P., Jackson, P., Massey, D., Robson, B. T., Thrift, N. J. and Wilson, A. G. 1987. Horizons and opportunities in research. *Area* 19, 266–72.

Hall, T. 2014. Making their own futures? Research change and diversity amongst contemporary British human geographers. *Geographical Journal* 180, 39–51.

Hall, T., Toms, P., McGuinness, M., Parker, C. and Roberts, N. 2015. Where's the Geography department? The changing administrative place of Geography in UK higher education. *Area* 47, 56–64.

Halvorson, P. and Stave, B. M. 1978. A conversation with Brian J. L. Berry. *Journal of Urban History* 4, 209–38.

Hamilton, F. E. J. 1974. A view of spatial behaviour, industrial organizations, and decision-making. In F. E. I. Hamilton (ed.) *Spatial Perspectives on Industrial Organization and Decision-making*. London: John Wiley, 3–46.

Hammett, D. 2012. Tales from the road: Reflections on power and disciplining within the academy. *Environment and Planning A* 44, 445–457.

Hamnett, C. 1977. Non-explanation in urban geography: Throwing the baby out with the bath water. *Area* 9, 143–5.

Hamnett, C. 1991. The blind men and the elephant: The explanation of gentrification. *Transactions of the Institute of British Geographers* NS16, 259–7.

Hamnett, C. 1997. The sleep of reason? *Environment and Planning D: Society and Space* 16, 127–8.

Hamnett, C. 2003. Contemporary human geography: Fiddling while Rome burns? *Geoforum* 34, 1–3.

Hannah, M. and Strohmayer, U. 1992. Postmodernism (s)trained. *Annals of the Association of American Geographers* 82, 308–10.

Hansen, K. (ed.) 2010. *Millennium Cohort Study: First, second, third and fourth surveys – a guide to the datasets*. London: Institute for Education, University of London.

Hanson, S. 1992. Geography and feminism: Worlds in collision? *Annals of the Association of American Geographers* 82, 569–86.

Hanson, S. 1993. 'Never question the assumptions' and other scenes from the revolution. *Urban Geography* 14, 552–6.

Hanson, S. 2002. Connections. *Gender, Place and Culture* 9, 301–3.

Hanson, S. 2004. Who are 'we'? An important question for geography's future. *Annals of the Association of American Geographers* 94, 715–22.

Hanson, S. and Pratt, G. 1995. *Gender, Work and Space*. London: Routledge.

Hare, F. K. 1974. Geography and public policy: A Canadian view. *Transactions of the Institute of British Geographers* 63, 25–8.

Hare, F. K. 1977. Man's world and geographers: A secular sermon. In D. R. Deskins, G. Kish, J. D. Nystuen and G. Olsson (eds) *Geographic Humanism, Analysis and Social Action: Proceedings of symposia celebrating a half century of geography at Michigan*. Ann Arbor, MI: Michigan Geographical Publication No. 17, 259–73.

Haritaworn, J. 2007. Queer mixed race? Interrogating homonormativity through Thai interraciality. In K. Browne, J. Lim and G. Brown (eds) *Geographies of Sexualities: Theory, practices and politics*. Farnham and Burlington, VT: Ashgate, 101–12.

Haritaworn, J. 2007. Shifting positionalities: Reflections on a queer/trans of colour methodology. *Sociological Research Online* 13 (1). Available at: www.socresonline.org.uk/13/1/13.html

Haritaworn, J. 2009. Hybrid border crossers? Towards a radical socialisation of 'mixed race'. *Journal of Ethnic and Migration Studies* 35, 115–32.

Haritaworn, J. 2012. *The Biopolitics of Mixing: Thai multiracialities and haunted ascendancies*. Farnham: Ashgate.

Harland, R. 1987. *Superstructuralism: The philosophy of structuralism and post-structuralism*. London: Methuen.

Harley, J. B. 1989. Deconstructing the map. *Cartographica* 26, 1–20.

Harley, J. B. 1990. Cartography, ethics and social theory. *Cartographica* 27, 1–23.

Harley, J. B. 1992. Rereading the maps of the Columbian encounter. *Annals of the Association of American Geographers* 82, 522–42.

Harries, K. D. 1974. *The Geography of Crime and Justice.* New York: McGraw-Hill.

Harries, K. D. 1975. Rejoinder to Richard Peet: 'The geography of crime: a political critique'. *Professional Geographer* 27, 280–2.

Harries, K. D. 1976. Observations on radical versus liberal theories of crime causation. *Professional Geographer* 28, 100–13.

Harris, C. D. 1954. The market as a factor in the localization of industry in the United States. *Annals of the Association of American Geographers* 44, 315–48.

Harris, C. D. 1977. Edward Louis Ullman, 1912–1976. *Annals of the Association of American Geographers* 67, 595–600.

Harris, C. D. 1978. Patterns of cities. In J. D. Eyre (ed.) *A Man for All Regions: The contributions of Edward L. Ullman to geography.* Chapel Hill, NC: University of North Carolina, Department of Geography, Studies in Geography No. 11, 66–79.

Harris, C. D. 1990. Urban geography in the United States: The formative years. *Urban Geography* 11, 403–17.

Harris, C. D. 1992. Areal patterns of cities through time and space: Technology and culture ('The nature of cities' further considered). *Colloquium Geographicum* 22, 41–53.

Harris, C. D. 1997a. Geographers in the U. S. government in Washington, DC, during World War II. *Professional Geographer* 49, 245–56.

Harris, C. D. 1997b. 'The nature of cities' and urban geography in the last half century. *Urban Geography* 18, 15–35.

Harris, C. D. and Ullman, E. L. 1945. The nature of cities. *Annals of the American Academy of Political and Social Science* 242, 7–17.

Harris, P., Brunsdon, C. and Charlton, M. 2011. Geographically weighted principal components analysis. *International Journal of Geographical Information Science* 25, 1717–36.

Harris, R. C. 1971. Theory and synthesis in historical geography. *Canadian Geographer* 15, 157–72.

Harris, R. C. 1978. The historical mind and the practice of geography. In D. Ley and M. S. Samuels (eds) *Humanistic Geography: Problems and prospects.* Chicago: Maaroufa Press, 123–37.

Harris, R., Sleight, P. and Webber, R. 2005. *Geodemographics, GIS and Neighbourhood Targeting.* Chichester: John Wiley.

Harrison, P. 2002. The caesura: Remarks on Wittgenstein's interruption of theory, or, why practices elude representation. *Geoforum* 32, 487–503.

Harrison, R. T. and Livingstone, D. N. 1982. Understanding in geography: Structuring the subjective. In D. T. Herbert and R. J. Johnston (eds) *Geography and the Urban Environment*, 5. Chichester: John Wiley, 1–40.

Harrison Church, R. J. 1957. *West Africa: A study of the environment and man's use of it.* London: Longmans Green.

Hart, J. F. 1982a. The highest form of the geographer's art. *Annals of the Association of American Geographers* 72, 1–29.

Hart, J. F. 1982b. Comment in reply. Commentary on 'The highest form of the geographer's art'. *Annals of the Association of American Geographers* 72, 559.

Hart, J. F. 1990. Canons of good editorship. *Professional Geographer* 42, 354–8.

Hartshorne, R. 1927. Location as a factor in geography. *Annals of the Association of American Geographers* 17, 92–9.

Hartshorne, R. 1939. *The Nature of Geography.* Lancaster, PA: Association of American Geographers.

Hartshorne, R. 1948. On the mores of methodological discussion in American geography. *Annals of the Association of American Geographers* 38, 492–504.

Hartshorne, R. 1954a. Political geography. In P. E. James and C. F. Jones (eds) *American Geography: Inventory and prospect.* Syracuse: Syracuse University Press, 167–225.

Hartshorne, R. 1954b. Comment on 'Exceptionalism in geography'. *Annals of the Association of American Geographers* 44, 108–9.

Hartshorne, R. 1955. 'Exceptionalism in Geography' re-examined. *Annals of the Association of American Geographers* 45, 205–44.

Hartshorne, R. 1958. The concept of geography as a science of space from Kant and Humboldt to Hettner. *Annals of the Association of American Geographers* 48, 97–108.

Hartshorne, R. 1959. *Perspective on the Nature of Geography.* Chicago: Rand McNally.

Hartshorne, R. 1972. Review of Kant's concept of geography. *Canadian Geographer* 16, 77–9.

Hartshorne, R. 1979. Notes towards a bibliography of The Nature of Geography. *Annals of the Association of American Geographers* 69, 63–76.

Hartshorne, R. 1984. Meetings: Session 1983–84. *Geographical Journal* 150, 429.

Harvey, D. 1967a. Models of the evolution of spatial patterns in geography. In R. J. Chorley and P. Haggett (eds) *Models in Geography*. London: Methuen, 549–608.

Harvey, D. 1967b. Editorial introduction: The problem of theory construction in geography. *Journal of Regional Science* 7, 211–16.

Harvey, D. 1969a. *Explanation in Geography*. London: Edward Arnold.

Harvey, D. 1969b. Review of A. Pred, *Behavior and location: Part I*. *Geographical Review* 59, 312–14.

Harvey, D. 1969c. Conceptual and measurement problems in the cognitive-behavioral approach to location theory. In K. R. Cox and R. G. Golledge (eds) *Behavioral Problems in Geography: A symposium*. Northwestern University Studies in Geography 17, 35–68.

Harvey, D. 1970. Behavioral postulates and the construction of theory in human geography. *Geographica Polonica* 18, 27–46.

Harvey, D. 1972. Revolutionary and counter-revolutionary theory in geography and the problem of ghetto formation. *Antipode: A Radical Journal of Geography* 4(2), 1–13.

Harvey, D. 1973. *Social Justice and the City*. London: Edward Arnold.

Harvey, D. 1974a. A commentary on the comments. *Antipode: A Radical Journal of Geography* 4(2), 36–41.

Harvey, D. 1974b. Discussion with Brian Berry. *Antipode: A Radical Journal of Geography* 6(2), 145–8.

Harvey, D. 1974c. What kind of geography for what kind of public policy? *Transactions of the Institute of British Geographers* 63, 18–24.

Harvey, D. 1974d. Population, resources and the ideology of science. *Economic Geography* 50, 256–77.

Harvey, D. 1974e. Class-monopoly rent, finance capital and the urban revolution, *Regional Studies* 8, 239–55.

Harvey, D. 1975a. Class structure in a capitalist society and the theory of residential differentiation. In R. Peel, M. Chisholm and P. Haggett (eds) *Processes in Physical and Human Geography: Bristol essays*. London: Heinemann, 354–69.

Harvey, D. 1975b. The political economy of urbanization in advanced capitalist societies: The case of the United States. In G. Gappert and H. M. Rose (eds) *The Social Economy of Cities*. Beverly Hills: Sage, 119–63.

Harvey, D. 1975c. Review of B. J. L. Berry, *The Human Consequences of Urbanization*. *Annals of the Association of American Geographers* 65, 99–103.

Harvey, D. 1976. The Marxian theory of the state. *Antipode: A Radical Journal of Geography* 8(2), 80–9.

Harvey, D. 1978. The urban process under capitalism: A framework for analysis. *International Journal of Urban and Regional Research* 2, 101–32.

Harvey, D. 1979. Monument and myth. *Annals of the Association of American Geographers* 69, 362–81.

Harvey, D. 1982. *The Limits to Capital*. Oxford: Blackwell.

Harvey, D. 1984. On the history and present condition of geography: An historical materialist manifesto. *Professional Geographer* 36, 1–11.

Harvey, D. 1985. The world-systems theory trap. *Studies in Comparative International Development* 22, 42–7.

Harvey, D. 1985a. *The Urbanization of Capital*. Oxford: Basil Blackwell.

Harvey, D. 1985b. The geopolitics of capitalism. In D. Gregory and J. Urry (eds) *Social Relations and Spatial Structures*. London: Macmillan, 128–63.

Harvey, D. 1985c. *Consciousness and the Urban Experience*. Oxford: Basil Blackwell.

Harvey, D. 1987. Three myths in search of a reality in urban studies. *Environment and Planning D: Society and Space* 5, 367–76.

Harvey, D. 1989a. *The Condition of Postmodernity*. Oxford: Basil Blackwell.

Harvey, D. 1989b. From models to Marx: Notes on the project to 'remodel' contemporary geography. In B. Macmillan (ed.) *Remodelling Geography*. Oxford: Basil Blackwell, 211–16.

Harvey, D. 1989c. From managerialism to entrepreneurialism: The transformation of urban governance in late capitalism. *Geografiska Annaler* 71B, 3–17.

Harvey, D. 1990. Between space and time: Reflections on the geographical imagination. *Annals of the Association of American Geographers* 80, 418–34.

Harvey, D. 1992. Postmodern morality plays. *Antipode: A Radical Journal of Geography* 24, 300–26.

Harvey, D. 1993a. Class relations, social justice and the politics of difference. In M. Keith and S. Pile (eds) *Place and the Politics of Identity*. London: Routledge, 41–66.

Harvey, D. 1993b. From space to place and back again: Reflections on the condition of postmodernity. In J. Bird et al. (eds) *Mapping the Futures: Local cultures, global change*. London: Routledge, 3–29.

Harvey, D. 1994. The nature of environment: The dialectics of social and environmental change. In R. Miliband and L. Panitch (eds) *Real Problems, False Solutions: Socialist Register 1993*. London: The Merlin Press, 1–51.

Harvey, D. 1995. Militant particularism and global ambition: The conceptual politics of place, space, and environment in the work of Raymond Williams. *Social Text* 42, 69–98.

Harvey, D. 1996a. *Justice, Nature and the Geography of Difference.* Baltimore, MD: John Hopkins University Press.

Harvey, D. 1996b. Cities or urbanization. *City* 1–2, 38–61.

Harvey, D. 1999. *The Limits to Capital*, 2nd edn. London: Verso.

Harvey, D. 2000. *Spaces of Hope.* Berkeley, CA and Los Angeles: University of California Press.

Harvey, D. 2002. Memories and desires. In P. Gould and F. R. Pitts, (eds) *Geographical Voices: Fourteen autobiographical essays.* Syracuse: Syracuse University Press, 149–88.

Harvey, D. 2003. *The New Imperialism.* (Clarendon lectures in geography and environmental studies.) Oxford and New York: Oxford University Press.

Harvey, D. 2005. 'For a ruthless criticism of everything existing': Jim Blaut's contribution to geographical knowledge. *Antipode: A Radical Journal of Geography* 37, 927–35.

Harvey, D. 2005a. The sociological and geographical imaginations. *International Journal of Politics, Culture and Society* 18, 211–55.

Harvey, D. 2005b. *A Brief History of Neoliberalism.* Oxford and New York: Oxford University Press.

Harvey, D. 2005c. *The New Imperialism.* Oxford: Oxford University Press.

Harvey, D. 2006. The geographies of critical geography. *Transactions of the Institute of British Geographers* NS31, 409–12.

Harvey, D. 2009. *Cosmopolitanism and the Geographies of Freedom.* New York: Columbia University Press.

Harvey, D. 2011. The urban roots of financial crisis: Reclaiming the city for anti-capitalist struggle. *Socialist Register* 48, 1–35.

Harvey, D. 2012. *Rebel Cities: From the right to the city to the urban revolution.* London: Verso.

Harvey, D. 2014. *Seventeen Contradictions and the End of Capitalism.* London: Profile Books.

Harvey, D. and Scott, A. J. 1989. The practice of human geography: Theory and empirical specificity in the transition from Fordism to flexible accumulation. In B. Macmillan (ed.) *Remodelling Geography.* Oxford: Basil Blackwell, 217–29.

Harvey, F. and Wardenga, U. 2006. Richard Hartshorne's adaptation of Alfred Hettner's system of geography. *Journal of Historical Geography* 32, 422–40.

Harvey, M. E. and Holly, B. P. 1981. Paradigm, philosophy and geographic thought. In M. E. Harvey and B. P. Holly (eds) *Themes in Geographic Thought.* London: Croom Helm, 11–37.

Hay, A. M. 1978. Some problems in regional forecasting. In J. I. Clarke and J. Pelletier (eds) *Régions géographique et régions d'aménagement. Collection les hommes et les lettres*, 7. Lyon: Editions Hermes.

Hay, A. M. 1979a. Positivism in human geography: Response to critics. In D. T. Herbert and R. J. Johnston (eds) *Geography and the Urban Environment*, Vol. 2. London: John Wiley, 1–26.

Hay, A. M. 1979b. The geographical explanation of commodity flow. *Progress in Human Geography* 3, 1–12.

Hay, A. M. 1985a. Scientific method in geography. In R. J. Johnston (ed.) *The Future of Geography.* London: Methuen, 129–42.

Hay, A. M. 1985b. Statistical tests in the absence of samples: A comment. *Professional Geographer* 37, 334–8.

Hay, A. M. and Jophston, R. J. 1983. The study of process in quantitative human geography. *L'Espace Géographique* 12, 69–76.

Hayford, A. M. 1974. The geography of women: An historical introduction. *Antipode: A Radical Journal of Geography* 6, 1–19.

Haynes, R. M. 1975. Dimensional analysis: Some applications in human geography. *Geographical Analysis* 7, 51–68.

Haynes, R. M. 1978. A note on dimensions and relationships in human geography. *Geographical Analysis* 10, 288–92.

Haynes, R. M. 1982. An introduction to dimensional analysis for geographers. *CATMOG* 33, Norwich: Geo Books.

Hayter, R. and Watt, H. D. 1983. The geography of enterprise. *Progress in Human Geography* 7, 157–81.

Heffernan, M. 2000. Balancing visions: Comments on Gearoid O'Tuathail's Critical geopolitics. *Political Geography* 19, 347–52.

Heffernan, M. 2001. History, geography and the French national spaces: The question of Alsace-Lorraine, 1914–1918. *Space and Polity* 5, 27–48.

Heffernan, M. 2003. Histories of geography. In S. L. Holloway, S. P. Rice and G. Valentine (eds) *Key Concepts in Geography.* London: Sage, 3–22.

Held, D. 1980. *Introduction to Critical Theory: Horkheimer to Habermas.* London: Hutchinson.

Hendrikse, R. P. and Sidaway, J. D. 2010. Neoliberalism 3.0. *Environment and Planning A* 42, 2037–42.

Hepple, L. W. 1989. Destroying local Leviathans and designing landscapes of liberty? Public choice theory and the poll tax. *Transactions of the Institute of British Geographers* NS14, 387–399.

Hepple, L. W. 1992. Metaphor, geopolitical discourse and the military in South America. In T. J. Barnes and J. S. Duncan (eds) *Writing Worlds: Discourse, text and metaphor in the representation of landscape.* London: Routledge, 136–54.

Hepple, L. W. 2008. Geography and the pragmatic tradition: The threefold engagement. *Geoforum* 39, 1530–41.

Herberet, D. T. and Johnston, R. J. 1978. Geography and the urban environment. In D. T. Herbert and R. J. Johnston (eds) *Geography and the Urban Environment*, Vol. 1. London: John Wiley, 1–29.

Herbert, S. 2000. For ethnography. *Progress in Human Geography* 24, 550–68.

Herbertson, A. J. 1905. The major natural regions, *Geographical Journal* 25, 300–10.

Herod, A. 1999. Reflections on interviewing foreign elites: Praxis, positionality, validity, and the cult of the insider. *Geoforum* 30, 313–27.

Herod, A. 2001. *Labor Geographies: Workers and the landscapes of capitalism.* New York: Guilford.

Herod, A. 2009. *Geographies of Globalization.* Chichester: Wiley-Blackwell.

Herod, A. and Wright, A. (eds) 2002. *Geographies of Power: Placing scale.* Oxford: Blackwell.

Herod, A., Ó Tuathail, G. and Roberts, S. (eds) 1998. *An Unruly World? Globalization, Governance and Geography.* London and New York: Routledge.

Hewitt, K. (ed.) 1983. *Interpretations of Calamity.* London: George Allen & Unwin.

Hill, M. R. 1982. Positivism: A 'hidden' philosophy in geography. In M. E. Harvey and B. P. Holly (eds) *Themes in Geographic Thought.* London: Croom Helm, 38–60.

Hinchliffe, S. 1996. Technology, power, and space – the means and ends of geographies of technology. *Environment and Planning D: Society and Space* 14, 659–82.

Hinchliffe, S. 2007. *Geographies of Nature: Societies, environments, ecologies.* London: Sage.

Hinshaw, R. E. 2006. *Living with Nature's Extremes: The life of Gilbert Fowler White.* Boulder, CO: Johnson Books.

Hix, S., Johnston, R. J. and McLean, I. 2010. *Choosing an Electoral System.* London: The British Academy.

Hodder, I. and Orton, C. 1976. *Spatial Analysis in Archaeology.* Cambridge: Cambridge University Press.

Hodgart, R. L. 1978. Optimizing access to public services: A review of problems, models, and methods of locating central facilities. *Progress in Human Geography* 2, 17–48.

Hodges, A. 2014 [1983]. *Alan Turing: The enigma.* London: Vintage.

Hoggart, K., Lees, L. and Davies, A. (eds) 2002. *Researching Human Geography.* London: Arnold.

Hohn, U. 1994. The bomber's Baedeker – target book for strategic bombing in the economic warfare against German towns 1943-45. *GeoJournal* 34, 213–30.

Hollowat, J. 2000. Institutional geographies of the new age movement. *Geoforum* 31, 553–65.

Holloway, L. and Hubbard, P. 2001. *People and Place: The extraordinary geographies of everyday life.* Harlow: Prentice Hall.

Holloway, S. L. and Valentine, G. (eds) 2000. *Children's Geographies: Playing, living, learning.* London and New York: Routledge.

Holmes, J. H. 2002. Geography's emerging cross-disciplinary links: Process, causes, outcomes and challenges. *Australian Geographical Studies* 40, 2–20.

Holt-Jensen, A. 1981. *Geography: Its history and concepts.* London: Harper & Row.

Holt-Jensen, A. 1988. *Geography: Its history and concepts*, 2nd edn. London: Harper & Row.

Holt-Jensen, A. 2009. *Geography: History and concepts: A student's guide*, 4th edn. London: Sage.

Hook, J. C. 1955. Areal differentiation of the density of the rural farm population in the northeastern United States. *Annals of the Association of American Geographers* 45, 189–90.

Hooper, B. 2009. Los Angeles school of post-modern urbanism. In R. Kitchin and N. Thrift (eds) *International Encyclopedia of Human Geography.* Amsterdam: Elsevier, 293–7.

Hooson, D. J. M. 1981. Carl O. Sauer. In B. W. Blouet (ed.) *The Origins of Academic Geography in the United States.* Hamden, CT: Archon Books, 165–74.

Hooson, D. J. M. 1984. The Soviet Union. In R. J. Johnston and P. Claval (eds) *Geography since the Second World War: An international survey.* London: Croom Helm, 79–106.

Hooson, D. J. M. 2002. Geography in Russia: Glories and disappointments. In G. S. Dunbar (ed.) *Geography: Discipline, profession and subject since 1870.* Dordrecht: Kluwer, 225–44.

Hopkins, P. and Noble, G. 2009. Masculinities in place: Situated identities, relations and intersectionality. *Social and Cultural Geography* 10, 811–19.

Hopkins, P. and Jackson, P. 2013. Researching masculinities and the future of the WGSG. *Area* 45, 9–10.

House, J. W. 1973. Geographers, decision takers and policy makers. In M. Chisholm and B. Rodgers (eds) *Studies in Human Geography*. London: Heinemann, 272–305.

Howell, P. 2009. Sexuality. In R. Kitchen and N. Thrift (eds) *International Encyclopedia of Human Geography*. Amsterdam: Elsevier, 59–63.

Howitt, R. 2002. Scale and the other: Levinas and geography. *Geoforum* 33, 299–313.

Hoyningen-Huene, P. 1993. *Reconstructing Scientific Revolutions*. Chicago: University of Chicago Press.

Hsu, J.-Y. and Sidaway, J. D. 2009. Commentary: In-between sessions at the AAG. *Environment and Planning A* 41, 2288–92.

Hubbard, P. 2000. Desire/disgust: Moral geographies of heterosexuality. *Progress in Human Geography* 24, 191–217.

Hubbard, P. and Kitchin, R. 2007. Battleground geographies and conspiracy theories: A response to Johnston (2006). *Transactions of the Institute of British Geographers* 32, 428–34.

Hubbard, P., Kitchin, R. and Valentine, G. (eds) 2004. *Key Thinkers on Space and Place*. London: Sage.

Hubbard, P., Kitchin, R. and Valentine, G. (eds) 2008. *Key Texts in Human Geography*. London, Thousand Oaks, CA and New Delhi and Singapore: Sage.

Hubbard, P., Kitchin, R., Bartley, B. and Fuller, D. 2002. *Thinking Geographically: Space, theory and contemporary human geography*. London and New York: Continuum.

Huckle, J. 1985. Geography and schooling. In R. J. Johnston (ed.) *The Future of Geography*. London: Methuen, 291–306.

Hudson, R. 1983. The question of theory in political geography: Outlines for a critical theory approach. In N. Kliot and S. Waterman (eds) *Pluralism and Political Geography*. London: Croom Helm, 39–55.

Hudson, R. 1988. Uneven development in capitalist societies. *Transactions of the Institute of British Geographers* NS13, 484–96.

Hudson, R. 2003. Geographers and the regional problem. In R. J. Johnston and M. Williams (eds) *A Century of British Geography*. Oxford: Oxford University Press for the British Academy, 583–602.

Hudson, R. 2012. Critical political economy and material transformation. *New Political Economy* 17, 373–97.

Huggett, R. J. 1980. *Systems Analysis in Geography*. Oxford: Oxford University Press.

Huggett, R. J. 1994. *Geoecology: An evolutionary approach*. London: Routledge.

Huggett, R. J. and Thomas, R. W. 1980. *Modelling in Geography*. London: Harper & Row.

Hughes, T. P. 2000. *Rescuing Prometheus: Four monumental projects that changed the modern world*. New York: Vintage.

Hugill, P. J. and Foote, K. E. 1994. Foreword: Culture and geography: Thirty years of advance. In K. E. Foote, P. J. Hugill, K. Mathewson and J. M. Smith (eds) *Re-reading Cultural Geography*. Austin, TX: University of Texas Press, 9–26.

Huntington, E. 1915. *Civilization and Climate*. New Haven, CT: Yale University Press.

Huntington, E. 1945. *Mainsprings of Civilization*. New York: John Wiley.

Huntington, E., Williams, F. E. and Van Valkenburg, S. 1933. *Economic and Social Geography*. New York: John Wiley.

Hyndman, J. 2001. Towards a feminist geopolitics. *Canadian Geographer* 45, 210–22.

Imrie, R. F. 1996. Ableist geographies, disablist spaces: Towards a reconstruction of Golledge's 'geography and the disabled'. *Transactions of the Institute of British Geographers* NS21, 397–403.

Imrie, R. F. 2000. Disability and discourses of mobility and movement. *Environment and Planning A* 32, 1641–56.

Isard, W. 1956a. *Location and Space Economy*. New York: John Wiley.

Isard, W. 1956b. Regional science, the concept of region, and regional structure. *Papers and Proceedings, Regional Science Association* 2, 13–39.

Isard, W. 1960. *Methods of Regional Analysis: An introduction in regional science*. New York: John Wiley.

Isard, W. 1975. *An Introduction to Regional Science*. Englewood Cliffs, NJ: Prentice-Hall.

Isard, W. 2003. *History of Regional Science and the Regional Science Association International: The beginnings and early history*. Berlin: Springer.

Isard, W. with Smith, T. and Isard, P., Tung, T. H., Dacey, M. 1969. *General Theory: Social, political, economic and regional with particular reference to decision-making analysis*. Cambridge, MA: MIT Press.

Isserman, A. M. 1995. The history, status, and future of regional science: An American perspective. *International Regional Science Review* 17, 249–96.

Jackson, P. 1984. Social disorganization and moral order in the city. *Transactions of the Institute of British Geographers* NS9, 168–80.

Jackson, P. 1985. Urban ethnography. *Progress in Human Geography* 9, 157–76.

Jackson, P. 1988. Definitions of the situation. In J. Eyles and D. M. Smith (eds) *Qualitative Methods in Human Geography*. Cambridge: Polity Press, 49–74.

Jackson, P. 1989. *Maps of Meaning: An introduction to cultural geography*. London and Boston, MA: Unwin Hyman.

Jackson, P. 1991a. Mapping meanings – a cultural critique of locality studies. *Environment and Planning A* 23, 215–28.

Jackson, P. 1991b. The cultural politics of masculinity: Towards a social geography. *Transactions of the Institute of British Geographers* NS16, 199–213.

Jackson, P. 1993a. Changing ourselves: A geography of position. In R. J. Johnston (ed.) *The Challenge for Geography. A changing world: A changing discipline*. Oxford: Blackwell Publishers, 198–214.

Jackson, P. 1993b. Berkeley and beyond: Broadening the horizons of cultural geography. *Annals of the Association of American Geographers* 83, 519–20.

Jackson, P. 1999. Postmodern urbanism and the ethnographic void. *Urban Geography* 20, 400–2.

Jackson, P. and Jacobs, J. M. 1996. Postcolonialism and the politics of race. *Environment and Planning D: Society and Space* 14, 1–3.

Jackson, P. and Penrose, J. 1993. Introduction: Placing 'race' and nation. In P. Jackson and J. Penrose (eds) *Constructions of Race, Place and Nation*. London: UCL Press, 1–26.

Jackson, P. and Smith, S. J. 1981. Introduction. In P. Jackson and S. J. Smith (eds) *Social Interaction and Ethnic Segregation*. London: Academic Press, 1–18.

Jackson, P. and Smith, S. J. 1984. *Exploring Social Geography*. London: George Allen & Unwin.

Jackson, P., Smith, S. J. and Johnston, R. J. 1988. An equal opportunities policy for the IBG. *Area* 20, 279–80.

Jacobs, J. 1994a. Negotiating the heart: Heritage, development and identity in postimperial London. *Environment and Planning D: Society and Space* 12, 751–72.

Jacobs, J. 1994b. Earth honouring: Western desires and indigenous knowledges. In A. Blunt and G. Rose (eds) *Writing Women and Space*. New York: Guilford, 169–96.

Jacobs, J. 1996. *Edge of Empire: Postcolonialism and the city*. London: Routledge.

Jacobs, J. 2001. Postcolonial geography. In N. J. Smelser and P. B. Baltes (eds) *International Encyclopedia of the Social and Behavioral Sciences*. Oxford: Elsevier Science, 11838–41.

Jacobs, J. and Jackson, P. 1996. Postcolonialism and the politics of race. *Environment and Planning D: Society and Space* 14, 1–4.

Jacobs, J. and Lees, L. 2013. Defensible space on the move: Revisiting the urban geography of Alice Coleman. *International Journal of Urban and Regional Research* 37, 1559–83.

James, P. E. 1942. *Latin America*. London: Cassell.

James, P. E. 1952. Toward a further understanding of the regional concept. *Annals of the Association of American Geographers* 42, 195–222.

James, P. E. 1954. Introduction: The field of geography. In P. E. James and C. F. Jones (eds) *American Geography: Inventory and prospect*. Syracuse: Syracuse University Press, 2–18.

James, P. E. 1965. The President's session. *Professional Geographer* 17(4), 35–7.

James, P. E. 1972. *All Possible Worlds: A history of geographical ideas*. Indianapolis, IN: Odyssey Press.

James, P. E. and Jones, C. F. (eds) 1954. *American Geography: Inventory and prospect*. Syracuse: Syracuse University Press.

James, P. E. and Martin, G. J. 1978. *The Association of American Geographers: The first seventy-five years, 1904–1979*. Washington, DC: The Association of American Geographers.

James, P. E. and Martin, G. J. 1981. *All Possible Worlds: A history of geographical ideas*, 2nd edn. New York: John Wiley.

James, P. E. and Mather, E. C. 1977. The role of periodic field conferences in the development of geographical ideas in the United States. *Geographical Review* 67, 446–61.

Janelle, D. G. 1968. Central-place development in a time-space framework. *Professional Geographer* 20, 5–10.

Janelle, D. G. 1969. Spatial reorganization: A model and concept. *Annals of the Association of American Geographers* 59, 348–64.

Janelle, D. G. 1973. Measuring human extensibility in a shrinking world. *Journal of Geography* 72, 8–15.

Jayne, M., Valentine, G. and Holloway, S. L. 2011. What use are units? Critical geographies of alcohol policy. *Antipode: A Radical Journal of Geography* 44, 828–46.

Jazeel, T. 2014. Subaltern geographies: Geographical knowledge and postcolonial strategy. *Singapore Journal of Tropical Geography* 35, 88–103.

Jenkins, A. 1995. The impact of Research Assessment Exercises on teaching in selected geography departments in England and Wales. *Geography* 80, 367–74.

Jenkins, A. and Smith, P. 1993. Expansion, efficiency and teaching quality: The experience of British geography departments, 1986–1991. *Transactions of the Institute of British Geographers* NS18, 500–15.

Johannesson, G. T. and Bærenholddt, J. O. 2009. Actor-network theory/network geographies. In R. Kitchin and N. Thrift (eds) *International Encyclopedia of Human Geography*. Amsterdam: Elsevier, 15–19.

Johnson, J. H. and Pooley, C. G. (eds) 1982. *The Structure of Nineteenth-Century Cities*. London: Croom Helm.

Johnson, L. C. 1987. (Un)realist perspectives: Patriarchy and feminist challenges in geography. *Antipode: A Radical Journal of Geography* 19, 210–15.

Johnson, L. C. 1989. Geography, planning and gender. *New Zealand Geographer* 45, 85–91.

Johnson, L. C. 1994. What future for feminist geography? *Gender, Place and Culture* 1, 103–13.

Johnson, L. C. 2009. Feminism/feminist geography. In R. Kitchen and N. Thrift (eds) *International Encyclopedia of Human Geography*. Amsterdam: Elsevier, 44–58.

Johnson, L. C. 2012. Feminist geography 30 years on: They came, they saw but did they conquer? *Geographical Research* 50, 345–55.

Johnston, R. J. 1969. Urban geography in New Zealand 1945–1969. *New Zealand Geographer* 25, 121–35.

Johnston, R. J. 1971. *Urban Residential Patterns: An introductory review*. London: G. Bell & Sons.

Johnston, R. J. 1974. Continually changing human geography revisited: David Harvey: *Social Justice and the City. New Zealand Geographer* 30, 180–92.

Johnston, R. J. 1975. But some are more equal . . . who gets what where, and how, in New Zealand. *Geography* 60, 255–68.

Johnston, R. J. 1976a. Observations on accounting procedures and urban-size policies. *Environment and Planning A* 8, 327–40.

Johnston, R. J. 1976b. *The World Trade System: Some enquiries into its spatial structure*. London: G. Bell & Sons.

Johnston, R. J. 1976c. Anarchy, conspiracy and apathy: The three conditions of geography. *Area* 8, 1–3.

Johnston, R. J. 1978a. Paradigms and revolutions or evolution: Observations on human geography since the Second World War. *Progress in Human Geography* 2, 189–206.

Johnston, R. J. 1978b. *Multivariate Statistical Analysis in Geography: A primer on the general linear model*. London: Longman.

Johnston, R. J. 1978c. *Political, Electoral and Spatial Systems*. London: Oxford University Press.

Johnston, R. J. 1979. *Geography and Geographers: Anglo-American human geography since 1945*, 1st edn. London: Edward Arnold.

Johnston, R. J. 1980a. *City and Society*. London: Penguin.

Johnston, R. J. 1980b. On the nature of explanation in human geography. *Transactions of the Institute of British Geographers* NS5, 402–12.

Johnston, R. J. 1981a. Applied geography, quantitative analysis and ideology. *Applied Geography* 1, 213–19.

Johnston, R. J. 1981b. Paradigms, revolutions, schools of thought and anarchy: Reflections on the recent history of Anglo-American human geography. In B. W. Blouet (ed.) *The Origins of Academic Geography in the United States*. Hamden, CT: Archon Books, 303–18.

Johnston, R. J. 1982a. On ecological analysis and spatial autocorrelation. In L. le Rouzic (ed.) *L'autocorrelation spatiale*. Reims: Travaux de l'Institut de Géographie, 3–16.

Johnston, R. J. 1982b. *Geography and the State*. London: Macmillan.

Johnston, R. J. 1982c. On the nature of human geography. *Transactions of the Institute of British Geographers* NS7, 123–5.

Johnston, R. J. 1983a. *Philosophy and Human Geography: An introduction to contemporary approaches*, 1st edn. London: Edward Arnold.

Johnston, R. J. 1983b. Resource analysis, resource management and the integration of human and physical geography. *Progress in Physical Geography* 7, 127–46.

Johnston, R. J. 1983c. On geography and the history of geography. *History of Geography Newsletter* 3, 1–7.

Johnston, R. J. 1984a. The region in twentieth century British geography. *History of Geography Newsletter* 4, 26–35.

Johnston, R. J. 1984b. Quantitative ecological analysis in human geography: An evaluation of four problem areas. In G. Bahrenberg, M. Fischer and P. Nijkamp (eds) *Recent Developments in Spatial Data Analysis*. Aldershot: Gower, 131–44.

Johnston, R. J. 1984c. *Residential Segregation, The State and Constitutional Conflict in American Urban Areas*. London: Academic Press.

Johnston, R. J. 1984d. The world is our oyster. *Transactions of the Institute of British Geographers* NS9, 443–59.

Johnston, R. J. 1984e. A foundling floundering in World Three. In M. Billinge, D. Gregory and R. Martin (eds) *Recollections of a Revolution*. London: Macmillan, 39–56.

Johnston, R. J. (ed.) 1985a. *The Future of Geography*. London: Methuen.

Johnston, R. J. 1985b. *The Geography of English Politics: The 1983 general election*. London: Croom Helm.

Johnston, R. J. 1985c. Places matter. *Irish Geography* 18, 58–63.

Johnston, R. J. 1985d. To the ends of the earth. In R. J. Johnston (ed.) *The Future of Geography*. London: Methuen, 326–38.

Johnston, R. J. 1986a. *On Human Geography*. Oxford: Basil Blackwell.

Johnston, R. J. 1986b. Four fixations and the quest for unity in geography. *Transactions of the Institute of British Geographers* NS11, 449–53.

Johnston, R. J. 1986c. *Philosophy and Human Geography: An introduction to contemporary approaches*, 2nd edn. London: Edward Arnold.

Johnston, R. J. 1986d. Placing politics. *Political Geography Quarterly* 5, s63–s78.

Johnston, R. J. 1986d. The neighbourhood effect revisited: Spatial science or political regionalism. *Environment and Planning D: Society and Space* 4, 41–56.

Johnston, R. J. 1989. *Environmental Problems: Nature, economy and state*. London: Belhaven Press.

Johnston, R. J. 1990a. The challenge for regional geography: Some proposals for research frontiers. In R. J. Johnston, J. Hauer and G. A. Hoekveld (eds) *The Challenge of Regional Geography*. London: Routledge, 124–41.

Johnston, R. J. 1990b. Some misconceptions about conceptual issues. *Tijdschrift voor Economische en Sociale Geografie* 81, 14–18.

Johnston, R. J. 1991. *A Question of Place: Exploring the practice of human geography*. Oxford: Blackwell.

Johnston, R. J. 1992. The rise and decline of the corporate-welfare state: A comparative analysis in global context. In P. J. Taylor (ed.) *Political Geography of the Twentieth Century: A global analysis*. London: Belhaven Press, 115–70.

Johnston, R. J. 1993a. Removing the blindfold after the game is over: The financial outcomes of the 1992 Research Assessment Exercise. *Journal of Geography in Higher Education* 17, 174–80.

Johnston, R. J. 1993b. The geographer's degrees of freedom: Wreford Watson, postwar progress in human geography and the future of scholarship in UK geography. *Progress in Human Geography* 17, 319–32.

Johnston, R. J. 1993c. A voice in the wilderness. *Geography* 78, 204–7.

Johnston, R. J. 1993d. Meet the challenge: Make the change. In R. J. Johnston (ed.) *The Challenge for Geography: A changing world: A changing discipline*. Oxford: Blackwell, 151–80.

Johnston, R. J. 1993e. Real political geography. *Political Geography* 12, 473–80.

Johnston, R. J. (ed.) 1993f. *The Challenge for Geography: A changing world: A changing discipline*. Oxford: Blackwell.

Johnston, R. J. 1994a. Resources, student: Staff ratios and teaching quality in British higher education: Some speculations aroused by Jenkins and Smith. *Transactions of the Institute of British Geographers* NS19, 359–65.

Johnston, R. J. 1994b. The 'quality industry' in British higher education and the AAG's publications. *Professional Geographer* 46, 491–7.

Johnston, R. J. 1994c. On spatial analysis, place and realism. *Urban Geography* 15, 290–5.

Johnston, R. J. 1994d. One world, millions of places: The end of History and the ascendancy of Geography. *Political Geography* 13, 111–22.

Johnston, R. J. 1995a. Geographical research, geography and geographers in the changing British university system. *Progress in Human Geography* 19, 355–71.

Johnston, R. J. 1995b. The business of British geography. In A. D. Cliff, P. R. Gould, A. G. Hoare and N. J. Thrift (eds) *Diffusing Geography: Essays for Peter Haggett*. Oxford: Blackwell, 317–41.

Johnston, R. J. 1996a. And now it's all over was it worth all the effort? *Journal of Geography in Higher Education* 20, 159–65.

Johnston, R. J. 1996b. The expansion and fragmentation of geography in higher education. In R. J. Huggett, M. Robinson and Douglas, I. (eds) *Companion Encyclopedia of Geography*. London: Routledge, 794–817.

Johnston, R. J. 1996c. Jean Gottmann: French regional and political geographer extraordinaire. *Progress in Human Geography* 20, 183–93.

Johnston, R. J. 1996d. *Nature, State and Economy: The political economy of environmental problems*. Chichester: John Wiley.

Johnston, R. J. 1996e. Academic tribes and territories: The realpolitik of opening up the social sciences. *Environment and Planning A* 28.

Johnston, R. J. 1996f. Quality in research, quality in teaching and quality in debate: A response to Graham Gibbs. *Quality in Higher Education* 2, 165–70.

Johnston, R. J. 1997a. Where's my bit gone? Reflections on *Rediscovering Geography. Urban Geography* 18, 353–9.

Johnston, R. J. 1997b. Australian geography seen from afar: Through a glass darkly. *Australian Geographer* 28, 29–37.

Johnston, R. J. 1998. Fragmentation around a defended core: The territoriality of geography. *Geographical Journal* 164, 139–47.

Johnston, R. J. 1999. Classics in human geography revisited. *Progress in Human Geography* 23, 253–66.

Johnston, R. J. 2000a. On disciplinary history and textbooks: Or where has spatial analysis gone. *Australian Geographical Studies* 38, 125–37.

Johnston, R. J. 2000b. Intellectual respectability and disciplinary transformation? Radical geography and the institutionalisation of geography in the USA since 1945. *Environment and Planning A* 32, 971–90.

Johnston, R. J. 2000c. City-regions and a federal Europe: Robert Dickinson and post-World War II reconstruction. *Geopolitics* 5, 153–76.

Johnston, R. J. 2002a. Robert E Dickinson and the growth of urban geography: An evaluation. *Urban Geography* 22, 702–36.

Johnston, R. J. 2002b. Reflections on Nigel Thrift's optimism: Political strategies to implement his vision. *Geoforum* 33, 421–5.

Johnston, R. J. 2003a. Geography: A different sort of discipline? *Transaction of the Institute of British Geographers* NS29, 133–41.

Johnston, R. J. 2003b. The institutionalisation of geography as an academic discipline. In R. J. Johnston and M. Williams (eds) *A Century of British Geography.* Oxford: Oxford University Press for the British Academy, 45–90.

Johnston, R. J. 2003c. Order in space: Geography as a discipline in distance. In R. J. Johnston and M. Williams (eds) *A Century of British Geography.* Oxford: Oxford University Press for the British Academy, 303–46.

Johnston, R. J. 2003d. Regionalization and classification. In K. Kempf-Leonard, J. Heckman, G. King and P. Tracy (eds) *Encyclopaedia of Social Measurement.* New York: Academic Press, 337–50.

Johnston, R. J. 2004a. Communications technology and the production of geographical knowledge. In S. D. Brunn, S. L. Cutter and J. W. Harrington Jr. (eds) *Geography and Technology.* Boston, MA: Kluwer, 17–36.

Johnston, R. J. 2004b. Geography – coming apart at the seams. In N. Castree, A. Rogers and D. Sherman (eds) *Questioning Geography: Fundamental debates.* Oxford: Blackwell, 9–25.

Johnston, R. J. 2004c. Institutions and disciplinary fortunes: Two moments in the history of UK geography in the 1960s – 1. *Progress in Human Geography* 28, 57–78.

Johnston, R. J. 2004d. Territory and territoriality in a globalizing world. *Ekistics* 70, 64–70.

Johnston, R. J. 2004e. Disciplinary change and career paths. In R. Lee and D. M. Smith (eds) *Geographies and Moralities: International perspectives on justice, development and place.* Oxford: Blackwell , 265–83.

Johnston, R. J. 2005. Learning our history from our pioneers: UK academic geographers in the *Oxford Dictionary of National Biography. Progress in Human Geography* 29, 651–67.

Johnston, R. J. 2006. The politics of changing human geography's agenda: Textbooks and the representation of increasing diversity. *Transactions of the Institute of British Geographers* NS31, 286–303.

Johnston, R. J. 2007a. Author's response. *Progress in Human Geography* 31, 47–52.

Johnston, R. J. 2007b. On duplicitous battleground conspiracies. *Transactions of the Institute of British Geographers* 32, 435–8.

Johnston, R. J. 2008. Emrys Jones. In *Proceedings of the British Academy 153: Biographical Memoirs of Fellows VII.* Oxford: Oxford University Press for the British Academy, 243–92.

Johnston, R. J. 2009a. The extent of influence: An alternative approach to identifying dominant contributors to a discipline's literature. *Scientometrics* 78, 409–20.

Johnston, R. J. 2009b. On *Geographic* and geography. *New Zealand Geographer* 65, 167–70.

Johnston, R. J. 2009c. On geography, *Geography* and geographical magazines. *Geography* 94, 207–14.

Johnston, R. J. 2009d. Popular geographies and geographical imaginations: Contemporary English-language geographical magazines. *GeoJournal* 74, 347–62.

Johnston, R. J. 2010a. Book review essay: What every human geography student needs to read to know the discipline? *Progress in Human Geography* 34, 528–35.

Johnston, R. J. 2010b. Leslie Curry (1922–2009): The scholar, the teacher and the climatologist. *Progress in Human Geography* 34, 387–98.

Johnston, R. J. 2011. Promoting geography (or part of it) – yet again? *Professional Geographer* 63, 325–331.

Johnston, R. J. 2012. Seats, votes and the spatial organisation of elections revisited. In G. Gudgin and P. J. Taylor, *Seats, Votes and the Spatial Organisation of Elections*. Colchester: ECPR Press, ix–xxxix.

Johnston, R. J. 2013. Review essay. Geographical societies, academics and publics: Reading *Civic Discipline: Geography in America, 1860–1890*. *Geographical Journal* 179, 87–91.

Johnston, R. J. and Brack, E. V. 1983. Appointment and promotion in the academic labour market: A preliminary survey of British University Departments of Geography. *Transactions of the Institute of British Geographers* NS8, 100–11.

Johnston, R. J. and Claval, P. (eds) 1984. *Geography Since the Second World War: An international survey*. London: Croom Helm.

Johnston, R. J. and Gregory, S. 1984. The United Kingdom. In R. J. Johnston and P. Claval (eds) *Geography Since the Second World War: An international survey*. London: Croom Helm, 107–31.

Johnston, R. J. and Pattie, C. J. 1990. The regional impact of Thatcherism: Attitudes and votes in Great Britain in the 1980s. *Regional Studies* 24, 479–93.

Johnston, R. J. and Pattie, C. J. 2000. Ecological inference and entropy-maximizing: An alternative estimation procedure for split-ticket voting. *Political Analysis* 8, 333–45.

Johnston, R. J. and Pattie, C. J. 2001. On geographers and ecological inference. *Annals of the Association of American Geographers* 91, 281–2.

Johnston, R. J. and Pattie, C. J. 2003. Evaluating an entropy-maximizing solution to the ecological inference problem: Split-ticket voting in New Zealand 1999. *Geographical Analysis* 35, 1–23.

Johnston, R. J. and Pattie, C. J. 2006. *Putting Voters in their Place: Geography and elections in Great Britain*. Oxford: Oxford University Press.

Johnston, R. J. and Sidaway, J. D. 2004. *Geography and Geographers: Anglo-American human geography since 1945*, 6th edn. London: Arnold.

Johnston, R. J. and Sidaway, J. D. 2015. Have the human geographical can(n)ons fallen silent; or were they never primed? *Journal of Historical Geography*, 49, 49–60.

Johnston, R. J. and Taylor, P. J. (eds) 1986. *A World in Crisis? Geographical Perspectives*. Oxford: Basil Blackwell.

Johnston, R. J. and Taylor, P. J. (eds) 1989. *A World in Crisis? Geographical Perspectives*, 2nd edn. Oxford: Basil Blackwell.

Johnston, R. J. and Thrift, N. J. 1993. Ringing the changes: The intellectual history of *Environment and Planning A*. *Environment and Planning A* Anniversary Issue, 14–21.

Johnston, R. J., Jones, K. and Gould, M. 1995. Department size and research in English Universities: Inter-university variations. *Quality in Higher Education* 1, 41–7.

Johnston, R. J., Hauer, J. and Hoekveld, G. A. (eds) 1990. *Regional Geography: Current developments and future prospects*. London: Routledge.

Johnston, R. J., Poulsen, M. F. and Forrest, J. 2009. Measuring ethnic residential segregation: Putting some more geography in. *Urban Geography* 30, 91–109.

Johnston, R. J., Poulsen, M. F. and Forrest, J. 2010. Moving on from indices, refocusing on mix: On measuring and understanding ethnic patterns of residential segregation. *Journal of Ethnic and Migration Studies* 36, 697–706.

Johnston, R. J., Taylor, P. J. and O'Loughlin, I. 1987. The geography of violence and premature death. In Vayrynen, R. (ed.) *The Quest for Peace*. London: Sage, 241–59.

Johnston, R. J., Taylor, P. J. and Watts, M. J. (eds) 1995. *Geographies of Global Change: Remapping the world in the late twentieth century*. Oxford: Blackwell.

Johnston, R. J., Taylor, P. J. and Watts, M. J. (eds) 2002. *Geographies of Global Change: Remapping the world*, 2nd edn. Oxford: Blackwell.

Johnston, R. J., Harris, R., Jones, K. and Manley, D. 2014. A response to Gorard. *Psychology of Education Review* 38 (Autumn), 5–8.

Johnston, R. J., Jones, K., Haggett, P. and Dodds, K. 2010. Leslie Wilson Hepple (1947–2007). In H. Lorimer and C. W. J. Withers (eds) *Geographers: Biobibliographical studies*, Vol. 29. London and New York: Continuum, 73-96.

Johnston, R. J., Pattie, C. J., Dorling, D. F. L. and Rossiter, D. J. 2001. *From Votes to Seats: The operation of the UK electoral system since 1945*. Manchester: Manchester University Press.

Johnston, R. J., Fairbrother, M., Hayes, D., Hoare, T. and Jones, K. 2008. The Cold War and geography's quantitative revolution: Some messy reflections on Barnes' geographical underworld. *Geoforum* 39, 1802–6.

Johnston, R. J., Manley, D., Jones, K., Harris, R. and Hoare, A. G. 2015. University admissions and the prediction of degree performance: An analysis in the light of changes to the English schools' examination system. *Higher Education Quarterly*. doi: 10.1111/hequ.12067

Johnston, R. J., Harris, R., Jones, K., Manley, D., Sabel, C. E. and Wang, W. W. 2014a. Mutual misunderstanding and avoidance, misrepresentations and disciplinary politics: Spatial science and quantitative analysis in (United Kingdom) geographical curricula. *Dialogues in Human Geography* 4, 3–25.

Johnston, R. J., Harris, R., Jones, K., Manley, D., Sabel, C. E. and Wang, W. W. 2014b. One step forwards but two steps back to the proper appreciation of spatial science. *Dialogues in Human Geography* 4, 59–69.

Johnston, R. J., Hepple, L. W., Hoare, A. G., Jones, K. and Plummer, P. 2003. Contemporary fiddling in human geography while Rome burns: Has quantitative analysis been largely abandoned – and should it? *Geoforum* 34, 157–61.

Johnston, R. J., Propper, C., Burgess, S., Sarker, R., Bolster, A. and Jones, K. 2005a. Spatial scale and the neighbourhood effect: Multinomial models of voting at two recent British general elections. *British Journal of Political Science* 35, 487–514.

Johnston, R. J., Propper, C., Sarker, R., Jones, K., Bolster, A. and Burgess, S. 2005b. Neighbourhood social capital and neighbourhood effects. *Environment and Planning A* 37, 1143–59.

Jonas, A. E. G. 1988. A new regional geography of localities? *Area* 20, 101–10.

Jonas, A. E. G. 2012. Region and place: Regionalism in question. *Progress in Human Geography* 36, 263–72.

Jones, A. 1998. (Re)producing gender cultures: Theorizing gender in investment banking recruitment. *Geoforum* 29, 451–74.

Jones, A. 1999. Dialectics and difference: Against Harvey's dialectical 'post-Marxism'. *Progress in Human Geography* 23, 529–55.

Jones, A. 2009. Marxism/Marxist geographies II. In R. Kitchin and N. Thrift (eds) *International Encyclopedia of Human Geography*. Amsterdam: Elsevier, 474.

Jones, E. 1956. Cause and effect in human geography. *Annals of the Association of American Geographers* 46, 369–77.

Jones, E. 1980. Social geography. In E. H. Brown (ed.) *Geography Yesterday and Tomorrow*. Oxford: Oxford University Press, 251–62.

Jones III, J. P. and Casetti, E. (eds) 1992. *Applications of the Expansion Method*. London: Routledge.

Jones III, J. P. and Dixon D. P. 2004, Guest editorial. What next? *Environment and Planning A* 36, 381–90.

Jones III, J. P. and Hanham, R. Q. 1995. Contingency, realism and the expansion method. *Geographical Analysis* 4, 185–207.

Jones III, J. P., Nast, H. and Roberts, S. (eds) 1997. *Thresholds in Feminist Geography*. Lanham, MD: Rowman & Littlefield.

Jones, K. 1984. Graphical methods for exploring relationships. In G. Bahrenberg, M. M. Fischer and P. Nijkamp (eds) *Recent Developments in Spatial Data Analysis*. Aldershot: Gower, 215–30.

Jones, K. 1991. Specifying and estimating multi-level models for geographical research. *Transactions of the Institute of British Geographers* NS16, 148–60.

Jones, K. 1997. Multilevel approaches to modeling contextuality: From nuisance to substance in the analysis of voting behaviour. In G. P. Westert and R. N. Verhoeef (eds) *Places and People: Multilevel modeling in geographical research. Nederlandse Geografische Studies no. 227*. Utrecht: Royal Dutch Geographical Society and Faculty of Geosciences, Utrecht University, 19–43.

Jones, K. 2010. The practice of quantitative methods. In B. Somekh and C. Lewin (eds) *Research Methods in the Social Sciences*, 2nd edn. London: Sage, 201–11.

Jones, K. and Almond, S. 1992. Moving out of the linear rut: The possibilities of generalized additive models. *Transactions of the Institute of British Geographers* NS17, 434–47.

Jones, K. and Bullen, N. 1994. Contextual models of urban house prices: A comparison of fixed- and random-coefficient models developed by expansion. *Economic Geography* 70, 252–72.

Jones, K. and Duncan, C. 1996. People and places: The multilevel model as a general framework for the quantitative analysis of geographical data. In P. A. Longley and M. Batty (eds) *Spatial Analysis: Geographical modelling in a GIS environment*. New York: John Wiley, 1–16.

Jones, K. and Duncan, C. 1998. Modelling context and heterogeneity: Applying multilevel models. In E. Scarbrough and E. Tanenbaum (eds) *Research Strategies in the Social Sciences: A guide to new approaches*, Volume 9. Oxford: Oxford University Press, 94–123.

Jones, K. and Wrigley, N. 1995. Generalized additive models, graphical diagnostics and logistic regression. *Geographical Analysis* 27, 1–21.

Jones, K., Johnston, R. J. and Pattie, C. J. 1992. People, places and regions: Exploring the use of multi-level modelling in the analysis of electoral data. *British Journal of Political Science* 22, 343–80.

Jones, L. V., Lindley, G. and Coggeshall, P. E. (eds) 1982. *An Assessment of Research-Doctorate Programs in the United States: Social and behavioral sciences*. Washington, DC: National Academy Press.

Joseph, A. E. and Phillips, D. R. 1984. *Accessibility and Utilization: Perspectives on health care delivery*. London: Harper & Row.

Kahneman, D. 2012. *Thinking Fast and Slow*. London: Penguin.

Kain, R. and Delano-Smith, C. 2003. Geography displayed: Maps and mapping. In R. J. Johnston and M. Williams (eds) *A Century of British Geography*. Oxford: Oxford University Press for the British Academy, 371–427.

Kansky, K. J. 1963. *Structure of Transportation Networks*. Chicago: University of Chicago, Department of Geography, Research Paper 84.

Kaolan, R. 2012. *The Revenge of Geography: What the map tells us about coming conflicts and the battle against fate*. New York: Random House

Kariya, P. 1993. The Department of Indian Affairs and Northern Development: The culture-building process within an institution. In J. S. Duncan and D. Ley (eds) *Place/Culture/Representation*. London: Routledge, 187–204.

Kasbarian, J. A. 1996. Mapping Edward Said: Geography, identity, and the politics of location. *Environment and Planning D: Society and Space* 14, 529–57.

Kasperson, R. E. 1971. The post-behavioral revolution in geography. *British Columbia Geographical Series* 12, 5–20.

Kates, R. W. 1962. *Hazard and Choice Perception in Flood Plain Management*. Chicago: University of Chicago, Department of Geography, Research Paper 78.

Kates, R. W. 1972. Review of *Perspectives on Resource Management*. *Annals of the Association of American Geographers* 62, 519–20.

Kates, R. W. 1987. The human environment: The road not taken, the road still beckoning. *Annals of the Association of American Geographers* 77, 525–34.

Kates, R. W. 1995. Labnotes from the Jeremiad experiment: Hope for a sustainable transition. *Annals of the Association of American Geographers* 85, 623–40.

Kates, R. W. and Burton, I. (eds) 1986a. *Geography Resources and Environment* (2 vols). Chicago: University of Chicago Press.

Kates, R. W. and Burton, I. 1986b. Introduction. In R. W. Kates and I. Burton (eds) *Geography, Resources and Environment, Vol. 1. Selected writings of Gilbert F. White*. Chicago: University of Chicago Press, xi–xiv.

Katz, C. and Monk, J. (eds) 1993. *Full Circles: Geographies of women over the life course*. London and New York: Routledge.

Kearn, L., Hawkins, R., Al-Hindi, K. F., Moss, P. 2014, A collective biography of joy in academic practice. *Social and Cultural Geography* 15, 834–51.

Kearns, G. 1984. Closed space and political practice: Frederick Jackson Turner and Halford Mackinder. *Environment and Planning D: Society and Space* 2, 23–34.

Kearns, G. 2009. *Geopolitics and Empire: The legacy of Halford Mackinder*. New York: Oxford University Press.

Keighren, I. M. 2006. Bringing geography to the book: Charting the reception of *Influences of Geographic Environment*. *Transactions of the Institute of British Geographers* 31, 525–40.

Keighren, I. M. 2010. *Bringing Geography to Book: Ellen Semple and the reception of geographical knowledge*. London: I.B. Taurus.

Keighren, I. M., Abrahamsson, C. and Della Dora, V. 2012. On canonical geographies. *Dialogues in Human Geography* 2, 296–312.

Keighren, I. M. (ed.) 2015. Teaching the history of geography: Current challenges and future directions. *Progress in Human Geography*. doi: 10.1177/0309132515575940

Kelly, A. and Burrows, R. 2011. Measuring the value of sociology? Some notes on performative metricization in the contemporary academy. *Sociological Review* 59, 130–50.

Keltie, J. S. 1886. *Report of the Proceedings of the Royal Geographical Society in Reference to the Improvement of Geographical Education*. London: John Murray.

Keltie, J. S. 1908. *Applied Geography: A preliminary sketch*. London: G. Philip.

Keylock, C. J. 2003. Mark Melton's geomorphology and geography's quantitative revolution. *Transactions of the Institute of British Geographers* NS28, 142–57.

Kimble, G. H. T. 1951. The inadequacy of the regional concept. In L. D. Stamp and S. W. Wooldridge (eds) *London Essays in Geography: Rodwell Jones memorial volume*. London: Longmans Green, 151–74.

Kindon, S., Pain, R. and Kesby, M. (eds) 2007. *Participatory Action Research Approaches and Methods: Connecting people*. London: Routledge.

King, G. 1997. *A Solution to the Ecological Inference Problem: Reconstructing individual behavior from aggregate data*. Princeton, NJ: Princeton University Press.

King, L. J. 1960. A note on theory and reality. *Professional Geographer* 12(3), 4–6.

King, L. J. 1961. A multivariate analysis of the spacing of urban settlement in the United States. *Annals of the Association of American Geographers* 51, 222–33.

King, L. J. 1962. A quantitative expression of the pattern of urban settlement sin selected parts of the United States. *Tijdschrift voor Economische en Sociale Geografie* 53, 1–7.

King, L. J. 1969a. *Statistical Analysis in Geography*. Englewood Cliffs, NJ: Prentice-Hall.

King, L. J. 1969b. The analysis of spatial form and relationship to geographic theory. *Annals of the Association of American Geographers* 59, 573–95.

King, L. J. 1976. Alternatives to a positive economic geography. *Annals of the Association of American Geographers* 66, 293–308.

King, L. J. 1979a. Areal associations and regressions. *Annals of the Association of American Geographers* 69, 124–8.

King, L. J. 1979b. The seventies: Disillusionment and consolidation. *Annals of the Association of American Geographers* 69, 155–7.

King, L. J. 1993. Spatial science and the institutionalization of geography as a social science. *Urban Geography* 14, 538–51.

King, L. J. (ed.) 2008. *North American Explorations: Ten memoirs of geographers from down under*. Victoria, BC: Trafford Publishing.

King, L. J. and Clark, G. L. 1978. Government policy and regional development. *Progress in Human Geography* 2, 1–16.

King, R., Connell, J. and White, P. (eds) 1995. *Writing Across Worlds: Literature and migration*. London: Routledge.

Kingsbury, P. and Pile, S. 2014. Introduction. The unconscious transference, drives, repetition and other things tied to geography. In P. Kingsbury and S. Pile (eds) *Psychoanalytic Geographies*. Farnham: Ashgate, 1–38.

Kirk, W. 1951. Historical geography and the concept of the behavioural environment. *Indian Geographical Journal* 25, 152–60.

Kirk, W. 1963. Problems of geography. *Geography* 48, 357–71.

Kirk, W. 1978. The road from Mandalay: Towards a geographical philosophy. *Transactions of the Institute of British Geographers* NS3, 381–94.

Kish, G. and Ward, R. 1981. A survival package for geography and other endangered disciplines. *Newsletter Association of American Geographers*, 16, 8,14.

Kitchin, R. 1996. Increasing the integrity of cognitive mapping research: Appraising conceptual schemata of environment-behaviour interaction. *Progress in Human Geography* 20, 56–84.

Kitchin, R. 2013. Big data and human geography: Opportunities, challenges and risks. *Dialogues in Human Geography* 3, 262–7.

Kitchin, R. and Dodge, M. 2007. Rethinking maps. *Progress in Human Geography* 31, 331–44.

Kitchin, R. and Dodge, M. 2011. *Code/Space: Software and everyday life*. Cambridge, MA: MIT Press.

Kitchin, R. and Kneale, J. 2001. Science fiction or future fact? Exploring imaginative geographies of the new millennium. *Progress in Human Geography* 25, 19–36.

Kitchin, R. and Lysacht, K. 2003. Heterosexism and geographies of everyday life in Belfast, Northern Ireland. *Environment and Planning A* 35, 489–510.

Knopp, L. 1990. Some theoretical implications of gay involvement in an urban land market. *Political Geography Quarterly* 9, 337–52.

Knopp, L. 1992. Sexuality and the spatial dynamics of capitalism. *Environment and Planning D: Society and Space* 10, 651–69.

Knopp, L. 2007. On the relationship between queer and feminist geographies. *Professional Geographer* 59, 47–55.

Knopp, L. and Lauria, M. 1987. Gender relations and social relations. *Antipode: A Radical Journal of Geography* 19, 48–53.

Knos, D. S. 1968. The distribution of land values in Topeka, Kansas. In B. J. L. Berry and D. F. Marble (eds) *Spatial Analysis*. Englewood Cliffs, NJ: Prentice-Hall, 269–89.

Knox, P. L. 1975. *Social Well-being: A spatial perspective*. London: Oxford University Press.

Knox, P. L. 1987. The social production of the built environment: Architects, architecture and the post-modern city. *Progress in Human Geography* 11, 354–78.

Knox, P. L. and Agnew, J. A. 1989. *The Geography of the World-Economy*. London: Edward Arnold.

Knox, P. L. and Taylor, P. J. (eds) 1995. *World Cities in a World-System*. Cambridge: Cambridge University Press.

Knox, P. L., Bartels, E. H., Bohland, J. R., Holcomb, B. and Johnston, R. J. 1988. *The United States: A contemporary human geography*. London: Longman.

Kobayashi, A. and Mackenzie, S. (eds) 1989. *Remaking Human Geography*. Boston, MA: Unwin Hyman.

Kobayshi, A. and Peake, L. 2000. Racism out of place: Thoughts on whiteness and antiracist geography in the new millennium. *Annals of the Association of American Geographers* 90, 392–403.

Koelsch, W. A. 1988. Geography at Clark: The first fifty years, 1921–1971. In J. E. Harmon and T. J. Rickard (eds) *Geography in New England*. New Britain, CT: New England-St Lawrence Valley Geographical Society, 40–8.

Koelsch, W. A. 2002. Academic geography, American style: An institutional perspective. In G. S. Dunbar (ed.) *Geography: Discipline, profession and subject since 1870*. Dordrecht: Kluwer, 245–80.

Kofman, E. 1988. Is there a cultural geography beyond the fragments? *Area* 20, 85–7.

Kofman, E. and England, K. 1997. Guest editorial. Citizenship and international migration: taking account of gender, sexuality, and 'race'. *Environment and Planning A* 29, 191–3.

Kofman, E. and Peake, L. 1990. Into the 1990s: A gendered agenda for political geography. *Political Geography Quarterly* 9, 313–36.

Kollmorgen, W. N. 1979. Kollmorgen as a bureaucrat. *Annals of the Association of American Geographers* 69, 77–89.

Kong, L. 1990. Geography and religion: Trends and prospects. *Progress in Human Geography* 14, 355–71.

Kong, L. 2007. The promises and prospects of geography in higher education. *Journal of Geography in Higher Education* 31, 13–17.

Kost, K. 1989. The conception of politics in political geography and geopolitics in Germany 1920–1950. *Political Geography Quarterly* 8, 369–86.

Kroeber, A. L. 1952. *The Nature of Culture*. Chicago: University of Chicago Press.

Krumbein, W. C. and Graybill, F. A. 1965. *An Introduction to Statistical Models in Geology*. New York: McGraw Hill.

Kuhn, T. S. 1957. *The Copernican Revolution: Planetary astronomy in the development of western thought*. Cambridge MA: Harvard University Press.

Kuhn, T. S. 1962. *The Structure of Scientific Revolutions*. Chicago: University of Chicago Press.

Kuhn, T. S. 1969. Comment on the relations of science and art. *Comparative Studies in Society and History* 11, 403–12.

Kuhn, T. S. 1970a. *The Structure of Scientific Resolutions*, 2nd edn. Chicago: University of Chicago Press.

Kuhn, T. S. 1970b. Logic of discovery or psychology of research? In I. Lakatos and A. Musgrave (eds) *Criticism and the Growth of Knowledge*. Cambridge: Cambridge University Press, 1–23.

Kuhn, T. S. 1970c. Reflections on my critics. In I. Lakatos and A. Musgrave (eds) *Criticism and the Growth of Knowledge*. Cambridge: Cambridge University Press, 231–78.

Kuhn, T. S. 1977. Second thoughts on paradigms. In F. Suppe (ed.) *The Structure of Scientific Theories*. Urbana, IL: University of Illinois Press, 459–82, plus discussion 500–17.

Kuhn, T. S. 1991. The natural and the human sciences. In D. J. Hiley, R. Bohman and R. Shusterman (eds) *The Interpretative Turn: Philosophy, science, culture*. Ithaca: Cornell University Press, 17–24.

Kwan, M.-P. 2002. Feminist visualization: Re-envisioning GIS as a method in feminist geographic research. *Annals of the Association of American Geographers* 92, 645–61.

Kwan, M.-P. 2004. Beyond difference: From canonical geography to hybrid geographies. *Annals of the Association of American Geographers* 94, 756–63.

Kwan, M.-P. 2012. The uncertain geographic concept problem. *Annals of the Association of American Geographers* 102, 958–68.

Kwan, M.-P. (ed.) 2014. *Geographies of Health, Disease and Well-being*. London and New York Routledge.

Kwan, M.-P. and Schwanen, T. 2009a. Critical quantitative geographies. *Environment and Planning A* 41, 261–4.

Kwan, M.-P. and Schwanen, T. 2009b. Quantitative revolution 2: The critical (re)turn. *Professional Geographer* 61, 283–91.

Labedz, L. 1977. Anarchism. In A. Bullock and O. Stallybrass (eds) *The Fontana Dictionary of Modern Thought*. London: William Collins, 22.

Lacoste, Y. 1973. An illustration of geographical warfare: Bombing of the dikes on the Red River, North Vietnam. *Antipode: A Radical Journal of Geography* 5, 1–13.

Lakatos, I. 1978a. Falsification and the methodology of scientific research programmes. In J. Worrall and G. Currie (eds) *The Methodology of Scientific Research Programmes: Philosophical papers, Vol. 1*. Cambridge: Cambridge University Press, 8–101.

Lakatos, I. 1978b. History of science and its rational reconstructions. In J. Worrall and G. Currie (eds) *The Methodology of Scientific Research Programmes: Philosophical Papers, Vol. 1*. Cambridge: Cambridge University Press, 102–38.

Landström, C., Whatmore, S. J., Lane, S. N., Odoni, N. A., Ward, N. and Brady, S. 2011. Coproducing flood risk knowledge: Redistributing expertise in critical 'participatory modelling'. *Environment and Planning A* 43, 1617–33.

Langton, J. 1972. Potentialities and problems of adapting a systems approach to the study of change in human geography. In C. Board, R. J. Chorley, P. Haggett and D. R. Stoddart (eds) *Progress in Geography, Vol. 4*. London: Edward Arnold, 125–79.

Langton, J. 1984. The industrial revolution and the regional geography of England. *Transactions of the Institute of British Geographers* NS9, 145–67.

Laponce, J. A. 1980. Political science: An import-export analysis of journals and footnotes. *Political Studies* 28, 401–19.

Larkin, R. P. and Peters, G. L. 1993. *Biographical Dictionary of Geography*. Westport, CT: Greenwood Press.

Larner, W. 2011. C-change: Geographies of crisis? *Dialogues in Human Geography* 1, 319–35.

Lash, S. and Urry, J. 1987. *The End of Organized Capitalism*. Cambridge: Polity Press.

Latour, B. 1987. *Science in Action: How to follow scientists and engineers around society*. Cambridge, MA: Harvard University Press.

Latour, B. 1999. *Pandora's Hope: Essays on the reality of science studies*. Cambridge, MA: Harvard University Press.

Latour, B. and Woolgar, S. 1979. *Laboratory Life: The construction of scientific facts*. Beverly Hills: Sage.

Lauria, M. and Knopp, L. 1985. Toward an analysis of the role of gay communities in urban renaissance. *Urban Geography* 6, 152–69.

Laurie, N., Dwyer, C., Holloway, S. and Smith, F. 1999. *Geographies of New Femininities*. Harlow: Longman.

Lavalle, P., McConnell, H. and Brown, R. G. 1967. Certain aspects of the expansion of quantitative methodology in American geography. *Annals of the Association of American Geographers* 57, 423–36.

Law, J. 1976. Theories and methods in the sociology of science: an interpretative approach. In G. Lemaine et al., *Perspectives on the Emergence of Scientific Disciplines*. The Hague: Mouton, 221–31.

Law, J. and Hassard, J. (eds) 1999. *Actor Network Theory and After*. Oxford and Keele: Blackwell and The Sociological Review.

Law, R. 1999. Beyond 'women and transport': Towards new geographies of gender and daily mobility. *Progress in Human Geography* 23, 567–88.

Lawson, V. 2007. Geographies of care and responsibility. *Annals of the Association of American Geographers* 97, 1–11.

Lawson, V. and Staeheli, L. A. 1991. On critical realism, geography, and arcane sects! *Professional Geographer* 43, 231–3.

Lawton, R. and Miller, E. W. 2001. Two economic geography texts. *Progress in Human Geography* 25, 303–9.

Leach, B. 1974. Race, problems and geography. *Transactions of the Institute of British Geographers* 63, 41–7.

Leach, E. R. 1974. *Lévi-Strauss*. London: William Collins.

Lee, R. 1984. Process and region in the A-level syllabus. *Geography* 69, 97–107.

Lee, R. 1985. The future of the region: Regional geography as education for transformation. In R. King (ed.) *Geographical Futures*. Sheffield: Geographical Association, 77–91.

Lee, R. and Philo, C. 2009. Welfare geography. In R. Kitchin and N. Thrift (eds) *International Encyclopedia of Human Geography*. Amsterdam: Elsevier, 224–9.

Lee, R. and Wills, J. (eds) 1997. *Geographies of Economies*. London: Arnold.

Lee, R., Castree, N., Kitchin, R., Lawson, V., Paasi, A., Philo, C., Radcliffe, S., Roberts, S. M. and Withers, C. 2014. Prologue: The vital requirement of reflexivity. In Lee, R. et al. (eds) *The SAGE Handbook of Human Geography*. Thousand Oaks, CA and London, New Delhi, Singapore: Sage, Vol. 1, ix–x.

Lee, Y. 1975. A rejoinder to 'The geography of crime: A political critique'. *Professional Geographer* 27, 284–5.

Lees, L. 2000. A reappraisal of gentrification: Towards a 'geography of gentrification'. *Progress in Human Geography* 24, 389–408.

Lees, L. 2012. The geography of gentrification: Thinking through comparative urbanism. *Progress in Human Geography* 36, 155–71.

Lees, L., Slater, T. and Wyly, E. (eds) 2010. *The Gentrification Reader*. Abingdon and New York: Routledge.

Leighley, J. 1937. Some comments on contemporary geographic methods. *Annals of the Association of American Geographers* 27, 125–41.

Leighley, J. 1955. What has happened to physical geography? *Annals of the Association of American Geographers* 45, 309–18.

Lemaine, G., MacLeod, R., Mulkay, M. and Weingart, P. 1976. Introduction: Problems in the emergence of new disciplines. In G. Lemaine et al. (eds) *Perspectives on the Emergence of Scientific Disciplines*. The Hague: Mouton, 1–24.

Leonard, S. 1982. Urban managerialism: A period of transition. *Progress in Human Geography* 6, 190–215.

Leszczynski, A. 2014. On the neo in neogeography. *Annals of the Association of American Geographers* 104, 60–79.

Lewis, G. M. 1966. Regional ideas and reality in the Cis-Rocky Mountain West. *Transactions of the Institute of British Geographers* 38, 135–50.

Lewis, G. M. 1968. Levels of living in the Northeastern United States c. 1960: A new approach to regional geography. *Transactions of the Institute of British Geographers* 45, 11–37.

Lewis, G. M. (ed.) 1998. *Cartographic Encounters: Perspectives on native American mapmaking and map use*. Chicago: University of Chicago Press.

Lewis, M. W. and Wigen, K. 1997. The myth of continents: A critique of metageography. Berkeley, CA: University of California Press.

Lewis, P. W. 1965. Three related problems in the formulation of laws in geography. *Professional Geographer* 17(5), 24–7.

Lewis, P. W. 1977. *Maps and Statistics*. London: Methuen.

Lewthwaite, G. R. 1966. Environmentalism and determinism: A search for clarification. *Annals of the Association of American Geographers* 56, 1–23.

Ley, D. 1974. *The Black Inner City as Frontier Outpost*. Washington, DC: Association of American Geographers.

Ley, D. 1977a. The personality of a geographical fact. *Professional Geographer* 29, 8–13.

Ley, D. 1977b. Social geography and the taken-for-granted world. *Transactions of the Institute of British Geographers* NS2, 498–512.

Ley, D. 1978. Social geography and social action. In D. Ley and M. S. Samuels (eds) *Humanistic Geography: Problems and prospects*. Chicago: Maaroufa Press, 41–57.

Ley, D. 1980. *Geography Without Man: A humanistic critique*. Oxford Research Paper 24, School of Geography, Oxford: University of Oxford.

Ley, D. 1981. Behavioral geography and the philosophies of meaning. In K. R. Cox and R. G. Golledge (eds) *Behavioral Problems in Geography Revisited*. London: Methuen, 209–30.

Ley, D. 1983. *A Social Geography of the City*. New York: Harper & Row.

Ley, D. 1989. Fragmentation, coherence, and the limits to theory in human geography. In A. Kobayashi and S. Mackenzie (eds) *Remaking Human Geography*. Boston, MA: Unwin Hyman, 227–44.

Ley, D. 1993. Postmodernism, or the cultural logic of advanced industrial capital. *Tijdschrift voor Economische en Sociale Geografie* 84, 171–4.

Ley, D. and Duncan, J. S. 1993. Epilogue. In J. S. Duncan and D. Ley (eds) *Place/Culture/Representation*. London: Routledge, 329–36.

Ley, D. and Samuels, M. S. 1978. Introduction: Contexts of modern humanism in geography. In D. Ley and M. S. Samuels (eds) *Humanistic Geography: Prospects and problems*. Chicago: Maaroufa Press, 1–18.

Leyshon, A. 1995. Missing words: Whatever happened to the geography of poverty? *Environment and Planning A* 27, 1021–5.

Leyshon, A., Matless, D. and Revill, G. 1995. The place of music. *Transactions of the Institute of British Geographers* NS20, 423–33.

Lichtenberger, E. 1984. The German-speaking countries. In R. J. Johnston and P. Claval (eds) *Geography Since the Second World War: An international survey*. London: Croom Helm, 156–84.

Lichtenberger, E. 1997. Harris and Ullman's 'The nature of cities': The paper's historical context and its impact for further research. *Urban Geography* 18, 7–14.

Lipietz, A. 1993. From Althusserianism to 'regulation theory'. In E. A. Kaplan and M. Sprinker (eds) *The Althusserian Legacy*. London and New York: Verso, 99–138.

Little, J., Peake, L. and Richardson, P. (eds) 1988. *Women in Cities*. London: Macmillan.

Liu, X. and Zhan, F. B. 2011. Placement of graduates in PhD-granting departments as a measure of productivity of doctoral geography programs in the United States: 1960–2008. *Professional Geographer* 64, 1–16.

Liu, X. and Zhan, F. B. 2012. Productivity of doctoral graduate placement among PhD–granting geography programs in the United States: 1960–2010. *Professional Geographer* 64, 475–90.

Liverman, D. M. 2009. The geopolitics of climate change: Avoiding determinism, fostering sustainable development. *Climate Change* 96, 7–11.

Livingstone, D. N. 1984. Natural theory and neo-Lamarckism: The changing context of nineteenth century geography in the United States and Great Britain. *Annals of the Association of American Geographers* 74, 9–28.

Livingstone, D. N. 1992. *The Geographical Tradition: Episodes in the history of a contested enterprise*. Oxford: Blackwell Publishers.

Livingstone, D. N. 1995. Geographical traditions. *Transactions of the Institute of British Geographers* NS20, 420–2.

Livingstone, D. N. 2000a. Putting geography in its place. *Australian Geographical Studies*, 39, 1–9.

Livingstone, D. N. 2000b. Making space for science. *Erdkunde* 54, 285–96.

Livingstone, D. N. 2002. *Science, Space and Hermeneutics*. Heidelberg: Department of Geography, University of Heidelberg.

Livingstone, D. N. 2003a. British geography, 1500–1900: An imprecise review. In R. J. Johnston and M. Williams (eds) *A Century of British Geography*. Oxford: Oxford University Press, 11–44.

Livingstone, D. N. 2003b. *Putting Science in its Place: Geographies of scientific knowledge*. Chicago: University of Chicago Press.

Livingstone, D. N. 2005. Science, text and space: Thoughts on the geography of reading. *Transactions of the Institute of British Geographers* 30, 391–401.

Livinstone, D. N. 2007. Johnston, R. J. 1979: *Geography and Geographers: Anglo–American human geography since 1945*. London: Edward Arnold. Commentary 1. *Progress in Human Geography* 31, 43–5.

Livingstone, D. N. and Harrison, R.T. 1981. Immanuel Kant, subjectivism, and human geography: A preliminary investigation. *Transactions of the Institute of British Geographers* NS6, 359–74.

Livingstone, D. N. and Withers, C. W. J. (eds) 1999. *Geography and Enlightenment*. Chicago: University of Chicago Press.

Lloyd, C. D., Pawlowsky–Glahn, V. and Egozcue, J. J. 2012. Compositional data analysis in population studies. *Annals of the Association of American Geographers* 102, 1–16.

Lobben, A., Lawrence, M. and Pickett, R. 2014. The map effect. *Annals of the Association of American Geographers* 104, 96–113.

Longhurst, R. 1995. Viewpoint – The body and geography. *Gender, Place & Culture: A Journal of Feminist Geography* 2, 97–106.

Longhurst, R. 1997. (Dis)embodied geographies. *Progress in Human Geography* 21, 486–501.

Longhurst, R. 2000. 'Corporeographies' of pregnancy: 'Bikini babes'. *Environment and Planning D: Society and Space* 18, 453–72.

Longley, P. A. 1995. GIS and planning for businesses and services. *Environment and Planning B: Planning and Design* 22, 127–9.

Longley, P. A. 1998. Foundations. In P. A. Longley, S. M. Brooks, R. MacDonnell and B. Macmillan (eds) *Geocomputation: A primer*. Chichester: John Wiley, 3–16.

Longley, P. A. 2000. Spatial analysis in the new millennium. *Annals of the Association of American Geographers* 90, 157–65.

Longley, P. A. 2002. Geographical Information Systems: Will developments in urban remote sensing and GIS lead to 'better' urban geography? *Progress in Human Geography* 26, 240–51.

Longley, P. A. 2004. Geographical Information Systems: On modelling and representation. *Progress in Human Geography* 28.

Longley, P. A. 2012. Geodemographics and the practice of geographical information science. Special issue in honor of Michael Goodchild, *International Journal of Geographic Information Science* 26, 2227–37.

Longley, P. A. and Mesev, V. 2000. On the measurement and generalisation of urban form. *Environment and Planning A* 32, 473–88.

Longley, P. A., Webber, R. and Lloyd, D. 2007. The quantitative analysis of family names: Historic migration and the present day neighborhood structure of Middlesbrough, England. *Annals of the Association of American Geographers* 97, 31–48.

Longley, P. A., Goodchild, M. F., Maguire, D. J. and Rhind, D. W. (eds) 1999a. *Geographical Information Systems: Principles, techniques, management and applications* (2 vols). New York: Wiley.

Longley, P. A., Goodchild, M. F., Maguire, D. J. and Rhind, D. W. (eds) 1999b. Introduction. In P. A. Longley, M. F. Goodchild, D. J. Maguire and D. W. Rhind (eds) *Geographical Information Systems: Principles, techniques, management and applications* (2 vols). New York: Wiley, 1–20.

Longley, P. A., Goodchild, M. F., Maguire, D. J. and Rhind, D. W. 2001. *Geographic Information Systems and Science*, 1st edn. Chichester: John Wiley.

Longley, P. A., Goodchild, M. F., Maguire, D. J. and Rhind, D. W. (eds) 2005. *Geographical Information Systems and Science: Principles, techniques, management and applications*, 2nd edn (abridged). Harlow, NJ: Wiley.

Longley, P. A., Goodchild, M. F., Maguire, D. J. and Rhind, D. W. 2011. *Geographic Information Systems and Science*, 3rd edn. Chichester: John Wiley.

Longley, P. A., Goodchild, M. F., Maguire, D. J. and Rhind, D. W. 2016. *Geographic Information Systems and Science*, 4th edn. Hoboken, NJ: John Wiley.

Loomis, J. M., Montello, D. R. and Klatzky, R. L. 2010. Reginald G. Golledge. *Progress in Human Geography* 34, 678–90.

Lorimer, H. 2003. Telling small stories: Spaces of knowledge and the practice of geography. *Transactions of the Institute of British Geographers* NS28, 197–217.

Lorimer, H. 2005. Cultural geography: The busyness of being 'more-than-representational'. *Progress in Human Geography* 29, 83–94.

Lorimer, H. and Spedding, N. 2002. Excavating geography's hidden spaces. *Area* 34, 294–302.

Lorimer, H. and Withers, C. 2007. Geographers: Lives, works, possibilities. In H. Lorimer and C. Withers (eds) *Geographers: Biobibliographical Studies, Vol. 26.* London and New York: Continuum, 1–5.

Lorimer, H. and Withers, C. W. J. (eds) 2014. *Geographers: Biobibliographical Studies, Vol. 33.* London: Bloomsbury Academic.

Lösch, A. 1954. *The Economics of Location.* New Haven, CT: Yale University Press.

Louden, R. B. 2014. The last frontier: The importance of Kant's Geography. *Environment and Planning D: Society and Space* 32, 1–16.

Lovering, J. 1987. Militarism, capitalism and the nation-state: Towards a realist synthesis. *Environment and Planning D: Society and Space* 5, 283–302.

Lowe, M. S. and Short, J. R. 1990. Progressive human geography. *Progress in Human Geography* 14, 1–11.

Lowenthal, D. (ed.) 1965. *George Perkins Marsh: Man and nature.* Cambridge, MA: Harvard University Press.

Lowenthal, D. 1958. *George Perkins Marsh: Versatile Vermonter.* New York: Columbia University Press.

Lowenthal, D. 1961. Geography, experience, and imagination: Towards a geographical epistemology. *Annals of the Association of American Geographers* 51, 241–60.

Lowenthal, D. 1968. The American scene. *Geographical Review* 48, 61–88.

Lowenthal, D. 1975. Past time, present place: Landscape and memory. *Geographical Review* 65, 1–36.

Lowenthal, D. 1985. *The Past is a Foreign Country.* Cambridge: Cambridge University Press.

Lowenthal, D. 2001. *George Perkins Marsh: Prophet of conservation.* Seattle, WA: University of Washington Press.

Lowenthal, D. 2014. Essay: Albion's other islets: Offshore, overseas, out of sorts. *Geographical Reivew* 104, 101–8.

Lowenthal, D. and Bowden, M. J. (eds) 1975. *Geographies of the Mind: Essays in historical geosophy in honor of John Kirkland Wright.* New York: Oxford University Press.

Lowenthal, D. and Prince, H. C. 1965. English landscape tastes. *Geographical Review* 55, 186–222.

Lowenthal, D., Burton, I., Cooley, R. and Mikesell, M. 1973. Report of the AAG task force on environmental quality. *Professional Geographer* 25, 39–46.

Luke, T. W. 2000. The discipline as disciplinary normalization: Networks of research. In R. Sil and E. M. Doherty (eds) *Beyond Boundaries? Disciplines, paradigms, and theoretical integration in international studies.* Albany: State University of New York Press, 207–29.

Lukermann, F. 1958. Towards a more geographic economic geography. *Professional Geographer* 10(1), 2–10.

Lukermann, F. 1960a. On explanation, model, and prediction. *Professional Geographer* 12(1), 1–2.

Lukermann, F. 1960b. The geography of cement? *Professional Geographer* 12(4), 1–6.

Lukermann, F. 1961. The role of theory in geographical inquiry. *Professional Geographer* 13(2), 1–6.

Lukermann, F. 1964a. *Geography Among the Sciences.* Minneapolis, MN: Kalamata.

Lukermann, F. 1964b. Geography as a formal intellectual discipline and the way in which it contributes to knowledge. *Canadian Geographer* 8, 167–72.

Lukermann, F. 1965. Geography: De facto or de jure. *Journal of the Minnesota Academy of Science* 32, 189–96.

Lukermann, F. 1990. The nature of geography: Post hoc, ergo propter hoc? In J. N. Entrikin and S. D. Brunn (eds) *Reflections on Richard Hartshorne's The Nature of Geography.* Washington, DC: Association of American Geographers, 53–68.

Lutz, C. A. and Collins, J. L. 1993. *Reading National Geographic.* Chicago: University of Chicago Press.

Lynch, K. 1960. *The Image of the City.* Cambridge, MA: MIT Press.

Lynch, M. 2012. Notes on Kuhn and Latour. *Social Studies of Science* 42, 449–55.

Mabogunje, A. K. 1977. In search of spatial order: Geography and the new programme of urbanization in Nigeria. In D. R. Deskins et al. (eds) *Geographic Humanism, Analysis and Social Action: A half century of geography at Michigan.* Ann Arbor, MI: Michigan Geography Publications No. 17, 347–76.

MacAllister, I., Pattie, C. J., Tunstall, H., Dorling, D. F. L. and Rossiter, D. J. 2001. Class dealignment and the neighbourhood effect: Miller revisited. *British Journal of Political Science* 31, 41–60.

MacDonald, N., Chester, D., Sangster, H., Todd, B. and Hooke, J. 2012. The significance of Gilbert F. White's 1945 paper 'Human adjustment to floods' in the development of risk and hazard management. *Progress in Physical Geography* 35, 125–33.

MacEachren, A. 1995. *How Maps Work*. New York: Guilford.

MacGill, S. M. 1983. The Q-controversy: Issues and nonissues. *Environment and Planning B: Planning and Design* 10, 371–80.

Machamer, P. and Osbeck, L. 2003. Scientific normativity as non-epistemic: A hidden Kuhnian legacy. *Social Epistemology* 17, 3–11.

Mackinder, H. J. 1887. On the scope and methods of geography. *Proceedings, Royal Geographical Society and Monthly Record of Geography* 9, 141–74.

Mackinder, H. J. 1890. The physical basis of political geography. *Scottish Geographical Magazine* 6, 78–84.

Mackinder, H. J. 1904. The geographical pivot of history. *Geographical Journal* 23, 421–37.

Mackinder, H. J. 1919. *Democratic Ideals and Reality: A study in the politics of reconstruction*. London: Constable.

MacKinnon, D. 2009. Regional Geography II. In R. Kitchin and N. Thrift (eds) *International Encyclopedia of Human Geography*. Amsterdam: Elsevier, 228–35.

MacLean, K. 2011. G. G. Chisholm, A. G. Ogilvie and the 1912 America transcontinental excursion. *Scottish Geographical Journal* 127, 231–48.

MacLeod, G. 1997. Globalizing Parisian thought-waves: Recent advances in the study of social regulation, politics, discourse and space. *Progress in Human Geography* 21, 530–3.

MacLeod, G. 1998. In what sense a region? Place, hybridity, symbolic shape, and institutional formation in (post-)modern Scotland. *Political Geography* 17, 622–63.

MacLeod, G. and Goodwin, M. 1999. Space, scale and state strategy: Towards a reinterpretation of contemporary urban and regional governance. *Progress in Human Geography* 23, 503–28.

MacLeod, G. and Holden, A. 2009. Regulation. In R. Kitchin and N. Thrift (eds) *International Encyclopedia of Human Geography*. Amsterdam: Elsevier, 309–13.

MacLeod, G. and Jones, M. 2001. Renewing the geography of regions. *Environment and Planning D: Society and Space* 19, 695–99.

Macmillan, B. 1989a. Quantitative theory construction in human geography. In B. Macmillan (ed.) *Remodelling Geography*. Oxford: Basil Blackwell, 89–107.

Macmillan, B. 1989b. Modelling through: An afterword to *Remodelling Geography*. In B. Macmillan (ed.) *Remodelling Geography*. Oxford: Basil Blackwell, 291–313.

Macmillan, B. 1998c. Epilogue. In P. Longley, S. M. Brooks, R. MacDonnell and B. Macmillan (eds) *Geocomputation: A primer*. Chichester: John Wiley, 257–64.

Maddrell, A. 2006. Revisiting the region: The 'ordinary' and 'exceptional' regions in the work of Hilda Ormsby: 1917–1940. *Environment and Planning A* 38, 1739–1752.

Maddrell, A. 2008. The 'map girls': British women geographers' war work, shifting gender boundaries and reflections on the history of geography. *Transactions of the Institute of British Geographers*, NS33, 127–48.

Maddrell, A. 2009. *Complex Locations: Women's geographical work in the UK 1850–1970*. Oxford and Malden, MA: Wiley-Blackwell.

Madge, C. 1993. Boundary disputes: Comments on Sidaway (1992). *Area*, 25, 294–9.

Made, C. 1997. An ode to geography. In Women and Geography Study Group (eds) *Feminist Geographies: Explorations in diversity and difference*. Harlow: Longman, 32–3.

Magee, B. 1975. *Popper*. London: William Collins.

Maguire, D. J. 1989. The Domesday interactive videodisc system in geography teaching. *Journal of Geography in Higher Education* 13, 55–68.

Maguire, D. J., Goodchild, M. F. and Rhind, D. W. (eds) 1991. *Geographical Information Systems*. London: Longman.

Mahtiani, M. 2001. Racial ReMappings: The potential of paradoxical space. *Gender, Place and Culture* 8, 299–305.

Mahtani, M. 2006. Challenging the ivory tower: Proposing antiracist geographies in the academy. *Gender, Place and Culture* 13, 21–5.

Mair, A. 1986. Thomas Kuhn and understanding geography. *Progress in Human Geography* 10, 345–70.

Manion, T. and Whitelegg, J. 1979. Radical geography and Marxism. *Area* 11, 122–4.

Mann, M. 1996. Neither nation-state nor globalism. *Environment and Planning A* 28.

Manners, I. R. and Mikesell, M. W. (eds) 1974. *Perspectives on Environment*. Washington, DC: Commission on College Geography, Association of American Geographers.

Mao, L. 2014. The geography, structure, and evolution of the GIS research community in the US: A network analysis from 1992 to 2011. *Transactions in GIS* 18, 704–17.

Marble, D. F. and Peuquet, D. F. 1993. The computer and geography: Ten years later. *Professional Geographer* 45, 446–8.

Marchand, B. 1978. A dialectical approach in geography. *Geographical Analysis* 10, 105–19.

Marcus, M. G. 1979. Coming full circle: Physical geography in the twentieth century. *Annals of the Association of American Geographers* 69, 521–32.

Marginson, S. 2004. Higher education. In R. Manne (ed.) *The Howard Years*. Melbourne: Black Inc., 216–44.

Mark, D. M., Chrisman, N., Frank, A. U., McHaffie, P. and Pickles, J. 1996. *The GIS History Project*. Available at: www.geog.buffalo.edu/ncgia/gishist

Mark, D. M., Freksa, C., Hirtle, S. C., Lloyd, R. and Tverske, B. 1999. Cognitive models of geographic space. *International Journal of Geographical Information Science* 13, 747–74.

Markusen, A. 1999. Fuzzy concepts, scanty evidence, policy distance: The case for rigour and policy relevance in critical regional studies. *Regional Studies* 33, 868–94.

Marsh, G. P. 1864. *Man and Nature, or Physical Geography as Modified by Human Action*. New York: Charles Scribner.

Marshall, J. U. 1985. Geography as a scientific enterprise. In R. J. Johnston (ed.) *The Future of Geography*. London: Methuen, 113–28.

Marston, S. Jones, J. P. and Woodward, K. 2005. Human geography without scale. *Transactions of the Institute of British Geographers* NS30, 416–32.

Martin, A. F. 1951. The necessity for determinism. *Transactions and Papers, Institute of British Geographers* 17, 1–12.

Martin, D. 2002. Output areas for 2001. In P. Rees, D. Martin and P. Williamson (eds) *The Census Data System*. Chichester: John Wiley, 37–46.

Martin, D., Nolan, A. and Tranmer, M. 2001. The application of zone-design methodology in the 2001 UK census. *Environment and Planning A* 33, 1949–62.

Martin, G. J. 1980. *The Life and Thought of Isaiah Bowman*. Hamden, CT: Archon Books.

Martin, G. J. 1981. Ontography and Davisian physiography. In B. W. Blouet (ed.) *The Origins of Academic Geography in the United States*. Hamden, CT: Archon Books, 279–90.

Martin, G. J. 1988. On Whittlesey, Bowman and Harvard. *Annals of the Association of American Geographers* 78, 152–8.

Martin, G. J. 1990. *The Nature of Geography* and the Schaefer-Hartshorne debate. In J. N. Entrikin and S. D. Brunn (eds) *Reflections on Richard Hartshorne's* The Nature of Geography. Washington, DC: Association of American Geographers, 69–88.

Martin, G. J. 2005. *All Possible Worlds: A history of geographical ideas*, 4th edn. New York: Oxford University Press.

Martin, G. J. 1998. The emergence and development of geographic thought in New England. *Economic Geography* (special issue), 1–13.

Martin, G. J. and James, P. E. 1993. *All Possible Worlds: A history of geographical ideas*, 3rd edn. New York: Wiley.

Martin, G. J. 2013. Maynard Weston Dow (1929–2011) and 'Geographers on Film'. *Annals of the Association of American Geographers* 103, 1–4.

Martin, G. J. 2015. *American Geography and Geographers: Towards geographic science*. Oxford and New York: Oxford University Press.

Martin, G. J. and James, P. E. 1993. *All Possible Worlds: A history of geographical ideas*, 3rd edn. New York: John Wiley.

Martin, R. L. 1999. The 'new geographical turn' in economics: Some critical reflections. *Cambridge Journal of Economics* 23, 65–92.

Martin, R. L. 2000. In memory of maps. *Transactions of the Institute of British Geographers* NS25, 3–6.

Martin, R. L. 2001a. The geographer as social critic – getting indignant about income inequality. *Transactions of the Institute of British Geographers* NS26, 267–72.

Martin, R. L. 2001b. Geography and public policy: The case of the missing agenda. *Progress in Human Geography* 25, 189–210.

Martin, R. L. 2002. A geography for policy or a policy for geography? A response to Dorling and Shaw. *Progress in Human Geography* 26, 642–4.

Martin, R. L. and Oeppen, J. E. 1975. The identification of regional forecasting models using space-time correlation functions. *Transactions of the Institute of British Geographers* 66, 95–118.

Martin, R. L. and Sunley, P. J. 2001. Rethinking the 'economic in economic geography: Broadening our vision or losing our focus? *Antipode: A Radical Journal of Geography* 33, 148–61.

Mason, K., Brown, G. and Pickerill, J. 2013. Epistemologies of participation, or, what do critical human geographers know that's of any use? *Antipode: A Radical Journal of Geography* 45, 252–5.

Massam, B. H. 1976. *Location and Space in Social Administration*. London: Edward Arnold.

Massey, D. 1975. Behavioral research. *Area* 7, 201–3.

Massey, D. 1984a. *Spatial Divisions of Labour: Social structures and the geography of production*. London: Macmillan.

Massey, D. 1984b. Introduction: Geography matters. In D. Massey and J. Allen (eds) *Geography Matters! A reader*. Cambridge: Cambridge University Press, 1–11.

Massey, D. 1985. New directions in space. In D. Gregory and J. Urry (eds) *Social Relations and Spatial Structures*. London: Macmillan, 9–19.

Massey, D. 1991. Flexible sexism. *Environment and Planning D: Society and Space* 9, 31–58.

Massey, D. 1992. Politics and space/time. *New Left Review* 196, 65–84.

Massey, D. 1995. *Spatial Divisions of Labour: Social structures and the geography of production*, 2nd edn. London: Macmillan.

Massey, D. 2000. Practising political relevance. *Transaction of the Institute of British Geographers* NS25, 131–3.

Massey, D. 2001. Geography on the agenda. *Progress in Human Geography* 25, 5–17.

Massey, D. 2002. Geography, policy and politics: A response to Dorling and Shaw. *Progress in Human Geography* 26, 645–6.

Massey, D. 2005. *For Space*. London, Thousand Oaks, CA and New Delhi: Sage.

Massey, D. and Meegan, R. A. 1979. The geography of industrial reorganization. *Progress in Planning* 10, 155–237.

Massey, D. and Meegan, R. A. 1982. *The Anatomy of Job Loss*. London: Methuen.

Massey, D. and Meegan, R. A. 1985. Introduction: The debate. In D. Massey and R. Meegan (eds) *Politics and Method: Contrasting studies in industrial geography*. London: Methuen, 1–12.

Masaterman, M. 1970. The nature of a paradigm. In I. Lakatos and A. Musgrave (eds) *Criticism and the Growth of Knowledge*. London: Cambridge University Press, 59–90.

Mather, J. R. and Sanderson, M. 1996. *The Genius of C. Warren Thornthwaite, Climatologist–Geographer*. Norman, OK: University of Oklahoma Press.

Mather, P. M. 1999. *Computer Processing of Remotely-sensed Images: An introduction*, 2nd edn. Chichester: John Wiley.

Mather, P. M. and Koch, M. 2010. *Computer Processing of Remotely-sensed Images: An introduction*, 4th edn. Chichester: John Wiley.

Mathewson, K. (ed.) 1993. *Culture, Form and Place: Essays in cultural and historical geography*. Baton Rouge: Louisiana State University, Department of Geography and Anthropology, *Geoscience and Man* publication, no. 32.

Mathewson, K. and Wisner, B. 2005. Introduction: The geographical and political vision of J M Blaut. *Antipode* 37, 900–10.

Matless, D. 1992. An occasion for geography: Landscape, representation and Foucault's corpus. *Environment and Planning D: Society and Space* 10, 41–56.

Matless, D. 1995. Effects of history. *Transactions of the Institute of British Geographers* NS20, 405–9.

Matthews, H. and Livinstone, I. 1996. Geography and lifelong learning. *Journal of Geography in Higher Education* 20, 5–10.

Matthews, J. A. and Herbert, D. T. (eds) 2004. *Unifying Geography: Common heritage, shared future*. London: Routledge.

Mattingly, D. and Falconer Al-Hindi, K. 1995. Should women count? A context for the debate. *Professional Geographer* 47, 427–35.

May, J. 1996a. 'A little taste of something more exotic': The imaginative geographies of everyday life. *Geography* 81, 57–64.

May, J. 1996b. Globalization and the politics of place: Place and identity in an inner London neighbourhood. *Transactions of the Institute of British Geographers* NS21, 194–215.

May, J. A. 1970. *Kant's Concept of Geography: And its relation to recent geographical thought*. Toronto: Department of Geography, University of Toronto, Research Publication 4.

May, J. A 1972. A reply to Professor Hartshorne. *Canadian Geographer* 16, 79–81.

Mayer, H. M. 1954. Urban geography. In P. E. James and C. F. Jones (eds) *American Geography: Inventory and prospect*. Syracuse: Syracuse University Press, 142–66.

Mayer, H. M. and Kohn, C. F. (eds) 1959. *Readings in Urban Geography*. Chicago: University of Chicago Press.

Mayer, T. 1989. Consensus and invisibility – the representation of women in human-geography textbooks. *Professional Geographer* 41(4), 397–409.

Mayhew, R. J. 2011a. Cosmographers, explorers, cartographers, chorographers: Defining, inscribing and practicing early modern geography, c. 1450–1850. In J. A. Agnew and J. S. Duncan (eds) *The Wiley-Blackwell Companion to Human Geography*. Chichester: Wiley–Blackwell, 23–49.

Mayhew, R. J. 2011b. Geography's genealogies. In J. A. Agnew and D. N. Livingstone (eds) *The Sage Handbook of Geographical Knowledge*. London, Thousand Oaks, CA and New Delhi and Singapore: Sage, 21–38.

Mayhew, R. J. 2014. *Malthus: The life and legacies of an untimely prophet.* Cambridge MA: Harvard University Press.

McCarty, H. H. 1940. *The Geographic Basis of American Economic Life.* New York: Harper & Brothers.

McCarty, H. H. 1952. *McCarty on McCarthy: The spatial distribution of the McCarthy vote, 1952.* Unpublished paper, Department of Geography, State University of Iowa, Iowa City.

McCarty, H. H. 1953. An approach to a theory of economic geography. *Annals of the Association of American Geographers* 43, 183–4.

McCarty, H. H. 1954. An approach to a theory of economic geography. *Economic Geography* 30, 95–101.

McCarty, H. H. 1958. Science, measurement, and area analysis. *Economic Geography* 34, facing page 283.

McCarty, H. H. 1979. Geography at Iowa. *Annals of the Association of American Geographers* 69, 121–4.

McCarty, H. H. and Lindberg, J. B. 1966. *A Preface to Economic Geography.* Englewood Cliffs, NJ: Prentice-Hall.

McCarty, H. H., Hook, J. C. and Knos, D. S. 1956. *The Measurement of Association in Industrial Geography.* Iowa City: Department of Geography, State University of Iowa.

McCormack, D. P. 2003. An event of geographical ethics in spaces of affect. *Transaction of the Institute of British Geographers* 28, 488–507.

McDaniel, R. and Eliot Hurst, M. E. 1968. *A Systems Analytic Approach to Economic Geography.* Washington, DC: Commission on College Geography, Publication 8, Association of American Geographers.

McDowell, L. 1986a. Feminist geography. In R. J. Johnston, D. Gregory and D. M. Smith (eds) *The Dictionary of Human Geography.* Oxford: Basil Blackwell, 151–2.

McDowell, L. 1986b. Beyond patriarchy: A class-based explanation of women's subordination. *Antipode: A Radical Journal of Geography* 18, 311–21.

McDowell, L. 1989. Women, gender and the organisation of space. In D. Gregory and R. Walford (eds) *Horizons in Human Geography.* London: Macmillan, 136–51.

McDowell, L. 1990. Sex and power in academia. *Area* 22, 323–32.

McDowell, L. 1991. Life without father and Ford: The new gender order of post-fordism. *Transactions of the Institute of British Geographers* NS 16, 400–19.

McDowell, L. 1991. The baby and the bath water: Diversity, deconstruction and feminist theory in geography. *Geoforum* 22, 122–33.

McDowell, L. 1992. Valid games? *Professional Geographer* 44, 212–15.

McDowell, L. 1993a. Space, place and gender relations: Part I. Feminist empiricism and the geography of social relations. *Progress in Human Geography* 17, 157–79.

McDowell, L. 1993b. Space, place and gender relations: Part II. Identity, difference, feminist geometries and geographies. *Progress in Human Geography* 17, 305–18.

McDowell, L. 1994. Polyphony and pedagogic authority. *Area* 26, 241–8.

McDowell, L. 1995. Bodywork: Heterosexual gender performances in city workplaces. In D. Bell and G. Valentine (eds) *Mapping Desire: Geographies of sexualities.* London and New York: Routledge, 75–98.

McDowell, L. 1999. Feminist geography. In L. McDowell and J. P. Sharp (eds) *A Feminist Glossary of Human Geography.* London: Arnold, 90–3.

McDowell, L. 2000. Economy, culture, difference and justice. In I. Cook, D. Crouch, S. Naylor and J. R. Ryan (eds) *Cultural Turns/Geographical Turns: Perspectives on cultural geography.* Harlow: Prentice Hall, 182–95.

McDowell, L. 2002. Masculine discourse and dissonances: Strutting 'lads', protest masculinity and domestic respectability. *Environment and Planning D: Society and Space* 20, 97–119.

McDowell, L. and Court, G. 1994. Missing subjects: Gender, power and sexuality in merchant banking. *Economic Geography* 70, 229–51.

McDowell, L. and Massey, D. 1984. A woman's place? In D. Massey and J. Allen (eds) *Geography Matters!* Cambridge: Cambridge University Press, 128–47.

McDowell, L. and Sharp, J. P. (eds) 1997. *Space, Gender, Knowledge: Feminist readings.* London: Arnold.

McDowell, L. and Sharp, J. P. 1999. *A Feminist Glossary of Human Geography.* London: Arnold.

McEwan, C. 1998. Cutting power lines within the palace? Countering paternity and eurocentrism in the 'geographical tradition'. *Transactions of the Institute of British Geographers* NS 23, 371–84.

McGuiness, M. 1999. Geography matters? Whiteness and contemporary geography. *Area* 32, 225–30.

McKendrick, J. 1995. The provincial geographical socities in Britain, 1884–1914. In M. Bell, R. A. Butlin and M. Heffernan (eds) *Geography and Imperialism: 1820–1940.* Manchester: Manchester University Press.

McKinney, W. M. 1968. Carey, Spencer, and modern geography. *Professional Geographer* 20, 103–6.

McTaggart, W. D. 1974. Structuralism and universalism in geography: Reflections on contributions by H. C. Brookfield. *Australian Geographer* 12, 510–16.

Mead, W. R. 1963. The adoption of other lands: Experiences in a Finnish context. *Geography* 48, 241–54.

Mead, W. R. 1980. Regional geography. In E. H. Brown (ed.) *Geography Yesterday and Tomorrow*. Oxford: Oxford University Press, 292–302.

Mead, W. R. 1993. *An Experience of Finland*. London: Hurst & Company.

Mead, W. R. 2002. *A Celebration of Norway*. London: Hurst & Company.

Mead, W. R. 2004. Obituary: Torsten Hägerstrand 1916–2004. *Geographical Journal* 170, 279.

Mead, W. R. 2007. *Adopting Finland*. Helsinki: Hakapaino Oy.

Meadows, D. H., Meadows, D. L., Randers, J. and Behrens III, W. W. 1972. *The Limits to Growth*. New York: Universal Books.

Meeus, B., Schuermans, N. and de Maesschalck, F. 2011. Is there a world beyond academic geography? A reply to Ben Derudder. *Area* 43, 113–14.

Meinig, D. W. 1972. American wests: Preface to a geographical introduction. *Annals of the Association of American Geographers* 62, 159–84.

Meinig, D. W. 1978. The continuous shaping of America: A prospectus for geographers and historians. *American Historical Review* 83, 1186–217.

Meinig, D. W. 1983. Geography as an art. *Transactions of the Institute of British Geographers* NS8, 314–28.

Meinig, D. W. 1986. *The Shaping of America: A geographical perspective on 500 years of history*. Vol. 1: *Atlantic America, 1492–1800*. New Haven, CT: Yale University Press.

Meinig, D. W. 1986–2004. *The Shaping of America: A geographical perspective on 500 years of history. Vols. 1–4.* New Haven, CT: Yale University Press.

Meinig, D. W. 1989. The historical geography imperative. *Annals of the Association of American Geographers* 79, 79–87.

Meinig, D. W. 1993. *The Shaping of America: A geographical perspective on 500 years of history*. Vol. 2. *Continental America 1800–1867*. New Haven, CT: Yale University Press.

Meinig, D. W. 2002. The life of learning. In P. Gould and F. R. Pitts (eds) *Geographical Voices: Fourteen autobiographical essays*. Syracuse: Syracuse University Press, 189–210.

Mercer, D. C. 1977. *Conflict and Consensus in Human Geography*. Clayton, Victoria, Australia: Monash Publications in Geography No. 17.

Mercer, D. C. 1984. Unmasking technocratic geography. In M. Billinge, D. Gregory and R. Martin (eds) *Recollections of a Revolution*. London: Macmillan, 153–99.

Mercer, D. C. and Powell, J. M. 1972. *Phenomenology and Related Non-positivistic Viewpoints in the Social Sciences*. Clayton, Victoria, Australia: Monash Publications in Geography, No. 1.

Merrifield, A. 1995. Situated knowledge through exploration: Reflections on Bunge's 'geographical expeditions'. *Antipode: A Radical Journal of Geography* 27, 49–70.

Merriman, P., Jones, M., Olsson, G., Sheppard, E., Thrift, N. J. and Tuan, Y.-F. 2012. Space and spatiality in theory. *Dialogues in Human Geography* 2, 3–22.

Mesev, V. 2003. *Remotely Sensed Cities*. London: Taylor & Francis.

Meyer, D. R. 1972. Geographical population data: Statistical description not statistical inference. *Professional Geographer* 24, 26–8.

Mialet, H. 2012. Where would STS be without Latour? What would be missing? *Social Studies of Science* 42, 456–61.

Michie, J. and Cooper, C. (eds) 2015. *Why the Social Sciences Matter*. Basingstoke: Palgrave-Macmillan.

Middleton, N. and Thomas, D. S. G. 1994. *Desertification: Exploding the myth*. Chichester: John Wiley.

Mikesell, M. W. (ed.) 1973. *Geographers Abroad: Essays on the prospects of research in foreign areas*. Chicago: Department of Geography, University of Chicago, Research Paper 152.

Mikesell, M. W. 1967. Geographical perspectives in anthropology. *Annals of the Association of American Geographers* 57, 617–34.

Mikesell, M. W. 1969. The borderlands of geography as a social science. In M. Sherif and C. W. Sherif (eds) *Interdisciplinary Relationships in the Social Sciences*. Chicago: Aldine Publishing Company, 227–48.

Mikesell, M. W. 1974. Geography as the study of environment: An assessment of some old and new commitments. In I. R. Manners and M. W. Mikesell (eds) *Perspectives on Environment*, Washington, DC: Commission on College Geography, Association of American Geographers, 1–23.

Mikesell, M. W. 1978. Tradition and innovation in cultural geography. *Annals of the Association of American Geographers* 68, 1–16.

Mikesell, M. W. 1981. Continuity and change. In B. W. Blouet (ed.) *The Origins of Academic Geography in the United States*. Hamden, CT: Archon Books, 1–15.

Mikesell, M. W. 1984. North America. In R. J. Johnston and P. Claval (eds) *Geography Since the Second World War: An international survey.* London: Croom Helm, 185–213.

Mikesell, M. W. 1999. Afterword: New interests, unsolved problems and persisting tasks. In K. E. Foote, P. J. Hugill, K. Mathewson and J. M. Smith (eds) *Re-reading Cultural Geography.* Austin, TX: University of Texas Press, 437–44.

Miller, D. and Hashmi, S. H. (eds) 2001. *Boundaries and Justice: Diverse ethical perspectives.* Princeton, NJ: Princeton University Press.

Miller, E. W. (ed.) 1993. *The American Society for Professional Geographers.* Papers presented on the occasion of the fiftieth anniversary of its founding. Washington, DC: Association of American Geographers, Occasional Publication 3.

Miller, H. J. 2004. Tobler's first law and spatial analysis. *Annals of the Association of American Geographers* 94, 284–9.

Miller, H. J. and Wentz, E. 2003. Geographic representation and spatial analysis in geographic information systems. *Annals of the Association of American Geographers* 93, 574–94.

Mills, C. W. 1959. *The Sociological Imagination.* New York: Oxford University Press.

Minca, C. 2000. Guest editorial: Venetian geographical praxis. *Environment and Planning D: Society and Space* 18, 285–9.

Minca, C. 2001. Acknowledgements. In C. Minca (ed.) *Postmodern Geography: Theory and praxis.* Oxford: Blackwell, xii–xiii.

Minca, C. 2003. Guest editorial: Critical peripheries. *Environment and Planning D: Society and Space* 21, 160–8.

Minca, C. 2009. Postmodernism/postmodern geography. In R. Kitchin and N. Thrift (eds) *International Encyclopedia of Human Geography.* Amsterdam: Elsevier, 363–72.

Mitchell, B. and Draper, D. 1982. *Relevance and Ethics in Geography.* London: Longman.

Mitchell, D. 1995. There's no such thing as culture: Towards a reconceptualization of the idea of culture in geography. *Transactions of the Institute of British Geographers* NS20, 102–16.

Mitchell, D. 2003. Dead labour and the political economy of landscape – California living, California dying. In K. Anderson, M. Domosh, S. Pile and N. Thrift (eds) *The Handbook of Cultural Geography.* London, Thousand Oaks, CA and New Delhi: Sage, 233–48.

Mitchell, D. 2004. Geography in an age of extremes: A blueprint for a geography of justice. *Annals of the Association of American Geographers* 94, 764–70.

Mitchell, D. 2009. People's geography. In R. Kitchin and N. Thrift (eds) *International Encyclopedia of Human Geography.* Amsterdam: Elsevier, 116–19.

Mitchell, R., Martin, D. and Foody, G. M. 1998. Unmixing aggregate data: Estimating the social composition of enumeration districts. *Environment and Planning A* 30, 1929–42.

Miyares, I. M. and McGlade, M. S. 1994. Specialization in 'jobs in geography' 1980–1993. *Professional Geographer* 46, 170–7.

Mohammad, R. 1999. Marginalisation, Islamism and the production of the 'Other's Other'. *Gender, Place and Culture* 6, 221–40.

Mohammad, R. and Sidaway, J. D. 2012. Reflections on affect: A meta-commentary occasioned by Pile (2010) and subsequent exchanges. *Transactions of the Institute of British Geographers* NS37, 655–7.

Mohan, G. 1994. Destruction of the con: Geography and the commodification of knowledge. *Area* 26, 387–90.

Mohan, J. 2000. Geographies of welfare and social exclusion. *Progress in Human Geography* 24, 291–300.

Mohan, J. 2003. Geography and social policy: Spatial divisions of welfare. *Progress in Human Geography* 27, 363–74.

Monk, J. 1998. The women were always welcome at Clark. *Economic Geography* (special issue), 14–30.

Monk, J. 2015. Spaces and flows. In S. C. Aitken and G. Valentine (eds) *Approaches to Human Geography: Philosophies, theories, people and practices,* 2nd edn. London, Thousand Oaks, CA and New Delhi and Singapore: Sage, 272–8.

Monk, J. and Hanson, S. 1982. On not excluding half of the human in human geography. *Professional Geographer* 34, 11–23.

Monmonier, M. S. 1993. What a friend we have in GIS. *Professional Geographer* 45, 448–50.

Monmonier, M. S. 1996. *How to Lie With Maps,* 2nd edn. Chicago: University of Chicago Press.

Montefiore, A. G. and Williams, W. M. 1955. Determinism and possibilism. *Geographical Studies* 2, 1–11.

Moodie, D. W. and Lehr, J. C. 1976. Fact and theory in historical geography. *Professional Geographer* 28, 132–6.

Moos, A. I. and Dear, M. J. 1986. Structuration theory in urban analysis: 1. Theoretical exegesis. *Environment and Planning A* 18, 231–52.

Moran, W. 2000. Exceptionalism in the Antipodes. *Progress in Human Geography* 24, 429–38.

Moreton, B. 2009. *To Serve God and Wal-Mart: The making of Christian free enterprise*. Cambridge MA: Harvard University Press.

Morgan, M. A. 1967. Hardware models in geography. In R. J. Chorley and P. Haggett (eds) *Models in Geography*. London: Methuen, 727–74.

Morgan, W. B. and Moss, R. P. 1965. Geography and ecology: The concept of the community and its relationship to environment. *Annals of the Association of American Geographers* 55, 339–50.

Morin, K. M. 2009. Feminist groups within geography. In R. Kitchen and N. Thrift (eds) *International Encyclopedia of Human Geography*. Amsterdam: Elsevier, 64–70.

Morin, K. M. 2011. *Civic Discipline: Geography in America, 1860–1890*. Aldershot: Ashgate.

Morin, K. M. and Rothenberg, T. Y. 2011. Our theories, ourselves: Hierarchies of place and status in the U.S. academy. *ACME: An International E-Journal for Critical Geographies* 10, 58–68.

Morrell, J. 1990. Professionalisation. In R. Olby (ed.) *Companion to the History of Modern Science*. London: Routledge, 980–9.

Morrill, R. L. 1965. Migration and the growth of urban settlement. Lund Studies in Geography, Series B. 24, Lund: C. W. K. Gleerup.

Morrill, R. L. 1968. Waves of spatial diffusion. *Journal of Regional Science* 8, 1–18.

Morrill, R. L. 1969. Geography and the transformation of society. *Antipode: A Radical Journal of Geography* 1(1), 6–9.

Morrill, R. L. 1970a. *The Spatial Organization of Society*, 2nd edn. Belmont, CA: Wadsworth.

Morrill, R. L. 1970b. Geography and the transformation of society: Part II. *Antipode: A Radical Journal of Geography* 2(1), 4–10.

Morrill, R. L. 1973. Ideal and reality in reapportionment. *Annals of the Association of American Geographers* 63, 463–77.

Morrill, R. L. 1974. Review of D. Harvey, *Social Justice and the City*. *Annals of the Association of American Geographers* 64, 475–7.

Morrill, R. L. 1978. Geography as spatial interaction. In J. D. Eyre (ed.) *A Man for all Regions: The contributions of Edward L. Ullman to geography*. Chapel Hill, NC: University of North Carolina, Department of Geography, Studies in Geography no. 11, 16–29.

Morrill, R. L. 1980. Productivity of American Ph.D-granting Departments of Geography. *Professional Geographer* 32, 85–9.

Morrill, R. L. 1981. *Political Redistricting*. Washington, DC: Resource Publications in Geography, Association of American Geographers.

Morrill, R. L. 1984. Recollections of the 'Quantitative Revolution's' early years: The University of Washington 1955–65. In M. Billinge, D. Gregory and R. Martin (eds) *Recollections of a Revolution*. London: Macmillan, 57–72.

Morrill, R. L. 1985. Some important geographic questions. *Professional Geographer* 37, 263–70.

Morrill, R. L. 1993. Geography, spatial analysis and social science. *Urban Geography* 14, 442–6.

Morrill, R. L. 1994. Response to Johnston. *Urban Geography*, 15, 296.

Morrill, R. L. 2002. Pausing for breath. In P. R. Gould and F. R. Pitts (eds) *Geographical Voices: Fourteen autobiographical essays*. Syracuse: Syracuse University Press, 211–36.

Morrill, R. L. 2005. Hägerstrand and the 'quantitative revolution': A personal appreciation. *Progress in Human Geography* 28, 328–32.

Morrill, R. L. and Dormtzer, J. 1979. *The Spatial Order: An introduction to modern geography*. North Scituate, RI: Duxbury.

Morrill, R. L. and Garrison, W. L. 1960. Projections of interregional patterns of trade in wheat and flour. *Economic Geography* 36, 116–26.

Morrill, R. L. and Wohlenberg, E. H. 1971. *The Geography of Poverty in the United States*. New York: McGraw-Hill.

Morton, P. A. 2011. The uses and abuses of human geography. *Journal of Architecture* 16, 805–20.

Moss, P. (ed.) 2001. *Placing Autobiography in Geography*. Syracuse: Syracuse University Press.

Moss, P. and Falconer Al-Hindi, K. 2008. An introduction: Feminisms, geographies, knowledges. In P. Moss and K. Falconer Al-Hindi (eds) *Feminisms in Geography: Rethinking space, place and knowledges*. Lanham, MA: Rowan & Littlefield, 1–27.

Moss, P., Berg, L. D. and Desbiens, C. 2002. The political economy of publishing in geography. *ACME: An International Journal for Critical Geographies* 7, 1–7.

Moss, R. P. 1970. Authority and charisma: Criteria of validity in geographical method. *South African Geographical Journal* 52, 13–37.

Moss, R. P. 1977. Deductive strategies in geographical generalization. *Progress in Physical Geography* 1, 23–39.

Moss, R. P. and Morgan, W. B. 1967. The concept of the community: Some applications in geographical research. *Transactions of the Institute of British Geographers* 41, 21–32.

mrs kinpaisby 2008. Taking stock of participatory geographies: Envisioning the communiversity. *Transactions of the Institute of British Geographers* NS33, 292–9.

Muir, R. 1975. *Modern Political Geography*. London: Macmillan.

Muir, R. 1978. Radical geography or a new orthodoxy? *Area* 10, 322–7.

Muir, R. 1979. Radical geography and Marxism. *Area* 11, 126–7.

Mulkay, M. J. 1975. Three models of scientific development. *Sociological Review* 23, 509–26.

Mulkay, M. J. 1976. Methodology in the sociology of science: Some reflections on the study of radio astronomy. In G. Lemaine *et al.* (eds) *Perspectives in the Emergence of Scientific Disciplines*. The Hague: Mouton, 207–20.

Mulkay, M. J. 1978. Consensus in science. *Social Science Information* 17, 107–22.

Mulkay, M. J., Gilbert, G. N. and Woolgar, S. 1975. Problem areas and research networks in science. *Sociology* 9, 187–203.

Müller-Wille, C. 1978. The forgotten heritage: Christaller's antecedents. In B. J. L. Berry (ed.) *The Nature of Change in Geographical Ideas*. DeKalb, IL: Northern Illinois University Press, 37–64.

Mumford, L. 1956. Prospect. In W. L. Thomas (ed.) *Man's Role in Changing the Face of the Earth*. Chicago: University of Chicago Press, 1141–52.

Murdie, R. A. 1969. *Factorial Ecology of Metropolitan Toronto 1951–1961*. Chicago: University of Chicago, Department of Geography, Research Paper 116.

Murdoch, J. 1998. The spaces of actor-network theory. *Geoforum* 29, 357–74.

Murdoch, J. 2006. *Post-Structuralist Geography*. London, Thousand Oaks, CA and New Delhi: Sage.

Murphy, A. B. 1991. Regions as social constructs: The gap between theory and practice. *Progress in Human Geography* 15, 22–35.

Murphy, A. B. 2004. Centennial forum: Where we have come from and where we are going. *Annals of the Association of American Geographers* 94, 701–2.

Murphy, A. B. 2007. Geography's place in higher education in the United States. *Journal of Geography in Higher Education* 31, 121–41

Murphy, A. B. 2013. Advancing geographical understanding: Why engaging grand regional narratives matters. *Dialogues in Human Geography* 3, 131–49 (with subsequent discussion and response).

Myrdal, G. 1957. *Economic Theory and Underdeveloped Regions*. London: Duckworth.

Nagar, R., Lawson, V., McDowell, L. and Hanson, S. 2002. Locating globalization: Feminist (re)readings of subjects and spaces of globalization. *Economic Geography* 78, 257–84.

Nairn, T. 1977. *The Break-up of Britain: Crisis and neo-nationalism*. London: New Left Books.

Nash, C. 1996. Reclaiming vision: Looking at landscape and the body. *Gender, Place and Culture* 3, 149–69.

Nash, C. 2000. Performativity in practice: Some recent work in cultural geography. *Progress in Human Geography* 24, 653–64.

Nast, H. 1994. Opening remarks on 'Women in the field'. *Professional Geographer* 46, 54–66.

Nast, H. 2002. Special Issue: Queer patriarchies, queer racisms, international. Prologue: Crosscurrents. *Antipode* 34, 835–44.

National Academy of Sciences-National Research Council (NAS-NRC) 1965. *The Science of Geography*. Washington, DC: NAS-NRC.

National Academy of Sciences-National Research Council (NAS-NRC) 1997. *Rediscovering Geography: New relevance for science and society*. Washington, DC: National Research Council.

National Research Council (NRC) 2010. *A Data–Based Assessment of Research–Doctorate Programs in the United States*. Washington, DC: National Academies Press.

National Research Council (NRC) 2011. *Understanding the Changing Planet: Strategic directions for the geographical sciences*. Washington, DC: The National Academies Press.

Natter, W. 2003. Geopolitics in Germany, 1919–45. In J. Agnew, K. Mitchell and G. Toal (eds) *A Companion to Political Geography*. Oxford: Blackwell, 187–203.

Nayak, A. 2003. Last of the 'real Geordies'? White masculinities and the subcultural response to deindustrialisation. *Environment and Planning D: Society and Space* 21, 7–25.

Nayak, A. 2010. Race, affect, and emotion: Young people, racism, and graffiti in the postcolonial English suburbs. *Environment and Planning A* 42, 2370–92.

Nayak, A. and Jeffrey, A. 2011. *Geographical Thought: An introduction to ideas in human geography*. London: Pearson.

Neft, D. 1966. *Statistical Analysis for Areal Distributions.* Philadelphia, PA: Monograph 2, Regional Science Research Institute.

Nelson, L. and Seager, J. 2005. *A Companion to Feminist Geography.* Malden, MA and Oxford: Blackwell.

Nelson, T. A. 2012. Trends in spatial statistics. *Professional Geographer* 64, 83–94.

Newman, D. 1996. Writing together separately: Critical discourse and the problems of cross-ethnic co-authorship. *Area* 28, 1–12.

Newman, D. and Portugali, J. 1987. Israeli–Palestinian relations as reflected in the scientific literature. *Progress in Human Geography* 11, 315–32.

Newman, J. L. 1973. The use of the term 'hypothesis' in geography. *Annals of the Association of American Geographers* 63, 22–7.

Newman, O. 1972. *Defensible Space: Crime prevention through urban design.* New York: Macmillan.

Nicholls, W. J. 2011. The Los Angeles school: Difference, politics, city. *International Journal of Urban and Regional Research* 35, 189–206.

Nickles, T. (ed.) 2003. *Thomas Kuhn.* Cambridge: Cambridge University Press.

Nijman, J. 2000. The paradigmatic city. *Annals of the Association of American Geographers* 90, 135–45.

Nystuen, J. D. 1968 [1963]. Identification of some fundamental spatial concepts. *Papers of the Michigan Academy of Science, Arts, and Letters,* 48, 373–84. Reprinted in B. J. L. Berry and D. F. Marble (eds) *Spatial Analysis: A reader in statistical geography.* Englewood Cliffs, NJ: Prentice-Hall, 35–41.

Nystuen, J. D. 1984. Comment on 'Artificial intelligence and its applicability to geographical problem solving'. *Professional Geographer* 36, 358–9.

Ó hUallacháin, B. and Leslie, T. F. 2013. Spatial pattern and order in sunbelt retailing: Shopping in Phoenix in the twenty-first century. *Professional Geographer* 65, 396–420.

O'Kelly, M. 1999. Introduction to the thirtieth anniversary special issue. *Geographical Analysis* 30, 311–17.

O'Loughlin, J. 2002. The electoral geography of Weimar Germany: Exploratory spatial data analyses (ESDA) of Protestant support for the Nazi party. *Political Analysis* 10, 217–43.

O'Riordan, T. 1971a. Environmental management. In C. Board, R. J. Chorley, P. Haggett and D. R. Stoddart (eds) *Progress in Geography,* Vol. 3. London: Edward Arnold, 173–231.

O'Riordan, T. 1971b. *Perspectives in Resource Management.* London: Pion.

O'Riordan, T. 1976. *Environmentalism.* London: Pion.

O'Riordan, T. 1981. *Environmentalism,* 2nd edn. London: Pion.

O'Sullivan, D. 2014, Commentary. Don't panic! The need for change and for curricular pluralism. *Dialogues in Human Geography* 4, 39–44.

O'Sullivan, D. and Unwin, D. J. 2003. *Geographic Information Analysis.* New York: John Wiley.

Ó Tuathail, G. 1992. Foreign policy and the hyperreal: The Reagan administration and the 'scripting' of South Africa'. In T. J. Barnes and J. S. Duncan (eds) *Writing Worlds: Discourse, text and metaphor in the representation of landscape.* London: Routledge, 155–75.

Ó Tuathail, G. 1994. (Dis)placing geopolitics: Writing on the maps of global politics. *Environment and Planning D: Society and Space* 12, 525–46.

Ó Tuathail, G. 1996. *Critical Geopolitics.* London and New York: Routledge.

Ó Tuathail, G. 2000. Dis/placing the geo-politics which one cannot not want. *Political Geography* 19, 385–96.

Ó Tuathail, G. 2003. 'Just out looking for a fight': American affect and the invasion of Iraq. *Antipode: A Radical Journal of Geography* 35, 856–70.

Ó Tuathail, G. and Dalby, S. 1994. Critical geopolitics: Unfolding spaces for thought in geography and global politics. *Environment and Planning D: Society and Space* 12, 513–14.

Ó Tuathail, G. and Dalby, S. 1998. Introduction. Rethinking geopolitics: Towards a critical geopolitics. In G. Ó Tuathail and S. Dalby (eds) *Rethinking Geopolitics.* London and New York: Routledge, 1–15.

Öberg, S. 2005. Hägerstrand and the remaking of Sweden. *Progress in Human Geography* 29, 340–9.

Obermeyer, N. J. 1994. GIS: A new profession? *Professional Geographer* 46, 498–503.

Odum, H. W. and Moore, H. E. 1938. *American Regionalism: A cultural–historical approach to national integration.* New York: H. Holt & Company.

Olds, K. 2001. Practices for 'process geographies': A view from within and outside the periphery. *Environment and Planning D: Society and Space* 19, 127–36.

Ollman, B. 1993. *Dialectical Investigations.* London and New York: Routledge.

Olsson, G. 1965. *Distance and Human Interaction: A review and bibliography.* Philadelphia, PA: Regional Science Research Institute, Bibliography Series Number 2.

Olsson, G. 1969. Inference problems in locational analysis. In K. R. Cox and R. G. Golledge (eds) *Behavioral Problems in Geography: A symposium*. Evanston, IL: Northwestern University Studies in Geography 17, 14–34.

Olsson, G. 1978. Of ambiguity or far cries from a memorializing mamafesta. In D. Ley and M. S. Samuels (eds) *Humanistic Geography*. London: Croom Helm, 109–20.

Olsson, G. 1979. Social science and human action or on hitting your head against the ceiling of language. In S. Gale and G. Olsson (eds) *Philosophy in Geography*. Dordrecht: Reidel, 287–308.

Olsson, G. 1980. *Birds in Egg/Eggs in Bird*. London: Pion.

Olsson, G. 1982. -/-. In P. R. Gould and G. Olsson (eds) *A Search for Common Ground*. London: Pion, 223–31.

Olsson, G. 1991. *Lines of Power/Limits of Language*. Minneapolis, MN: University of Minnesota Press.

Olsson, G. 1992. Lines of power. In T. J. Barnes and J. S. Duncan (eds) *Writing Worlds: Discourse, text and metaphor in the representation of landscape*. London: Routledge, 86–96.

Olsson, G. 2007. *Abysmal: A critique of cartographic reason*. Chicago: University of Chicago Press.

Openshaw, S. 1984a. The modifiable areal unit problem. *CATMOG* 38, Norwich: Geo Books.

Openshaw, S. 1984b. Ecological fallacies and the analysis of areal census data. *Environment and Planning A* 16, 17–32.

Openshaw, S. 1986. *Nuclear Power: Siting and safety*. London: Routledge & Kegan Paul.

Openshaw, S. 1989. Computer modelling in human geography. In B. Macmillan (ed.) *Remodelling Geography*. Oxford: Basil Blackwell, 70–88.

Openshaw, S. 1991. A view on the GIS crisis in geography: Or, using GIS to put Humpty-Dumpty back together again. *Environment and Planning A* 23, 621–8.

Openshaw, S. 1992. Further thoughts on geography and GIS: A reply. *Environment and Planning A* 24, 463–6.

Openshaw, S. 1994. Computational human geography: Towards a research agenda. *Environment and Planning A* 26, 499–505.

Openshaw, S. 1995. Human systems modelling as a new grand challenge area in science: What has happened to the science in social science? *Environment and Planning A* 27, 159–64.

Openshaw, S. 1996. Fuzzy logic as a new scientific paradigm for doing geography. *Environment and Planning A* 28, 761–8.

Openshaw, S. 1998. Building automated geographical analysis and explanation machines. In P. Longley, S. Brooks, R. MacDonnell and B. Macmillan (eds) *Geocomputation: A primer*. Chichester: John Wiley, 95–116.

Openshaw, S. and Goddard, J. B. 1987. Some implications of the commodification of information and the emerging information economy for applied geographical analysis in the United Kingdom. *Environment and Planning A* 19, 1423–40.

Openshaw, S. and Openshaw, C. 1997. *Artificial Intelligence in Geography*. Chichester: John Wiley.

Openshaw, S. and Rao, L. 1995. Algorithms for reengineering 1991 census geography. *Environment and Planning A* 27, 425–46.

Openshaw, S. and Taylor, P. J. 1979. A million or so correlation coefficients: Three experiments on the modifiable areal unit problem. In N. Wrigley (ed.) *Statistical Applications in the Spatial Sciences*. London: Pion, 127–44.

Openshaw, S. and Taylor, P. J. 1981. The modifiable areal unit problem. In N. Wrigley and R. J. Bennett, (eds) *Quantitative Geography: A British view*. London: Routledge & Kegan Paul, 60–70.

Openshaw, S. and Turton, I. 2001. Using a geographical explanations machine to explore spatial factors relating to primary school performance. *Geographical and Environmental Modelling* 5, 85–101.

Openshaw, S., Blake, M. and Wymer, C. 1995. Using neurocomputing methods to classify Britain's residential areas. In P. F. Fisher (ed.) *Innovations in GIS 2*. London: Taylor & Francis, 97–112.

Openshaw, S., Carver, S. and Fernie, J. 1989. *Britain's Nuclear Waste: Siting and safety*. London: Belhaven Press.

Openshaw, S., Charlton, M., Craft, A. W. and Birch, J. 1988. Investigation of leukaemia clusters by use of a geographical analysis machine. *Lancet*, 6 February, 272–3.

Openshaw, S., Steadman, P. and Greene, O. 1983. *Doomsday: Britain after nuclear attack*. Oxford: Basil Blackwell.

Openshaw, S., Wymer, C. and Charlton, M. 1986. A geographical information and mapping system for the BBC Domesday optical discs. *Transactions of the Institute of British Geographers* NS11, 296–304.

Openshaw, S., Wymer, C. and Craft, A. W. 1988. A Mark I geographical analysis machine for the automated analysis of point data sets. *International Journal of Geographical Information Systems* 1, 335–58.

Ord, J. K. and Getis, A. 1995. Local spatial autocorrelation statistics: Distributional issues and an application. *Geographical Analysis* 27, 286–96.

Orme, A. R. 2002. Shifting paradigms in geomorphology: The fate of research ideas in an educational context. *Geomorphology* 47, 325–42.

Östh, J., Clark, W. A. V. and Malmberg, B. 2015. Measuring the scale of segregation using k-nearest neighbour aggregates. *Geographical Analysis* 47, 34–49.

Owens, P. L. 1984. Rural leisure and recreation research: A retrospective evaluation. *Progress in Human Geography* 8, 157–88.

Ozouf-Marignier, M.-V. and Robic, M.-C. 1999. The Tableau is alive and well . . . reactions to the *Tableau de la Géographie de la France* of Paul Vidal de la Blache. In A. Buttimer, S. D. Brunn and U. Wardenga (eds) *Text and Image: Social construction of regional knowledges*. Leipzig: Leibniz Institute for Regional Geography, Beiträge zur regionalen Geographie 49, 54–66.

Paasi, A. 1991. Deconstructing regions: Notes on the scales of social life. *Environment and Planning A* 23, 239–56.

Paasi, A. 2005. Globalisation, academic capitalism, and the uneven geographies of international journal publishing spaces. *Environment and Planning A* 37, 769–89.

Pacione, M. (ed.) 1999. *Applied Geography: Principles and practice*. London: Routledge.

Pacione, M. 1990a. Conceptual issues in applied urban geography. *Tijdschrift voor Economische en Sociale Geografie* 81, 3–13.

Pacione, M. 1990b. On the dangers of misinterpretation. *Tijdschrift voor Economische en Sociale Geografie* 81, 26–8.

Pacione, M. 2014. *Scottish Geography: A historiography*. Glasgow: Royal Scottish Geographical Society.

Pagliara, F. and Wilson, A. G. 2010. The state-of-the-art in building residential location models. In F. Pagliara, J. Preston and D. Simmons (eds) *Residential Location Choice*. New York: Springer, 1–20.

Pahl, R. E. 1965. Trends in social geography. In R. J. Chorley and P. Haggett (eds) *Frontiers in Geographical Teaching*. London: Methuen, 81–100.

Pahl, R. E. 1969. Urban social theory and research. *Environment and Planning* 1, 143–54. (Reprinted in R. E. Pahl 1970, *Whose City?* London: Longman, 209–25.)

Pahl, R. E. 1975. *Whose city? And Other Essays*, 2nd edn. Harmondsworth: Penguin Books.

Pahl, R. E. 1979. Socio-political factors in resource allocation. In D. T. Herbert and D. M. Smith (eds) *Social Problems and the City: Geographical perspectives*. Oxford: Oxford University Press, 33–46.

Pain, R. 1991. Space, sexual violence and social control: Integrating geographical and feminist analyses of women's fear of crime. *Progress in Human Geography* 15, 415–31.

Pain, R. 2003. Social geography: On action-orientated research. *Progress in Human Geography* 27, 649–57, 677–85.

Pain, R., Kesby, M. and Askins, K. 2011. Geographies of impact: Power, participation and potential. *Area* 43, 183–88.

Painter, J. 2000. Pierre Bourdieu. In M. Crang and N. Thrift (eds) *Thinking Space*. New York and London: Routledge, 239–59.

Painter, J. 2002. The rise of the workfare state. In R. J. Johnston, P. J. Taylor and M. J. Watts (eds) *Geographies of Global Change: Remapping the world*. Oxford: Blackwell Publishers, 158–73.

Painter, J. and Philo, C. 1995. Spaces of citizenship: An introduction. *Political Geography* 14, 107–20.

Palfreyman, D. and Tapper, T. 2014. *Reshaping the University: The rise of the regulated market in higher education*. Oxford: Oxford University Press.

Palm, R. 1979. Financial and real estate institutions in the housing market. In D. T. Herbert and R. J. Johnston (eds) *Geography and the Urban Environment*, Vol. 2. Chichester: John Wiley, 83–124.

Palm, R. 2003. Textbooks that moved generations. *Progress in Human Geography* 27, 515–18.

Palm, R. and Pred, A. R. 1978. The status of American women: A time-geographic view. In D. Lanegran and R. Palm (eds) *Invitation to Geography*, 2nd edn. New York: McGraw-Hill, 99–109.

Papageorggiou, G. J. (ed.) 1976. *Mathematical Land Use Theory*. Lexington, MA: D. C. Heath.

Papageorgiou, G. J. 1969. Description of a basis necessary to the analysis of spatial systems. *Geographical Analysis* 1, 213–15.

Pappenberger, F., Cloke, H. L., Parker, D. J., Wetterhall, F., Richardson, D. S. and Theilen, J. 2015. The monetary benefit of early flood warnings in Europe. *Environmental Science and Policy* 51, 278–91.

Park, C. P. 1994. *Sacred Worlds: An introduction to geography and religion*. London and New York: Routledge.

Parker, G. 1985. *Western Geopolitical Thought in the Twentieth Century*. London: Croom Helm.

Parker, W. H. 1982. *Mackinder: Geography as an aid to statecraft*. Oxford: The Clarendon Press.

Parkes, D. N. and Thrift, N. J. 1980. *Times, Spaces and Places*. Chichester: John Wiley.

Parkinson, S., Searle, B. A., Smith, S. J., Stoakes, A. and Wood, G. 2009. Mortgage equity withdrawal in Australia and Britain: Towards a wealth–fare state? *European Journal of Housing Policy* 9, 363–87.

Parr, D. A. and Lu, Y. 2010. The landscape of GIScience publications 1997–2007: An empirical investigation with latent semantic analysis. *Transactions in GIS* 14, 689–708.

Parry, M. and Duncan, R. 1995. *The Economic Implications of Climate Change*. London: Earthscan.

Parsons, J. J. 1977. Geography as exploration and discovery. *Annals of the Association of American Geographers* 67, 1–16.

Paterson, J. H. 1974. Writing regional geography. In C. Board, R. J. Chorley, P. Haggett and D. R. Stoddart (eds) *Progress in Geography*, Vol. 6. London: Edward Arnold, 1–26.

Paterson, J. L. 1985. *David Harvey's Geography*. London: Croom Helm.

Patmore, J. A. 1970. *Land and Leisure*. Newton Abbott: David & Charles.

Patmore, J. A. 1983. *Recreation and Resources: Leisure patterns and leisure places*. Oxford: Basil Blackwell.

Patterson, T. C. 1986. The last sixty years: Toward a social history of Americanist archaeology in the United States. *American Anthropologist* 88, 7–26.

Pattie, C. J. and Johnston, R. J. 1995. 'Its not like that round here': Region, economic evaluations and voting at the 1992 British general election. *European Journal of Political Research* 28, 1–32.

Pattie, C. J. and Johnston, R. J. 2000. 'People who talk together vote together': An exploration of contextual effects in Great Britain. *Annals of the Association of American Geographers* 90, 41–66.

Pattison, W. D. 1964. The four traditions of geography. *Journal of Geography* 63, 211–16.

Pawson, E. J. 1999. Postcolonial New Zealand? In K. Anderson and F. Gale (eds) *Cultural Geographies*. Melbourne: Longman, 25–50.

Pawson, E. J. 2009. Oxbridge geographies. In R. Kitchin and N. Thrift (ed.) *International Encyclopedia of Human Geography*. Oxford: Elsevier, 56–62.

Pawson, E. J. 2011. Creating public spaces for geography in New Zealand: Towards an assessment of the contributions of Kenneth Cumberland. *New Zealand Geographer* 67, 102–15.

Peach, C. 2002. Social geography: New religions and ethnoburbs – contrasts with cultural geography. *Progress in Human Geography* 26, 252–60.

Peach, C. 2003. Geography and the fragmented city. In R. J. Johnston and M. Williams (eds) *A Century of British Geography*. Oxford: Oxford University Press for the British Academy, 563–82.

Peach, C. and Smith, S. J. 1981. Introduction. In C. Peach, V. Robinson and S. J. Smith (eds) *Ethnic Segregation in Cities*. London: Croom Helm, 9–24.

Peake, L. 1993. Race and Sexuality: Challenging the patriarchal structuring of urban social space. *Environment and Planning D: Society and Space* 11, 415–32.

Peake, L. 1994. 'Proper words in proper places . . .' Or, of young Turks and old turkeys. *Canadian Geographer* 38, 204–6.

Peake, L. and Kobayashi, A. 2002. Policies and practices for an antiracist geography at the millennium. *Professional Geographer* 54, 50–61.

Peake, L. and Sheppard, E. 2014. The emergence of radical/critical geography in North America. *ACME: An International E-Journal for Critical Geographies* 13, 305–27.

Peck, J. 1996. *Work-place: The social regulation of labor markets*. New York: Guilford.

Peck, J. 1999. Grey geography? *Transactions of the Institute of British Geographers* NS24, 131–5.

Peck, J. 2000. Jumping in, joining up and getting on. *Transactions of the Institute of British Geographers* NS25, 255–8.

Peck, J. 2001. *Workfare States*. New York: Guilford.

Peck, J. 2013. Making space for labour. In D. Featherstone and J. Painter (eds) *Spatial Politics: Essays for Doreen Massey*. Oxford: Wiley-Blackwell, 99–114.

Peck, J. and Tickell, A. 1992. Local modes of social regulation? Regulation theory, Thatcherism and regional development. *Geoforum* 23, 347–64.

Peck, J. and Tickell, A. 1994. 'Jungle law breaks out': Neoliberalism and global–local disorder. *Area* 26, 317–26.

Peck, J. and Tickell, A. 1995. The social regulation of uneven development: 'Regulatory deficit', England's South East, and the collapse of Thatcherism. *Environment and Planning A* 27, 15–40.

Peck, J. and Tickell, A. 1996. The return of the Manchester men: Men's words and men's deeds in the remaking of the local state. *Transactions of the Institute of British Geographers* 21, 595–616.

Peck, J. and Tickell, A. 2002. Neoliberalizing space. *Antipode: A Radical Journal of Geography* 34, 380–404.

Peck, J. and Tickell, A. 2012. Apparitions of neoliberalism: Revisiting 'Jungle law breaks out'. *Area* 44, 245–9.

Peet, R. (ed.) 1978. *Radical Geography: Alternative viewpoints on contemporary social issues*. London: Methuen.

Peet, R. 1971. Poor, hungry America. *Professional Geographer* 23, 99–104.

Peet, R. 1975a. Inequality and poverty: A Marxist-geographic theory. *Annals of the Association of American Geographers* 65, 564–71.

Peet, R. 1975b. The geography of crime: A political critique. *Professional Geographer* 27, 277–80.

Peet, R. 1976. Further comments on the geography of crime. *Professional Geographer* 28, 96–100.

Peet, R. 1977. The development of radical geography in the United States. *Progress in Human Geography* 1, 240–63.

Peet, R. 1979. Societal contradiction and Marxist geography. *Annals of the Association of American Geographers* 69, 164–9.

Peet, R. 1980. The transition from feudalism to capitalism. In R. Peet (ed.) *An Introduction to Marxist Theories of Underdevelopment*. Canberra. Canberra: Australian National University, Research School of Pacific Studies, Department of Human Geography, 51–74.

Peet, R. 1985a. The social origins of environmental determinism. *Annals of the Association of American Geographers* 75, 309–33.

Peet, R. 1985b. Radical geography in the United States: A personal history. *Antipode: A Radical Journal of Geography* 17, 1–7.

Peet, R. 1989. World capitalism and the destruction of regional cultures. In R. J. Johnston and P. J. Taylor (eds) *A World in Crisis?* Oxford: Basil Blackwell, 175–99.

Peet, R. 1991. *Global Capitalism: Theories of societal development*. London: Routledge.

Peet, R. 1993. Reading Fukuyama: Politics at the end of history. *Political Geography* 12, 64–78.

Peet, R. 1996a. A sign taken for history: Daniel Shays' memorial in Peterham, Massachusetts. *Annals of the Association of American Geographers* 86, 21–43.

Peet, R. 1996b: Discursive idealism in the 'landscape-as-text' school. *Professional Geographer* 48, 96–8.

Peet, R. 1998. *Modern Geographical Thought*. Oxford and Malden, MA: Blackwell.

Peet, R. 2005. Bio-gaze: Review essays on *Key Thinkers on Space and Place. Environment and Planning A* 37, 165–7.

Peet, R. and Lyons, J. V. 1981. Marxism: Dialectical materialism, social formation and the geographic relations. In M. E. Harvey and B. P. Holly (eds) *Themes in Geographic Thought*. London: Croom Helm, 187–205.

Peet, R. and Thrift, N. J. 1989. Political economy and human geography. In R. Peet and N. J. Thrift (eds) *New Models in Geography, Vol. 1*. London: Unwin Hyman, 3–27.

Peet, R. and Watts, M. (eds) 1996. *Liberation Ecologies: Nature, development and social movements*. London and New York: Routledge.

Peet, R. and Watts, M. (eds) 2004. *Liberation Ecologies*, 2nd edn. London and New York: Routledge.

Peet, R., Robbins, R. and Watts, M. (eds) 2011. *Global Political Ecology*. Abingdon and New York: Routledge.

Peltier, L. C. 1954. Geomorphology. In P. E. James and C. F. Jones (eds) *American Geography: Inventory and prospect*. Syracuse: Syracuse University Press, 362–81.

Penning-Rowsell, E. C. 1981. Fluctuating fortunes in gauging landscape value. *Progress in Human Geography* 5, 25–41.

Penning-Rowsell, E. C. and Pardoe, J. 2012. Who benefits and who loses from flood risk reduction? *Environment and Planning C: Government and Policy* 30, 448–66.

Penrose, J. and Jackson, P. 1993. Identity and the politics of difference. In P. Jackson and J. Penrose (eds) *Constructions of Race, Place and Nation*. London: UCL Press, 202–10.

Pepper, D. 1984. *The Roots of Modern Environmentalism*. London: Croom Helm.

Pepper, D. 1996. *Modern Environmentalism: An introduction*. London: Routledge.

Pepper, D. and Jenkins, A. (eds) 1985. *The Geography of Peace and War*. Oxford: Basil Blackwell.

Pepper, D. and Jenkins, A. 1983. A call to arms: Geography and peace studies. *Area* 15, 202–8.

Pepper, S. C. 1942. *World Hypotheses*. Berkeley, CA: University of California Press.

Perry, P. J. 1969. H. C. Darby and historical geography: A survey and review. *Geographische Zeitschrift* 57, 161–77.

Perry, P. J. 1979. Beyond Domesday. *Progress in Human Geography* 3, 407–16.

Persson, O. and Ellegård, E. 2011. Torsten Hägerstrand in the citation time web. *Professional Geographer* 64, 1–12.

Petch, J. R. and Haines-Young, R. H. 1980. The challenge of critical rationalism for methodology in physical geography. *Progress in Physical Geography* 4, 63–78.

Peter, L. and Hull, R. 1969. *The Peter Principle*. London: Bantam Books.

Phelps N. A. and Tewdwr-Jones, M. 2008. If geography is anything, maybe it's planning alter ego? Reflections on policy relevance in two disciplines concerned with place and space. *Transactions of the Institute of British Geographers* 33, 566–84.

Philbrick, A. K. 1957. Principles of areal functional organization in regional human geography. *Economic Geography* 33, 299–366.

Phillips, M. and Unwin, T. 1985. British historical geography: Places and people. *Area* 17, 155–64.

Phillips, R. 2006. Book review: *Spaces of Masculinities*, edited by Bettina van Hoven and Kathrin Horschelmann. *Progress in Human Geography* 30, 550–2.

Phillips, R. 2010. The impact agenda and geographies of curiosity. *Transactions of the Institute of British Geographers* 35, 447–52.

Philo, C. (comp.) 1991. *New Words, New Worlds: Reconceptualising social and cultural geography* – conference proceedings. Aberystwyth: Cambrian Printers.

Philo, C. 1992. Foucault's geography. *Environment and Planning D: Society and Space* 10, 137–61.

Philo, C. 1994. Political geography and everything: Invited notes on 'transpolitical geography'. *Geoforum* 25, 525–32.

Philo, C. 1995. Animals, geography and the city: Notes on inclusions and exclusions. *Environment and Planning D: Society and Space* 13, 655–81.

Philo, C. 1998. Reading Drumlin: Academic geography and a student geographical magazine. *Progress in Human Geography* 22, 344–67.

Philo, C. 2000. More words, more worlds: Reflections on the 'cultural turn' and human geography. In I. Cook, D. Crouch, S. Naylor and J. R. Ryan (eds) *Cultural Turns/Geographical Turns: Perspectives on cultural geography*. Harlow: Prentice Hall, 26–53.

Philo, C. 2005. Spacing lives and lively spaces: Partial remarks on Sarah Whatmore's Hybrid Geographies. *Antipode: A Radical Journal of Geography* 37, 824–33.

Philo, C. (ed.) 2008. *Theory and Methods: Critical essays in human geography*. Aldershot: Ashgate.

Philo, C. 2009. Cultural turn. In R. Kitchin and N. Thrift (eds) *International Encyclopedia of Human Geography*. Amsterdam: Elsevier, 442–50.

Philo, C. 2012. A 'new Foucault' with lively implications – or 'the crawfish advances sideways'. *Transactions of the Institute of British Geographers* NS37, 496–514.

Philo, C. and Wilbert, C. (eds) 2000. *Animal Spaces, Beastly Places: New geographies of human–animal relations*. London: Routledge.

Phipps, A. G. 2001. Empirical applications of structuration theory. *Geografiska Annaler: Series B, Human Geography* 83, 189–204.

Pickerill, J. and Chatterton, P. 2006. Notes towards autonomous geographies: Creation, resistance and self-management as survival tactics. *Progress in Human Geography* 6, 730–46.

Pickering, A. 2012. The world since Kuhn. *Social Studies of Science* 42, 467–73.

Pickles, J. 1985. *Phenomenology, Science and Geography: Spatiality and the human sciences*. Cambridge: Cambridge University Press.

Pickles, J. 1986. *Geography and Humanism*. CATMOG 44, Norwich: Geo Books.

Pickles, J. 1988. From fact-world to life-world: The phenomenological method and social science. In J. Eyles and D. M. Smith (eds) *Qualitative Methods in Human Geography*. Cambridge: Polity Press, 233–54.

Pickles, J. 1992. Texts, hermeneutics and propaganda. In T. J. Barnes and J. S. Duncan (eds) *Writing Worlds: Discourse, text and metaphor in the representation of landscape*. London: Routledge, 193–230.

Pickles, J. 1993. Discourse on method and the history of discipline: Reflections on Dobson's 1983 automated geography. *Professional Geographer* 45, 451–5.

Pickles, J. (ed.) 1995a. *Ground Truth: The social implications of geographical information systems*. New York: The Guilford Press.

Pickles, J. 1995b. Representations in an electronic age: Geography, GIS and democracy. In J. Pickles (ed.) *Ground Truth: The social implications of geographical information systems*. New York: The Guilford Press, 1–30.

Pickles, J. 2004. *A History of Spaces: Cartographic reason, mapping and the geo-coded world*. London and New York: Routledge.

Pickles, J. 2007. Radical thought-in-action: Gunnar Olsson's critique of cartographic reason. *Geografiska Annaler: Series B, Human Geography* 89, 394–97.

Pickles, J. and Smith, A. (eds) 1998. *Theorising Transition: The political economy of post-Communist transformations*. London and New York: Routledge.

Pickvance, C. 1976. Housing, reproduction of capital and reproduction of labour power: Some recent French work. *Antipode: A Radical Journal of Geography* 8, 58–68.

Pile, S. 1991. Practising interpretative geography. *Transactions of the Institute of British Geographers* NS16, 458–69.

Pile, S. 1993. Human agency and human geography revisited: A critique of 'new models' of the self. *Transaction of the Institute of British Geographers* NS18, 122–39.

Pile, S. 1994. Masculinism, the use of dualistic epistemologies and third spaces. *Antipode: A Radical Journal of Geography* 26, 255–77.

Pile, S. 1996. *The Body and the City. Psychoanalysis, space and subjectivity*. London: Routledge.

Pile, S. 2010. Emotions and affect in recent human geography. *Transactions of the Institute of British Geographers* 35, 5–20.

Pile, S. and Keith, M. 1997. *Geographies of Resistance*. London: Routledge.

Pile, S. and Rose, G. 1992. All or nothing – politics and critique in the modernism-postmodernism debate. *Environment and Planning D: Society and Space* 10, 123–36.

Pile, S. and Thrift, N. (eds) 1995. *Mapping the Subject: Geographies of cultural transformation*. London and New York: Routledge.

Pinch, S. P. 1985. *Cities and Services: The geography of collective consumption*. London: Routledge & Kegan Paul.

Pinch, S. P. 1997. *Worlds of Welfare*. London: Routledge.

Pinch, S. P. 1998. Knowledge communities, spatial theory and social policy. *Social Policy and Administration*, 32, 104–9.

Pipkin, J. S. 1981. Cognitive behavioral geography and repetitive travel. In K. R. Cox and R. G. Golledge (eds) *Behavioral Problems in Geography Revisited*. London: Methuen, 145–80.

Pirie, G. H. 1976. Thoughts on revealed and spatial behaviour. *Environment and Planning A* 8, 947–55.

Pitman, A. J. 2005. On the role of geography in earth system science. *Geoforum* 36, 137–48.

Pitts, F. R. 1965. A graph theoretic approach to historical geography. *Professional Geographer* 17(5), 15–20.

Platt, R. H. 1986. Floods and man: A geographer's agenda. In R. W. Kates and I. Burton (eds) *Geography, Resources and Environment, Vol. 2: Themes from the work of Gilbert F. White*. Chicago: University of Chicago Press, 28–68.

Ploszajska, T. 1999. *Geographical Education, Empire and Citizenship: Geography teaching and learning in English schools, 1870–1944*. Historical Geography Research Series, 35. London: Royal Geographical Society (with the Institute of British Geographers).

Plummer, P. 2001. Vague theories, sophisticated techniques and poor data. *Environment and Planning A* 33, 761–4.

Plummer, P. and Sheppard, E. S. 2001. Must emancipatory economic geography be qualitative? *Antipode* 33, 194–9.

Pocock, D. C. D. 1983. The paradox of humanistic geography. *Area* 15, 355–8.

Pocock, D. C. D. and Hudson, R. 1978. *Images of the Urban Environment*. London: Macmillan.

Poiker, T. K. 1983. The shining armor of the white knight. *Professional Geographer* 35, 348–9.

Pollard, J. S., Henry, N., Bryson, J. and Daniels, P. 2000. Shades of grey? Geography and policy. *Transactions of the Institute of British Geographers* NS25, 243–8.

Pooler, J. A. 1977. The origins of the spatial tradition in geography: An interpretation. *Ontario Geography* 11, 56–83.

Popper, K. R. 1959. *The Logic of Scientific Discovery*. London: Hutchinson.

Popper, K. R. 1967. Replies to my critics. In P. A. Schipp (ed.) *The Philosophy of Karl Popper*, Vol. 2. La Salle, IN: Open Court Publishing Company, 961–97.

Popper, K. R. 1970. Normal science and its dangers. In I. Lakatos and A. Musgrave (eds) *Criticism and the Growth of Knowledge*. London: Cambridge University Press, 51–8.

Porteous, J. D. 1977. *Environment and Behavior*. Reading, MA: Addison-Wesley.

Porteous, J. D. 1985. Literature and Humanist Geography. *Area* 17, 117–22.

Porteous, J. D. 1986. Bodyscape: The body-landscape metaphor. *Canadian Geographer* 30, 2–19.

Porteous, J. D. 1988. Topocide: The annihilation of place. In J. Eyles and D. M. Smith (eds) *Qualitative Methods in Human Geography*. Cambridge: Polity Press, 75–93.

Porter, J. R. 2011. Context, location, and space: The continued development of our geo-sociological imaginations. *American Sociologist* 42, 288–302.

Porter, P. W. 1978. Geography as human ecology. Human geography: Coming of age. *American Behavioral Scientist* 22, 15–40.

Porter, P. W. and Lukermann, F. 1975. The geography of utopia. In D. Lowenthal and M. J. Bowden (eds) *Geography of the Mind: Essays in historical geosophy*. New York: Oxford University Press, 197–224.

Poulsen, M. F., Johnston, R. J. and Forrest, J. 2011. Using local statistics and neighbourhood classifications to portray ethnic residential segregation: A London example. *Environment and Planning B* 38, 636–58.

Powell, J. M. 1970. *The Public Lands of Australia Felix: Settlement and land appraisal in Victoria 1834–1891*. Melbourne: Oxford University Press.

Powell, J. M. 1971. Utopia, millennium and the cooperative ideal: A behavioral matrix in the settlement process. *Australian Geographer* 11, 606–18.

Powell, J. M. 1972. *Images of Australia*. Clayton, Victoria, Australia: Monash University Publications in Geography No. 3.

Powell, J. M. 1977. *Mirrors of the New World: Images and image-makers in the settlement process.* Folkestone: Dawson.

Powell, J. M. 1980a. Thomas Griffith Taylor 1880–1963. In T. W. Freeman and P. Pinchemel (eds) *Geographers: Biobibliographical studies, Vol. 5.* London: Mansell, 141–54.

Powell, J. M. 1980b. The haunting of Saloman's house: Geography and the limits of science. *Australian Geographer* 14, 327–41.

Powell, J. M. 1981. Editorial comment: 'Professional' geography into the eighties? *Australian Geographical Studies* 19, 228–30.

Powell, M. and Boyne, G. A. 2001. The spatial strategy of equality and the spatial division of welfare. *Social Policy and Administration* 35, 181–94.

Powell, R. C. 2011. Echoes of the new geography? History and philosophy of geography I. *Progress in Human Geography* 36, 518–26.

Powell, R. C. 2015. Notes on a geographical canon? Measures, models and scholarly enterprise. *Journal of Historical Geography,* 49, 2–8.

Power, M. 1998. The dissemination of development. *Environment and Planning D: Society and Space* 16, 577–98.

Power, M. and Sidaway, J. D. 2004. The degeneration of tropical geography. *Annals of the Association of American Geographers* 94, 585–601.

Pratt, G. 1989. Quantitative techniques and humanistic–historical materialist perspectives. In A. Kobayashi and S. Mackenzie (eds) *Remaking Human Geography.* Boston, MA: Unwin Hyman, 101–15.

Pratt, G. 1992. Spatial metaphors and speaking positions. *Environment and Planning D: Society and Space* 10, 241–4.

Pratt, G. 1999. From registered nurse to registered nanny: Discursive geographies of Filipina domestic workers in Vancouver, BC. *Economic Geography* 75, 215–36.

Pratt, G. 2000. Feminist geographies. In R. J. Johnston, D. Gregory, G. Pratt and M. Watts (eds) *The Dictionary of Human Geography,* 4th edn. Oxford: Blackwell, 259–62.

Pratt, G. 2009. Critical realism/critical realist geographies. In R. Kitchin and N. Thrift (eds) *International Encyclopedia of Human Geography.* Amsterdam: Elsevier, 379–84.

Pratt, G. and Hanson, S. 1988. Gender, class and space. *Environment and Planning D: Society and Space* 6, 15–35.

Pratt, G. and Hanson, S. 1994. Geography and the construction of difference. *Gender, Place and Culture* 1, 5–29.

Pred, A. 2005. Hägerstrand matters: life(-path) and death matters – some touching remarks. *Progress in Human Geography* 29, 328–32.

Pred, A. R. 1965a. The concentration of high value-added manufacturing. *Economic Geography* 41, 108–32.

Pred, A. R. 1965b. Industrialization, initial advantage, and American metropolitan growth. *Geographical Review* 55, 158–85.

Pred, A. R. 1967. *Behavior and Location: Foundations for a geographic and dynamic location theory.* Part I. Lund: C. W. K. Gleerup.

Pred, A. R. 1969. *Behavior and Location: Foundations for a geographic and dynamic location theory.* Part II. Lund: C. W. K. Gleerup.

Pred, A. R. 1973. Urbanization, domestic planning problems and Swedish geographic research. In C. Board, R. J. Chorley, P. Haggett and D. R. Stoddart (eds) *Progress in Geography, Vol. 5.* London: Edward Arnold, 1–77.

Pred, A. R. 1977a. The choreography of existence: Comments on Hägerstrand's time geography and its usefulness. *Economic Geography* 53, 207–21.

Pred, A. R. 1977b. *City-Systems in Advanced Economies.* London: Hutchinson.

Pred, A. R. 1979. The academic past through a time-geographic looking glass. *Annals of the Association of American Geographers* 69, 175–80.

Pred, A. R. 1981a. Production, family, and free-time projects: A time-geographic perspective on the individual and societal change in nineteenth century US cities. *Journal of Historical Geography* 7, 3–36.

Pred, A. R. 1981b. Of paths and projects: Individual behavior and its societal context. In K. R. Cox and R. G. Golledge (eds) *Behavioral Problems in Geography Revisited.* London: Methuen, 231–55.

Pred, A. R. 1984a. From here and now to there and then: Some notes on diffusions, defusions, and disillusions. In M. Billinge, D. Gregory and R. Martin (eds) *Recollections of a Revolution.* London: Macmillan, 86–103.

Pred, A. R. 1984b. Structuration, biography formation, and knowledge: Observations on port growth during the late mercantile period. *Environment and Planning D: Society and Space* 2, 251–76.

Pred, A. R. 1984c. Place as historically contingent process: Structuration and the time geography of becoming places. *Annals of the Association of American Geographers* 74, 279–97.

Pred, A. R. 1985. The social becomes the spatial and the spatial becomes the social. In D. Gregory and J. Urry (eds) *Social Relations and Spatial Structures.* London: Macmillan, 336–75.

Pred, A. R. 1986. *Becoming Places, Practice and Structure: The emergence and aftermath of enclosures in the plains villages of southwestern Skane.* Oxford: Polity Press.

Pred, A. R. 1988. Lost words as reflections of lost worlds. In R. G. Golledge, H. Couclelis and P. R. Gould (eds) *A Ground for Common Search.* Santa Barbara, CA: The Santa Barbara Geographical Press, 138–47.

Pred, A. R. 1989. The locally spoken word and local struggles. *Environment and Planning D: Society and Space* 7, 211–34.

Pred, A. R. 1990. In other wor(l)ds: Fragmented and integrated observations on gendered languages, gendered spaces and local transformation. *Antipode: A Radical Journal of Geography* 22, 33–52.

Pred, A. R. 1992. Straw men build straw houses. *Annals of the Association of American Geographers* 82, 305–8.

Pred, A. R. 1996. Interfusions: Consumption, identity and the practices and power relations of everyday life. *Environment and Planning A* 28, 11–24.

Pred, A. R. and Kibel, B. M. 1970. An application of gaming simulation to a general model of economic locational processes. *Economic Geography* 46, 136–56.

Pred, A. R. and Palm, R. 1978. The status of American women: A time-geographic view. In D. A. Lanegran and R. Palm (eds) *An Invitation to Geography,* 2nd edn. New York: McGraw-Hill, 99–109.

Preticeille, E. 1976. Urban planning: The contradiction of capitalist urbanization. *Antipode: A Radical Journal of Geography* 8, 69–76.

Price, M. and Lewis, M. 1993a. The reinvention of cultural geography. *Annals of the Association of American Geographers* 83, 1–17.

Price, M. and Lewis, M. 1993b. Reply: On reading cultural geography. *Annals of the Association of American Geographers* 83, 520–2.

Price, P. L. 2010. At the crossroads: Critical race theory and critical geographies of race. *Progress in Human Geography* 34, 147–74.

Prince, H. C. 1961–2. The geographical imagination. *Landscape* 11, 21–5.

Prince, H. C. 1971a. Real, imagined and abstract worlds of the past. In C. Board, R. J. Chorley, P. Haggett and D. R. Stoddart (eds) *Progress in Geography, Vol. 3.* London: Edward Arnold, 1–86.

Prince, H. C. 1971b. America! America? Views on a pot melting 1. Questions of social relevance. *Area* 3, 150–3.

Prince, H. C. 1979. About half Marx for the transition from feudalism to capitalism. *Area* 11, 47–51.

Prince, H. C. 2000. *Geographers Engaged in Historical Geography in British Higher Education 1931–1991.* Historical Geography Research Series, 36. London: Royal Geographical Society (with The Institute of British Geographers).

Proctor, J. D. 1998a. Ethics in geography: Giving moral form to the geographical imagination. *Area* 30, 8–18.

Proctor, J. D. 1998b. The social construction of nature: Relativist accusations, pragmatist and critical realist responses. *Annals of the Association of American Geographers* 88, 352–76.

Proctor, J. D. and Smith, D. M. (eds) 1999. *Geography and Ethics: Journeys in a moral terrain.* London: Routledge.

Proudfoot, M. J. 1937. City retail structure. *Economic Geography,* 13, 425–8.

Pruitt, E. L. 1979. The Office of Naval Research and geography. *Annals of the Association of American Geographers* 69, 103–8.

Prunty, M. C. 1979. Clark in the early 1940s. *Annals of the Association of America Geographers* 69, 42–5.

Puar, J. K., Rushbrook, D. and Schein, L. 2003. Sexuality and space: Queering geographies of globalization. *Environment and Planning D: Society and Space* 21, 383–7.

Pudup, M. B. 1988. Arguments within regional geography. *Progress in Human Geography* 12, 369–90.

Pulido, L. 2002. Reflections on a white discipline. *Professional Geographer* 54, 42–9.

Purcell, M. 2003. Islands of practice and the Marston/Brenner debate: Toward a more synthetic critical human geography. *Progress in Human Geography* 27, 317–32.

Quaini, M. 1982. *Geography and Marxism.* Oxford: Blackwell.

Radcliffe, S. A. 1990. Ethnicity, patriarchy and incorporation into the nation: Female migrants as domestic servants in Peru. *Environment and Planning D: Society and Space* 8, 379–93.

Radcliffe, S. A. 1994. (Re)presenting post-colonial women: Authority, difference and feminism. *Area* 26, 25–32.

Radcliffe, S. A. 1996. Gendered nations: Nostalgia, development and territory in Ecuador. *Gender, Place and Culture* 3, 5–21.

Radcliffe, S. A. and Westwood, S. (eds) 1996. *Remaking the Nation: Place, identity and politics in Latin America.* London: Routledge.

Radcliffe, S. A., 2005. Development and geography: Towards a postcolonial development geography? *Progress in Human Geography* 29, 291–98.

Radford, J. P. 1981. The social geography of the nineteenth century US city. In D. T. Herbert and R. J. Johnston (eds) *Geography and the Urban Environment 4.* Chichester: John Wiley, 257–93.

Rana, S. and Joliveau, T. 2009. NeoGraphy: An extension of mainstream geography for everyone made by everyone. *Journal of Location Based Services* 3, 75–81.

Ratzel, F. 1882–91. *Anthropogeographie*. Stuttgart: J. Engelghorn.

Ratzel, F. 1897. *Politische Geografie, oder die Geographie der Staaten, des Verkhers, und der Krieges*. Munich: R. Oldenbourg.

Ravenstien, E. G. 1885. The laws of migration. *Journal of the Royal Statistical Society* 48, 167–235.

Rawstron, E. M. 1958. Three principles of industrial location. *Transactions of the Institute of British Geographers* 25, 135–42.

Rawstron, E. M. 2002. Textbooks that moved generations. *Progress in Human Geography* 26, 831–6.

Ray, D. M., Villeneuve, P. Y. and Roberge, R. A. 1974. Functional prerequisites, spatial diffusion, and allometric growth. *Economic Geography* 50, 341–51.

Rees, J. 1985. *Natural Resources: Allocation economics, and policy*. London: Methuen.

Rees, P. H. 1986. Developments in the modelling of spatial populations. In R. I. Woods and P. H. Rees (eds) *Population Structures and Models: Developments in spatial demography*. London: Allen & Unwin, 97–125.

Rees, P. H. and Wilson, A. G. 1977. *Spatial Population Analysis*. London: Edward Arnold.

Reitsma, F. 2013. Revisiting the 'Is GIScience a science?' debate (or is it quite possibly scientific gerry-mandering?). *International Journal of Geographic Information Science* 27, 211–21.

Relph, E. 1970. An inquiry into the relations between phenomenology and geography. *Canadian Geographer* 14, 193–201.

Relph, E. 1976. *Place and Placelessness*. London: Pion.

Relph, E. 1977. Humanism, phenomenology, and geography. *Annals of the Association of American Geographers* 67, 177–9.

Relph, E. 1981a. Phenomenology. In M. E. Harvey and B. P. Holly (eds) *Themes in Geographic Thought*. London: Croom Helm, 99–114.

Relph, E. 1981b. *Rational Landscapes and Humanistic Geography*. London: Croom Helm.

Renfrew, A. C. 1981. Space, time and man. *Transactions of the Institute of British Geographers* NS6, 257–78.

Reynolds, R. B. 1956. Statistical methods in geographical research. *Geographical Review* 46, 129–32.

Rhind, D. W. 1981. Geographical information systems in Britain. In N. Wrigley and R. J. Bennett (eds) *Quantitative Geography*. London: Routledge & Kegan Paul, 17–35.

Rhind, D. W. 1986. Remote sensing, digital mapping and GIS: The creation of government policy in the UK. *Environment and Planning C: Government and Policy* 4, 91–100.

Rhind, D. W. 1989. Computing, academic geography, and the world outside. In B. Macmillan (ed.) *Remodelling Geography*. Oxford: Basil Blackwell, 177–90.

Rhind, D. W. 1996. Differential research funding – a comment on Smith. *Area* 28, 96–7.

Rhind, D. W. 2003. The geographical underpinning of society and its radical transformation. In R. J. Johnston and M. Williams (eds) *A Century of British Geography*. Oxford: Oxford University Press for the British Academy, 429–62.

Rhind, D. W. and Adams, T. A. 1980. Recent developments in surveying and mapping. In E. H. Brown (ed.) *Geography, Yesterday and Tomorrow*. Oxford: Oxford University Press, 181–99.

Rhind, D. W. and Hudson, R. 1981. *Land Use*. London: Methuen.

Rhind, D. W. and Mounsey, H. 1989. The Chorley Committee and 'Handling geographical information'. *Environment and Planning A* 21, 571–86.

Richards, K. and Wrigley, N. 1996. Geography in the United Kingdom 1992–1996. *Geographical Journal* 162, 41–62.

Richards, K., Batty, M., Edwards, K., Findlay, A., Foody, G., Frostick, L., Jones, K., Lee, R., Livingstone, D., Marsden, T., Petts, J., Philo, C., Simon, D. and Thomas, D. 2009. The nature of publishing and assessment in geography and environmental studies: Evidence from the Research Assessment Exercise 2008. *Area* 41, 231–43.

Richardson, D. 2003. Federal funding for geography education. *AAG Newsletter* 38 (June), 2–4.

Richardson, H. W. 1973. *The Economics of Urban Size*. Farnborough: Saxon House.

Riddell, J. B. 1970. *The Spatial Dynamics of Modernization in Sierra Leone*. Evanston, IL: Northwestern University Press.

Rieser, R. 1973. The territorial illusion and the behavioural sink: Critical notes on behavioural geography. *Antipode: A Radical Journal of Geography* 5(3), 52–7.

Ritter, K. 1817–59. *Die Erdkunde*, 19 vols. Berlin: G. Reimer.

Robbins, P. 2004. *Political Ecology: A critical introduction*. Oxford: Blackwell.

Roberts, S. 2000. Realizing critical geographies of the university. *Antipode: A Radical Journal of Geography* 32, 230–44.

Robinson, A. H. 1954. Geographical cartography. In P. E. James and C. F. Jones (eds) *American Geography: Inventory and prospects*. New York: Syracuse University Press, 553–77.

Robinson, A. H. 1956. The necessity of weighting values in correlation analysis of area data. *Annals of the Association of American Geographers* 46, 233–6.

Robinson, A. H. 1961. On perks and pokes. *Economic Geography* 37, 181–3.

Robinson, A. H. 1962. Mapping the correspondence of isarithmic maps. *Annals of the Association of American Geographers* 52, 414–25.

Robinson, A. H. and Bryson, R. A. 1957. A method for describing quantitatively the correspondence of geographical distributions. *Annals of the Association of American Geographers* 47, 379–91.

Robinson, A. H., Lindberg, J. B. and Brinkman, L. W. 1961. A correlation and regression analysis applied to rural farm densities in the Great Plains. *Annals of the Association of American Geographers* 51, 211–21.

Robinson, G. M. 1991. An appreciation of James Wreford Watson with a bibliography of his work. In G. M. Robinson (ed.) *A Social Geography of Canada*. Toronto: Dundurn Press, 492–506.

Robinson, G. M. 1998. *Methods and Techniques in Human Geography*. New York: John Wiley.

Robinson, J. 2003a. Postcolonialising geography: Tactics and pitfalls. *Singapore Journal of Tropical Geography* 24, 273–89.

Robinson, J. 2003b. Political geography in a postcolonial context. *Political Geography* 22, 647–52.

Robinson, J. L. 1976. A new look at the four traditions of geography. *Journal of Geography* 75, 520–30.

Robinson, M. E. 1982. Representation, misrepresentation, and 'uncritical rhetoric'. *Professional Geographer* 34, 224–6.

Robinson, W. S. 1950. Ecological correlation and the behavior of individuals. *American Sociological Review* 15, 351–57.

Robson, B. T. 1969. *Urban Analysis*. Cambridge: Cambridge University Press.

Robson, B. T. 1972. The corridors of geography. *Area* 4, 213–14.

Robson, B. T. 1982. Introduction. In B. T. Robson and J. Rees (eds) *Geographical Agenda for a Changing world*. London: Social Science Research Council, 1–6.

Robson, B. T. 1984. A pleasant pain. In M. Billinge, D. Gregory and R. Martin (eds) *Recollections of a Revolution*. London: Macmillan, 104–6.

Roche, M. 2011. New Zealand geography, biography and autobiography. *New Zealand Geographer* 67, 73–8.

Rocheleau, D. 1995. Maps, numbers, text and context: Mixing methods in feminist political ecology. *Professional Geographer* 47, 458–66.

Rodaway, P. 1994. *Sensuous Geographies: Body, sense and place*. London and New York, Routledge.

Roder, W. 1961. Attitudes and knowledge on the Topeka flood plain. In G. F. White (ed.) *Papers on Flood Problems*. Chicago: University of Chicago, Department of Geography, Research Paper 70, 62–83.

Rodríguez-Pose, A. 2001. Killing economic geography with a 'cultural turn' overdose. *Antipode: A Radical Journal of Geography* 33, 176–82.

Rodríguez-Pose, A. 2004. On English as a way to preserve geographical diversity. *Progress in Human Geography* 28.

Rogers, A. 2006. Review of *Geography and Geographers: Anglo–American human geography since 1945*, 6th edn. *Progress in Human Geography* 30, 546–8.

Rogers, A., Bear, C., Hunt, M., Mills, S. and Sandover, R. 2014. Intervention: The impact agenda and human geography in UK higher education. *ACME: An International E-Journal for Critical Geographies* 13, 1–9.

Rose, C. 1980. Human geography as text interpretation. In A. Buttimer and D. Seamon (eds) *The Human Experience of Space and Place*. London: Croom Helm, 123–34.

Rose, C. 1981. Wilhelm Dilthey's philosophy of human understanding. In D. R. Stoddart (ed.) *Geography, Ideology and Social Concern*. Oxford: Basil Blackwell, 99–133.

Rose, C. 1987. The problem of reference and geographic structuration. *Environment and Planning D: Society and Space* 5, 93–112.

Rose, D. 1987. Home ownership, subsistence, and historical change: The mining district of West Cornwall in the late nineteenth century. In N. J. Thrift and P. Williams (eds) *Class and Space: The making of urban society*. London: Routledge, 108–53.

Rose, D. and Pevalin, D. J. (eds) 2003. *A Researcher's Guide to the National Statistics Socio-Economic Classification*. London: Sage.

Rose, G. 1989. Locality-studies and waged labour – an historical critique. *Transactions of the Institute of British Geographers* NS14, 317–28.

Rose, G. 1993a. *Feminism and Geography*. Cambridge: Polity Press.

Rose, G. 1993b. Progress in geography and gender: Or something else. *Progress in Human Geography* 17, 531–7.

Rose, G. 1994. The cultural politics of place: Local representation and oppositional discourse in two films. *Transactions of the Institute of British Geographers* NS19, 46–60.

Rose, G. 1995. Tradition and paternity: Same difference? *Transactions of the Institute of British Geographers* NS20, 414–16.

Rose, G. 1997. Situating knowledges: Positionality, reflexivities and other tactics. *Progress in Human Geography* 21, 305–20.

Rose, G. 2001. *Visual Methodologies: An introduction to the interpretation of visual materials*. London: Sage.

Rose, G., Gregson, N., Foord, J., Bowlby, S., Dwyer, C., Hollowat, S. Laurie, N., Maddrell, A. and Skelton, T. 1997. Introduction. In Women and Geography Study Group (eds) *Feminist Geographies: Explorations in diversity and difference*. Harlow: Longman, 1–12.

Rose, J. K. 1936. Corn yield and climate in the Corn Belt. *Geographical Review* 26, 88–102.

Rossiter, D. J. and Johnston, R. J. 1981. Program GROUP: The identification of all possible solutions to a constituency-delimitation problem. *Environment and Planning A* 13, 231–8.

Rothenberg, T. Y. 2007. *Presenting America's World: Strategies of innocence in National Geographic Magazine, 1888–1945*. Aldershot: Ashgate.

Rothstein, J. 1958. *Communication, Organization and Science*. Colorado: Falcon's Wing Press.

Rouhani, F. 2007. Religion, identity and activism: Queer Muslim diasporic identities. In K. Browne, J. Lim and G. Brown (eds) *Geographies of Sexualities*. Farnham and Burlington, VT: Ashgate, 169–180.

Routledge, P. 2009. Activist geographies. In R. Kitchin and N. Thrift (eds) *International Encyclopedia of Human Geography*. Amsterdam: Elsevier, 7–14.

Rowles, G. D. 1978. Reflections on experiential field work. In D. Ley and M. S. Samuels (eds) *Humanistic Geography: Problems and prospects*. London: Croom Helm, 173–93.

Rowntree, L., Foote, K. E. and Domosh, M. 1989. Cultural geography. In G. L. Gaile and C. J. Willmott (eds) *Geography in America*. Columbus, OH: Merrill, 209–17.

Ruming, K. 2009. Following the actors: Mobilising an actor–network theory methodology in geography. *Australian Geographer* 40, 451–69.

Rupke, N. A. 2005. *Alexander von Humboldt: A metabiography*. Frankfurt am Main: Peter Lang.

Rushton, G. 1969. Analysis of spatial behavior by revealed space preference. *Annals of the Association of American Geographers* 59, 391–400.

Rushton, G. 1979. On behavioral and perception geography. *Annals of the Association of American Geographers* 69, 463–4.

Russell, A. T. 1997. A question of interaction: Using logistic regression to examine geographic effects on British voting behaviour. In C. J. Pattie *et al.* (eds) *British Elections and Parties Review*, Vol. 7. London: Frank Cass, 91–109.

Russell, C. A. 1977. *Chemists by Profession: The origins and rise of the Royal Institute of Chemistry*. Milton Keynes: The Open University Press.

Rusu, M. S. (2012) Multi-paradigmacity, scattered cumulativity, multi-localized ignorance: The tumultuous condition of sociological knowledge. *Revista de Cercetare si Interventie Sociola* 39, 187–203.

Rycroft, S. and Cosgrove, D. 1999. Regional knowledge for pedagogy and planning: Dudley Stamp and the Land Utilisation Survey. In A. Buttimer, S. D. Brunn and U. Wardenga (eds) *Text and Image: Social construction of regional knowledges*. Leipzig: Leibniz Institute for Regional Geography, Beiträge zur regionalen Geographie 49, 122–9.

Saarinen, T. F. 1979. Commentary: Critique of Bunting-Guelke paper. *Annals of the Association of American Geographers* 69, 464–8.

Sachs, A. 2007. *The Humboldt Current: A European explorer and his American disciples*. Oxford: Oxford University Press.

Sack, R. D. 1972. Geography, geometry and explanation. *Annals of the Association of American Geographers* 62, 61–78.

Sack, R. D. 1973a. Comment in reply. *Annals of the Association of American Geographers* 63, 568–9.

Sack, R. D. 1973b. A concept of physical space in geography. *Geographical Analysis* 5, 16–34.

Sack, R. D. 1974a. The spatial separatist theme in geography. *Economic Geography* 50, 1–19.

Sack, R. D. 1974b. Chorology and spatial analysis. *Annals of the Association of American Geographers* 64, 439–52.

Sack, R. D. 1981. *Conceptions of Space in Social Thought*. London: Macmillan.

Sack, R. D. 1983. Human territoriality: A theory. *Annals of the Association of American Geographers* 73, 55–74.

Sack, R. D. 1986. *Human Territoriality: Its theory and history*. Cambridge: Cambridge University Press.

Sack, R. D. 1997. *Homo Geographicus: A framework for action, awareness and moral concern*. Baltimore, MD: Johns Hopkins University Press.

Sack, R. D. 2001. The geographic problematic: Empirical issues. *Norsk Geografisk Tidsskrift* 55, 107–16.

Said, E. 1978. *Orientalism*. New York: Harper.

Saldanha, A. 2010. Skin, affect, aggregation: Guattarian variations on Fanon. *Environment and Planning A* 42, 2410–27.

Samers, M. and Sidaway, J. D. 2000. Guest editorial: Exclusions, inclusions and occlusions in 'Anglo-American geography': Reflections on Minca's 'Venetian geographical praxis'. *Environment and Planning D* 18, 663–6.

Samuels, M. S. 1978. Existentialism and human geography. In D. Ley and M. S. Samuels (eds) *Humanistic Geography: Problems and prospects*. Chicago: Maaroufa Press, 22–40.

Sandbach, F. 1980. *Environment, Ideology and Policy*. Oxford: Blackwell.

Sanders, R. 2006. Social justice and women of colour in geography: Philosophical musings, trying again. *Gender, Place and Culture* 13, 49–55.

Sanderson, M. 1988. *Griffith Taylor: Antarctic scientist and pioneer geographer*. Ottawa: Carleton University Press.

Santos, M. 1974. Geography, Marxism and underdevelopment. *Antipode: A Radical Journal of Geography* 6(3), 1–9.

Sarre, P. 1987. Realism in practice. *Area* 19, 3–10.

Sarre, P., Phillips, D. and Skellington, R. 1989. *Ethnic Minority Housing: Explanations and policies*. Aldershot: Avebury.

Sauer, C. O. 1918. Geography and the gerrymander. *American Political Science Review* 12, 403–26.

Sauer, C. O. 1925. The morphology of landscape. *University of California Publications in Geography* 2, 19–54.

Sauer, C. O. 1941. Foreword to Historical Geography. *Annals of the Association of American Geographers* 31, 1–24.

Sauer, C. O. 1956a. The education of a geographer. *Annals of the Association of American Geographers* 46, 287–99.

Sauer, C. O. 1956b. The agency of man on earth. In W. L. Thomas (ed.) *Man's Role in Changing the Face of the Earth*. Chicago: University of Chicago Press, 49–69.

Sauer, C. O. 1956c. Retrospect. In W. L. Thomas (ed.) *Man's Role in Changing the Face of the Earth*. Chicago: University of Chicago Press, 1131–5.

Saunders, A. 2010. Literary geography: Reforging the connections. *Progress in Human Geography* 34, 436–52.

Saunders, P. and Williams, P. R. 1986. The new conservatism: Some thoughts on recent and future developments in urban studies. *Environment and Planning D: Society and Space* 4, 393–9.

Saunders, P. and Williams, P. R. 1987. For an emancipated social science. *Environment and Planning D: Society and Space* 5, 427–30.

Sayer, A. 1979. Epistemology and conceptions of people and nature in geography. *Geoforum* 10, 19–44.

Sayer, A. 1981. Defensible values in geography. In D. T. Herbert and R. J. Johnston (eds) *Geography and the Urban Environment*, Vol. 4. Chichester: John Wiley, 29–56.

Sayer, A. 1982. Explanation in economic geography. *Progress in Human Geography* 6, 68–88.

Sayer, A. 1983. Notes on geography and the relationship between people and nature. In T. Cannon, M. Forbes and J. Mackie (eds) *Society and Nature: Socialist perspectives on the relationship between human and physical geography*. London: Union of Socialist Geographers, 47–57.

Sayer, A. 1984. *Method in Social Science: A realist approach*. London: Hutchinson.

Sayer, A. 1985a. Realism and Geography. In R. J. Johnston (ed.) *The Future of Geography*. London: Methuen, 159–73.

Sayer, A. 1985b. The difference that space makes. In D. Gregory and J. Urry (eds) *Social Relations and Spatial Structures*. London: Macmillan, 49–66.

Sayer, A. 1987. Hard work and its alternatives. *Environment and Planning D: Society and Space* 5, 395–9.

Sayer, A. 1989a. The new regional geography and problems of narrative. *Environment and Planning D: Society and Space* 7, 253–76.

Sayer, A. 1989b. On the dialogue between humanism and historical materialism in geography. In A. Kobayashi and S. Mackenzie (eds) *Remaking Human Geography*. Boston, MA: Unwin Hyman, 206–26.

Sayer, A. 1992a. *Method in Social Science: A realist approach*, 2nd edn. London: Routledge.

Sayer, A. 1992b. What's left to do? A reply to Hadjimichalis and Smith. *Antipode: A Radical Journal of Geography* 24, 214–17.

Sayer, A. 1995. *Radical Political Economy: A critique*. Oxford: Blackwell Publishers.

Sayer, A. 2000. Critical and uncritical turns. In I. Cook, D. Crouch, S. Naylor and J. R. Ryan (eds) *Cultural Turns/ Geographical Turns: Perspectives on cultural geography*. Harlow: Prentice Hall, 166–81.

Sayer, A. 2001. For a critical cultural political economy. *Antipode: A Radical Journal of Geography* 33, 687–708.

Sayer, A. 2013. Looking forward to new realist debates. *Dialogues in Human Geography* 3, 22–5.

Sayer, A. 2015. *Why We Can't Afford the Rich*. Bristol: Policy Press.

Sayer, A. and Morgan, K. 1985. A modern industry in a declining region: Links between method, theory and policy. In D. Massey and R. Meegan (eds) *Politics and Method: Contrasting studies in industrial geography*. London: Methuen, 144–68.

Sayer, A. and Storper, M. 1997. Ethics unbound: For a normative turn in social theory. *Environment and Planning D: Society and Space* 15, 1–17.

Sayer, A. and Walker, R. A. 1992. *The New Social Economy: Reworking the division of labour*. Oxford: Blackwell Publishers.

Sayer, D. 2015. *Rank Hypocrisies: The insult of the REF*. London: Sage.

Scargill, D. I. 1976. The RGS and the foundations of geography at Oxford. *Geographical Journal* 142, 438–61.

Schaefer, F. K. 1953. Exceptionalism in geography: A methodological examination. *Annals of the Association of American Geographers* 43, 226–49.

Schelhaas, B. and Hönsch, I. 2002. History of German geography: Worldwide reputation and strategies of nationalisation and institutionalisation. In G. S. Dunbar (ed.) *Geography: Discipline, profession and subject since 1870*. Dordrecht: Kluwer, 9–44.

Schlemper, M. B., Adams, J. K. and Solem, M. 2014. Geographers in business, government and nonprofit organizations: Skills, challenges, and professional identities. *Professional Geographer* 66, 480–92.

Schoenberger, E. 1992. Self-criticism and self-awareness in research: A reply to Linda McDowell. *Professional Geographer* 44, 215–18.

Schoenberger, E. 1993. On knowing what to know. *Environment and Planning A* 25, 1225–8.

Schoenberger, E. 1997. *The Cultural Crisis of the Firm*. Oxford: Blackwell Publishers.

Schoenberger, E. 2001. Interdisciplinarity and social power. *Progress in Human Geography* 25, 365–82.

Schuermans, N., Meeus, B. and De Maeschalck, F. 2010. Is there a world beyond the Web of Science? Publication practices outside the heartland of academic geography. *Area* 42, 417–24.

Schulten, S. 2001. *The Geographical Imagination in America, 1880–1950*. Chicago: University of Chicago Press.

Schuurman, N. 1999. An interview with Michael Goodchild, January 6, 1998, Santa Barbara, California. *Environment and Planning D: Society and Space* 17, 3–15.

Schuurman, N. 2000. Trouble in the heartland: GIS and its critics in the 1990s. *Progress in Human Geography* 24, 569–90.

Schuurman, N. 2006. Formalization matters: Critical GIS and ontology research. *Annals of the Association of American Geographers* 96, 726–39.

Schwanen, T. and Kwan, M.-P. 2009. 'Doing' critical geographies with numbers. *Professional Geographer* 61, 459–64.

Scott, A. J. 1982. The meaning and social origins of discourse on the spatial foundations of society. In P. R. Gould and G. Olsson (eds) *A Search for Common Ground*. London: Pion, 141–56.

Scott, A. J. 1985. Location processes, urbanization, and territorial development: An exploratory essay. *Environment and Planning A* 17, 479–501.

Scott, A. J. 1986. Industrialization and urbanization: A geographical agenda. *Annals of the Association of American Geographers* 76, 25–37.

Scott, A. J. 1988. *Metropolis*. Los Angeles, CA: University of California Press.

Scott, A. J. 2000. Economic geography: The great half-century. In G. L. Clark, M. P. Feldmann and M. S. Gertler (eds) *The Oxford Handbook of Economic Geography*. Oxford: Oxford University Press, 19–43.

Scott, A. J. and Cooke, P. N. 1988. The new geography and sociology of production. *Environment and Planning D: Society and Space* 6, 241–4.

Scott, A. J. and Storper, M. (eds) 1985. *Production, Work, Territory*. Boston, MA: George Allen & Unwin.

Seager, J. and Enloe, C. 2011. *The Real State of America Atlas*. London: University of California Press.

Seager, J. and Olson, A. 1986. *Women in the World: An international atlas*. London: Pan.

Seager, J., Reed, C. and Stott, P. 1995. *The State of the Environment Atlas*, 2nd edn. London: Pan.

Seamon, D. 1984. Phenomenology and environment-behavior research. In G. T. Moore and E. Zube (eds) *Advances in Environment, Behavior and Design*. New York: Plenum, 3–36.

Seamon, D. 1984. The question of reliable knowledge: The irony and tragedy of positivist research. *Professional Geographer* 36, 216–18.

Seamon, D. and Sowers, J. 2009. Existentialism/existential geography. In R. Kitchin and N. Thrift (eds) *International Encyclopedia of Human Geography*. Amsterdam: Elsevier, 666–71.

Seemann, J. 2015. VII Provocations for effective teaching in the history of geography. *Progress in Human Geography*. doi: 10.1177/0309132515575940

Self, A., Thomas, J. and Randall, C. (ONS). 2012. *Measuring National Well-Being: Life in the UK*, 2012. Office of National Statistics, UK. Available at: www.ons.gov.uk/ons/dcp171766_287415.pdf

Semple, E. C. 1911. *Influences of Geographical Environment*. New York: Henry Holt.

Shannon, G. W. and Dever, G. E. A. 1974. *Health Care Delivery: Spatial perspectives*. New York: McGraw-Hill.

Sharp, J. P. 1993. Publishing American identity: Popular geopolitics, myth, and *The Reader's Digest*. *Political Geography* 12, 491–504.

Sharp, J. P. 2000. Remasculinising geo-politics? Comments on Gearoid O'Tuathail's *Critical Geopolitics*. *Political Geography* 19, 361–4.

Sharp, J. P. 2009. *Geographies of Postcolonialism*. London: Sage.

Sharp, J. P., Routledge, P., Philo, C. and Paddison, R. (eds) 2000. *Entanglements of Power: Geographies of domination/resistance*. London: Routledge.

Sharrock, W. and Read, R. 2002. *Kuhn: Philosopher of scientific revolution*. Cambridge: Polity Press.

Shaw, W. 2013. Auto-ethnography and autobiography in geographical research. *Geoforum* 46, 1–4.

Shelton, N. J. 2005. The future for new geographers. *Area* 37, 110–14.

Sheppard, E. S. 1979. Gravity parameter estimation. *Geographical Analysis* 11, 120–33.

Sheppard, E. S. 1993. Automated geography: What kind of geography for what kind of society? *Professional Geographer* 45, 457–60.

Sheppard, E. S. 1995. Dissenting from spatial analysis. *Urban Geography* 16, 283–303.

Sheppard, E. S. and Barnes, T. J. (eds) 2000. *A Companion to Economic Geography*. Oxford: Blackwell.

Sheppard, E. S. and Barnes, T. J. 1990. *The Capitalist Space Economy: Geographical analysis after Ricardo, Marx and Sraffa*. London and Boston: Unwin Hyman.

Short, J. R. 1984. *The Urban Arena: Capital, state and community in contemporary Britain*. London: Macmillan.

Short, J. R. 2002. The disturbing case of the concentration of power in human geography. *Area* 34, 323–4.

Short, J. R., Boniche, A., Kim, Y. and Li, P. L. 2001. Cultural globalization, global English, and geography journals. *Professional Geographer* 53, 1–11.

Shevky, E. and Bell, W. 1955. *Social Area Analysis*. Stanford, CA: Stanford University Press.

Sibley, D. 1995. *Geographies of Exclusion*. London and New York: Routledge.

Sibley, D. 1998. Sensations and spatial science: Gratification and anxiety in the production of ordered landscapes. *Environment and Planning A* 30, 235–46.

Sibley, D. 2000. Placing anxieties. In I. Cook, D. Crouch, S. Naylor and J. R. Ryan (eds) *Cultural Turns/Geographical Turns: Perspectives on cultural geography*. Harlow: Prentice Hall, 359–69.

Sidaway, J. D. 1992. In other worlds: On the politics of research by First World geographers in the Third World. *Area* 24, 403–8.

Sidaway, J. D. 1997. The production of British geography. *Transactions of the Institute of British Geographers* NS22, 488–504.

Sidaway, J. D. 2000a. Post-colonial geographies: An exploratory essay. *Progress in Human Geography* 24, 591–612.

Sidaway, J. D. 2000b. Recontextualising positionality: Geographic research and academic fields of power. *Antipode: A Radical Journal of Geography* 32, 260–70.

Sidaway, J. D. and Johnston, R. J. 2007. Geography in higher education in the UK. *Journal of Geography in Higher Education* 31, 57–80.

Sidaway, J. D., Woon, C. Y. and Jacobs, J. M. 2014. Planetary postcolonialism. *Singapore Journal of Tropical Geography* 35, 4–21.

Siddall, W. R. 1961. Two kinds of geography. *Economic Geography* 36, facing page 189.

Simandan, D. 2002. On what it takes to be a good geographer. *Area* 34, 284–93.

Simmons, I. G. 1993. *Interpreting Nature: Cultural constructions of the environment*. London: Routledge.

Simmons, I. G. 1997. *Humanity and Environment: A cultural ecology*. London: Addison, Wesley Longman.

Simon, H. A. 1957. *Models of Man: Social and rational*. New York: John Wiley.

Simonsen, K. 2002. Global–local ambivalence. *Transactions of the Institute of British Geographers* NS27, 391–4.

Sinclair, J. C. and Kissling, C. C. 1971. A network analysis approach to fruit distribution planning. In R. J. Johnston and J. M. Soons (eds) *Proceedings, Sixth New Zealand Geography Conference*, Christchurch, August 1970, Vol. 1. Christchurch: New Zealand Geographical Society, 131–6.

Singleton, A. D. and Spielman, S. E. 2014. The past, present and future of geodemographic research in the United States and the United Kingdom. *Professional Geographer* 66, 558–67.

Sismondo, S. 2012. Fifty years of *The Structure of Scientific Revolutions*, twenty–five of *Science in Action*. *Social Studies of Science* 42, 415–19.

Skupin, A. 2014. Making a Mark: A computational and visual analysis of one researcher's intellectual domain. *International Journal of Geographical Information Science* 28, 1209–32.

Slater, D. 1973. Geography and underdevelopment – 1. *Antipode: A Radical Journal of Geography* 5(3), 21–53.

Slater, D. 1977. Geography and underdevelopment – 2. *Antipode: A Radical Journal of Geography* 9, 1–31.

Slater, T. 2012. Commentary: Impacted geographers: A response to Pain, Kesby and Askins. *Area* 44, 117–19.

Slater, T. R. 1988. Redbrick academic geography. *Geographical Journal* 154, 169–80.

Smailes, A. E. 1946. The urban mesh of England and Wales. *Transactions and Papers, Institute of British Geographers* 11, 85–101.

Smallman-Raynor, M. and Cliff, A. D. 1990. Aquired Immune Deficiency Syndrome (AIDS): Literature, geographical origins and global patterns. *Progress in Human Geography* 14, 157–213.

Smallman-Raynor, M., Cliff, A. D. and Haggett, P. 1992. *Atlas of AIDS*. Oxford: Blackwell Reference.

Smallman-Raynor, M., Cliff, A. D. and Haggett, P. 2004. *World Atlas of Epidemic Diseases*. London: Hodder.

Smith, C. T. 1965. Historical geography: Current trends and prospects. In R. J. Chorley and P. Haggett (eds) *Frontiers in Geographical Teaching*. London: Methuen, 118–43.

Smith, D. M. 1971. America! America? Views on a pot melting. 2. Radical geography – the next revolution? *Area* 3, 153–7.

Smith, D. M. 1973a. Alternative 'relevant' professional roles. *Area* 5, 1–4.

Smith, D. M. 1973b. *The Geography of Social Well-Being in the United States*. New York: McGraw-Hill.

Smith, D. M. 1975. On the concept of welfare. *Area* 7, 33–6.

Smith, D. M. 1977. *Human Geography: A welfare approach*. London: Edward Arnold.

Smith, D. M. 1979. *Where the Grass is Greener: Living in an unequal world*. London: Penguin.

Smith, D. M. 1981. *Industrial Location: An economic geographical analysis*, 2nd edn. New York: John Wiley.

Smith, D. M. 1984. Recollections of a random variable. In M. Billinge, D. Gregory and R. Martin (eds) *Recollections of a Revolution*. London: Macmillan, 117–33.

Smith, D. M. 1985. The 'new blood' scheme and its application to geography. *Area* 17, 237–43.

Smith, D. M. 1986. UGC research ratings: Pass or fail? *Area* 18, 247–9.

Smith, D. M. 1988a. Towards an interpretative human geography. In J. Eyles and D. M. Smith (eds) *Qualitative Methods in Human Geography*. Cambridge: Polity Press, 255–67.

Smith, D. M. 1988b. On academic performance. *Area* 20, 3–13.

Smith, D. M. 1994a. *Geography and Social Justice*. Oxford: Blackwell Publishers.

Smith, D. M. 1994b. On professional responsibility to distant others. *Area* 26, 359–67.

Smith, D. M. 1995. Against differential research funding. *Area* 27, 79–83.

Smith, D. M. 1996. Reply to Rhind – value for money, the continuing debate. *Area* 28, 97–101.

Smith, D. M. 1997a. Geography and ethics: A moral turn? *Progress in Human Geography* 21, 583–90.

Smith, D. M. 1997b. Back to the good life: Towards an enlarged conception of social justice. *Environment and Planning D: Society and Space* 15, 19–35.

Smith, D. M. 1998. How far should we care? On the spatial scope of beneficence. *Progress in Human Geography* 22, 15–38.

Smith, D. M. 2000. *Moral Geographies: Ethics in a world of difference*. Edinburgh: Edinburgh University Press.

Smith, G. D. and Winchester, H. P. M. 1998. Negotiating space: Alternative masculinities at the work/home boundary. *Australian Geographer* 29, 327–39.

Smith, J. M. 1996. Geographical rhetoric: Modes and tropes of appeal. *Annals of the Association of American Geographers* 86, 1–20.

Smith, J. M. 2001. Constructing the nation: Eighteenth-century geographies for children. *Mosaic* 34, 133–48.

Smith, J. M. 2004. Unlawful relations and verbal inflation. *Annals of the Association of American Geographers* 94, 294–9.

Smith, J. M. 2009. Humanism/humanistic geography. In R. Kitchin and N. Thrift (eds) *International Encyclopedia of Human Geography*. Amsterdam: Elsevier, 239–50.

Smith, N. 1979a. Geography, science and post-positivist modes of explanation. *Progress in Human Geography* 3, 365–83.

Smith, N. 1979b. Toward a theory of gentrification: A back to the city movement by capital, not people. *Journal of the American Planning Association* 45, 538–48.

Smith, N. 1984. *Uneven Development: Nature, capital and the production of space*. Oxford: Basil Blackwell.

Smith, N. 1986. On the necessity of uneven development. *International Journal of Urban and Regional Research* 10, 87–104.

Smith, N. 1987a. 'Academic war over the field of geography': The elimination of geography at Harvard, 1947–1951. *Annals of the Association of American Geographers* 77, 155–72.

Smith, N. 1987b. Danger of the empirical turn: The CURS initiative. *Antipode: A Radical Journal of Geography* 19, 59–68.

Smith, N. 1987c. Rascal concepts, minimalizing discourse, and the politics of geography. *Environment and Planning D: Society and Space* 5, 377–83.

Smith, N. 1990. Geography as museum: Private history and conservative idealism in *The Nature of Geography*. In J. N. Entrikin and S. D. Brunn (eds) *Reflections on Richard Hartshorne's* The Nature of Geography. Washington, DC: Association of American Geographers, 89–120.

Smith, N. 1991. What's left? A lot's left. *Antipode: A Radical Journal of Geography* 23, 406–18.

Smith, N. 1992. History and philosophy of geography: Real wars, theory wars. *Progress in Human Geography* 16, 257–71.

Smith, N. 1994. Shaking loose the colonies: Isaiah Bowman and the 'de-colonization' of the British Empire. In A. Godlewska and N. Smith (eds) *Geography and Empire*. Oxford: Blackwell, 270–99.

Smith, N. 2001. Marxism and geography in the Anglophone world. *Geographische Revue* 3, 5–22.

Smith, N. 2003. *American Empire: Roosevelt's geographer and the prelude to globalization*. Berkeley, CA: University of California Press.

Smith, N. 2005a. Neo–critical geography, or, the flat pluralist world of business class. *Antipode: A Radical Journal of Geography* 37, 887–97.

Smith, N. 2005b. *Endgame of Globalization*. Abingdon, Oxford and New York: Routledge:

Smith, R. G. 2003. Baudrillard's non-representational theory: Burn the signs and journey without maps. *Environment and Planning D: Society and Space* 21, 67–84.

Smith, R. G. and Doel, M. A. 2001. Baudrillard unwound: The duplicity of post-Marxism and deconstruction. *Environment and Planning D: Society and Space* 19, 137–59.

Smith, S. J. 1984. Practising humanistic geography. *Annals of the Association of American Geographers* 74, 353–74.

Smith, S. J. 1986. *Crime, Space and Society*. Cambridge: Cambridge University Press.

Smith, S. J. 1988. Constructing local knowledge: The analysis of self in everyday life. In J. Eyles and D. M. Smith (eds) *Qualitative Methods in Human Geography*. Cambridge: Polity Press, 17–38.

Smith, S. J. 1989. Society, space and citizenship: A human geography for the 'new times'? *Transactions of the Institute of British Geographers* NS14, 144–56.

Smith, S. J. 2008. Owner-occupation: At home with a hybrid of money and materials. *Environment and Planning A* 40, 520–35.

Smith, S. J. 2015. Owner occupation: At home in a spatial financial paradox. *International Journal of Housing Policy* 15, 61–83.

Smith, S. J. and Searle, B. A. 2010. Housing futures: A role for derivatives. In S. J. Smith and B. A. Searle (eds) *The Blackwell Companion to the Economics of Housing*. Oxford: Wiley-Blackwell.

Smith, S. J., Munro, M. and Christie, H. 2006. Performing (housing) markets. *Urban Studies* 43, 81–98.

Smith, S. J., Searle, B. A. and Cooke, N. 2008. Rethinking the risks of home ownership. *Journal of Social Policy* 38, 83–102.

Smith, T. R. 1984. Artificial intelligence and its applicability to geographical problem solving. *Professional Geographer* 36, 147–58.

Smith, T. R., Clark, W. A. V. and Cotton, J. W. 1984. Deriving and testing production system models of sequential decision-making behavior. *Geographical Analysis* 16, 191–222.

Smith, W. 1949. *An Economic Geography of Great Britain*. London: Methuen.

Soffer, A. and Minghi, J. 1986. Israel's security landscapes: The impact of military considerations on land uses. *Professional Geographer* 38, 28–41.

Soja, E. W. 1968. *The Geography of Modernization in Kenya*. Syracuse: Syracuse University Press.

Soja, E. W. 1980. The socio-spatial dialectic. *Annals of the Association of American Geographers* 70, 207–25.

Soja, E. W. 1985. The spatiality of social life: Towards a transformative retheorization. In D. Gregory and J. Urry (eds) *Social Relations and Spatial Structures*. London: Macmillan, 90–127.

Soja, E. W. 1989. *Postmodern Geographies: The reassertion of space in critical social theory*. London: Verso.

Soja, E. W. 2001a. Postmodernism in geography. In N. J. Smelser and P. B. Baltes (eds) *International Encyclopedia of the Social and Behavioral Sciences*. Oxford: Elsevier Science, 11860–5.

Soja, E. W. 2001b. Afterword. In C. Minca (ed.) *Postmodern Geography: Theory and praxis*. Oxford: Blackwell, 282–94.

Soja, E. W. and Hooper, B. 1993. The spaces that difference makes: Some notes on the geographical margins of the new cultural politics. In M. Keith and S. Pile (eds) *Place and the Politics of Identity*. London: Routledge, 183–205.

Sparke, M. 1994a. Negotiating national action: Free trade, constitutional debate and the gendered geopolitics of Canada. *Political Geography* 15, 615–39.

Sparke, M. 1994b. Writing on patriarchal missiles: The chauvinism of the 'Gulf War' and the limits of critique. *Environment and Planning A* 26, 1061–89.

Sparke, M. 2013. From global dispossession to local repossession: Towards a wordly cultural geography of occupy activism. In N. C. Johnson, R. H. Schein and J. Winders (eds) *The Wiley-Blackwell Companion to Cultural Geography*. Chichester: Wiley-Blackwell, 387–408.

Spate, O. H. K. 1957. How determined is possibilism? *Geographical Studies* 4, 3–12.

Spate, O. H. K. 1960a. Quantity and quality in geography. *Annals of the Association of America Geographers* 50, 477–94.

Spate, O. H. K. 1960b. Lord Kelvin rides again. *Economic Geography* 36, facing page 1.

Spate, O. H. K. 1963. Letter to the editor. *Geography* 48, 206.

Spate, O. H. K. 1989. Foreword. In F. W. Boal and D. N. Livingstone (eds) *The Behavioural Environment*. London: Routledge, xvii–xx.

Spencer, H. 1892. *A System of Synthetic Philosophy, Vol. 1. First principles*, 4th edn. New York: Appleton.

Spencer, J. E. 1979. A geographer west of the Sierra Nevada. *Annals of the Association of American Geographers* 69, 46–52.

Speth, W. W. 1999. *How it Came to Be: Carl O. Sauer, Franz Boas and the meaning of anthropogeography*. Berkeley, CA: University of California Press.

Springer, S. 2013. Anarchism! What geography still ought to be. *Antipode: A Radical Journal of Geography* 44, 1605–24.

Springer, S. 2014. Why a radical geography must be anarchist. *Dialogues in Human Geography* 3, 249–70.

Staeheli, L. A. and Cope, M. 1994. Empowering women's citizenship. *Political Geography* 13, 443–60.

Staeheli, L. A. and Nagar, R. 2002. Feminists talking across worlds. *Gender, Place and Culture* 9, 167–72.

Staeheli, L. A. and Mitchell, D. 2005. The complex politics of relevance in geography. *Annals of the Association of American Geographers* 95, 357–72.

Stamp, L. D. 1934. Planning the land for the future. *Science* 80, 510–12.

Stamp, L. D. 1946a. *The Land of Britain*. London: Longman.

Stamp, L. D. 1946b. The place of science in town and country planning. *Advancement of Science* 3, 337–48.

Stamp, L. D. 1948. Applied geography. In L. D. Stamp and S. W. Wooldridge (eds) *London Essays in Geography: Rodwell Jones memorial volume*. London: Longmans, Green, 1–18.

Stamp, L. D. 1949. The planning of land use. *Advancement of Science* 6, 224–32.

Stamp, L. D. 1960. *Applied Geography*. Harmondsworth: Penguin Books.

Stamp, L. D. 1966. Ten years on. *Transactions of the Institute of British Geographers* 40, 11–20.

Stamp, L. D. and Beaver, S. H. 1947. *The British Isles*. London: Longman.

Stannard, K. 2003. Commentary. Earth to academia: On the need to reconnect university and school geography. *Area* 35, 316–22.

Steel, R. W. (ed.) 1987. *British Geography 1918–1945*. Cambridge: Cambridge University Press.

Steel, R. W. 1974. The Third World: Geography in practice. *Geography* 59, 189–207.

Steel, R. W. 1982. Regional geography in practice. *Geography* 67, 2–8.

Stegmuller, W. 1976. *The Structure and Dynamics of Theories*. New York: Springer-Verlag.

Steinitz, C. 2014. The beginnings of geographical information systems: A personal historical perspective. *Planning Perspectives* 29, 239–54.

Stevens, A. 1921. *Applied Geography*. Glasgow: Blackie.

Stewart, J. Q. 1945. *Coasts, Waves and Weather*. Boston, MA: Ginn & Co.

Stewart, J. Q. 1947. Empirical mathematical rules concerning the distribution and equilibrium of population. *Geographical Review* 37, 461–85.

Stewart, J. Q. 1956. The development of social physics. *American Journal of Physics* 18, 239–53.

Stewart, J. Q. and Warntz, W. 1958. Macrogeography and social science. *Geographical Review* 48, 167–84.

Stewart, J. Q. and Warntz, W. 1959. Physics of population distribution. *Journal of Regional Science* 1, 99–123.

Stiell, B. and England, K. 1997. Domestic distinctions: Constructing difference among paid domestic workers in Toronto. *Gender, Place and Culture* 4, 339–59.

Stimson, R. J. 2012. You don't need sight to have vision: Reginald G. Golledge was a giant in analytical human geography. *Annals of the Association of American Geographers* 102, 234–43.

Stoddart, D. R. 1965. Geography and the ecological approach: The ecosystem as a geographic principle and method. *Geography* 50, 242–51.

Stoddart, D. R. 1966. Darwin's impact on geography. *Annals of the Association of American Geographers* 56, 683–98.

Stoddart, D. R. 1967a. Growth and structure of geography. *Transactions of the Institute of British Geographers* 41, 1–19.

Stoddart, D. R. 1967b. Organism and ecosystem as geographic models. In R. J. Chorley and P. Haggett (eds) *Models in Geography*. London: Methuen, 511–47.

Stoddart, D. R. 1975a. The RGS and the foundations of geography at Cambridge. *Geographical Journal* 141, 216–39.

Stoddart, D. R. 1975b. Kropotkin, Reclus and relevant geography. *Area* 7, 188–90.

Stoddart, D. R. 1977. The paradigm concept and the history of geography. Abstract of a paper for the conference of the International Geographical Union Commission on the History of Geographic Thought, Edinburgh.

Stoddart, D. R. 1981a. The paradigm concept and the history of geography. In D. R. Stoddart (ed.) *Geography, Ideology and Social Concern*. Oxford: Blackwell, 70–80.

Stoddart, D. R. 1981b. Ideas and interpretation in the history of geography. In D. R. Stoddart (ed.) *Geography, Ideology and Social Concern*. Oxford: Blackwell, 1–7.

Stoddart, D. R. 1986. *On Geography: And its history*. Oxford: Basil Blackwell.

Stoddart, D. R. 1987. To claim the high ground: Geography for the end of the century. *Transactions of the Institute of British Geographers* NS12, 327–36.

Stoddart, D. R. 1990. Epilogue: Homage to Richard Hartshorne. In J. N. Entrikin and S. D. Brunn (eds) *Reflections on Richard Hartshorne's The Nature of Geography*. Washington, DC: Association of American Geographers, 163–6.

Stoddart, D. R. 1991. Do we need a feminist historiography of geography – and if we do, what should it be? *Transactions of the Institute of British Geographers* NS16, 484–7.

Stoddart, D. R. 1997a. Richard J. Chorley and modern geomorphology. In D. R. Stoddart (ed.) *Process and Form in Geomorphology*. London: Routledge, 383–99.

Stoddart, D. R. 1997b. Carl Sauer: Geomorphologist. In D. R. Stoddart (ed.) *Process and Form in Geomorphology*. London: Routledge, 340–79.

Stoddart, D. R. 2001. Be of good cheer, my weary readers, for I have espied land. *Atoll Research Bulletin* 494, 234–72.

Stone, K. H. 1979. Geography's wartime service. *Annals of the Association of American Geographers* 69, 89–97.

Storey, D. W. 2001. *Territory: The claiming of space*. Harlow: Longman.

Storper, M. 1987. The post-Enlightenment challenge to Marxist urban studies. *Environment and Planning D: Society and Space* 5, 418–26.

Storper, M. 2001. The poverty of radical theory today: From the false promises of Marxism to the mirage of the cultural turn. *International Journal of Urban and Regional Studies* 25, 155–79.

Stouffer, S. A. 1940. Intervening opportunities: A theory relating mobility and distance. *American Sociological Review* 5, 845–67.

Strange, C. and Bashford, A. 2008. *Griffith Taylor: Visionary environmentalist explorer*. Toronto: University of Toronto Press.

Subramanian, S. V., Jones, K., Kaddour, A. and Krieger, N. 2008. Revisiting Robinson: The perils of individualistic and ecologic fallacy. *International Journal of Epidemiology* 38, 342–60.

Sui, D. Z. 1994. GIS and urban studies: Positivism, post-positivism and beyond. *Urban Geography* 15, 258–78.

Sui, D. Z. 1999. Postmodern urbanism disrobed: Or why postmodern urbanism is a dead end for urban geography. *Urban Geography* 20, 403–11.

Sui, D. Z. 2004a. GIS, cartography, and the 'third culture': Geographic imaginations in the computer age. *Professional Geographer* 56, 62–72.

Sui, D. Z. 2004b. Tobler's first law of geography: A big idea for a small world? *Annals of the Association of American Geographers* 94, 269–77.

Sui, D. Z. (ed.) 2011. Focus: Discussions on NRC report's section: Strategic directions for the geographical sciences. *Professional Geographer* 63, 305–42.

Sui, D. Z. 2012. Looking through Hägerstrand's dual vistas: Towards a unifying framework for time geography. *Journal of Transport Geography* 23, 5–16.

Sui, D. Z., Fotheringham, A. S., Anselin, L., O'Loughlin, J. and King, G. 2000. Book review forum on *A Solution to the Ecological Problem* by Gary King. *Annals of the Association of American Geographers* 90, 579–606.

Sullivan, R. 2011. *Geography Speaks: Performative aspects of geography*. Farnham: Ashgate.

Summerfield, M. A. 1983. Population, samples and statistical inference in geography. *Professional Geographer* 35, 143–8.

Suppe, F. 1977a. The search for philosophic understanding of scientific theories. In F. Suppe (ed.) *The Structure of Scientific Theories*. Urbana, IL: University of Illinois Press, 3–233.

Suppe, F. 1977b. Exemplars, theories and disciplinary matrices. In F. Suppe (ed.) *The Structure of Scientific Theories*. Urbana, IL: University of Illinois Press, 473–99.

Suppe, F. 1977c. Afterword – 1977. In F. Suppe (ed.) *The Structure of Scientific Theories*. Urbana, IL: University of Illinois Press, 617–730.

Sviatlovsky, E. E. and Eels, W. C. 1937. The centrographical method and regional analysis. *Geographical Review* 27, 240–54.

Swanton, D. 2010. Sorting bodies: Race, affect, and everyday multiculture in a mill town in northern England. *Environment and Planning A* 42, 2332–50.

Swartz, D. 1997. *Culture and Power: The sociology of Pierre Bourdieu*. London and Chicago: University of Chicago Press.

Symankski, R. 1994. Why we should fear postmodernists. *Annals of the Association of American Geographers* 84, 301–4.

Symanski, R. 2002. *Geography Inside Out*. Syracuse: Syracuse University Press.

Symanski, R. 2007. The world's foremost living cultural geographer. In R. Symanski, *Irreverent Essays on Geographers*. Irvine, CA: Estrilda, 193–210.

Symanski, R. and Agnew, J. A. 1981. *Order and Skepticism: Human geography and the dialectic of science*. Washington, DC: Association of American Geographers.

Symanski, I. R. and Pickard, J. 1996. Rules by which we are judged. *Progress in Human Geography* 20, 175–82.

Taaffe, E. J. 1970. *Geography*. Englewood Cliffs, NJ: Prentice-Hall.

Taaffe, E. J. 1974. The spatial view in context. *Annals of the Association of American Geographers* 64, 1–16.

Taaffe, E. J. 1979. In the Chicago area. *Annals of the Association of American Geographers* 69, 133–8.

Taaffe, E. J. 1993. Spatial analysis: Development and outlook. *Urban Geography* 14, 422–33.

Taaffe, E. J., Morrill, R. L. and Gould, P. R. 1963. Transport expansion in underdeveloped countries: A comparative analysis. *Geographical Review* 53, 503–29.

Tambolo, L. 2015. A tale of three theories: Feyerabend and Popper on progress and the aim of science. *Studies in the History and Philosophy of Science* 51, 33–41.

Tarrant, J. R. 1968. Computers in geography. *Institute of British Geographers, Newsletter* 6, 11–25.

Tatham, G. 1953. Environmentalism and possibilism. In G. Taylor (ed.) *Geography in the Twentieth Century*. London: Methuen, 128–64.

Taylor, P. J. 1971. Distance decay curves and distance transformations. *Geographical Analysis* 3, 221–38.

Taylor, P. J. 1976. An interpretation of the quantification debate in British geography. *Transactions of the Institute of British Geographers* NS1, 129–42.

Taylor, P. J. 1977. *Quantitative Methods in Geography*. New York: Harper & Row.

Taylor, P. J. 1978. Political geography. *Progress in Human Geography* 2, 53–62.

Taylor, P. J. 1979. 'Difficult-to-let', 'difficult-to-live-in', and sometimes 'difficult-to-get-out-of': An essay on the provision of council housing. *Environment and Planning A* 11, 1305–20.

Taylor, P. J. 1982. A materialist framework for political geography. *Transactions of the Institute of British Geographers* NS7, 15–34.

Taylor, P. J. 1985a. The value of a geographical perspective. In R. J. Johnston (ed.) *The Future of Geography*. London: Methuen, 92–110.

Taylor, P. J. 1985b. The geography of elections. In M. Pacione (ed.) *Progress in Political Geography*. London: Croom Helm, 243–72.

Taylor, P. J. 1988. History's dialogue: An exemplification from political geography. *Progress in Human Geography* 12, 1–14.

Taylor, P. J. 1990a. Journeyman editor. *Professional Geographer* 42, 359–60.

Taylor, P. J. 1990b. Editorial comment: GKS. *Political Geography Quarterly* 9, 211–12.

Taylor, P. J. 1990c. *Britain and the Cold War: 1945 as geopolitical transition*. London: Belhaven Press.

Taylor, P. J. 1993. Full circle, or new meaning for the global? In R. J. Johnston (ed.) *The Challenge for Geography: A changing world: A changing discipline*. Oxford: Blackwell Publishers, 181–97.

Taylor, P. J. 1994. The state as container: Territoriality in the modern world-system. *Progress in Human Geography* 18, 151–62.

Taylor, P. J. 1995. Beyond containers: Internationality, interstateness, interterritoriality. *Progress in Human Geography* 19, 1–15.

Taylor, P. J. 1996. Embedded statism and the social sciences: Opening up to new spaces. *Environment and Planning A* 28, 1917–28.

Taylor, P. J. 1999. Spaces, places and Macy's: Place-space tensions in the political geography of modernities. *Progress in Human Geography* 23, 7–26.

Taylor, P. J. 2003. Radical political geographies. In J. Agnew, K. Mitchell and G. Toal (eds) *A Companion to Political Geography*. Oxford: Blackwell, 47–58.

Taylor, P. J. and Goddard, J. (eds) 1974. Geography and statistics: An introduction. *Statistician* 23, 149–56.

Taylor, P. J. and Gudgin, G. 1976. A statistical theory of electoral redistricting. *Environment and Planning A* 8, 43–58.

Taylor, P. J. and Johnston, R. J. 1979. *Geography of Elections*. Harmondsworth: Penguin Books.

Taylor, P. J. and Overton, M. 1991. Further thoughts on geography and GIS. *Environment and Planning A* 23, 1087–94.

Taylor, T. G. 1927. *Environment and Race*. London: Oxford University Press.

Taylor, T. G. 1937. *Environment, Race and Migration*. Toronto: University of Toronto Press.

Taylor, T. G. 1957. Introduction: The scope of the volume. In G. Taylor (ed.) *Geography in the Twentieth Century: A study of growth, fields, techniques, aims and trends*. London: Methuen, 1–27.

Taylor, T. G. 1958. *Journeyman Taylor: The education of a scientist*. London: Robert Hale.

Tesch, R. 1990. *Qualitative Research: Analysis types and research tools*. Brighton: Falmer Press.

Tewdwr-Jones, M., Phelps, N. A. and Freestone, R. 2014. *The Planning Imagination: Peter Hall and the study of urban and regional planning*. London: Routledge.

Thien, D. 2005. After or beyond feeling? A consideration of affect and emotion in geography. *Area* 37, 450–6.

Thien, D. 2009. Feminist methodologies. In R. Kitchin and N. Thrift (eds) *International Encyclopedia of Human Geography, Vol. 1*. Oxford: Elsevier, 71–8.

Thoman, R. S. 1965. Some comments on *The Science of Geography*. *Professional Geographer* 17(6), 8–10.

Thomas, E. N. 1960. Areal associations between population growth and selected factors in the Chicago urbanized area. *Economic Geography* 36, 158–70.

Thomas, E. N. and Anderson, D. L. 1965. Additional comments on weighting values in correlation analysis of areal data. *Annals of the Association of American Geographers* 55, 492–505.

Thomas, R. W. 1982. *Information Statistics in Geography*. Norwich: Geo Books.

Thomas, R. W. 1992. *Geomedical Systems: Intervention and control*. Routledge: London.

Thomas, W. L. Jr. (ed.) 1956. *Man's Role in Changing the Face of the Earth*. Chicago: University of Chicago Press.

Thompson, D'Arcy W. 1917. *On Growth and Form*, 1st edn. Cambridge: Cambridge University Press.

Thompson, J. H., Sufrin, S. C., Gould, P. R., Buck, M. A. 1962. Toward a geography of economic health: The case of New York state. *Annals of the Association of American Geographers* 52, 1–20.

Thorne, C. R. (ed.) 1993. University Funding Council Research Selectivity Exercise, 1992: Implications for higher education in geography. *Journal of Geography in Higher Education* 17, 167–99.

Thornes, J. B. 1989a. Geomorphology and grass roots models. In B. Macmillan (ed.) *Remodelling Geography*. Oxford: Basil Blackwell, 3–21.

Thornes, J. B. 1989b. Environmental systems. In M. J. Clark, K. J. Gregory and A. M. Gurnell (eds) *Horizons in Physical Geography*. London: Macmillan, 27–46.

Thrall, G. I. 1985. Scientific geography. *Area* 17, 254.

Thrall, G. I. 1986. Reply to Felix Driver and Christopher Philo. *Area* 18, 162–3.

Thrift, N. J. and Olds, K. 1996. Refiguring the economic in economic geography. *Progress in Human Geography* 20, 311–37.

Thrift, N. J. 1977. An introduction to time geography. *CATMOG* 13, Norwich: Geo Books.

Thrift, N. J. 1979. Unemployment in the inner city: Urban problem or structural imperative? A review of the British experience. In D. T. Herbert and R. J. Johnston (eds) *Geography and Urban Environment: Progress in Research and Applications, Vol. 2*. London: John Wiley, 125–226.

Thrift, N. J. 1981. Behavioural geography. In N. Wrigley and R. J. Bennett (eds) *Quantitative Geography*. London: Routledge & Kegan Paul, 352–65.

Thrift, N. J. 1983. On the determination of social action in space and time. *Environment and Planning D: Society and Space* 1, 23–57.

Thrift, N. J. 1987. No perfect symmetry. *Environment and Planning D: Society and Space* 5, 400–7.

Thrift, N. J. 1990. For a new regional geography 1. *Progress in Human Geography* 14, 272–9.

Thrift, N. J. 1991. For a new regional geography 2. *Progress in Human Geography* 15, 456–565.

Thrift, N. J. 1993. For a new regional geography 3. *Progress in Human Geography* 17, 92–100.

Thrift, N. J. 1994. Taking aim at the heart of the region. In D. Gregory, R. Martin and G. Smith (eds) *Human Geography: Society, space and social science*. London: Macmillan, 200–31.

Thrift, N. J. 2000. Pandora's box? Cultural geographies of economies. In G. L. Clark, M. P. Feldmann and M. S. Gertler (eds) *The Oxford Handbook of Economic Geography*. Oxford: Oxford University Press, 689–704.

Thrift, N. J. 2002. The future of geography. *Geoforum* 33, 291–8.

Thrift, N. J. 2005. Hägerstrand and social theory. *Progress in Human Geography* 29, 337–40.

Thrift, N. J. 2008. *Non-representational Theory: Space/politics/affect*. Abingdon and New York: Routledge.

Thrift, N. J. and Dewsbury, J.-D. 2000. Dead geographies – and how to make them live. *Environment and Planning D: Society and Space* 18, 411–32.

Thrift, N. J. and Olds, K. 1996. Refiguring the economic in economic geography. *Progress in Human Geography* 20, 311–17.

Thrift, N. J. and Pred, A. R. 1981. Time geography: A new beginning. *Progress in Human Geography* 5, 277–86.

Tickell, A. and Peck, J. 1992. Accumulation, regulation and the geographies of post-Fordism: Missing links in regulationist research. *Progress in Human Geography* 16, 190–218.

Tickell, A. and Peck, J. 1996. The return of the Manchester Men: Men's words and men's deeds in the remaking of the local state. *Transactions of the Institute of British Geographers* NS21, 595–616.

Tiefelsdof, M. 2003. Misspecifications in interaction model distance decay relations: A spatial structure effect. *Journal of Geographical Systems* 5, 25–50.

Timmermans, H. J. P. and Golledge, R. G. 1990. Applications of behavioural research on spatial problems: II Preference and choice. *Progress in Human Geography* 14, 311–54.

Timmermans, H. J. P., Aremze, T. and Joh, C.-H. 2001. Analysing space-time behaviour: New approaches to old problems. *Progress in Human Geography* 26, 175–90.

Timmins, N. 1995. *The Five Giants: A biography of the welfare state*. London: HarperCollins.

Timms, D. 1965. Quantitative techniques in urban social geography. In R. J. Chorley and P. Haggett (eds) *Frontiers in Geographical Teaching*. London: Methuen, 239–65.

Timms, D. 1971. *The Urban Mosaic: Towards a theory of residential differentiation*. Cambridge: Cambridge University Press.

Tivers, J. 1978. How the other half lives: The geographical study of women. *Area* 10, 302–6.

Tobler, W. R. 1959. Automation and cartography. *Geographical Review* 49, 526–34.

Tobler, W. R. 1970. A computer movie simulating urban growth in the Detroit region. *Economic Geography* 46, 234–40.

Tobler, W. R. 1995. Migration: Ravenstein, Thornthwaite and beyond. *Urban Geography* 16, 327–43.

Tobler, W. R. 2002. Ma vie: Growing up in America and Europe. In P. Gould and F. R. Pitts (eds) *Geographical Voices: Fourteen autobiographical essays*. Syracuse: Syracuse University Press, 293–322.

Tocalis, T. R. 1978. Changing theoretical foundations of the gravity concept of human interaction. In B. J. L. Berry (ed.) *The Nature of Change in Geographical Ideas*. DeKalb, IL: Northern Illinois University Press, 65–124.

Tolia-Kelly, D. P. 2006. Affect – an ethnocentric encounter? Exploring the 'universalist' imperative of emotional/affectual geographies. *Area* 38, 213–17.

Tomlinson, R. F. 1989. Geographic information systems and geographers in the 1990s. *Canadian Geographer* 33, 290–8.

Tomlinson, R. F. 2007. *Thinking about GIS*, 3rd edn. Redlands: ESRI Press.

Tomlinson, R. F. 2009. Changing the face of geography: GIS and the IGU. *ArcNews Online*. Available at: www.esri.com/news/arcnews/spring09articles/changing-the-face.html

Toulmin, S. E. 1970. Does the distinction between normal and revolutionary science hold water? In I. Lakatos and A. Musgrave (eds) *Criticism and the Growth of Knowledge*. London: Cambridge University Press, 39–48.

Trewartha, G. T. 1973. Comments on geography and public policy. *Professional Geographer* 25, 78–9.

Trudgill, S. T. 1990. *Barriers to a Better Environment*. London: Belhaven Press.

Tso, B. and Mather, P. M. 2001. *Classification Methods for Remotely Sensed Data*. London: Taylor & Francis.

Tuan, Y.-F. 1968. *The Hydrological Cycle and the Wisdom of God: A theme in geoteleology*. Toronto: University of Toronto Press.

Tuan, Y.-F. 1971. Geography, phenomenology, and the study of human nature. *Canadian Geographer* 15, 181–92.

Tuan, Y.-F. 1974a. Space and place: Humanistic perspectives. In C. Board, R. J. Chorley, P. Haggett and D. R. Stoddart (eds) *Progress in Geography*, Vol. 6. London: Edward Arnold, 211–52.

Tuan, Y.-F. 1974b. *Topophilia: A study of environmental perception, attitudes and values*. Englewood Cliffs, NJ: Prentice-Hall.

Tuan, Y.-F. 1975a. Images and mental maps. *Annals of the Association of American Geographers* 65, 205–13.

Tuan, Y.-F. 1975b. Place: An experiential perspective. *Geographical Review* 65, 151–65.

Tuan, Y.-F. 1976. Humanistic geography. *Annals of the Association of American Geographers* 66, 266–76.

Tuan, Y.-F. 1977. *Space and Place*. London: Edward Arnold.

Tuan, Y.-F. 1978. Literature and geography: Implications for geographical research. In D. Ley and M. S. Samuels (eds) *Humanistic Geography: Prospects and problems*. Chicago: Maaroufa Press, 194–206.

Tuan, Y.-F. 1979. *Landscapes of Fear*. Oxford: Basil Blackwell.

Tuan, Y.-F. 1982. *Segmented Worlds and Self*. Minneapolis, MN: University of Minnesota Press.

Tuan, Y.-F. 1984. *Dominance and Affection*. New Haven, CT: Yale University Press.

Tuan, Y.-F. 1999. *Who am I? An autobiography of emotion, mind and spirit*. Madison, WI: The University of Wisconsin Press.

Tuan, Y.-F. 2002a. *Dear Colleague: Common and uncommon observations*. Minneapolis, MN and London: University of Minnesota Press.

Tuan, Y.-F. 2002b. A life of learning. In P. Gould and F. R. Pitts (eds) *Geographical Voices: Fourteen autobiographical essays*. Syracuse: Syracuse University Press, 323–40.

Tullock, G. 1976. *The Vote Motive*. London: Institute of Economic Affairs.

Turner, A. 2006. *Introduction to Neogeography*. Sebastopol, CA: O'Reilly.

Turner, B. L. II. 1989. The specialist-synthesis approach to the revival of geography: The case of cultural ecology. *Annals of the Association of American Geographers* 79, 88–100.

Turner, B. L. II. 2002. Contested identities: Human-environment geography and disciplinary implications in a restructuring academy. *Annals of the Association of American Geographers* 92, 52–74.

Turner, B. L. II. 2005. Geography's profile in public debate: 'Inside the beltway' and the national academies. *Professional Geographer* 57, 462–7.

Turner, B. L. II and Meyer, W. B. 1985. The use of citation indices in comparing geography programs: An exploratory study. *Professional Geographer* 37, 271–8.

Turner, B. L. II and Varlyguin, D. 1995. Foreign-area expertise in U.S. geography: An assessment of capacity based on foreign-area dissertations, 1977–1991. *Professional Geographer* 47, 308–14.

Turner, B. L. II, Clark, W. C., Kates, R. W., Richards, J. F., Matthews, J. T. and Meyer, W. B. (eds) 1990. *The Earth as Transformed by Human Action: Global and regional changes in the biosphere over the past 300 years*. Cambridge: Cambridge University Press.

Turner, R. K. 1993. Sustainability: Principles and practice. In R. K. Turner (ed.) *Sustainable Environmental Economics and Management: Principles and practice*. London: Belhaven Press, 3–36.

Ullman, E. L. 1941. A theory of location for cities. *American Journal of Sociology* 46, 853–64.

Ullman, E. L. 1953. Human geography and area research. *Annals of the Association of American Geographers* 43, 54–66.

Ullman, E. L. 1954a. Geography as spatial interaction. In *Interregional Linkages: Proceedings of the Western Committee on Regional Economic Analysis*. Berkeley, CA: University of California Press, 63–71. Reprinted in R. R. Boyce (ed.) 1980, 13–27.

Ullman, E. L. 1954b. Transportation geography. In P. E. James and C. F. Jones (eds) *American Geography: Inventory and prospect*. Syracuse: Syracuse University Press, 310–33.

Ullman, E. L. 1956. The role of transportation and the bases for interaction. In W. L. Thomas (ed.) *Man's Role in Changing the Face of the Earth*. Chicago: University of Chicago Press, 862–80.

Ullman, E. L. 1959 [1957]. *American Commodity Flow: A geographical interpretation of rail and water traffic based on principles of spatial interchange*. Seattle, WA: University of Washington Press.

Ullman, E. L. 1962. Presidential address: The nature of cities reconsidered. *Papers in Regional Science* 9, 7–23.

Unstead, J. F. 1933. A system of regional geography. *Geography* 18, 175–87.

Unwin, A. 1996. Exploratory spatial analysis and local statistics. *Computing and Statistics* 11, 387–400.

Unwin, D. J. 1981. *Introductory Spatial Analysis*. London: Methuen.

Uunwin, D. J. 1994. Cartography, VisC and GIS. *Progress in Human Geography* 18, 516–22.

Unwin, D. J. 2005. Fiddling on a different planet. *Geoforum* 36, 681–84.

Unwin, D. J. and Dawson, J. A. 1985. *Computer Programming for Geographers*. London: Longman.

Unwin, D. J. and Hepple, L. W. 1974. The statistical analysis of spatial series. *Statistician* 23, 211–28.

Unwin, T. 1992. *The Place of Geography*. Harlow: Longman.

Upton, G. J. G. and Fingleton, B. 1985. *Spatial Data Analysis by Example, Vol. 1: Point pattern and quantitative data*. Chichester: John Wiley.

Upton, G. J. G. and Fingleton, B. 1989. *Spatial Data Analysis by Example, Vol. 2: Categorical and directional data*. Chichester: John Wiley.

Urlich-Cloher, D. U. 1972. Migrations of the North Island Maoris 1800–1840: A systems view of migration. *New Zealand Geographer* 28, 23–35.

Urlich-Cloher, D. U. 1975. A perspective on Australian urbanization. In J. M. Powell and M. Williams (eds) *Australian Space, Australian Time: Geographical perspectives*. Melbourne: Oxford University Press, 104–59.

Urry, J. 1985. Space, time and the study of the social. In H. Newby *et al.* (eds) *Restructuring Capital*. London: Macmillan, 21–40.

Urry, J. 1986. Locality research: The case of Lancaster. *Regional Studies* 20, 233–42.

Urry, J. 1987. Society, space and locality. *Environment and Planning D: Society and Space* 5, 435–44.

Vaiou, D. 2003. Guest editorial. Radical debate between 'local' and 'international': A view from the periphery. *Environment and Planning D: Society and Space* 21, 133–7.

Vale, T. R. 2002. From Clements and Davis to Gould and Botkin: Ideals of progress in physical geography. In R. D. Sack (ed.) *Progress: Geographical essays*. Baltimore, MD: Johns Hopkins University Press, 1–21.

Valentine, G. 1993a. (Hetero)sexing space: Lesbian perceptions and experiences of everyday spaces. *Environment and Planning D: Society and Space* 11, 395–413.

Valentine, G. 1993b. Negotiating and maintaining multiple sexual identities. *Transactions of the Institute of British Geographers* NS18, 237–43.

Valentine, G. 1998. 'Sticks and stones may break my bones': A personal geography of harassment. *Antipode: A Radical Journal of Geography* 30, 305–32.

Valentine, G. 2003. Geography and ethics: in pursuit of social justice. *Progress in Human Geography* 27, 375–80.

Valentine, G. 2005. Geography and ethics: Moral geographies? Ethical commitment in research and teaching. *Progress in Human Geography* 29, 483–87.

Van den Daele, W. and Weingart, P. 1976. Resistance and receptivity of science to external direction: The emergence of new disciplines under the impact of science policy. In G. Lemaine *et al.* (eds) *Perspectives on the Emergence of Scientific Disciplines*. The Hague: Mouton, 247–75.

Van der Laan, L. and Piersma, A. 1982. The image of man: Paradigmatic cornerstone in human geography. *Annals of the Association of American Geographers* 73, 411–26.

Van der Wusten, H. and O'Loughlin, J. 1986. Claiming new territory for a stable peace: How geography can contribute. *Professional Geographer* 38, 18–27.

Van Hoven, B. and Hörschelmann, K. (eds) 2005. *Spaces of Masculinities*. London and New York: Routledge.

Van Paassen, C. 1981. The philosophy of geography: From Vidal to Hägerstrand. In A. Pred and G. Tornquist (eds) *Space and Time in Geography*. Lund: C. W. K. Gleerup, 17–29.

Vance, J. E. 1970. *The Merchant's World*. Englewood Cliffs, NJ: Prentice-Hall.

Vance, J. E. 1978. Geography and the study of cities. Human geography: Coming of age. *American Behavioral Scientist* 22, 131–49.

Velikonja, J. 1994. Geography at the University of Washington. *Yearbook of the Association of Pacific Coast Geographers*, 56.

Vidal de la Blache, P. 1903. *Tableau de la Géographie de la France*. Paris: Hachette.

von Bertalanffy, L. 1950. An outline of general systems theory. *British Journal of the Philosophy of Science* 1, 134–65.

von Humboldt, A. 1845–1862. *Kosmos, Entwurf einer physichien Weltbescheibung*. Stuttgart: J. C. Gotta Verlag.

Wagner, P. L. 1976. Reflections on a radical geography. *Antipode: A Radical Journal of Geography* 8(3), 83–5.

Wainwright, E., Barker, J., Ansell, N., Buckingham, S., Hemming, P. and Smith, F. 2014. Geographers out of place: Institutions, (inter)disciplinarity and identity. *Area* 46, 410–17.

Wainwright, J. 2010. On Gramsci's 'conceptions of the world'. *Transactions of the Institute of British Geographers* 35, 507–21.

Wainwright, J. 2013. *Geopiracy: Oaxaca, militant empiricism, and geographical thought*. New York: Palgrave Macmillan.

Wakefield, J., Quinn, M. and Raab, G. (eds) 2001. Disease clusters and ecological studies. Special issue, *Journal of the Royal Statistical Society Series A: Statistics in Society* 164, 1–207.

Walby, S. 1986. *Patriarchy at Work*. Cambridge: Polity Press.

Walford, R. 2001. *Geography in British Schools 1850–2000: Making a world of difference*. London: Woburn Press.

Walker, R. A. 1981a. A theory of suburbanization. In M. J. Dear and A. J. Scott (eds) *Urbanization and Urban Planning in Capitalist Societies*. London: Methuen, 383–430.

Walker, R. A. 1981b. Left-wing libertarianism, an academic disorder: A reply to David Sibley. *Professional Geographer* 33, 5–9.

Walker, R. A. 1989a. What's left to do? *Antipode* 21, 133–65.

Walker, R. A. 1989b. Geography from the left. In G. L. Gaile and C. J. Willmott (eds) *Geography in America*. Columbus, OH: Merrill Publishing Company, 619–51.

Wallace, I. 1989. *The Global Economic System*. London: Unwin Hyman.

Walmsley, D. J. 1972. *Systems Theory: A framework for human geographical enquiry*. Research School of Pacific Studies. Department of Human Geography Publication HG/7, Canberra: Australian National University.

Walmsley, D. J. 1974. Positivism and phenomenology in human geography. *Canadian Geographer* 18, 95–107.

Walmsley, D. J. and Sorensen, A. D. 1980. What Marx for the radicals? An antipodean viewpoint. *Area* 12, 137–41.

Ward, D. 1971. *Cities and Immigrants: A geography of change in nineteenth-century America*. New York: Oxford University Press.

Ward, K. 2006. Geography and public policy: Towards public geographies. *Progress in Human Geography* 30, 495–503.

Ward, K. 2007. 'Public intellectuals', geography, its representations and publics. *Geoforum* 38, 1058–64.

Ward, K. 2001. The re-interpretation of urban politics: Three authors, four papers and the 'shibboleth of regulation theory'. *Transactions of the Institute of British Geographers* NS26, 127–33.

Ward, K., Blunt, A., Norcup, J., Sidaway, J. D., Withers, C. W. J. and Maddrell, A. 2010. Book Review Forum: Complex locations: Women's geographical work in the UK 1850–1970 – by Avril Maddrell. *Area* 42, 394–400.

Ward, K., Johnston, R. J., Richards, K., Gandy, M., Taylor, Z., Paasi, A., Fox, R., Serje, M., Yeung, H.W.C., Barnes, T., Blunt, A. and McDowell, L. 2009. The future of research monographs: An international set of perspectives. *Progress in Human Geography* 33, 101–26.

Ward, M. D. and Gleditsch, K. S. 2002. Location, location, location: An MCMC approach to modelling the spatial context of war and peace. *Political Analysis* 10, 244–60.

Wardenga, U. 1999. Constructing regional knowledge in German geography: The Central Commission on the Regional Geography of Germany, 1882–1941. In A. Buttimer, S. D. Brunn and U. Wardenga (eds) *Text and Image: Social construction of regional knowledges*. Leipzig: Leibniz Institute for Regional Geography, Beiträge zur regionalen Geographie 49, 77–84.

Warf, B. 1986. Ideology, everyday life and emancipatory phenomenology. *Antipode: A Radical Journal of Geography* 18, 268–83.

Warf, B. 1988. The resurrection of local uniqueness. In R. G. Golledge, H. Couclelis and P. R. Gould (eds) *A Ground for Common Search*. Santa Barbara, CA: The Santa Barbara Geographical Press, 51–62.

Warf, B. 1993. Postmodernism and the localities debate: Ontological questions and epistemological implications. *Tijdschrift voor Economische en Sociale Geografie* 84, 162–8.

Warf, B. 1995. Separated at birth? Regional science and social theory. *International Regional Science Review* 17, 185–94.

Warf, B. 2006. Humanistic geography. In B. Warf (ed.) *Encyclopedia of Human Geography*. London: Thousand Oaks, CA and New Delhi: Sage, 233–6.

Warf, B. and Arias, S. (eds) 2009. *The Spatial Turn: Interdisciplinary perspectives*. London: Routledge.

Warntz, W. 1959a. *Toward a Geography of Price*. Philadelphia, PA: University of Pennsylvania Press.

Warntz, W. 1959b. Geography at mid-twentieth century. *World Politics* 11, 442–54.

Warntz, W. 1959c. Progress in economic geography. In P. E. James (ed.) *New Viewpoints in Geography*. Washington, DC: National Council for the Social Studies, 54–75.

Warntz, W. 1968. Letter to the editor. *Professional Geographer* 20, 357.

Warntz, W. 1984. Trajectories and coordinates. In M. Billinge, D. Gregory and R. Martin (eds) *Recollections of a Revolution*. London: Macmillan, 134–52.

Waterman, S. 1985. Not just the milk and honey – now a way of life: Israeli human geography since the six-day war. *Progress in Human Geography* 9, 194–234.

Waterman, S. and Kliot, N. 1990. The political impact on writing the geography of Palestine-Israel. *Progress in Human Geography* 14, 237–60.

Waterstone, M. 2002. A radical journal of geography or a journal of radical geography? *Antipode* 34, 662–6.

Watkins, J. W. N. 1970. Against normal science. In I. Lakatos and A. Musgrave (eds) *Criticism and the Growth of Knowledge*. London: Cambridge University Press, 25–38.

Watson, J. D. 1968. *The Double Helix: A personal account of the discovery of the structure of DNA*. London: Weidenfeld & Nicolson.

Watson, J. W. 1953. The sociological aspects of geography. In G. Taylor (ed.) *Geography in the Twentieth Century*. London: Methuen, 453–99.

Watson, J. W. 1955. Geography: A discipline in distance. *Scottish Geographical Magazine* 71, 1–13.

Watson, J. W. 1983. The soul of geography. *Transactions of the Institute of British Geographers* NS8, 385–99.

Watson, M. K. 1978. The scale problem in human geography. *Geografiska Annaler* 60B, 36–47.

Watts, M. 1988. Deconstructing determinism. *Antipode: A Radical Journal of Geography* 20, 142–68.

Watts, M. 2000. *Struggles over Geography: Violence, freedom and development at the millennium*. Heidelberg: Department of Geography, University of Heidelberg.

Weaver, J. C. 1943. Climatic relations of American barley production. *Geographical Review* 33, 569–88.

Weaver, J. C. 1954. Crop-combination regions in the Middle West. *Geographical Review* 44, 175–200.

Webber, M. J. 1972. *Impact of Uncertainty on Location*. Canberra: Australian National University Press.

Webber, M. J. 1977. Pedagogy again: What is entropy? *Annals of the Association of American Geographers* 67, 254–66.

Webber, M. M. 1964. The urban place and the non-place urban realm. In M. M. Webber, J. W. Dyckman, D. L. Foley, A. Z. Guttenberg and C. Bauer Wurster (eds) *Explorations into Urban Structure*. Philadelphia, PA: University of Pennsylvania Press, 79–153.

Weightman, B. 1981. Gay bars as private places. *Landscape* 24, 9–17.

Werritty, A. G. 2010. D'Arcy Thompson's 'On growth and form' and the rediscovery of geometry within the geographic tradition. *Scottish Geographical Journal* 126, 231–57.

Wescoat, J. L. 1992. Common themes in the work of Gilbert White and John Dewey: A pragmatic appraisal. *Annals of the Association of American Geographers* 82, 587–607.

Western, J. S. 1978. Knowing one's place: 'The coloured people' and the Group Areas Act in Cape Town. In D. Ley and M. S. Samuels (eds) *Humanistic Geography: Problems and prospects*. Chicago: Maaroufa Press, 297–318.

Whatmore, S. 1997. Dissecting the autonomous self: Hybrid cartographies for a relational ethics. *Environment and Planning D: Society and Space* 15, 37–54.

Whatmore, S. 1999. Hybrid geographies: Rethinking the human in human geography. In D. Massey, J. Allen and P. Sarre (eds) *Human Geography Today*. Cambridge: Polity Press, 281–307.

Whatmore, S. 2002. *Hybrid Geographies: Natures, cultures, spaces*. London, Thousand Oaks, CA and New Delhi: Sage.

Whatmore, S. 2005. *Hybrid Geographies*: Author's responses and reflections. *Antipode* 37, 842–5.

Wheeler, J. O. 1998. Mappophobia in geography? 1980–1996. *Urban Geography* 19, 1–5.

Wheeler, J. O. 2002a. Assessing the role of spatial analysis in urban geography in the 1960s. *Urban Geography* 22, 549–58.

Wheeler, J. O. 2002b. From urban economic to social/cultural geography, 1980–2001. *Urban Geography* 22, 97–102.

Wheeler, P. B. 1982. Revolutions, research programmes and human geography. *Area* 14, 1–6.

White, G. F. 1945. *Human Adjustment to Floods*. Chicago: University of Chicago, Department of Geography, Research Paper 29.

White, G. F. 1972. Geography and public policy. *Professional Geographer* 24, 101–4.

White, G. F. 1973. Natural hazards research. In R. J. Chorley (ed.) *Directions in Geography*. London: Methuen, 193–216.

White, G. F. 1985. Geographers in a perilously changing world. *Annals of the Association of American Geographers* 75, 10–16.

White, G. F. 2002. Autobiographical essay. In P. Gould and F. R. Pitts, (eds) *Geographical Voices: Fourteen autobiographical essays*. Syracuse: Syracuse University Press, 341–64.

White, P. E. 1985. On the use of creative literature in migration study. *Area* 17, 277–83.

White, P. E. 1995. Geography, literature and migration. In R. King, J. Connell and P. White (eds) *Writing Across Worlds: Literature and migration*. London: Routledge, 1–19.

Whitehand, J. W. R. 1970. Innovation diffusion in an academic discipline: The case of the 'new' geography. *Area* 2, 19–30.

Whitehand, J. W. R. 1984. The impact of geographical journals: A look at ISI data. *Area* 16, 185–7.

Whitehand, J. W. R. 1985. Contributors to the recent development and influence of human geography: What citation analysis suggests. *Transactions of the Institute of British Geographers* NS10, 222–3.

Whitehand, J. W. R. 2002. Contributors to the recent development and influence of human geography: What citation analysis suggests. *Progress in Human Geography* 26, 511–19.

Whitehand, J. W. R. and Edmondson, P. M. 1977. Europe and America: The reorientation in geographical communication in the post-war period. *Professional Geographer* 29, 278–82.

Whitehand, J. W. R. and Patten, J. H. C. (eds) 1977. Change in the town. *Transaction of the Institute of British Geographers* NS2, 257–416.

Whittlesey, D. 1954. The regional concept and the regional method. In P. E. James and C. F. Jones (eds) *American Geography: Inventory and prospect*. Syracuse: Syracuse University Press, 19–68.

Wilbanks, T. J. 1995. Employment trends in geography: Introduction. *Professional Geographer* 47, 315–17.

Wilbanks, T. J. and Libbee, M. 1979. Avoiding the demise of geography in the United States. *Professional Geographer* 31, 1–7.

Wilbert, C. 2009. Animal geographies. In R. Kitchen and N. Thrift (eds) *International Encyclopedia of Human Geography*. Amsterdam: Elsevier, 122–6.

Williams, M. 1983. The apple of my eye: Carl Sauer and historical geography. *Journal of Historical Geography* 9, 1–28.

Williams, M. 1989. *Americans and their Forests: A historical geography*. Cambridge: Cambridge University Press.

Williams, M. 2003. *Deforesting the Earth: From prehistory to global crisis*. Chicago: University of Chicago Press.

Williams, M. 2009. Berkeley school. In R. Kitchin and N. Thrift (eds) *International Encyclopedia of Human Geography*. Amsterdam: Elsevier, 300–4.

Williams, M. 2012. The creation of humanized landscapes. In R. J. Johnston and M. Williams (eds) *A Century of British Geography*. Oxford: Oxford University Press, 196–212.

Williams, M. 2014. *To Pass on a Good Earth: The Life and work of Carl Sauer*. Charlottesville, VA: University of Virginia Press.

Williams, W. M. and Herbert, D. T. 1962. The social geography of Newcastle-under-Lyme. *North Staffordshire Journal of Field Studies* 2, 108–26.

Williamson, P., Birkin, M. and Rees, P. H. 1998. The estimation of population microdata by using data from small area statistics and samples of anonymised records. *Environment and Planning A* 30, 785–816.

Wills, J. 2000. Political economy II: The politics and geography of capitalism. *Progress in Human Geography* 24, 641–52.

Wills, J. 2002. Political economy III: Neoliberal chickens, Seattle and geography. *Progress in Human Geography* 26, 90–100.

Wills, J. 2014. Engaging. In R. Lee, N. Castree, V. Kitchin, A. Paasi, C. Philo, S. Radcliffe, S. M. Roberts and C. Withers (eds) *The SAGE Handbook of Human Geography*. London: Sage, 363–80.

Wills, J. and Peck, J. 2002. Progress or retreat? Antipode and the radical geographical project. *Antipode: A Radical Journal of Geography* 34, 667–71.

Wilson, A. C. and Henry, M. 2011. Evolving disciplinary imaginaries: Mapping the *New Zealand Geographer*, 1945–1969. *New Zealand Geographer* 67, 116–25.

Wilson, A. G. 1967. A statistical theory of spatial distribution models. *Transportation Research* 1, 253–69.

Wilson, A. G. 1970. *Entropy in Urban and Regional Modelling*. London: Pion.

Wilson, A. G. 1974. *Urban and Regional Models in Geography and Planning*. London: John Wiley.

Wilson, A. G. 1976a. Catastrophe theory and urban modelling: An application to modal choice. *Environment and Planning A* 8, 351–46.

Wilson, A. G. 1976b. Retailers' profits and consumers' welfare in a spatial interaction shopping model. In I. Masser (ed.) *Theory and Practice in Regional Science*. London: Pion, 42–57.

Wilson, A. G. 1978. *Mathematical Education for Geographers*. Department of Geography, University of Leeds, Discussion Paper 211, Leeds

Wilson, A. G. 1981a. *Geography and the Environment: Systems Analytical Methods*. Chichester: John Wiley.

Wilson, A. G. 1981b. *Catastrophe Theory and Bifurcation: Applications to urban and regional systems*. London: Croom Helm.

Wilson, A. G. 1984a. One man's quantitative geography: Frameworks, evaluations, uses and prospects. In M. Billinge, D. Gregory and R. L. Martin (eds) *Recollections of a Revolution: Geography as spatial science*. London: Macmillan, 200–26.

Wilson, A. G. 1984b. Making urban models more realistic: Some strategies for future research. *Environment and Planning A* 16, 1419–32.

Wilson, A. G. 1989a. Mathematical models and geographic theory. In D. Gregory and R. Walford (eds) *Horizons in Human Geography*. London: Macmillan, 29–47.

Wilson, A. G. 1989b. Classics, modelling and critical theory: Human geography as structured pluralism. In B. Macmillan (ed.) *Remodelling Geography*. Oxford: Basil Blackwell, 61–9.

Wilson, A. G. 2000. *Complex Spatial Systems: The modelling foundations of urban and regional analysis*. Chichester: John Wiley.

Wilson, A. G. 2011. *Catastrophe Theory and Bifurcation: Applications to urban and regional systems*. Routledge Revivals reprint. London: Routledge.

Wilson, A. G. and Bennett, R. J. 1985. *Mathematical Methods in Human Geography and Planning*. Chichester: John Wiley.

Wilson, A. G. and Kirkby, M. J. 1975. *Mathematics for Geographers and Planners*. Oxford: Oxford University Press.

Wilson, A. G., Rees, P. H. and Leigh, C. 1977. *Models of Cities and Regions*. London: John Wiley.

Wilson, E. O. 1998. Consilience among the great branches of learning. *Daedalus* 127, 131–50.

Wilton, R. and Evans, J. 2009. Disability and chronic illness. In R. Kitchen and N. Thrift (eds) *International Encyclopedia of Human Geography*. Amsterdam: Elsevier, 205–10.

Winter, C. 1997. Ethnocentric bias in geography textbooks: A framework for reconstruction. In D. Tilbury and M. Williams (eds) *Teaching and Learning in Geography*. London: Routledge, 180–8.

Wise, M. J. 1968. *Sir Dudley Stamp: His life and times*. Land use and resources: studies in applied geography. London: Institute of British Geographers, Special Publication 1, 261–70.

Wise, M. J. 1975. A university teacher of geography. *Transactions of the Institute of British Geographers* 66, 1–16.

Wise, M. J. 1977. On progress and geography. *Progress in Human Geography* 1, 1–11.

Wise, M. J. 1986. The Scott Keltie Report 1885 and the teaching of geography in Great Britain. *Geographical Journal* 152, 367–82.

Wisner, B. 1970. Introduction: On radical methodology. *Antipode: A Radical Journal of Geography* 2, 1–3.

Withers, C. W. J. 2001. *Geography, Science and National Identity: Scotland since 1520*. Cambridge: Cambridge University Press.

Withers, C. W. J. 2002. Constructing 'the geographical archive'. *Area* 34, 303–11.

Withers, C. W. J. 2007. *Placing the Enlightenment: Thinking geographically about the age of reason*. Chicago: University of Chicago Press.

Withers, C. W. J. 2009. Place and the 'spatial turn' in geography and in history. *Journal of the History of Ideas* 70, 637–58.

Withers, C. W. J. 2010. *Geography and Science in Britain, 1831–1939: A study of the British Association for the Advancement of Science*. Manchester: Manchester University Press.

Withers, C. W. J. 2011. Geography's narratives and intellectual history. In J. A. Agnew and D. N. Livingstone (eds) *The Sage Handbook of Geographical Knowledge*. London, Thousand Oaks, CA and New Delhi and Singapore: Sage, 39–50.

Withers, C. W. J. and Mayhew, R. J. 2002. Rethinking 'disciplinary' history: Geography in British universities, c.1580–1887. *Transactions of the Institute of British Geographers* NS27, 11–29.

Wolch, J. and Dear, M. J. (eds) 1989. *The Power of Geography: How territory shapes social life*. Boston, MA: Unwin Hyman.

Woldenberg, M. J. and Berry, B. J. L. 1967. Rivers and central places: Analogous systems? *Journal of Regional Science* 7, 129–40.

Wolf, A. 2002. *Does Education Matter? Myths about education and economic growth*. London: Penguin.

Wolf, L. G. 1976. Comments on the Harries–Peet controversy. *Professional Geographer* 28, 196–8.

Wolpert, J. 1964. The decision process in spatial context. *Annals of the Association of American Geographers* 54, 337–58.

Wolpert, J. 1965. Behavioral aspects of the decision to migrate. *Papers and Proceedings Regional Science Association* 15, 159–72.

Wolpert, J. 1967. Distance and directional bias in inter-urban migratory streams. *Annals of the Association of American Geographers* 57, 605–16.

Wolpert, J. 1970. Departures from the usual environment in locational analysis. *Annals of the Association of American Geographers* 60, 220–9.

Wolpert, J., Dear, M. J. and Crawford, R. 1975. Satellite mental health facilities. *Annals of the Association of American Geographers* 65, 24–35.

Women and Geography Study Group-IBG (WGSG). 1984. *Geography and Gender: An introduction to feminist geography*. London: Hutchinson.

Women and Geography Study Group-IBG (WGSG). 1997. *Feminist Geographers: Explorations in diversity and difference*. Harlow: Longman.

Woods, M. and Gardner, G. 2011. Applied policy research and critical human geography: Some reflection on swimming in murky waters. *Dialogues in Human Geography* 2, 198–214.

Woods, R. I. 1982. *Theoretical Population Geography*. London: Longman.

Woods, R. I. and Rees, P. H. (eds) 1986. *Population Structures and Models: Developments in spatial demography*. London: Allen & Unwin.

Woodward, K., Dixon, D. P. and Jones, J. P. III. 2009. Poststructuralism/poststructuralist geographies. In R. Kitchin and N. Thrift (eds) *International Encyclopedia of Human Geography*. Amsterdam: Elsevier, 396–407.

Wooldridge, S. W. 1956. *The Geographer as Scientist*. London: Thomas Nelson.

Wooldridge, S. W. and East, W. G. 1958. *The Spirit and Purpose of Geography*. London: Hutchinson.

Wright, D. J. 2012. Theory and application in a post-GISystems world. *International Journal of Geographical Information Science* 26, 2197–2209.

Wright, J. K. 1925. *The Geographical Lore of the Time of the Crusades: A study in the history of medieval science and tradition in western Europe*. New York: American Geographical Society.

Wright, J. K. 1947. Terrae incognitae: The place of imagination in geography. *Annals of the Association of American Geographers* 37, 1–15.

Wright, J. K. 1952. *Geography in the Making: The American Geographical Society, 1851–1951*. New York: American Geographical Society.

Wright, J. K. 1962. Miss Semple's *Influences of Geographic Environment*: Notes towards a bibliography. *Geographical Review* 52, 346–61.

Wright, R. and Koch, N. 2009. Ivy League and geography in the US. In R. Kitchin and N. Thrift (eds) *International Encyclopedia of Human Geography*. Amsterdam: Elsevier, 616–21.

Wrigley, E. A. 1965. Changes in the philosophy of geography. In R. J. Chorley and P. Haggett (eds) *Frontiers in Geographical Teaching*. London: Methuen, 3–24.

Wrigley, N. 1973. The use of percentages in geographical research. *Area* 5, 183–6.

Wrigley, N. 1976. Introduction to the use of logit models in geography. *CATMOG* 10, Norwich: Geo Books.

Wrigley, N. 1983. Quantitative methods: On data and diagnostics. *Progress in Human Geography* 7, 567–77.

Wrigley, N. 1984. Quantitative methods: Diagnostics revisited. *Progress in Human Geography* 8, 525–35.

Wrigley, N. 1985. *Categorical Data Analysis for Geographers and Environmental Scientists*. London: Longman.

Wrigley, N. 1995. Revisiting the modifiable areal unit problem and the ecological fallacy. In A. D. Cliff, P. R. Gould, A. G. Hoare and N. J. Thrift (eds) *Diffusing Geography: Essays for Peter Haggett*. Oxford: Blackwell Publishers, 49–71.

Wrigley, N. 2002. 'Food deserts' in British cities: Policy context and research priorities. *Urban Studies* 39, 2029–40.

Wrigley, N. 2013. Towards a policy engaged retail geography. In C. Garrocho (ed.) *Advances in Commercial Geography: Prospects, methods and applications*. Zinacantepec, MX: El Colegio Mexiquense, 59–93.

Wrigley, N. and Bennett, R. J. (eds) 1981. *Quantitative Geography: A British view*. London: Routledge & Kegan Paul.

Wrigley, N. and Longley, P. A. 1984. Discrete choice modelling in urban analysis. In D. T. Herbert and R. J. Johnston (eds) *Geography and the Urban Environment*, Vol. 6. Chichester: John Wiley, 45–94.

Wrigley, N. and Matthews, S. 1986. Citation classics and citation levels in geography. *Area* 18, 185–94.

Wylie, J. 2005. A single day's walking: Narrating self and landscape on the South West Coast Path. *Transactions of the Institute of British Geographers* NS30, 234–47.

Wyly, E. 2009. Strategic positivism. *Professional Geographer* 61, 310–22.

Wyly, E. 2011. Positively radical. *International Journal of Urban and Regional Research* 35, 889–912.

Wyly, E. 2014. The new quantitative revolution. *Dialogues in Human Geography* 4, 26–38.

Wyly, E., Moos, M., Foxcroft, H. and Kabahizi, E. 2008. Subprime mortgage segmentation in the American urban system. *Tijdschrift voor Economische en Sociale Geografie* 99, 3–23.

Wyly, E., Moos, M., Hammel, D. and Kabahizi, E. 2009. Cartographies of race and class: Mapping the class–monopoly rents of American subprime mortgage capital. *International Journal of Urban and Regional Research* 33, 332–54.

Yang, C., Goodchild, M., Huang, Q., Nebert, D., Raskin, R., Xu, Y. Bambacus, M. and Fay, D. 2011. Spatial cloud computing: How can the geospatial sciences use and help shape cloud computing? *International Journal of Digital Earth* 4, 305–29.

Yeates, M. H. 2001. Yesterday as tomorrow's song: The contribution of the 1960s 'Chicago School' to urban geography. *Urban Geography* 22, 514–29.

Yeung, H.W.C. 1997. Critical realism and realist research in human geography: A method or a philosophy in search of a method? *Progress in Human Geography* 21, 51–74.

Yeung, H.W.C. 2002. Deciphering citations. *Environment and Planning A* 34, 2093–2101.

Young, I. M. 1990. *Justice and the Politics of Difference*. Princeton, NJ: Princeton University Press.

Yule, G. U. and Kendall, M. G. 1950. *An Introduction to the Theory of Statistics*. London: Griffin.

Zelinsky, W. 1970. Beyond the exponentials: The role of geography in the great transition. *Economic Geography* 46, 499–535.

Zelinsky, W. 1973a. *The Cultural Geography of the United States*. Englewood Cliffs, NJ: Prentice-Hall.

Zelinsky, W. 1973b. Women in geography: A brief factual report. *Professional Geographer* 25, 151–65.

Zelinsky, W. 1974. Selfward bound? Personal preference patterns and the changing map of American society. *Economic Geography* 50, 144–79.

Zelinsky, W. 1975. The demigod's dilemma. *Annals of the Association of American Geographers* 65, 123–43.

Zelinsky, W. 1978. Introduction. Human geography: Coming of age. *American Behavioral Scientist* 22, 5–13.

Zelinsky, W., Monk, J. and Hanson, S. 1982. Women and geography: A review and prospectus. *Progress in Human Geography* 6, 317–66.

Zimmerer, K. S. 2010. Retrospective on nature–society geography: Tracing trajectories (1911–2010) and reflecting on translations. *Annals of the Association of American Geographers* 100, 1076–94.

Zimmermann, E. W. 1972. *World Resources and Industries*, 3rd edn. New York: Harper & Row.

Zipf, G. K. 1949. *Human Behavior and the Principle of Least Effort*. Cambridge, MA: Addison-Wesley.

Zolnik, E. J. 2009. Context in human geography: A multilevel approach to study human–environment interactions. *Professional Geographer* 61, 336–49.

Author index

Subject index